FRONTIERS OF MATERIALS RESEARCH /
ELECTRONIC AND OPTICAL MATERIALS

C-MRS INTERNATIONAL 1990 SYMPOSIA PROCEEDINGS

Editors-in-Chief:

Dongsheng YAN
Academia Sinica

Changxu SHI
National Natural Science Foundation
of China (NSFC)

Hengde LI
Department of Materials Science & Engineering
Tsinghua University

Volume 1
Frontiers of Materials Research / Electronic and Optical Materials
(Edited by: Meiying KONG and Liji HUANG)

Volume 2
Advanced Structural Materials
(Edited by: Yafang HAN)

Volume 3
Polymers and Biomaterials
(Edited by: Hanbao FENG, Yafang HAN and Liji HUANG)

Volume 4
Thin Films and Beam-Solid Interactions
(Edited by: Liji HUANG)

Volume 5
Mechanical Properties / Materials Design
(Edited by: Boqun WU)

FRONTIERS OF MATERIALS RESEARCH / ELECTRONIC AND OPTICAL MATERIALS

PROCEEDINGS OF THE SYMPOSIA N: FRONTIERS OF MATERIALS RESEARCH, A: HIGH T_c SUPERCONDUCTORS, AND D: OPTOELECTRONIC MATERIALS AND FUNCTIONAL CRYSTALS OF THE C-MRS INTERNATIONAL 1990 CONFERENCE

BEIJING, CHINA, 18 – 22 JUNE 1990

Volume editors:

Meiying KONG
Institute of Semiconductors
Chinese Academy of Science

Liji HUANG
Department of Materials Science & Engineering
Tsinghua University

VOLUME 1

1991
NORTH-HOLLAND
AMSTERDAM • NEW YORK • OXFORD • TOKYO

Elsevier Science Publishers
P.O. Box 211
1000 AE Amsterdam
The Netherlands

Sole distributors for the U.S.A. and Canada:

Elsevier Science Publishing Company, Inc.
655 Avenue of the Americas
New York, N.Y. 10010
U.S.A.

Library of Congress Cataloging-in-Publication Data

C-MRS International Conference (1990 : Peking, China)
Frontiers of materials research--electronic and optical materials : proceedings of the symposia N, frontiers of materials research, A, high Tc superconductors, and D, optoelectronic materials and functional crystals of the C-MRS International 1990 Conference, Beijing, China, 18-22 June 1990 / volume editors, Meiying Kong, Liji Huang.
p. cm. -- (C-MRS international 1990 symposia proceedings ; v. 1)
Includes index.
ISBN 0-444-88825-X
1. Materials--Congresses. 2. Materials--Research--Congresses. 3. High temperture superconductors--Congresses. 4. Optoelectronics--Materials--Congresses. I. Kong, Meiying. II. Huang, Liji. III. Chinese Materials Research Society. IV. Title. V. Series.
TA401.3.C2 1990a
620.1'12972--dc20 90-28528
CIP

ISBN: 0 444 88825 X (Volume 1)
0 444 89016 5 (5 Volume set)

Printed in The Netherlands

PREFACE

This volume is one of the five-volume proceedings of C-MRS International'90 Conference held in Beijing, China on June 18 to 22. The Conference was organized into 14 symposia (through Symposium A to N) covering research areas of high Tc superconductors(A), advanced composite materials(B), high performance ceramics(C), optoelectronic materials and functional crystals(D), advanced engineering plastics(E), functional polymers(F), intermetallic compounds and high temperature materials(G), laser and particle-beam interactions with solids(H), thin films(I), fatigue and fracture behavior in materials(J), recent advances of tribomaterials(K), biomedical materials(L), microstructures and materials design(M), and frontiers of materials research(N). Over 100 scientists from 18 countries abroad and more than 600 scientists from different regions of China attended this conference. Altogether 630 papers were presented at 63 oral sessions including 3 plenary sessions and 4 poster sessions. In the plenary sessions 14 famous scientists from 8 countries presented their recent achievements, they are: L.E. Cross(U.S.A.), K.C. Taylor(U.S.A.), R.P.H. Chang(U.S.A.), J.D. Dow(U.S.A.), P. Hagenmuller(France), R.W. Cahn(U.K.), E. Fitzer (F.R.G.), R.J. Brook(F.R.G.), M. Doyama(Japan), L.A. Utracki(Canada), A. Wagendrisel(Austria), J.S. Williams(Australia), D. Feng(China), T. Ko(China) and Chuangtian Chen(China).

This volume contains papers presented at the following three symposia: Symposium N on "Frontiers of Materials Reseach", Symposium A on "High Tc Superconductors" and Symposium D on "Optoelectronic Materials and Functional Crystals". The broad range of subjects was covered in about 100 theoretical and technical presentations that attracted many of the C-MRS International'90 Conference attendees.

The development of high Tc superconductor materials will explore the applications widely in high speed electronics. The understanding and controling of crystal growth processes present a major challenge to materials scientists (many workers have achieved significient advances in technology). Optoelectronic materials, functional crystals, devices and processing technology are becoming mature, and the fruits of the reseach will play a strong role in the near future.

The works of this volume were refereed and carefully selected from a variety of papers presented at the conference, which contains important new research results in these rapidly growing fields of high Tc superconductors, optoelectronic materials, functional crystals and devices and processing technologies. It also involves the frontiers of materials research. The present volume is divided into three parts covering the main topics of the three symposia.

Editorial Board

ACKNOWLEDGEMENTS

The organizers wish to thank the many participants and particularly the authors whose work made this volume an outstanding one. We are indebted to authors and referees whose efforts made rapid publication of the proceedings possible. We also acknowledge the invited speakers who had given the excellent overview of currently important topics. Thanks particularly go to the session chairs for keeping the discussions relevant. They are:

YAN Dongsheng, SHI Changxu, LI Hengde, GAN Zizhao, ZHAO Zhongxian, GAN Fuxi, JIANG Minhua, KONG Meiying and Charles W. TU.

Special acknowledgements are due to Ms. Mary Carpenter and Dr. Albert E.M.J. Fischer of Elsevier Science Publishers; their efficient work helped to speed up the publications.

The C-MRS International'90 was financially supported by China Association for Science and Technology(CAST), National Natural Science Foundation of China(NSFC), Chinese Academy of Science(Academia Sinica) and National Advanced Materials Committee of China(NAMCC).

COMMITTEES OF C-MRS INTERNATIONAL'90 CONFERENCE

INTERNATIONAL ADVISORY COMMITTEE

CAHN Robert W. (U.K)
CARLSSON Janne (SWEDEN)
CHANG R.P.H. (U.S.A.)
CHU C.W. (U.S.A.)
GYULAI Jozsef (HUNGARY)
HASIGUTI R.R. (JAPAN)
KE Jun (CHINA)
LIN Lanying (CHINA)
NAKABAYASHI Nobuo (JAPAN)
PETERSON Marshall B. (U.S.A.)
PETZOW Gunter (F.R.G.)
PLATE N. (U.S.S.R.)
QIAN Renyuan (CHINA)
RAO C.N.R. (INDIA)
RIMINI E.R. (ITALY)
ROY Rustum (U.S.A.)
SHI Changxu (CHINA)
SIFFERT P. (FRANCE)
TAYLOR Kathleen C. (U.S.A.)
UTRACKI L.A. (CANADA)
WAGENDRISTEL A. (AUSTRIA)
WILLIAMS J.S. (AUSTRALIA)
YAN Dongsheng (CHINA)

CONFERENCE CHAIRMEN

YAN Dongsheng SHI Changxu

LOCAL ORGANIZING COMMITTEE

YAN Dongsheng (Chairman)
WANG Fosong (Chairman)
KE Jun
LI Hengde
LI Yiyi
LIN Lanying
LIN Zhenshen
QIAN Renyuan
SHI Changxu
WU Jiaxiang
WU Renjie
XIAO Jimei
YAN Minggao
YUAN Haibo
ZENG Hanmin
ZHAO Zhongxian
ZHANG Ruisheng
ZHANG Xiaowen
ZHU Hesun

CONTENTS

PART 1: FRONTIERS OF MATERIALS RESEARCH

PART 2: HIGH Tc SUPERCONDUCTORS

PART 3: OPTOELECTRONIC MATERIALS AND FUNCTIONAL CRYSTALS

AUTHOR AND SUBJECT INDICES

PART 1: FRONTIERS OF MATERIALS RESEARCH

Frontiers of Materials Research/
Electronic and Optical Materials
M. Kong, L. Huang (editors)

A BRIEF INTRODUCTION TO THE SCIENCE POLICY OF THE CHINESE ACADEMY OF SCIENCES

Guangzhao ZHOU

The Chinese Academy of Sciences, Beijing, China

It is really a great pleasure and honour to be invited to give a keynote speech at the C-MRS International'90. I don't think I can make any scientific contribution to this conference, as I am not a material scientist. Therefore, I have decided to talk about the science policy of the Chinese Academy of Sciences, which might be of interest to our friends from abroad, who have made great efforts in the past to develop international scientific collaborations with China.

There are some key factors one has to consider in the process of science policy making, namely,

1. Trends and perspectives in the world scientific development;
2. Basic conditions and economic needs of the country;
3. Research environment that is favourable to encourage initiativity, creativity, collaboration and information exchange.

I will elaborate a bit on the first two aspects, and the third point will be mentioned here and there in between.

I.TRENDS AND PERSPECTIVES OF THE WORLD SCIENTIFIC DEVELOPMENT

The contemporary science and technology is undergoing profound changes which could be characterized by the following features:

1. Interdisciplinary and multidisciplinary research have become the main trend of the development of science and technology. Since 60s, due to the development of research method and broadening field of vision, there has been a strong inclination towards the convergence of vision, there has been a strong inclination towards the convergence of different disciplines in research. Chemical and physical processes, which are the basis of the human endeavor in exploring the origin of life, are pulling chemists, physicists and life scientists together in their common scientific pursuit; the formation and movement of the earth crust has brought mechanics scientists into the realm of earth sciences; the study on chemical molecular reaction dynamics is inseparable from the development of atomic and molecular physics and laser science; the origin of celestial body and space-time principles are the common research subjects of astronomy, mathematics and theoretical physics; And furthermore, the social and economical development is giving great impetus to the advancement of some comprehensive sciences like information science, materials science, space science and technology, oceanology, etc. All these tell us that the disciplinary interactions are the main trend for development of modern science.

2. The severe impact on nature by human activities, which is changing and damaging the ecological environment. Soil erosion, deforestation, desertification, water and air pollution, the increase of CO_2, ozone depletion, etc. have become important issues of world concern, imposing enormous challenges to science and engineering, concern, to our Administration, and to the world community of nations. And the environmental science is now widely accepted as a most important discipline by the Governments and scientific communities of various countries. When we recognize the economic and social development is becoming

more dependent on science and technology, the contemporary research on natural sciences is by no means something purely initiated by some individual scientists but rather an indispensable activity in the modern society.

3. The cycle from scientific discovery to application is becoming shorter. This is because of the increasingly active two-way flow of research personnel, information and technologies between the increased financial input to basic research from the industry forced by the long-term market-demand. The "human capital", transfer, has speeded up the process from scientific results to helped maintain the vitality of the basic research team, which in turn has become a necessary condition for building up a crack team on basic research.

4. Collective effort is becoming one of the key factors in many successful scientific thrusts. Quite a number of the scientific discoveries during the Mid-1900s were mainly achieved by individual scientists through their persistent and painstaking research and observations. However, it has become increasingly evident that collective effort by various scientific communities is playing a more and more important role today, particularly in areas of "big science" and oriented basic research with important application prospects. Taking example of high energy physics experiments exploring deep-level mater structure, one research group would usually have hundreds of scientists from various countries, and to build one detector would take massive amount of funds and a time-span of several years. This tendency is starting also in life science, e.g., construction of human gene bank will need a cooperative and coordinated effort of the world scientists costing several billions of dollars. It is no exaggeration to say that a good scientist in the 21st century should also be a capable organizer and an active and popular public figure. It is, to a large extent, caused by the science modernization and socialization.

When I have touched on some points of the world scientific development trend, I would like to turn to the situation in China, which has a significant bearing on science-policy-making in this country. The predictability and continuity of resources are essential prerequisites for the productivity of science policy.

II. BASIC FACTS ABOUT CHINA

It has been proved by the forty years' history of China's scientific and technological development that China can not simply copy the model of the developed countries in their development of science and technology. We should select a road suitable to our own conditions by proceeding from China's actual situation and following the inherent law governing the development of science and technology. In this context, we should bear in mind the following three major aspects:

1. A Huge Population:
   - 1989 1.16 billion
   - 2000 1.36 billion
   - 2030 >1.56 billion
2. Relatively Poor Natural Resources:
   - Arable land per capita 1.36mu(1989)
   - Forest coverage 11.5%

   Total water resource: 2.9 trillion m^3; 2730 m_3 per capita, ranking the 88th place in the world

   - lack of iron, copper, phosphorus, potassium etc. minerals. It is estimated that China's mineral resources per capita is only half of the world average level.
3. Present situation about economy, education and social development:

   1989 GNP 1285 yuan per capita;
   Illiteracy 0.22 billion;
   hospital beds one per 104 people;
   college graduates one per 104 people;
   rural population 80%;
   serious air and water pollution;

Therefore, to develop eduction, science and technology is the most urgent task of China.

4. Problems and Challenges Facing Chinese Scientists:

The above-mentioned facts about China have presented us a clear although not very encouraging picture. It is suggested that if the Chinese scientists are to make further contribution to the country's economy, they will have to deal with the following issues:

- Control of population growth;
- Preservation and controlled use of land, water, forest and other natural resources;
- Protection of environment and ecosystems;
- Promotion of economic growth through technological innovation;
- Development and dissemination of scientific knowledge;

The Chinese Academy of Sciences is a national comprehensive research centre and is therefore placed with the historical task to actively participate in the effort to solve these problems. We will continue to carry out the policy of "deploying the bulk of its research force on the main battlefield of serving the national economy and at the same time maintaining a crack research team for basic research and high-tech tracking". To achieve this aim, we have to introduce a new system pertaining to the overall operation of the Academy's work, which could help us to deal with different types of work more efficiently.

The main problems of the old system are:

1. lack of organic conncection of research institutes with both universities and enterprises;

2. equalitarianism in distribution and universal employment insurance;

3. the inbreeding personnel system and the bureaucratic style of work.

All these drawbacks make the institutes overstaffed, less efficient and not competitive.

The essential points of the reform consist of the following:

1. to implement different funding and administrative policies according to the different nature of research work.

2. to expand the scope of autonomy of research institutes and lessen the government administrative interferences;

3. to abolish step by step the old distribution and personnel system by introducing marketing and competition mechanism;

4. to open the institutes to the whole country and the world; to encourage talent flow and information exchanges; to strengthen the lateral ties among institutes, universities and enterprises;

5. to keep a rational proportion of crack force for basic research while emphasizing the service to the national economic development.

Here, I would like to give a brief account on the new policies adopted for basic sciences, environmental sciences and engineering sciences.

Fundamental Sciences (basic)

China is a large country. It will be impossible for her to build up its own capability for developing science and technology without a proper emphasis on fundamental sciences. We have altogether 15,000 scientists who are engaged in basic research. And China's overall investment to basic research is still rather low and its per capita investment is about 100 times lower than that in some of the developed countries. While voice is being raised to call for more governmental support, we have to be fully aware that no great expectation is to be met in the foreseeable future. Therefore, only a restricted target could be set for basic research in terms of financial, manpower, instrumental and institutional input. For a proper use of the limited funds and equipment and for a better show in the world competition, we should break down traditional compartmentalization in China and encourage a new mechanism which will allow for openness, mobility, joint entities, cooperation and competition. Only in this way can we pull together outstanding scientists of the country to tackle with our national key projects, provide more opportunities for good young scientists to display their talents and create an atmosphere of free interactions among scientists. And only in this way can we have more interactions between industries and research institutions. It is gratifying to note that in the recent years this new system has been adopted by our Academy and a number of laboratories, in which permanent staff should be less than half and the projects are approved and financed through a peer review system.

Environmental Sciences

Time and again, the Chinese environmental scientists have been calling for a national attention that China is facing with a serious and widely-spread environmental pollution and ecological crisis. The increasing human impact on nature is found everywhere because of the modernization drive including township enterprise, which has caused environmental degradation and pollution. The area of soil erosion has increased by one third with annual solid loss amounting to 5 billion tons; forest coverage has been decreasing and the annual consumption of timber has exceeded the production by 100 million m^3; the area of grassland degradation has reached 770 million mu. the area of desertification is expanding annually by 1500 km^2 and is now covering 11.4% of the total area of the country; about 150 million people are drinking polluted water; and acid rain is also a headache for us particularly in the southwest part of China. China in the long run, will be frustrated by the contradicting situation of too large a population, limited natural resources and ever-increasing environmental problem. Only the advancement of science can find appropriate ways in solving these contradictions. For a systematic and persistent observation, data-collecting and comprehensive study, strong governmental support is given to create a network of field observation stations and comprehensive research centers. In return, scientists put a lot of efforts to establish model counties guided by scientific and ecological considerations, try their best to predict nature disasters and give scientific advice on government policy-making.

Engineering Sciences

Engineering science is a bridge which makes possible the transformation of research results into production force. Compared with other areas, work on engineering sciences used to be a weak link in China. As determined by its own nature, the development of engineering sciences will be largely oriented towards the needs of the economy. Since we are not likely to get enough research funds from the government, we have in recent years made some adjustments of the operation mechanism over the academy's work. Different operation work. As to the engineering sciences, it is believed we should rely both on governmental support and industrial investment. Recently an academy industry partnership has been established. The Academy has signed a number of cooperative agreements with many provinces covering quite a range of industries such as petroleum, automobile, metallurgy and chemical industries etc. Taking the partnership with the Hunan Province alone, more than 40 research institutes of the academy have established cooperative relationship with nearly 100 enterprises in 12 cities and prefectures involving 14 sectors of industry. And up to now 24 items of technology transfer are completed and already put to production, resulting an annual income increase of 60 million yuan. Another good example is the Academy's collaboration with more than 30 enterprises of the petroleum industry, which conducted 162 items of research projects. Another new phenomenon is that the Academy has sent a number of scientists to take up posts as deputy county magistrates or deputy mayors to help the rural areas to build up local scientific and technological strength. The present research institutions in China were established according to scientific disciplines and no major adjustment has been made so far. However, the trend of the development of modern science is inclined towards an interdisciplinary pattern with disciplines crossing with each other. Therefore, the Academy is considering to set up some R&D centers to adapt itself with this development trend. The idea is meant to create an optimum working environment, attracting good younger scientists and exploring a new model of research institution. CAS also serves as an incubator of high-tech firms which, as we all know, is a special organizational form emerged from the development of high technology. At present different types of incubators are under discussion for development is science parks or various high-tech development regions in China including within our Academy. As a matter of fact, the Academy has established more than 300 companies for developing different types of technologies. Some of them are Academy invested in the form of interest-free loans and some of the companies are invested by the research institutes. About 8,000 scientists have walked out of their research labs to get engaged in the development work, and about 1,100 types of products have been developed. Among them about 450 products are classified as new products of the country. Before I finish my talk, I would like to brief you about the situation of international cooperation. We believe that science by its nature has no national boundary and international collaboration is essential for the healthy development of world science.

International Cooperation Since 1978, the Chinese Academy of Sciences has established quite a complete network with its counterparts in the world. The Academy has up to now signed over 70 agreements on scientific and technological cooperation with its counterparts in more than 50 countries and regions. These formal agreements have made it possible for both sides to conduct various kinds of scientific collaborations including exchange of visits, holding of seminars and conferences, conducting joint research projects, etc. And on the basis of these interactions, a free and more extensive cooperative mode between individual scientists has been created. Since 1980, CAS's exchange volume with outsiders in terms of number of visits has increased by three folds, reaching 6800 person termes in 1988. Although some interference has been made because of the "June 4th Event" last year, CAS is still committed to the belief that science is international. During the four months after the June Event, CAS had sent abroad about 1840 scientists. The majority of the above-mentioned agreements are still active. Many of the substantial joint projects are going well, such as the joint survey on glaciology by Chinese and German scientists, discussion about joint scientists, China/Canada dinosaur project, Sino/French joint research and expedition to the Western Kunlun and the Karakunlun Mountain and Sino/US High Energy Physics Meeting. And more projects are underway like the signing of the Sino/Japanese Cooperative Program on Atmosphere-Land Surface Processes Experiment at Heihe River Basin between Lanzhou Institute for Plateau Atmospheric Physics, CAS and Disaster Prevention Research Institute Kyoto University, China/Germany Cooperative Program for Experimental Physics to be supported by Volkswagen Foundation, etc. And moreover, about 20 million US dollars of the World Bank Loan has been confirmed to CAS which will be used to update our scientific equipments and raise the capabilities of the CAS laboratories in training graduate students. Materials science is one of the most important, rapid developing and fruitful field of research. It attracts scientists from many different disciplines and has formed the backbone of high technology and modern civilization. The CAS will do its best to support the development of materials science. Finally, I wish all the participants of this conference a great success in their scientific pursuits. Thank you.

Frontiers of Materials Research/
Electronic and Optical Materials
M. Kong, L. Huang (editors)

RECENT ADVANCE OF FOCUSED ION BEAM TECHNOLOGY IN JAPAN

Susumu NAMBA

Research Center for Extreme Materials and
Faculty of Engineering Science
Osaka University
Toyonaka, Osaka 560, Japan

Recent advance of focused ion beam technology in Japan is reviewed. Various focused ion beam systems which generate fine focused ion beams at energies ranging from 100eV up to 200keV have been developed, and various maskless processing techniques have been developed. Present FIB technology finds interesting applications in local modification such as circuit testing and modification in addition to various maskless processing.

1. INTRODUCTION

Since first work on focused ion beam technology was reported in 1974 by Hughes Research Laboratories group[1], focused ion beam technology has made significant progress both in FIB optics systems and microfabrication technology. In Japan, research in focused beam technology is very active, and new focused ion beam systems and various maskless fabrication techniques have developed. These include ion beam assisted etching and deposition and recent nanostructure fabrications. Low energy FIB, which is of potential importance for low damage processing but was generally difficult to form because of increasing chromatic aberration at low energy, has already been developed[2,3].

This paper will discuss recent activity on FIB research in Japan.

2. ACTIVITY IN JAPAN

Figure 1 shows the number of FIB systems used in Japan. This clearly reflects the activity and its growth of FIB works in Japan. Since the first machine was developed in our laboratory in 1981, there was a rapid increase and now, 79 machines are in use in many institutes. At the initial stage of development, high energy and high vacuum FIB systems were fabricated. These systems have been used mainly for maskless ion implantation and lithography.

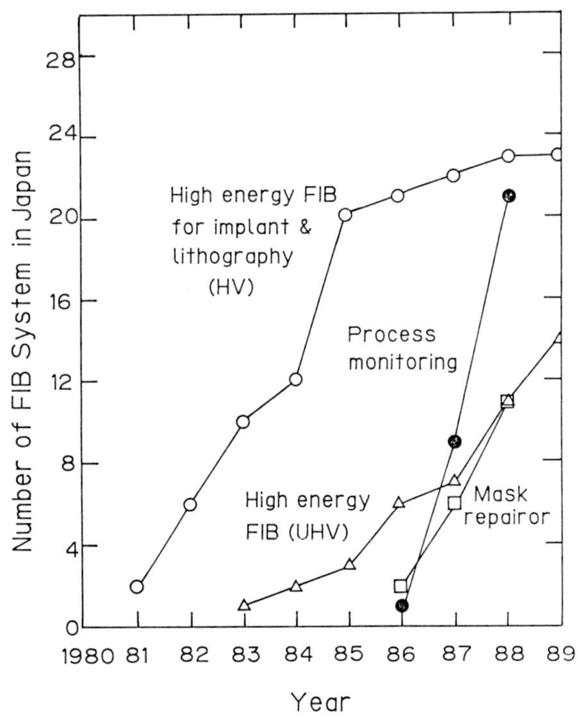

Fig. 1 Number of FIB systems used in Japan

This statistics, however, show a tendency that interest is growing in the applications for

Table 1 Focused ion beam systems

Energy (keV)	Beam spot (nm)	Current (pA)	Remarks	Reference
25	200	0.1	He field ion source	4
150	100	100	Computer controlled X-Y stage	5
30-200	60	50(200keV)	Two focusing mode	6
0.1-35	100	30(1keV)	Low energy FIB	3
200	<80	100	Highest beam energy	7
20		50-170	Mask repair	8
100	100-200	30	High mass resolution ($M/dM \geq 63$)	10
35	<100	100	SIMS with multi-channel detector	9

process monitoring, mask repair and ultra high vacuum systems. Many of the UHV system are designed to use for fabrications of III-V compound devices or quantum effect devices and combine with molecular beam epitaxy system.

For the past few years, many FIB systems have been developed and important improvements have been achieved. Table 1 summarizes various FIB systems developed for various direct microprocessing such as scanning ion beam lithography, direct etching and deposition, and total vacuum processing. FIB is also attractive for microbeam analysis and several systems was developed[9,11].

Most FIB systems have beam energy ranging from 30 to 200keV and form focused beams with submicron or smaller diameter and current density ranging from 0.1 to over $1A/cm^2$.

For various applications, it is necessary to vary beam energy in wide range. However, focusing characteristics of electron optics generally becomes poor at low energy. The beam spot size is determined by chromatic and spherical aberration. For FIB systems which use liquid metal ion sources, a chromatic aberration determines the beam spot size because of relatively large energy spread of liquid metal ion sources (2-10eV in FWHM), and becomes large at a low acceleration energy. This makes it difficult to obtain submicron beam spot in wide energy range, say 50 to 200keV. This problem has been solved using two focusing mode[6].

Some ion sources generate ions with a mass close to each other. To separate these ions, very high mass resolution is required but high mass resolution is difficult to achieve without degrading focusing performance, because a deflection chromatic aberration increases for large beam deflection. Recently, FIB system with high mass resolution has been developed which incorporates a four stage EXB, achromatic mass filter[10].

Present FIB systems use liquid metal ion sources and generate various metal ions. However, gaseous FIB for light mass ions such as H and He beam is important for various processing such as lithography, ion beam assisted etching and deposition, and microbeam analysis. H and He focused ion beams are expected to be focused nanometer size because H or He field ionization sources has low energy spread and high brightness, and they induce less damage compared to heavy ions such as Si or Ga. Unfortunately, it has generally been difficult to obtain stable operation. Figure 2 shows an example of He FIB system which has recently been developed[4]. For the beam alignment, gimbal assembly is used. The optics system is composed of two electrostatic lenses. The probe radius of this system is shown in Fig. 3 as a function of beam half angle. A few nanometer sized beam can be expected to be generated which is difficult to generate by using liquid metal ion sources. As He field ion source has very low energy spread and small source size, the beam size is determined by spherical aberration. This is quite different from conventional liquid metal ion source systems.

Using focused ion beams, various maskless fabrication techniques such as scanning ion beam lithography, ion beam assisted etching and deposition, and implantation have been investigated. All of these works make use of advantages of focused ion beam technology, that is, flexibility, ease in lateral doping profiles and capability of direct submicron structures.

Scanning ion beam lithogrphy[12-14] is one of promising applications of FIB. Ion beam suffers negligible scattering compared with electron beams and consequently shows negligible proximity effects. This is important for future VLSI fabrication.

Maskless etching, deposition and ion implantation realizes process simplification. Mask and circuit repair and testing, or circuit reconstruction and adjustment[15-17] are getting important because of increasing complexity and scale of integration of VLSI devices. These local adjustment technology is one of important application and are difficult with techniques other than focused ion beams.

Many device applications of maskless implantations have been reported including fabrication of submicron channel MOS and EPROM[18,19]. All of these applications make use of advantages of focused ion beams, i.e., capability of maskless

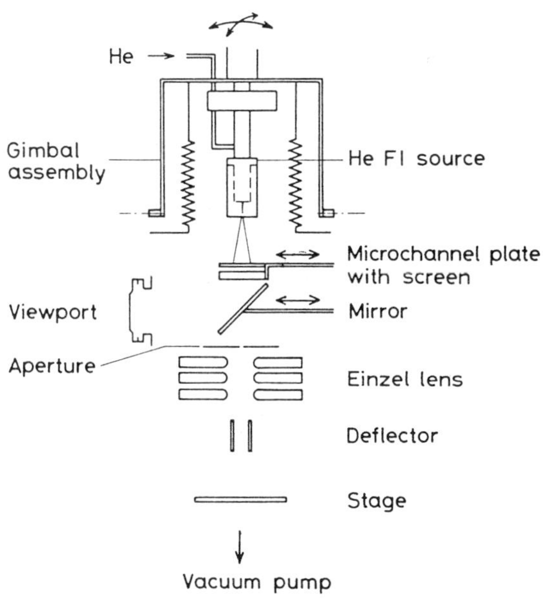

Fig.2 A schematic representation of the optical column for He FIB.

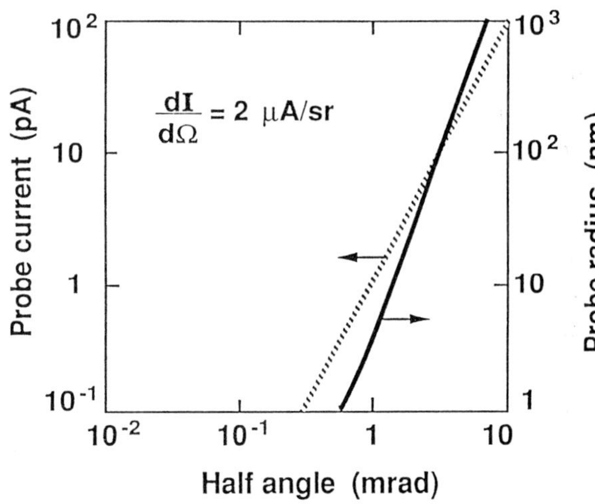

Fig. 3 Simulated probe radius and probe current as a function of a beam half angle.

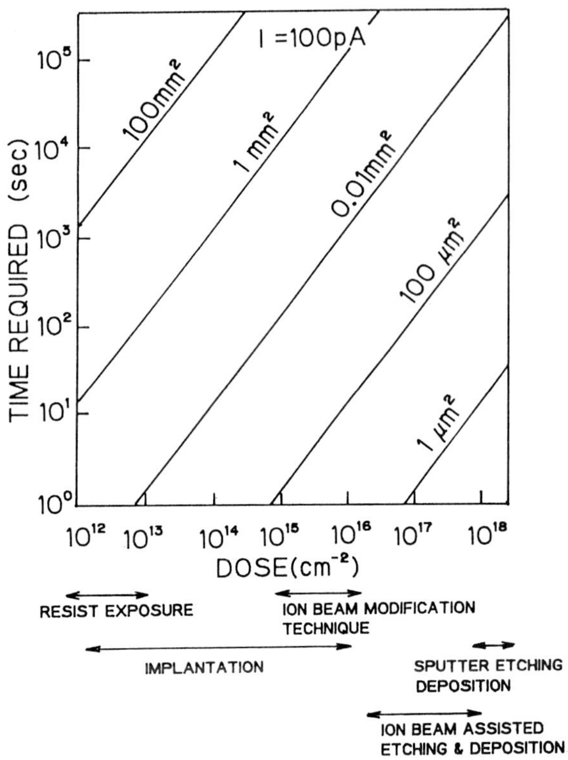

Fig. 4 Processing time required for various treatment area as a function of the dose. The beam current is assumed to be 100pA.

fine line formation and ease of change in local doping concentration.

There is an increasing interest in applications of FIB processing to nanofabrication or fabrication of various quantum effect devices. Here, fabrication techniques which can define submicron or smaller lateral dimensions with high reliability and flexibility without inducing surface or interface contamination are required. FIB techniques fulfill these requirements.

3. APPLICATIONS OF FIB

Figure 4 shows the time required to process various treatment area as a function of dose using FIB with a current intensity of 100pA. The doses required for various processing steps are also shown. Most focused ion beams give very high current density of $1A/cm^2$ or higher, but a total current intensity is low and ranges from 10pA to several nA. Ion beam lithography is one of promising application because very low dose is enough to delineate fine patters in resist layer. The high current intensity makes FIB processing very attractive for processing or modifying local area. For example, sputter etching of an area of $1m^2$ takes only about 10 sec, while the same
to process process takes about 1000 sec by conventional ion beam system which has a current density of $100A/cm^2$. Therefore, FIB find very important applications for mask repair, VLSI circuit testing and repair and many interesting works have been done recently. However, the low current intensity limits applications of focused ion beams. To obtain high current, liquid metal ion sources with low energy spreads are desirable. Shaped beam systems may be another promising alternative[20]. Using this technique, more than one order of magnitudes larger current intensity than the present focused ion beam system will be obtained.

3.1 Ion beam lithography

Ion beam lithography is new potential means to delineate patterns with deep submicron or smaller dimensions. It provides means with very high resolution and high sensitivity. Pattern delineation using mask has also been investigated but thin membrane masks are very fragile and require advanced microfabrication technology. Focused ion beam lithography removes this difficulty.

Figure 5 shows an example of 260nm thick PMMA resist pattern delineated by 25keV focused He ion beam[21]. In this example, two narrow lines adjacent to large pad pattern are separately delineated leaving 140 and 70nm gaps in-between. It can also be noticed that the line width is unchanged between regions adjacent and separated to the large pad. The same resist pattern

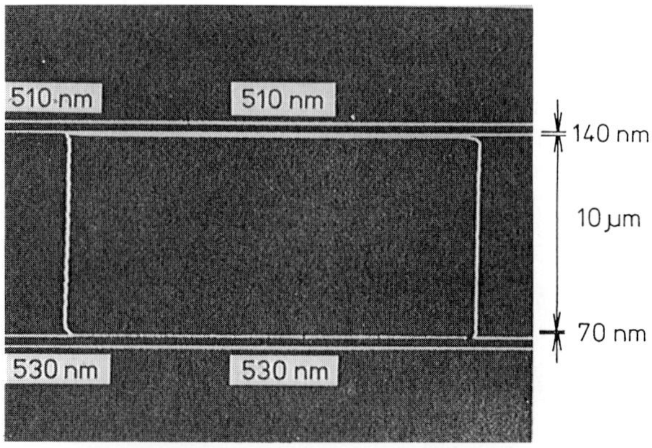

Fig. 5 PMMA pattern delineated by 25 keV He FIB.

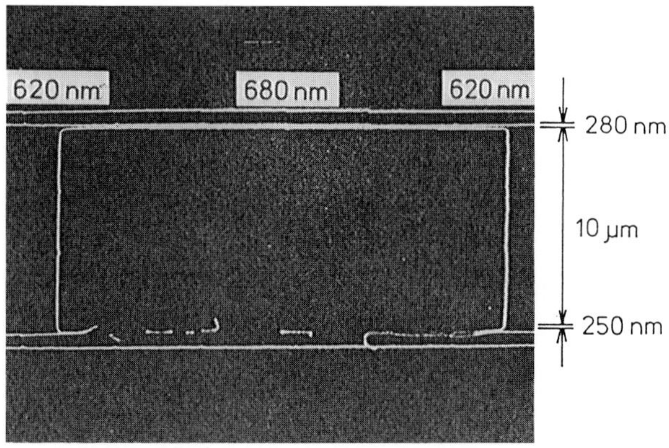

Fig. 6 PMMA pattern delineated by 30 keV electron beam

delineated by 30 keV electron beam is shown in Fig. 6 for comparison. For electron beam exposure, lines separated by 280nm were delineated although the line width was broader at the region adjacent to the pad and lines separated 250nm could not be resolved. These results clearly indicate that ion beam lithography suffers very low proximity effect and has very high resolution compared with electron beams.

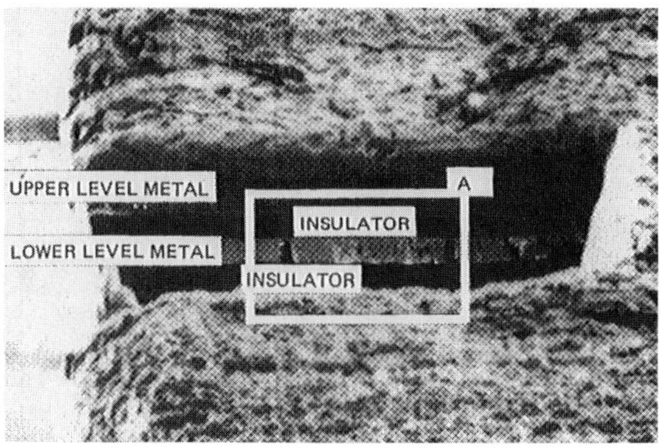

(a)

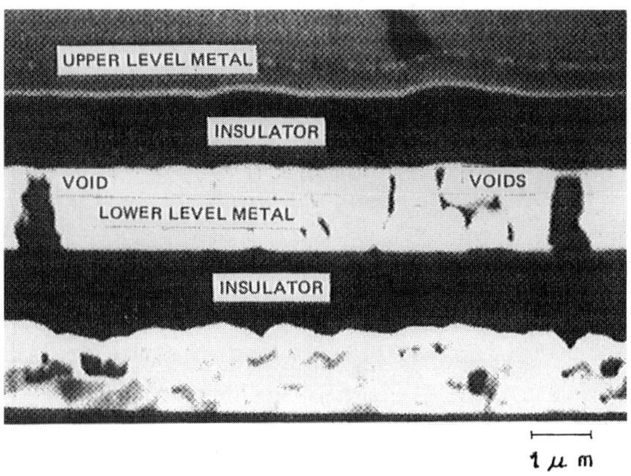

(b)

Fig. 7 Voids induced by electromigration and observed by Ga FIB.

3.2 Circuit surgery and micromachining

Figure 7 shows an example of failure analysis of VLSI memory devices[16]. The device was subjected to high temperature bias test and failed electrically. The failure occurred in the lower level aluminum lines due to electromigration. By other means such as metallurgical microscope, for example, no anomalies will be observed. However, it is possible to observe the lower level layers using focused ion beams as shown in Fig. 7. In the present example, the sample was first etched using Ga focused ion beams to form the grove and then the cross-section of the etched grooves were observed in-situ by seconodary electron image. Figure 7(a) shows the observed secondary electron image and Figure 7(b) is the image of the area A in Fig 7(a) observed at higher magnification. In this example, 100nm

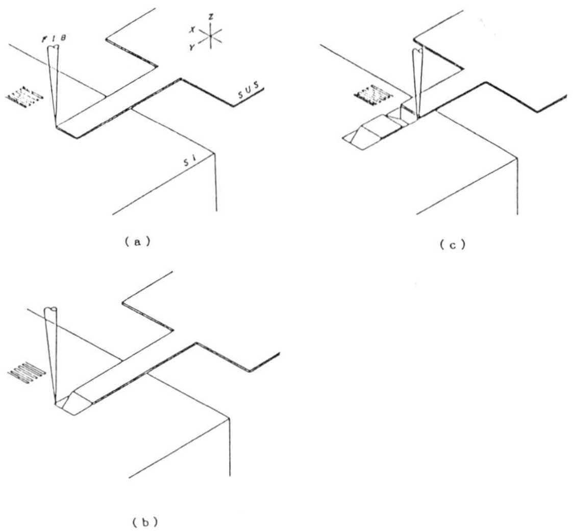

Fig. 8 Transplantation of devices using FIB.

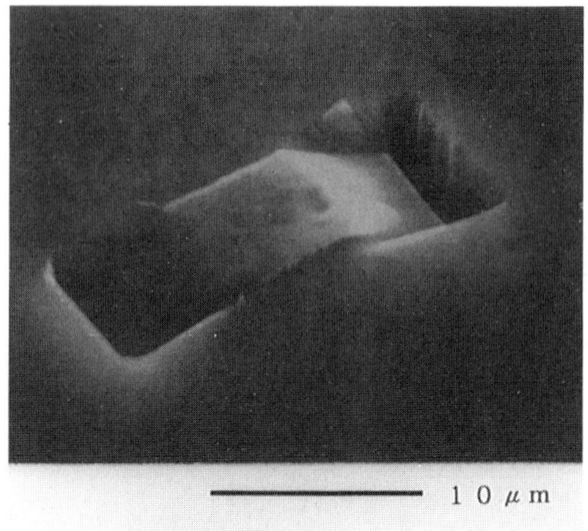

Fig. 9 Scanning electron microscope of transplanted dummy device.

or smaller voids formed by electromigration are detected as shown in Fig. 7(b). At present days, fine focused beams are available and a resolution better than 30nm will be possible.

Figure 8 shows another interesting application of local modification performed by focused ion beams[17]. Focused ion beam is used to transplant devices for circuit repair or assembly by using sputtering, redeposition and ion beam assisted deposition. In Fig. 8, a dummy device is transplanted to demonstrate the feasibility. In Fig. 8(a), a dummy device, which is 2m thick stainless steel strip is first transferred to the desired position on the Si substrate using piezoelectric manipulator. Next, the top sidewall of the dummy device is fixed to the Si substrate by redeposition as shown in Fig. 8(b) and finally, the dummy device is cut out by sputter etching and fixed the bottom side-wall by redeposition as shown in Fig. 8(c). The cutting and deposition may be also performed by using ion beam assisted etching and deposition if higher processing speed is necessary.

Figure 9 shows the scanning electron microscope image of transplanted stainless steel dummy tip on Si substrate. The redeposition formed slopes toward the top edges of the dummy device and good fixation to the substrate.

3.3 Nanofabrication

Nanofabrication using focused ion beam is one of interesting subjects. This is because FIB has many advantages for applications to nanofabrication. The advantages are as follows. First, the nanofabrication process using FIB is flexible because no mask step is necessary. This is especially important for exploratory nanostructure study. Second, clean and contamination free process is possible by combining various maskless process techniques. Third, nanostructures with a dimension smaller than 0.1m are easily fabricated.

Techniques applicable for nanostructure fabrication are doping, lithography, etching, disordering, damage effect and in-situ process techniques. For example, ballistic point contact, quantum wires and dots were fabricated by using these techniques[22-25].

Figure 10 shows GaAs gratings with very short period which were fabricated by using FIB[26]. The short period grating is very important for surface layer superlattice devices, quantum

(a) (b)

Fig. 10 Grating on GaAs surface formed by radiation enhanced etching (a) and enlarged view (b). 100keV FIB Ga was irradiated at a dose of $2.2X10^8$/cm.

wires and other quantum devices. Here, these gratings are fabricated by making use of radiation enhanced etching. The fabrication process is as follows. First, 100keV Ga FIB was irradiated to induce radiation damage with the grating pattern and next, the irradiated region was removed by etching using HCl and H_2O_2 etchant. Gratings with a period of 41nm were hard to form because of the spreading of the beam but those with 63nm were formed as shown in Fig. 10. No characteristics of the grating has been measured yet but it can expect that the residual damage should be few because the damaged region is selectively removed by this process. This is very attractive point for nanometer device fabrication.

4. Summary

Recent works on focused ion beam research in Japan were reviewed. It seems that FIB finds important applications to local modification such as circuit repair, modification and testing. This is because FIB has very high current density and fine focused probe. Nanofabrication is also an important application of FIB. This is because it realizes very simple means for nanofabrication. This was already demonstrated by fabricating ballistic point contact devices or GaAs gratings with 60nm period gratings on GaAs surface. Current intensity is still low and much improvement is required though this is still difficult.

REFERENCES

1. Seliger, R.L., Ward, J.W., Wang, V. and Kubena, R.L., Appl. Phys. Lett. **34** (1979) 310.
2. D. H. Narum and R. Fabian Pease: J. Vac. Sci. Technol. **B6** (1988) 966.
3. H. Kasahara, H. Sawaragi, R. Aihara, K. Gamo and S. Namba: J. Vac. Sci. Technol. **B6** (1988) 974.
4. T. Itakura, K. Horiuchi, S. Yamamoto: Microelectronic Eng. 3 (1985) 153.
5. H. Sawaragi, H. Kasahara, R. Aihara and M.H. Shearer: J. Vac. Sci. Technol. **B6** (1988) 958.
6. R. Aihara, H. Sawaragi, H. Morimoto, H. Hosono, Y. Sasaki, T. Kato and M. H. Shearer: J. Vac. Sci. Technol. **B6** (1988) 245.
7. T. Shiokawa, P. H. Kim, K. Toyoda, S. Namba, K. Gamo, R. Aihara and N. Anazawa: Jpn. J. Appl. Phys. 24 (1985) L566.
8. M. Yamamoto, M. Sato, H. Kyoogoku and K. Aita: Proc. of SPIE, Electron Beam, X-ray and Ion Beam Techniques for Submicrometer Lithog-

raphy, Vol.632 (1986) p.99.
9. Y. Nihei, H. Satoh and M. Owari: J. Vac. Sci. Technol. **B6** (1988) 915.
10. Y. Kawanami, T. Ishitani and K. Umemura: Nucl. Inst. Methods **B37/38** (1989) 240.
11. R. Levi-setti, Y.L. Wang and G. Crow: J. Phys. 45 (1984) C9-197.
12. 3. T. Shiokawa, Y. Aoyagi, Pil Hyon Kim, K. Toyoda and S. Namba: Jpn. J. Appl. Phys. **23** (1984) L232.
13. 4. Y. Sasaki, H. Morimoto, H. Onodera, M. Komaru, M. Kobiki, Y. Mitsui, O. Ishihara and T. Kato: Proc. Int. Symp. GaAs and Related Compounds, Inst. Phys. Conf. Ser. No.79 (1986) 487.
14. 5. Shinji Matsui, Yoshikatsu Kojima and Yukinori Ochiai: Appl. Phys. Lett. **53** (1988) 868.
15. K. Saito, H. Onoda, H. Morimoto, T. Katayama, Y. Watakabe and Kato: J. Vac. Sci. Technol. **B6** (1988) 1032.
16. K. Nikawa, K. Nasu, M. Murase, T. Kaito, T. Adachi and S. Inoue: Proc. 1989 International Reliability Physics Symposium 1989 p.16.
17. Tsuyoshi Ohnish, Yoshimi Kawanami and Tohru Ishitani: Jpn. J. Appl. Phys. **29** (1990) 188.
18. Shoji Shukuri, Yasuo Wada, Masao Tamura, Hiroo Masuda and Tohru Ishitanni: Extended Abstracts of 16th Conf. Solid State Devices and Materials (Jpn. Soc. Appl. Phys., 1984) p.91.
19. S. Shukuri, Y. Wada, T. Hagiwara, K. Komori and M. Tamura, Extended Abstracts of 18th Conf. Solid State Devices and Materials (Jpn. Soc. Appl. Phys., 1986) p.327.
20.T. Itakura, K. Horiuchi and S. Yamamoto: K. Horiuchi, T. Itakura and H. Ishikawa, J. Vac. Sci. Technol. B6 (1988) 241.
21. H. N. Slingerland, J. H. Bohlander, K. D. v. d. Mast and E. Koets: Microelectronic Eng. 5 (1986) 155.
22. Yoshihiro Hirayama and Tadashi Saku: Jpn. J. Appl. Phys. **29** (1990) L368.
23. Toshio Hiramoto, Kazuhiko Hirakawa, Yasuhiro Iye and Toshiaki Ikoma: Appl. Phys. Lett. **54** (1989) 2103.
24. Y. Hirayama and H. Okamoto: J. Vac. Sci. Technol. **B6** (1988) 1018.
25. F. Wakaya, T. Kakuta, Y. Takagaki, Y. Yuba, S. Takaoka, K. Murase, T. Shiokawa, K. Gamo and S. Namba: J. Vac. Sci. Technol. (to be published).
26. T. Shiokawa, K. Ishibashi, P.H. Kim, Y. Aoyagi, K. Toyoda and S. Namba: 1990 Intern. MicroProcess Conference (July 16-19, 1990, Makuhari, Japan) (to be published in Jpn. J. Appl. Phys.).

Frontiers of Materials Research/
Electronic and Optical Materials
M. Kong, L. Huang (editors)

NEW DEVELOPMENTS IN INTERMETALLIC COMPOUNDS IN WEST GERMANY

Gerhard SAUTHOFF

Max-Planck-Institut für Eisenforschung GmbH, Max-Planck-Straße 1, D-4000 Düsseldorf

In West Germany several alloy systems which are based on intermetallics are regarded as promising for new materials developments, and respective developments have been initiated. The present work which is centered on γ-TiAl, Ti_5Si_3-based alloys, Mg_2Si-Al, NiAl-Cr, Al_3Nb-NiAl and Laves phase-based alloys is overviewed, and the mechnical properties - strength, ductility and/or toughness - are described.

1. INTRODUCTION

In view of their outstanding mechanical properties intermetallic phases are regarded as highly promising for applications as structural materials at high temperatures (e.g. [1,2,3], and various materials developments on the basis of intermetallic phases are under way, or are being started (e.g. [4,5,7]).

After 1985 research programs were established in West Germany for promoting the fundamental research on intermetallic phases and for initiating new materials developments. The primary aim of the German activities has been the screening of various intermetallic phases to explore the possibilities for new materials developments. The objective has been and still is the development of new lightweight materials on one hand where the low density is the first prerequisite, and of high-temperature high-strength materials with applications above 1000 °C on the other hand where the high-temperature strength and the creep strength should be as high as possible. The recent work has led to new projects which are to make use of promising candidate phases, and some of these new developments are overviewed in the following. It has to be noted that most of the respective materials contain more than one phase, i.e. the particular intermetallic phase is combined with another metallic or intermetallic phase to have a good combination of strength and toughness. These two-phase or multiphase materials are called intermetallic alloys in this overview.

2. LIGHTWEIGHT ALLOYS

2.1. TiAl

As in other countries the γ-TiAl phase has been selected for materials developments because of its low density. Various projects are under way or beeing started to develop TiAl-based materials for specific applications. The present state is illustrated by Figure 1. Here the temperature dependence of strength and fracture strain is shown for a cast TiAl alloy with 45 at.% Al (prepared by skull melting) after various heat treatments [8,9]. Obviously the aging treatments of this alloy which is two-phase with Ti_3Al as second phase reduce both the strength and ductility. Fracture toughness (K_{IC}) is between 12 and 23 MN $m^{-3/2}$ [9,10,11]. The further work aims at increasing the strength by varying the composition, alloying with further elements and thermomechanical treatments. Alloying with Cr not only increases the hardness, but also the oxidation resistance [8,9].

Other work on TiAl is directed at studying the effects of processing on the microstructure with respect to alloy optimization [12]. Additions of third elements affect the microstructure in a sensitive way, and Mn can be used

a)

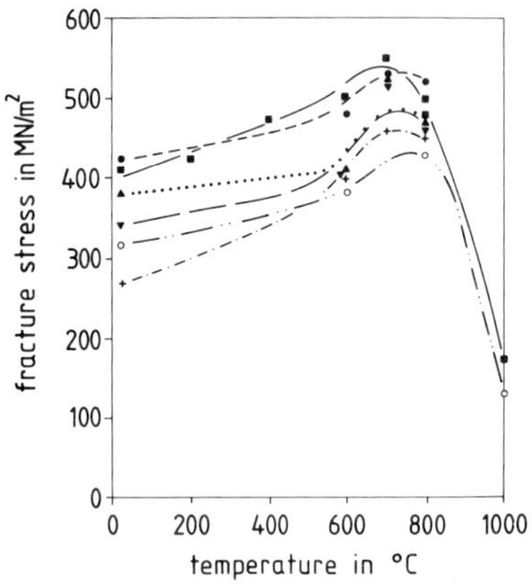

b)

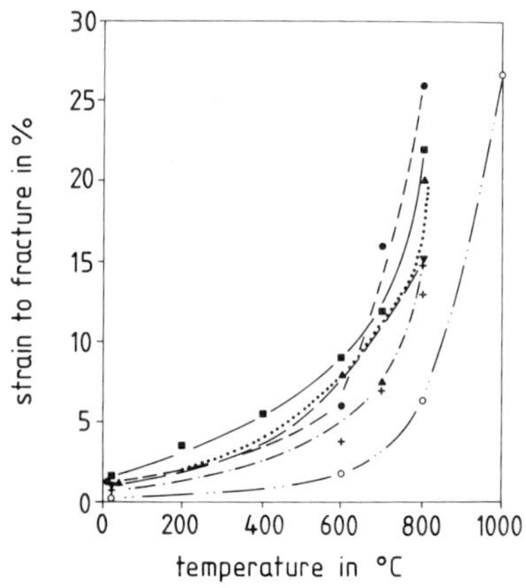

FIGURE 1a and b
Fracture stress (a) and strain to fracture (b) (both in tension with 1.7 x $10^{-7}s^{-1}$ deformation rate in argon) [8,9]:

■ as cast
▲ 168 h at 600°C in air,
▼ 336 h at 650°C in air,
\+ 168 h at 700°C in air,
• 3 h at 1000°C in argon,
○ 168 h at 1000°C in argon.

besides Nb for improving the ductiliy [13]. Hot working of TiAl induces dynamic recrystallization by which the microstructure - including texture - and the mechanical behaviour can be controlled [14].

2.2. Ti_5Si_3 - Ti - Al

Ti_5Si_3 is very appealing because of its low density and its very high melting temperature of 2130 °C, but this phase is very brittle, too. There is the possibility to combine this phase with another less brittle phase in order to obtain a two-phase intermetallic alloy with relieved brittleness. Indeed Ti_5Si_3 forms a eutectic equilibrium not only with Ti [15] but also with the three titanium aluminides [16]. Ti_3Al - Ti_5Si_3 has been studied with respect to the effects of composition and processing on microstructure and porperties, and it has been found that such alloys show promising hightemperature mechanical properties [17,18]. Representative data are shown in Figure 2, and the fracture toughness of such alloys is between 8 and 13 $MNm^{-3/2}$ [18]. Now the effect of Nb additions on the alloy properties is studied [20].

Other Ti_5Si_3-base alloys are studied for obtaining good combinations of strength and toughness by applying special alloy preparation methods [21]. A directionally solidified eutectic Ti_5Si_3-Ti alloy shows some deformability in compression even at room temperature, whereas single-phase Ti_5Si_3 can be deformed plastically only at 1000 °C or above. A Ti_5Si_3-Al_3Nb alloy with 8 wt.% Al (prepared by hot isostatic pressing of ground Ti_5Si_3 and Al powder) is characterized by a creep strength which is lower than that of Ti_5Si_3 by an order of magnitude [21] corresponding to a rule of mixtures.

2.3. Mg_2 Si-Al

Mg_2Si with cubic C1 structure has a density of only 2 Mg/m^3, and its melting temperature is nearly 1100 °C. Thus it is regarded as promising candidate phase for lightweight mate-a)

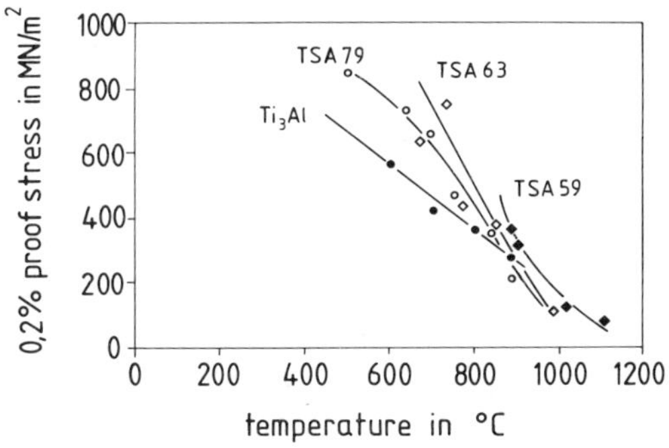

b)

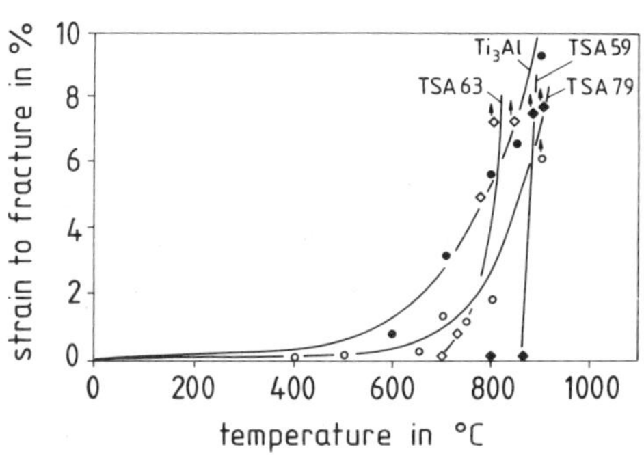

FIGURE 2a and b
Strength (a) and strain to fracture in bending (b) as a function of temperature for the Ti_3Al-Ti_5Si_3 alloys TSA 59 (29 at.% Al, 6.5 at.% Si), TSA 63 (23.5. at.% Al, 6.5 at.% Si) and TSA 79 (20.5 at.% Al, 8.5 at.% Si)[18] in comparison to those for Ti_3Al [19].

a)

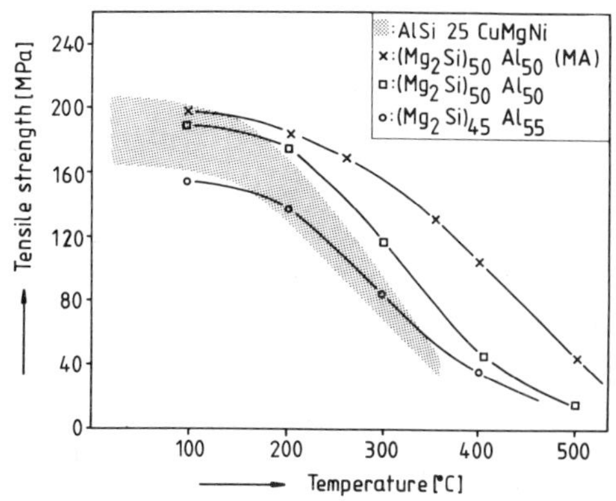

b)

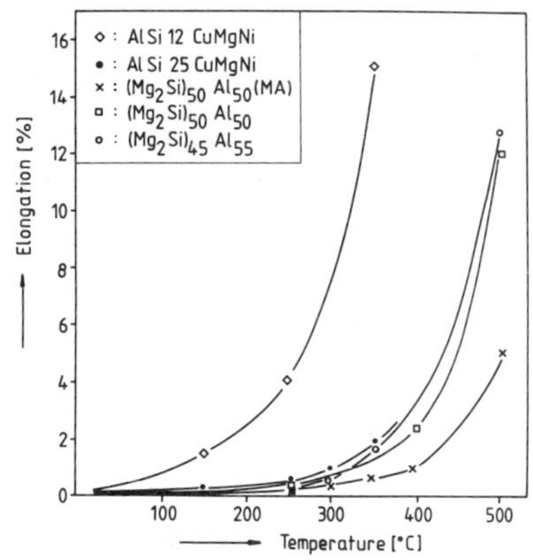

FIGURE 3a and b
Tensile strength (a) and tensile fracture strain (b) as a function of temperature for the two-phase alloys $(Mg_2Si)_{50}Al_{50}$ (mechanically alloyed), $(Mg_2Si)_{50}Al_{50}$ Ccast) and $(Mg_2Si)_{45}$ Al_{45} (cast) in comparison to the commercial cast Al alloy AlSi25CuMgNi [23].

rials, and it has been selected for a materials development which aims at applications in car engines [22]. Since the single-phase Mg_2Si is too brittle the development is centered on two-phase Mg_2Si-Al alloys. Strength and deformability of these alloys is illustrated by Figure 3 according to which these alloys indeed show a significantly higher strength than commercial Al alloys at 300 °C and above, but there is also a sensible reduction of deformability [23].

3. HIGH-TEMPERATURE HIGH-STRENGTH ALLOYS

New materials are to be developed for applications above 1000 °C. Cand-idate phases should contain Al for providing the necessary oxidation resistance and for having a not too high density, and the melting temperature should be sufficiently high for having a high creep strength which scales with the diffusion coefficient [24]. NiAl with B2 structure has an appreciably higher melting temperature, a lower density and a better oxidation resistance than Ni_3Al. However, it could not be "ductilized" at room temperature, and it shows a poor creep strength in spite of the high melting temperature [24]. In order to increase its strength NiAl is combined with other phases as shown in the following.

3.1. NiAl-Cr

NiAl forms eutectics with Cr, Mo, W which may be used to produce composite materials by directional solidification [25], and indeed the NiAl-Cr composite shows a promising high-temperature strength [26]. Such alloys are now studied intensively in order to develop high-temperature high-strength alloys [27,28]. For the preparation of alloys, both powder metallurgy (PM) and ingot metallurgy - in particular directional solidification - is used. Figure 4 shows compressive strength data for PM NiAl-Cr alloys according to which the strength increases with increasing volume fraction of the α-Cr-phase in these two-phase alloys. An even higher strength

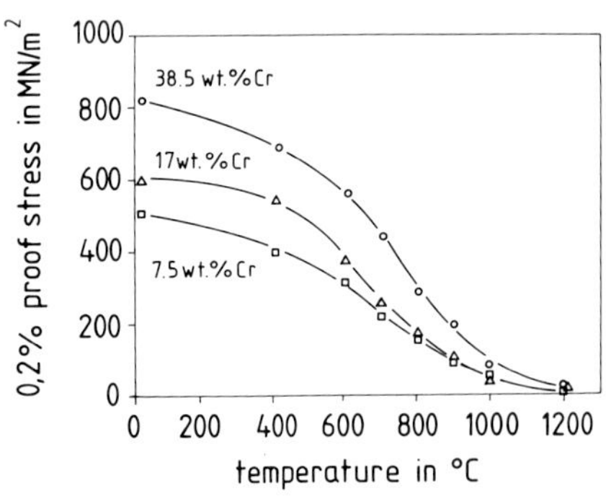

FIGURE 4
Temperature dependence of strength (in compression) for two-phase PM NiAl-Cr alloys with 7.5 wt.% Cr, 17 wt.% Cr, and 38.5 wt.% Cr (eutectic composition), respectively, as HIPed [27,28].

is obtained with cast alloys with much coarser grain structure. The fracture toughness K_{IC} is between 8 and 10 MN $m^{-3/2}$ depending on processing, i.e. the low toughness can be improved by optimizing the alloy with respect to composition and processing.

3.2. Al_3Nb-NiAl

Al_3Nb melts at 1680 °C [29], i.e. the melting temperature is much higher than that of Ni_3Al and even higher than that of NiAl. The DO_{22} structure of Al_3Nb is derived from the $L1_2$ structure of Ni_3Al by further ordering with tetragonal distortion [24]. Thus Al_3Nb should be significantly stronger than both nickel aluminides NiAl and Ni_3Al. However, it is rather brittle and it was regarded as not useful for high-temperature applications [30].

Recently Al_3Nb and Al_3Nb-base alloys have been studied in detail, and possibilities for materials developments have been found. Al_3Nb can be alloyed with the lightweight DO_{22} phase Al_3Ti to form single-phase ternary alloys $Al_3(Nb,Ti)$, and Al_3Nb can be alloyed with NiAl to form two-phase Al_3Nb-NiAl alloys. Figures 5 and 6 show the temperature dependence of strength of the resulting alloys, always compared with the respective binary phases. The binary Al_3Nb shows the highest strength at temperatures above 1000 °C, but there is practically no plasticity at lower temperatures. The alloying with the other phases extends the range of deformability to significantly lower temperatures whereas the high-temperature strength is affected only to a small degree [31]. In a separate study of NiAl-Fe alloys [33] it was shown that strength and in particular creep strength can be described with a rule of mixtures as long as the phase distribution is sufficiently coarse whereas very fine phase distributions result in an extra threshold stress for creep (see [24]).

The oxidation behaviour of these Al_3Nb-based alloys has been studied, too, in some detail,

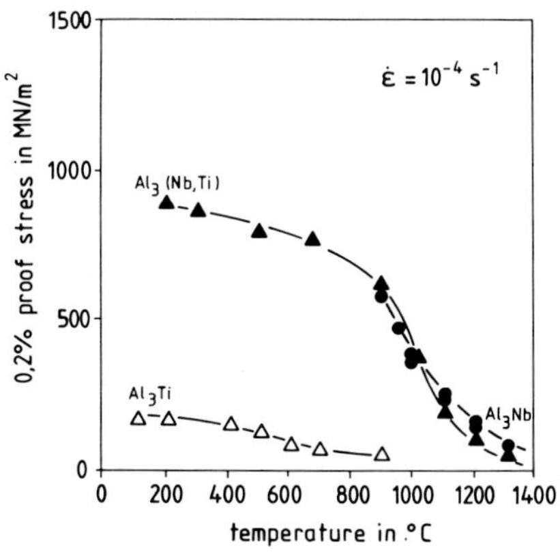

FIGURE 5
Temperature dependence of the 0.2 % proof stress ($10^{-4}s^{-1}$ strain rate in compression) for stoichiometric Al_3Nb, $Al_3(Nb_{0.75}Ti_{0.25})$ [24], and Al_3Ti[32], all with DO_{22} structure.

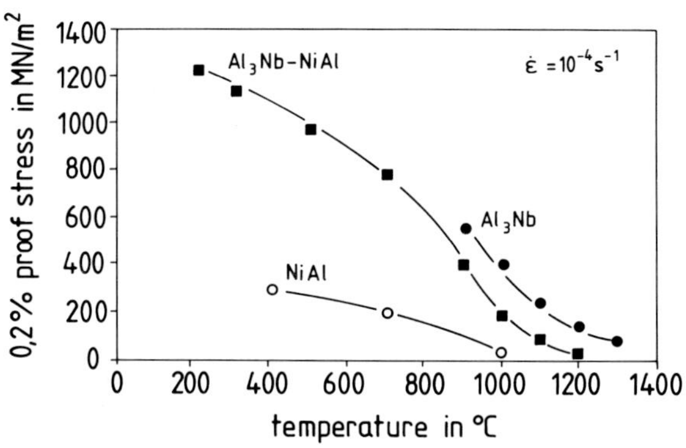

FIGURE 6
Temperature dependence of the 0.2 % proof stress in compression ($10^{-4}s^{-1}$ strain rate) for stoichiometric B2-NiAl , DO_{22} Al_3Nb and three-phase Al_3Nb-NiAl (with 22 vol.% NiAl and 4 vol.% C14-NbNiAl) [24].

and it was found that the good oxidation resistance of the binary NiAl can nearly be reached by proper adjustment of the volume fractions of the respective phases [34].

3.3. NbNiAl-NiAl

The high-temperature strength of NiAl with Cr or Al_3Nb only reaches or slightly surpasses that of the superalloys. For high-temperature strength above that of the superalloys, one has to make use of still stronger intermetallic phases. NbNiAl is a ternary Laves phase with C14 structure [35]. It is significantly stronger than Al_3Nb, but also more brittle, and specimens of good quality can be prepared only by powder metallurgy [36]. However, NbNiAl forms a stable equilibrium with NiAl, and the resulting NbNiAl-NiAl alloys can be produced by vacuum induction melting. The brittleness of these two-phase alloys is reduced to a tolerable level, and thus these alloys have been studied in some detail with respect to eventual high-temperature applications.

Figure 7 shows the strength of NbNiAl-NiAl alloys at 1200 °C as a function of the volume fraction of Laves phase which forms a continuous skeleton in these alloys, i.e. the Laves phase plays the role of the matrix even in the NbNiAl-poor alloys. The bending strength does not increase with the NbNiAl volume fraction as the compressive strength at the higher volume fractions which may be due to alloy defects - pores and microcracks. The fracture toughness of these alloys at room temperature is 2-4 $MNm^{-3/2}$, and the brittle-to-ductile transition temperature is between about 500 °C and 700 °C depending on composition and processing [36].

In the transition-temperature range cracks in the Laves phase are stopped at the Laves phase-NiAl interface by producing a plastic zone in the NiAl phase [36, 37]. Other and still harder ternary Laves phases with Al are NbFeAl, TaFeAl, TaNiAl [35], which form stable equilibria with the respective B2 aluminides FeAl and

NiAl, too. Figure 8 shows the high temperature strength of such phases and of such tow-phase alloys. The resulting strength-temperature range is indeed above that of the superalloys.

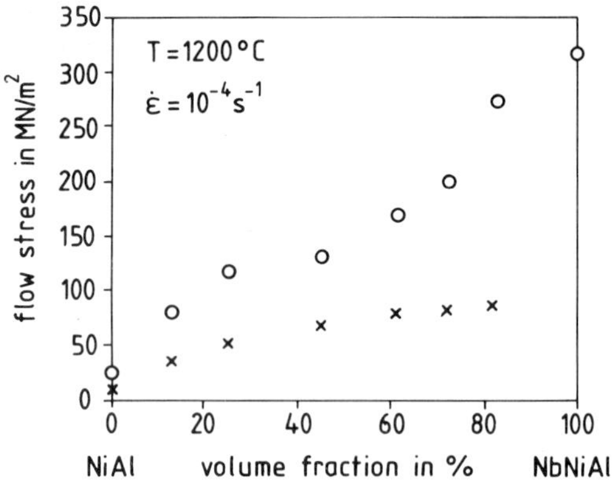

FIGURE 7
Flow stress of two-phase NbNiAl-NiAl as a function of the volume fraction of C14-NbNiAl at 1200 °C in compression (o: 0.2 % proof stress) and in bending (x) [24].

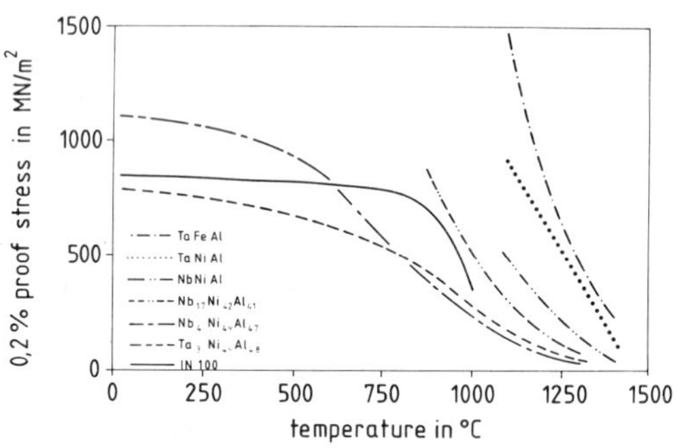

FIGURE 8
Temperature dependence of the 0.2 % proof stress in compression ($10^{-4}s^{-1}$ strain rate) for the Laves phases (single phase) TaFeAl, TaNiAl, NbNiAl and the alloys $Nb_{17}Ni_{42}Al_{41}$ (NbNiAl-39 vol.% NiAl), $Nb_4Ni_{49}Al_{47}$ (NbNiAl-87 vol.% NiAl), $Ta_3Ni_{49}Al_{48}$ (NiAl-12 vol.% TaNiAl) compared with the superalloy IN 100 (tension data) [36].

However, these alloys compare less favourably with the superalloys with respect to long-term behaviour, i.e. creep strength. Figure 9 shows the limited creep strength range of some of these alloys, and it can be seen that the weak NiAl is strenghtened significantly by the Laves phases. However, the NbNiAl-NiAl data do not follow a simple rule of mixtures, i.e., the creep strength depends sensitively on the phase distribution which is produced by the alloy processing. Correspondingly the creep strenght of the eutectic NiAl-Cr alloy compares favourably with NbNiAl-NiAl in spite of the very low creep resistance of pure α-Cr (about 2 MN/m² [38]) compared with that of the Laves phases. For the optimization of the Laves phase-based alloys, phase composition and phase distributions are varied to obtain higher creep strengths with still tolerable brittleness. From the presented results it has been concluded that there are good prospects for developing intermetallic alloys with high creep strengths at temperatures above the application temperatures of superalloys.

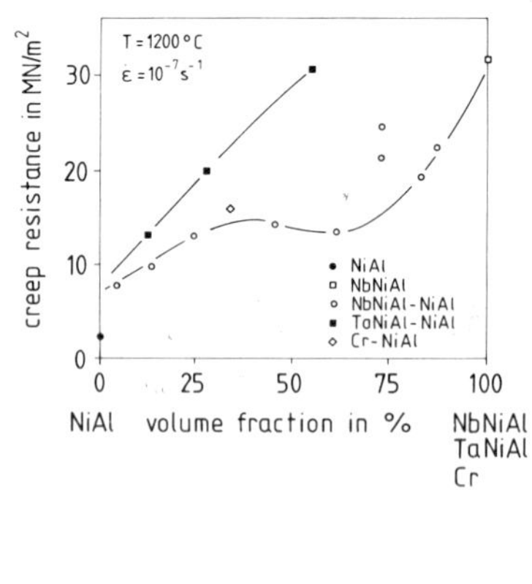

FIGURE 9
Creep resistance (stress for $10^{-7}s^{-1}$ secondary creep rate) at 1200 °C as a function of volume fraction of second phase phase for NiAl-NbNiAl, NiAl-TaNiAl and NiAl-Cr [36].

ACKNOWLEDGEMENTS

The financial support by the Bundesministerium für Forschung und Technologie (BMFT) and the Deutsche Forschungsgemeinschaft (DFG) is gratefully acknowledged.

REFERENCES

1. P.R. Munroe and I. Baker, Metals and Materials 4 (1988) 435.

2. J.D. Destefani, Advanced Materials and Processes 2/89 (1989) 37.

3. G. Sauthoff, Z. Metallkde. 80 (1989) 337-344.

4. High-Temperature Ordered Intermetallic Alloys, eds. C.C. Koch, C.T. Liu, and N.S. Stoloff (Mat. Res. Soc., Pittsburgh, 1985).

5. High-Temperature Ordered Intermetallic Alloys II, eds. N.S. Stoloff, C.C. Koch C.T. Liu, and O. Izumumi (Mat. Res. Soc., Pittsburgh, 1987).

6. High-Temperature Ordered Intermetallic Alloys III, eds. C.T. Liu, A.I. Taub, N.S. Stoloff, and C.C. Koch (Mat. Res. Soc., Pittsburgh, 1989).

7. G. Sauthoff, Z. Metallkde. 77 (1986) 654.

8. T. Kremser, G. Frommeyer, and K.-H. Matucha, Mechanische Eigenschaften schmelzmetallurgisch hergestellter TiAl-Basislegierungen, poster presented at the DGM 1989 Annual Meeting in Karlsruhe.

9. T. Kremser and G. Frommeyer, to be published in Z. Metallkde. 81 (1990).

10. S. Reuss and H. Vehoff, Bruchzähigkeiten an ausgewählten intermetallischen Phasen, poster presented at the DGM 1989 Annual Meeting in Karlsruhe.

11. S. Reuss and H. Vehoff, Scripta Met. 24 (1990) 1021.

12. C. Hartig, S. Chen, P.A. Beaven, and H. Fukutomi, Mechanical Properties of Powder Metallurgy and Ingot Metallurgy Intermetallic TiAl Alloys, in: Proc. Sixth World Conf. on Titanium, Part II, eds. P. Lacombe, R. Tricot, and G. Beranger (Les Editions de Physique, Les Ulis Cedex, France, 1989) pp. 1021-1026.

13. B. Dogan, R. Wagner, and P.A. Beaven, Fracture Behaviour of TiAl(Mn), presentation at the 1989 TMS Fall Meeting in Indianapolis, to be published in Scripta Met. 24 (1990).

14. H. Fukutomi, C. Hartig, and H. Mecking, Z. Metallkde. 81 (1990) 272.

15. F.W. Crossman and A.S. Yue, Metall. Trans. 2 (1971) 1545.

16. O. Schob, H. Nowotny, and F. Benesovsky, Planseeberichte f. Pulvermetallurgie 10 (1962) 65.

17. J.S. Wu, P.A. Beaven, R. Wagner, C. Hartig, and J. Seeger, Microstructure and Mechanical Properties of Dual Phase Alloys Consisting of the Intermetallic Phases $Ti_3(Al,Si)$ and $Ti_5(Si,Al)_3$, in ref. 6, pp. 761-766.

18. P.A. Beaven, J.S. Wu, B. Dogan, C. Hartig, J. Seeger and R. Wagner, Entwicklung von intermetallischen Verbindungen für neue Hochtemperaturwerkstoffe, in GKSS-Jahresbericht (Geesthacht, 1989) pp. 49-61.

19. H.A. Lipsitt, D. Shechtman, and R.E. Schafrick, Metall. Trans. 11 A (1980) 1369.

20. J.S. Wu, D. Chen, P.A. Beaven, R. Wagner, and J. Seeger, Microstructure and Properties of Two-Phase Alloys Based on $(Ti,Nb)_3(Al,Si)$ and $(Ti,Nb)_5(Si,Al)_3$, this volume.

21. G. Frommeyer, R. Rosenkranz, and C. Lüdecke, Z. Metallkde. 81 (1990) 307.

22. K.H. Matucha and E. Schmid, Warmfeste Leichtbauwerkstoffe auf Basis Intermetallischer Phasen, Presentation at the DGM Symposium "Intermetallische Phasen" 24.-25.11.1988 in Bad Nauheim.

23. K. von Oldenburg, G. Frommeyer, E. Schmid, and W. Henning, Microstructures and Mechanical Properties of As Cast and Mechanically Alloyed Mg_2Si-Al Alloys, in: Advanced Aluminium and Magnesium Alloys, eds. T. Khan and G. Effenberg (ASM Europe, 1990), in print.

24. G. Sauthoff, Mechanical Properties of Intermetallics at High Tem-peratures, in: High Temperature Aluminides and Intermetallics, eds. C.T. Liu et al. (ASM/TMS, Metals Park, 1990), in print.

25. H.E. Cline, J.L. Walter, E. Lifshin, and R.R. Russel, Metall. Trans. 2, (1971) 189.

26. J.L. Walter and H.E. Cline, Metall. Trans. 1 (1970) 1221.

27. G. Frommeyer, W. Kowalski, and C. Lüdecke, Plastizität und Festigkeit intermetallischer NiAl-Legierungen, poster presented at the DGM 1989 Annual Meeting in Karlsruhe.

28. W. Kowalski and G. Frommeyer, to be published in Z. Metallkde. 81 (1990).

29. R.L. Fleischer, J. Mat. Sci. 22 (1987) 2281.

30. D. Shechtman and L.A. Jacobson, Metall. Trans. 6A (1975) 1325.

31. C.-P. Reip and G. Sauthoff, Verformungsverhalten der intermetallischen Phase Al_3Nb, poster presented at the DGM 1989 Annual Meeting in Karslruhe.

32. M. Yamaguchi, Y. Umakoshi and T. Yamane, Phil. Mag. A55 (1987) 301.

33. J. Klöwer, Untersuchung des Kriechverhaltens gerichtet erstarrter Eisen-Nickel-Aluminium-Legierungen, Dr. Ing. thesis (RWTH Aachen, 1989).

34. M. Steinhorst, Oxidationsverhalten des Aluminids $NbAl_3$ und mehrphasiger Legierungen dem System Nb-Ni-Al, Dr. Ing. thesis (Universität Dortmund, 1989).

35. P. Villars and L.D. Calvert, Pearson's Handbook of Crystallographic Data for Intermetallic Phases (ASM, Metals Park, 1985).

36. L. Machon and G. Sauthoff, Verformungsverhalten zweiphasiger intermetallischer Legierungen der Laves-Phase NbNiAl und der B2-Phase NiAl, poster presented at the DGM 1989 Annual Meeting in Karlsruhe.

37. W. Wunderlich, L. Machon and G. Sauthoff, Versetzungsanalyse an Rißspitzen an der Phasengrenzfläche NiAl/NbNiAl in intermetallischen Ni-Al-Nb-Legierungen, presentation at the DPG 1990 spring meeting in Regensburg.

38. H.J. Frost and M.F. Ashby, Deformation Mechanism Maps (The Plasticity and Creep of Metals and Ceramics) (Pergamon Press, Oxford, 1982), ch. 5.

Frontiers of Materials Research/
Electronic and Optical Materials
M. Kong, L. Huang (editors)
Elsevier Science Publishers B.V., 1991

NON STOICHIOMETRY IN PEROVSKITE-TYPE OXIDES. INFLUENCE OF THE ELECTRONIC CONFIGURATION ON FORMATION OF EXTENDED DEFECTS

P. HAGENMULLER, M. POUCHARD and J.C. GRENIER

Laboratoire de Chimie du Solide du CNRS, Université de BORDEAUX I, 351 cours de la Libération, 33405 TALENCE Cedex (France)

In equilibrium conditions the defects appearing in perovskite-type and related oxides tend to align so as to minimize the free energy. For oxygen vacancies the extended defect structure depends largely on the electronic configuration of the B cations as shown on hand of appropriate examples. In the particular case of the superconducting copper oxides, the electronic delocalization mechanism is closely related to the type of layer structure.

After the failure of BERTHOLLET to refute PROUST'S law for solid crystals, the presence in the metallic phase diagrams of wide homogeneity domains shown by KURNAKOV's school and corresponding to monotonous variations of physical properties was largely considered before 1930 as a specific feature of metallurgy. Progressively accepted, at first for the palladium hydrides but as an extension of the classical metallic systems, later for $Fe_{1-x}O$, the concept of non stoichiometry asserted its fundamental character thanks to development of X-ray diffraction which pointed out the fundamental structural analogy between intermetallic and ionocovalent compounds. Its real birth certificate was actually the publication of SCHOTTKY and WAGNER in 1930 who introduced notions of statistical thermodynamics to account for the defects in real crystals |1|

It is convenient to distinguish in non-stoichiometric compounds SCHOTTKY defects, which represent either cation or anion vacancies and FRENKEL defects of interstitial nature, generally cationic by reason of size. The free energy change at constant volume ΔG results from a compensation between the increase of internal energy ΔH required by the formation of the defects and increasing entropy resulting from disorder enhancement ($\Delta G = \Delta H - T\Delta S$). The practical consequence is that the number of defects has to increase with temperature and can only theoretically be zero at 0 K.

If we consider N ion pairs, w_S being the energy required to remove either an anion or a cation from the bulk to the surface, one may show on hand of a thermodynamical treatment that the number of SCHOTTKY defects is :

$$n_S \simeq N \exp. (-w_S/2kT)$$

In a similar way the number of FRENKEL defects appears to be :

$$n_F \simeq NN^* \exp. (-w_F/2kT),$$

where N^* is the number of interstitial sites available and w_F the transfer energy .

Whereas compounds with highly electronegative anions (e.g. fluorides) do not have generally a measurable homogeneity range, in oxides all types of ionic defects may occur, sometimes simultaneously. Frequent occurence of electronic localization and existence of various oxidation states may be the main reasons of this diversity.

A notion which is quite important for understanding the behavior of oxides is the concept of defect pairing |2|. If for instance

w_{vv} is formally the interaction energy of two anion vacancies (w_{vv} is negative when the defects attract each other) one may show that there is a critical temperature $T_c = -w_{vv}/2k$ above which the vacancies are randomly distributed, but below which they tend to order. If the defects are sufficiently numerous and $|w_{vv}|$ large enough, for $T < T_c$ two possibilities may occur :

a) formation of a unique ordered phase,

b) separation into two distinct phases.

The structural mechanism is actually formation of local clusters (w_{vv} involving several neighboring cations) which, if sufficiently large, will nucleate a novel "microphase" with less anions. A second phase will appear for a critical composition representing the maximum number of vacancies the structure is able to accomodate. Both phases may be related in fact by an intergrowth phenomenon if they have structural analogies.

When the vacancy clusters develop enough to give rise to a microphase with rearrangement of the remaining anions, formation of "extended defects" is observed. Appearance of shear planes (WADSLEY defects) as in the MAGNELI W_nO_{3n-1} or W_nO_{3n-2} oxides, or alignment of the point defects in fluorite-type vacancy oxides are classical examples.

As far as ΔH decreases due to ordering of the defects into regular arrays and may largely compensate for decreasing configuration entropy, particularly at low temperature, a sequence of ordered phases will be often more stable than a single phase with wide compositional variations. To understand the defect association mechanism choice of appropriate formulations and annealing processus will be useful as well as electron diffraction and HREM characterization.

Non stoichiometry of oxides of cubic perovskite-type and related structures is of particular interest due to the diversity of the crystallographic features, but also to their impact on many different properties (e.g. magnetic, electronic, dielectric, optical, catalytic) |3|.

It is well known that cubic symmetry for an ABO_3 perovskite corresponds to a so-called GOLDSCHMIIDT tolerance factor $t = \frac{r_A + r_O}{\sqrt{2}(r_B + r_O)}$ equal to 1 or slightly smaller. If t decreases enough (smaller A or bigger B cation) cation displacements and tilting of corner sharing BO_6 octahedra leading to orthorhombic symmetry ($GdFeO_3$-type structure) may occur. Conversely $t > 1$ gives rise to hexagonal or rhombohedral perovskites with constitution of blocks of face sharing octahedra, but due to small B-B distances, factors such as B-B bonding with eventual electronic delocalization and B-O covalency become of major importance for the defect formation |4| |5|.

Several types of non-stoichiometry can be encountered : A-site vacancies, B-site vacancies, oxygen deficiency (or even sometimes excess).

1) <u>Cation vacancies</u>

A-vacancies are relatively easy to achieve, as a result of the high coordination number (C.N. = 12). Some tungsten bronzes constitute a classical example of A-cation deficient oxides. The perovskite structure appears with an A_xWO_3 formulation either for small values of x or for very large ones, x being then close to 1. While the former oxides, which exhibit structural distortions have a semi-conducting behavior, the latter ones are metallic. Screening by the conduction electrons in the last family makes uneasy the formation of a long range order and the

A-cations are more or less randomly distributed. On the contrary the insulating $Cu_{0.50}TaO_3$ phase is characterized by a complex copper ordering |6| |7|. A somewhat different example is that of $La_{0.33}NbO_3$ where the lanthanum atoms occupy randomly one A-plane out of two, the second one being empty (Fig. 1). However, lithium is able by intercalation reaction to fill partly the vacant sites, occupying probably preferentially the empty planes by electrostatic reasons |8| |13|. Cation ordering in A-defect perovskites seems to be favored either by absence or by localization of d-electrons on the B-cations.

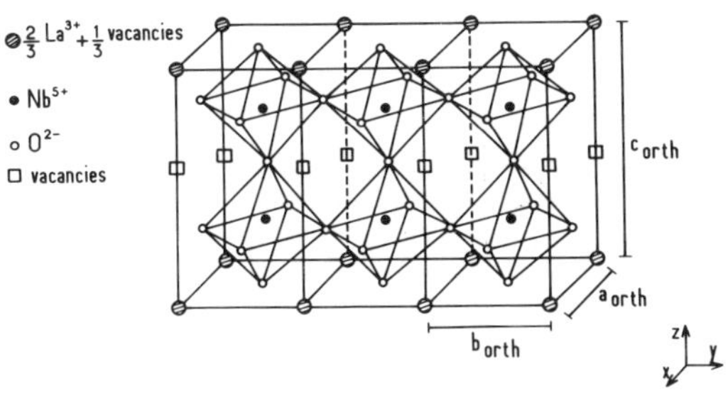

Fig. 1 Structure of $La_{0.33}NbO_3$.

WO_3 may be considered as a lower limit of the A-deficient cubic perovskite phase for small x values. At low temperature it has in fact a triclinic distortion. The symmetry increases by insertion of A-cations or by temperature raising.

B-site vacancies appear often in hexagonal perovskites, resulting from the strong B-B repulsions through the common faces, especially when B is large and highly charged. Due to the electrostatic character of this phenomenon B-vacancies are generally ordered occupying privileged sites of the hexagonal unit cell |11| |12|. As B-B repulsion is strongly attenuated in corner sharing octahedra, few examples of B-cation vacancies have been reported for cubic perovskites. Ordering seems to be favored either by redox equilibrium between the remaining B-cations (e.g. in $Ba_2Sm_{2/3}UO_6$ or $Ba_2Ce_{3/4}SbO_6$) or by cluster formation ($Ba_5Ta_4O_{15}$) |9| |10|.

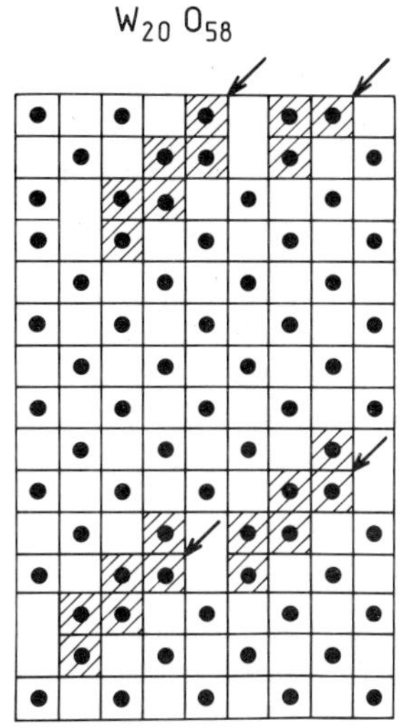

Fig. 2 Formation of equidistant shear planes in $W_{20}O_{58}$ resulting from the equilibrium obtained after partial oxygen evolution.

2) Oxygen deficiency

In the BO_3 oxides of ReO_3-type related structure such as MoO_3 or WO_3 the oxygen vacancies are stabilized by formation of common edges leading easily to shear planes when they are numerous enough (Fig. 2). Such a defect ordering mechanism is excluded, however, for ABO_3 perovskites as a result of the constant A/B ratio. The oxygen vacancies form clusters and, when present in sufficiently large number, especially at low temperature, give rise to extended defects. The nature of the resulting supercell depends on the electronic configuration of the B-cation.

A well characterized defect phase is

brownmillerite where the generated supercell is indeed bigger than the related cubic perovskite cell ($a \simeq 2a_c, b \simeq \sqrt{2}a_c, c \simeq 4a_c$). Planes of perovskite structure alternate along the c-axis with planes where thanks to the missing oxygen atoms tetrahedra grouped two by two share a common corner (Fig. 3). The B cations are slightly shifted in these tetrahedra planes. For $Ca_2Fe_2O_5$ the oxygen vacancies result in a succession of FeO_6 and FeO_4 sheets (otot) |14| |15|. Such a structure occurs only indeed as far iron (III) is able to accomodate in both octahedral and tetrahedral sites. Within the isostructural Ca_2FeCoO_5 iron will occupy the tetrahedral positions, whereas low spin cobalt (III) fills the octahedral sites of the perovskite sheets |16|. On the contrary one may expect that in Ca_2FeAlO_5 the small aluminium atoms are in the tetrahedral sites.

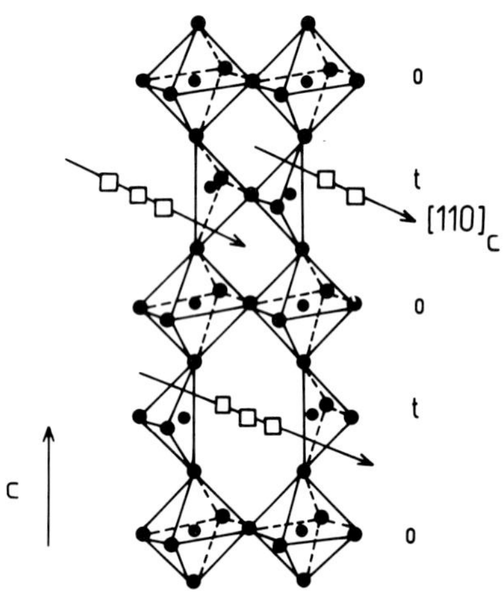

Fig. 3 Brownmillerite-type structure.

$Ca_2Mn_2O_5$ and $Sr_2Mn_2O_5$ despite similar formulation have a structure which differs from the brownmillerite one. As a result of the high spin configuration of manganese(III) (i.e. $t_{2g}^3 d_{z^2}^1$) leading to a Jahn-Teller effect, all coordination polyhedra are square based pyramids |17|, |22|. The structural relation to the perovskite cell is : $a = \sqrt{2}a_c$, $b = 2\sqrt{2}a_c$, $c = a_c$ (Fig. 4).

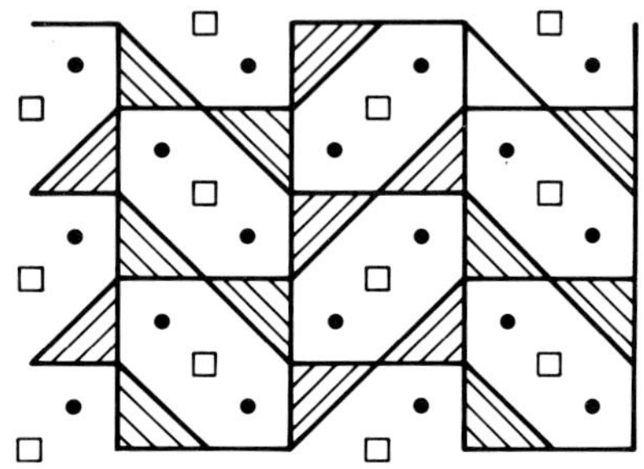

Fig. 4 Idealized representation of the structure of $Sr_2Mn_2O_5$ and $Ca_2Co_2O_5$ showing the formation of square based pyramids thanks to ordered oxygen vacancies.

$Ca_2Co_2O_5$ shows a structure similar to that of $Ca_2Mn_2O_5$, but the doubling of the a and c parameters is probably due to different orientations of the fourfold axes of the square pyramids |18|. On the contrary homologous $Sr_2Co_2O_5$ seems to have an oxygen vacancy 2H-type structure below 900°C and a brownmillerite-type lattice above. Under oxygen pressure this high temperature variety gives rise progressively to a defect perovskite structure by filling of the oxygen vacancies |19|. Differences in electronic configuration of cobalt account certainly for the diversity of the coordination polyhedra. The 2H-structure is probably stabilized for instance by electronic delocalization in a partly filled band along the c-axis.

$SrTiO_{2.5}$ and $SrVO_{2.5}$ have a cubic unit cell close to that of the perovskite structure, but the titanium and vanadium atoms seem to exhibit a trigonal bipyramid surrounding. According to the authors the defects are randomly distributed in the lattice or show only small tendency to order. This feature results indeed from the itinerant character of the d-electrons of titanium and vanadium leading to metallic behavior |20| |21|. Neither $CaCrO_{2.5}$ nor $SrCrO_{2.5}$ have been reported : the stability of the $3d^3$ configuration in an octahedral surrounding does not allow a lower coordination to occur. On the other hand d-orbitals are not extended enough to permit an electronic delocalization able to overcome this feature.

$La_2Ni_2O_5$ has been recently reported after reduction by hydrogen of $LaNiO_3$ around 550°C and a brownmillerite structure suggested on hand of a X-ray diffraction pattern |23|. As tetrahedral surrounding of nickel (II) is actually unlikely, successions of octahedral and square planar coordination entities (the latter may be with low spin configuration of Ni^{2+}) have been proposed as a credible alternative (Fig. 5) |24|.

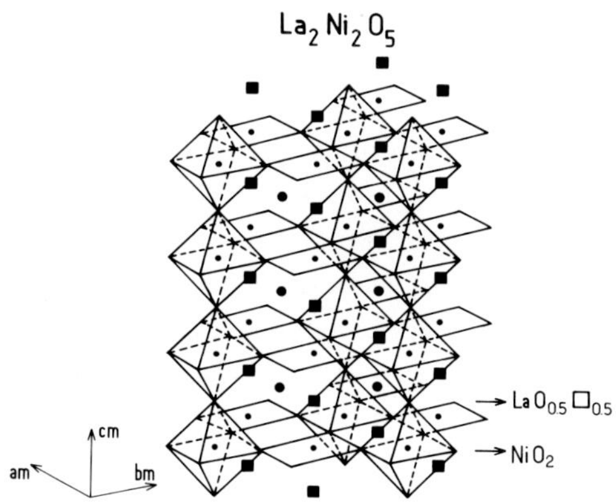

Fig. 5 Structure of $La_2Ni_2O_5$ showing simultaneous presence of high spin Ni^{2+} in octahedral sites and low spin Ni^{2+} in square planar sites.

Many **phases with a composition intermediate between ABO_3 and $A_2B_2O_5$** have been prepared and investigated. The structures depend indeed on the formulation and the elaboration process. For an ABO_{3-x} compound with x small and prepared at high temperature one may expect a random distribution of point defects with local appearance of coordination number five sites especially for close shell cations such as Fe^{3+} |25|. As Mn^{3+} $(3d^4)$ is subjected to Jahn-Teller distortion, whereas Mn^{4+} $(3d^3)$ is isotropic, the $CaMnO_{3-x}$ system leads to several ordered phases with formation of a superlattice containing both slightly distorted MnO_6 octahedra and MnO_5 square pyramids so as to accomodate Mn^{4+} and Mn^{3+}, e.g. $CaMnO_{2.80}$, $CaMnO_{2.75}$ and $CaMnO_{2.67}$ with an increasing proportion of pyramids. HREM shows that intermediate compositions are characterized by microphase intergrowth (Fig. 6) |26| |27|.

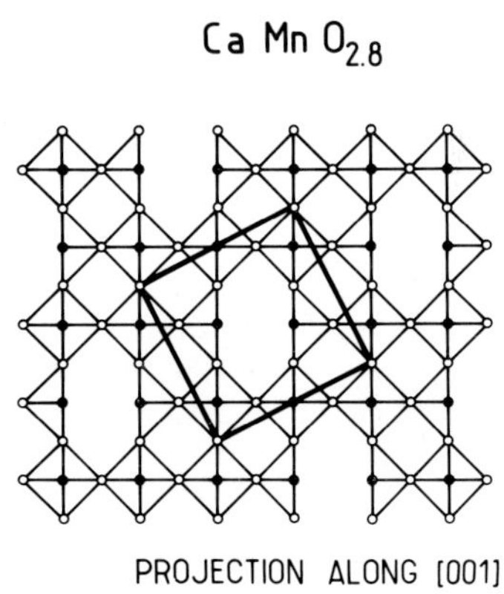

Fig. 6 Idealized representation of the manganese-oxygen framework in $CaMnO_{2.8}$.

An extensive research based on the feature that Fe^{3+} is able to adapt indifferently to both octahedral and tetrahedral coordinations has been carried out on the systems

$Ca_xLa_{1-x}FeO_{3-y}$ and $CaTi_{1-x}Fe_xO_{3-y}$ with $y = \frac{x}{2}$ in order to preserve the oxidation state III of iron, i.e. with compositions intermediate either between $LaFeO_3$ (perovskite) and $Ca_2Fe_2O_5$ (brownmillerite) or between $CaTiO_3$ and $Ca_2Fe_2O_5$. A homologous series of $A_nB_nO_{3n-1}$ phases (A = Ca, La ; B = Fe, Ti) have been characterized by X-ray diffraction, physical measurements like Mössbauer spectroscopy and HREM |28| |29| |34|. The structure of these phases consists usually of a succession along the c-axis of n-1 perovskite-type sheets constituted by BO_6 octahedra (o layers) and a corner sharing sheet of BO_4 tetrahedra (t layers). $n=\infty$ corresponds to the perovskite structure (oooo) and n=2 to the brownmillerite (otot).

Other members of the series are $Ca_2LaFe_3O_8$, $Ca_3Fe_2TiO_8$ (n=3) and $Ca_4Fe_2Ti_2O_{11}$ (n=4) with layer sequences ootoot and oootooot respectively (Fig. 7).

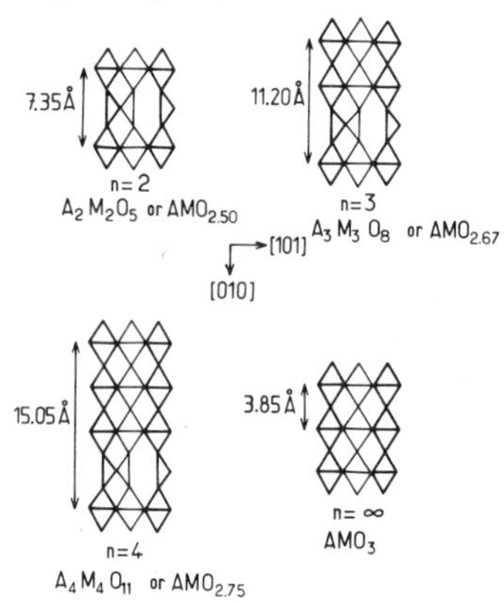

Fig. 7 Succession of layers with corner sharing octahedra of tetrahedra in $A_nB_nO_{3n-1}$ phases for n=2, 3, 4 and

Intergrowth phases have been reported. If they have a relevant composition and if the corresponding slabs are thin enough a regular succession of the previous motives may be detected by HREM. $Ca_5Fe_4TiO_{13}$ and $Ca_4YFe_5O_{13}$ (n=2.5) contain for instance alternatively n=2 (otot) and n=3 (ootoot) sheets |30|. $Ca_7Fe_6TiO_{18}$ (n=2.33) should be characterized by an ototoot sequence, but many stacking faults actually appear due to the complexity of the expected packing (Fig. 8).

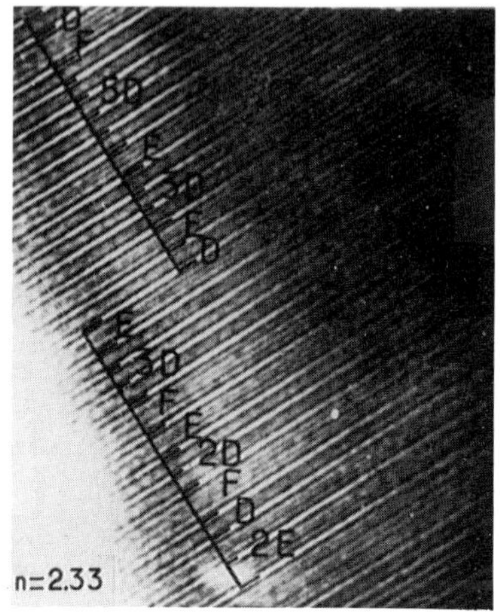

Fig. 8 TEM picture of the $Ca_7Fe_6TiO_{18}$ phase (n=2.33). The expected ototoot sequence (E-stacking) is mostly replaced by adjacent ootot (n=2.5 : D-stacking) and otototoot (n=2.25 : F-stacking) sequences.

Isolated point defects occur only in iron oxides for a very small number of oxygen defects. As their number increases they result in the formation of tetrahedra rather than bipyramids as already expected 10 years ago from conductivity vs. oxygen pressure measurements (86). The tetrahedra align easily and for a rising proportion of oxygen vacancies these files may form planes, which in turns tend to order.

But when Fe^{4+} is present, for example in $Ca_xLa_{1-x}FeO_{3-y}$ samples ($0.25 \lesssim y < 0.5$) prepared in air at 1400°C, whereas X-ray diffraction seems to characterize a cubic perovskite-type structure (a ≃ 3.85 A), electronic diffraction and HREM show actual formation of three dimensionally oriented intergrown microdomains of brownmillerite or $Ca_2LaFe_3O_8$ (grenierite) bulk compositions. The dimension of the individual domains, the volume of which is about $10^6 Å^3$, tends to decrease with rising oxygen content (i.e. lower annealing temperature), probably as a consequence of increasing formation of $Fe^{4+}(3d^4)$ with coordination five on the domain walls |31| |32| |33|. Similar accomodation of mixed valency phenomena in twin plane boundaries seems to be related to the superconductivity mechanism in $Ba_2YCu_3O_{7-\delta}$ as shown above.

An extensive theoretical investigation has confirmed the overwhelming formation of octahedra and tetrahedra in the ABO_{3-y} compounds containing iron (III) in agreement with Mössbauer resonance results. It is based on the hypothesis that the coordination number of the A cations should be between 12 (perovskite) and 10 (brownmillerite), so that the oxygen vacancies affecting A and as a consequence B are aligned. The vanishing number of C.N.5 sites especially for $y > 0.25$ which results from these realistic assumptions agrees with the experimental data and has been later confirmed by HREM (Fig. 9) |35|.

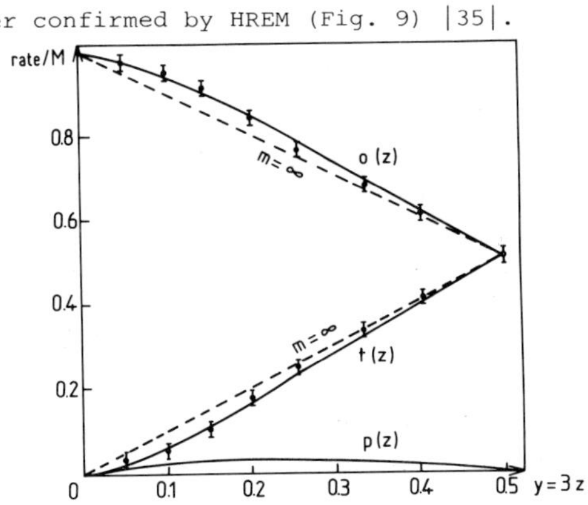

Fig. 9 Relative number of tetrahedral, pyramidal and octahedral sites in a statistic model supposing vacancy alignment in iron (III) perovskites with 0 y 0.5 deviation from stoichiometry. If the files are infinite the terminal trigonal bipyramids tend to disappear.

In the $BaFeO_{3-y}$ system the phases obtained are either of brownmillerite type or hexagonal perovkite polytypes. As a consequence of the oxygen vacancies face sharing octahedra should be largely replaced by corner sharing tetrahedra. Tetrahedra formation would be probably favored by high covalency of the Fe-O bonds. In contrast the vacancies seem to lead to edge sharing trigonal bipyramids in the $BaMnO_{3-y}$ system, which has been investigated by heating above 1150°C the 2H-type $BaMnO_3$ where the MnO_6 octahedra form common face files |36| |37| |38|. This feature results obviously from the tendency of the obtained Mn^{3+} to give rise to fivefold coordination |39| |40| |41|. In the 12H-type oxide of ideal composition $BaCoO_{2.67}$ two CoO_4 tetrahedra containing mainly high spin $Co^{4+}(3d^5)$ share corners with four face sharing CoO_6 octahedra with low spin configuration Co^{3+} (Fig. 10) |42| |43|.

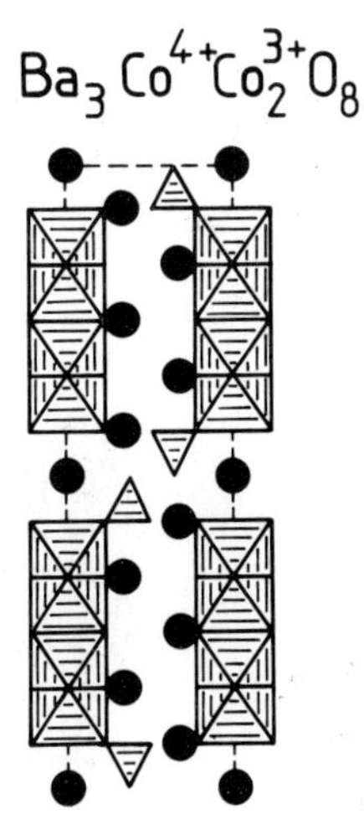

Fig. 10 $BaCoO_{2.67}$ showing two tetrahedra containing high spin Co^{4+} connected to blocks of four face sharing octahedra with low spin Co^{3+}.

Intergrowth phenomena and phase mixtures strongly influenced by the preparation conditions make such studies particularly complex even utilizing systematically HREM. No oxidized phase derived from $La_2Ni_2O_5$ has been found except perovskite-type $LaNiO_3$.

The coordination preference resulting from the vacancy ordering is less obvious for copper oxides. For instance in the tetragonal cubic perovskite derived phase $La_3Ba_3Cu_6O_{14+\delta}$, which contains simultaneously copper (II) and (III), squares, square planar pyramids and elongated octahedra seem to coexist |44| |80|.

Anion-excess non stoichiometry is not as common as oxygen deficiency in perovskite oxides due to the difficulty to introduce additional oxygen atoms into a compact structure. Some apparent anion surplus may result, however, from A vacancies and eventually B-deficiencies |45| |46|.

Large oxygen excess is generally the consequence of an $A_nB_nO_{3n+2}$ stacking between cubic perovskite slabs and A layers. Several terms of this homologous phase series have been found ranging from n=1 to n=6. $La_2Ti_2O_7$ and $Ca_2Nb_2O_7$ for example constitute n=4 members but chemically they can be considered as resulting from the oxidation of $LaTiO_3$ and $CaNbO_3$ (Fig. 11) |47|.

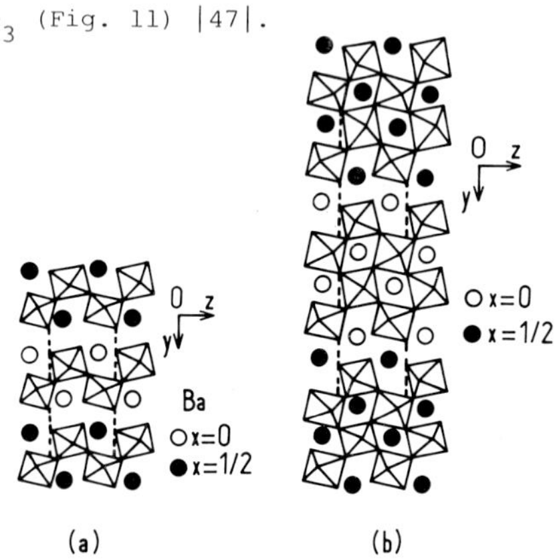

Fig. 11 $A_nB_nO_{3n+2}$ (n=2 and 4 terms) phases showing an excess of oxygen thanks to intercalation of A-layers between perovskite slabs.

Ordered intergrowth has often been found for the $A_nB_nO_{3n+2}$ families after prolonged annealing, extending over large values of the c-parameter, for instance between the n=4 and n=5 members detected in the (Na, Ca) NbO_{3+x} system |47|.

3) Structures related to perovskite-type

The most common family is characterized by alternative stacking of perovskite ABO_3 and rock salt AO layers. The A_2BO_4 composition corresponds generally to the K_2NiF_4-type structure. More complex stacking showing apparent oxygen excess can be detected, for instance in intergrowth phases such as $La_3Ni_2O_7$ or $La_4Ni_3O_{10}$ |48|.

Oxygen deficiency leads to structures analogous to those of the perovskite phases. Reduction of Ca_2MnO_4 gives rise for instance to a $Ca_2MnO_{3.5}$ phase containing $Mn^{3+}O_5$ square pyramids, whereas $Ca_2FeO_{3.5}$ derives from the brownmillerite-type structure |16| |17|.

An alternative to the perovskite structure of ABO_3 oxides is the defect pyrochlore structure ($A_2B_2O_6\square$) which may appear when the A or B cations have a high polarizability (e.g. for $Tl_2B_2O_6$ with B = Nb, Ta or U or $A_2Sb_2O_6$ with A = Na, K, Ag or $Pb_2Ru_2O_6$) |49| |50| |92|.

Aurivillius phases containing $(Bi_2O_2)^{2+}$ layers with edge sharing square pyramids separated by perovskite-type slabs have been found with the composition $(Bi_2O_2)^{2+}(A_{n-1}B_nO_{3n+1})^{2-}$ (B = Cr, Fe, Ti). They may show ordered or disordered intergrowth |51| |88|.

As a consequence of the superconducting behavior detected for some of them above liquid nitrogen temperature, perovskite derived copper oxides have been extensively investigated since 1986 |52|. It is difficult to attribute unambiguously coordination polyhedra to the various oxidation states and spin configurations of copper, as most of them have a metallic conductivity, but as a rule Cu^{+} has a linear coordination (like in Cu_2O or in delafossite-type $CuFeO_2$), Cu^{2+} has a preference for elongated octahedra or square based pyramids (like in $La_2CuO_{4-\delta}$), Cu^{3+} for squares or square pyramids as a consequence of the Jahn-Teller effect (involving $3d^9$ or $3d^8$ low spin configurations) |54| |55| |74| |75|. Metallic $BaLa_4Cu_5O_{13}$ for instance which contains formally 3 Cu^{2+} and 2 Cu^{3+} has an ideal structure with 3 CuO_6 elongated octahedra and 2 CuO_5 square based pyramids (Fig. 12) |61|.

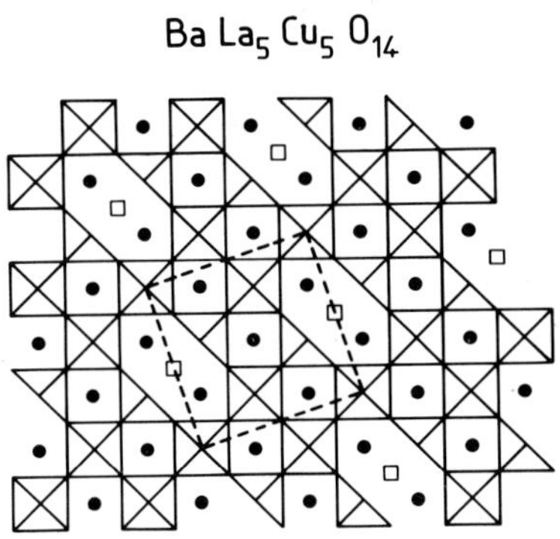

Fig. 12 Structure of $BaLa_4Cu_5O_{13}$ with 3 $Cu^{2+}O_6$ octahedra and 2 $Cu^{3+}O_5$ square pyramids in each unit cell.

Tetragonal $Ba_2YCu_3O_6$ contains one Cu^{+} in coordination 2 and two Cu^{2+} in coordination 5 with a long coaxial distance, while the oxygen richer $Ba_2YCu_3O_7$ has formally two $(Cu^{2+}O_5)$ square pyramids and one $(Cu^{3+}O_4)$ planar group Fig. 13 (a) and (b) show that the addition of O_1 has transformed the isolated CuO_4^2 linear units into chains formed by corner sharing $CuO_4^2O_1^2$ distorted squares (Cu_1-O_4 = 1.84 Å, Cu_1-O_1 = 1.94 A). Fig. 14 gives evidence of the similarity between $Ba_2YCu_3O_7$ and the n = 3 term of the $A_nFe_nO_{3n-1}$ series where due to the isotropic Fe^{3+} high spin configuration the oxygen vacancies order to give rise to tetrahedra layers. In $Ba_2YCu_3O_7$ the vacancies lead to a succession of square based pyramids and squares parallel to b-axis. Hence the $Ba_2YCu_3O_7$ structure is orthorhombic and leads to twinned single crystals resulting from alternating orientation of the square chains along a- and b-axes. Random occupancy of the O_1 and O_5 sites results in tetragonal symmetry (Fig. 13 (c)) |62| |63| |64| |65| |66| |67| |68| |87| |96|.

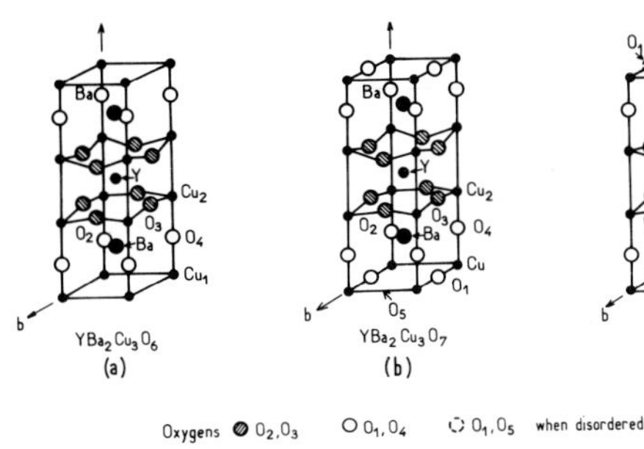

Fig. 13 Structures of tetragonal $YBa_2Cu_3O_6$ (a) and orthorhombic $YBa_2Cu_3O_7$ (b). The (c) structure represents the tetragonal disordered structure.

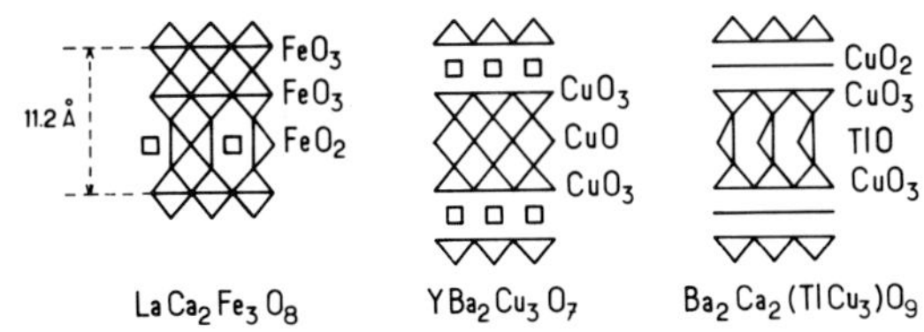

Fig. 14 Similarities of the framework structures of $LaCa_2Fe_3O_8$, $YBa_2Cu_3O_7$ and $Ba_2Ca_2(TlCu_3)O_9$. In the latter phase distorted "TlO" layers appear between the tops of the square based Cu-pyramids, analogous to the tetrahedra layers between the octahedra sheets in $LaCa_2Fe_3O_8$. Besides corner sharing square planar CuO_4 groups form intermediate copper-oxygen layers between the pyramid bases.

$La_{2-x}Sr_xCuO_4$ and $La_{2-x}Ba_xCuO_4$ are superconducting for small values of x close to metal-insulator transition. T_c is in fact small (i.e. in the 20-40 K range) but these materials have a historical importance as they were the first discovered high T_c oxides, giving rise afterwards to many investigations |56| |57| |58| |59| |60|. Possible superconducting behavior has been recently suggested for the homologous $La_{2-x}Sr_xNiO_4$ series (x 0.2) |91|.

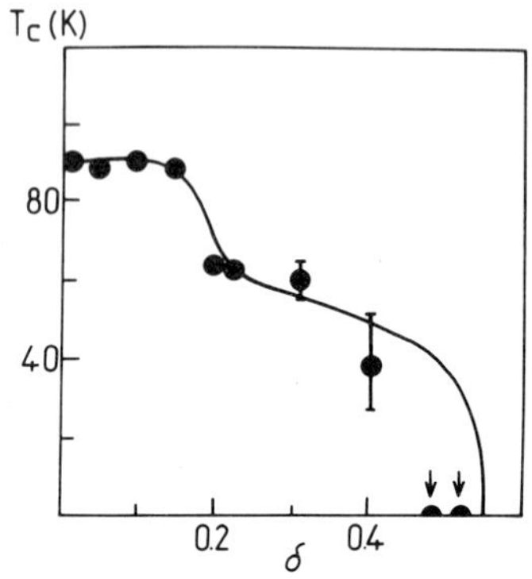

Fig. 15 Variation of T_c with for $YBa_2Cu_3O_{7-}$ determined from the resistivity drops.

$Ba_2YCu_3O_{7-\delta}$ has a superconducting behavior which depends on δ, as far it is orthorhombic, with T_c often exceeding 90 K. The decreasing values of T_c (Fig. 15) at rising may be related to cutting of the square chains resulting from oxygen departure. The plateau observed in the $0.2 < \delta < 0.4$ range which is also reflected by magnetic and electric measurements may be attributed to structural factors so far not elucidated |96|. The T_c breakdown noticed at $\delta \simeq 0.6$ coincides with vanishing orthorhombic distortion. Negative values of δ which correspond to formation of octahedra by filling of the O_5 sites lead to quick collapse of superconductivity.

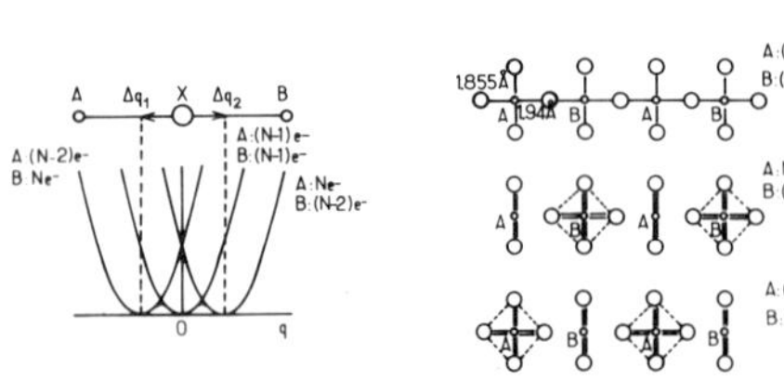

Fig. 16 Potential energy variation for N, N-1 and N-2 electron cations in identical sites. Electron pair transfer in a planar square chain.

As illustrated by Fig. 16 high T_c superconductivity observed in these materials has been attributed to disproportionation without activation energy of 2 $Cu^{2+} \rightleftarrows Cu^{+} + Cu^{3+}$ type in the chains, which would account for the presence of only one copper site within such files (Cu_1 in Fig. 13) but of two Cu-O distances |84|. XPS and Auger spectroscopy seem to confirm this interpretation in showing the simultaneous presence of Cu^{+} and Cu^{2+} with enrichment in Cu^{+} at decreasing temperature. The fact that Cu^{3+} has not been detected should be explained by a $Cu^{3+} + O^{2-} \rightleftarrows Cu^{2+} + O^{-}$ equilibrium with formation of very mobile oxygen holes in the basic square planes containing Cu_2. They could either bridge two Cu^{2+} ions or form temporarily O_2^{2-} dimers |93| |94|. The shape of the UPS spectra of oxygen agrees with such hypothesas, they are consistent with the strong covalent character of the Cu-O bonds. This interpretation would account for the p-type conductivity. Intermediate pairing in a peroxide O_2^{2-} anion can be compared with occurence of S_2^{2-} ions in many strongly covalent 3d-sulfides |82| |83|. Structural determinations as well as physical characterizations are consistent with the proposed model, though the exact conduction mechanism is still questioned. Electronic delocalization excludes oversimplified explanations (Fig. 16). Actually ten years ago the isothermal electrical conductivity mechanism with varying p_{O_2} in $LaCa_2Fe_3O_{8+\delta}$ had already been attributed to transitory formation of O_2^{2-} anions leading to transform ation of tetrahedra into octahedra |85| |86|. The superconductivity of analogous $La_{3-x}Ba_{3+x}Cu_6O_{14+\delta}$ phases has probably a similar explanation, as far T_c seems to be enhanced by increasing othorhombic distortion (i.e. occupation of the O_1 site at the expense of O_5) |80| |81|.

The proposed mechanism can be likely extended to the (Ba, K) (Bi, Pb) O_3 perovskites involving simultaneously Bi^{3+}, Bi^{5+} and Pb^{4+} |57| |95|. The 3D-character of these materials which distinguishes them from the previous layer-type superconducting copper oxides explains probably that T_c is much lower.

Recently several new bismuth and thallium cuprates containing apparently Bi_2O_2 (or Tl_2O_2) sheets analogous to those of the Aurivillius phases or modulated metal-oxygen layers exhibited T_c values attaining 125 K |76| |77| |78| |79|. Evidence has been given of structural similarities with previously investigated ordered defect oxides (Fig. 14 and 17).

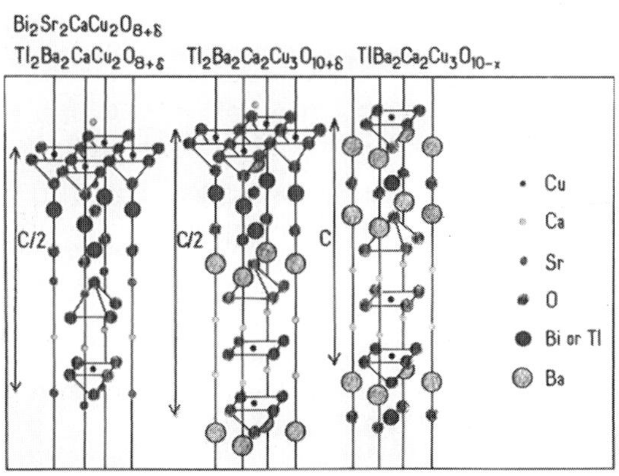

Fig. 17 Layer structures proposed for some recent bismuth or thallium syperconducting oxides. For most of these phases the oxygen positions are still controversial.

The layer structures, as far as they will be determined in detail, the present disagreements being lifted, should allow to formulate the hypothesis that superconductivity results simultaneously from an equilibrium involving Cu^+ and Cu^{3+} as well as Bi^{3+} and Bi^{5+} (or Tl^+ and Tl^{3+}). Presence of Pb^{2+} seems to enhance slightly the superconductivity of the thallium oxides |72|. Rising number of copper-oxygen planes apparently increases also T_c but multiplication of the Cu-O layers makes more difficult growing single crystals with homogeneous stacking and leads quickly to saturation of such possibilities |69| |70| |71| |72| |73| |89| |90| |96|.

The occurence of two disproportionation mechanisms in the bismuth and thallium cuprates, as far it would be confirmed, could account for the larger resistivity dropping domain observed at decreasing temperature for the concerned oxides.

The crystal chemical developments in this area, where phonon-electron interactions are essential, are delayed by the difficulty to obtain pure phases, temperature stable and with a crystal quality sufficient to allow precise structural determinations, at least by TEM.

Despite complexity of the conduction mechanism the present copper oxides are narrowly related to the defect perovskite-like oxides previously investigated. The type of defects differs indeed with the formulation, the prevailing factor being the electronic configuration of the B cation. The fact that the only 3d-element leading unambiguously to superconductivity is copper and that in the superconducting oxides substitution of copper by homologous cations leads quickly to vanishing of the phenomenon may be attributed to a relatively unique feature, most likely the mentioned activation energy-free disproportionation mechanism favored by the crystal structure. Such a coincidence explains probably why so few superconducting oxide series have been found and why most of the layer type materials concerned have strong structural similarities |87|.

The recent breakthrough in the field of

superconducting copper oxides emphasizes in any case two general rules in solid state chemistry : the necessity never to renounce to prepare new materials and the importance not to separate new phenomena from their scientific background.

REFERENCES

1. W. SCHOTTKY and C. WAGNER, Z. phys. Chem., B 11 (1930) 163.
2. J.S. ANDERSON, Proc. R. Soc. A 185 (1946) 69.
3. C.N.R. RAO and G.V. SUBBARAO, Physica Status Solidi, A 1 (1970) 597.
4. L. KATZ and R. WARD, Inorg. Chem. 3 (1964) 205.
5. J.B. GOODENOUGH and J. LONGO, Landolt-Bornstein Tabellen, III-4a, Springer V. (1970).
6. J. LONGO and A.W. SLEIGHT, Mat. Res. Bull. 10 (1975) 1273.
7. H. VINCENT, B. BOCHU, J. AUBERT, J.C. JOUBERT and M. MAREZIO, J. Sol. St. Chem. 10 (197) 1273.
8. P.N. IVER and A.J. SMITH, Acta Cryst. 23 (1967) 740.
9. G. RAUSER and S. KAMMLER-SACK, J. Sol. St. Chem. 33 (1980) 135.
10. J. SHANNON and L. KATZ, Acta Cryst. B26 (1970) 102.
11. C. CALVO, N.N. NG and B. CHAMBERLAND, Inorg. Chem. 17 (1978) 699.
12. K.R. POEPPELMEIER, A.J. JACOBSON and J. LONGO, Mat. Res. Bull. 15 (1980) 339 and 16 (1981) 359.
13. A. NADIRI, G. LE FLEM and C. DELMAS, J. Sol. St. Chem. 73 (1988) 338.
14. J. BERGGREN, Acta Chem. Scand., 25 (1971) 3616.
15. J.C. GRENIER, M. POUCHARD and P. HAGENMULLER, Structure and Bonding, Springer, 47 (1981)
16. K. VIDYASAGAR, J. GOPALAKRISHNAN and C.N.R. RAO, Inorg. Chem. (1984).
17. K.R. POEPPELMEIER, M.E. LEONOWICZ and J.M. LONGO, J. Sol. St. Chem. 44 (1982) 89.
18. C.N.R. RAO, J. GOPALAKRISHNAN and K. VIDYASAGAR, Ind. J. Chem. 23 A (1984) 275.
19. J.C. GRENIER, S. GHODBANE, G. DEMAZEAU, M. POUCHARD and P. HAGENMULLER, Mat. Res. Bull. 14 (1979) 831.
20. M.A. ALARIO-FRANCO and M.V. REGI, Nature, 270 (1977) 706.
21. B.C. TOFIELD, Nature, 272 (1978) 713.
22. V. CAIGNAERT, N. NGUYEN, M. HERVIEU and B. RAVEAU, Mat. Res. Bull., 20 (1985) 479.
23. M. CRESPIN, P. LEVITZ and L. GATINEAU, J. Chem. Soc. Faraday II, 79 (1983) 1181.
24. C.N.R. RAO, J. GOPALAKRISHNAN and k. VIDYASAGAR, Ind. J. Chem., 23A (1984) 265.
25. B. TOFIELD, C. GREAVES and B.E.F. FENDER, Mat. Res. Bull., 10 (1975) 737.
26. A. RELLER, D.A. JEFFERSON, J.M. THOMAS and M.K. UPPAL, J. Phys. Chem., 87 (1983) 913.
27. A. RELLER, J.M. THOMAS, D.A. JEFFERSON and M.K. UPPAL, Proc. Roy. Soc. (1984).
28. J.C. GRENIER, M. POUCHARD and P. HAGENMULLER, Structure and Bonding, Springer, 47 (1981) 1.
29. J.C. GRENIER, J. DARRIET, M. POUCHARD and P. HAGENMULLER, Mat. Res. Bull., 11 (1976) 1279.
30. Y. BANDO, Y. SEKIKAWA, H. NAKAMARA and Y. MATSUI, Acta Cryst., A37 (1981) 723.
31. M.A. ALARIO-FRANCO, M.J.R. HENCHE, M.V. REGI, J.M.G. CALBET, J.C. GRENIER, A. WATTIAUX and P. HAGENMULLER, J. Sol. St. Chem., 46 (1983) 23.
32. M.A. ALARIO-FRANCO, J.M.G. CALBET, M.V. REGI and J.C. GRENIER, J. Sol. St. Chem., 49 (1983) 219.
33. J.M.G. CALBET, M.V. REGI, M.A. ALARIO-FRANCO and J.C. GRENIER, Mat. Res. Bull., 18 (1983) 285.
34. J.C. GRENIER, M. POUCHARD, P. HAGENMULLER, G. SCHIFFMACHER and P. CARO, J. Phys. 38C7 (1977) 84.
35. S. KOMORNICKI, J.C. GRENIER, M. POUCHARD and P. HAGENMULLER, Nouv. J. Chimie, 5 n° 3 (1981) 161.
36. S. MORI, J. Phys. Soc. Jap., 28 (1970) 44.
37. M. ZANNE and Ch. GLEITZER, Bull. Soc. Chim. Fr., (1971) 1567.
38. A. JACOBSON, Acta Cryst., B32 (1976) 1087.
39. T. NEGAS and R.S. ROTH, J. Sol. St. Chem., 3 (1971) 323.
40. A. J. JACOBSON and A.J.W. HORROX, Acta Cryst., B 32 (1976) 1003.
41. J.L. HUTCHISON and A.J. JACOBSON, J. Sol. St. Chem., 20 (1977) 417.
42. M. ZANNE, A. COURTOIS and Ch. GLEITZER, Bull. Soc. Chim. Fr., (1972) 4470.
43. A.J. JACOBSON and J.L. HUTCHISON, J. Sol. St. Chem., 35 (1980) 334.
44. L. ER-RAKHO, C. MICHEL, J. PROVOST and B. RAVEAU, J. Sol. St. Chem., 37 (1981) 151.
45. B.C. TOFIELD and W.R. SCOTT, J. Sol. St. Chem., 10 (1974) 183.
46. N. RAMADAN, J. GOPALAKRISHNAN and M.V.C. DE SASTRI, J. Inorg. and Nucl. Chem., 40 (1978) 1453.
47. R. PORTIER, A. CARPY, M. FAYARD and J. GALY, J. Phys. Status Solidi, A 30 (1975) 683.
48. J. DRENNEN, C.P. TAVARES and B.C.H. STEELE, Mat. Res. Bull., 17 (1981) 621.
49. M.A. SUBRAMANIAN, G. ARAVAMUDAN and G.V. SUBBARAO, Prog. Sol. St. Chem., 15 (1983) 55.
50. Y. PIFFARD and M. TOURNOUX, Acta Cryst., 35 B (1979) 1450.
51. J.L. HUTCHISON, J.S. ANDERSON and C.N.R. RAO, Proc. Roy. Soc. London, 355 (1977) 301.

52. J.G. BEDNORZ and A. MÜLLER, Z. Phys., 864 (1986) 187.
53. A.W. SLEIGHT et al., Sol. St. Commun., 17 (1975) 27.
54. J. BEILLE et al., C.R.Acad. Sc., 18 (1987) 304.
55. S.M. FINE, M. GREENBLATT, S. SIMIZU and S.A. FRIEDBERG, ACS symposium series, 10 (1987).
56. P. DAY et al., J. Phys. Sol. St., C20 (1987) 1429.
57 - R. CAVA et al., Phys. Rev. Lett., 58 (1987) 408.
58. C.W. CHU et al., Phys. Rev. Lett., 58 (198) 405.
59. R. MOHAN, P. GANGULY and C.N.R. RAO, Phase Transitions, 10 (1987) 107.
60. S. UCHIDA et al., Jap. J. Appl. Chem., 28 (1987) 4.
61. C. MICHEL, L. ER-RAKHO and B. RAVEAU, Mat. Res. Bull., 20 (1985) 667.
62. M.K. WU, Phys. Rev. Lett., 58 (1987) 908.
63. C.N.R. RAO et al., Nature, 326 (1987) 856.
64. W.F. DAVID et al., Nature, 327 (1987) 310.
65. P. BORDET et al., Nature, 327 (1987) 687 and Progress in HT superconductivity, 7 (1988) 76, World Scient. Publ.
66. J.D. JORGENSEN et al., Phys. Rev., 835 (1987) 975.
67. R.J. CAVA, Phys. Rev. Lett., 58 (1987) 405 and Nature, 58 (1987) 408.
68. C.N.R. RAO, Progress in HT superconductivity, 7 (1988) 1, World Scient. Publ.
69. C.N.R. RAO, Progress in HT Superconductivity, 15 (1987) 103, World Scient. Publ.
70. C. MICHEL et al., Z. Phys., B68 (1987) 421.
71. H. MAEDA, Y. TANAKA, M. FUKUTOMI and T. ASANO, Jap. J. appl. Phys., 27 (1988) 288.
72. M.A. SUBRAMANIAN et al., Science, 239 (1988) 1015, 242 (1988) 249.
73. J.M. TARASCON et al., Phys. Rev. B. 37 (1988) 9382.
74. B. RAVEAU et al., Progr. in HT superconductivity, 2 (1988) 115, World Scient. Publ.
75. C.N.R. RAO and P. GANGULY, Current Sciences, 56 (1987) 47.
76. C.W. CHU et al., Phys. Rev. Lett., (1988), in press.
77. J.M. TARASCON et al., Phys. Rev., B36 (1987) 226.
78. Z.Z. SHENG and A.M. HERMANN, Nature, 32 (1988) 555.
79. J.M. TARASCON et al., Eur. J. Sol. St. Chem. 27 (1990) 81.
80. W.F. DAVID et al., Nature, 328 (1987) 328.
81. C.U. SEGRE et al., Nature, 329 (1987) 227.
82. D.D. SARMA and C.N.R. RAO, J. Phys. Sol. St. C20 (1987) 659.
83. D.D. SARMA et al., Phys. Rev. B 36 (1987) 2371.
84. M. POUCHARD, J.C. GRENIER et J.P. DOUMERC, C.R.Acad. Sc., 305 (1987) 571.
85. H. ODA, Y. YAMAGUCHI, H. TAKEI and H. WATANABE, J. Phys. Soc. Japan, 41(1) (1977) 101.
86. J.P. BONNET, J.C. GRENIER, M. ONILLON, M. POUCHARD and P. HAGENMULLER, Mat. Res. Bull., 14 (1979) 67.
87. M. POUCHARD, J.C. GRENIER, J.P. DOUMERC, G. DEMAZEAU, J.P. CHAMINADE, A. WATTIAUX, P. DORDOR and P. HAGENMULLER, Proc. MRS meeting, high T_c oxide symposium, San Diego (1989).
88. B. AURIVILLIUS, Arkiv Kemi, 1 (1949) 499 ; 2 (1950) 519.
89. C.N.R. RAO and B. RAVEAU, Acc. Chem. Res. 22-3 (1989) 106.
90. C.N.R. RAO, R. VIJARAYARAGHAVAN, L. GANAPATHI and S.V. BHAT, J. Sol. St. Chem. 79 (1989) 177.
91. Z. KAKOL, J. SPALEK and J.M. HONIG, J. Sol. St. Chem. 79 (1989) 288.
92. J.M. LONGO, P. RACCAH and J.B. GOODENOUGH, Mat. Res. Bull. 4 (1969) 190.
93. B.K. CHAKRAVERTY, D.D. SARMA and C.N.R. RAO, Physica C 156 (1988) 413.
94. Y.GUO, J. LANGLOIS, W.A. GODDARD, Science 239 (1988) 896.
95. R.J. CAVA et al. Nature 332 (1988) 814.
96. B. RAVEAU, M. HERVIEU, C. MICHEL, D. GROULT and J. PROVOST, Eur. J. Sol. St. Chem. 27 (1990) 25.

Frontiers of Materials Research/
Electronic and Optical Materials
M. Kong, L. Huang (editors)
Elsevier Science Publishers B.V., 1991

PROGRESS IN POLYMER PROCESSING

L.A. UTRACKI

NRCC/IMRI, 75 De Mortagne, Boucherville, Que., Canada, J4B 6Y4

The plastics industry shows undiminishing expansion. On the global scale it is the fastest growing industry. Even when, at the turn of the century, the consumption of commodity resins in the advanced countries levels off, the large capital investments in the developing countries will more than compensate for the local slowdown. Production of commodity polymers during the next five years is expected to grow at an annual rate of: PVC 2.8%, PS 3.2%, LDPE 3.2%, HDPE 4.7% and LLDPE 9%. For the engineering resins the rates are higher: POM 5.1%, PA 6.1%, MPPE 10.1%, PC 10.4% and Polyesters 13.5%. The most rapidly expanding are the multiphase polymeric materials: alloys, blends and composites whose growth rate will be about 14% per annum. In recent years the plastics industry experienced evolutionary rather than revolutionary growth. This is apparent for all factors. There are no new outstanding polymers but a continuous improvement and maturing of both production and utilization of the 5-10 years old resins, such as PEEK or LCP. In commodity resins the advances are achieved in the type of catalysts which allow synthesis of newer and better grades (e.g. flexomers). Similarly the processing equipment undergoes a gradual improvement, increased automatization and loop control as well as growth of size; injection of 100 kg per shot is a reality. The true advances are made in compounding and die design. For example, using the lubricated die one can extrude materials, e.g. UHMWPE, by the process which more resembles metal forming than classical plastics extrusion.

1. INTRODUCTION

World production of plastics in 1995 is projected at 76 million metric tons with the annual growth rate, AGR, throughout the decade of 3.7% [1]; this volume is sufficient to form a ring 1.6 m in diameter all around the earth per year. The expected AGR of polymer alloys and blends is 12%, and that of composites 16%. In 1987, already 21% of polymers were used in blends and 29% in composites and filled plastics [2]. It is not difficult to calculate that if these tendencies persist, by 1995 all manufactured resins will be used in multi-phase polymeric systems. It should be noted that this trend is particularly strong for the fastest growing segment of the polymer industry, the engineering resins. There are two factors which will moderate this tendency: (i) a need for single-phase polymers in some applications (e.g. PTFE, UHMWPE, light-sensitive polymers, etc.), and (ii) use of polymer alloys and blends as matrices for composites, reinforcements and foams.

2. FLOW-INDUCED MORPHOLOGY

Flow plays a dominant role in the structure of processed material. Compounding or mixing must break the dispersed phase (aggregates) and randomly distribute the fragments within the matrix. Thus the compounding equipment is designed to provide dispersing and distributive stress/strain fields. It is well known that the same compound may process well on one processing line but poorly on another of the same design. Apparently, a small variation in screw and/or die wear, a seemingly insignificant variation in the runners may lead to variation in flow-induced morphology (FIM) and property changes in the finished product. It is useful to consider FIM in relation to an individual particle and than to the whole system.

I. FIM-particle ought be discussed in terms of the orientation, deformation and migration.

I.1 Orientation is limited to anisometric (fiber or platelet) particles. The effect of irrotational (extensional) field is particularly pronounced.

On starting, the particle aligns in the direction of the extensional force and then preserves this orientation during the flow. Experimentally, the orientation angle may show ± 5° variations. In the shear field all particles rotate. The rotation is periodic: a rapid change followed by a nearly stationary flow [3].

I.2 Deformation affects both liquid drops and solid particles (especially those with high aspect ratio). Again, the irrotational field is more powerful in deforming liquid drops [2]:

I.3 Migration of solid particles toward the low stress in the capillary center was reported [3]. Such a hydrodynamic effect differs from the wall exclusion one, in which the solid particles are expelled away from the wall zone. The shear field is also known to be able to fractionate a polydisperse mixture of particles, placing the smaller particles in the higher stress field [4]. These observations are in accord with known "laziness of nature," otherwise known as the principle of energy minimization. Thus, it is of no surprise that during flow of a two-phase system the low viscosity component migrates toward the high stress domain, e.g. lubricating the flow through pipes, runners, dies, capillaries, etc. [5]. The migration is a kinetic process whose rate depends on the stress gradient, as well as the relative viscosity and normal stress of the components.

II FIM-system strongly depends on concentration. Already at $\phi >> 0.02$ the solid particles preferentially combine into doublets while the liquid drops undergo a continuous process of dispersion and coalescence. With an increase of concentration, invariably the yield stress appears leading to the plug flow. This in turn may be responsible for several morphological features: skin-core, absence of extrudate swell (in systems containing solid particles or in blends with large values of the viscosity ratio, λ), jetting into the mold, poor weld lines, etc. In polymer blends the morphology strongly depends on concentration (ϕ) and stress field. While at $\phi \leq 0.12$ generally only drops are observed, at the higher range, say up to $\phi \cong 0.35$, fibers are apparent. In the middle concentration range, near the phase conversion, (dependent on λ), the two phases are co-continuous.

3. COMPOUNDING

Compounding is a four stage process involving (i) preparation of ingredients (drying, sizing, heating, etc.), (ii) premixing (dry blending, homogenization, breakage of agglomerates, fluxing, etc.), (iii) melt-mixing in dispersive and distributive fields (usually with degasing), and (iv) chopping, e.g. granulation, pelletization or dicing.

3.1 Compounding Equipment

The melt compounders are either batch or continuous type. The former requires lower investment cost, is more labor-intensive, has low output and poor batch-to-batch reproducibility. Recent developments in process control and automatization eliminated some of these traditional disadvantages.

To the batch mixers belong e.g. the two-roll mills, internal and kinetic energy mixers. The two-roll mill primary function is dispersion in the high shear and extensional stress field, adjustable during mixing. The distributive action is provided by continuous cutting and folding of the mixed slab. Modern mills can be programmed to perform automatically these operations. Like two-roll mills the internal mixers were originally developed for the rubber industry. Today they find use in mixing not only rubbers but a diversity of plastics formulations e.g. PVC, ABS, EVAc, PP, PE, engineering resins, master batches, blends, filled and reinforced systems, etc. Some of the internal mixers are fully automatic with capacity from 0.5

to 300 liters [6]. There is a wide variety of rotor designs. Mixing is carried out under controlled ram pressure.

All the batch mixers require additional equipment either to pelletize the compound or process it into the final product. With exception of the kinetic energy mixer compounders are difficult to scale-up. Some, theoretically based, guidance for internal mixer scaling is available [7].

The continuous mixers require high capital investment, but are easily robotizable, have high output, and can be run with a statistical quality loop control (QC) providing a high quality, reproducible compound. To this group belong: (i) twin-shaft continuous mixers, (ii) single screw extruders, (iii) twin-screw extruders, and (iv) speciality mixers.

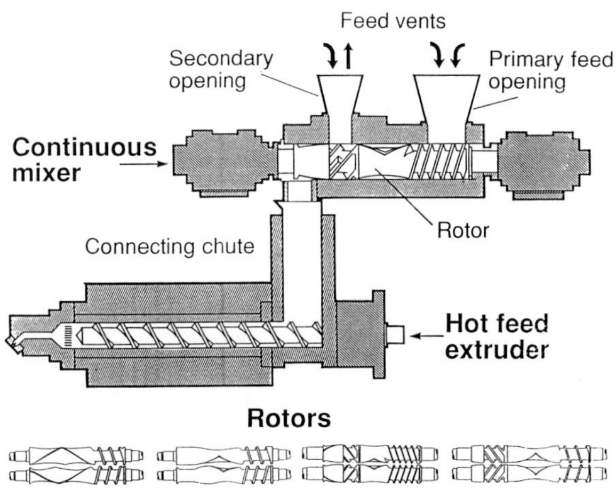

Fig. 1 Twin-shaft continuous mixer compounding line. Four standard rotors for (from left) temperature sensitive materials, filled systems, devolatilization, and short residence time materials (courtesy Farrel Co.) [22].

The twin-shaft continuous mixer (CM) or continuous internal mixer (see Fig. 1) combines the high intensity mixing with continuous operation. The first non-intermeshing, counter-rotating mixer was developed nearly 30 years ago. Since that time the machine underwent several modifications. The unit provides solid conveying, melting and mixing functions. Similarly to the batch mixers, CM's also require a separate forming unit, usually in-line melt extruder. The rotors (single or double stage) are designed with two principal elements: the solid conveying screw and two mixing wings twisted in opposite directions. An adjustable outlet gate is used to control the pressure and residence time. The relative angular positions of the wings are adjustable providing an added control element of the process. The material is continuously pumped from one rotor chamber to another. Thus, the maximum stress, randomization and total mixing time can be independently adjusted providing a wide variety of mixing conditions. The distribution of residence time is relatively narrow with a tail at longer times [8-10].

The single screw extruder is an inexpensive, difficult to scale up, poor mixer with broad residence time distribution and relatively long residence time [11]. There are two exceptions to this rule: (i) speciality mixing extruders and (ii) modified mixing extruders. To the first category belong e.g. Buss Kneader or Barmag extruders (neither inexpensive nor poor), to the second the standard single screw extruders equipped with special mixing screws, mixing and add-on devices.

There is a wide variety of modifications of a single screw extruder to improve its mixing capabilities. Diverse screw designs with mixing elements and/or mixing heads (Fig. 2) are available from most screw manufacturing companies. Some of the mixing heads can be provided as add-ons either to be affixed to the end of the screw or operated independently with its own drive [12-16]. Both the Barmag and the Cavity Transfer Mixer (CTM) require machining of both the screw and the barrel add-ons. By contrast, the University of Twente (UT) mixing

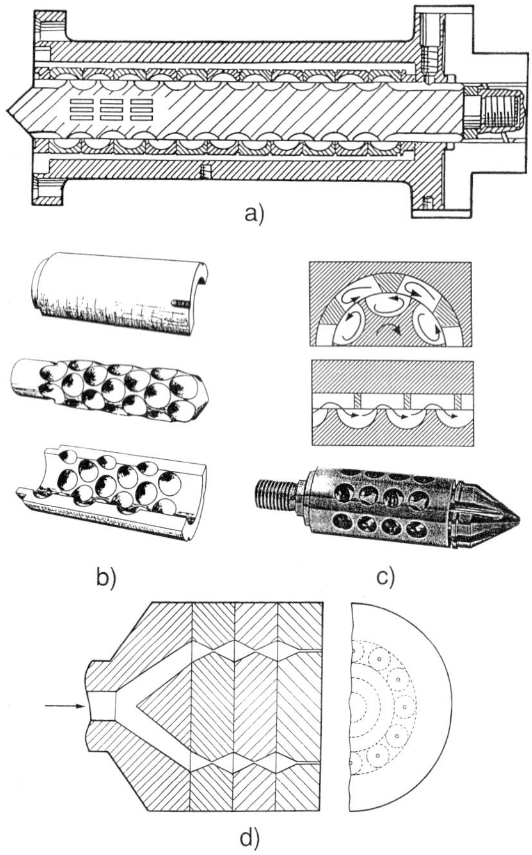

Fig. 2 Single screw extruder mixing heads: (a) Barmag, (b) CTM, (c) University of Twente, and (d) Showa Denko convergent/divergent homogenizer.

head uses a smooth barrel [17]. All three mixers are primarily distributive. In case of UT, a sleeve (with holes) placed between the barrel and the screw plays the role of the dented barrel required for the two preceding heads. The sleeve slowly rotates under influence of the drag force.

A special add-on mixer for homogenization of liquid mixtures, blends and alloys was developed by Suzaka (see Fig. 2d). The mixer operates on the principle of flow through consecutive convergent and divergent channels where the drops are elongated and broken into smaller ones. Significant improvement of mechanical properties was reported for PE/PP blends homogenized by the mixer [18].

The most popular compounders are twin-screw extruders. There is a wide variety of twin-screw extruders available on the market (Fig. 3): co- and counter-rotating, partially or not at all, intermeshing (lengthwise or crosswise) fully, with parallel or conical screws, with solid, segmental and clam-shell barrels, etc. [19]. In addition, there is a possibility of creating hundreds screw designs and configurations making selection of equipment difficult. Several existing rules may simplify this task.

In general, the lower the degree of intermeshing, the broader the distribution of the relaxation times and the lower dispersive mixing is achieved. At the limit, the non-intermeshing twin-screw extruders behave as a single screw one with enhanced distributive mixing. Thus, the main application of low intermeshing twin-screw extruders is for dispersing reinforcements, mixing blends with low and equal viscosity ingredients, etc.

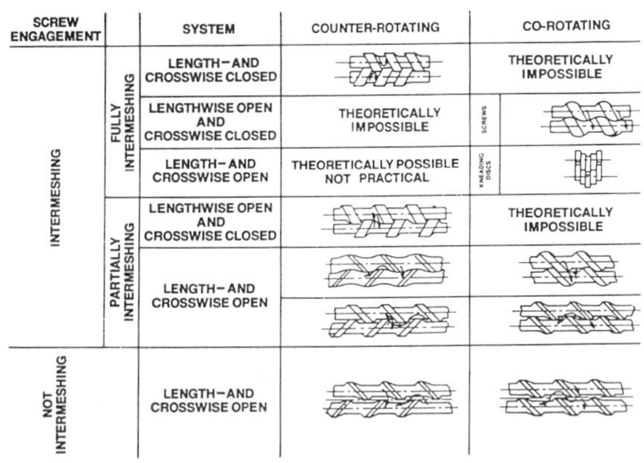

Fig. 3 Classification of twin-screw extruders .

By contrast, the intermeshing twin-screw machines are characterized by relatively narrow residence time distribution. They provide a high and controllable degree of dispersive and distributive mixing leading to homogeneous melt. In addition, here the melting is largely independent of the metal/melt friction coefficients (controlling the process in single screw extruders); there is a low pressure, the positive and controllable conveying of material, good venting, self wiping action, surge-free discharge and possibility of multiple feeding.

The intermeshing counter-rotating extruders generate higher shear and extensional stresses in the gap than the co-rotating ones. In addition, there is a different flow pattern in these two types with a higher flux of radial flow in the former. Thus the co-rotating intermeshing twin screw provides more uniform, controllable stress field. These machines are used for reactive processing, compounding of alloys and blends, for color masterbatches and fiber reinforced composition. Co-rotating extruders with low degree of intermeshing are also used for the latter application. The quality of the product critically depends on selection of screws and its configuration. Optimization of screws and operating condition is still an art [20-24].

Recent developments in process control place emphasis on use of the critical product parameters to control the process (e.g. rheological properties, chemical composition, degree of dispersion, color) instead of guarding the processing parameters constant as the older controllers used to do [25-27]. For large volume production ($\dot{Q}$ > 20 t/h) the twin-screw compounders outperform the single-screw extruders.

3.2 Compounding Industry

The annual growth rate in the compounding industry has been significantly higher (16 to 18%) than that of the plastics (3-5%) [28, 29]. The growth rate of compounded polymer alloys and color masterbatches has been the highest (18.5%), while reinforced and filled plastics is lower but still impressive 10%.

In the USA, nearly half of all plastics is being compounded by independent compounders. There is a great diversity of compounding equipment, with 44% being more than 15 years old and a growing tendency for preferential purchase of twin screw extruders and continuous mixers, with increasing emphasis on microprocessor control of the mixing and statistical quality control of the product.

The economy of compounding depends on several parameters : (i) productivity of the compounding line, Pr, expressed as a product of the volumetric output, $\dot{Q}$, and equipment availability, A:

$$Pr = \dot{Q}A \qquad (1)$$

where A depends on reliability of the machine, lifetime of wearing parts, availability of spare parts, ease of maintenance and speed of change-over from one to another formulation (strongly dependent on the residence time distribution), (ii) low amount of low-quality products and rejects, (iii) low cost-to-throughput ratio, (iv) low energy consumption and (v) low personnel cost (automatization, quality control, self-cleaning capability, robotization, etc.).

There are twelve criteria for economic, high-quality compounding [31]: 1. Sound fundamental design principles of the compounding machine. 2. Flexibility of operation. 3. High screw torque. 4. Appropriate selection of the screw elements. 5. A balanced combination of the high shear dispersive and low shear distributive screw elements. 6. Low pressure at the die. 7. Reliability of components and parts. 8. Long lifetime of compounders' parts. 9. Reliable feeding/dosing equipment. 10. Reliable granulation method. 11. Automation of the process. 12. On-line quality inspection and closed-loop control.

The importance of closed-loop line control is easily understood - it usually reduces the amount of recyclable products from 20-30% to 1-3%.

4. EXTRUSION

Mathematical description of extrusion of the two-phase systems starts with the same set of equations describing conservation of mass, momentum and energy for a single phase system [32]. The crucial step involves selection of the constitutive equation describing the stress-deformation dependencies. Computations provide pressure, stress and velocity fields which in turn can be used for calculation of bubble/drop deformation or solid particle motion. So far only severely simplified cases have been treated. Two recent volumes on mixing and extruding are worth study [33, 34]. The extruder provides the following functions: solids conveying, plasticating or melting and melt conveying or pumping. The multiphase nature of the two-phase systems becomes important in the latter two.Assuming a non-isothermal power-law flow in a co-rotating twin-screw extruder with heat flux through the barrel, one can compute the temperature, stress, pressure and velocity profiles [35]. The numerical analysis for each screw section starts with assumption of Newtonian flow, from which the "length" of drag and pressure flow are calculated. Next, at the incremental step the local pressures, temperatures, leakage flow and power consumption are determined, then the computation loops to correct for leakage flow and the absolute pressure. Once the correct value of the absolute pressure is obtained the next screw element is computed in analogous manner. Excellent predictive ability of the method was reported; the melt temperatures in ZSK300 under diverse processing conditions were predicted with a difference of ± 3°C from the measured values, whereas the specific energies were predicted with a difference of ± 3%. Surprisingly, as shown in Fig. 4, the simple schematic worked better for the kneading disks sections than for the conveying screw elements [35].

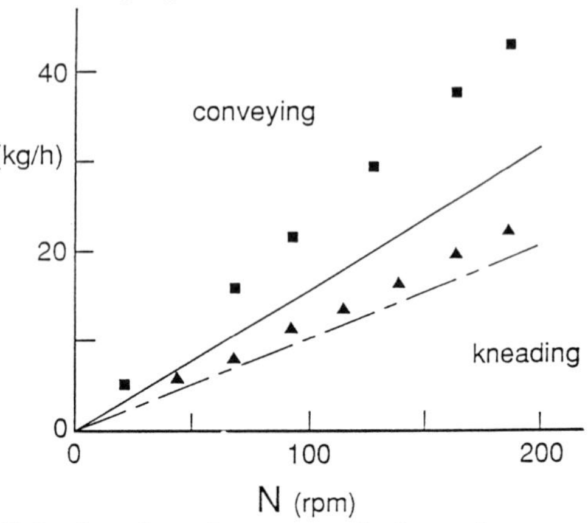

Fig. 4 Calculated and measured throughput vs. screw speed for conveying and kneading elements of a twin-screw extruder [35].

4.1 Blends Extrusion

It is difficult to generalize blend extrusion. Such blends as ABS, HIPS, Noryl, Bayblend, etc. can be processed as single phase polymers under the condition that processing parameters specified by the manufacturer are strictly adhered to. However, blends are generally more sensitive to process variables (e.g. stress, temperature, pressure, residence time) and need more sophisticated line control than the single-phase polymers [2, 36-39]. The key is stability of the morphology.

4.2 Extrusion of Particulate-filled Systems

Devolatilization during extrusion of particulate-filled polymer is required. Thus, the screws used for this application are of two-stage type. Owing to abrasive wear the screw, the barrel and the die should be hardened and the wear systematically measured [33, 40, 41].

Flow of highly filled thermoplastics is characterized by the yield stress leading to plug flow in the extruder and through the die, which in turn results in small, if any, extrudate swell [42-46]. Processing these materials requires relatively small flight depth which means either small single screw extruder, a mixing extruder or a twin-screw extruder. Orientation of the anisometric particles is controlled by flow convergence or divergence [46-48].

Fiber breaking during processing of fiber-filled thermoplastics (attrition) originates in flow vorticity and occurs at any filler concentration [49, 50]. By contrast, platelet breaking is primarily due to the mechanical platelet-platelet contact and occurs only above the dilute range of concentration $\phi \geq p^2$ [50]. The higher the stress and the longer the residence time, the higher the degree of attrition. In Fig. 5 the effect of concentration on particle attrition is illustrated.

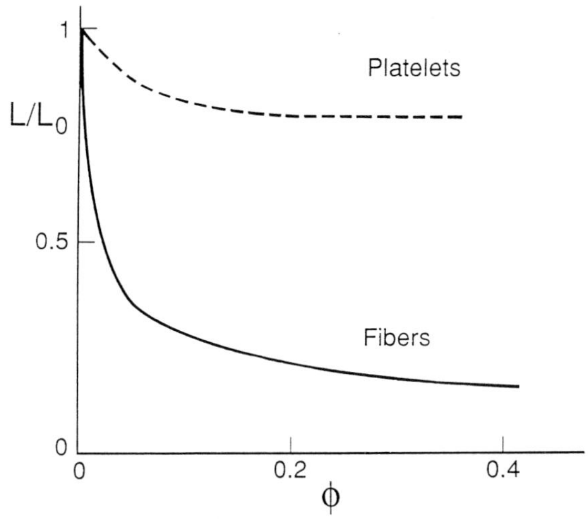

Fig. 5 Concentration dependence of glass fiber and mica flakes attrition during 2 min compounding with PP in an internal mixer at 200°C [50].

An interesting solution to the abrasive wear and particle attrition can be use of a short, non-contacting cantilever screw and roller die [51]. Increasingly, the compounding and forming of reinforced systems is combined into a single operation. The automatization, closed-loop control as well as new developments in side (or bottom) feeding of the reinforcements to molten thermoplastics are responsible for the trend. Several machine manufacturers developed the latter devices [52, 53]. The side feed introduces the chopped glass fiber (GF) at a point where there is no melt pressure. It uses a short side, starve-fed twin-screw extruder and a vent. The main extruder screw is of the three-stage type, allowing for further devolatilization and a two-stage GF mixing. The advantage of the side feed is the rapid filler incorporation into the melt which reduces attrition and machine wear.

5. INJECTION MOLDING

Each year the injection molding units are becoming larger and more numerous. It is estimated that worldwide there are over 50,000 injection molding plants, consuming annually about $4 billion worth of processing equipment (or 40% of total), manufactured by 400 companies. The injection molding machines with 90,000 kN clamping force, shot size up to 90 liters from two injection units are available. Two giant multiprocessing machines are being built for injection and compression molding of thermoplastics based composites for automotive and housing industries [54].

However, the biggest changes in the field are primarily due to robotization, automation and control. Robots (e.g. 20t, 7 axes unit from Battenfeld) perform multiple tasks at a speed of several m/s and payload capacity of several hundred kilograms. Larger to 1000 t robots are also available (e.g. from ATM Automation). These can handle up to 100 programs and operate with 100 μm precision at speeds up to 1 m/s. Robotization and computerization leads to automatized computer integrated manufacturing (CIM) where flow of material (just-in-time delivery), control of the injection molding machine, statistical process control, parts handling and documentation are integrated for maximum efficiency and economy. The integration includes peripheral operations such as tool transport, handling, changing and preheating facilities, transport of raw materials and finished goods. In near future the system will also include the data base for technical and managerial information to cover the whole spectrum of operations from mold part design to the product sale.

Computer aided design and manufacture (CAD/CAM) is reaching its maturity and universal acceptance. As indicated in Fig. 6 the

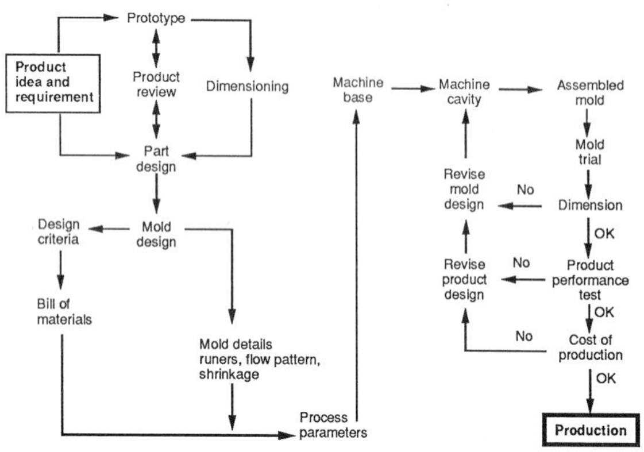

Fig. 6 Schematic representation of CAD/CAM mold and part design.

process starts with definition of requirements for the plastic part and selection of material. Next,

the part is designed and the computer analyzes its performance in the intended environment as well as verifies if the design respects the accepted criteria. The runners are then positioned and melt flow in the cavity computed with particular attention given to the weld lines, air entrapment as well as pressure (P) and temperature (T) profile. In the last stage the holding pressure and mold shrinkage are computed. After each stage the computer program makes it possible to go back in order to optimize the product molding and performance. The final stage generates the set of suggested process parameters and TC type for mold cutting [39]. Optimization of processing parameters is usually done by operator, although automatic, self-optimization programs are being developed [55].

Different control systems are available. These usually incorporate: user-friendly programming, integrated quality control (surveys basic molding parameters, makes statistical control, rejects defects), molding system (automatic adjustment of molding parameters to preserve molding quality), filling and feeding control (automatic adjustment of shot volume and hold-on time) and PVT program (check the holding stage, as far as P, T and time is concerned to minimize stresses and shrinkage in the finished product) [91].

The final part of automation of the injection molding line is the mold change. The programmable quick mold changing (QMC) systems are available either from the injection molding manufacturers or from independent houses [56-59].

6. OTHER PROCESSING METHODS

The extrusion and injection molding hardly exhaust the processing methods. However, the aim of this review is not to provide a complete list of processing methods, but rather to touch on those considered more important.

Some blends have been specifically designed to be used in fiber spinning or blow molding. The extensional flow deforms the dispersed phase into fibrillas or platelets, enhancing the processing and/or performance. In the first case better line stability, better surface finish, zero thermal shrinkage and improved fiber strength have been observed [60]. In the second case improved barrier properties were achieved [61].

There is a large potential for use of metal-forming methods with blends and reinforced thermoplastics. Hydrostatic extrusion, rolling, forging, etc. have been successfully used for neat polymers [62-64]. The preliminary results on rolling of mica-filled polypropylene [65] and rotational extrusion of glass-fiber-filled polypropylene [66] are quite promising. Good control of orientation and improvement of mechanical properties were demonstrated.

There is a growing trend for replacing the advanced composites based on thermosetting resins by those containing thermoplastics [67].

REFERENCES

1. Anon. Chem. Market. Rep., p. 3, 23, Nov. 23 (1987).
2. Utracki, L.A., "Polymer Alloys and Blends," Hanser Publ., Munich (1989).
3. Goldsmith, H. L., Mason, S. G., in "Rheology, Theory and Applications," F. R. Eirich, Ed., Vol. 4, Academic Press, New York (1967).
4. Utracki, L.A., Ch. 15 in "Rheological Measurements," Collyer, A. A. and Clegg, D. W. Eds., Elsevier Appl. Sci., London (1988).
5. Han, C. D., "Multiphase Flow in Polymer Processing," Academic Press, New York (1981).
6. Anon. "Moriyama Dispersion Mixers," Moriyama Corp. Japan.
7. Manas-Zloczower, I., Tadmor, Z., Rubber Chem. Technol., **57**, 48 (1984).
8. Hold, P., Adv. Polym. Technol., **4**, 281 (1984).
9. Valsamis, L. N., Canedo, E. L., Donoian, G. S., Farrell Corp., (1989).
10. Bull. Nos. 234, 238, 239, Farrell Corp. (1988, 1989).
11. Hold, P., Adv. Polym. Technol., **2**, 197 (1982).
12. Renk, P., U. S. Pat. Nos. 4,128,342, Dec. 5 (1978); 4,253,771, March 3 (1981).
13. Anon., "Barmag 3DD Mixer," Barmag A. G., Inf. Serv. No. 30 (1989).
14. Gale, G. M., Plast. Comp., p. 70, May-June (1985).
15. Hindmarch, R. S., Gale, G.M., Elastomeric, 114(8), 20 (1982).
16. Alzner, B.G., Allen, W. F., Csongor, D., S.P.E. Techn. Pap., **32**, 883 (1986).
17. Housz, I., Polym. Process. Soc., European Meeting, Kerkrade, 30 Oct. - 2 Nov. (1989).
18. Suzaka, Y., U. S. Pat. No. 4,334,783, June 15 (1982).
19. Eise, K., Herrmann, H., Werner, H., Burkhardt, U., Adv. Plast. Technol., **1**, 1 (1981).
20. Anon., "High Performance Twin-screw Compounding System," Werner & Pfleiderer GmbH, Techn. Inf. Bull. Nos. 21E.8910A.1000, 05-098/2, 05-085/1, 05-100/2 (1989).
21. Anon., "Bitruders BT", Reifenhäuser Techn. Inf. Bull. No. 0/3.2-8903 (1989).
22. Anon., "A No-Nonsense Guide to Extrusion Systems for Continuous Compounding," Welding Eng. Inc., Techn. Bull (1989).
23. Anon., "MPC/V, APV Chemical Machinery," (Baker Perkins), Techn. Inf. (1989).
24. Anon., "Twin-Screw Continuous Mixers," ComacPlast, Techn. Inf. (1989).
25. Anon., "Bex 2-65/BEC 2000", Battenfeld Extrusiontechnik Techn. Inf. (1989).
26. Ultsch, S., Fritz, H.-G., Polym. Proc. Soc., Annual Meeting, Kyoto, Apr. 11-14 (1989).
27. Anon., "Production of Silane Crosslinkable PE-tubes", Werner & Pfleiderer Techn. Inf. 05-075/1 (1989).
28. Anon., Plast. Comp., pg. 63, May-June (1985).
29. Brewer, D., Plast. Comp., pg. 12, Jan-Feb. (1987).
30. Bigg, D. M., in "Science and Technology of Polymer Processing," Suh, N. P., Sung, N.-H., Eds., MIT Press, Cambridge, MA (1979).
31. Herrmann, H., Kunststoffe, **78**, 876 (1988).
32. Bird, R. B., Armstrong, R. C., Hassager, O., "Dynamics of Polymeric Liquids," 2nd edition, J. Wiley and Sons, New York (1987).
33. Rauwendaal, C., "Polymer Extrusion," Hanser Publ., Munich (1986).
34. Cheremisinoff, N. P., "Polymer Mixing and Extrusion Technology," M. Dekker Inc., New York (1987).
35. Elemans, P. H. M. , PhD thesis, Techn. Univ. Eindhoven (1989).
36. Golba, J. C., Seeger, G. T., S.P.E. RETEC, Huron, OH., March 10-11 (1986).

37. Anon., "Bayer Thermoplastics Processing Extrusion," Bayer Techn. Ring Folder "Thermoplastics" No. E63-861/848736 (1989).

38. Menges, G., Makromol. Chem., Macromol. Symp., **23**, 13 (1989).

39. Michaeli, W., Angew. Chem., Int. Ed., **28**, 644 (1989); World Plast. Rubb. Technol., **2**, 187 (1990).

40. Menning, G., Voltz, P., Kunststoffe, **70**, 385 (1980).

41. Schuele, H., Fritz, H.-G., Kunststoffe, **73**, 603 (1983).

42. Utracki, L.A., Fisa, B., Polym. Compos., **3**, 193 (1982).

43. Utracki, L.A., Favis, B. D., Fisa, B., Polym. Compos., **5**, 227 (1984).

44. Utracki, L.A., Rubber Chem. Technol., **57**, 507 (1984).

45. Utracki, L.A., Polym. Compos., **7**, 274 (1986).

46. Fisa, B., Utracki, L.A., Polym. Compos., **5**, 36 (1984).

47. Freundlich, R. A., Brit. Pat. Appl. No. 2,055,680, March 11 (1981).

48. Goettler, L.A., Lambright, A. J., U. S. Pat. Nos. 4,056,591, 4,057,610 Nov. (1977).

49. Czarnecki, L., White, J. L., J. Appl. Polym. Sci., **25**, 1217 (1980).

50. Fisa, B., Polym. Compos., **6**, 232 (1985).

51. Hagiware, K., Nakagawa, K., U. S. Pat No. 4,304,539, Dec. 8 (1981).

52. Hulbert, W. H., S.P.E. RETEC, Huron, OH. March 10-11 (1986).

53. Anon., "ZSK 40 Alloying, Reinforcing and Colouring", Werner & Pfleiderer Techn. Inf. No. 05 085/1-1.5 (1989).

54. Anon., Modern. Plast. Int., K89 Newsletters, Nov. 3, p. 4 (1989).

55. Fillmann, W., World Plast. Rubb. Technol., **2**, 139 (1990).

56. Anon., "Sandretto Series Eight Injection Molding Machines," Techn. Inf. (1989).

57. Anon., "Rapidomat Automatic Mold Changer," Arburg Techn. Inf. (1989).

58. Anon., "Enerpac QMC," Techn. Inf. (1989).

59. Anon., "PS20E5ASE", Nissei Techn. Inf (1989).

60. Utracki, L.A., S.P.E., NATEC, Bal Harbour, FA, Oct. 25-27, 1982; Polym. Eng. Sci., **23**, 602 (1983).

61. Anon., "SELAR Barrier Resins," DuPont Techn. Inf. E 73973 (1984).

62. Cifferri, A., Ward, I. M. (Eds.), "Ultra-high Modulus Polymers," Appl. Sci. Publ., London (1979).

63. Zachariades, A. E., Porter, R. S. (Eds.), "The Strength and Stiffness of Polymers," M. Dekker, New York (1983).

64. Ward, I. M., "Mechanical Properties of Solid Polymers," 2nd ed; J. Wiley and Sons, New York (1983).

65. Lee, I., Turner, S., Woodhams, R. T., Polym. Compos., **3**, 212 (1982).

66. Zachariades, A. E., Chung., B., S.P.E. Techn. Pap., **32**, 965 (1986).

67. Utracki, L.A. (Ed.), "Panel Discussion on Thermoplastics vs. Thermosets in Advanced Composites," Polym. Compos., **8**, 437 (1987).

Frontiers of Materials Research/
Electronic and Optical Materials
M. Kong, L. Huang (editors)

AUTOMOBILE EXHAUST EMISSION CONTROL BY CATALYSTS

Kathleen C. Taylor

Physical Chemistry Department, General Motors Research Laboratories, Warren, Michigan 48090 USA

This paper presents a short history of the use of catalysts for controlling automobile exhaust emissions in the United States from 1975 through 1989. The evolution of catalyst materials over that time period is reviewed.

1. INTRODUCTION

Both the noble metal catalysts and the emission control systems in automobiles have undergone several changes during the last decade in response to tightening emission control requirements, while at the same time meeting the demands for vehicle performance. Today the composition of the catalyst has been tuned to meet the need for high activity and durability; engine control systems have been developed that meet the catalyst operating requirements. Research on emission control catalysts has provided understanding of the catalyst formulation - catalytic chemistry relationships. The evolution of exhaust emission control catalysts from 1975 through 1989 will be reviewed here. For additional information on automobile exhaust catalysts and their application the reader is referred to several review articles[1,2,3] and many publications of the Society of Automotive Engineers[4].

2. HISTORY

2.1. 1975 to 1980

The passage of the 1970 Clean Air Act by the U.S. Congress placed stringent limits on emissions of carbon monoxide (CO), hydrocarbons (HC), and nitrogen oxides (NO_x) from automobiles. An abbreviated version of these requirements is listed in Table 1. To meet the 1975 HC and CO requirements, oxidation catalysts were placed on all vehicles. These catalysts were composed of platinum and palladium supported on alumina. What this catalyst

Table 1. Passenger car regulatory exhaust emission control requirements in the United States. Emissions expressed in grams per mile as determined by the 1975 Federal Test Procedure.

Model Year	HC	CO	NO_x
Federal			
1975	1.5	15	3.1
1977	1.5	15	2.0
1981	0.41	7	1.0
1983	0.41	3.4	1.0
California			
1975	0.9	9	2.0
1977	0.41	9	1.5
1980	0.39(non CH_4)	9	1.0
1981	0.39	7	0.7
1989	0.39	7	0.4
1993	0.25	3.4	0.4

does is to promote the complete oxidation of CO and HC to carbon dioxide and water at catalyst temperatures above 300°C under net oxidizing (lean air-fuel ratios) engine operating conditions. Nitrogen oxides are not reduced by the catalyst under lean operating conditions. NO_x control was achieved by exhaust gas recirculation which lowers the amount of NO_x formed in the engine.

Exhaust emission control was a totally new application of catalysts so many decisions had to be made associated with the selection of materials, including noble metal availability, catalyst performance, and long term durability of the catalyst. Lead had to be removed from gasoline, and cars had to be modified to operate on 91 octane lead-free fuel.

The introduction of catalysts on cars also raised some concerns. For example, the regulators were concerned that high levels of sulfuric acid would be formed over the catalyst and accumulate near busy freeways. A large field study conducted by General Motors demonstrated that sulfuric acid accumulation was not a problem[5]. Another concern was that the catalyst would be poisoned by inappropriate use of leaded fuel.

Stricter exhaust emission standards for HC and CO in the late 1970's were met using new low-density catalyst supports (pellet type) which were more resistant to poisoning, and for some vehicles air was added to the exhaust ahead of the catalyst to enhance the oxidation reactions.

2.2. 1981 to 1989

Stricter standards for NO_x starting in 1981 meant that NO_x had to be controlled with the catalyst. Several approaches were used. One was a dual-catalyst system (reduction catalyst followed by oxidation catalyst) whereby the engine was operated slightly rich. The first catalyst promoted the reduction of NO_x to nitrogen. Air was then added to the exhaust between the two catalysts so the second catalyst completed the oxidation of CO and HC.

The approach which is now widely used for controlling all three pollutants is the current 3-way catalyst. Use of single 3-way catalyst became feasible with improved A/F control. This catalyst uses rhodium together with platinum to promote the simultaneous reduction of NO_x to nitrogen and the oxidation of CO and HC. The vehicle is operated at the stoichiometrically balanced air/fuel ratio because this is where the Pt/Rh catalyst best promotes all three reactions (Table 2).

Table 2. Overall reactions catalyzed by the 3-way catalyst.

$$CO + 1/2O_2 \longrightarrow CO_2$$
$$\text{hydrocarbons} + O_2 \longrightarrow H_2O + CO_2$$
$$NO_x + CO\ \&\ H_2 \longrightarrow N_2 + CO_2 + H_2O$$

To maintain the exhaust gas at the stoichiometrically balanced composition, a closed-loop control system which makes use of an oxygen sensor and microprocessor was introduced on all cars. The property of rhodium to simultaneously react HC, CO, and NO_x is shown by the high conversions in the shaded area in Figure 1. The conversion of NO_x is high under reducing conditions, and the conversions of CO and HC are high under oxidizing conditions. Close to the stoichiometrically balanced composition there is a narrow region where the conversion of all 3 pollutants is simultaneously high. Cerium is added to catalyst support of the 3-way catalyst to enhance both the activity and durability. Current understanding of the role of cerium is dealt with in a later section of this paper.

Automobile exhaust catalysts account for a significant percentage of the demand for the platinum group metals in the western world. In 1988 autocatalysts represented 73% of the

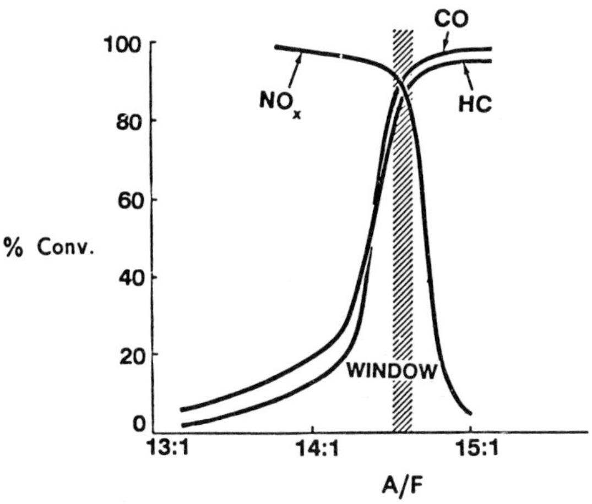

FIGURE 1
Three-way catalyst operating window

total demand for rhodium, 32% of the total demand for platinum, and 6% of the total demand for palladium. The demand for platinum for autocatalysts was 1,325,000 oz t[6]. The amount of rhodium used in autocatalysts in relation to the amount of platinum is approximately twice the naturally occurring mine ratio for the upper deposit mines[2].

Currently the U.S. Congress is debating stricter vehicle emissions standards. We see no radically new technology to further reduce emissions in the near future. Effort is being placed on ways to bring the catalyst up to full activity sooner following the start of the vehicle since the CO/HC emissions released during the first several minutes following the start of the cold engine are a significant fraction of the emissions for the test cycle[7].

3. CATALYST SUPPORTS AND CATALYST PREPARATION

The supports for automotive catalysts are of two general types: alumina pellets and monoliths. Most monoliths are composed of cordierite ($2MgO \bullet 2Al_2O_3 \bullet 5SiO_2$) formed in a honeycomb structure with 400 channel openings per sq. in. Typical monolith volumes are 170 cu. in. for a dual monolith converter and 110 cu. in. or 85 cu. in. for a single monolith converter. A diagram of a monolith type converter is shown in Figure 2. Metal foil

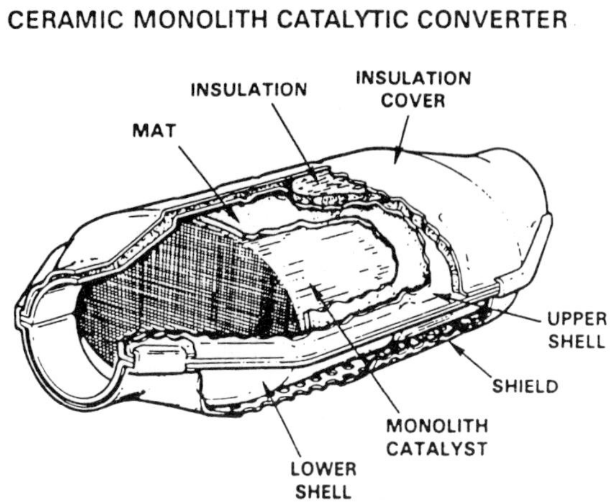

FIGURE 2
Ceramic monolith catalytic converter

monoliths are gaining wider acceptance, especially for highly engineered applications requiring complex geometries. The thinner wall thickness of the metal foil monolith compared with the ceramic (50 μm vs 150 μm) can mean larger channel openings if the number of channels and monolith volume are held constant, or more channel openings and more internal support area for noble metals, if the number of channels is increased while the monolith frontal area and volume are held constant.

Studies of the influence of monolith support design on catalyst light-off (defined as 50% conversion), steady-state conversion, and pressure drop on converter performance have demonstrated that cell density, monolith length, and frontal area are important design factors[8]. A three-dimensional model of a monolithic catalytic converter has been developed for use in analyzing the transient thermal and conversion characteristics during warm up and thermal cycling[9,10]. These models are used to assess simultaneously the combined design objectives for structural reliability and conversion efficiency.

An alumina washcoat is applied to the surface of the monolith as a thin slurry to form a 30-50 micron thick, high surface area (approximately 100 m^2/g) coating. Alumina pellets used in pellet type catalytic converters are generally 1/8 inch in diameter and are composed of thermally stable transition alumina phases. Thermal stabilization is achieved by the addition of rare earth oxides such as lanthanum[11,12]. The stabilization mechanism is a strong surface interaction between the alumina and a stable cubic perovskite compound, $LaAlO_3$, which inhibits the structural transition to corundum[11]. Alkaline earth oxides such as barium and strantium can also enhance the stability of alumina[12,13].

The noble metals are applied to the pellets by spraying the pellets with a dilute solution of the metals. The noble metals are applied to the monolith by dipping the alumina coated monolith into a dilute solution of salts of the noble metals. Noble metal use is typically in the range 0.05-0.08 oz t platinum and 0.005-0.009 oz t rhodium per converter.

4. CURRENT RESEARCH

Three areas of recent research on three-way catalysts will be reviewed briefly. These are

(1) catalyst promotion by ceria,

(2) durability/deactivation of catalysts, and

(3) control/prevention of hydrogen sulfide formation.

4.1. Catalyst promotion by ceria

The impact of ceria on the performance of three-way catalysts has been shown to have multiple functions which were not initially understood in detail. Ceria was first used at the 2 wt% level in the washcoat of 3-way catalysts. The ceria content of the washcoat of current 3-way catalysts is 10 to 30 wt%. One explanation which has been given for ceria's beneficial effect is that ceria stores oxygen by undergoing oxidation (benefiting NO reduction to N_2) during excursions to lean A/F, and this oxygen is then available for reaction with CO during excursions to rich A/F (benefiting CO conversion). This simple view of the role of ceria could not fully explain the magnitude of the CO reaction observed in laboratory experiments which lead to the conclusion that ceria was also promoting the water-gas shift reaction ($CO + H_2O \rightarrow CO_2 + H_2$)[14,15]. In recent studies of the kinetics of CO oxidation and NO reduction over ceria containing rhodium/alumina catalysts it has been demonstrated that ceria directly promotes the CO-O_2 reaction as well. Ceria favorably alters the oxidation of CO over rhodium by suppressing the inhibition of the CO-O_2 reaction by CO, by decreasing the sensitivity of the reaction rate to the gas-phase oxygen concentration, and decreasing the apparent activation energy under moderately oxidizing or net-reducing conditions[16]. The effect on CO_2 production of the addition of various levels of ceria to a 0.014 wt% rhodium/alumina catalyst is shown in Figure 3. Likewise ceria addition to a low-loaded alumina supported rhodium catalyst enhanced the NO reduction activity at low temperatures by decreasing the apparent activation energy for the NO-CO reaction and by shifting to positive order the dependence of the rate on NO partial pressure[17].

The influence of ceria in increasing the oxygen storage capacity of platinum and rhodium containing catalysts has been demonstrated in a study which compared the nitric oxide decomposition activity of Pt/alumina, Pt/ceria, Rh/alumina, and Rh/ceria[18]. The alumina supported platinum and rhodium cata-

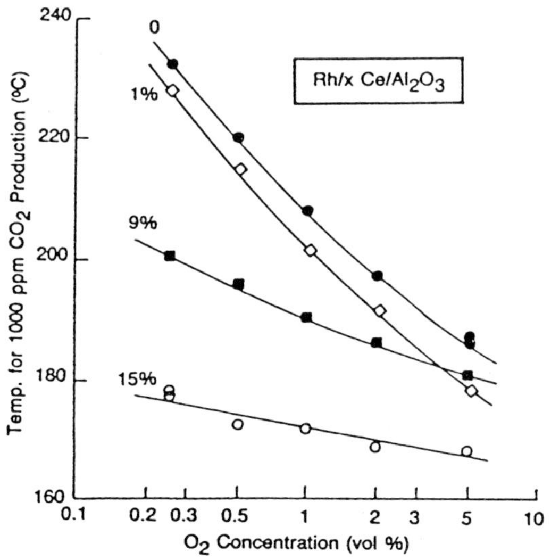

FIGURE 3
Temperature for 1000 ppm CO_2 production over 0.014% Rh/Al_2O_3 containing variable levels of Ce as a function of the O_2 concentration in the feed. The feedstream concentration of CO was held constant at 2 vol%

lysts are active for the decomposition of NO, but are slowly deactivated by the accumulation of oxygen derived from NO decomposition. The Pt/ceria and Rh/ceria catalysts were shown to deactivate more slowly than the alumina supported counterparts, due to a greater oxygen storage capacity of the ceria support than the alumina support[18]. The transient NO decomposition over Rh/alumina and Rh/ceria at 300°C following reduction of the catalyst at 600°C is shown in Figure 4. In other studies the noble metal has been shown to play a role in facilitating the reduction of ceria which is important because oxygen storage capacity is regenerated at a lower temperature[13]. Using temperature programmed reduction rhodium was shown to lower the temperature of reduction of the surface oxide on ceria to 350°C compared with a reduction temperature of 500°C for ceria without noble metal present[13].

An important role of ceria as a promoter of rhodium containing exhaust catalysts may be the maintenance of high rhodium dispersions.

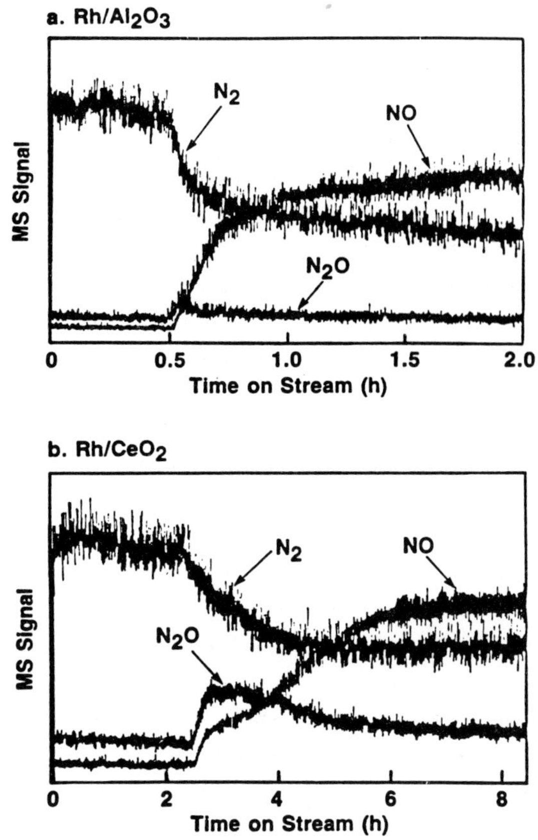

FIGURE 4
Transient NO decomposition over Rh/Al_2O_3 and Rh/CeO_2 at 300°C

An investigation of the promotional effects of ceria on rhodium catalysts by infrared spectroscopy which made use of the distinctive adsorbed states of CO on rhodium showed that the total integrated absorbance of the carbonyl absorbates did not decline significantly during a series of sintering experiments[19]. A high state of dispersion was maintained in the presence of CO induced sintering because the rhodium presumably coalesces to form 2-dimensional rhodium surfaces[19]. In contrast, sintered rhodium particles on non-ceria catalysts are 3-dimensional and larger than those of the ceria promoted catalyst[19,20].

4.2. Durability/deactivation of catalysts. Exhaust catalyst must be able to withstand relatively high temperatures during use. While 350°C to 500°C is typical, higher temperatures are possible at high vehicle speeds and loads. Some vehicles are being equipped with small additional catalysts up close to the engine to enhance the removal of cold-start emissions. These small converters see temperatures perhaps 100°C higher than converters in the traditional under-floor location. For very high temperatures (above 1000°C) the monolith may be damaged due to thermal shock or melting and the alumina washcoat is susceptible to loss of surface area. However, at temperatures as low as 700°C there can be gradual loss in the dispersion of platinum and rhodium catalyst particles supported on alumina[21,22,23]. High dispersion is important because metal buried within large metal particles is not available for promoting catalytic reactions.

Numerous studies have been done which examine changes in the morphology and activity of platinum and rhodium containing catalysts as a result of extended use in exhaust as well as changes produced as a result of controlled exposures in laboratory experiments. The laboratory studies while important for understanding the influence of specific conditions on catalyst performance do not capture completely the varied and dynamic conditions experienced by a working catalyst. The surface state of the catalyst is dependent on reaction conditions and undergo a continuous restructuring in response to the changing oxidizing and reducing conditions[24]. An analytical electron microscopy study of two vehicle-aged automotive exhaust catalysts having dissimilar activities revealed that while the microstructure of the supports were found to be similar, the average metal particle size on the poorer catalyst (21 nm) was larger than on the better catalyst (16 nm)[25]. For both catalysts these sizes represent an increase in particle size compared with the fresh catalyst (8 nm)[25]. A high-resolution electron microscopy study of a rhodium/ceria/alumina catalyst aged in air for 24 hours at 950°C revealed that platinum particles grew to ten times their size on the fresh catalyst while the rhodium-rhodium oxide particles remained disperse[24].

Rhodium supported on alumina oxidized at 700° to 800°C is converted to a hard to reduce form which appears to dissolve into the alumina support[22,23,26,27]. Rhodium/alumina converted to this hard to reduce form has been reported to exhibit a five- to ten-fold decrease in CO oxidation activity[26]. A similar high temperature reduction in hydrogen can restore the rhodium to the surface[22,23,26]. High temperature oxidation of Rh/alumina containing exhaust catalysts is believed to be detrimental since strong reducing conditions which reverse the deactivation are infrequent.

Deposits of contaminants which have been identified on vehicle-aged exhaust catalysts are P, Pb, Zn, Ca, and Si[25]. These elements were present as dense, noncrystalline deposits and often appeared as a shell less than 10 um deep at the outermost surface of the catalyst[25]. The major source of P, Ca, and Zn is engine oil; the source of Pb is fuel additives. The specific chemical effect of poisoning is difficult to assess even qualitatively[25]. Fortunately, with the removal of lead from fuel in the U.S., catalyst poisoning is probably no longer a major cause of catalyst deactivation.

4.3. Control/prevention of hydrogen sulfide formation

The sulfur present in gasoline results in engine-out emissions of sulfur dioxide which depend on the sulfur content of the gasoline. A typical exhaust gas concentration of sulfur dioxide is 20 ppm. This SO_2 may be oxidized over the catalyst under lean exhaust conditions and stored on the catalyst as aluminum sulfate and cerium sulfate. The use of high concentrations of cerium in the 3-way catalyst washcoat has increased the amount of sulfur stored on the catalyst. This stored sulfur can be a source of high instantaneous emissions of hydrogen sulfide under net reducing vehicle operating conditions. During momentary vehicle operation at net reducing A/F the stored sulfate is rapidly reduced to hydrogen sulfide, and levels of hydrogen sulfide above the odor threshold can be emitted from the catalyst. Several studies have been undertaken to understand the detailed mechanism of hydrogen sulfide formation[28,29,30].

Exhaust catalysts have been modified to control and prevent hydrogen sulfide formation. One method is to reduce the release of hydrogen sulfide by adding nickel oxide to the catalyst. Nickel oxide acts as a getter by forming a stable sulfide under rich exhaust conditions. Any hydrogen sulfide formed quickly reacts with the nickel oxide and is not released from the catalyst. The sulfur storage capacity of the nickel oxide is regenerated by release of the sulfur as SO_2 under net oxidizing conditions. Another way to prevent hydrogen sulfide formation is to modify the catalyst support to lower its capacity for storing sulfate[29]. For example, the amount of sulfur oxides stored on the catalyst can be reduced by increasing the crystallite diameter of the ceria or by modifying the surface hydroxyl-groups of the support[29].

REFERENCES

1. Kathleen C. Taylor, Automobile Catalytic Converters (Springer-Verlag, Berlin, 1984).
2. M.L. Church, B.J. Cooper, and P.J. Willson, SAE Paper 890815 Feb. 27-Mar. 3, 1989.
3. J.T. Kummer, Progress in Energy and Combustion Science, 6(1980)177.
4. Society of Automotive Engineers, 400 Commonwealth Drive, Warrendale, PA 15096, USA.
5. S. Cadle, D.P. Chock, P.R. Monson, and J.M. Heuss, Air. Poll. Contr. Assoc. Journal 27(1977) 33.
6. Platinum 1989 (Johnson Matthey, London, 1989).
7. W.A. Whittenberger and J.E. Kubsh, SAE Paper 900503, Feb. 26-Mar. 2, 1990.
8. J.P. Day and L.S. Socha, SAE Paper 881590 Oct. 10-13, 1988.
9. D.K.S. Chen, E.J. Bissett, S.H. Oh, and D.L. Van Ostrom, SAE Paper 880282 Feb. 29-Mar. 4, 1988.
10. D.K.S. Chen and C.E. Cole, SAE Paper 890798 Feb. 27-Mar. 3, 1989.
11. F. Oudet, P. Courtine, and A. Vejux, J. Catal. 114(1988)112.
13. B. Harrison, A.F. Diwell, and C. Hallett, Platinum Metals Reviews 32(2) (1988) 73.
14. J.C. Schlatter and P.J. Mitchell, Ind. Chem. Prod. Res. Develop. 19(1980)288.
15. R.K. Herz, J. Kiela, and J. Sell, Ind. Chem. Prod. Res. Develop. 22(1983)381.
16. S.H. Oh and C.C. Eickel, J. Catal. 112(1988)543.
17. S.H. Oh, "Effects of Cerium Addition on the CO-NO Reaction Kinetics over Alumina-Supported Rhodium Catalysts", General Motors Research Publication, GMR-6874, November, 1989. J. Catal. in print.
18. B.K. Cho and C.J. Stock, "Transient Study of NO Decomposition over Supported Noble Metal Catalysts", General Motors Research Publication GMR-5497. August, 1986. Presented at the AIChE Meeting, Miami Beach, FL, November, 1986.
19. R. Dictor and S. Roberts, J. Phys. Chem. 93(1989)5846.
20. R. Dictor and S. Roberts, J. Phys. Chem. 93(1989)2526.
21. D.D. Beck and C.J. Carr, J. Catal. 110(1988) 285.

22. D.D. Beck and C.J. Carr, "A Study of Thermal Aging of Rhodium/Alumina Using Temperature Programmed Desorption Spectroscopy of Deuterium and Carbon Monoxide" General Motors Research Publication, GMR-6421, September, 1988. J. Phys. Chem. in print.

23. D.D. Beck and C.J. Carr, "Effects of In Situ Thermal Aging on the Carbon Monoxide Oxidation Activity of Rhodium/Alumina", General Motors Research Publication, GMR-6676, May, 1989. J. Catal. in print.

24. B. Engler, E. Koberstein, and P. Schubert, Applied Catal. 48(1989)71.

25. S. Kim and M.J. D'Aniello, Jr., Applied Catal. 56(1989)23.

26. C. Wong and R.W. McCabe, J. Catal. 119(1989)47.

27. J.G. Chen, M.L. Colaianni, J.T. Yates, and G.B. Fisher, "Thermal Behavior of a Rh/Al_2O_3 Model Catalyst: The Disappearance of Surface Rh Upon Heating", General Motors Research Publication, GMR-6830, October, 1989.

28. T.J. Truex, H. Windawi, and P.C. Ellgen, SAE Paper 872162 Nov. 2-5, 1987.

29. E.S. Lox, B.H. Engler, and E. Koberstein, SAE Paper 890795 Feb. 25-Mar. 3, 1989.

30. I. Gottberg, E. Hogberg, and K. Weber, SAE Paper 890491 Feb. 27-Mar. 3, 1989.

Frontiers of Materials Research/
Electronic and Optical Materials
M. Kong, L. Huang (editors)

ORDERED INTERMETALLIC COMPOUNDS, 1916-1990

Robert W. CAHN

Department of Materials Science and Metallurgy, Cambridge University, Pembroke Street, Cambridge CB2 3QZ, England

The first ordered intermetallic compound was identified in Russia in 1916, although the concept of long-range order was only clarified by x-ray diffraction, in Sweden, ten years later. Initially, such materials were studied only by chemists trying to understand the basis of combination between metals; later, metallurgists began to be interested in their properties, first magnetic ones and later mechanical properties. The first conference on such compounds,concerned with mechanical properties, was not held until 1960; now, there are apt to be several conferences devoted to intermetallics in any one year. — The paper outlines some of the generalizations which can be made about the variation of the strength and ductility of intermetallic compounds as a function of temperature, with a sideways glance at the behaviour of such alloys when they are plastically annealed (those that are capable of this) and then annealed. The role of varying degrees of long-range ordered is emphasized. More recently, the beneficial effect of combining a 'monolithic' intermetallic phase with another to form an intimate two-phase structure has been established, and this will be exemplified. Finally, an attempt is made to look into the future.

1. INTRODUCTION AND HISTORICAL OUTLINE

Chemical combination among metals has been studied for a long time. In the early days, chemists had much difficulty in understanding how metals could combine at all among themselves, often in several different proportions, since this seemed often to be at variance with elementary notions of valency. An early classic on the subject was a book published by two Italian chemists in 1918[1]. The British metallurgist C.H. Desch wrote a monograph on intermetallic compounds in 1914[2] (a fateful year when the name 'physical metallurgy' was first proposed, and the chemical bias of metallurgy shifted towards a physical bias), and he declared that "attempts to form a theory of the constitution of intermetallic compounds...have been comparatively unsuccessful." When he returned to the subject 20 years later[3], he felt much more cheerful about the state of the theory, because in the meantime Hume-Rothery had introduced the concepts of atomic size factor, electron/ atom ratio and electronegativity (a concept also associated with the name of Linus Pauling) into the discussion of intermetallic phases, or compounds. Today, a battery of sophisticated physical concepts is used to make sense of both binary and multi-element intermetallic compounds[4,5] and they are no longer regarded as chemical anomalies.

The early authors had no idea that intermetallic phases (as distinct from terminal solid solutions) were, as a rule, atomically ordered. This unawareness caused much confusion, especially to the eminent Russian metallurgist Kurnakov, who just before the Revolution discovered some puzzling anomalies in the physical and chemical properties of a series of Cu-Au alloys"; at certain compositions only, after slow cooling, the electrical resistivity was anomalously low and so was liability to chemical attack by acids[6]. Kurnakov tied himself into knots seeking to understand this, trying to reconcile the formation of a continuous series of solid solutions between Cu and Au with these well defined anomalies. The fact that alloys quenched from a high temperature did not show the anomalies made the matter even more puzzling. It was not till Johansson and Linde[7], in 1925, applied x-ray diffraction to the problem that the existence of long-range order (LRO), also named "superlattices", was at length discovered. Subsequently, the extraordinarily wide distribution of LRO was established by numerous x-ray diffraction analysts who studied thousands of different intermetallic phases and found a great variety of superlattice types. Some of these researches (e.g., the study of the Heusler-type superlattice in Fe_3Al, by Bradley and Jay[8]) achieved classic status.

Kurnakov's early findings were seen to imply that solid phases such as Cu_3Au and CuAu were disordered at high temperatures and only became ordered on slow cooling. Critical temperatures were discovered and related to the physical theory of critical phenomena; in due course, first-order and second-order ordering transitions were distinguished, and later linked with the postwar theory of spinodal transformations[9]. This is one of several areas where physicists' theory and metallurgists' interpretations are still separate, even today. Metallurgists are still apt to find illumination in the brilliant, but now outdated, interpretation of long-range ordering and critical temperatures by Bragg and Williams[10] (introduced to the metallurgical community in 1948 in Cottrell's famous textbook, "Theoretical Structural Metallurgy"[11]), but they are also beginning to come to terms with more precise modern approaches to the same issues, especially the cluster variation method[12], in which interactions between more than two atoms at a time are taken into account.— In 1936, Sykes in England discovered antiphase domains[13,14], and correctly interpreted them as the consequence of the independent nucleation of long-range order in many different locations in a disordered crystal. Only recently has it been clearly recognized that the presence of antiphase domains in an ordered intermetallic compound is evidence that the compound has a critical ordering temperature below the freezing temperature[15]. Sykes's work was done entirely by x-ray diffraction, but in recent decades research on antiphase domains and other defects in superlattices (including domains with differently oriented tetragonal or rhombohedral axes) has been done mostly by transmission electron microscopy, including high-resolution methods, and some remarkable insights have resulted[16]. Even the humble technique of optical microscopy can give useful information: thus, a study of ordering in the rhombohedral phase CuPt, using polarised light, has shown how domains can form at grain boundaries and at defects such as scratches[17]. The conclusions from both optical and electron microscopy concerning domain morphologies were pulled together in a brilliant review by Tanner and Leamy[18].

Some alloys have a tendency to long-range order but a low value of the critical temperature, T_c (typically ≈150°C), so that ordering is necessarily very sluggish. These include Cu_3Zn, Cu_3Al, NiFe and Au_3Cu, and the ordering has been studied most effectively by scanning calorimetry[19]. A notable study of NiFe has shown that LRO can be achieved by annealing the alloy while it is being irradiated to enhance diffusion rates[20], and subsequently it was discovered that certain massive meteorites, which are known to have cooled in space at a rate of the order of 1°C in a million years and then broken up on terrestrial impact, have had enough time to form long-range ordered NiFe. — There is a substantial body of both experiment and theory concerning the isothermal ordering kinetics of alloys originally quenched from high temperature to preserve disorder, and their relation to diffusivity; both are reviewed by Cahn[21].

The early concerns with understanding the solid-state chemistry of intermetallic compounds have developed today into a courageous effort to *predict*, either from first principles of by semi-empirical methods, the crystal structures of binary compounds, especially simple ones like AB, AB_2, AB_3, and slightly less simple ones such as $A_{17}B_2$ and AB_{13}. A particularly successful semi-empirical approach to structure prediction has been pioneered by Pettifor[22,23;24] this method of creating "structure maps" involves threading all the elements in the periodic table along a "string" which winds its way back and forth along the groups of the table; each element has a "Mendeleev number' marking its place on the string (rather like a gene along a DNA molecule) and the structure map is made by plotting Mendeleev numbers of the two constituents along two orthogonal axes. Compounds with similar pairs of Menedeleev numbers have similar ordered crystal structures. This is only one, albeit the most successful, of several approaches to the construction of structure maps; another such approach, based on size factors, e/A ratios and electronegativity, is due to Villars[25].— Another very influential semi-empirical approach has been Miedema's method of estimating formation energies of compounds: the fruits of this method have now been published in a recent book in which the cohesion of transition metal intermetallic compounds, in the form of their formation energies, are calculated[26].

Rigorous first-principles methods for calculating crystal structures, lattice spacings and cohesive properties have gradually improved, until now the "local density functional" approach[27,23], using modern supercomputers, is used routinely and successfully for such calculations. There are in fact several rival theorists' approaches, the nature and uses of which are well set out in a recent NATO Advanced Study Institute[28]. One can see how far the theoretical understanding of intermetallic compounds has advanced in the past decade by comparing this volume with another published in 1980[29]

In this brief historical overview of intermetallic compounds (nowadays simply called "intermetallics"), nothing has been said yet about mechanical properties, which are responsible for the present-day revival of interest in this category of alloy. We turn to this next.

2. MECHANICAL PROPERTIES OF INTERMETALLICS

The study of the mechanical behaviour of ordered alloys began in earnest in the mid-1950s, mainly at the GE Research Laboratory in Schenectady (Westbrook) and at Birmingham University (Cahn and collaborators). For a number of years, attention was focused on those phases which could be disordered by heating above T_c and quenching (Cahn[15] has proposed the term 'reversibly ordered' for such phases, in distinction to phases ordered up to the melting-point, which he termed 'permanently ordered'). The results of these early studies were summarised in 1966 by Stoloff and Davies[30] and by Cahn[31], in 1967 by Lawley[32] and in 1971 by Stoloff[33]. The last of these is the most comprehensive and offers a useful mix of experiment and theory. Compounds which were most studied in those days include Cu_3Au, Ni_3Fe, Ni_3Mn, FeCo, CuZn, NiAl, Fe_3Al and Mg_3Cd, and the effect of short-range (local) order was examined as well as that of long-range order.

Antiphase domain size has a pronounced effect on flow stress for a given degree of order (S and domain size can be varied independently); flow stress peaks for a small domain size, typically 50Å.

Short-range order (or alternatively highly imperfect LRO, as in Cu_3Al) enhances the flow stress. Increasing amounts of long-range order first increase the flow stress (up to $S \approx 0.5$, where S is the Bragg-Williams order parameter), then decrease it again as order increases further; this is quite generally true, both in alloys quenched out with different degrees of order and in those tested at high temperatures (just above and below T_c). It is clear[33,34] that this peak is due to a changeover, near $S = 0.5$, from a regime of ndividual dislocations moving through a short-range ordered region and disrupting it, to paired dislocations connected by a ribbon of stacking-fault (so-called superdislocations) as order builds up. Other features that affect the flow stress, such as the existence of a field of directional short-range order within the stress field surrounding a dislocation, a kind of "SRO cloud", are discussed by Cahn[31] and Stoloff[33]. — Very recently, Jang and Koch[35] have shown that this important rule applies also to a permanently ordered compound, Ni_3Al. Such an alloy cannot be disordered by quenching from a high temperature or even by melt-spinning, but Jang and Koch showed that this alloy can be progressively disordered by intensive ball-milling, and they measured the microhardness of powder particles of the alloy as a function of the order parameter: there was a very sharp hardness peak at $S = 0.5$.

Ordered alloys work-harden much more rapidly than disordered ones, and there is a body of theory relating to this: the notion, still not quite firmly established, of jogs in a moving dislocation trailing a tube of antiphase domain boundary, is an important part of this.

The Hall-Petch slope and intercept of alloys such as FeCo in polycrystalline form are very sensitive to the state of order, and so is the ductile-brittle transition in those alloys where this can be a problem. Thus, FeCo is brittle at room temperature when fully ordered, but suddenly becomes ductile when S falls below 0.2.

Creep of intermetallics has never been studied in as much detail as its central importance deserves. It is clear from both experiment and theory that the steady-state creep rate is proportional to the first power of the self-diffusivity (though it has never been clear which constituent diffusivity is rate-determining!) Early studies such as that by Lawley et al. on Fe_3Al[36], and recent ones such as that by Whittenberger on NiAl[37], make this clear. Because self-diffusivity is strongly reduced by the presence of LRO in B2 (CsCl-type) superlattices, but little affected for $L1_2$ (Cu_3Au-type) superlattices[21], it has become clear that B2 intermetallics and those derived therefrom, such as the Heusler phase Ni_2TiAl, are a much better bet for achieving high creep resistance than are $L1_2$ intermetallics. An important special case of this is the discovery that epitaxial two-phase alloys, incorporating both B2 and Heusler phases, have a remarkably high creep resistance[38].

Recovery and recrystallization of plastically deformed intermetallics (mostly this work was done with reversibly ordered alloys) shows that the presence of LRO slows down grain-boundary mobility by a very large factor, of 100-300 ×. Accordingly, a cold-worked intermetallic (this applies notably to Ni_3Al) can preserve the extra strength deriving from the cold-work up to quite a high temperature, so long as this is below T_c . The work in this field has recently been reviewed by Cahn[39]. Before recrystallization starts, the alloy can either soften or harden (recover or strain-age harden). It has recently become clear that which of these happens depends mainly on the initial state of the alloy before deformation (ordered or disordered) , on the degree of deformation and on how far the anneal temperature is below T_c ; the hardening or softening at this stage is linked to changes in the order parameter during annealing rather than to dislocation rearrangements[40].

Such are some of the general principles established, mostly with reversibly ordered model intermetallics, often more than twenty years ago, and they are in some danger of being forgotten now that attention is devoted chiefly to permanently ordered materials and one or two other intermetallics specially suited to high-temperature structural use...the current criterion that determines the availability of men and money.

3. HIGH-TEMPERATURE INTERMETALLICS

In an article devoted mainly to a historical overview, it is appropriate to be brief about the current spate of research, because this cannot yet be viewed with any sort of historical detachment. This Section, then, will merely mention some of the key researches and cite a few reviews judged by the author to be of particular value.

Four high-temperature intermetallics received a good deal of attention during the 1970s and some even earlier: Fe_3Al, NiAl, Ni_3Al and Ti_3Al, and I shall briefly survey these.

Fe_3Al:— The U.S. Navy made a precocious attempt to develop Fe_3Al and ternary alloys, especially with molybdenum, based thereon, in the 1950s, but at that time the brittleness problem at ambient temperature, the besetting difficulty with most otherwise useful intermetallics, proved insuperable. It is only recently that alloys between the Fe_3Al and FeAl compositions, with Mo, Cr, TiB_2 and other additions, have been taken up again at Oak Ridge National Laboratory[41,42]. Although these alloys will not be useful much above 600°C and are not of exceptional strength, yet their high corrosion resistance to many agents will ensure their use in applications such as those involving coal combustion. It cannot be emphasized too strongly that the uses of intermetallics will *not* be restricted to aerospace!

NiAl:— Research on the mechanical properties of NiAl goes back to pioneering work by Ball and Smallman in the 1960s[43], following on from even earlier work by Rachinger and Cottrell, in 1956,[44] which aimed at predicting the slip systems of B2 intermetallics, which in turn affects the multiplicity of independent slip systems and thus (via the von Mises principle requiring 5 such independent slip systems for possible ductility in a polycrystal) also aimed at predicting which B2 alloys might be ductile. (It is ironic that a recent study casts doubt on the general applicability of von Mises' criterion[45]). Rachinger and Cottrell's theoretical predictive method, centred on the magnitude of the ordering energy, led to many subsequent attempts to improve upon such prediction, up to a paper by Munroe and Baker in 1989[46]. NiAl and its ternary-alloyed derivatives in general have suffered from brittleness in polycrystals, but very recently, in a study with careful attention to purity, it has been found that the accurately stoichiometric NiAl alloy shows several per cent ductility[47]; this may be linked with a mysterious tendency of stoichiometric alloys to take up less interstitial impurity than non-stoichiometric ones[48]. The present state of knowledge on B2 intermetallics, with special emphasis on NiAl, has been very well reviewed by Baker and Munroe[49].

Since Baker and Munroe's report was written, several studies have shown very clearly that NiAl alloyed to produce a multiphase microstructure, (B2 + $L1_2$) or (B2 + f.c.c.) or (B2 + $L1_2$ + f.c.c.) or even (B2 + $L2_1$[i.e., Heusler phase] + $L1_2$), can be reasonably ductile at room temperature while preserving good high-temperature strength[50-53], provided the microstructure is correctly controlled. — These very recent studies have reinforced a general principle which is rapidly becoming accepted by practitioners in this field: *Good mechanical properties, including ductility, are much more likely to be achieved if a multiphase microstructure based on an intermetallic matrix is produced. This principle includes metal matrix composites based on intermetallics.*

Ni_3Al: — An enormous body of research followed on from Aoki and Izumi's discovery[54] that small amounts of boron dopant ductilises Ni_3Al. It soon became clear that this only works provided the nickel content is 76at.% or above, and much attention was devoted to the mechanism of the boron ductilisation, the effects of grain size and shape, and similar issues. There is no space here even to summarize the great mass of research in the past decade: a recent review of Ni_3Al by Stoloff[55] is a good source of information. — Chinese metallurgists have made a good contribution to this research (e.g.[56,57]).

Special mention should be made of an intriguing recent hypothesis[58], backed by a good deal of evidemce, to the effect that linked segregation of B and Ni to grain boundaries in Ni_3Alcan generate a thin layer of disordered material in this region and thereby enhance ductility. The varied evidence and opinions about this hypothesis are discussed in a recent survey by Cahn[59]; the most recent publication on the phenomenon however casts doubt on its very existence[60].

The main problem with Ni_3Al as a matrix phase is its limited creep resistance, clearly linked with the relatively high self-diffusivity of this kind of phase, but nevertheless a series of highly alloyed "advanced aluminides" created by Liu and his colleagues at Oak Ridge have found many terrestrial uses (e.g.[55]).

<u>Ti_3Al</u>: — Here again, a huge body of research has been undertaken, though some of it remains classified. The research spread from the U.S. Air Force to other countries, and there is much work in progress in Japan and several European countries. More recently, research has spread to TiAl and even more lately to the very brittle compound $TiAl_3$. The multiphase principle applies with special force to these difficult phases, and useful properties imperatively require ternary and quaternary alloying. Research has been divided into two large categories: a great deal of work on the dislocation mechanisms operating, especially in Ti_3Al, and work on alloying and processing. Particular interest attaches to the creation of powders and composites by forms of self-sustaining high-temperature synthesis, especially the XD™ process[61-63]; this approach has also prove useful with nickel aluminides.The reader is referred to a few recent review articles[64-67]. The Fleischer/Dimiduk/Lipsitt review[64] also incorporates a discussion of the incipient research on "exotic" intermetallics, novel phases with unfamiliar crystal structures, which owes much to Fleischer's advocacy. This is a fascinating topic which, again, there is no space to discuss here.

4. THE FUTURE

Someone who has worked on intermetallics for 35 years may be forgiven for feeling some curiosity as to the future course of research on, and applications of, this family of materials. Only fools and saints indulge in prophecy, and the author lays no claim to sanctity. Nevertheless, here are a few tentative guesses.

1) Useful intermetallics will all be multiphase; often, most or all the constituent phases of alloys with good properties — especially, those with good ductility — will themselves be intermetallic compounds.

2) Many uses of intermetallics — perhaps, in tonnage terms, most such uses — will be terrestrial.

3) "Grain boundary engineering" — i.e., the control of composition and state of order at grain boundaries — will be vital for achieving reasonable ductility in intermetallic polycrystals.

4) Novel processing methods will have to be used to achieve good microstructures and properties, particularly forms of self-sustaining high-temperature synthesis such as the XD™ process.

5) Numerous issues in pure science will continue to appear and lead to plenty of interesting research in physical metallurgy...issues such as the possible existence of disordered grain

boundary zones in doped intermetallics, the role of departures from stoichiometry on ductility, the exact relation between creep and self-diffusion and the relationship between the self-diffusivities of the components of an intermetallic. The problem will be to persuade those who are responsible for research finance that such research is worth doing — but that has always been a problem!

6) First-principles quantum-mechanical calculations and the use of structure maps will play an increasing role in the search for new and improved intermetallics.

REFERENCES

1. Michele Giua and Clara Giua-Lollini, Chemical Combination among Metals, transl. G. Wooding Robinson (J.& A. Churchill, London, 1918).
2. C.H. Desch, Intermetallic Compounds (Longmans Green, London, 1914).
3. C.H. Desch, The Chemistry of Solids (Cornell University Press, Ithaca, NY, 1934). pp 153-160.
4. K. Girgis, Structure of intermetallic compounds, in: Physical Metallurgy, eds. R.W. Cahn and P. Haasen (North-Holland, Amsterdam, 1983) pp 219-270.
5. H. Schäfer, Ann. Rev. Mat. Sci. 15 (1985) 1-41.
6. N. Kurnakov, S. Zemczuzny and M. Zasedelev, J. Inst. Metals 15 (1916) 305.
7. C.H. Johansson and J.O. Linde, Ann. Physik 78 (1925) 439.
8. A.J. Bradley and A.H. Jay, Proc. Roy. Soc. [Lond.] A 136 (1932) 210.
9. W. A. Soffa and D.E. Laughlin, Acta Metall. 37 (1989) 3019.
10. W.L. Bragg and E.J. Williams, Proc. Roy. Soc. [Lond.] A 145 (1934) 699.
11. A.H. Cottrell, Theoretical Structural Metallurgy (Edward Arnold, London, 1948).
12. D. de Fontaine, The cluster variation method and the calculation of alloy phase diagrams, in: Alloy Phase Stability, eds. G.M. Stocks and A. Gonis (Dordrecht Academic Publishers, Dordrecht, Netherlands, 1989), pp. 177-204.
13. C. Sykes and H. Evans, J. Inst. Metals 58 (1936) 255.
14. C. Sykes and F.W. Jones, Proc. Roy. Soc. [Lond.] A 145 (1936) 699.
15. R.W. Cahn, P.A. Siemers, E.L. Hall, J.E. Geiger and P. Bardhan. Acta Metall. 35 (1987) 2737, 2753.
16. G. vanTendeloo, Electron microscopy of ordering in alloys, in: Alloy Phase Stability, eds. G.M. Stocks and A. Gonis (Dordrecht Academic Publishers, Dordrecht, Netherlands, 1989) pp. 75-100.
17. R.S. Irani and R.W. Cahn, J.Mater. Sci. 8 (1973) 1453.
18. L.E. Tanner and H.J. Leamy, The microstructure of order-disorder transitions, in: Order-Disorder Transformations in Alloys, ed. H. Warlimont (Springer, Heidelberg, 1974) pp. 180-239.
19. L. Clarebrough, M.E. Hargreaves and M.H. Loretto, Proc. Roy. Soc. [Lond.] 257A (1960) 338, 363; idem, 261A (1961) 500.
20. L. Néel, J. Paulevé, R. Pauthenet, J. Langier and D. Dautreppe, J. Appl. Phys. 35 (1964) 873.
21. R.W. Cahn, Ordering kinetics and diffusion in some $L1_2$ compounds, in: Phase Transitions in Condensed Systems: Experiments and Theory, eds. G.S. Cargill, F. Spaepen and K.-N. Tu (Mat. Res. Soc. Symp. Proc., vol. 57, 1987) pp. 385-404.
22. D.G. Pettifor, J. Phys. C 19 (1986) 285.
23. D.G. Pettifor, It's all very well in practice but what about theory?, in: New Materials for the Next Century, ed. G.A.D. Briggs (Blackwell, Oxford, 1990), in press.
24. D.G. Pettifor, Quantum mechanics in alloy design, in: Alloy Phase Stability, as ref. 16, pp. 329-350.
25. P. Villars, K. Mathis and F. Hulliger, Environment classification and structural stability maps, in: Cohesion and Structure, Vol. 2,

eds. F.R. de Boer and D.G. Pettifor (North-Holland, Amsterdam, 1989) pp. 1-103.

26. F.R. de Boer, R. Boom, W.C.M. Mattens, A.R. Miedema and A.K. Niessen, Cohesion in Metals: Transition Metal Alloys [Cohesion and Structure, Vol. 1] (Elsevier, Amsterdam, 1988).

27. R.O. Jones and O. Gunnarsson, Rev. Mod. Phys. 61 (1989) 689.

28. A.J. Freeman, Structural stability of intermetallic compounds: a computational metallurgical approach, in: Alloy Phase Stability, G.M. Stocks and A. Gonis (eds.), (Kluwer Academic Publishers, Dordrecht, 1989), pp. 365-376, and other articles therein.

29. L.H. Bennett, ed., Theory of Alloy Phase Formation (The Metallurgical Society of AIME, New York, 1980).

30. N.S. Stoloff and R.G. Davies, Prog. Mat. Sci. 13 (1966) 1.

31. R.W. Cahn, Correlation of local order with mechanical properties, in: Local Atomic Arrangements Studied by X-ray Diffraction, eds. J.B. Cohen and J.E. Hilliard (Gordeon and Breach, New York, 1966) pp. 179-214.

32. A. Lawley, in: Intermetallic Compounds, ed. J.H. Westbrook (Wiley, New York, 1967) pp. 464-490.

33. N.S. Stoloff, Intermetallic compounds and ordered phases, in: Strengthening Methods in Crystals, eds. A. Kelly and R.B. Nicholson (Elsevier, Amsterdam, 1971) pp.193-259.

34. R. Moine, J.P. Eymery and P. Grosbras, phys. stat. sol. (b) 46 (1971) 177.

35. J.S.C. Jang and C.C, Koch, J. Mater. Research 5 (1990) 498.

36. A. Lawley, J.A. Coll and R.W. Cahn, Trans. AIME 218 (1960) 166.

37. J.D. Whittenberger, J. Mater. Sci. 23 (1987) 274.

38. R.S. Polvani, W.S. Tzeng and P.R. Strutt, Met. Trans. 7A (1976) 23, 31.

39. R.W. Cahn, Recovery, strain-age-hardening and recrystallization in deformed intermetallics, in: High Temperature Aluminides and Intermetallics, eds. S.H. Whang, C.T. Liu, D.P. Pope and J.O. Stiegler (TMS-AIME, Warrendale, PA, 1990), in press.

40. R.W. Cahn, M. Takeyama, J.A. Horton and C.T. Liu, submitted to J. Mater. Research, June 1990.

41. C.G. McKamey and J.A. Horton, Met. Trans. 20A (1989) 751.

42. C.G. McKamey, J.A. Horton and C.T. Liu, MRS Symp. Proc. 81 (1987) 321.

43. A. Ball and R.E. Smallman, Acta Metall. 14 (1966) 1517.

44. W.A. Rachinger and A.H. Cottrell, Acta Metall. 4 (1956) 109.

45. R.L. Fleischer, Acta Metall. 35 (1987) 2129.

46. P.R. Munroe and I. Baker, Scripta Metall. 23 (1989) 495.

47. K.H. Hahn and K. Vedula, Scripta Metall. 23 (1989) 7.

48. T.G. Nieh, J. Wadsworth and C.T. Liu, Scripta Metall. 22 (1988) 1409.

49. I. Baker and P.R. Munroe, A review of the properties of B2 compounds, in: High Temperature Aluminides and Intermetallics, eds. S.H. Whang et al. (TMS-AIME, Warrendale, 1990), in press.

50. D.R. Pank, M.V. Nathal and D.A. Koss, J.Mater. Research 5 (1990) 942.

51. K. Ishida, R. Kainuma, N. Ueno and T. Nishizawa, submitted to Met. Trans. (1990).

52. R. Yang, J.A. Leake and R.W. Cahn, to be presented at the MRS Symposium on High-Temperature Ordered Intermetallics, Boston, November 1990.

53. T, Khan, P. Caron and S. Naka, Mechanical behavior of Ni_3Al-based intermetallics and the need for designing multiphase alloys, in: High Temperature Aluminides and Intermetallics, eds. S.H. Whang et al. (TMS-AIME, Warrendale, PA, 1990) in press.

54. K. Aoki and O. Izumi, J. Japan Inst. Met. 43 (1979) 1190.

55. N.S. Stoloff, Internat. Mater. Reviews 34 (1989) 153.

56. Z. Jishan et al., MRS Symp. Proc. Vol. 133 (1989) 549.

57. Z. Yizhang et al., MRS Symp. Proc. Vol. 133 (1989) 555.

58. I. Baker and E.M. Schulson, Scripta Metall. 23 (1989) 1883.

59. R.W. Cahn, Mat. Sci. & Techn 6 (1990) in press.

60. S.S. Brenner and M.-J. Hua, Scripta Metall. 24 (1990) 671.

61. L. Christodoulou, P.A. Parrish and C.R. Crowe, XD™ titanium aluminide composites, in: High Temperature, High Performance Composites, eds. F.D. Lemkey et al. (MRS Symp. Proc. Vol. 120, 1989).

62. D.E. Larsen, M.L. Adams, S.L. Kampe, L. Christodoulou and J.D. Bryant, Scripta Metall. 24 (1990) 851.

63. A.K. Kuruvilla, K.S.Prasad, V.V. Bhanuprasad and Y.R. Mahajan, Scripta Metall. 24 (1990) 873.

64. R.L. Fleischer, D.M. Dimiduk and H.A. Lipsitt, Ann. Rev. Mat. Sci. 19 (1989) 231.

65. Y.W. Kim, JOM 41/7 (1989) 24.

66. M. Strangwood and A. Vassel, Harwell (U.K.) Report R 13198 (1989).

67. M. Yamaguchi and Y. Shirai, Deformation behavior of single and poly-crystal Al_3Ti and Al_3Ti with terney alloying additions, in: Dispersion Strengthened Aluminum Alloys (TMS-AIME, Warrendale,PA, 1988).

Frontiers of Materials Research/
Electronic and Optical Materials
M. Kong, L. Huang (editors)

COMPUTER APPLICATIONS TO MATERIALS SCIENCE

Masao DOYAMA

Department of Materials Engineering, The Nishi Tokyo University
Uenohara, Yamanashi 409-01 JAPAN

As examples of calculations, atomic configurations of dislocation cores,using molecular dynamics, rapid quenching of liquid, tensile deformation of amorphous and crystalline iron, shear plastic deformation of copper small crystals, lattice vibrations are shown.

1. INTRODUCTION

The properties of materials are determined not only by the components or compositions but also by the microscopic and macroscopic arrangement of atoms and molecules. Crystalline materials usually consist of poly-crystals and are rarely single crystals. Silicon waffers or jewels are typical examples for single crystals. The crystallographic directions of grains in poly-crystals are usually at random but sometimes are not randomly distributed. Silicon steels for transformers are made so that [100] directions are almost aligned in the rolling direction. Practical engineering materials often have more than one phase.

Recently computers have been rapidly developing. The computing speed has been very much improving and the capacity of memory has been enlarging. Computers are used not only for camputation but also for simulations, data base, and artificial intelligence. Structural bodies have been calculated by mechanics and plasticity, ships and air-planes by fluid mechanics, simulations are used and huge wind tunnels or water tanks are used as the last means. Computers are used for the design of electronic circuits.

Engineering materials are very complicated. It is not easy to treat plactical materials rigorously.

Atomic and molecular approach has been performed from the first principle. Concerning metals, cohesive energies, elastic constants, electronic structures of pure crystalline metals and of intermetallic compounds have been calculated. Random alloys are difficult to be treated. Using interatomic potentials molecular dynamics are applied to calculate the dynamic behavior of materials.

2. AB INITIO CALCULATIONS

The simplest calculations are on elements. Eighty three out of one hundred and three elements are metals. Metals usually have crystalline forms. In metals, the valence electrons are not tightly bound to an ion but rather freely move around in a crystal as conduction electrons. There are the Coulomb interactions between electron-electron, ion-ion, ion-electron. Treating the interactions is one of the typical many body problems which are difficult to solve. 10^{24} electrons and 10^{22} atomic nuclei cannot be rigorously treated even by a supercomputer. Each atom is identical so that the unit region for one atomic nucleus is defined as Wigner-Seitz cell. Near the nucleus, the electric field for an electron to see is treated as a muffin-tin (MT) potential. Electrons outside the core are often treated as the sum of plane waves, spherical waves, etc. Moreover Bloch's theorem can be used. Using the Wigner-Seitz cells and Bloch's theorem the number of particles to be treated can be reduced to the order of tens. When the Wigner-Seitz cell, the charge of the nucleus(atomic number) and the number of electrons(usually equal to the atomic number) are given, the relation between wave number k and energy ε, i.e. the band struc-tures are calculated for pure metals. Intermetallic compounds can be similarly treated. Calculations for pure metals have been performed including cohesive energies, elastic constants, lattice constants, Fermi surfaces and Fermi energies, etc. Linear band theory removes the energy dependence on the ground state wave functions. The behavior of the wave function in the muffin-tin sphere can be expressed in a practical ener-gy region as a linear combination of two wave functions. The Linear Augumented Plane Wave method is de-veloped from the APW method and the Lin-ear Muf fin-Tin Orbital Method is from the KKR method. The OPW(Orthogonal Plane Wave) method has chang-ed to the pseudopotential method. The NCPP (Norm Conserving Pseudo Potential)method, inner shell electrons are not treated straight-forward, but the valence electrons are in relief. The ASW (Augumented Spherical Wave)method and the linear regorous celluar(LRC) method close tothe LMTO method have been developed.

3. ALLOYS AND MATERIALS

For most engineering mate-rials, the periodic boundary conditions cannot be used. Inusual ma-terials, different kind of atoms are randomly occupied the lattice points. The most materials such as partically ordered alloys, random alloys, amorphous metals and alloys, liquid metals and alloys are not periodic. Random alloy have been treated by the Coherent Potential Approximation Method. These materials are treated as a single phase. Real materials usually have multi-phases. There are two kind of boundaries of two phases, a coherent boundary and an incoherent boundary. Usual materials also consist of poly-crystals. The crystallographic directions of grains are usually at random. Materials often have multi-phases. Not many calculations and simulations of physical properties of poly-crystals have been performed. For metallic materials, the composi-tions of elements are of course important but the atomic arrangement, phases, crystal structures, size of grains and the distribution of crystal orientation, etc. are also important. Practical materials are complicated and many atoms have to be treated.

4. ATOMIC STATIC MECHANICS

An atom in a condenced matter interacts with surrounding atoms. The atom is pushed by the re-sultant force exerted by the surrounding atoms until the equilibrium is achieved. It is not easy to find a real equilibrium but it often finds metastable configurations. A thermally activated process also cannot be considered. The molecular dynamics is better in this sence.

The resultant force from surrounding atoms within a radius rc is cal-culated. It is im-portant to know how to calculate the interaction between atoms. The simplest is a two body inter-action which only depends upon the distance be-tween atoms. Three body interac-tions have been also discussed. The interactioncan be calculated from the first principle, but this depends on the atomic arrangement surrounding the atom. Each time a huge calculation is required. Even in pure elements the determination of the interac-tion potential is not easy. The interaction be-tween different species are much more compli-cated.

5. MOLECULAR DYNAMICS

The calculation of a resultant force exerting on an atom is the same as that in statics. Posi-tions and velocities of all atoms in the system at time t = 0 are given from the resultant force the accerelations, velocities, can be calculated by solving Newton's equation. The differential equations are reducedto difference equations. From the accerelations, velocities and positions of all atoms in thesystem at time t these values at $t + \Delta t$ can becalculated.

The first work of molecular dynamics has been performed by Vineyard's group [1, 2] to show how atoms in a small crystal are displaced as a fuction of time after a momentum is given to an atom in the crystal. They showed that the atom given a momentum pushes the next atoms and the displaced atom pushes another next atom, leaving a vacancy at the original positions of the atom given the momentum.

Gibson et al [3] calculated the threshold energies in copper and they obtained the lowest energy in <100>, then <110> and the highest in <111>. Ergynsoy et al [4,5] calculated the displacement during radiation damage in a small iron crystal. They have clearly shown the channel ing. Velocity auto-correlation function, vibrational spectrum as dynamical properties, diffusion coefficient, and viscousity as transport properties are calculated. Radiation damage of KCl has been calculated [6].

As the vibrational frequencies of atoms in solids are of the order of $10^{13}sec^{-1}$, the time step should be order of 10^{-15} sec. The total time is of the order of 10^{-9} sec, even 1 million steps are taken. Semidynamical approaches to treat cascade have been performed [7,8,9].

This method is applied to the motion of atoms in molten rare gas [10] and metals, alkaline halides[11], silicide glass [11] using a peridic boundary condiction and the statistical properti es such as pair distribution function, the diffu -sion of atoms, structure factor, the equations of state, thermodynamic properties, thermal expansion coefficient, enthalpy, constant volume heat capacity, constant pressure heat, etc. were obtained. The atomic configuration in amorphous metals have been calculated [12,13,14]. The statistics of Voronoi polyhedra have been studied [15]. ZrF_4-BaF_2 glass [16,17] and metasilicate $R_2O \cdot SiO_2$[18] has been also treated.

Ultra fine particles have been studied experimentally and theoretically. The total number of particles in ultra fine particles are not very large so the regorous calculations are possible but the calculations of dynamical properties require large computations.

6. MONTE CARLO METHOD

Using random numbers, the interatomic interactions with a probability elements and creation of a cannonical distribution have been performed.

7. INTERACTION POTENTIALS

The interaction potential between atoms can be calculated from the first principle but phenomen ological potentials are often used. Electron core-core repulsion is often given by the Born-Mayer potential[19]. Morse potential[20] has a repulsive and attractive term. Girifalco and Weizer [21] determined the constants from the lattice parameter, elastic constant and sublimation energy. This is good for the uniform expansion or contraction. The present author proposed to use the formation energy of a vacancy instead of the sublimation energy [22]. Lenard-Jones potential represents the interaction between rare gas atoms. In Buckingham potential the attactive term between rare gas atoms is due to the attrac tion between the electric dipoles and the repulsive term is the core electron repulsion which i s exponential and represented by the Born-Mayer potential. Ionic potential [23,24]: The interaction between ions are given by ionic potential. Polynomials are used also to represent the interactions [25].

A major problem is that the potential determined from a perfect crystal cannot be applied to imperfect lattice. Interaction potentials have been discussed[26,27,28]. The embedded-atom method [29,30] derived from local density-functional theory has been applied to impurities, surface and other crystalline defects. In a disordered lattice electrons are redistributed. When the positions of atoms are given, the wave functions of electrons which give the lowest energy of the system are calculated by introducing a virtual time, atoms are moved toward the force and the wave functions of electrons are recalculated. This is an ab initio molecular dynamics method [31,32]. This method will be greatly used in the future, but so far not many applications are published. Atomic structures, phonon spectra and electronic properties in amorphous silicon [33] the atomic structure of Ge(100) [34,35] and twist boundary of $\Sigma = 5(100)$ [36] have been calculated by this method.

8. OTHERS

The phase diagrams of the electronic and statistical-mechanical theory of sp-bonded alloys have been calculated [37].

When the σ phase appears in nickel base refractory alloys, the alloy becomes brittle. PHACOMP (Phase COMPutation)[38] is developed for the prediction of σ phase formation. Yamazaki et al [39] developed an alloy design program based on experiments and theory. Yukawa and Morinaga [40] use the cluster theory to devel ope nickel base refractory alloys and titanium alloys.

The calculations of lattice constants and for bidden bands in binary mixed crystls of semiconductors such as $Al_xGa_{1-x}As$ have been calculated. Self consistent calculations of heterojunctions [41],and other many calculations haveb een performed. Calculations on quantum devices, devices, process simulations, process design of ULSI, and carrier transport have been also performed.

The properties of organic compounds depend on the properties of constituent molecules. Proper ties of many bases are known.

Computer applications extend to ceramics, glasses, polymers, organic materials, drugs, proteins, bio-materials, chemical reaction path, transport phnomena, chemical kinetics, organic reaction mechanism, luminous reactions, quantum devices, devices and process simulations, refining,purification, solidification, crystal growth, sintering, synthesis design, etc.

9. EXAMPLES OF COMPUTER APPLICATIONS TO MATERIALS SCIENCE

There are vast number of computer applications to materials science. In this paper the studies close to the authors work are shown.

9.1 Statics

It has been shown that if four vacancies coagurated on a (111) plane in the nearest neighbor position (Fig. 1) it may collaps but if they form a tetragonal void, it does not collapse[42].

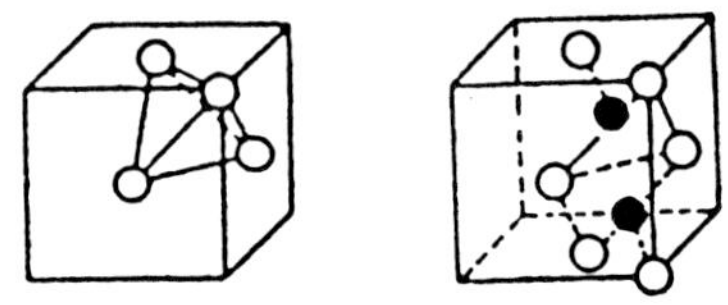

Fig.1: (a)Most stable quadri-vacancy, (b) Two atoms above and below a rhombic quadri-vacancy relaxed forming an elementary stacking fault.

9.2 Dislocations

The atomic arranmgement near the center of an edge dislocation in a face centered cubic lattice, copper was calculated. Burgers vector is 1/2[110]a (a = lattice constant). Figure 2 [43] is the projection to (111). It was believed that the core was melted or hollow when this calculation was performed. It was also shown by computer calculations that the complete dislocations (edge and screw) in copper split into two

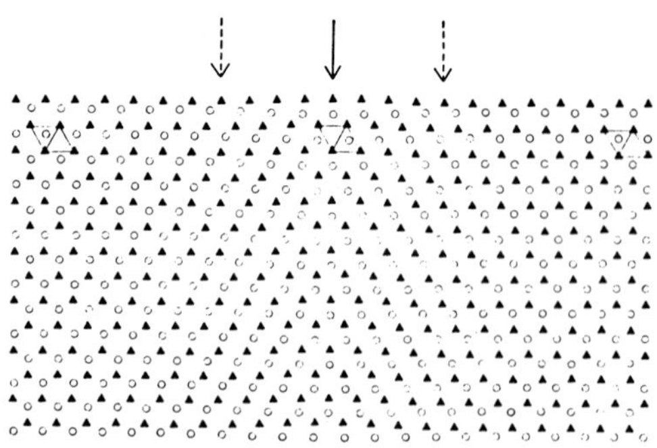

DISSOCIATED EDGE DISLOCATION

Fig. 2: Position of atoms in two {111} planes, one above (triangles) and one below(circles) the slip plane of the dissociated edge dislocation.

partial dislocations connected with stacking fault without activation energy. According to the first order elastic theory, the strain energy within the radius r is given by

$$E = A \log r + B = A \log (r/r_0)$$

However, constant B or r_0 cannot be determined by the first order elastic theory. By atomic calculations, B or r_0 can be determined. r_0 was determined to be 1.3 A and 1.0 A [44] for complete edge dislocation and complete screw dislocation in copper, respectively. J.J. Gilman[45] criticized our calculations in Comments on Solid State Physics saying that unrealistic "two-body potentials" cannot be used and "the man(Doyama) and the machine are well advised to seek other employment". After one quarter of a century, it has been cleared that his criticizm was not adequate at all. Many calculations have been performed and much new information has been obtained. The present author was definitely discouraged to continue the work for a while. These calculations have been applied to dislocations in a body centered lattice by Vitek et al [46,47] and have shown that (a/2)[111]screw dislocations split on three equivalent planes and that because of this the yield stress in iron is high. Yamaguchi et al [48] applied the calculations to intermetallic compounds. Jindo-Masuda [49] calculated the core structures of (a/2)<111> screw dislocation in α-iron under stress using the electron theory. The core structure of dislocations in silicon has been calculated using Keating potential[50,51]. The kink and jogs are calculated [52]. The electronic states and the structure of the core have been calculated self-consistently [53]. The core structure of an dislocation in an ionic solid has been calculated by Kurosawa [54].

9.3 Molecular dynamics

9.3.1 Rapid quenching

The atomic configurations in amorphous metals have been calculated. The Voronoi polyhedra have been statistically studied. Quenching of liquid metals can be simulated by taking away the kinetic energy of atoms. It should be remembered that the total time elapsed is at most of the order of 10^{-9} sec, so that the quenching speed is much faster than the experintal quenching speed. Figure 3 shows the radial distribution function when the temperture is lowered [55]. The second peak of the "frozen" radial distribution function has a shoulder and this is typical for amorphous metals.

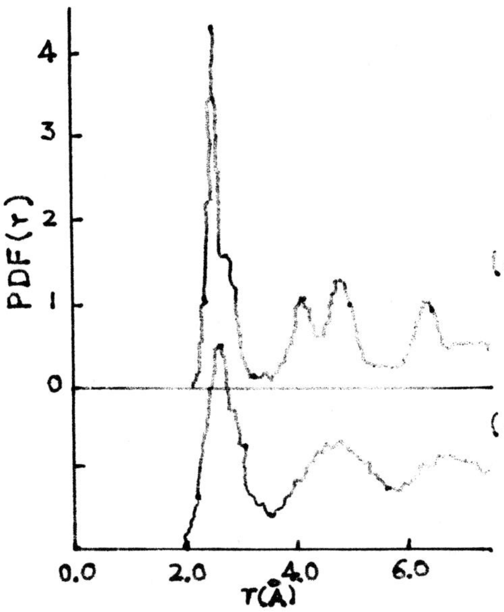

Fig. 3: Two-body distribution function. Quenched from Liquid (a)(n=0.0756), (d) (n=0.0848). n is the atomic dencity = atom/A^3.

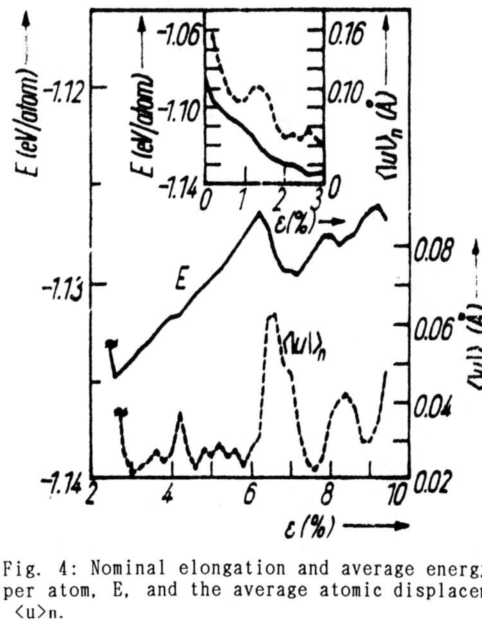

Fig. 4: Nominal elongation and average energy per atom, E, and the average atomic displacement <u>n.

9.3.2. Tensile deformation

A small amorphous iron specimen was pulled by the method of molecular dynamics [56]. The relation between energy and nominal elongation is plotted in Fig. 4. In some region of elongation strain energy-elongation relation is linear and in some region the energy is relaxed suddenly. In the linear part the displacement of atoms are just expected from the elastic theory(Fig. 5). In the relaxed region, the displacements of atoms above and below a plane about 45degree to the stretching direction are roughlyopposite, and a slip is observed (Fig. 6).

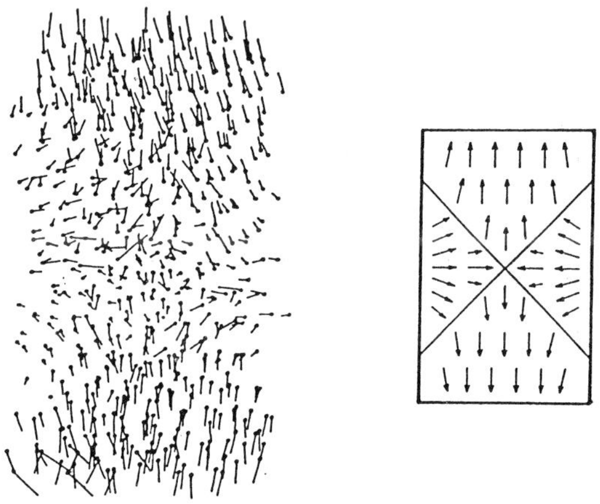

Fig. 5: The displacements of atoms in the elastic region (nominal elongation 2.6%→6.2%) are projected on (y, z).

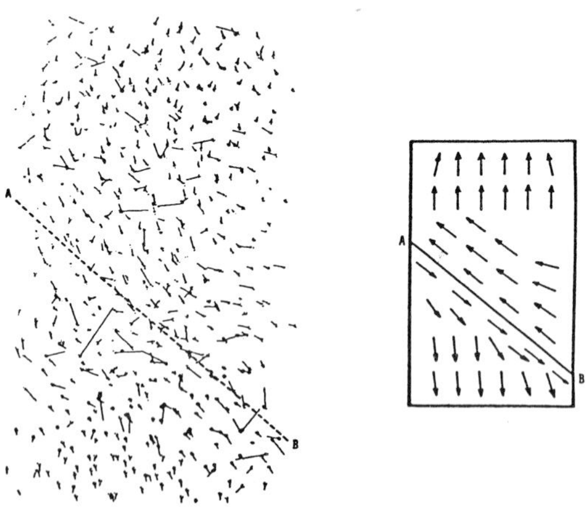

Fig. 6: The displacements during plastic deformation (nominal elongation 66.2%→7.5%) are plotted on (y, z).

An iron whisker crystals with a small crack was pulled in a super computer using the method of molecular dynamics. Figures 7～10 are the projection to {112}[57]. A hexagonal structure was observed only under a tensilestress. The whisker was completely broken into two pieces(Fig. 10). The stress-strain curve is plotted in Fig. 11.

9.3.3. Shear plastic deformation

To see the motion of dislocation more clearly a small crystalline piece of copper was deformed in a supercomputer[58]. A displacement was given in the upper half plane of a rectangular parallelpiped of a small copper crystal [Fig. 12～17]. Partial edge dislocation was first introcuced with a stacking fault followed. Then the other partial dislocation was created at the surface

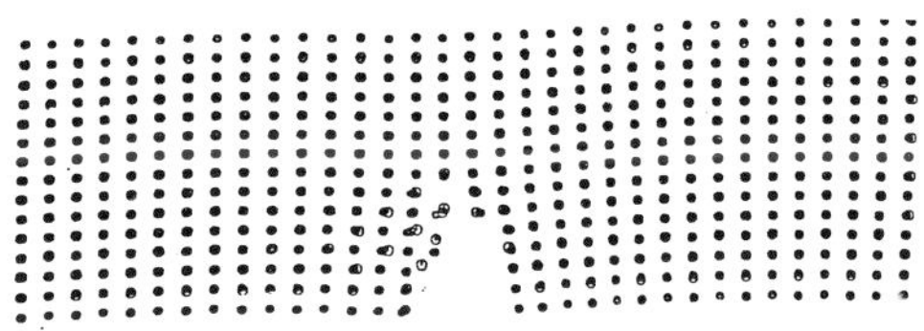

Fig. 7: Tensile specimen A. Nominal elongation 6.4%.

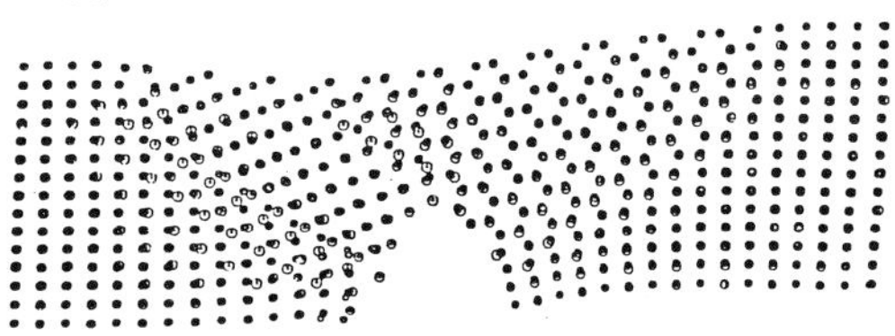

Fig. 8: Tensile specimen A. Twins are observed at A.

Fig. 9: Tensile specimen A. Necking is observed.

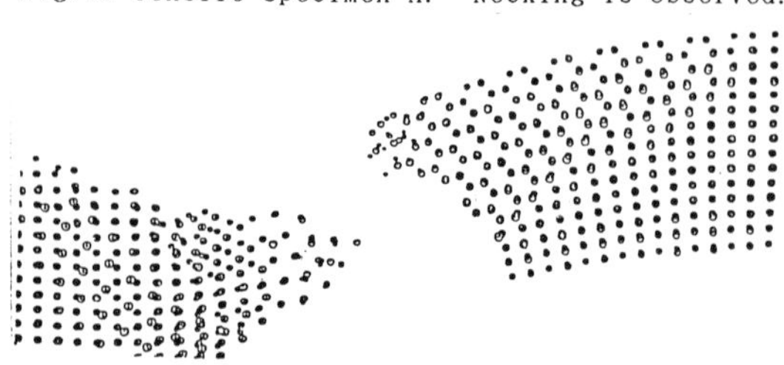

Fig. 10: Tensile specimen A. Just after fracture.

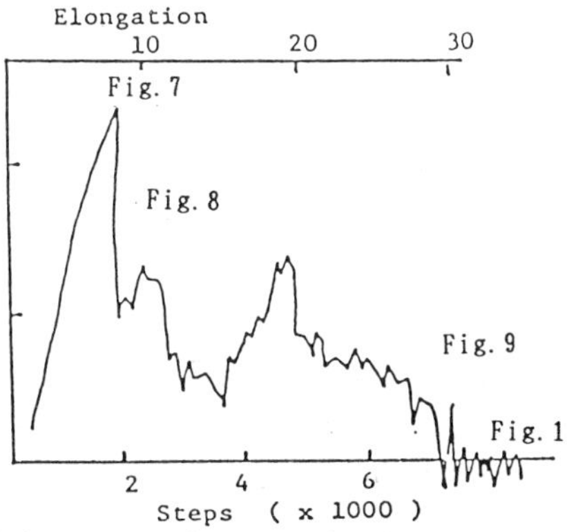

Fig. 11: Stress-strain curve for Specimen A pulled in [111]direction. The numbers represent figures of atomic configuration.

Fig.12: Displacement 0.4d.

Fig.15: Displacement 0.4d.

Fig.13: Displacement 1.2d.

Fig.16: Displacement 1.2d.

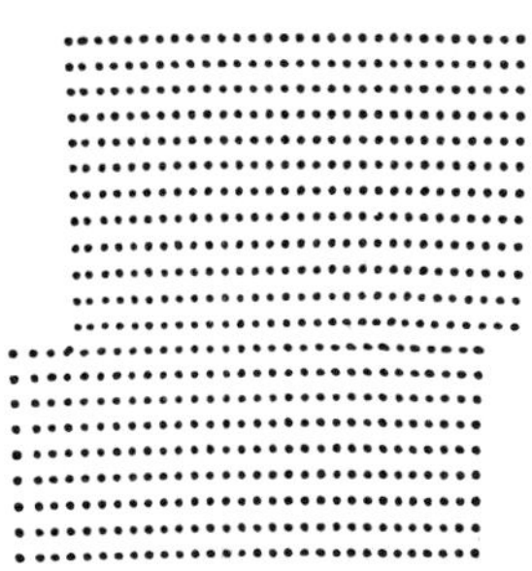

Fig.14: Displacement 2d.

Fig.17: Displacement 2d.

with a lower stress because it is pulled by a stacking fault. The pair of the split dislocations moved toward the other surface and the plastic deformation of one atomic plane was completed. The stress to introduce a partial dislocation in a crystal is higher than that to introduce the second partial disloation. The second dislocation was created with a lower stress than that for the first dislocation. The stress to introduce a dislocation is lower in the crystal with a crack than in that without a crack (Fig. 19).

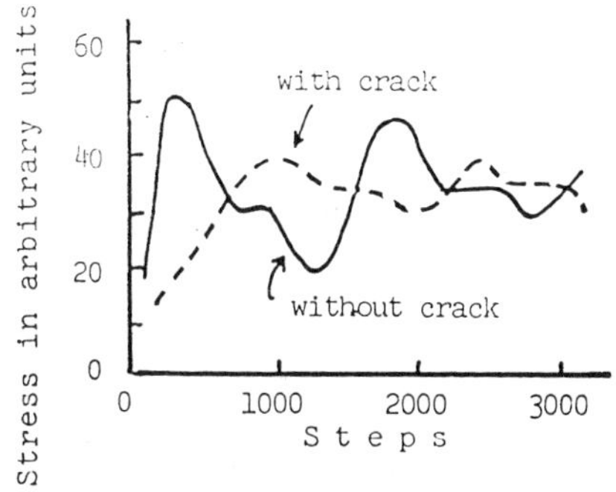

Fig.19: Shear stress-strain curve for the small copper crystal.

9.3.4. Lattice vibrations

The phonon dispersion relation inpure copper using a Morse potential has been calculated[59]. The phonon frequency the phonon density of states for the nearestneighbor atoms, 2nd, 3rd, 4th nearest neighboratoms of a vacancy in copper have been also calculated [60]. The nearest neighbor of a vacancy has high density at lower vibrational frequency. The atomic structure and the relation between the phenon frequency the phenon density of states at a grain boundary in iron. Figure 20 is the atomic structures and the relation between the frequencies and density of states at the grain boudary in iron [61,62].

9.3.5. Interaction between defects

The interactions between defects are important, particularly the interaction between vacancies and defects. Vacancies play an important role in diffusion of atoms in crystals. The interaction between vacancy-vacancy, vacancy-impurity and

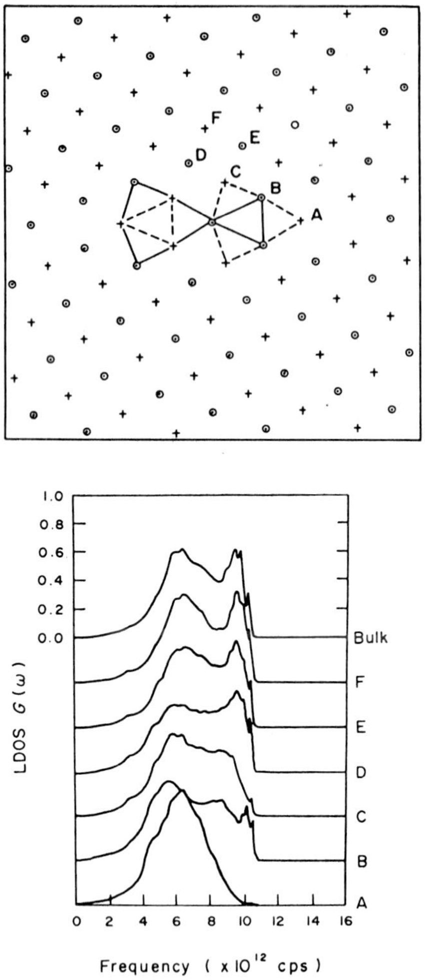

Fig. 20: Atomic configuration and the the relation between phonon frequency and the local density of states near $\Sigma=5$ grain boundary in $\alpha-$Fe

impurity -impurity have been caculated by use of pseudopotential [63].

10. CONCLUSIONS

The above are examples on atomic scale. To treat a macroscopic problems, finite element method is often used. In linear finite element method, the displacement is proportional to stress. In the non-linear finite element method, the displacement and stress relationship is given. In some cases as the deformation proceeds the net size of finite elements is changed in the region where the displacement is large. It is,however, assumed that the stress-strain relation is held. As the element becomes smaller to the order of grain size, the relation should be changed.

The speed of computers are becoming faster and faster, and the memory capacity is becoming larger and larger and the capacity of computers are being improved. There are, however, many problems which cannot be calculated straightforwardly, "grafting a bamboo on a tree" is some times necessary. Originality to handle these problems is important. If we emphasize too much on the calculations from the first principle, complicated problems such as those involving materials cannot be calculated.

The author expresses his thanks to his associates and students who have worked this problem with him. He also acknowledges the support by an Grant-in-Aid of the Ministry of Education, Science and Culture.

References

1. G. H. Vineyard, Discussions Faraday Soc. 31 (1961) 7.
2. G. H. Vineyard, J. Phys Soc. Japan, 18 Suppl. III (1963) 144.
3. J. B. Gobson, A. N. Gol and M.Milgram and G.H. Vineyard, Phys. Rev. 120 (1960) 1229.
4. C.Erginsoy,G.H.Vineyard snd A.Englert, Phys. Rev. 133 (1964) A595.
5. C.Erginsoy,G.H.Vineyard snd A.Shimizu, Phys. Rev. 139 (1965) A118.
6. McC. Torrens and L.T. Chadderton, Phys. Rev. 159 (1967) 671.
7. M.Yoshida, J. Phys. Soc. Japan 16 (1961) 44.
8. J.R.Beeler, Jr. and D.G.Besco, J. Appl. Phys. 34 (1963) 2873.
9. J.R.Beeler, Jr., Phys. Rev. 150 (1966) 470.
10. A.Rahman, Phys. Rev. 45 (1966) 2585.
11. K. Kawamura and I.Okada, Nippon kinzokugakkai Kaiho 16 (1977) 834 (in Japanes) , I. Okaga, ibid, 23 (1984) 600.
12. J.L.Finney, Proc. Roy. Soc. London, A319 (1970) 479.
13. R.Yamamoto, K. Haga, H. Shibuta and M.Doyama, J. Phys. F8 (1978) L179.
14. M.Tanaka, J. Pys. Soc. Jpn 55 (1986) 3108. ibid, 55 (1986) 3428.

15. R.Yamamoto, T. Mihara and M. Doyama, phys. stat. sol. (a) 50 (1978) 165.
16. I.Yasui, H.Inoue, J.Non-Cryst. Solids, 71 (1985) 39.
17. H.Inoue, H. Hasegawa, I.Yasui, Phys. Chem. Glasses, 26 (1985) 74.
18. "Calculational Materials Science" eds. M.Doyama and R.Yamamoto ed. (Kaibundo, Tokyo, 1987) (in Japanese)
19. M.Born and E.Mayer, Z. Physik 75 (1932) 1.
20. P.M. Morse, Phys. Rev. 34 (1929) 57.
21. L.A.Girifalco and V.G.Weizer, Phys. Rev. 114 (1959) 687.
22. M.Doyama and R.M.J.Cotterill, "Lattice Defects and Their Interactions" ed. R. R. Hasiguti (Gorden and Breach, New York, 1967). pp.79.
23. F.G.Fumi, M.P.Tosi, J. Phys. Chem. Solids 25 (1964) 31.
24. M.P.Tosi, F.G.Fumi, J. Phys. Chem. Solids 25 (1964) 45.
25. R.A.Johnson and W.D.Wilson, "Interaction Potentials and Simulation of Lattice Defects", eds. P.C.Gehlen, J.R.Beeler, Jr. and R.I. Jaffee, (Plenum Press, New York,1972).pp.301.
26. "Interaction Potentials and Simulation of Lattice Defects", eds. P.C.Gehlen,J.R.Beeler, Jr. and R.I.Jaffee, (Plenum Press, New York, 1972).
27. I.M.Torrens, "Interatomic Potentials", (Academic Press, New York, 1972).
28. J.K.Lee ed. "Interatomic Potentials and Crystalline Defects,Metallurgical Society of AIME, New York, 1981.
29. M.S.Daw and M.I.Baskes, Pys. Rev. Lett. 50 (1983) 1285.
30. M.S.Daw and M.I.Baskes, Pys. Rev. B29 (1984) 6443-6453.
31. R.Car and M.Parrinello, Phys. Rev. Lett. 55 (1985) 2471.
32. M.C.Payne, J.D.Joanno, D.C.Allan, M.P.Teter, D.H.Vanderbilt, Phys. Rev. Lett. 56 (1986) 2656.
33. R.Car and M.Parrinello, Phys. Rev. Lett. 60 (1988) 204-207.
34. M.Needels,M.C.Payne, J.D.Joannopoulos, Phys. Rev. Lett. 58 (1987) 1765-1768.
35. M.Needels,M.C.Payne, J.D.Joannopoulos, Phys. Rev. B (1988-I) 5543-5546 .
36. M.C.Payne,P.D.Bristowe, J.D.Joannopoulos, Phys. Rev. Lett. 58 (1987) 1348-1364,
37. J.Hafner,From Hamiltonians to Phase Diagrams, Springer-Verlag, Berlin, 1987.
38. W.J.Boesch and J.S.Slaney, Metal Progress, 86(1964) 109.
39. H.Harada, K.Ohno, T.Yamagata, T.Yokokawa and M.Yamazaki,"Superalloys", ed. S. Reichman, D.N.Duhl, G.Maurer, S.Antolovich and C.Lund, The Metallurgical Society, 1988)
40. H.Esaki, M.Morinaga, N.Yukawa, H.Adachi, Phil. Mag. A 53(1986) 709.
41. W.E.Pikett,S.G.Louie, and M.L.Cohen: Phys. Rev. B 17 (1987) 815.
42. M. Doyama and J, S. Koehler, Phys. Rev. 134 (1964) A522-A529.
43. R.M.J.Cotterill and M.Doyama, Phys. Rev. 145 (1966) 465-478.
44. M.Doyama and R.M.J.Cotterill, Phys. Rev. 150 (1966) 448-455.
45. J.J.Gilman, Comments on Solid State Physics 2 (1968-70) 37-39.
46. V.Vitek, Proc. Roy. Soc., London 352A (1976) 109.
47. V.Vitek, R.C.Perrin and D. K.Bowen, Phil.Mag. 21 (1970) 1049.
48. M.Yamaguchi, V.Vitek, and D.P.Pope, Phil. Mag, 43A (1981) 1027 (1981). ibid, 43A (1981) 1265., M.Yamaguchi,V.Paidar, D.Pope, V.Viket, ibid, 45A (1982) 867.ibid, 45A(1982) 882.
49. K.Masuda, Phil. Mag. 43B(1981) 1.
50. M.Heggie and R. Jones, Microscopy of Semi-conducting Materials, Inst. Phys. Conf. Ser. 67 (1983) 45.
51. S.Marklund, Solid State Comm. 50(1984) 185; 54 (1985) 555.
52. K.Masuda(jindo), K.Kojima, T.Hoshino, Jap. J. Appl. Phys. 22 (1983) 1240.
53. H.Veth and H.Teichler, Phil. Mag. 49B(1984) 371.
54. T. Kurosawa, J. Phys. Soc. Japan 19(1964) 2096.
55. R.Yamamoto,T.Mihara, K.Taira and M.Doyama, Physics Letters 70A (1979) 41.
56. R.Yamamoto,H.Matsuoka and M.Doyama, physica status solidi (a)51(1979) 163.
57. M.Doyama and R.Yamamoto, Dislocation in Solids eds. H.Suzuki and T.Ninomiya, K.Sumino,S.Takeuchi, University of Tokyo Press, 1985.pp.85.
58. M.Doyama and R.Yamamoto, Materials Science Forum 37 (1989) 77.
59. F.W.de Wett, R.M.J.Cotterill and M.Doyama, Physics Letters 23(1966) 309.
60. R.Yamamoto, Haga, T.Mihara, M.Doyama, J. Phys. F8(1979) L179.
61. M.Hashimoto, Y.Ishida, S.Wakayama,R.Yamamoto and M.Doyama, Acta Met. 32(1984) 13.
62. S.Wakayama, M.Hashimoto,Y.Ishida, R.Yamamoto and M.Doyama, Acta Met. 32(1984) 21.
63. O.Takai, R.Yamamoto, M.Doyama and Y. Hisamatsu: Physical Rev. B10 (1974) 3113.

Frontiers of Materials Research/
Electronic and Optical Materials
M. Kong, L. Huang (editors)

TRIBOLOGY OF CERAMICS -TRIBOCHEMISTRY AND LOW FRICTION-

Koji KATO

Mechanical Engineering Department, Tohoku University
Sendai, 980, Japan

Ceramics can offer low friction in air and in water if frictional conditions are properly selected and designed. The tribochemical reaction at the interface is the key to have low friction. Recent representative results are analyzed to understand this situation and pressure-velocity map is introduced for the control of low friction.

1. INTRODUCTION

Ceramics are hard, brittle and inert chemically. The melting temperature is high and the thermal conductivity is small. Ceramics of such mechanical and chemical properties are expected to be good materials for tribological elements.

The question is why those properties could be good for low friction ? How we can expect the potentials of ceramics as good materials for tribological usage?

The purpose of this paper is to discuss on this point with recent published results and introduce the water lubrication of ceramics as an attractive direction of ceramic tribology. Friction of diamond against ceramics and tribochemical reaction are also introduced to help the understanding.

2. FRICTION COEFFICIENT IN UNLUBRICATED SLIDING

Friction coefficients of Si_3N_4 , SiC, Al_2O_3 , ZrO_2 and diamond distribute in the range of about 0.3-0.9 in vacuum, inert gases or dry air[1-7]. In many cases of these observations, friction was performed between similar materials and the frictional pair was pin-on-disk.

Against these values of friction coefficient such as 0.3-0.9, the representative values of metals in unlubricated sliding may be supposed as 0.3-1.5 in air[8,9] .

It is common, in addition, that the friction coefficient larger than 2 is observed in vacuum for metals. Friction coefficient larger than 30 can be observed with gold[9].

If we compare the friction coefficient of ceramics with those of metals, ceramics could be assumed better than metals to avoid large friction coefficient.

But the value of 0.3-0.9 is not small enough to use ceramics for popular machines and devices in air or vacuum without lubricants.

3. FRICTION COEFFICIENTS OF DIAMOND/DIAMOND PAIR AND DIAMOND/CERAMIC PAIR IN AIR

Friction of diamond against itself is an exception among ceramics which gives the friction coefficient of 0.05 in air without lubricants[1] . Diamond gives such low friction coefficient when it slides against other ceramics in air without positive supply of lubricants.

The pair of diamond pin/Si_3N_4 disk showed the friction coefficient of 0.05-0.06 with a pin of 0.005mm radius under the load of 0.1-1.0 N[10] , and the friction coefficient changed by the direction of sliding against the direction of hot-press direction of Si_3N_4.

The friction coefficient of a diamond pin/Si_3N_4 plane pair was 0.04 at 10% room humidity. The generation of hydrated oxide layer was supposed to be responsible to reduce friction [11].

In the case of a pair of diamond pin/SiC disk, the friction coefficient changed from 0.05 to 0.12 during repeated friction cycles [12]. When a SiC ball slided on a diamond film coated on SiC, the friction coefficient was about 0.08 under the load of 4-40 N [13]. A thin film of graphite formed at the real contact area was assumed important to generate such a small friction coefficient.

It is evident from those observations that a diamond/ceramic pair can promise the friction coefficient of about 0.04 if sliding conditions such as surface roughness, load, humidity and material combination would be properly selected.

The essential points for low friction suggested from these observations would be smooth surface of diamond, a light load, adsorbed water and tribological generation of surface film such as hydrated silicon oxide and/or graphite.

A smooth surface and a light load may mean the same contact situation which does not cause the abrasive friction.

The tribological generation of a surface film means that gases in air such as oxygen, hydrogen and water vapor are special lubricants for diamond and ceramics.

4. FRICTION COEFFICIENT OF CERAMIC/CERAMIC PAIR IN AIR AND IN WATER

Smooth surfaces, good conforming of contact surfaces and the third body or film of low shearing strength at the contact in interface are reasonable factors which give low friction. The best situation is to keep these factors during the process of sliding. Diamond/diamond and diamond/ceramic could be good pairs of this meaning.

Some ceramic/ceramic pairs are possible to respond to such requests under certain frictional conditions;

(1) Smooth surfaces are generated as a result of wear.
(2) Good conforming of contact surfaces is kept because of the high hardness and low thermal expansion of ceramics.
(3) The third body or film is generated at the contact interface during sliding which reduces friction.

4.1 Si_3N_4/Si_3N_4 Pair

It was shown with a Si_3N_4/Si_3N_4 pair that Si_3N_4 does oxidize in air at room temperature under pure rolling contact. Rough surface became smooth and shiny by generating SiO_2 films as wear particles [14]. The wear particle size as a film increased drastically by the introduction of a small amount of slippage at the interface and the wear rate increased [15].

Such oxidation of Si_3N_4 caused by friction was called tribochemical reaction and the reduction of wear by humidity was explained from this viewpoint [16]. Tribochemical reaction of Si_3N_4 become important for friction in water where friction coefficient was reduced below 0.002 [17]. Contact surfaces became smooth by tribochemical wear where the tribochemical reaction was assumed as follows.

$$Si_3N_4 + 6H_2O \rightarrow 3SiO_2 + 4\ NH_3$$
$$SiO_2 + 2H_2O \rightarrow Si(OH)_4$$

This mean that SiO_2 was generated by friction in water and it dissolved in the water much more rapidly than quiescent amorphous SiO_2 particles.

Generation of NH_3 was confirmed experimentally for the case of wet grinding of silicon nitride powder[18]. Generation of SiO_2 and NH_3 was confirmed also for the sliding in air, and it was shown that the amount of SiO_2 and NH_3 increased with humidity [19].

Si_3N_4 surfaces in contact become smooth by such tribochemical wear and good conforming of contact surfaces is attained. As a result, a thin water film may be formed at the contact interface in water and gives very low friction coefficient in the range of 0.001-0.01 . High hardness of ceramics would be important in this case to keep thin and uniform water film.

4.2 SiC/SiC Pair

It was once shown that SiC/SiC pair kept high friction coefficient such as 0.26 in water and did not give such a value as Si_3N_4[17].

But it was shown later that SiC/SiC pair gave the friction coefficient of 0.02 in water if the running-in process would be carefully carried out [21]. Although graphite was partly formed on wear surface and it lubricating property was expected, it was not the main reason for the low friction coefficient of 0.02. Smoothness of the wear surface and sliding velocity were important factors for low friction. It means that water film at the contact interface may generate hydrodynamic lubrication.

Similar result was obtained in other observation [20].

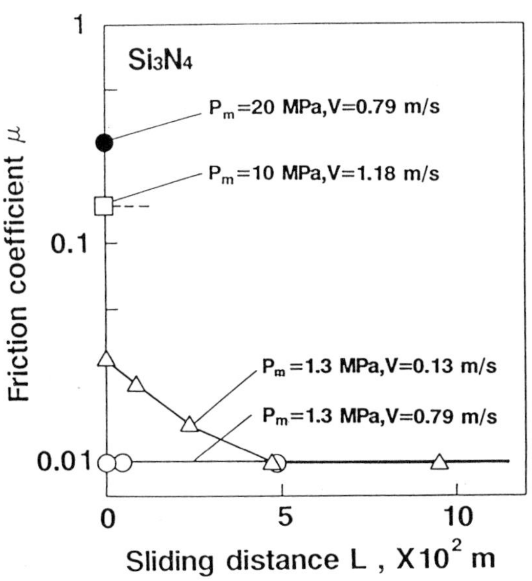

FIGURE 1
The relationship between friction coefficient and sliding distance for Si_3N_4

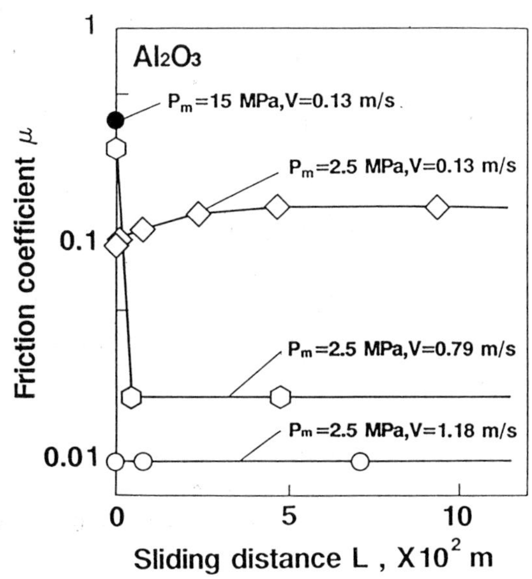

FIGURE 2
The relationship between friction coefficient and sliding distance for Al_2O_3

SiC/SiC pair showed the friction coefficient less than 0.1 in air when humidity was over 50% and the style of friction was fretting.

These results mean that SiC/SiC could be a good material for low friction in air and water.

5. P-V MAP FOR CERAMICS IN WATER

It is shown above that Si_3N_4 and SiC are well lubricated by water and give very low friction coefficients.

It suggests that ceramics can be widely used in wet circumstances as structural materials and water may be used as a good and popular lubricant.

Under this situation, pressure-velocity map of ceramics was introduced to give the general view on water lubricated ceramics [22]

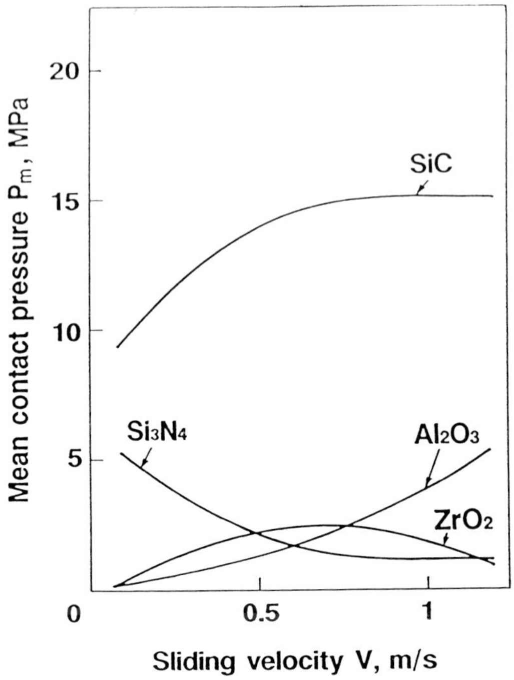

FIGURE 3
The pressure-velocity map for $\mu < 0.01$

Friction type was cylinder on cylinder and the apparent contact area was 200mm^2. The mean contact pressure Pm was changed from 0.25MPa to 20MPa and the sliding velocity from 0.13m/s to 1.18m/s.

The relationship between friction coefficient and sliding distance was observed at first under several loads and sliding velocities. Figs. 1 and 2 show examples observed with Si_3N_4 and Al_2O_3.

Similar observations were carried out with SiC and ZrO_2, and the (Pm, V) combinations to keep the friction coefficient below 0.01 were summarized. Fig.3 shows the obtained results for each ceramic. In the figure, μ is larger than 0.01 above the solid line and μ is smaller than 0.01 below the solid line in each ceramic. It is obvious from Fig.3 that SiC has the wide range for the friction coefficients below 0.01, and SiC and ZrO_2 has also the low friction range.

6. CONCLUSION

Ceramics has a large potential to give low friction in air and in water.

Smooth surface generated by tribochemical wear, good conforming of hard two surfaces in sliding and tribochemical products are main functions to have low friction.

This paper described only the side of friction of ceramics. Other paper [23] would help the understanding on wear of ceramics.

REFERENCES

1. F.P. Bowden and D. Tabor, The Friction and Lubrication of Solids, Part 2, Oxford, Clarendon Press, (1964)169.

2. D.H. Buckley, The influence of the atomic nature of crystalline materials on friction, ASLE Trans., (1967)134.

3. T.E. Fischer and H. Tomizawa, Interaction of tribochemistry and microfracture in the friction and wear of silicon nitride, ASME, Wear of Materials, Vancouver, (1985)22.

4. H. Ishigaki, I. Kawaguchi, M. Iwasa and Y. Toibana, Friction and wear of hot pressed silicon nitride, ASME, Wear of Materials, Vancouver, (1985)13.

5. I. Shimura and Y. Mizutani, Wear of ceramics in high temperature, 29th Proc. JSLE Conf., (1985)173.

6. S. Sasaki, The effects of water on friction and wear of ceramics, J. JSLE, 33(1988) 620.

7. C.S. Yust and F.J. Carignan, Observation on the sliding wear of ceramics, ASLE Trans., 28(2)(1985)245.

8. F.P. Bowden and D. Tabor, The Friction and Lubrication of Solids, Part 1, Oxford, Clarendon Press, (1954)151.

9. T. Kayaba and K. Kato, Friction properties of surface films in air and in vacuum, Wear, 47 (1978)93.

10. H. Ishigaki, I. Kawaguchi, M. Iwasa and Y. Toibana, Friction and wear of hot pressed silicon nitride and other ceramics, Wear of Materials, ASME(1985)13.

11. P. Kapsa and Y. Enomoto, Sliding damage on hot-pressed and sintered silicon nitride caused by a diamond tip under controlled humidity, Wear, 127(1988)65.

12. M. Kohzaki, S. Noda, H. Doi and O. Kamigaito, Tribology of Niobium-coated SiC ceramics and the effects of high energy ion irradiation, Wear, 131(1989)341.

13. S. Jahanmir, D.E. Deckman, L.K. Ives, A. Feldman and E. Farabaugh, Tribological characteristics of synthesize diamond films on silicon carbide, Wear, 133(1989)73.

14. M. Akazawa, K. Kato and K. Umeya, Wear properties of silicon nitride in rolling contact, Proc. JSLE Int. Conf., Tokyo (1985)191. and, Wear, 110(1986)285.

15. M. Akazawa and K. Kato, Wear properties of silicon nitride in rolling/sliding contact, Wear of Materials, ASME,(1987)267. and, Wear,124(1988)123.

16. T.E. Fischer and H. Tomizawa, Interaction of tribochemistry and microstructure in the friction and wear of silicon nitride, Wear of Materials, ASME, (1985)22. and, Wear, 105,(1985)29.

17. H. Tomizawa and T.E. Fischer, Friction and wear of silicon nitride and silicon carbide in water : Hydrodynamic lubrication at low sliding speed obtained by tribochemical wear, ASLE Trans., 30, 1(1987)41.

18. Y. Kanno, K. Suzuki and Y. Kuwahara, NH_3 formation caused by the presence of H_2O in the wet grinding of sialon nitride powder, Yogyo-Kyokai-Shi, 91(8), (1983)386.

19. K. Kawamura, Y. Imada, F. Honda and K. Nakajima, The effect of atmosphere on wear of conductive ceramics(Si_3N_4+TiN), Proc.,33rd JSLE Conf.,(1989) 49.

20. S. Sasaki, The effects of the surrounding atmosphere on the friction and wear of alumina, zirconia, silicon carbide and silicon nitride, Wear, 134(1989)185.

21. S. Hotta and Y. Mizutani, Effect of water and organic solvents on the friction of SiC, Proc. 32nd JSLE Conf., (1988)1.

22. K. Kato, H.C. Wong, N. Umehara and K. Nii, The fundamental friction properties of ceramics in water, Proc. JAST Conf., Tokyo, May(1990)329.

23. K. Kato, Tribology of ceramics, Wear, 136 (1990)117.

Frontiers of Materials Research/
Electronic and Optical Materials
M. Kong, L. Huang (editors)

FIBRE REINFORCED COMPOSITES -- A SOLUTION FOR MATERIALS SCIENCE, ENERGY SAVING AND POLLUTION DECREASE IN THE ENVIRONMENT.

Erich FITZER

Institut fur Chemische Technik, University of Karlsruhe, FRG

I. NATURE AS A MODEL FOR ANISOTROPIC MATERIALS

All biologically grown material in nature is ANISOTROPIC, that means it has different values of mechanical properties in the various directions. In such a way, nature is able to concentrate and limit strength and stiffness of the material to those directions, where these mechanical properties are really needed. Best models are the various grasstypes in a meadow or the structure of trees. The economic principle of nature is to achieve the maximum in strength and stiffness with a minimum of material used and thus to save living material and energy.

The solution to achieve specific anisotropy is the fibrous structure of all biomaterials, for instance the fibres in muscles, the fibrilles in cellulose, the use of reinforcement ribs in the wings of insects and so on.

We recognize in FIGs 1a and 1b [1] the analogous building principle in biomaterial and in modern composites and can conclude, that these new materials are not a human invention, but only a copy of the natural model.

Novel and a human invention was the successful development of suitable light weight reinforcement fibres, which are not only strong such as the well known glass fibres but also stiff. Furthermore, suitable

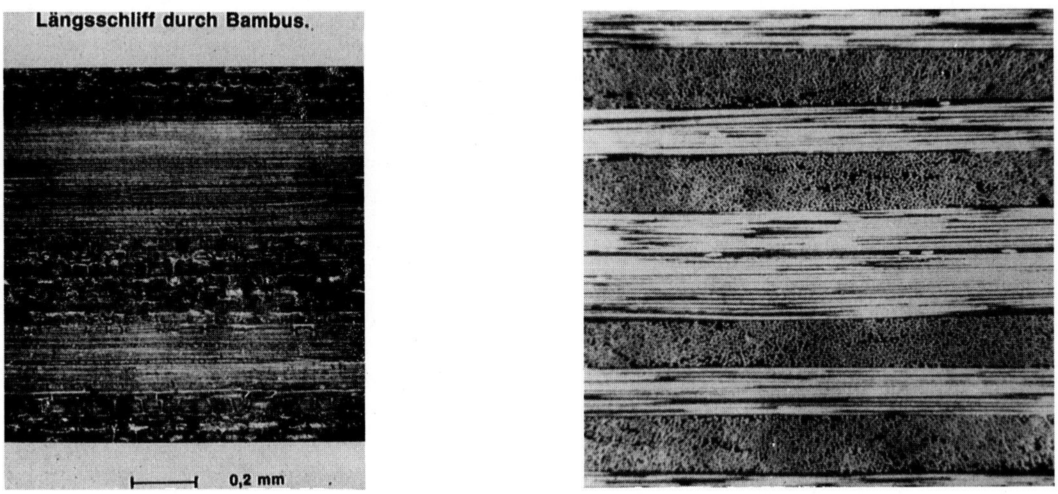

FIGURE 1

Cross sectional microstructures: a. Bamboo; b. Two-dimensional reinforced composite formulations of synthetic resins had to be developed for this special application as matrix in high performance composites.

It was a military development by the US and lateron also by the British airforce. Only after NASA had joined this classified material development for application in the space shuttle program, open literature became available on these superior new class of materials, surpassing conventional metals, see FIG 2 [2] and FIG 3 [3].

The term "Advanced Composites" became popular for such fibre reinforced polymers, indicating that it was a decisive progress if compared with the conventional fibre glass polymers without any stiffness properties.

II. THE REINFORCEMENT FIBRES AND MATRICES

2.1 The development of fibres with high specific stiffness

Boron fibres were the first selection of the US airforce. in the 50s as most promising candidate for light weight fibres with high strength and stiffness, suitable for reinforcement of polymers, to be applied in future airplanes.

A new technology for their fabrication was developed, the chemical vapor deposition on thin tungsten wires as substrates, resulting in thick monofilaments with more than 100 um diameter, having rather the character of thin wires than of fibres or filaments, with very difficult handling and enormous prices, see FIG 4 [1].

Therefore chemists started to develop filament fibres from the neighbouring element carbon, which is also light and has a high melting point, and from which fibres have been demonstrated 100 years earlier as filaments for electric light bulbs. The high stiffness of the first new carbon fibre type, which was rayon based, has been achieved by hot working, similar to the processing of metal wires, but only at much higher temperatures around 3000 K.

Real technical success with carbon fibres was achieved only after polyacrylonitrile (PAN) has been introduced as precursor fibre.

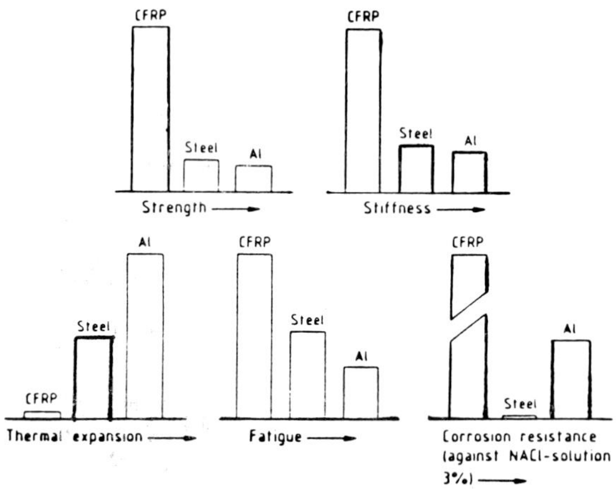

203 225

FIGURE 2

Properties of advanced composites compared with conventional materials: a. specific strength and modulus; b. further superior properties (arbitrary units) and some fields of non aerospace application.

FIGURE 3

US space shuttle in fabrication at Rocketyne Int. with CFRP doors for the loading room [3].

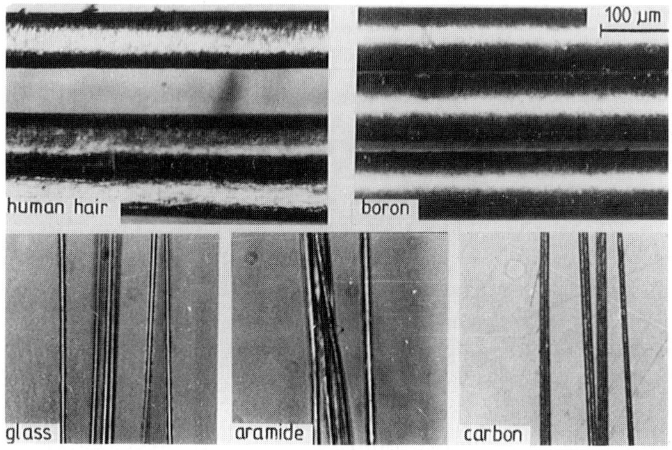

FIGURE 4

REM micrographs of reinforcement fibres compared with human hairs.

The former high-temperature forming process became superfluous upon introduction of the preferred orientation in the process step before thermal decomposition of the polyacrylonitrile fibres to the carbon fibre (British airforce. contribution). The PAN based carbon fibres, which can be handled in tows of some thousands monofilaments like glass fibres, have completely replaced boron fibres in their function of strengthening and stiffening polymer composites.

There are two types of carbon fibres, the low heat treated type "HT" with highest strength and only moderate stiffness and the high heat treated type "HM" with highest stiffness but only moderate strength, see FIGs 5a and 5b.

During last years, it became known, that

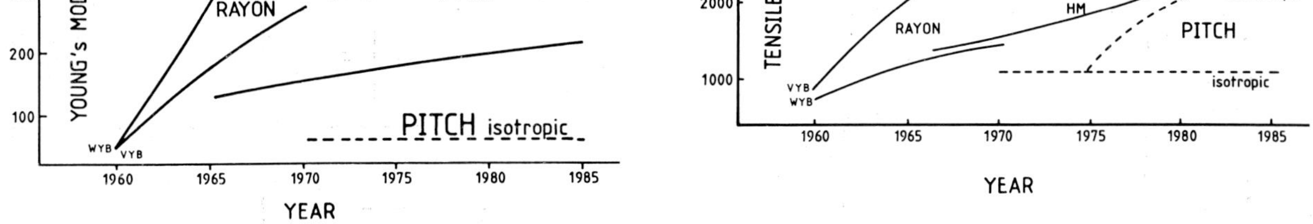

FIGURE 5

Development of properties of carbon fibres based on the various precursor fibres during last decades: a. Young's modulus; b. tensile strength.

this "HM" type with extremely high modulus can be fabricated beat from pitch via the so called mesophase route [4,5].

The enormous advantage of the "HT" type carbon fibres is the high value of strain to failure, which ranges in the order of nearly 2%. Modern design and engineering prefer the high strain to failure. 2% is the value which would be appreciated by the engineers. It can only be achieved and eventually surpassed by further improvement of the tensile strength of the fibre, because no carbon fibre type shows plastic deformation.

Table 1 [6] gives a compilation of today's commercially available types of reinforcement fibres. This table also contains the organic polyaramid fibres and the improved glass fibre qualities.

For better understanding of the alternatives for reinforcement of polymers as the most important matrix material in advanced composites so far, Table 2 offers a qualitative compilation of the respective typical properties:

Glass fibres are the most economic ones with good strength but lacking stiffness and highest density.

Table 1. Typical strength and stiffness values (state 1990 according to companies' data)

	Strength MPa	Modulus GPa	Strain %
PAN based			
standard HT			
Torayca T300	3530	230	1.5
Celion G40 700	3790	238	1.6
IM types			
Torayca T800	5490	294	1.9
Celion G50 300	5000	300	1.7
HM types			
Torayca M46	4210	436	1.0
Celion G50 300	2500	360	0.8
MP pitch based			
Amoca P25	1400	160	0.9
P55	1900	380	0.5
P100	2370	758	0.32
Dupont G E35	2900	262	1.0
E55	3200	393	0.75
105	3100	717	0.5

Polyaramide fibres are the lightest ones with moderate strength and stiffness, but high sensitivity against compression.

Carbon fibres are superior in most properties with the exception of impact behavior.

In order to improve this disadvantage, in modern composite technology, often hybrid reinforcement is applied with the goal of toughening, especially. against impact damage.

2.2 Carbon fibres, the future challenge for polymer composites

The specific mechanical properties of all actual reinforcement fibres are compiled in FIG 6.

For advanced composites with polymer matrix, there is no alternative to the superior carbon fibre types, as they have been achieved today. The question arises whether the present status is the final one already for carbon fibres or even further improvement can be expected in future.

Table 3 gives elastic numbers of the graphite lattice. Although carbon fibres do not consist of graphite, the C11 number can be taken for the bounding strength in the planar polyaromatic planes of the fibre lattice

Table 2. Qualitative comparison of today's reinforcement fibres

	Static Load						Dynamic Load				
	STRENGTH			MODULUS							
	Tensile	Flexural	Compressive	Tensile	Flexural	Compressive	IMPACT	FATIGUE	ILSS	Temperature Resistivity	DENSITY
GLASS	+	o	+	–	–	–	+	–	+	o	–
ARAMIDE	o	–	–	o	o	o	o	o	o	–	+
CARBON	+	+	o	+	+	+	–	+	+	+	o

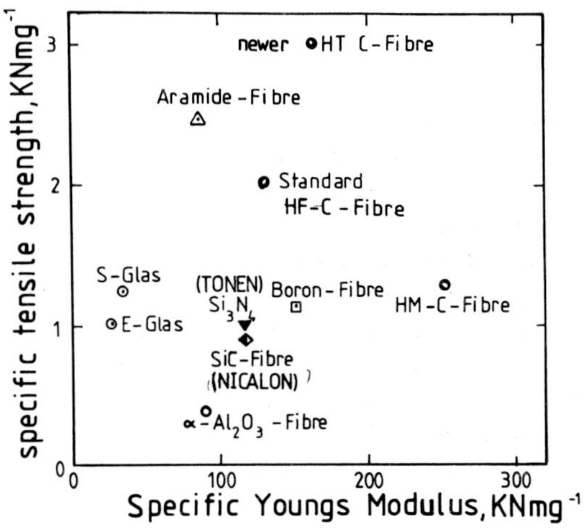

FIGURE 6

Specific mechanical properties of all newer reinforcement fibres (modulus versus strength).

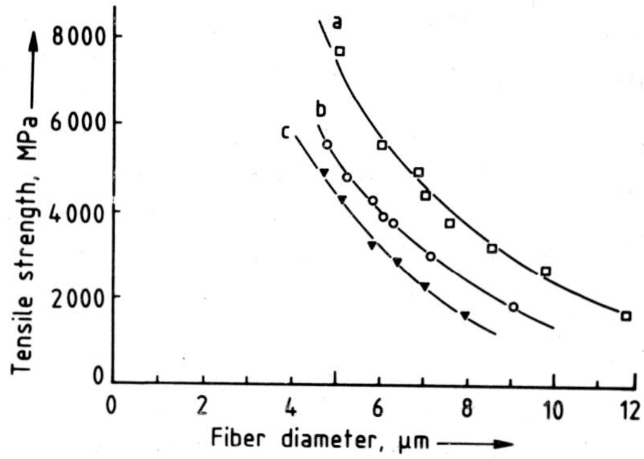

Tensile strength of monofilaments (carbon fibers) as a function of fiber diameter. a) Hercules AS4 (Hercules); b) Torayca T300 (Toray); c) Torayca M40 (Toray)

FIGURE 7

Monofilament tensile strength, depending on the fibre diameter.

Table 3. Elastic numbers of the graphite lattice

ELASTIC CONSTANT:	
a-direction	C_{11}=1060 GPa Young's modulus
c-direction	C_{33}=36.5 GPa
parallel to the planes	C_{44}=4.5 GPa Shear modulus
[-C-C]-DISTANCE:	
a-driection	0.142 nm
c-direction	0.335 nm

and shows that up to 80% of the theoretical maximum stiffness can be achieved already in the Ultra HM type C fibres.

Theoretical strength is achieved upto 20% based on the modulus in the HT types. These both percentages of theoretical maximum stiffness and strength data are unique, if compared with all other technical materials. In HM type carbon fibres however, the achieved strength is only a small percentage of the theoretical one.

There remains the probability for further improvement in strength, for the low modulus types as well as in combination with higher and also very high stiffness.

FIG 7 shows, that the superior strength improvement during last years has been achieved by reduction of the diameter of the monofilaments to 6 um. There is no doubt that further reduction of the fibre thickness will further improve the strength, but it is doubtful, whether such strength increase can be translated to strength increase of the composite because of the buckling problem in the fibre, embedded in the matrix - the buckling in compression, see FIG 8.

Short fibres have no buckling problem. However the reinforcement effect in the composite is minor, especially it is non isotropic. Composites with short fibres are therefore not in the projected line for new energy saving ANISOTROPIC materials as announced in the introduction.

2.3 Polymer matrices for advanced composites

So far as the polymer matrices are concerned, epoxy resins are the standard thermosetting polymers generally used.

These standard resins are bisphenol A diglycidil esters and are applied in liquid form together with a curing agent, mostly aromatic diamines. They have to be cured under carefully controlled conditions for some hours at 120 centigrades. For the aerospace application, so called tetra functional modifications of epoxies are used, which show improved strength and resistance against heat damage, however also less toughness and the need of higher curing temperatures around 180 centigrades.

The unsaturated polyesters, the unique matrix for the broad fibre glass application, are very easy and very fast to cure, even by light only. They are not applied for advanced composites, neither for the reinforcement with carbon fibres, nor even with the high quality glass fibres.

There are two developments in the matrix technology: the first one comprises the use of higher application temperatures by replacing of the epoxies with bismaleinimides. In such a way, the upper temperature limit can be extended up to 250 centigrades, however again with acceptance of higher brittleness. Today's application is below 5% of that of epoxies; the second tendency in changing the polymer basis of the matrix is the introduction of high temperature thermoplastics. The

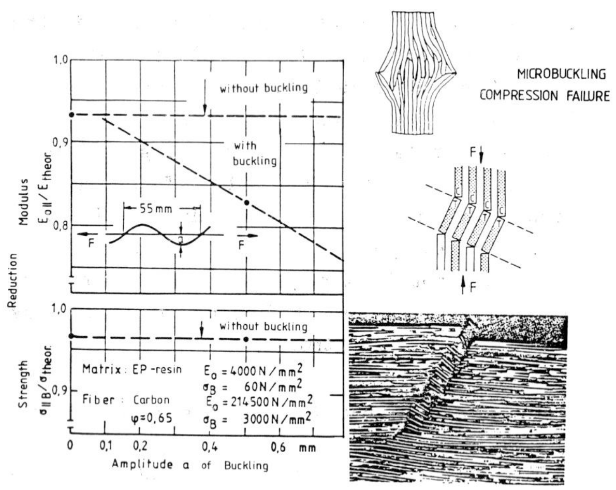

FIGURE 8

Buckling of thin carbon fibres under compression stresses.

enormous increase of polymers application in the past was generally due to the very easy and fast formidably of thermoplastics. The same advantage is expected in simplifying of the composite technology and by this means broadening of the application field of advanced composites.

Precondition for advanced fibre reinforced thermoplastics is the stability of these polymers as solids at higher temperatures, at least above 250 centigrades. High toughness is expected. Polyphenylen sulfides, polysulphones and first of all polyetheretherketons are promising candidates. Most critical parameter is the cost situation. FIG 13 will give an idea about the costs of these new matrix polymers.

2.4 Carbon fibres for reinforcement of others than polymer matrices

For a composite, chemical and physical compatibility is precondition.

The physical compatibility is mainly controlled by the thermal expansion. Carbon fibres have ANISOTROPIC thermal expansion coefficients with values in direction parallel with the fibre axis around zero, perpendicular between 10 and 20. 10 to the minus 6/K. Polymers have CTE values between 30 and 60. 10 to the minus 6/K.

Thermal treatment during curing is performed at about 200 centigrades. Consequence: In cooled composites, the carbon fibres are under residual compression stresses. This is advantageous in polymer matrix and improves even the flexural strength of the composite at cryogenic temperature [11].

Carbon fibres are chemically compatible with and can be used in carbon and graphite matrix. Such carbon-carbon composites are extremely resistant against thermal shock, because of the similarity in expansion behavior.

The most dangerous problem for carbon/carbon composites is the limitation of all high temperature applications to non-oxidizing environment.

All development work for protection of C/C composites against oxidation worldwide are still in mature state. Only limited success for short time applications up to 1600 centigrades has been achieved so far.

The long time application of advanced composites at high temperatures in oxidizing environment is the only case, where the carbon fibres have to be replaced by carbide or oxide fibres.

Carbon fibres can further be used for cement at room temperatures. Also this application, although very promising, is still in first stage of scientific research.

There are numerous activities to use C fibres to reinforce metals and refractory oxides with carbon fibres.

Both, physical and chemical compatibility are unsatisfactory in these two cases. Protective coatings have to be used as interface. Furthermore, ANISOTROPIC carbon fibres will embrittle the ductile metals. The chances, to use large quantities of carbon fibres for metal reinforcement are much minor than those of carbon fibres in polymers.

Similar situation can be expected for carbon fibre reinforcement of refractory oxides. Also in that case, chemical incompatibility with the matrix at higher temperatures makes protective coatings absolutely necessary.

In both cases, metals and refractory oxides, reinforcement fibres made of SiC and refractory oxides would be much more suitable than carbon fibres.

From mechanical viewpoint, those fibre materials can never reach the high quality level of carbon fibres, at least today and in near future [12]. An example of a new SiC reinforced refractory oxide will be presented in the part 4 of this paper.

III. THE CHALLENGE FOR CARBON FIBRE COMPOSITES

3.1 Design, fabrication and costs of advanced composites

Polymer composites consist of a volume fraction of about 60% fibres and 40% matrix.

In a unidirectional (UD) composite, where all fibres are arranged parallel to each other, the tensile strength of the composite in fibre direction is composed by the strength of the fibre part and that of the matrix part. This relation is controlled by the rule of mixture, that means in the case of high strength fibres, the bulk strength is in first approximation only proportional to the volume fraction of the fibres.

Good adhesion between fibre and matrix is precondition. Mostly 90% and more of translation of the fibre strength into the composite are achieved in modern fibre-matrix combinations. It means, that with a volume fraction of 60% of a modern HT fibre with 4000 MPA strength, a bulk strength of 2000 MPA can be expected. Taking into account an average bulk density of 1,9, the specific bulk strength will achieve 1160 m. That means, a composite strip of such length will stay undamaged when hanging under its own weight in the magnetic field of the earth. For more comparison, a high strength steel bar has a specific strength of 250 m only. With the newest HT fibre qualities (laboratory scale only) the 10 fold specific strength of that of high strength steel could be achieved.

In case, composites are designed for unique reinforcement in one direction only, the UD composites can be used. In case, some stresses will occur in other directions too, it must be considered, that in perpendicular direction of the UD composite, the strength of the matrix is the weakest part in the periodical geometrical sequence of fibre-matrix.

The bulk strength perpendicular to the fibre direction can be even lower, if the adhesion between fibre and matrix is lower than that of the intrinsic strength of the pure matrix. Another consequence of insufficient adhesion is the low shear strength in direction parallel with the fibre. Also the low shear resistance within the fibre itself can contribute to lower results. Especially HM type fibres are most disadvantageous in this respect. The most important consequence of the anisotropy of UD composites is the strong

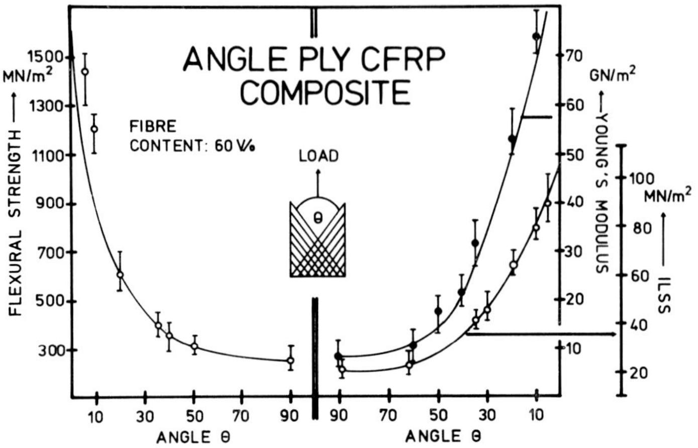

FIGURE 9

Angle dependence of the reinforcement effect from direction in an UD laminate.

angle dependence of the composite strength on fibre direction, for instance a 20% strength decrease in case of only 5% angle deviation from that of the fibre direction. In order to compensate the angle dependence of the testing direction from the fibre direction, composites are designed with different fibre directions, see FIGs 9, 10, 11 [1,10]. In such a way, the designer is able to tailor the mechanical properties of the composites.

The practical realization of optimized design of the composites, concerning the fibre directions, can be achieved by two different fabrication methods: the prepreg method, where thin preformed UD sheet are combined to an optimized laminate or the wet winding process, in which also different winding patterns are achieved fully automatically with computer controlled winding equipment.

In all cases of thermosetting matrices, the prepolymer, plastically deformable during the fabrication step of the green composite, must be cured in a subsequent process step at elevated temperatures up to 200 centigrades

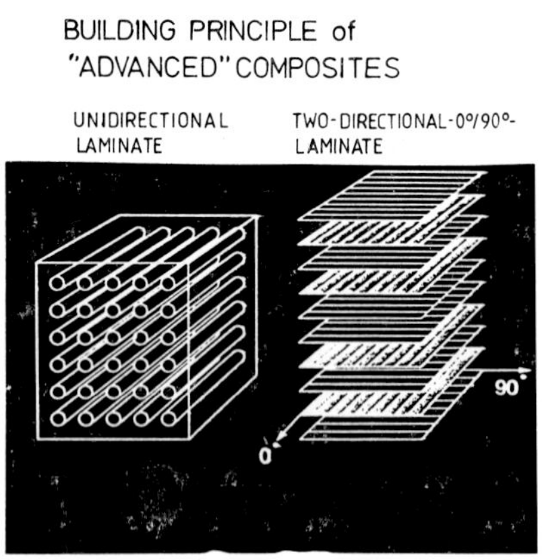

FIGURE 10

UD - laminate and two directional layer - laminate.

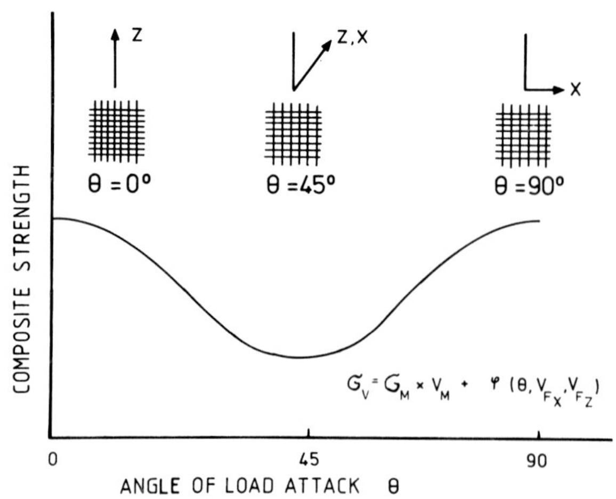

FIGURE 11

Composite strength in a 0/45 degree composite.

in an autoclave. This curing is one of the most expensive cost factors in composite technology. To give some idea of the costs: Fibre costs 15%, resin costs 10%, prepreg fabrication 15%, processing of the composite 50%.

Cost reduction can be achieved by the application of thermoplastic polymers as matrix. Concentrated research on the low cost fabrication of high-temperature thermoplastics will contribute to an eminent cost reduction for the composite technology.

Most critical parameter in the cost situation is the fabrication technology. Here thermoplastic can open a drastic cost saving. However, also fibre and matrix costs have to be considered. These both parts of the total cost for a composite increase considerably with improved qualities as shown in FIG 12 for the new polymers an in FIG 13 for the new developed fibre types [13].

The processing costs will further decrease with progressing automation. For this aspect, a large fabrication volume of same or similar structural types of the composites is precondition. Such a consideration is dominated by the field of application and thus by the market.

3.2 Application, controlled by cost/benefit considerations.

Nonmilitary applications of advanced composites today are based on cost benefit calculations only. The ratio cost benefit is favorable in case of fast moving structural parts according to the formula on kinetic energy, where energy saving is proportional to weight saving and inverse to the square of the speed. For missiles, aircrafts and racing cars, it is substantial. Also fast moving parts in machinery profit in the respect. Although automatives to be considered under same aspect, bring the economy of scale, based on steel technology and automation in production against the argument of weight saving. The cost/benefit ration is presented in FIG 14 [14]. As can be seen, aerospace is the most suitable field for application from economic viewpoint. The ratio is much less favorable for automatives.

The effective application is a direct mirror of these economic considerations as shown in FIG 15 [13].

Aerospace, already holding a fraction of more than a third, will increase much more that the other sectors. A fraction upto 65% is

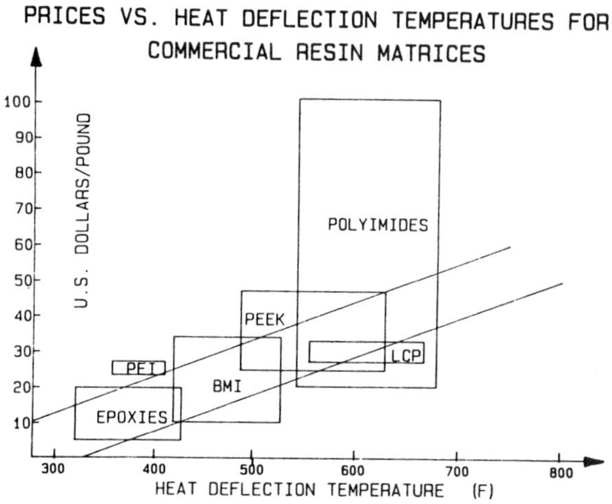

FIGURE 12

Prize increase for the new developed matrix polymers.

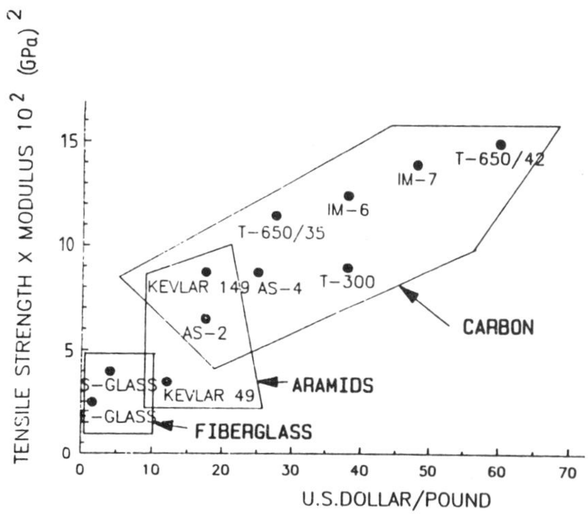

FIGURE 13

Prize development for the carbon fibres of the new generation.

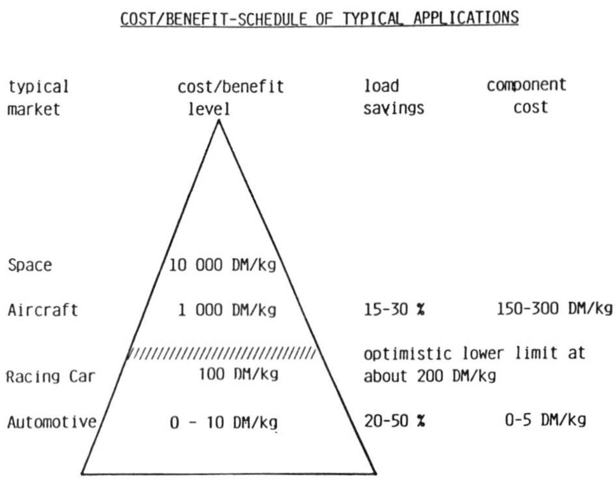

FIGURE 14

Cost/benefit ratio of some typical applications of CFRPs.

expected in 96, based on today's economic considerations.

3.3 Pollution and environment

However the situation is completely changed however, when the application of the advanced composites is considered from viewpoint of environment. Gasoline consumption by cars and trucks causes one of the main parts of air pollution in industrialized countries. The situation will became worse, when the Eastern Europe follows the standard of the western part in motorization. It will be catastrophic, as far as all development countries will follow. Only a drastic reduction of the specific gasoline consumption, additional to catalytic cleaning of combustion gases can save an acceptable atmosphere, not loaded by increasing dangerous ozone contents. There are experiments and demonstrations, that this goal can be achieved from technical viewpoint.

For instance some years ago, aerospatial has published in a press release, that a car has been designed, following the design principles as for an aircraft, for 5 passengers, 180 km/hr. max.speed and only 2-2,5 1/100km gasoline consumption.

Newest demonstration is given by a German company for city and cross country buses with weight and gasoline reduction in similar orders of magnitude, see FIG 16 [15].

The last example became successful because of the many traffic problems in all cities of the industrial world. Not successful was the French initiative because of the very strong

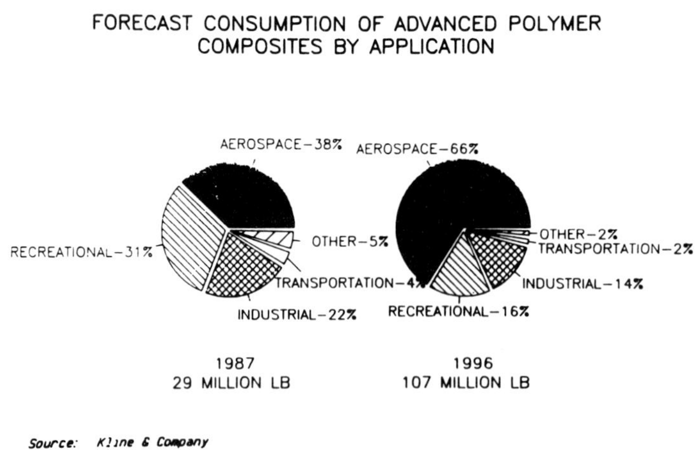

FIGURE 15

Splitting of CFRP application according to the various field in 1989 and forcast for 1996.

FIGURE 16

"Metroliner", a bus using advanced composites and crash test with a taxi, 40 km/hr speed.

competition of companies producing low cost street cars worldwide.

There is an other viewpoint for energy saving and improvement of the pollution problem: The energy need for production of the materials [16]. FIG 17 gives some information. As compared with conventional metals, advanced composites need only a small energy part. Also for production, energy consumption is combined with pollution, for instance in steel mills. The energy saving is equivalent with drastic reduction of pollution. All such technical information will not succeed without multilateral governmental regulations.

ENERGY REQUIREMENT IN kWh FOR THE PRODUCTION OF STEEL, ALUMINIUM, CFRP and PVC

Material	$\frac{kWh}{kg}$	$\frac{kWh}{l}$	$\frac{kWh}{l} / \frac{N}{mm^2}$
Al ($\rho = 2{,}8 \frac{g}{cm^3}$)	37,3	104,5	232,2
Fe ($\rho = 7{,}9 \frac{g}{cm^3}$)	6,8	53,6	26,8
CFRP, UD ($\rho = 1{,}5 \frac{g}{cm^3}$)	3,7	5,6	3,5
CFRP, 2D	3,7	5,6	6,9
PVC ($\rho = 1{,}4 \frac{g}{cm^3}$)	7,9	11,1	1110,0

194 600

FIGURE 17

Energy for production of various materials, including the raw materials.

Newer considerations of the antagonisme in our industrial world [17] have shown, that the former three - capital, work and consumption - must be supplemented by a forth - the environment, see FIG 18.

We have not to be prophets to expect a gigantic future for advanced composites.

IV. COMPOSITES FOR HIGH TEMPERATURE

For high temperature composites also, carbon fibres are of superior importance (see [18]). As mentioned before, they are chemically compatible with a carbon or graphite matrix. They are also physically compatible with a carbon or graphite matrix. They are also physically compatible with the matrix of synthetic carbon and especially of synthetic graphite, due to the porosity with slit formed micropores in such a matrix. This pore structure is the reason for the pseudoplastic fracture behavior of carbon/carbon composites. Two- or three-dimensionally reinforced carbon/carbon composites can even be nailed. FIG 19 shows a typical fracture surface of a 3 D - composite [19].

The high temperature strength is shown in FIG 20 [20].

It should be pointed out, that the three-dimensional reinforced composites do not follow the idea of anisotropy as discussed in the introduction. They utilize the outstanding refractoriness and especially the thermal shock tolerance of synthetic graphite, only in improved form as compared with the former granular finegrain graphite types. They are a special development for the military missile fabrication. The extra-ordinary properties can be utilized however also in the ANISOTROPIC version, for instant in aerospace [21].

According to a recent publication, the need for missile application is in order of

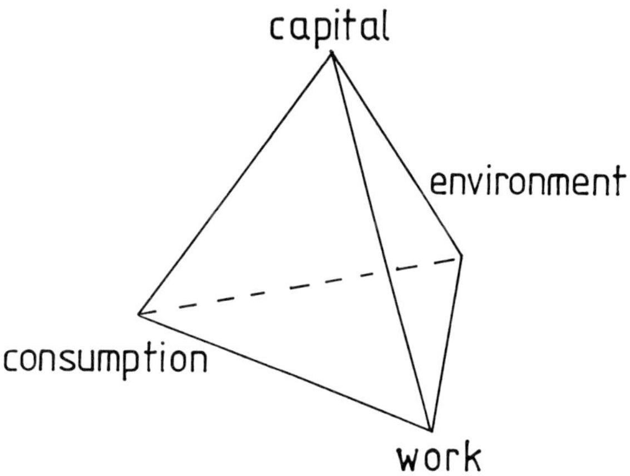

FIGURE 18

The antagonisme in the industrial world.

FIGURE 19

REM of a fracture surface of a 3 D - carbon/carbon composite.

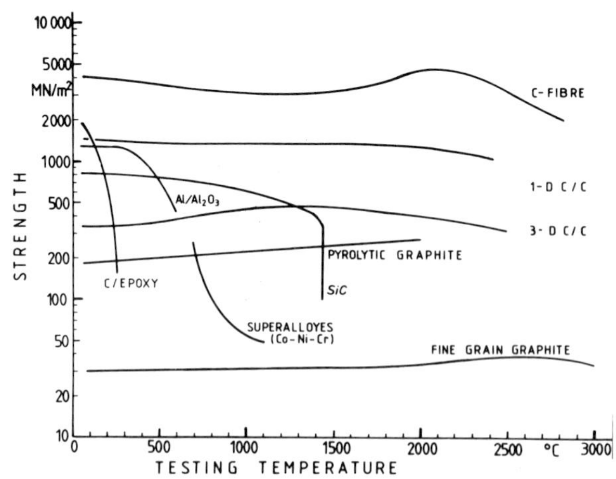

FIGURE 20

The high temperature strength of carbon materials.

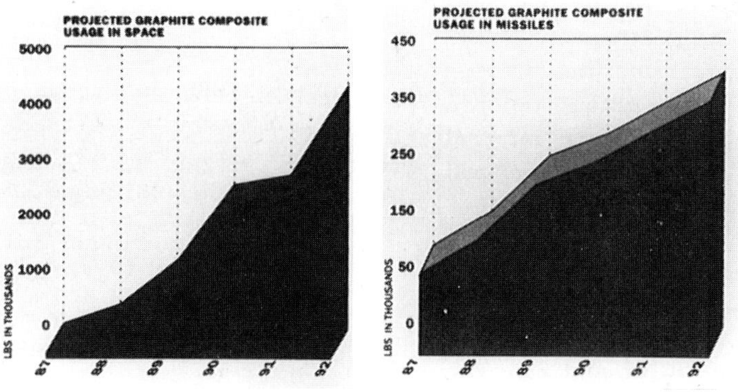

FIGURE 21

Projected carbon/carbon composite usage in USA.

magnitude less than that for aerospace, see FIG 21 [22]. Especially the very large future aerospace station must be designed from a material, resistant against damages by thermal cycles. It is achieved by a minimum in thermal expansion behavior. This is achieved by the anisotropy, following the same design principles as in C fibre reinforced polymers.

It has been mentioned already, that for long time applications in oxidizing environment, carbon fibres have to be replaced by oxidation resistant fibres, the commercial SiC or oxide based types, although their mechanical properties are worse, first of all their tendency to recrystallize at temperatures above 1000 centigrades, which transition from amorphous to crystalline state is combined with a decisive strength decrease [17].

As concerned the matrix material, there remains the alternative: SiC or refractory oxides.

Literature is available for both possibilities, for instance for Al-silicon carbride composites; their brittleness however cause severe limitations in applications [23].

Recently a new oxidation resistant composite has been published by us [24, 25, 26], which consists of a ZrO_2/SiO_2 matrix and continuous

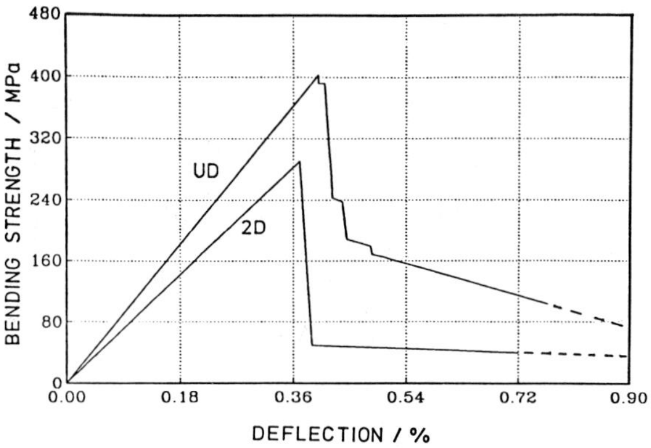

FIGURE 22

Stress/strain behavior of SiC reinforced UD and 2 D laminates with ZrO_2 modified SiO_2 matrix.

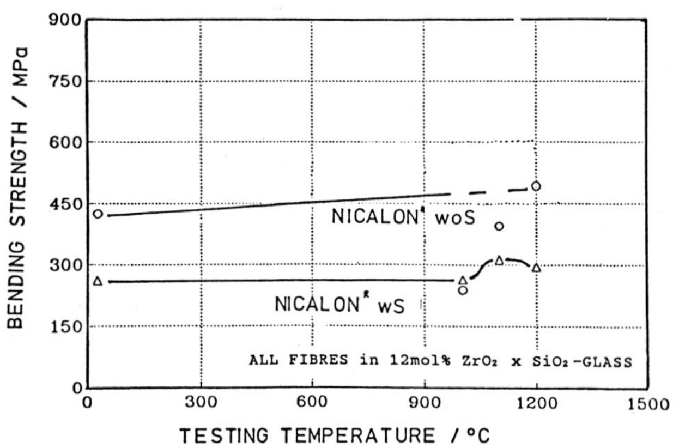

FIGURE 23

High temperature flexural short time strength of SiC reinforced ZrO_2/SiO_2 glass composites at high temperature.

SiC reinforcement fibres, based on polycarbosilane precursor. By the modification of the SiO_2 glass with 12 mol% ZrO_2, the thermal expansion coefficient of the matrix matches with that of the SiC fibres. Thus internal thermal stresses are avoided. A flexural strength up to 400 MPa at room temperature has been achieved with UD laminates, from which value a translation of the fibre strength up to 75% can be calculated. With temperature increase, a light strength increase is observed (see FIG 22 and 23). The composite is resistant against oxidation at least up to 1200 centigrades. Longtime test however have not yet been performed so far.

V. SUMMARY

The basic principle of advanced composites is their tailorable anisotropy. It is discussed, that the principle of anisotropy is the basic principle of natural growth and that there do not exist biomaterials not grown anisotropically.

Modern advanced composites follow the model of nature, since fibre materials with high stiffness became commercially available. Carbon fibres turned out to be the most suitable ones.

Today's preferred application field of CFRP's is the aircraft and space industry, which will expand most of all also in future. It is shown, that the technical use of these light, stiff and strong materials is limited so far by cost/benefit considerations only.

The advantageous energy saving in replacing of conventional metals by advanced composites is caused by the fabrication technology in general and in case of moving terestric structures during their application. The worldwide pollution problem by exhaust gases of automotives asks for international regulations. Advanced composites offer an effective way to contribute in solving this urgent environment problem.

Finally, the need of oxidation resistant high temperature light weight composites is discussed and a promising solution - SiC fibre reinforced ZrO_2/SiO_2 glass - si presented.

REFERENCES

1. E. FITZER, in Carbon Fibres and Filaments (Alvor, Kluwer Dordrecht, 1989) p. 3-42.

2. E. FITZER, in Caorbon Fibres and Their Composites (Springer, Berlin 1985) p. 3-45.

3. Bergmann, Fabrication at Rocketyne International (DVFLR, Braunschweig 1983)

4. E. FITZER and M. Heine, in Ullmann's Encyclopedia of Industrial Chemstry 1988, p. 42-63.

5. E. FITZER and M. Heine, in Composite Materials Series (Elsevier, Amsterdam 1988) p. 37-148.

6. E. FITZER and F. Kunkele, HTHP (1990) in press.

7. E. FITZER, in Ullmann's Encyclopdia of Industrial Chemstry 1988, p. 2-10.

8. R.W. Lang, H. Stutz, M. Heym and D. Nissen, Lecture at meeting of GDCh subgroup "Chemistry of Markov Moleculars", Bad Nauheim, 1986.

9. E. FITZER, CARBON, 27 (1989) 621

10. G. Menges, Proc. Techkeram 88, Wiesbaden, DEMAT Frankfurt 1988, p. 2.00-2.49.

11. E. FITZER, G. Hartwig and H. Jager, Proc. CARBON 86, Baden-Baden, DKG, 1986, p. 260-263

12. E. FITZER and W. Remmele, Proc. Techkeram 87, Weisbadden, DEMAT Frankfurt, 1987, p. 14.01-14.36.

13. R.A. Farth and C.H. Eckert, 33rd Int. SAMPE Symp. Anaheim (Elsvier, Amsterdam 1988) p. 369-380/

14. H. Boder, Proc. CARBON 86, Baden-Baden, DKG, 1986, p. 608-610.

15. B. Lee, Proc. Verbundwerk'90 Weisbaden, DEMAT Frankfurt, 1990, p. M04.

16. H. Fleckenstein, VDI-Z, 145 (1983) 123

17. R. Eckert, Chem. Ind. 110 (1987) 10.

18. E. FITZER, HT-HP, 18 (1986) 479

19. E. FITZER and M. Heym, in Processing and Uses of CFRP, (Baden-Baden, VDI-Verlag, 1981) p.85

20. E. FITZER, Pure and Appl. Chem. 60 (1988) 287

21. E. FITZER, in Silicon and Carbon (Elsvier, London, 1987) p. 203-237.

22. HERCULES prospect, 1990

23. E. FITZER, Proc. CARBON 86, Baden-Baden, DKG, 1986, p.643-645.

24. E. FITZER, Proc. Int. Conf. on Whisker-, Fibre-toughneed Ceramics Oak Ridge, 1988, p.165-192.

25. E. FITZER and A. Pfeiffer, HTHP (1990) in press.

26. E. FITZER and A. Pfeiffer, HTHP (1990) in press.

Frontiers of Materials Research/
Electronic and Optical Materials
M. Kong, L. Huang (editors)

CURRENT RESEARCH TRENDS ON STRUCTURAL CERAMICS

Jingkun GUO and Dongsheng YAN

Shanghai Institute of Ceramics, Chinese Academy of Sciences
Shanghai 200050, China

1. INTRODUCTION

As an art and technology, ceramics has a history of nearly 8,000 years in China[(1)]. The sherds of black pottery and grey pottery (5,190 ± 150 BC) unearthed from LUO Jiajiao in Zhejiang province have indicated that our ancestors at that times already knew the importance to choose raw materials and found the approaches to improve the thermal shock resistance and prevent cracks in the products. The application of clays with higher alumina content, the development of high temperature technology as well as the use of glaze, were the three main breakthroughs in the period from 1,500 - 300 BC in China and resulted in the appearance of neoporcelain. This is considered as the first stage of development in ceramics history, which was marked by the evolution from pottery to porcelain. The microstructure of porcelain ware was typified by a multiphased structure of mullite plus other crystalline phases and glassy phase. Such microstructural characteristics are the basic elements of traditional ceramics. The development of modern sciences promoted the study of ceramics as a sicence. Through the use of highly purified or chemically prepared starting materials, the improvement of ceramic processing technology, the development of more sophisticated instrumentation and methods for microstructural analysis, property studies and nondestructive evaluation, etc., the present day advanced ceramics or fine ceramics in a multitute of varieties have been evolved. They are, in general, characterized by a much finer microstructure in the micrometer scale and approaching single phased with high purity. They may be considered as the second stage of development in ceramics history. Upon analyzing current research trends on structural ceramics, it is probably worthwhile to note that we are entering night now into the third stage of development which may be represented by the following three main features: (1). the size of grains in the material has gradually been scaling down from micrometer to nanometer scale, (2). the microstructure has been fransferring from high purity, monophased to multiphased material but at a much more controlled level, and (3). with the progress of ceramic science and technology, people are gaining greater ability to tailor and design materials according to the properties needed in application. In this paper, it is intended to illustrate these issues by intorducing some of the works, especially carried out in our laboratory in recent years as examples.

2. NANOSCALED CERAMICS

Fig.1 shows some of the essential contents of research and property-application relationships of nanoscaled ceramics (ceramics with nanometer sized grains). While grain size reduces by an order of magnitude, the surface area of grains and the grain boundaries will increase by its square. When the grains are of a few nanometer in size, the grain boundaries will constitute an enormously large percentage of the whole volume (depending on how thick is the grain boundary structurally and/or electronically). These will lead to a series of drastic changes in such systems.

Firstly, for the fabrication of nanoscaled ceramics, nanoscaled powders will be needed along with other

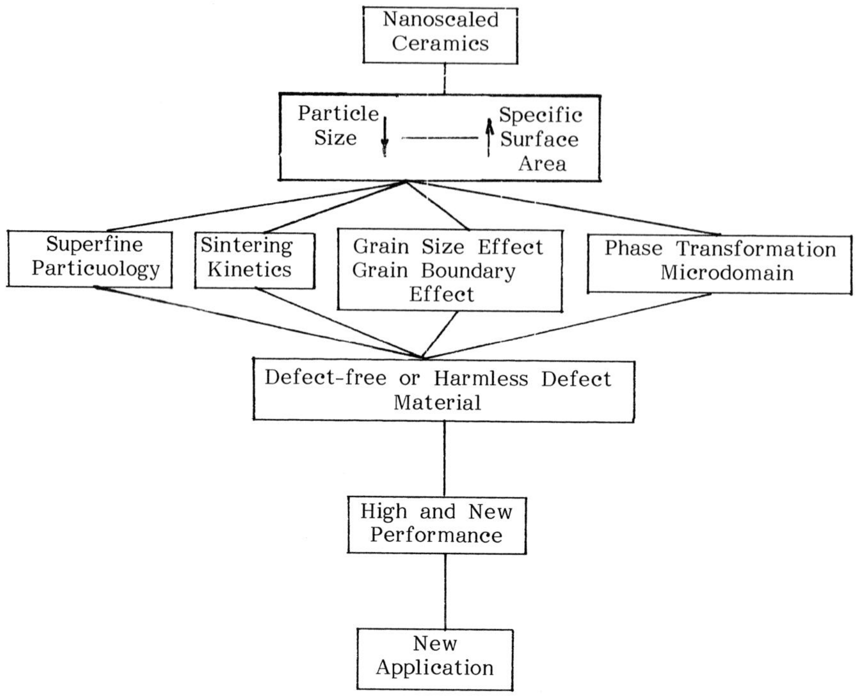

FIGURE 1
The basic research contents of nano-scaled ceramics

requirements, for instance, chemical purity, homogeneity, nanosized or desired size distribution of powders, etc. To meet with these qualifications, one will probably resort to various chemical routes as reported already in large amounts in recent literatures. One point of extreme importance that needs to be indicated is agglomeration which goes along with the extralarge surface areas of such powders. The mechanisms of such agglomeration, soft or hard, the consequences of agglomeration in both shaping of the powders into a green body and sintering into an applicable material as well as to optimize its microstructure and properties are complex problems to be faced and studied. The characterization of superfine powders is also a challenge.

Secondly, rate of sintering and densification of superfine powder compact will be dramatically enhanced by several orders of magnitude. Much higher driving force for diffusion, densification and chemical reaction stemming from the much higher surface energy of the system will present new contents for research.

Thirdly, the mechanical properties of the material will be greatly influenced by grain size and the microstructure. The solid line in Fig.2 shows what has been obtained for well processed material and the dotted extrapolation is what might be figured proposing that the defect size decreases simultaneously with grain size. The shape of extrapolating the curve is uncertain and will be a rich area for research. In addition, creep, plasticity (or superplasticity), and fracture toughness etc. will also be expected to show great diffe-

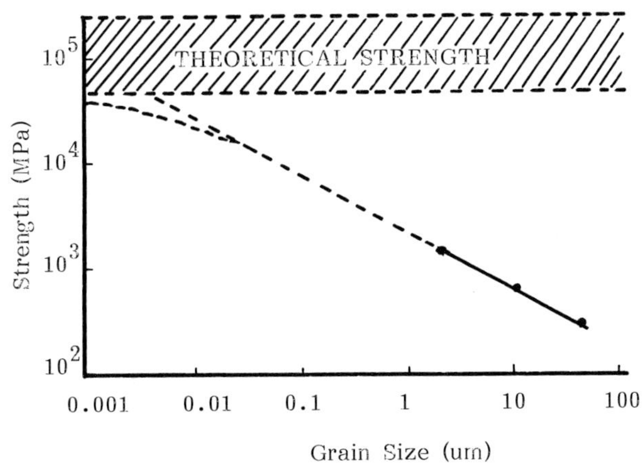

FIGURE 2
The sketch of relation between grain size and strength of ceramics

rences asking for new explanations.

For a doped or a multiphased material, the second phase will be distributed to a such large volume and yet very thin layer of grain boundaries in nanoscaled ceramics. The influence on the functional properties of such materials will be of great interest to study. Does nanoscaled material lend more room for enhancing or tailoring sensitivity, selectivity, and reliability? Processing science and technology will be of great importance.

Fourthly, what are the effects of grain refining down to nanoscale on phase transformation, on grain alignment for nonequiaxed material? Could we make use of these effects on transformation toughening or the critical current density of high Tc superconductors?

Fifthly, the pores and defects in a ceramic material usually have sizes of the same order of magnitude as those of the grains. Therefore, in this respect, these defects may do no harm on material's mechanical properties and can be refered to as "harmless" defects. One might say, he has made a "defect free" material.

Sixthly, when the grains of a material are degraded down to nanoscale, many of its properties may show an abrupt turn or jump at a certain stage. It will be interesting and rewarding to study how and why it happened.

And seventhly, these nanoscaled ceramics will have an unexpected dimension of new performances.

From those enumerated above, nanoscaled ceramics will call for new processing routes, will promote to develop new theories in ceramic sicence to explain the new phenomena and noval properties happening with grain refining and thus ceramic research may embark onto a new step of development. These materials may show unforeseen capability of performance. It would be attractive both from the practical or the theoretical viewpoints.

Can nanoscaled ceramics actually be realized?

The answer is, at least, essentially positive.

New processing methods can prepare ceramic powders with several tens or several nm in particle size. Using a 2.8 mol.% Y_2O_3 - ZrO_2 powder with particle size of ~20 nm, spherical in shape and loosely agglomerated can be formed by conventional methods and has good sinterability attaining a relative density of over 98% at 1250°C with grain size ~100 nm (see Fig.3)[2], which is 300-400°C lower than that for ordinary zirconia powders. Low temperature sintering is very effective to suppress grain growth attaining superfine grain structure. New sintering techniques, such as microwave sintering, that may allow heat generation and heat transfer and thus densification to be accomplished in much short time intervals —— fast sintering —— will even be more favorable to obtain superfine microstructure materials.

The development of nanoscaled ceramics will open up the area and potential of research in ceramic sicence and technology. Research on processing and synthesis, properties and performance will be gaining new grounds in scientific and technological contents. It can be anticipated that nanoscaled ceramics will be a new trend of research in the coming years. Ceramic materials with better and new properties and thus stretching their performance and application will have a bright future.

FIGURE 3
SEM picture of Y-TZP with grain size ~ 100 nm

3. MULTIPHASED CERAMICS

Advanced ceramics with monophased microstructure and much cleaner grain boundaries symbolizes the major development in the past few decades. This approach has been able to elevate the properties in significant measures than the multiphased conventional ceramic materials and to use them in more sophisticated applications. However, there are certainly limitations in this approach of advancement. One critical problem is brittleness and there are others. And there is a tendency, recently, to resort to new multiphased ceramics as indicated in the introductory part of this paper. A few illustration will be cited especially in relation with the aim to improve toughness.

3.1 Fibre (or whisker) reinforced ceramic composite

This has been a topic of research and development for some years. Both strength and toughness can be greatly improved by fibre (whisker) reinforcement[3]. Mechanisms for energy absorption and reinforcement have also been studied and proposed as due to debonding, crack deflection, crack branching, and fibre pull-out effects. C-fibre reinforced Si_3N_4 composite does not improve its mechanical strength due to the mismatch of Young's Moduli of the two materials, but still the fracture toughness of the composite is greatly improved[4]. SiC whisker reinforced TZP shows better high temperature strength (see Fig.4) as compared with Y-TZP material[5]. Both fibre (whisker) and particulate reinforcement come into play at higher temperatures when phase transformation loses its potential. For CMC material, interface study is of upmost importance. Fig.5 shows the interface in a SiC(w)/Si_3N_4 composite. There is just a couple of nanometer of amorphous layer in between these two phases which helps to improve the compatibility. Experiments indicated that 10-15 vol.% of SiC(w) gives the best strengthening and toughening effects[6].

The following criteria are favorable for achieving a successful CMC composite material:

(1). high temperature chemical compatibility between the reinforcement and the matrix materials,

(2). compatibility of thermal expansion coefficients, $\alpha_{(w)} \geqslant \alpha_{(m)}$,

(3). compatibility of the Young's Moduli,

$$E_{(w)} > E_{(m)}$$

(4). a favorable interface —— microstructure and micromechanism of the interface.

3.2 Particulate reinforced ceramic composite (PRCC)

Dispersion hardening in metal matrices has long been a successful approach in metal industries. Ceramic composites reinforced by particulate inclusion are relatively new and promising and we can learn from what has been achieved by metallurgists. There are natually basic differences with ceramic matrices. Both processing technology and mechanisms of reinforcement need to be intensively

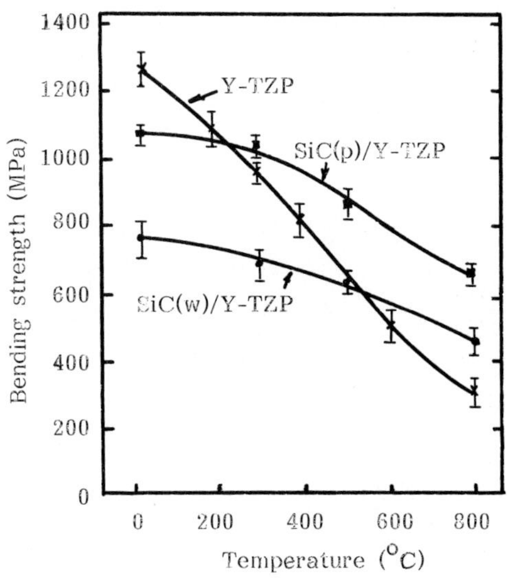

FIGURE 4
The comparison of high temperature strength of Y-TZP, SiC(w)/Y-TZP and SiC(p)/Y-TZP composites

studied. Factors of importance for strengthening and toughening in PRCC are complex. To name a few, the particle size and grain size of the particulate and matrix materials; the thermal expansion coefficients and Young's Moduli of the components in the system are items of great concern. One can roughly analyze their relative effects, but detailed studies are needed, although particulate reinforcement is or can be a fruitful approach. The following are a few experimental results obtained in our laboratory.

(1). SiC-TiC composite ceramics

TiC particulates were added to α-SiC powder and SiC-TiC composite ceramics were fabricated with their properties shown in Table 1 [7]. Dispersed TiC particulates in SiC matrix show a significant effect to enhance the fracture toughness of the material. Fine dispersed particulates also significantly increase its strength. Since there is almost a 50% difference in thermal expansion coefficients between TiC and SiC, strain fields were observed in SiC grain (Fig.6) Crack deflection and branching were also observed in the composite system (Fig.7).

(2). SiC-TiB_2 composite ceramics

The properties of SiC-TiB_2 composite ceramics are listed in Table 2 [8]. Apparently, there is an optimum value of ZrB_2 content in the composite for strengthening and toughening, which is about 10-15 v/o and there are also crack deflection and branching observed. More work will be done on this system.

(3). ZrO_2-SiC composite ceramics

By dispersing SiC particulate in 2.8 mol.% Y-TZP, it was shown in Fig.4 that the composite has better high temperature strength than SiC(w) reinforced Y-TZP [5]. Microstructural analysis indicated that the SiC particles are under stress

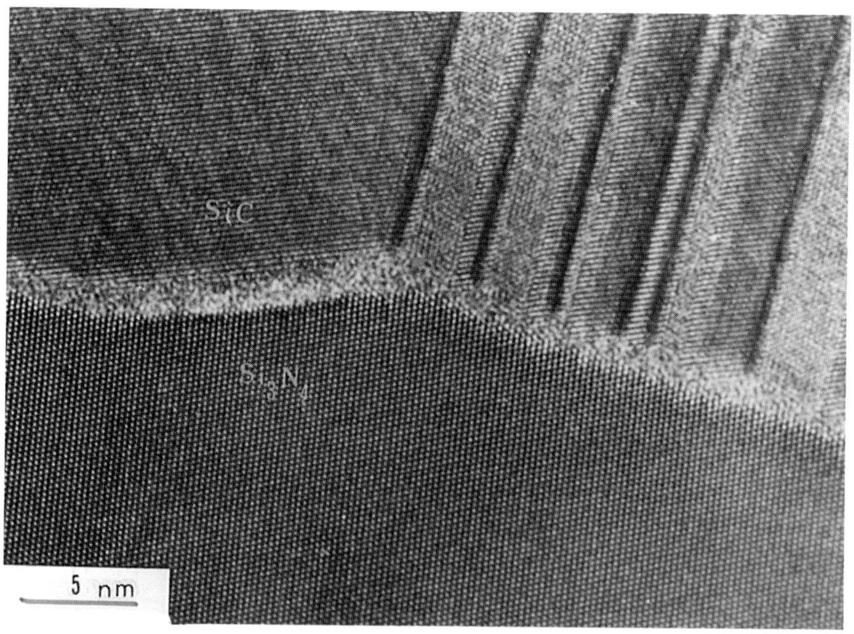

FIGURE 5
TEM picture of the interface structure in SiC(w)/Si_3N_4 composite

Table 1 The Mechanical Properties of SiC with Different TiC Content

Composite	Density (g/cm^3)	Flexural Strength (MPa)	Fracture Toughness ($MPa \cdot m^{\frac{1}{2}}$)
SiC-25 vol.%TiC*	3.65	418	6.2 (SENB)
SiC-25 vol.%TiC**	3.63	586	7.15 (ISB)
SiC-1%B_4C + 3%C	3.17	487	4.5 (SENB)

* TiC particle size 7.2 um ** TiC particle size 0.8 um

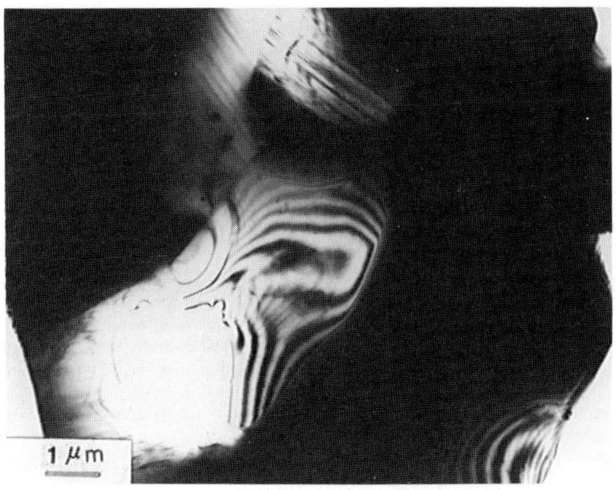

FIGURE 6
SiC-25 vol.% TiC multiphased ceramics, the stress loops were revealed in SiC grain

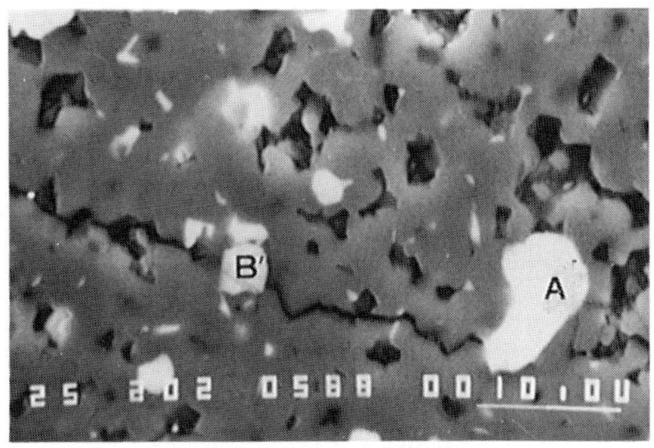

FIGURE 7
The crack deflection and branching in SiC-13 vol.% TiC, and A,B should be noted

Table 2 SiC Based Multiphased Ceramics with Different ZrB_2 Content

No.	Composition	Density (g/cm^3)	D/D_t (%)	Flexural Strength (MPa)	Fracture Toughness* ($MPa \cdot m^{\frac{1}{2}}$)
SZ21	SiC-21 vol.% ZrB_2	3.84	100	436	4.8
SZ15	SiC-15 vol.% ZrB_2	3.60	99	561	6.46
SZ10	SiC-10 vol.% ZrB_2	3.50	100	598	

* SENB method.

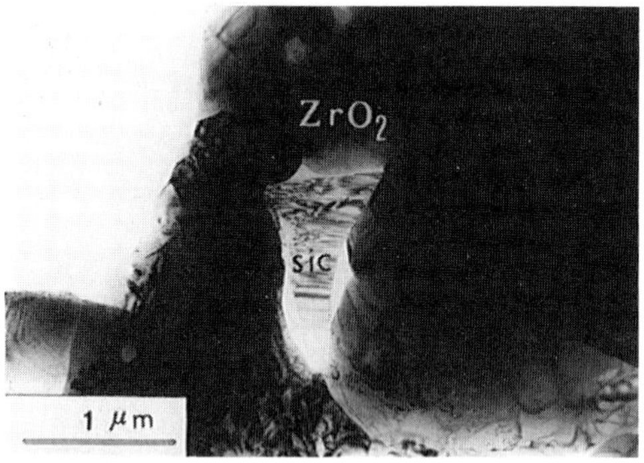

FIGURE 8
Y-TZP-SiC multiphased ceramics

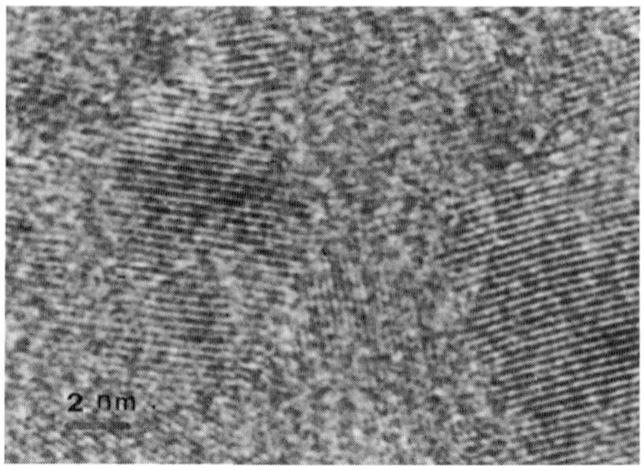

FIGURE 9
The crystallization of glassy phase in micro-area shown in Fig.8

(Fig.8) and the surrounding tetragonal zirconia grains are prematurelly transformed into monoclinic phase. The silica rich glassy phase at the grain boundaries contains a minor amount of Y_2O_3 diffused out apparently from the Y-TZP matrices. And there are micro-areas, a couple of nanometer in size, being recrystallized (Fig.9). The deflection of yttrium in tetragonal zirconia promotes its phase transformation and some microcrack formation at certain locations. This complex microstructure in superfine scale shows particularly effective strengthening effect especially at higher temperatures (Fig.4). Further work will be pursued in this class of particulate strengthened ceramic composites.

4. DESIGN AND TAILORING OF CERAMIC MATERIALS

Based on more thorough understanding of the phase compatibility relationships of complex systems as well as the understanding of the reaction kinetics and sequences of such complex systems, it is possible to tailor ceramic materials with double (or complex) major crystalline phases so that an optimization of properties can be achieved by compensating or reinforcing each other. α - β Sialon composite material design is one of such examples(9,10), compensating or combining the higher hardness and better thermal shock resistance of α-Sialon with higher strength properties of β-Sialon. Some of the results are shown in Table 3.

Microstructural tailoring of ceramic materials through processing is of no less importance and is of compensative nature together with compositional design. Usually, they are supportive one to the other. No good material can be obtained with poor microstructure. Compositional and processing design to obtain fibrous β-Si_3N_4 (or Sialon) and/or AlN polytypes with high aspect ratio (5-10) has been shown to give promising toughening results[11].

Grain boundary tailoring and engineering has been shown to be very significant to manipulate and upgrade the mechanical properties of Si_3N_4 ceramics. By using Y_2O_3 plus La_2O_3 as double additives for β-Si_3N_4 ceramics instead of Y_2O_3 plus Al_2O_3, the eutectic temperature of the glassy phase can be raised by at least 200°C. Proper heat treatment can further develop N-containing crystalline phases as well as refractory $LaYO_3$ phase at the grain boundaries. The high temperature strength can be maintained up to 1300°C (Fig.10)[12,13]. Further narrowing down the grain boundary width (specimen A & B with nominally 4 nm and 8 nm respectively) and processed by GPS, fracture toughness of the material is substantially improved at high temperatures beyond 1100°C (Fig.12). The high temperature (1300°C) creep behaviors of these materials are good. Specimen A with narrower grain boundary width (4 nm) has lower creep rate. A comparison of heat treated specimens A & B by creep experiments at 1300°C is shown in Fig.13. Apparently, materials with narrower grain boundary width have better creep resistance. The mechanisms of creep have also been studied. They are essentially controlled by diffusion plus visous flow. Heat treatment will favor diffusion as the major contribution due to recrystallization of the grain boundaries.

Table 3 The Properties of $\alpha+\beta$ -Sialon Multiphased Ceramics

Designation	α/β ratio	Density (g/cm^3)	Bending Strength (MPa) R.T.	Bending Strength (MPa) 1300°C	Fracture Toughness ($MPa \cdot m^{\frac{1}{2}}$)	Hardness (HR_4)	Thermal Conductivity ($W/m \cdot {}^oK$)	Thermal Expansion ($x10^6/{}^oC$)
2-2	——	3.26	700-750	540	5.7	93-94	9.1	3.87
8-2	——	3.26	750-800	620	6.8	93	11.2	3.66

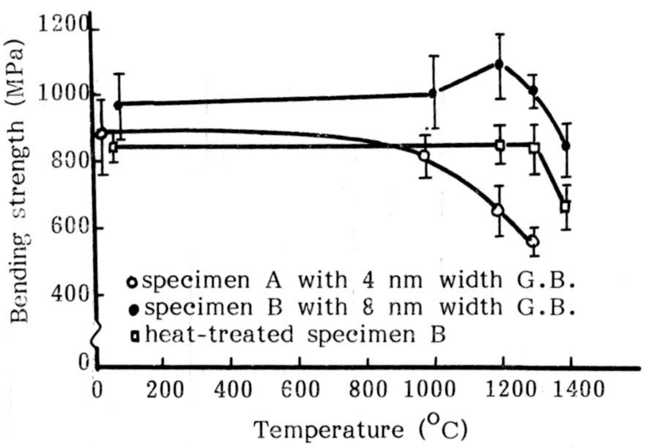

FIGURE 10
Bending strength vs temperature relationship of YL-1 and YLA-1

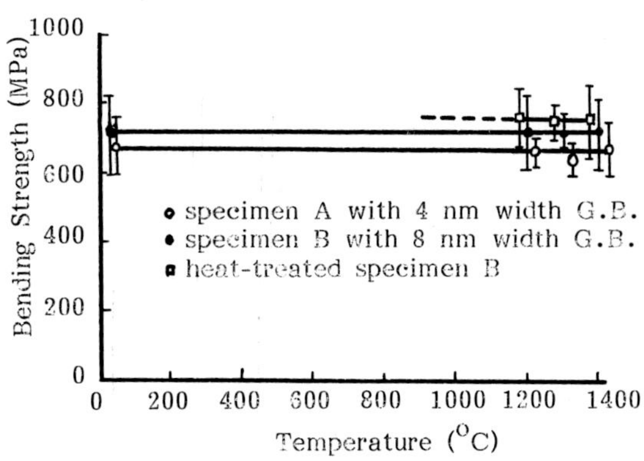

FIGURE 11
Bending strength vs temperature relationship of specimens with various G.B. width

the high temperature strength of these materials can be maintained up to 1400°C (Fig.11) and is further enhanced by heat treatment. The

5. CONCLUSION

It has been attempted to briefly outline the recent development trends on the research of

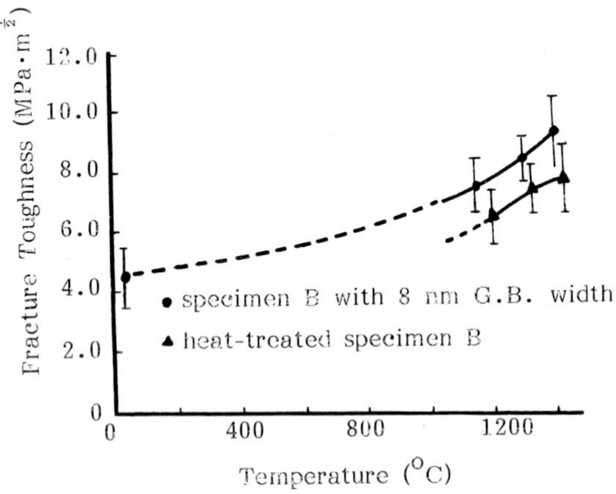

FIGURE 12
Fracture toughness vs temperature relationship of specimen B with and without heat treatment

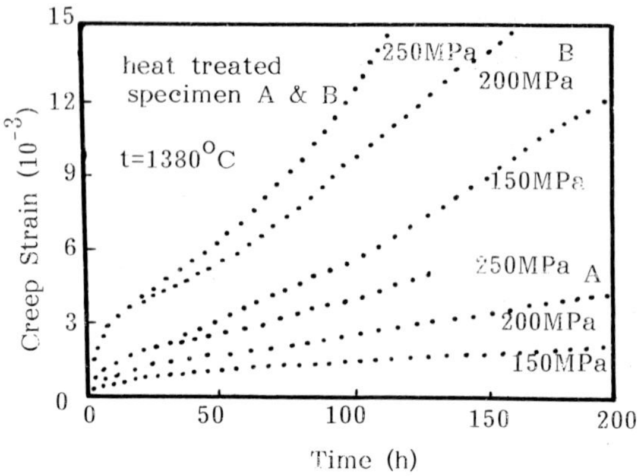

FIGURE 13
The creep behavior of YL-Si_3N_4 with various G.B. width (A 4 nm, B 8 nm)

structural ceramics.

1. Microstructure refining through processing to achieve nanoscaled ceramics by scaling down from grain size of micrometers. Noval properties can be anticipated.

2. Developing multiphased ceramic materials at a high level of control to improve especially the fracture toughness of the material. CMC with either fibre (or whisker) reinforcement or particulate dispersion are promising routes.

3. Compositional and microstructural tailoring of materials are approaches to call for serious consideration for property and performance improvement to meet the needs of application.

REFERENCES

1. T.S. YEN and F.K. ZHANG, The progress on Science and Technology of Chinese Ancient Ceramics, in: The studies of Chinese Ancient Ceramics, eds. Shanghai Institute of Ceramics (China, Beijing, Science Press, 1987) pp.1.

2. Y.R. XU and J.K. GUO, to be published.

3. J.K. GUO and T.S. YEN, Fibre Reinforced Ceramic Composite, in: Microstructure and Properties of Ceramic Materials, eds. T.S. YEN and J. A. PASK (China, Beijing, Science Press, 1984) pp.281-290.

4. J.K. GUO et al., J. Mater. Sci., 17 (1982) 3611-3616.

5. J.K. GUO et al., to be published.

6. H.R. ZHUANG et al., to be published.

7. D.L. JIANG, J.H. WANG, Y.L. LI and L.T. MA, Mater. Sci. Eng., 109(A) (1989) 401-406.

8. D.L. JIANG et al., J. Chinese Ceramic Soc., 18(2) (1990) 123.

9. W.Y. SUN, F.Y. WU and T.S. YEN, Material Letters, 6(1) (1987) 2.

10. L.J. DONG, Z.K. HUANG, W.Y. WU and L. P. HUANG, to be published.

11. H.R. ZHUANG, Z.K. HUANG et al., Private communication (1989).

12. Y.R. XU et al., KEXUE TONGBAO (Science Bulletin, China), 33(10) (1988) 832-836.

13. Y.R. XU, X.R. FU and D.S. YAN (T.S. YEN), Physica, 150(B) (1988) 276-282.

Frontiers of Materials Research/
Electronic and Optical Materials
M. Kong, L. Huang (editors)

HIGH Tc SUPERCONDUCTORS RESEARCH IN CHINA

Hu Suhui and Fang Junren*

Shanghai Institute of Metallurgy, Academia Sinica, 200050 Shanghai, China
* Changsha Research Institute of Mining and Metallurgy, 410012 Changsha, Hunan, China

A brief discription of some recent developments of high T_c ceramic superconductors research in China is given in this paper. Various processing method such as solid state reaction, melting, powder-in-tube etc. are being developed for preparing bulk ceramic materials and composite wire or tape. For fabricating high quality ceramic superconducting thin films, many methods are being used. The highest J_c (77K) value of bulk oxide material was obtained from YBCO fabricated by melt-textured method. Its' value at 2T and 7T were 23800 Acm^{-2} and 8600 Acm^{-2} respectively. The highest J_c (77K) value of YBCO thin film prepared by dc magnetron sputtering on $LaAlO_3$ substrate was $3.8\times10^6 Acm^{-2}$ at 0T and over $1\times10^6 Acm^{-2}$ at 1T. At present, several institutes can fabricate YBCO thin films with J_c (77K, 0T) value higher than $1\times10^6 Acm^{-2}$. Some topics about the material science and application of the high T_c ceramic superconductors are also presented in this paper.

The initial report by Bednorz and Müller[1] on the possible high T_c superconductivity in Ba-La-Cu-O system and the discovery of high T_c superconductor with zero resistance temperature above 90K[2,3] open up many exciting prospects for the applications of superconducting materials. In China, several institutes and universities have begun research work in this field and a National R & D Programme of HTS has been carried out since 1987, in which the main efforts are concentrated on the fabrication and properties of bulk materials, thin films and their applications, HTS physics and novel high T_c ceramics. The main advances recently are the preparation of high J_c YBCO bulk materials and high quality YBCO thin films. A high J_c (at 77K and 0T) of over $2\times10^4 Acm^{-2}$ (1μV/cm, dc measurement) was obtained by a melt-textured growth (MTG) process in $YBa_2Cu_3O_y$. J_c (77K) of $3.8\times10^6 Acm^{-2}$ at 0T and over $1\times10^6 Acm^{-2}$ at 1T were obtained for YBCO thin film prepared by dc magnetron sputtering method on $LaAlO_3$ substrate. Considerable work on the studies of microstructural characteristics, and superconducting properties have been carried out in order to improve the properties, especially the J_c, of the bulk oxide materials. Some groups have begun their research in developing HTS application. Here some recent advances of HTS are given and some of them will be described a little more in detail.

1. BULK MATERIALS

Although the high T_c ceramic superconductors show a high T_c, the critical current density J_c of the bulk sintered materials are very low because of the effects of such properties as weak-links and porosity. To enhance the J_c, many techniques are being developed to overcome these shortcomings. Among them, the most important is the melting process with controlled solidification, such as Melt-Texture-Growth (MTG)[4], Quench and Melt Growth (QMG)[5] and liquid phase processing method[6].

A group of GRI (General Research Institute for Nonferrous Metal) recently produced bulk YBCO material with an aligned structure exhibiting high J_c under magnetic field by using of MTG method[7]. The $YBa_2Cu_3O_y$ superconducting powder used was made by standard solid-state reaction and cold pressed the powder into sheet with dimensions of $1.5\times20\times60 mm^3$. The sheets were sintered in air at 950°C for 24h and furnace-cold. Then put the sheet-shape sample into a furnace with temperature gradient preheated in the temperature range of 1000-1150°C and maintained at this temperature for 5-30min. Cooled the sample at a rate of 0.5-5°C/h, maintained it at 925-980°C for 12h and then cooled down to room temperature at a rate of 1°C/min. Samples with dimensions of 0.07-0.3mm by 0.3-1mm by 10mm were cut along cleavage plane. After the samples were annealed in flowing oxygen in the temperature range of 400-600°C for about 100h, the silver leads were soldered on the sample with indium. X-ray diffraction pattern of cleavage plane in the MTG sample showed that the crystals were preferentially aligned and the cleavage plane is ab plane of 123 phase. When the MTG sample was ground into powder, the peaks of 211 phase could seen on the x-ray diffraction pattern. The SEM photographs taken from the fracture surface of the MTG sample showed that the sample has dense layer-structure with regular alignment, the crystals of 123 phase are plate-shaped with a thickness about 5-10μm and 1-2.5cm long. T_c of the MTG sample is 91.2K. When the applied field is perpendicular to the current but parallel to ab plane, the J_c (77K) are 23800 Acm^{-2} and 8600 Acm^{-2} at 2T and 7T respectively. The HVEM observation of the MTG sample revealed that the stacking faults, dislocations and dislocation loops exist in the vicinity of the 211 phase and CuO phase, small particles of the 211 phase and the CuO phase with an average grain size about 3μm and 0.5μm respectively are dispersed in the 123 phase. Twining bands cross perpendicularly. The twin spacing is about 0.01-0.1μm. These defects are the effective flux pinning center for enhancement the J_c of the MTG sample.

One group in NIN (Northwest Institute for Nonferrous Metal Research) developed a modified melt growth process[8] with which YBCO plate-shaped crystals were prepared from sintered Y-123 powders by melting at about 1100°C and slow cooling solidification with rate of 0.1-0.5^{o}C/min in a temperature gradient of 55°C/cm. XRD showed that the melted samples had well perfectly crystalline orientation with c-axis perpendicular to the growth direction. There is fewer grain boundaries, even in the c-axis direction, the boundaries disappeare in some regions. Hence, the supercurrent can pass sidways when it meet with any roadblock in the ab planes and the week-links density reduces significantly. The J_c (77K) of the melt grown YBCO sample achieved are over 12200 Acm^{-2} and 4000 Acm^{-2} at zero and 1T field respectively. Another group in NIN developed a novel melt process named Powder Melting Process (PMP)[9]. The 221 phase powder and Ba-Cu-O powders were used as the starting materials. Aligned YBCO crystals were obtained through the melting and unidirectional solidification of the precursor powders during peritectic reaction. The main advantage of the PMP is that the size of precursor powders could be easily controlled so that the favour microstructure of grain boundaries-free and well-distributed fine 211 particles could be obtained. J_c (77K, 1T) of a PMP sample with cross-section area of 0.227mm^2 is over 10460 Acm^{-2}. In the PMP samples, the week-links density reduced substantially and the fine dispersive 211 phase particles are the effective pinning centers.

IP (Institute of Physics), NIN and TU (Tsinghua University) prepared the polycrystalline $YBa_2Cu_3O_4$ superconducting sample using MTG method[10]. The melting temperature (1100-1200°C), melting time (3-6h), solidified temperature gradient (>55°C cm^{-1}) and cooling solidification rate (0.1-0.5^{o}C min^{-1}) are the key factors in preparing textured samples. SEM showed that the sample has a dense, layed structure with regular alignment of plate-shape crystals. The zero resistance temperature and onset of diamagnetism of the samples occur at

92 and 91K respectively. The highest J_c (77K, 0T) is 11000 Acm^{-2}. Comparing with the sintered samples, the J_c of MTG sample is enhanced and the field dependence of J_c is much reduced. This means that the limitation of the week-link problem in sintered materials can be reduced by MTG processing. The anisotropic field dependence of J_c seems to be small for the MTG samples.

Superconducting fibers of Bi-based system materials by means of the laser-heated pedestal growth method (LHPG) were prepared in TU[11]. The fibers with diameter 35-200μm were grown by two steps. Firstly, the source rode was pulled at the rate of about 2mm/min to become round, smooth and dense. Then, the fibers was grown at the rate of 0.04-1mm/min. The 2212 phase fibers after post-heat treatment at 820°C for 4h, the J_c (77K, 0T) value is 4200 Acm^{-2}. Whereas for 2223 phase (T_c=110K) fiber, J_c is only 2000 Acm^{-2} even if it was heat treated at 850°C for 48h. Using amorphous source material to fabricate the fibers, the highest J_c obtained is over 5000 Acm^{-2}. XRD analysis and SEM results revealed that the 2212 phase as-grown fibers have highly aligned morphology with the a-axis parallel to the growth direction, which is closely related to the high J_c. Hence, a short period of, or even without, post-growth heat treatment is favorable in LHPG method.

2. HIGH Tc OXIDE SUPERCONDUCTING COMPOSITE

In NIN, SIM (Shanghai Institute of Metallurgy) and GRI, YBCO/Ag or Ni composite tapes by using of powder rolling process or powder-in-tube processing were fabricated [12,13,14]. The J_c (77K, 0T) values of these tapes could reach to 6000-8000 Acm^{-2}. This J_c enhancement is not only caused by the crystals alignment but also by the size effect and increasing of the grain aspect ratio. Its field dependence is not improved in comparison with the sintered bulk materials.

Bi-base oxide/Ag composite tape were fabricated using powder-in-tube processing method by several groups in GRI, NIN, SIM and CRI (Changsha Research Institute of Mining and Metallurgy)[15,16,17,18]. The J_c (77K, 0T) obtained up to 5000-15000 Acm^{-2}. Experimental results show that the deformation-texture can not make the J_c values of both YBCO and Bi-based oxide improved remarkably. However, Bi-based oxide/Ag composite tapes have strong c-axis orientation perpendicular to the rolling direction, and its J_c at weak magnetic fields up to mT has been improved. So the flexible Bi-based oxide/Ag wires were used in some cases where the transport supercurrent is needed in weak magnetic fields, e.g. the flux transformer.

By using of powder-in-tube processing, Tl-Ba-Ca-Cu-O/Ag composite tape was fabricated in CRI[19]. The optimum J_c (77K, 0T) was measured to be 6441 Acm^{-2} under a criterion of 1μVcm^{-1}.

3. THIN FILMS

In order to fabricate high T_c superconducting thin films with high quality, many methods were adopted.

A group of NU (Nanjing University) deposited YBCO films using either dc and RF magnetron sputtering[20]. During sputtering, the temperature of the (100) $SrTiO_3$ substrate is 600-700°C and the argon and oxygen mixture gas pressure is 0.3 torr with 30% oxygen. The film thickness is typically 1000-3000Å. After sputtering, the YBCO films are plasma oxidized in situ at 480-520°C for 80-150min, during which time only oxygen is used, at a pressure of ≈1 torr. After oxidation, the sample is left in the chamber to cool to room temperature and is stored in the chamber for more than 4h. For the best processing conditions, T_c and J_c (77K, 0T) of the film are 91K and $3\times10^6 Acm^{-2}$, respectively. The x-ray diffraction pattern indicated that the c-axis is highly orientated and perpendicular to the film surface; SEM showed that grains of the film are of the order of submicrons in size. Films fabricated by magnetron sputtering and subsequent in situ plasma oxidation look very smooth and mirror-like. The roughness of the film surface is 105Å, while the roughness of the substrate surface is 125Å.

A group of IP deposited YBCO thin film on (100) $LaAlO_3$ substrate by dc magnetron sputtering[21]. The J_c (77K, 0T) is $3.8\times10^6Acm^{-2}$ and over $1\times10^6Acm^{-2}$ at 77K and 1T. With reducing the temperature down to 65K, J_c of the film at zero field is $5.8\times10^6Acm^{-2}$, even at 7T, its J_c is still as high as $1\times10^6Acm^{-2}$ [22]. The anisotropy behavior could be reduced through improving the film quality.

High T_c RE-Ba-Cu-O (RE=Y, Ho, Er, Gd and Sm) thin films were grown by reactive magnetron sputtering in PU (Peking University)[23]. Since bonbardment of a growing film by negative ions from the target surface is one of the problems of preparation of high quality $YBa_2Cu_3O_{y-\delta}$ thin films by dc or RF sputtering, a modified dc magnetron sputtering system was used. In the preparation of high T_c YBCO thin films, a substrate temperature, T_s, of 800°C and a pressure of 2×10^{-1} torr, with an argon/oxygen ratio of 10:1, were used. After in situ annealing at 400°C for 10 min, the films exhibited zero resistances at 90K and 87K for those deposited on (100) Y-stabilized ZrO_2 (YSZ) and (100) $SrTiO_3$ (STO) substrates, respectively. $1.4\times10^6Acm^{-2}$ of the J_c on (100) STO and $3\times10^5Acm^{-2}$ on YSZ at 77K, 0T have been obtained with the criterion of $2\times R_n/1000$.

At present, YBCO thin film with J_c (77K, 0T) over $1\times10^6Acm^{-2}$ could be steadily produced by using of different processing in several groups.

4. MATERIAL SCIENCE OF HTS

Large single crystals of BSCCO 2212 phase with high quality were grown successfully using directional solidification method in USTC (University of Science and Technology of China)[24]. In this method, a liquid alloy was used as coolant which has not been used to grow high T_c oxide superconducting single crystals up to date. Maximum single crystals $19\times3\times2mm^3$ were obtained and could be cleaved along the (001) plane. Back-reflection Lave method analysis revealed that the growth direction of the crystal is [100], and the crystals were indeed single crystals. The zero resistance temperature of the crystals annealed in oxygen is 81.5K.

A new phase with T_c>140K by ac susceptibility measurement in Pb and Sb doped Bi-Sr-Ca-Cu-O system was recently discovered by a group of USTC[25,26]. The specimens were prepared through solid-state reaction and had the nominal composition $Bi_{1.6}Sb_{0.3}Pb_{0.1}Sr_2Ca_2Cu_3O_y$. Ac susceptibility measurement showed that there is a higher superconducting transition in this sample at 140K but there is no superconducting transition at 110K in the initial measurement except a lower one at about 84K, only after five thermal cycles from 50K to room temperature in air, T_c is reduced to below 110K although it is stable in the first four cycles. By means of powder x-ray diffraction, electron diffraction and electron probe microanalysis, it was found that the new phase is $Bi_3Sb_{0.8}Pb_{0.2}Sr_{4.1}Ca_{3.9}CuO_{15+\delta}$ and belongs to monoclinic system with lattice parameters a=22.14, b=5.902, c=19.91Å and β=99.7°. If regard the Sb and Pb as the elements replacing the Bi, the chemical formular of this new phase could be $Bi_4Sr_4Ca_4CuO_{15+\delta}$, hence it can be called as 4441 phase. There may exist some connections between the new phase and 110K phase. The connections are considered to play an important role to the improvement of T_c.

In order to understand some relationship between the microstructure and J_c in high T_c superconductor, a group in SIC (Shanghai Institute of Ceramic) measured and compared the superconductivities of two $EuBa_2Cu_3O_{7-\delta}$ samples which have the same chemical composition but with different processing[27]. The T_c value of these two samples are nearly the same (90K) but the J_c value are quite different, 30 Acm^{-2} for sample S_1 and 300 Acm^{-2} for S_2. Hundreds of pictures were randomly taken by TEM from both samples. The microstructural information were analysed from TEM pictures using statistical method. From the microstructural difference between S_1 and S_2, the following conclusions are obtained. The distribution of grain size and the shape of grain might relate with J_c. The more grains with twin domains, the higher J_c is. The grain boundaries between parallel grains might be of benefit in enhancing J_c. The total thickness of the grain boundary of S_1 (J_c=30 Acm^{-2}) is much thicher than that of S_2 (J_c=300 Acm^{-2}). High density and low voids would benefit the J_c and the mechanical strength of the oxide material.

The oxygen behavior in high T_c ceramic superconductors was investigated in SIM[28]. In low frequency (f=1) internal friction (IF) measurements of $YBa_2Cu_3O_{7-x}$, two IF peaks located at 204^oC and 170^oC were observed and identified as stress-induced jump of oxygen atoms in the Cu-O basal planes of the orthorhombic and tetragonal phases YBCO. By analyzing the quantitative relationship between the 204^oC peak height and oxygen content (7-x), a criterion for the oxygen atomic movement in the basal plane can be expressed as "The Cu(1) atomic coordination can never exceed 4". The occupancies C_A, C_B and the potential barrier H_A, H_B of the oxygen atoms on sites $O_A(1/2,0,0)$ and O_B (0,1/2,0) were calculated based on the 204^oC IF peak height[29]. Oxygen diffusion process in YBCO was found to be of vacancy methanism and can take place only in the basal planes along (110) direction. Under the framework of one dimensional random walk process, the oxygen diffusivity can be expressed as

$$D = \frac{d^2}{4}\nu_o \frac{C_B}{C_A+C_B} e^{-H_B/kT} = D_o e^{-H_B/kT}$$

where

$$D_o = \frac{d^2}{4}\nu_o \frac{C_B}{C_A+C_B} \simeq 3.6\times10^{-4} cm^2/sec.$$

By ultilizing the IF data, the oxygen diffusion coefficient can be made out.

By using of thermogravimetrical analysis, a weight vs. temperature relationship of single 2223 phase $Bi_{1.6}Pb_{0.4}Sr_2Ca_2Cu_3O_x$ (T_c=104-108K) was obtained in SIM[30]. It is concluded that from room temperature to 800^oC the weight loss of the sample is only 0.056%, and the corresponding change of oxygen fraction per cell (assuming x=10) is 0.36%. This is quite different from Y-system with respect to the behavior of the change of oxygen content with temperature. In order to study the influence of oxygen on the superconductivity of 2223 phase, one 2223 phase specimen (T_c=104K) was annealed at different temperature in air for 10 h before quenching into liquid nitrogen. Experimental results showed that the R-T curves of the specimen quenched from 500-700^oC were almost the same but when T_q=800^oC, T_c=93K. If the specimen was annealed at 400^oC in flowing nitrogen for 12 h, its T_c=85K. When this annealed specimen was heated at 400^oC in air for only 2 min., its T_c rose to 102K and its R-T curve restored to the original situation for 12 h. Since the XRD pattern of the specimen subjected to the different heat treatment indicated not any change, the variation of its T_c is only dependent upon the change of oxygen content, even it is very small. The length of the lack of oxygen specimen during heating in air increases linearly with temperature, but departed from the linearity between 200^oC and 400^oC. This indicates that the oxygen replenishment results in reduction of the sample volume. It is concluded that the change of oxygen content, which brought about the T_c

variation, occurred in the entire sample rather than in interface between grains.

The kinetic behavior and mechanism of both in- and out-diffusion of oxygen in BPSCCO (2223) specimen was investigated in SIM by in situ resistivity measurements during isothermal annealing at 237.5-333°C in flowing oxygen and 387-455.5^{o}C in nitrogen[31]. It is concluded that the oxygen out-diffusion is surface reaction limited, following the diffusion equation of $dc/dt=-Kc^2$. c is the oxygen content. The surface barrier is 1.5 eV. The oxygen in-diffusion process is managed by a special diffusion mechanism in which the oxygen diffusion coefficient is variable with oxygen content and the motivation energy is 0.94 eV. For $Tl_2Ba_2Ca_2Cu_3O_x$, the diffusion process of oxygen is similar to that of the Bi-based system. The surface barrier is 2.0 eV and the motivation energy 1.6 eV. Both in- and out-diffusion of oxygen in Bi-system material were investigated using isothermal dilation measurements in flowing oxygen and nitrogen[32]. The oxygen in- and out-diffusion activation energy were calculated to be 0.87 eV and 1.4 eV respectively, which agree with that was gotten by the isothermal in situ resistivity measurements.

The infrared (IR) spectra of $YBa_2Cu_3O_x$ and $BiSrCaCu_2O_x$ was investigated in SIM[33,34]. The samples powder were prepared by conventional solid state reaction and mixed with KBr powder for IR measurements. Five peaks P_1 (636cm^{-1}), P_2 (590cm^{-1}), P_3 (550-580cm^{-1}), P_4 (410cm^{-1}) and P_5 (360cm^{-1}) were observed in the IR spectra of YBCO. P_1 and P_2 are assigned to the Cu-O asymmetrical stretching vibration in the one dimensional Cu-O chain along the b-axis between two Ba layers and their two dimensional Cu-O network between Ba and Y layers, respectively. P_3 is assigned to the asymmetrical stretching vibration mode of the Cu-O bond along the c-axis. P_4 and P_5 were assigned to the bending vibrations of Cu-O bond of the Cu-O planes between Ba-Ba layers and between Y-Ba layers, respectively.

Four peaks A_1 (610cm^{-1}), A_2 (475cm^{-1}), A_3 (450cm^{-1}) and A_4 (370cm^{-1}) were observed in the IR spectra of BSCCO. A_1 and A_2 were assigned to the stretching vibration of Cu-O and Bi-O of the Cu-O and Bi-O planes respectively. A_4 was attributed to the bending vibrations of Cu-O and Bi-O of the Cu-O and Bi-O planes, A_3 (450cm^{-1}) needs further investigation. It is suggested that IR spectroscopy can be an useful tool for material characterization of HTS including the characterization of thin films, wires and homogeneity of bulk material at different depths.

In SUST (Shanghai University of Science and Technology), the sample preparation, the effect of element substituted and the pressure (10-20 Kba) dependence of the T_c of YBCO and Bi-system oxide superconductors were investigated. Experimental results showed that T_c linearly increased with the increase of pressure applied for both materials. The microwave properties of YBCO at X- and W-band in the temperature range of 20-300K were characterized. The frequency dependence of $R_s \propto \omega^{\alpha}$ with $\alpha=2.32\pm0.5$ and at low temperature, e.g. 20K, its R_s value could be less than that of OFHC copper, indicating the possibility for the application of YBCO at W-band[35,36].

^{63}Cu NQR (Nuclear Quadrupole Resonance) spectra of $YBa_2Cu_3O_{7-\delta}$ with varying oxygen content was measured at room temperature under zero field in ECNU (East China Normal University)[37]. It is concluded that the line at 22.2 MHz belongs to the Cu(1) site, while the one at 31.2 MHz arises from the Cu(2) site. The line shapes and intensity of NQR spectra are strongly dependent on δ. There is an additional peak around 30 MHz which is related to the oxygen deficiency. Comparing with x-ray powder analysis, the NQR spectra analysis is more sensitive on varying oxygen content.

5. HIGH Tc OXIDE SUPERCONDUCTING DEVICES

A NIN group developed an YBCO superconducting magnetic shielding tube with a shielding coefficient of 10 and a capability of maximum shielding field of 4mT[38]. A superconducting magnetic flux transformer used in SQUID magnetometer was successfully made with sheathed silver BPSCCO wire in GRI[39]. When a signal of about $1\times10^{-6}Acm^{-2}$ produced by the magnet exciting coil, the SQUID detected a magnetic flux change of 10 ϕ_0 ($\phi_0=2.07\times10^{-15}$wb). The coupling coefficient is 0.44×10^{-3}. Another group of NIN made a flux transformer with BPSCCO/Ag wire [40]. A signal of 1.4×10^{-7}T was sensed by a SQUID with sensitivity of 1/7 ϕ_0.

Single hole RF-SQUIDs were made with high T_c bulk YBCO and BSCCO materials in NIMC (National Institute of Metrology)[47]. The device has been operated for over one year with more than 100 cycles from 77K to 300K for measuring the variations of the earth magnetic field, the biomagnetism and some properties of HTS. The equivalent flux noise of the SQUID at 77K is $5\times10^{-4}\phi_0/\sqrt{Hz}$ from 10 Hz to 100 Hz.

A variable temperature susceptibility measuring system using a high T_c RF-SQUID as a senser was operated in FU (Fudan University)[42]. The double-hole RF-SQUID used possesses flux resolution of $5\times10^{-4}\phi_0/\sqrt{Hz}$ in the range of 20-200 Hz. The susceptibility resolution of the system is $5\times10^{-8}emu/cm^3$.

A set of loop antenna made of BSCCO superconductor and copper with different loop diameter were fabricated by TU[43]. The optimum frequency is near 558 Hz and the relative gain enhancement of the superconducting BSCCO loop antenna is about 5.6 dB higher than that of the identically shaped copper antenna operating at 77K.

ACKNOWLEDGEMENTS

We would like to thank Professor L. Zhao and other colleagues for providing the valuable information. Most of the works are sponsored by the National Center for Research and Development on Superconductivity.

REFERENCES

1. J.A. Bednoze and K.A. Müller: Z. Phys. B. Condensed Matter, 64 (1986) 189.

2. M.K. Wu et al: Phys. Rev. Lett. 58 (1987) 908.

3. Z.X. Zhao et al: Kexue Tongbao, 32 (10) (1987) 661.

4. S. Jin et al: Appl. Phys. Lett. 51 (12) (1987) 21.

5. M. Murakami et al: Jpn. J. Appl. Phys. 28 (1989) L1125.

6. K. Salama et al: Appl. Phys. Lett. 54 (23) (1989) 2352.

7. H.T. Ren et al: ICEC-13 April 24-27, 1990, Beijing, China.

8. P.X. Zhang et al: ibid.

9. L. Zhou et al: to be published at ICMC '90 Conf.

10. C.G. Cui et al: ICEC-13 April 24-27, 1990, Beijing, China.

11. J.C. Zhang: ibid.

12. X.D. Tang et al: Proceedings of Beijing International Conf. on High Tc Superconductivity, Sept. 1989, Beijing, China.

13. Y.X. Fu et al: Preprint.

14. C.S. Zhao et al: Proceedings of Beijing International Conf. on High Tc Superconductivity, Sept. 1989, Beijing, China.

15. H.T. Ren et al: Communication of Science and Technology, No.1 (1990) Feb. 12.

16. P.X. Zhang et al: Proceedings of ICEC 13 Conf. 1990, April 24-29, Beijing, China.

17. E.Y. Zhou et al: Preprint.

18. J.R. Fang et al: Proceedings of ICEC 13 Conf. 1990, April 24-29, Beijing, China.

19. Z.W. Qi et al: ICEC-13 April 24-27, 1990, Beijing, China.

20. Z.J. Sun et al: Proceedings of Beijing

International Conf. on High Tc Superconductivity, Sept. 1989, Beijing, China.

21. Y.Z. Zhang et al: accepted by Modern Phys. Lett. B.

22. X.G. Qin et al: accepted by Phys. Rev. Lett.

23. S.Z. Wang et al: Proceedings of Beijing International Conf. on High Tc Superconductivity, Sept. 1989, Beijing, China.

24. G.D. Gu et al: Preprint.

25. I. Yang et al: Modern Physics Letters B 4 (5) (1990) 363.

26. T.Z. Cheng et al: Supercond. Sci. Technol. 3 (1990) 87.

27. X.Y. Song et al: Chinese Science Bulletin 35 (5) (1990) 375.

28. T.G. Chen et al: Chinese Phys. Lett. 5 (1988) 185.

29. X.M. Xie et al: Phys. Rev. B. 40 (1989) 4549.

30. T.W. Li et al: Proceedings of the Beijing International Conference on High Tc Superconductivity (1989) 435.

31. C.J. Zhou et al: Preprint.

32. T.G. Chen et al: Preprint.

33. T.S. Shi et al: Phys. Stat. Sol. (b) 148 (1988) 715.

34. Y.G. Zhao et al: J. Phys. Condens. Matter. 1 (1989) 8235.

35. P.J. Jin et al: Chin. Phys.- Supl. TR. 9 (1989) 193.

36. G.S. Bao et al: Preprint.

37. L.A. Zheng et al: Kexue Tongbao 35 (7) (1990) 7.

38. J.R. Wang et al: Proceedings ICMC '89, Los Angles 24-28, July, 1989.

39. H.T. Ren et al: ICEC 13 April 24-27, 1990, Beijing, China.

40. S.W. Wu et al: C-MRS Meeting, June 17-22, Beijing, China.

41. S.Q. Xue et al: ICEC 13, April 24-27, 1990, Beijing, China.

42. J.W. Qiu et al: ibid.

43. A.S. He et al: ibid.

Frontiers of Materials Research/
Electronic and Optical Materials
M. Kong, L. Huang (editors)

SOME FUNDAMENTAL PROBLEMS OF NANOCOMPOSITE MATERIALS *

Lishi WEN and Rongfang Huang

Institute of Metal Research, Academia Sinica, Shenyang 110015, China.

Nanocomposite materials are a class of new materials with many unusual physical and mechanical properties. Recent paper discusses their general fundamentals. The most outstanding feature of nanocomposites is low dimensionality of their composition components. As a measure of the spatial dimensionality of matter, characteristic length was introduced. Its critical value will be the threshold of macro－to－low dimension state transition of matter. As the results of low dimensionality, there are different quantum effects leading to variation of energy state and electronic structure of their components. Other basic features of them are the remarkable modification of the electronic structure by widespread interface interaction at electronic level and the great variety of microstructure ranging from high level ordered 3-dimensional periodic structure to stochastically dispersed medium of superfine particle. All these provide much more potential for tailoring property of materials than by formation of chemical compound or mixture including usual microcomposites. Elements of designing nanocomposites and their preparing by ion beam and plasma enhanced vapour deposition are discussed.

1. INTRODUCTION

Recently, artificial structure of nanometer scale shows ever more perspective for improving the property of materials and devices. Such structures can be multilayer or particulate and are called together nanocomposite materials or nanosolid. The efforts in this direction went back to the beginning of 70s, when Esaki and Tsu[1] shown their idea of semiconductor superlattice and Koehler [2] proposed that nanoscale multilayer structure consisting of two materials of widely different elastic constant should have a greatly enhanced resistance to brittle fracture.

As the experimental results, supermodulus effect was discovered in compositionally modulated multilayer films of Au-Ni and Cu-Pd systems by Hilliard et al.[3] at late 70s. Biaxial elastic modulus was shown to be the function of the modulation wavelength Λ and remarcable increase of 2-4 times was observed when Λ is close to a critical value near 2 nm depending on the system. Because the elastic moduli are usually structure insensitive, these results initiate great interest in mechanical property of artificial modulated structure. In the subsequent experimental works, similar results were obtained in Cu-Ni and Ag-Pd systems, but not in Cu-Au system [4-6]. Lehoczky[7,8] prepared Al／Cu and Al／Ag multilayer structures with yield stresses several times those of the pure component. Significant enhancement of hardness and yield strength was obtained for polycrystalline TiC/TiB_2 and Al/Al_xO_y multilayer films respectively[9,10]. Anomalously high microhardness of factor 2 for single crystal TiN /VN strained-layer superlattices (SLS's) by Sundgren et al. at $\Lambda = 5.2$ nm [11,12]. Increase of microhardness of TiN-based film from 17 GPa to 45 GPa was obtained by Wen et al. through refining the grain size to 20-30 nm by adding boron into TiN and bombarding it with low energy ion[13,14]. The best wear resistance was shown for TiC/TiB_2, TiN/TiB_2 and TiC/TiN by Holleck et al.[15,16] at $\Lambda = 20$-50 nm. Superplasticity of TiO_2 and CaF_2 was discovered by Gleiter et al.[17], when they decrease the grain size down to nanocrystalline. X-ray reflection mirror has been made with multilayer structure consisting of alternating elements of high and low scattering cross section[18].
The nanocomposites of superfine particle in different transparent media shown infrared emissivity considerably higher than that pre-

---

* Partially supported by Natural Science Foundation of China.

dicted by classical theory[1]. The most important events in process of their development are summarized in table 1.

Many models were proposed to explain the physical mechanism underlying the above mentioned phenomena. Each of them is useful for a special subject. In recent paper, we try, however, to take a new approach to the question, to discuss the general fundamentals, which must be helpful for an insight into the general nature of this new class of materials.

2. LOW DIMENSIONALITY OF MATTER

Why the structure of nanometer size plays such an important role in modifying property of materials? Nanometer scale is the macro-to-micro state transition region of matter. When the characteristic size of matter decreases to nanometer scale, its surface electron state prevails over bulk electron state and different quantum effects become significant, resulting in losing its macroscopic property. Matter in such a state is called low dimensional materials.

There are two types of effect of low dimensionality: surface effect and volume effect. The former appears when considerable part of atoms of materials are at the surface. For example, when the total number of atoms n_t in a superfine particle (of cubic form) is of order 10^9, the percentage of surface atoms of the particle N_s is negligible (Fig 1). When n_t is about 10^6, N_s is already more than 10%. When n_t decreases to 10^3, N_s becomes much more than that of bulk atoms. Surface effect of low dimensionality leads to the unique physical and chemical characteristics of low dimensional materials. For example, as the particle size of powder falls, increases its specific surface energy resulting in growth of grain boundary diffusion coefficient, decrease of activation energy of sintering and falling of sinpering temperature. All these are favorable for powder metallurgy and ceramic processes. On the other hand, as the particle size falls, aggregation of powder becomes ever more serious and fractal structure was formed, making much trouble for processing.

Volume effect appears when the total number of atoms in materials becomes much less than

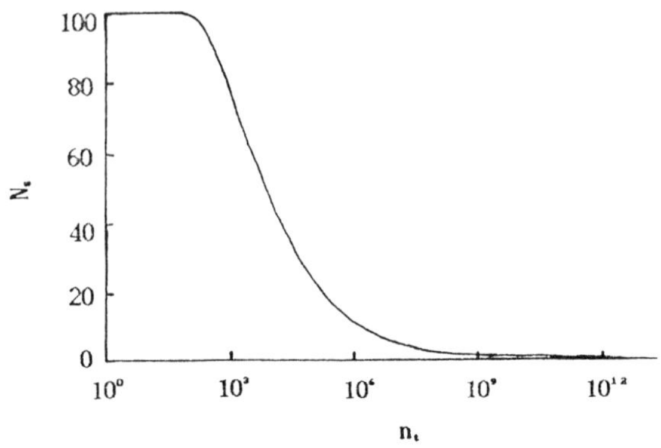

Fig 1 Surface effect of superfine particle

in the usual macroscopic bulk materials. For example, the gap between the neighboured sublevels of electron in energy band is given by

$$\delta = \frac{1}{N(0)\Omega} \tag{1}$$

where $N(0)$ is the state density at the Fermi level in unit volume, Ω is the volume of particle. When the total number of atoms in the superfine particle is much less than that in bulk materials, it becomes so-called small system. As the result, the traditional method of statistical physics could be inaccurate and its energy sublevel gap δ becames considerably higher than that of bulk. Therfore, the electron structure of low dimensional materials differs significantely from that of bulk materials (Fig 2)[20]. The electrical resistance of thin film (or wire) increases, when the thickness of the film (or the radius of the wire) decreases (Fig 3).

To describe the spatial dimensionality of matter, characteristic length Λ was introduced. Its critical value Λ_c is the threshold of macro-to-low dimensional state transition of matter. When the characteristic length of materials satisfies

$$\Lambda < \Lambda_c \tag{2}$$

they become low dimensional. The spatial di-

Table 1 The important events in development process of nanocomposites

Time	Author	Events	Reference
1970	L. Esaki and R. Tsu	Theory of semiconductor superlattice	1
1970	J. S. Koehler	Theory of strenthening materials by multilayer microstructure	2
1976	C. G. Grranqvist, R. A. Buhrman, J. Wyns and A. J. Sievers	Anomaly of infrared emissivity of superfine particle dispersed media	19
1977-1985	W. M. C. Yang, T. Tsakalakos and J. E. Hilliard	Supermodulus effect in Au/Ni, Cu/Pd, Cu/Ni and Ag/Pd systems	3-6
1978	S. L. Lehoczky	High yield strength of Al/Cu and Al/Ag multilayer films	7, 8
1978	R. W. Springer and D. S. Catlett	High yield strength of Al/Al_xO_y laminate	9
1981	E. Spiller	X-ray mirror by scattering cross section modulated structures	18
1982	B. A. Movchan, A. V. Demchishin, G. F. Badilenko, R. F. Bunshan, C. Sans, C. Deshpan and H. J. Doer	High yield strength of TiC/TiB_2 multilayer films	10
1986	M. Pomerantz, J. C. Slonczewski And E. Spiller	Interface magnetic coupling	21
1987	S. Torodova, U. Helmersson, S. A. Barnett, J. -E. Sundgren L. C. Markert and J. E. Greene	Superhigh hardness of superlattices of interstitial phases	11, 12
1987	J. Karch, R. Birringer and H. Gleiter	Superplasticity and toughening of ceramics	17
1987-1988	L. S. Wen, X. Z. Chen, Q. Q. Yang, Y. Q. Zheng, Y. Z. Chuang	Superhigh hardness of nanocrystalline Ti-B-N films	13, 14
1987-1989	H. Holleck and H. Schulz	Improved wear-resistance of nanocomposite multilayer TiC/TiB_2, TiN/TiB_2, TiC/TiN films	15, 16

mension of geometric graph is a pure geometric quantity, while the spatial dimension of materials is a physical quantity. Its critical value Λ_c depends, therefore, on the physical characteristics of materials and physical processes involved. For example, for the motion of electron in solids

$$\Lambda_c = l_e \tag{3}$$

where l_e is the free path of electron in the solid.

For superconducting processes,

$$\Lambda_c = \xi_0 \tag{4}$$

where ξ_0 is the correlation length of Kubo electron pair given by

$$\xi_0 = \frac{h\, v_f}{\pi \Delta(0)} \tag{5}$$

where v_f is Fermi velocity, $\Delta(0)$ is BCS energy gap at 0 K.

For interaction of electromagnetic waves with materials

$$\Lambda_c = K \delta_s \tag{6}$$

where K is constant, δ_s is the skin depth of

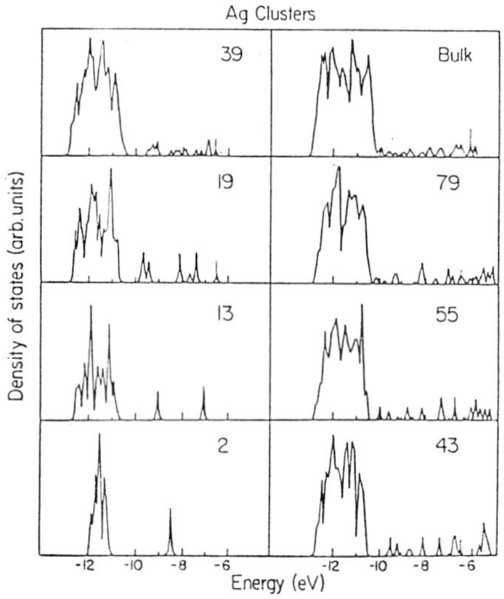

Fig 2 Dependence of electron structure of silver on its particle size

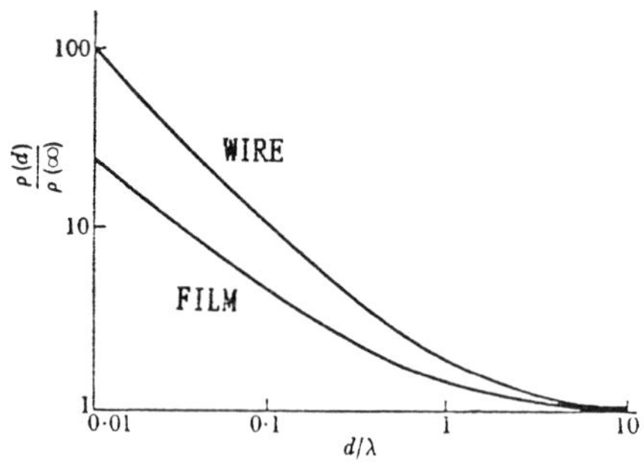

Fig 3 Electrical resistance of low dimensional materials

materials,

$$\delta_s = \frac{1}{\sqrt{\pi \nu \mu \sigma}} \qquad (7)$$

where ν is the frequancy of electromagnetic waves, μ is magnetic susceptbility of materials, σ is their electrical conductivity.

3. BASIC CHARACTERISTICS .OF NANOCOMPOSITE MATERIALS

Progress of different kinds of nanocomposite puts forward the need of studying their general fundamentals. The most outstanding feature of nanocomposites is low dimensionality of their composition components. Therefore, the components have electronic structure and energy state significantly different from that of bulk materials. That serves the physical ground in which nanocomposites differ from both chemical compound and mixture. The composition components of a chemical compound are limited in range of the atoms of chemical elements. In contrast, nanocomposite can be made of low dimensional components varying both chemical composition and characteristic size.

Another basic feature of nanocomposites is the remarkable modification of the electronic structure of materials by the widespread interface interactions at the electronic level between their composition components. Ordinary microcomposites are in nature a mixture of their components. They provide no obvious modification in the electronic structure of materials. Therefore, they behave in accordence with the rule of mixture. Their property is the linear combination of the initial property of their components with the volume percentage as the weighting factor. That determines the limit of the property improvement by microcomposites which could not be higher than that of the component of best property.

On the contrary, for nanocomposites, due to the volume effect of low dimensionality of the composition components, their initial electronic structure are already different from that of usual bulk materials. In addition, surface effect of low dimensionality leads to their much greater specific interface area, higher interface energy and lower activation energy of interface chemical reaction than in bulk materials. That results in formation of new electronic structures which are instable in usual materials, but could become stable in nanocomposites. Furthermore, due to the great interfacial area, such new electronic structures could be significant in quantity or even prevail over the bulk state. The enhanced interfacial interaction could also lead to new processess which occur seldom under usual conditions. For example, it is known that localized surface states play an important role in enhancement of surface magnetism for 3d transition metals. Magnetic coupling between ferromagnetic films separated by nonmagnetic thin film, has been observed experimentally by using ferromagnetic resonance[21]. Enhancement of magnetism at the transition metal/noble metal (or simple metal) interfaces has been predicted theoritically[21].

The third feature of nanocomposites is the great variety of their microstructure. Their composition components can be two-, one- or zero-dimensional. Homogeneous nanocomposites consist of one type of component.Heterogeneous nanocomposites consist of two or more types of component. They are multiphase materials.

The orderliness is one of the most important characteristics of microstructure in nanocomposites. It can be the periodic modulation of crystallographic structure or other active factors including chemical composition and its spatial distribution, interface adsorption and segregation, micromorphology, internal strain and its distribution, interface structure and defects, interface electron structure and band structure.

Such periodic modulation can be at most 3-dimensional, that is the modulation in three each to other mutually perpendicular spatial directions. Semiconductor superlattices are a typical example of nanocomposites with periodically modulated structure. Besides periodically modulated structure, there are also aperiodically modulated structures which are widely used for sensor and function transformation devices. Another limit case of microstructure is stochastically dispersed nanocomposite of superfine particle in different media. However, supercrystals of iron oxide in decalin with oleic acid has been observed experimentally[23].

The active factors in microstructure of nanocomposites for modifying the property of materials are listed in table 2.

It is evident now, that nanocomposite materials (or nanocomposites, nanosolids, nanostructure, nanophase, etc.) are a new kind of materials different from both chemical compound and mixture. Their composition components are low dimensional. The interface interaction between their components are strong and widespread. In addition to this, they have a great variety of different microstructures.

4. MATERIALS DESIGNING OF NANOCOMPOSITES

Analysis of fundamentals of nanocomposites revealed that the designing is of great importance for such a new class of high performance materials. The first step of materials designing is to outline an appropriate physical model. The physical and chemical processes determined by this model should be able to provide the necessary property of nanocomposites. The key point of this structure-property designing consist in the soundness of physical model and selection of characteristic length and its critical value.

The materials designing of nanocomposites can be theoretical, semiempirical and empirical. Theoretical representation of physical mechanism and its detailed mathematical analysis is important for sketching the first physical models and for making a new break through, when the old physical mechanism is already in serious disagreement with experimental data or is in principle not able to provide necessary property of materials. However, pure theoretical analysis is usually difficult due to the shortage of necessary physical and chemical data, especially, for low dimensional materials. On the other hand, pure empirical analysis is a profitable stage for accumulating the first experimental results and forming the idea of new physical mechanism and designing model. However, this process could be made more effective when the experimental results are correlated with theoretical representation.

For designing a special kind of materials, its physics and chemistry are the most important theoretical basis. The designing of superlattices is based on the laws of electron motion in solids. In the case of one-dimensional periodic modulation, quantum well has been obtained. The motion of electron was limited in two-dimensional quantum well. Two-dimensional periodic modulation makes the electron motion limited in one-dimensional quantum wire, while three-dimensional modulation limits the electron motion in zero-dimensional

Table 2 Effective factors modifying the property of materials and the types of nanocomposites

№	Effective factor	Type of nanocomposites
1	Periodic modulation of chemical composition	Semiconductor superlattices with compositional modulation and with doping modulation, metallic multilayer films with with supermodulus effect,high strength multilayer nanocomposite metal films, superhard multilayer nanocomposite films of interstitial phases
2	Periodic modulation of crystallographic structure and chemical bond	Intercalated compound
3	Periodic modulation of crystallographic defects	Strained layer superlattices, semiconductor superlattices with interface dislocation structure
4	Periodic modulation of or- and disorder	Superlattices of amorphous semiconductors
5	Periodic modulation of atomic scattering cross section	nanocomposite multilayer films for X-ray reflective mirror
6	Aperiodic modulation of different factors	Chemical conversion layer, nanocomposite films for sensors and function transformation devices
7	Homogeneous and heterogeneous nanocrystalline structure	High toughness nanocrystalline ceramics, high strength nanocrystalline metal, TiN based multiphase nanocrystalline superhard films
8	Stochastic dispersed media of superfine particle	Optical composite media of superfine particle, microwave absorbing coating of superfine powder, magnetofluid

quantum dot of nanometer scale. Variation of dimensionality of electron motion initiates a series of new characteristics of materials and shows a broad perpective for development of microelectronic and opto-electronic devices. Recently, strained layer superlattices show great potential of modifying optical property of materials. Band gap is the basic characteristics of materials determining their optical property. It is shown now that band gap of materials can be modified continuously with strained layer superlattices at constant chemical composition.

For designing nanocoposites with desirable magnetic property, it is important to take into consideration a series of interface phenomena and size effects, such as local structure of low temperature magnetization near surface and interfaces[24-26], magnetic hyperfine field oscillation[27], enhanced magnetic moments in monolayers and interfaces[28], magnetic interface anisotropy[29] and finite temperature size effects[30-32].

For mechanical property designing, Hall-Petch equation

$$\sigma_{lv} = \sigma_0 + k\, d^{-1/2} \qquad (8)$$

is important, where σ_{lv} is the lower yield strength of materials, σ_0 and k are constants and d is the grain size. The equation gives the correlation of yield strength with grain size. It is well known that refining grain size is also beneficial for toughening materials the other side,the hardness of materials is proportional to the yield strength. So, it is obvious, that decreasing the grain size of a thin film to nanometer scale should lead to improvement of both the hardness and toughness of the film. For example, when decreased the grain size of a TiN-based from 0.2-0.3 μm to 20-50 nm, its hardness increased more than two times and its toughness improved at the same time[13,14]. Our results can be understand when considering Hall-Petch equition. Recently, similar results were obtained with nanocomposite multilayer films of Ti_2N/TiN and Ti/TiN[32] prepared by hollow cathode discharge ion plating. Each kind of sample were made with a certain single layer thickness ranging from 10 to 100 nm. Results of measurement with supermicrohardness tester shown that as the thickness of single layer falls, increases the hardness of nanocomposites films (Fig 4). X-ray dif-

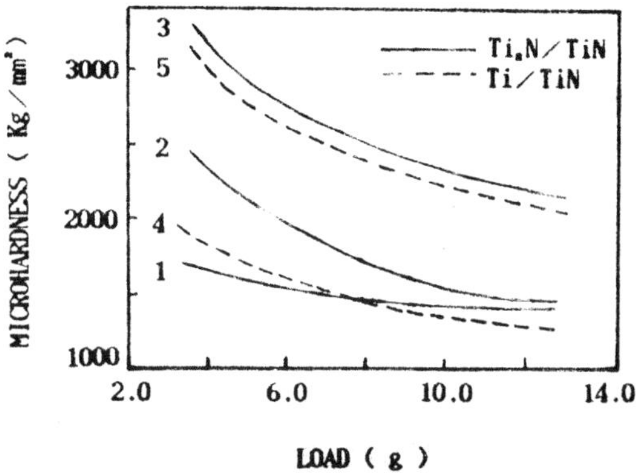

Fig 4 Microhardness of nanocomposite films, thickness of the single layer: 1 - 100 nm, 2 - 20 nm, 3 - 10 nm, 4 - 40 nm, 5 - 20 nm

fraction analysis revealed TiN, Ti_2N and Ti phases in the film (Fig 5). Results of transmission electron microscopy observation shown, that the grain size of the film decreases to less than 20 nm when the thickness of single layer falls down below 0.1μm (Fig 6).

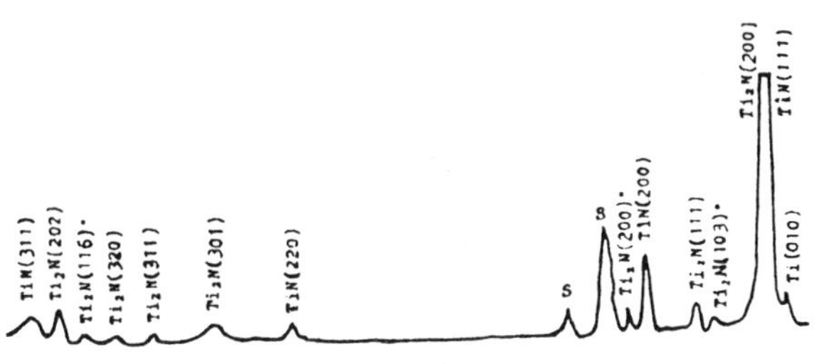

Fig 5 X-ray diffraction diagram of nanocomposite film

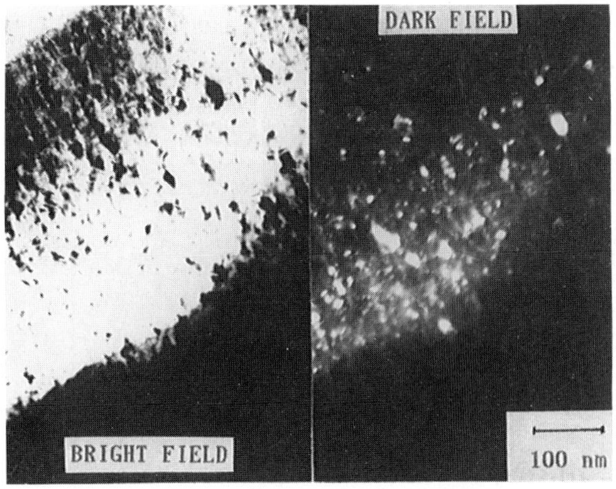

Fig 6 Microstructure of nanocomposite film.

One interesting factor which should be taken into consideration, is that due to the greatly enhanced interfacial area, the interface concentrition of detrimental trace elements shall be much lower than in bulk materials, even its total quantity is the same in both cases.

Interdisciplinarity is one of the most prominant feature of the materials designing of nanocomposites. So, the new physical theories are of great interest for successful designing nanocomposite materials. Theory of percolation studies property of inhomogeneous media which are represented as space filled in a disorded way with two (or more) different types of materials: conductor and insulator, opaque and transparent, superconducting and normal conductor, magnetic and nonmagnetic, solid and fluid. At first sight, theory of percolation is considered more useful for macroscopic systems. However, as it was shown by Deutscher[34], one of the earlier models for the study of percolation, a lattice with missing bonds in the tight binding approximation, was actually a quantum model. Based on quantum percolation, energy dissipation by free electron and other physical effects of nanocomposition, the percolation threshold of microwave has been obtained and a composite microwave absorbing film has been designed and prepareds[35]. Results of microwave absorptivity measurement gave an absorption frequency bandwidth of 3.6 GHz at 10 dB absoption level in 7-17 frequency band (Fig 7). That is of

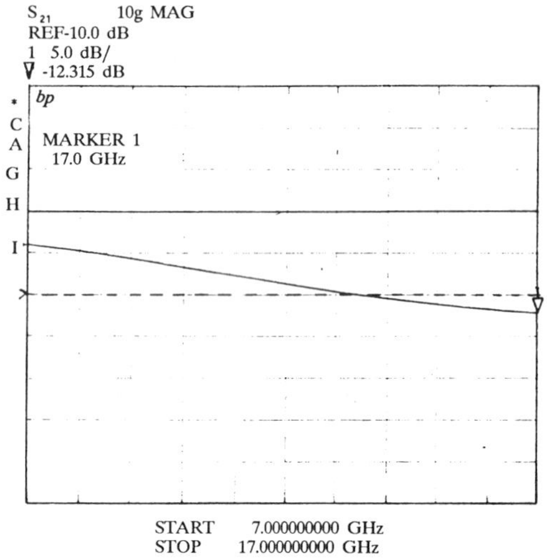

Fig 7 Microwave absorptivity of nanocomposite film

the same order as for the ordinary microwave absoption coating. While the thickness and the surface density of the nanocomposite film) together with the thickness and the weight of the deposition substrate are of one order lower than that of ordinary microwave absorbing coating.

Fractal dimension analysis could be an useful mathematical means for characterization of superfine powder, growth morphology and microstructure of thin film, and the percolation network of different property. The crux of the question seems to be introducing characteristic length and its critical value into fractal dimension calculation and connecting the variation of fractal dimension with evolution of physico-chemical mechanism underlying the change of property and performance of nanocomposite materials.

5. PREPARATION OF NANOCOMPOSITES

Preparation is by no means less important than designing. In this respect, vapour deposition enhanced by ion beam and electical discharge plasma is a kind of versatile prcesses for preparing both low dimensional materials (thin films, wiskers and superfine powders) and their nanocomposites.

Molecular beam epitaxy (MBE) and metalorganic chemical vapour deposition (MOCVD) have advantage over other methods for precisely layer by layer growing at the atomic scale. However, for more popular application, there is now a trend to use more inexpensive process.

Recently, vacuum evaporation with resistance and induction heating, and chemical vapour growth with arc plasma heating are more and more applied for preparing superfine powder. For preparation of multilayer nanocomposite films, vacuum evaporation by electron beam heating, magnetron sputtering and other physical vapour deposition and chemical vapour deposition processes enhanced by ion beam and electric discharge plasma, are widely applied.

Low energy ion beam enhancment attracts now more and more attention. It is well known that a kinetic energy of 1 eV is equivalent approximately to $1{,}16\times10^4$K. The interaction of a low energy (10^1-10^3 eV) ion beam with solid surface gives rise to the so-called spike effect, the nonlinear effects of collective agitations of all atoms within the spike region. Estimation based on different assumptions have shown that there is a maximum of temperature in the range from 10^3 to 10^4K, a pressure maximum of about 10 GPa and a lifetime of 10^{-11}s in the spike region[34]. Such highly nonequilibrium processes are especially favorable for different metastable or instable chemical compound and phases.

6. CONCLUSION

Nanocomposite materials have both the important scientific meaning and broad perspective for application. They make the modifying characteristics of materials possible by means of utilization of low dimensionality of matter and different physical effects of nanocomposition including modifying electron structure and energy state through interface interaction; disorder, orderliness and modulation of active factors in their microstructure and volume effects. The most important points in studying nanocomposites are the formulation of a rational physical model and materials designing, and processing for their preparing. It is still difficult to evaluate their potential for improving the property of materials at the present time. However, it is already evident now that the research and development of nanocomposites initiate a new great progress in materials science which will strongly promote forward the science, technology and economy of mankind.

REFERENCES

1. L. Esaki and R. Tsu, IBM J. Res. Dev., 14 (1970)61.
2. J. S. Koehler, Phys. Rev., B2(1970)547.
3. W. M. C. Yang, T.Tsakalakos and J. E. Hilliard, J. Appl. Phys., 48 (1977) 876.
4. G. E. Henein and J. E. Hilliard, J. Appl. Phys., 54 (1983) 728.
5. T. Tsakalakos and J. E. Hilliard, J. Appl. Phys., 54 (1983) 734.
6. D. Baral, J. B. Ketterson and J. E. Hilliard, J. Appl. Phys., 57 (1985) 1076.
7. S. L. Lehoczky, Phys. Rev. Lett., 41(1978) 1814.
8. S. L. Lehoczky, J. Appl. Phys., 49 (1978) 5479.
9. B. A. Movchan, A. V. Demchishin, G. F. Badilenko, R. F. Bunshan, C. Sans, C. Deshpan and H. J. Doerr, Thin Solid Films, 97 (1982)215.
10. R. W. Springer and D. S. Catlett, Thin Solid Films, 54(1978)197-205.
11. S. Todorova, U. Helmersson, S. A. Barnett, J. -E. Sundgren, J. E. Greene, IPAT 87, 6th Int. Conf.Ion and Plasma Asisted Techniques, Brighton, UK, May 1987, p248.
12. U. Helmersson, S. Torodova, S. A. Barnett, J. -E. Sundgren, L. C. Markert and J. E. Greene, J. Appl. Phys., 62(1987)481.
13. L. S. Wen, X. Z. Chen, Q. Q. Yang, Y. Q. Zheng, Y. Z. Chuang, Vacuum Science and Technology, 7 (1987) 31.
14. L. S. Wen, X. Z. Chen, Q. Q. Yang, Y. Q. Zheng, Y. Z. Chuang, 1st Int. Conf. Plasma Surface Engineering, Sept. 23 - 27, 1988, Garmisch-Partenkirchen, FRG.
15. H. Holleck and H. Schulz, Thin Solid Films, 153 (1987) 11.
16. H. Holleck, Metall, 43 (1989) 614.91
17. J. Karch, R. Birringer & H, Gleiter, Nature, 330 (1987) 556.
18. E. Spiller,in Low Energy X-Ray Diagnostics —1981, eds. D. Attwood and B. L. Henke, AIP Conf. Proc. № 75, AIP, New York, 1981, p.131.
19. C. G. Granqvist, R. A. Buhrman, J. Wyns and A. J. Sievers, Phys. Rev. Lett., 37 (1976) 625.
20. R. C. Baetzold, Surface Science, 106, (1981)243.
21. M. Pomerantz, J. C. Slonczewski and E. Spiller, J. Magnetism & Magnetic Materials, 54-57 (1986) 781.
22. C. L. Fu, A. J. Freeman, J. Magnetism & Magnetic Materils, 54-57 (1986) 777.
23. M. D. Bentzon, j, VAN Wonterghem, S. Morup , A. Th:len and C. J. W. Koch, Phil. Mag., B 60(1989)2,169.
24. D. Wang, A. J. Freeman and H. Krakauer, Phys. Rev., B 26(1982)1340.
25. J. Tersoff and Falicov, Phys. Rev., B 26 (1982)6186.
26. A. J. Freeman and R. H. Victora, in Magnetic properties of low-dimensional systems, ed. L. M. Falicov and J. L. Moran, Springer Verlag, 1986.
27. S. Ohnishi, M. Weinert and A. J. Freeman, Phys. Rev., B 30 (1984)36.
28. C. L. Fu, A. J. Freeman, T. Oguchi, Phys. Rev. lett., 54(1985)2700.
29. U. Gradmann, J. Magn. and Magn. Mat. 54-57 (1986)733.
30. U. Gradmann and Mueller, Phys. Stat. Sol. (1986)313.
31. U. Gradmann, R. Bergholz and E. Bergter, thin Solid Films, 126(1985)107.
32. L. M. Falicov and J. L. Moran-Lopez, Magnetic Properties of Low-dimensional Systems, Springer Verlag, 1986.
33. L. S. Wen, J. Gong, B. H. Yu, R. F. Huang, L. P. Guo, Paper presented at C-MRS Int.90 June, 1990, Beijing.
34. G. Deutscher, Applications of percolation, in Chance and Matter, eds. J. Souletie, J. Vannimenus and R. Stora, North-Holland Physics Publishing, Amsterdam, 1987.
35. L. S. Wen, R. F. Huang, J. Wu, H. Wang, R. J. Hong, J. Gong, Y. Q. Zheng, unpublished

work.
36. C. Weissmantel, Preparation, structure and properties of hard coatings on the basis of i-C and i-BN, in Thin films from free atoms and particles, ed. K. J. Klabunde, Acad. Press, Inc., Orlando, Florida, 1985.

Frontiers of Materials Research/
Electronic and Optical Materials
M. Kong, L. Huang (editors)
Elsevier Science Publishers B.V., 1991

OXYGEN-RELATED DONOR STATES IN CZOCHRALSKI SILICON+

Zhanguo WANG, Jiajun QIAN, Shouke WAN and Lanying LIN

Institute of Semiconductors, Chinese Academy of Sciences, P.O. Box 912, Beijing 100083, China

The behavior of oxygen-related donor traps in Czochralski silicon is briefly reviewed. The experimental data from recent electron nuclear double paramagnetic resonance studies indicate that no impurity or dopant is involved in the thermal donor defect as a substantial part, and that the core of the defect contains four oxygen atoms in C_{2v} symmetry occupying a silicon vacancy. This result shines some light on the nature of the thermal donor. However, the detailed formation kinetics and the origin of the electrical activity of this center are still open. For the new donor, we present a tentative structural model in which the chemical components of the new donor consist of neither the oxygen nor the carbon, but the Si self-interstitial dislocation loop itself resulting from oxygen precipitation during heat treatment process. This model may be viewed as the first step to approach the microstructure of the center. Since no spectroscopic data which is directly related to the new donor are available at present, therefore, the electronic states of the new donor are unknown.

1. INTRODUCTION

Silicon crystals grown by the Czochralski(Cz) method are saturated with oxygen at the melting temperature, but are supersaturated at lower temperature, i.e.,the temperature at which devices are processed.[1] The equilibrium solubility of oxygen at the melting point is about $10^{18}cm^{-3}$ and the oxygen atom is incorporated into the Si lattice on electrically inactive interstitial site. During subsequent heat treatment supersaturated oxygen will react with silicon to form SiOx clusters or precipitates which apparently act as donor traps. For Cz Si-wafers annealed at different temperatures, two kinds of oxygen related donor states can be formed. The thermal donors(TD) and the new donors(ND) are generated in the temperature ranges of 350-500 °C and 550-850 °C respectively.[2-5]

Extensive studies have been conducted mainly on the TD which have proved that they are double donors[6] which get shallower with annealing[7,8] and are usually associated with a series of oxygen clusters successively formed by the incorporation of the interstitial oxygen atoms into the oxygen aggregate. However, in spite of a tremendous amount of the experimental data gathered over the last 36 years no structural model for the TD could be established. The ND centers were observed by Capper et al.[4] in 1977. They differ from the TD, and are much less investigated. In the following, electronic properties and kinetic models of the TD centers will be reviewed briefly (Sec. 2). In Sec. 3, we will describe the main experimental results concerning the ND, including electronic states and a possible identification of a characteristic phtoluminescence (PL) line at 0.902 eV (K-line) related to the ND centers. Finally, on the basis of our transmi-

+This work was supported by NNSF of China, Grant No. 85267.

ssion electron microscopy (TEM) and high-resolution electron microscopy (HREM) observations, a tentative structural model of the ND is presented.

2. THERMAL DONORS

2.1. Electronic states of thermal donors

Since the discovery of the oxygen related TD effect 36 years ago,[2] many works have been devoted to the microscopic structure and growth mechanism for the TD. Early works to define a structure for this center were based on formation kinetics of the TD that were determined through resistivity measurements. The initial growth rate of increase in electron concentration is proportional to the fourth power of the oxygen concentration, $[Oi]^4$, and the maximum donor concentration is proportional to the third power of the oxygen concentration, $[Oi]^3$. In subsequent work the TD has been studied extensively by using various techniques such as: Hall effect,[8] Infrared absorption (IR),[9] Electron paramagnetic resonance(EPR),[10] Deep level transient spectroscopy(DLTS)[11] and Electron nuclear double resonance(ENDOR).[12] Hall measurements give two energy levels at 0.06 eV and 0.13 eV below the conduction band.[8] The compensation behavior and temperature dependence of these levels demonstrated that the TD is a double donor.[6] The donor activity of the oxygen related clusters is destroyed by annealing at temperature higher than 500°C with an activation energy of 2.5-2.8 eV.[7] Current transient spectra also show a double donor character of these centers (Ec-0.07 eV and Ec-0.15 eV) which correspond to the thermal ionization transitions of $TD^{0/+}$ and $TD^{+/++}$.[11] However, since DLTS is unable to resolve the energy levels of individual state of the TD species, the Ec- 0.07 eV and Ec-0.15 eV peaks represent the total response of the whole distribution of cluster sizes.

IR studies of electronic transitions associated with the TD show the presence of at least nine distinct double donor species, which have progressively shallower energy levels with nearly constant separation (~ 2 meV) due to the increased repulsive perturbation.[13,14] resulting from successive precipitation of oxygen at the core of the defect.

On the other hand, EPR spectrum labeled NL8 was observed following 450 °C heat treatment of Oxygen-rich silicon and has been identified as the TD,[12] and the defect has a C v symmetry. Despite efforts, no hyperfine interaction with O^{17} and Si^{29} nuclei has been detected by EPR. In order to determine the structure of the NL8 center, the ENDOR technique was performed, and the results showed that the different TD species identified in IR spectroscopy also show up in ENDOR. This means that the assignment of the NL8 centers as the TD is resonable.

2.2 Models of the TD formation

Based on the kinetic information mentioned above, Kaiser, Frisch and Reiss (KFR) considered that the TD is a Si_nO_x complex[15] which is built up in successive reaction steps, $Si_nO_x + O_i \rightarrow Si_nO_{x+i}$ (n, x integers). It was supposed that the donor activity stem from clusters containing four oxygen atoms, whose electrical activity is destroyed when a fifth oxygen atom is added to this cluster. This model is the most successful one of the many so far put forward, in the sense that it explains a number of features consistantly observed in the formation kinetics of the TD. Experimental work, however, has shown that it fails to explain two important facts. First, the measured rate of donor formation requires a diffusivity of oxygen which is several orders of magnitude higher than the accepted value. Secondly, the presence of at least nine electrically active species observed by the IR studies can not be interpreted using this model. In order to explain the observed rate of the TD generation, an accelerated diffusion of the oxygen interstitials is required. Recently, the evidence of enhanced oxygen diffusion (EOD) has been obtained from the out-diffusion profiles of oxygen in Cz-Si at 500 °C measured with SIMS[16] and from TEM study[17] of the oxygen precipitation kinetics at 480 °C. Several models explaining the EOD phenomenon have been proposed.[17-20] According to these models, the EOD occurs in the form of the fast diffusing molecular oxygen or oxygen atom with vacancy or Si-interstitial or Si-interstitial-oxygen interstitial pair. Helmreich and Sirtl (HS) model was rejected by an interesting experiment which shows a complete prevention of TD generation in Cz-Si at 450 °C by the formation of Chromium dicilicide ($CrSi_2$), a process which does provide a high supersaturation of vacancies.[21] A fast diffusing, gas-like molecular oxygen O_2 which has no bonds to the Si lattice, was suggested by Gosele and Tan (GT model). However, a linear dependence of $K_1 x N_{FDS}$ on N_0 recently obtained by Markerrich et al.[22] showed that only one oxygen atom is involved in the fast diffusing species (FDS), where K_1, N_{FDS} and N_0 are the rate constant, FDS concentration and the oxygen interstitial concentration respectively. This result rules out the molecular oxygen model.

Alternative kinetic models suggested by Ourmazd, Schroter and Bourret (OSB),[20] Borrenstein, Peak and Corbett (BPC)[23] as well as the Ylid model[24] are able to describe quantitatively the sequential formation of the TD, but the problem of the fast diffusing species involved in their formation still remains.

From IR measurements, Newman suggested that the TD could be in fact Si self-interstitial aggregates.[25] This suggestion is also supported by the recent findings of Bourret that the ribbon-like defects observed in annealed Cz-Si are small particles of the hexagonal silicon phase, formed by agglomeration of Si self-interstitials.[26] Following this idea, Mathiot[27] has shown that a good kinetic description of the TD growth is possible if one assume that the TD is formed by the agglomeration of the self-interstitials generated by the formation of the O_2 complex, which is the first stage of the oxygen precipiation; this Si self-interstitial agglomeration occurs around some small oxygen complexes which exist in the as-grown material, act as nucleation site, and are thus an essential constituent of the resulting TD. Such a model is able to account for the main features of the TD, such as the influence of an initial oxygen concentration or of carbon.

This confusing situation is changing only recently as the detailed structural imformation becomes available from ENDOR experiments. The preliminary ENDOR experiment was successfully established unambiguously for the first time the direct involvement of oxygen in the structure of the TD center. Further ENDOR studies on Boron and Phosphorus doped Cz-Si and O^{17} diffused Fz-Si showed that no impurity or dopant is involved in the defect as a substantial part, and that the core of the TD contains four oxygen atoms in C_{2v} symmetry occupying a silicon vacancy.

In summary, it is clear that even though the ENDOR study shines some light on the microstructure of the TD, however, an accurate description of the nature of the TD is still lacking, and even the detailed TD formation kinetics and the source of electrical activity of the TD center are still open.

3. EXPERIMENTAL DATA ON NEW DONORS

3.1. Previous work

The ND center was observed by Capper et al.[4] in 1977. Later on Kanamori et al.[5] published their systematic investigation of the effect of heat treatment in a wide temperature range upon the resistivity of Cz-Si. They could unambiguously distinguish a new donor different from the well-known TD, which occurs in the samples annealed at higher temperatures (600-850 ^{0}C) where the TD are completely annihilated. Contrary to the TD whose generation can be suppressed by a high carbon concentration, the ND formation, is, however, promoted by the presence of a large amount of carbon atoms, [28] and preannealing also enhanced the ND generation.

The ND is a single, shallow donor state and the activation energy of this center has been found to decrease continuously with the duration of annealing, reaching values as low as 19 meV. Some authors attributed this behavior to an energy distribution associated to interface states or the dangling bonds at the surface of SiO precipitates,[29,30] which are spatially distributed in the silicon lattice. In Lin et al's study, [31] three kinds of the ND with different activation energies (Ec-0.025 eV, Ec-0.05 eV and Ec-0.1 eV) were observed for as-grown N and P-type samples respectively, if the samples were subjected to a long enough time annealing until a saturated ND concentration was nearly reached. Contrary to the TD, no IR absorption bands related to the ND has been found up to now. But a sharp and strong PL line at 0.902 eV (K-line) is frequently observed in samples containing ND ceneters. [32,33] The fact that the intensity dependence of the K-line on the annealing time coincides with the concentration dependence expected for silicon self-interstitials emitted during the oxygen precipitation process led Tajima et al.[33] to the suggestion that this line should be correlated with Si-interstitials. As mentioned above, high carbon concentration in Cz-Si is proved to promote the ND generation, but the effect of carbon content in Si on the K-line has not been established previously. In Wang et al's work,[34] it was found that the intensity of the K-line decreases with the increasing carbon concentration, and it completely disappears when the carbon content reaches the critical value of about 1x10^{17} cm^{-3}, regardless of its conductivity type or very high ND's density. This not only tells of the quenching effect of high carbon concentration on the K-line, but also excludes the possibility of the K center just being the ND itself or related to carbon. On the other hand, however, for the samples of producing neither TD nor ND after two step thermal treatment, the K-line has been never observed, and this strongly indicates, from another point of view, the ND (at least, a part of it) should be a necessary component of the K center. On the basis of these experimental results and in combination with the high luminesent efficiency of the K-line, Wang et al proposed that the K-line is a neutral complex consisting of a shallow donor state and a strongly localized deep acceptor, and the K-line is emitted from excitons bound to an acceptor-like isoelectronic center. New experimental date obtained in present study further support this idea. We will return to this point later.

3.2. New results

I. Temperature dependence of the K-line

The normalized PL intensity of the K-line is plotted against 10^3 / T in the temperature range 4.2 to 77 K in fiqure 1. Thermal activation

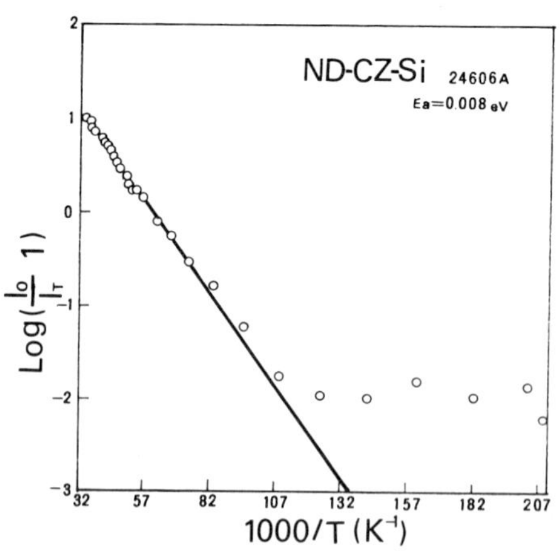

FIGURE 1

Temperature dependence of a normalized PL intensity of the K-line measured at 0.902 eV.

energy Ea, only 8 meV, is observed which is much smaller than the total binding energy of 253 meV expected from the K center (taking the excitonic gap as 1.1552 eV). This provides further support for the isoelectronic trap model of the K center proposed by Wang et al.[34] The smaller activation energy would correspond to the thermal release of one long range Coulomb potential bound particle from the isoelectronic center and the difference of the total binding energy of the K center and Ea is due to the particle that remains at the center.

II. TEM and HREM observations

The starting material is P-type B-doped Cz-Si wafers. The typical parameters are shown in Table 1. The wafers are preannealed at 450°C for 16 hrs and then annealed at 650°C for a few days under Ar atomsphere. The carrier concentration, oxygen and carbon content are measured by the conventional methods. The samples for TEM and HREM observation are first ground and polished, then thinned by Ar^+ at 4-5 KV up to an electronic transparency thickness. The electron diffraction patterns and the HREM images are observed by EM-420 and/or JEOL-200CX at 120 KV and 200 KV respectively.

The electrical measurements show that three kinds of samples A, B and C are turned into N-type after being annealed at 650 °C for a long enough time, whereas the resistivity of annealed sample D has no obvious change. According to IR measurements, the more interstitial oxygen [Oi] decreased([Oi]), the more ND generated. For the sample D, there is no oxygen precipitation, no ND produced(see Table 1). TEM results indicate that in the samples containing ND and with lower carbon concentration, several types of defects are observed such as: high density microdefects and dislocation dipoles as well as <110> oriented rod-like defects[35-37] with the density of 10^3 - 10^4 cm^{-3} and a length of a few microns etc., whereas dislocation dipoles and rod-like defects are not seen in sample of higher carbon concentration ($[Cs]>10^{17}cm^{-3}$). For the sample containing no ND, no such defects are generated (see Table 1). The TEM observed microdefects of high density and non-uniform distribution in samples A, B and C will assume to have black-white lobe contrast under certain diffraction conditions. Based on the spatial orientation of

TABLE 1. Some parameters of representative Cz-Si samples before and after annealing

	Before annealing				After annealing				
Sample No.	Conduc. type	P 10^{15} (cm^{-3})	Oi$^+$ 10^{17} (cm^{-3})	Cs 10^{17} (cm^{-3})	Conduc. type	ND 10^{15} (cm^{-3})	Oi 10^{17} (cm^{-3})	Rodlike defects (cm^{-2})	Small loops (cm^{-3})
A	P	1.81	9.3	0.1	N	2.6	7.2	10^6-10^7	10^{13}-10^{14}
B	P	0.05	6.7	0.1	N	0.5	2.4	10^4-10^5	10^{13}-10^{14}
C	P	1.5	6.6	3.8	N	2.2	4.4	none	10^{13}-10^{14}
D	P	2.0	4.6	0.4	P	none	0	none	none

$^+$According to ASTM F 121-81.

the dislocation dipoles in the samples as well as their projection length on the electron micrographs, the thickness of the thin film can be calculated, and thus the density of the microdefects is obtained to be about 10^{13}- 10^{14} cm^{-3}(fig.2). The HREM observations show that the microdefects have a size of 10 nm with a stress field expanding outwards,and the central portion of which has a plate-like precipitates[38,39]with a thickness of 1-2 atom layers. In the vicinity of the precipitate there is a pair of the extra half {111} silicon planes marked by arrow in fig. 3, and this obviously is an 60^0 dislocation loop. The diameter of the loops is of 2-7 nm with a {113} habit plane. A schematic structure of such dislocation loop is shown in figure 4, with the dislocation core D having dangling bonds.

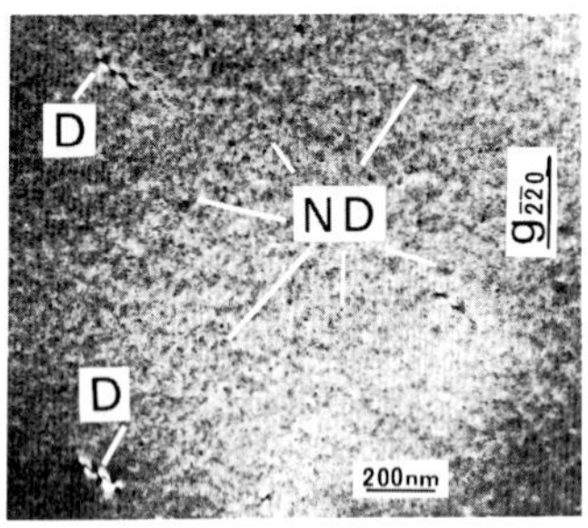

FIGURE 2
TEM micrograph of the ND and dislocation dipoles (D) in Cz-Si annealed at 650 ^{0}C for 5 days. $\vec{B}\sim[001]$, $\vec{g}\sim 220$, S > 0.

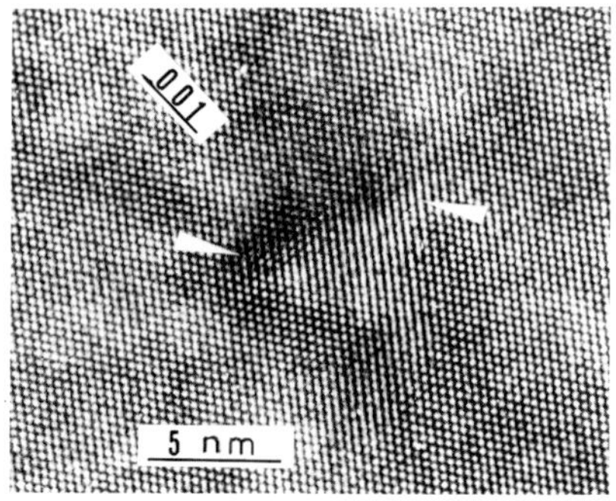

FIGURE 3
Lattice image of small interstitial dislocation loop. A pair of the extra half {111} silicon planes are denoted by arrow. Observation at 200 KV, [110] zone axis.

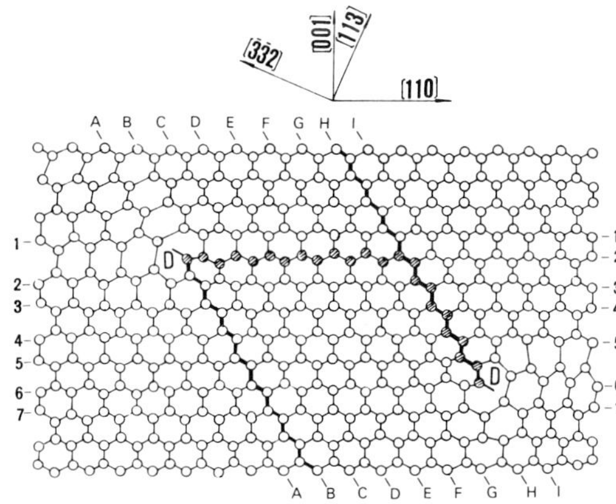

FIGURE 4
Schematic structrure of small dislocation loop, heavy lines denote a pair of the extra half {111} Si planes with the dangling bonds (D). The atoms with shadow are excessive Si atoms.

4. A NEW STRUCTURAL MODEL OF THE ND CENTER

As for the nature of the ND, Cazcarra and Zunino et al [38] correlated the generation of the ND with the reduction of the interstitial oxygen concentration and suggested that the Si_yO_x cluster could be related to the ND, while Gaworzewski et al.[28] proposed that C-O complexes are possibly the core of the ND. However, these models can not explain the origin of the ND electrical activity. Recently, several ND models have been suggested with the emphasis upon the interface states of the dangling bonds on the interface between the host silicon lattice and the oxygen precipitates.[29,30,39]Though such models can better explain the passivating action of the hydrogenation treatment on the ND and the origin of the ND electrical activity, yet it can not interpret why the density of the ND decreases in spite of the growth of SiO_2 precipitates and therefore the SiO_2 / Si interface at elevated temperature annealing. It is obvious that there is much work to do for revealing kinetic growth mechanism and an atomic structure of the ND center.

In this paper, a new structural and kinetic model of the ND is presented, in which the electrical activity of the ND does not originate from the oxygen atom itself but from the uncoordinated Si dangling bonds on the dislocation loops which result from the Si self-interstitials produced by oxygen precipitation during heat treatment process. Our new model is based on the following arguments:

4.1. Kinetic formation of the Si-interstitial -type 60^0 dislocation loop

It is generally accepted that the Si self-interstitials are emitted from growing oxygen precipitates. By this process excessive stress caused by a volume increase of about 100% during oxygen precipitation can be avoided.[39,40-43] Obviously,the Si self-interstitial concentration ($[N_{Si}]i$) is proportional to the volume of all oxygen precipitates already formed, and thus is proportional to the decreasing of the interstitial oxygen concentration ($\triangle[Oi]$) during the annealing process. However,if the $[N_{Si}]i$ reaches a certain level, they begin to agglomerate,which results in the formation of the interstitial-type dislocation cores. For silicon, each dangling bond already amounts to about 1 eV of free energy, thus the dislocation cores may be constructed with the least number of dangling bonds only if the dislocation line is in a <110> direction. For this reason, the linear configuration will be favored.[44] Along with increasing heat treatment time, newly generated Si self-interstitials will continuously agglomerate to the dislocation lines, then the latter can slip on {113} plane[45,46] under the action of the stress field to form the interstitial-type 60^0 dislocation loops with a {113} habit plane and a number of dangling bonds in their cores(fig. 4).

4.2. Relation between ND and dislocation loop

The experimental results show that the sample containing ND must have above mentioned dislocation loops with a certain density, conversely, such high density dislocation loops have never been observed in Si wafers containing no ND.

4.3. The origin of ND electrical activity

In homopolar covalent bonding Si crystal, the dangling bonds possess donor properties[47], thus the ND electrical activity is naturely considered to be related to the dangling bonds on the core of the Si self-interstitial type dislocation loop. The density of such defects with a diameter of 2 - 7 nm is 10^{13}- 10^{14} cm^{-3}. If the average diameter is about 5 nm, and Si atom spacing in <110> direction is $(\sqrt{2}/2)$ a (a=0.543 nm) then the number of the dangling bonds along the dislocation core of each loop will be about 40, thus all of them are $4x10^{14}$- $4x10^{15}$ cm^{-3}. This result is quite similar to the ND concentration obtained by Hall measurements.

4.4. Si interstitial emission and oxygen precipitation

Similarly, it can readily seen from fig. 4 that the atoms with shadow are extra Si atoms in comparison with perfect lattice surrounding the small dislocation loop which are transformed from Si self-interstitials caused by oxygen precipitates. For a loop with 40 dangling bonds (5 nm diamter), the excess Si atom are calculated to be about 300-500, which totally correspond to $(0.3\text{-}5)x10^{16}cm^{-3}$ extra Si atoms ($[N_{Si}]ex$). This is approximately equal to the $[N_{Si}]i$ generated by a volume increase resulting from (2.4-7.2)x $10^{17}cm^{-3}$oxygen precipitation in our case, if one takes into account the fact that the fraction of the stress not released by Si self-interstitial emission as well as other microdefects such as dipoles, rod-like defects etc. are also formed by Si interstitials during annealing at 650 ^{0}C.

4.5. Oxygen diffusion coefficient consideration

The formation of such a dislocation loop can approximately be considered as a consequence of oxygen diffusion and precipitation in a sphere with a radius of 100 nm, because the dislocation loop with 5 nm diameter of 10^{14} cm^{-3} corresponds to the average distance of 200 nm between the loops. According to the oxygen diffusion coefficient data at 650 ^{0}C,[11] the time calculated for oxygen atoms passing through 100nm is about 12 hrs. This value coincides with the intensity dependence of the K-line on the 650^{0}C annealing time (fig. 5). This intensity dependence was

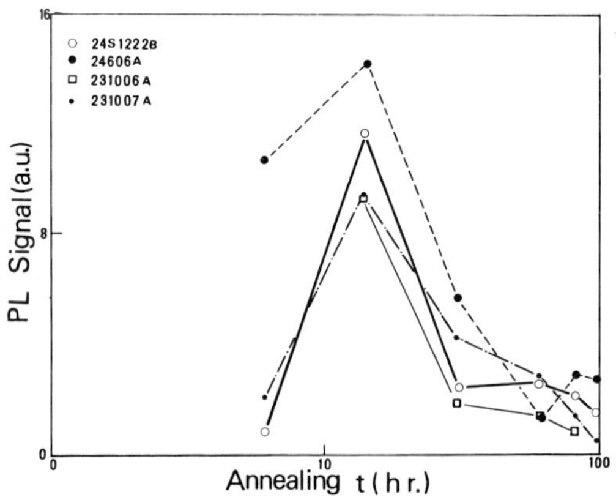

FIGURE 5

The intensity dependence of the 0.902 eV PL line on the annealing time in relation to the behavior of Si self-interstitials in Cz-Si.

suggested in relation to the behavior of Si self -interstitials emitted during the oxygen precipitation process.[33]

By means of the present model, the main experimental facts related to ND such as: kinetic formation, the origin of electrical activity,the effect of hydrogen passivation and carbon on ND, the inhomogeneity of ND distribution and the annihilation behavior of ND at higher temperature annealing (>900 ^{0}C) etc. can be better interpreted.

5. SUMMARY

We have briefly reviewed the behavior of oxygen related donor traps in Cz-Si. Even though recent ENDOR work casts a beam of light on the electronic structure of the TD, more careful experiments still need to be done in the forthcoming years.

For the ND, the confusing situation is just as the TD several years before. What limits our knowledge of or prevents an establishment of an accurate structural model of the Nd should be traced to the lack of available spectroscopic data concerning detailed structure information about the ND. Our model may be viewed as the step to approach the nature of the center.

Limitation on the length of this paper preclude an exhaustive discussion of this progress, and we apologize to the authors and readers for being selective and incomplete.

REFERENCES

1. J.C. Mikkelsen, in Oxygen, Carbon, Hydrogen and Nitrogen in crystalline silicon, eds. J.C. Mikkelsen,Jr. S.J. Pearton, J.W.Corbett and S.J. Pennycook (Materials Res. Soc. Pittsburgh, 1986) pp. 19

2. C.S. Fuller, N.B. Ditzenberger et al., Phys. Rev. 96 (1954) 833

3. W. Kaiser, Phys. Rev. 105 (1957) 1751

4. P. Capper, A.W. Jones et al., J. Appl.Phys. 48 (1977) 1646

5. A. Kanamori and M. Kanamori, J. Appl. Phys. 50 (1979) 8095

6. A.R. Bean and R.C. Newman, J. Phys. Chem. Solids, 33 (1972) 255

7. C.S. Fuller and R.A. Logan, J. Appl. Phys. 28 (1957) 1427

8. P. Gaworzewski and K. Schmalz, Phys. Stat. Sol. (a) 55 (1979) 699

9. B. Pajot, H. Compain et al., Physica, 117B/118B (1983) 110

10. S.H. Muller, S.M. Sprenger et al., Solid State Commun., 25 (1978) 987

11. M. Stavola, J.R. Patel et al., Appl. Phys. Lett., 42 (1983) 73

12. J. Michel, R.J. Niklas et al., Phys. Rev. Lett., 57 (1986) 611

13. J.W. Corbett, H.L. Frisch et al., Mater. Lett. et al., 2 (1984) 209

14. J.T. Borenstein, J.W. Corbett. J. Phys. C: Solid State Phys. 19 (1986) 2893

15. W. Kaiser, H.L Frish et al., Phys. Rev. 112 (1958) 1546

16. S.-J. Lee and P. Fellinger, Mater. Sci. Foum, 10-12 (1986) 1021

17. W. Bergholz, J.L. Hatchison et al., J. Appl. Phys. 58 (1985) 3419

18. H. Helmreich and E. Sirtl, Semicond. Silicon, (1977) 622

19. U. Gosele and T. Tan, Appl. Phys. A28 (1982) 79

20. A. Ourmazd, W. Schroter et al., J. Appl. Phys. 56 (1984) 1670

21. R.C. Newman, A.K. Tipping et al., J. Phys, C: Solid State Physics, 18 (1985) L861

22. V.P. Markevich, L.F. Makarenko et al., Mater. Sci. Foum, 38-41 (1989) 589

23. J.T. Borenstein, D. Peak et al., J. Mater. Res. 1 (1986) 527

24. L.C. Snyder and J.W. Corbett, in Pro. 13th. Intern. conf. on Defects in Semiconductors. Coronado, CA (1984) 693

25. R.C. Newman, Mater. Res. Soc. Symp. 59 (1986) 205

26. A. Bourret, Inst. Phys. Conf. Ser. 87 (1987) 39

27. D. Mathiot, Mater. Sci. Foum, 38-41(1989) 649

28. P. Gaworzewski and K. Schmalz, Phys. Status Solidi, A77 (1983) 571

29. A. Henry, J. L. Pautrat et al., J. Appl. Phys. 60 (1986) 3192

30. L. Zhong, J.X. Shi et al., in Semicond. Si. (Pennington, Electro-Chem. Soc. 1986) p813

31. L.Y. Lin, Z.G. Wang et al., in Pro. of Inter. conf. on Semicond. and IC's Tech. eds. X.Y. Wang and R.X. Mo (Woeld Scientific 1986) pp. 111

32. N.S. Minaey and V. Mudriyi, Phys. Status Solidi A68 (1981) 561

33. M. Tajima, U. Gosele et al., Appl. Phys. Lett. 43 (1983) 270

34. Z.G. Wang, S.K. Wan et al., in Proc. of Intern. Conf. on Mater. and Process. Charac. for VLSI, eds. X.F. Zong, Y.Y. Wang and and J. Chen, (world Scientific 1988) pp.92

35. K. Tempelhoff and F. Spiegelberg, in Semiconductor Silicon 1977, eds. H. Huff and E. Sirtl (Electro-Chem. Soc. Pennington, NJ, 1977) pp. 585

36. N. Yamamoto, P.M. Petroff et al., J. Appl. Phys. 54 (1983) 3475

37. K. Tempelhoff, R. Gleichmann et al., Phys. stat. Sol. (a), 56 (1979) 213

38. V. Cazcarra and P. Zunino, J. Appl. Phys. 51 (1980) 4206

39. K. Holzlein, G. Pensl et al., Appl. Phys. Lett. 48 (1986) 916

40. A. Bourrent, J. Thibault-Desseaux., J. Appl. Phys. 55 (1984) 825

41. P.E. Freeland, K.A. Jackson et al., Appl. Phys. Lett. 30 (1977) 31

42. S. Mahajan, G.A. Rozgony et al., Appl. Phys. Lett. 30 (1977) 73

43. J.R. Patel, in Semicond. Silicon 1981, eds. H.R. Huff, R.J. Kriegler and Y. Takeishi (Electro-Chem. Soc., Penninton, 1981) pp.189

44. T.Y. Tan, Phil. Mag. A44 (1981) 101

45. C.A. Ferreira Lima and A. Howie, Phil. Mag. 34 (1981) 1057

46. I.G. Salibury, J. Appl. Phys. 52 (1981) 1108

47. V. Heine, Phys. Rev. 146 (1966) 568

Frontiers of Materials Research/
Electronic and Optical Materials
M. Kong, L. Huang (editors)

ON THE BIONIC DESIGN OF COMPOSITE MATERIALS*

Benlian ZHOU**# and Changxu SHI**

**Institute of Metal Research, Academia Sinica and #International Center for Materials Physics, Academia Sinica, Shenyang 110015, China

Composite material is difficult to design owing to its complexity. Bio-bodies had been naturally selected and evolved for millions of years, their composite structures could be taken as reference. Goals to be reached in different stages of the development of composite materials are listed and the composite structural analyses of bio-materials in literature are reviewed. The last part of this paper deals with the bionic design of composite materials in light of regularities of the nature, such as morphological bionic simulation of composite processing and bionic healing of inner damage in composite materials as well as the methodology used in this work.

1. INTRODUCTION

Composite material often shows its high performance by taking advantages of both the reinforcement and matrix. Sometimes it performs as not only structural but also functional material. In other words, it reveals "hybrid superiority". But in practice, the structural design of composite material is very difficult owing to its complexity of structure and various factors affecting its performance. It is well known that the bio-bodies in the nature are compositely structured and had been naturally selected and evolved for millions of years, such that a great variety of naturally rational composite structures with hybrid superiority had been formed. All of these could be taken as our reference.

2. GOALS TO BE REACHED IN DIFFERENT STAGES OF THE DEVELOPMENT OF COMPOSITE MATERIALS[1]

1942-1960

Light weight and high rigidity is the goal pursued in this stage since the first composite of fibre reinforced resin fabricated in the United States.

1960-1975

High strength and high toughness form the goal of this stage for mechanical components mainly used in aerospace industry.

1975-present

Multi-functional properties consist in the major goal of composite studies in this stage. The so-called advanced composite materials nowadays are studied to meet the multi-functional requirement as well as high performance/price ratio.

3. THE COMPOSITE STRUCTURAL ANALYSIS OF BIO-MATERIALS

Many examples of bionic simulation can be found in material studies, such as the structure of early stage airplane is the mimic of birds and insects. The study of infra-red sensor had taken the hint of rattlesnake, and the presence of

*The major project supported by National Nature Science Foundation of China.

sonar was inspired by the instinctive function of dolphin. A special kind of artificial cobweb with very high strength has been worked out in the United States by bionic and genetic engineering studies.

The recent bionic studies on composite materials can be found in the "Journal of Japan Society of Composite Materials". A series of papers[2-6] under the topic "Bionic Design" were and will be published therein. Some brief descriptions of them are given here in this paper:

3.1. "Some hidden aspects of the structure of wood" by Akira Sakakibara[2]

The specific gravity of wood is 0.3 - 0.8; compressive strength is 1/3 - 2/3 of that of the tensile one; it behaves as a visco-elastic body and its creep behavior varies with water content etc. The so-called "A. T. wood" (compress-failed or stressed wood) can be taken as an example showing the stress induced structure of wood (Figure 1).

It is very interesting that not only the structure but also the chemical composition of wood varies with the internal stress in it. The lignin content of normal wood is about 25 - 30%, but that of compressed A.T. wood is more than 40%, while the cellulose content of tensioned one is about 1.5 ~ 2 times than that of the normal one.

3.2. "Shellfish and its hinge ligament" by katsumichi Ono[3]

The composite structure of shell of a kind of shellfish had been described at first in the paper as shown in Figure 2. Then the ligament structure was discussed in detail showing that the dry structure is as hard as resin while the wet one will be swollen as elastic as

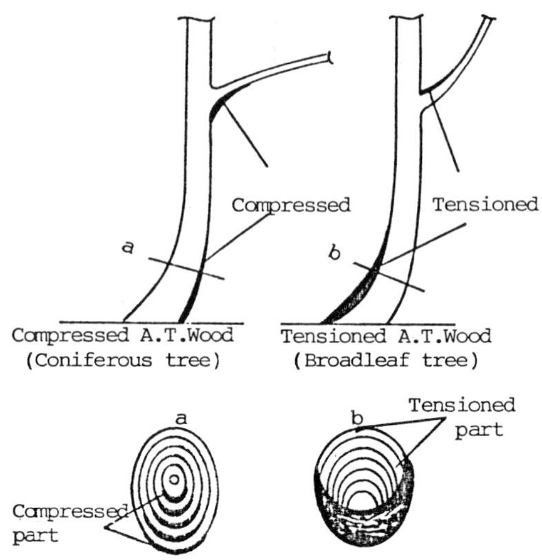

FIGURE 1
The forming position and cross-section of compressed A.T. wood (coniferous tree) and tensioned A.T. wood (broadleaf tree) (After A. Sakakibara[2])

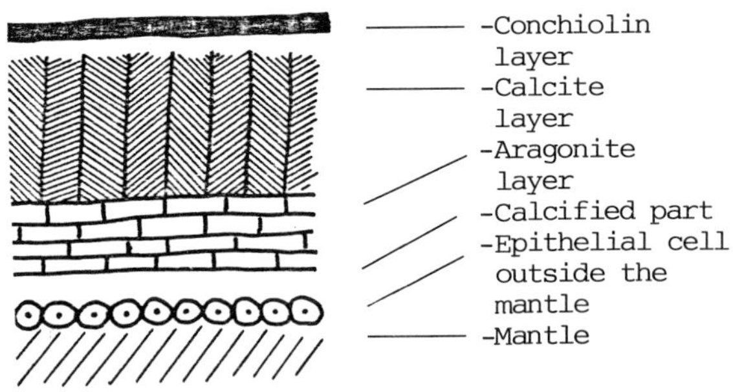

FIGURE 2
The structure of shell cross section. (After K. Ono[3])

sulphurated rubber. And the further analysis pointed out that the ligament is a kind of composite with protein matrix and short fibre reinforcement.

3.3. "The hard tissue of bio-bodies" by Tetsuya Tateishi[4]

The mechanical properties of dense

bone had been discussed especially in the aspect of its anisotropy, as shown in Figure 3. The anisotropy of elasticity, strength and toughness are similar.

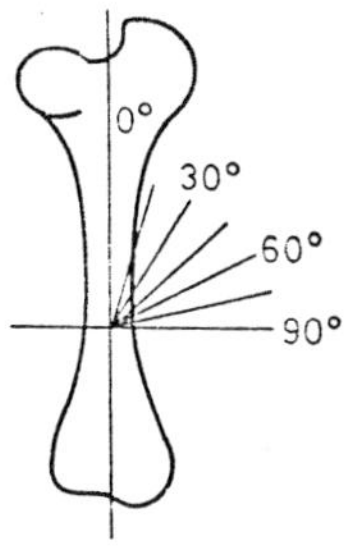

(a) Sampling position and direction of cow leg bone

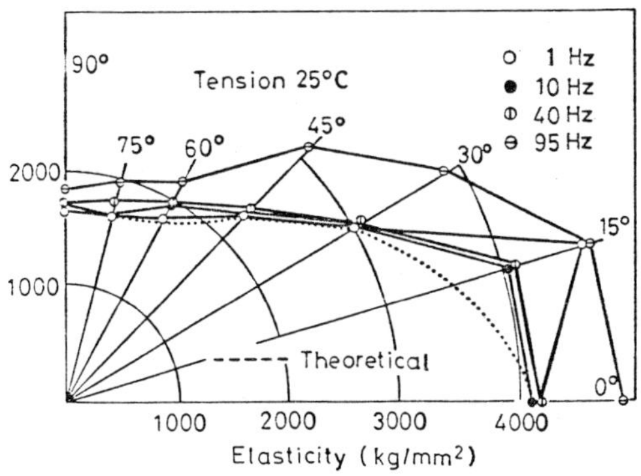

(b) Tensile elasticity

FIGURE 3
The directional dependence of the dynamic tensile elasticity of cow leg bone. (After T. Tateishi[4])

Another aspect discussed is the characteristics of spongy bone with visco-elastic behavior.

3.4. "Engineering application of skin Structure" by Shigeru Sato[5]

Human skin is a kind of substance with visco-elastic properties, six models have been suggested to discuss it, i.e. (a) Kelvin solid state model, (b) Maxwell fluid state model, (c) Three-element model, (d) Four-element model, (e) Brody model and (f) Gunner model as shown in Figure 4. The characteristics of skin can be taken as

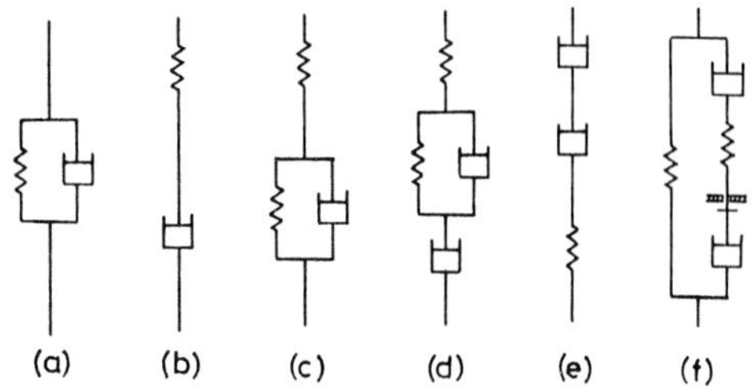

FIGURE 4
Different models of skin. (After S. Sato[5]) (a) Kelvin solid state model, (b) Maxwell fluid state model, (c) Three-element model, (d) Four-element model, (e) Brody model and (f) Gunner model

a reference for touching sensor studies.

After reading these papers and the similar one of Kozo Ikegami[6], it can be seen that all the authors are devoted themselves to the analyses of composite structures and properties of bio-materials, but the realistic bionic design has not yet been put into practice. This will be the main purpose which we are going to discuss in this paper.

4. THE BIONIC DESIGN OF COMPOSITE MATERIALS IN LIGHT OF REGULARITIES OF THE NATURE

Metallic and hybrid matrices composite materials are taken to be studied for the first step, and the morphology, structure, mechanism of injuries of bamboo, wood and bone cells are chosen to be simulated. Some meaningful initial results have been obtained and will be reported in other papers[7,8]. Certain part of them are described briefly as follows:

4.1 The Morphological Bionic Simulation of Reinforcement

It is known from botany and physiology of trees that almost all the fibre cells and stone cells are hollow, with multilayer and sometimes branch structured. In addition to meeting the need for nutriment transfer in the living body, many advantages are also connected with the hollowed structure than that of the solid one, if it is considered from material mechanics. The hollow one has better toughness and bending resistance. For example, a hollow fibre of 5μ outer and 3μ inner diameter has the same cross-section area as that of a solid fibre of 4μ diameter(Figure 5), but

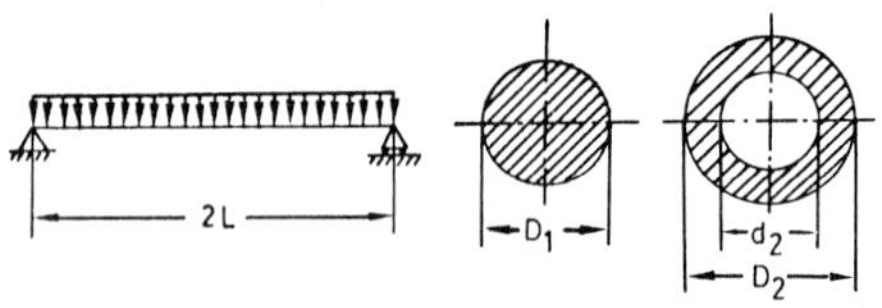

FIGURE 5
The cross section diagrams of hollow fibre and solid fibre of the same area

the bending strength of the former is 41% higher than that of the latter. It can be calculated from Equation (4) in the following derivations.

$$\sigma_{max} = M/Wz \quad (1)$$

$$Wz(D_1) = \pi D_1^3/32 \quad (2)$$

$$Wz(D_2,d_2) = (1-\alpha^4)*\pi D_2^3/32, \quad (\alpha = d_2/D_2) \quad (3)$$

where σ_{max}: maximum bending stress, M: moment, Wz: cross section modulus, D_1: diameter of the solid fibre, D_2,d_2 outer and inner diameters of the hollow fibre, and the relative difference $\Delta\sigma$ of maximum bending stress between that of the solid and hollow fibres is

$$\Delta\sigma = (\sigma_1 - \sigma_2)/\sigma_1 \quad (4)$$

Considering these analyses, hollow graphite fibres have been worked out in our laboratory by CVD method. The strength and flexibility of them are better than that of the solid carbon fibres. And the multilayer and branching techniques are also studied.

4.2 The Structural Bionic Simulation of Composite Processing[7]

As a first step of structural simulation, the bionic optimum design of bamboo simulating composite has been studied in our laboratory. The cross-section illustration of bamboo stem can

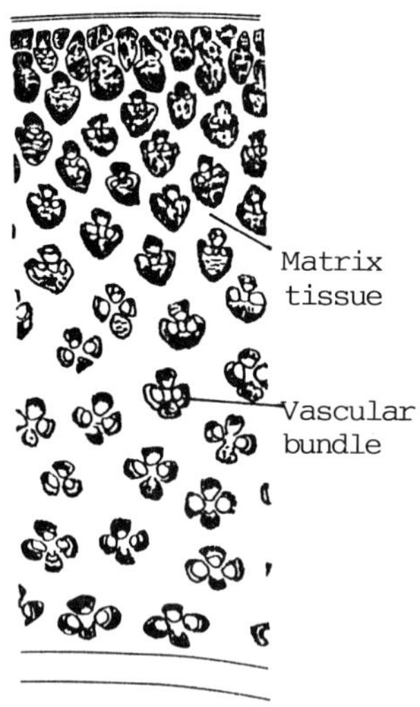

FIGURE 6
Cross section of bamboo stem

be found in any book of botany[9] as shown in Figure 6. The distribution of reinforcement -- vascular bundle is in-

TABLE 1
Results of Experimental Verification with Carbon/Epoxy Composite Materials

	Flexural Strength (MPa)					
Bionic Beam	214.3	359.8	258.7	240.6	217.1	345.4
Uniform Beam	133.2	188.5	130.8			
Increment (%)	60.9	90.9	102.9			
Aver. Incr. (%)	80.7					

homogeneous, it is dense in the outer layer(the so-called green part) and becomes dispersed gradually in the bulk (middle part), and changes to another kind of fine structure in the inner layer(yellow part). In-depth analysis and measurement have been carried out in our laboratory, the typical curve of

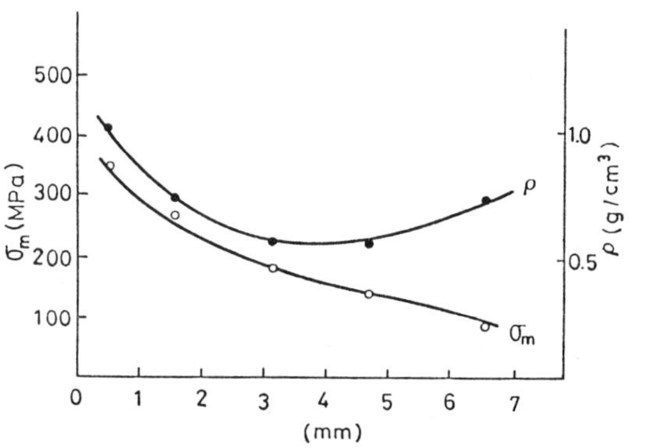

FIGURE 7
Dependence of strength and density of bamboo on the distance from the outer surface

strength and density is shown in Figure 7. The mathematical-mechanical analysis shows that this distribution forms a kind of very reasonable optimum structure. According to this structure, an optimum model of carbon fibre reinforced resin has been suggested. The experimental results show that the bending strength of composite with bamboo simulating structure is 81% higher in average than that of the homogeneous distributed structure of composite with the same amount of matrix and reinforcement. The highest one of the former is 103% higher than that of the latter as listed in Table 1.

4.3 The Bionic Healing of Inner Damage in Composite Materials[8]

Certain substance will be secreted in a plant or animal body through internal secretion system to cause filling, healing or local regenerating when the body is injured. This process gives us a hint to explore the way of healing inner damage in composite materials. From the view point of dissipation structure of physics, the essence of healing is the exchanging process of substance and energy of an open system with environment. This can be simulated in our case. Tentatively, aluminum sample was used, and it will be described in detail elsewhere[8]. An inner flaw of 0.1 - 0.4 mm was prepared by wire cutting and boundary brazing in an aluminum sample

FIGURE 8
Sample with a cutting seam (left);
Sample with an inner flaw (right)

(20x15mm) as shown in Figure 8. The spatial position of flaw was determined by CT technique and X-ray non-destructive testing. CT images along both axial and radial directions of the flawed sample are shown in Figure 9. After

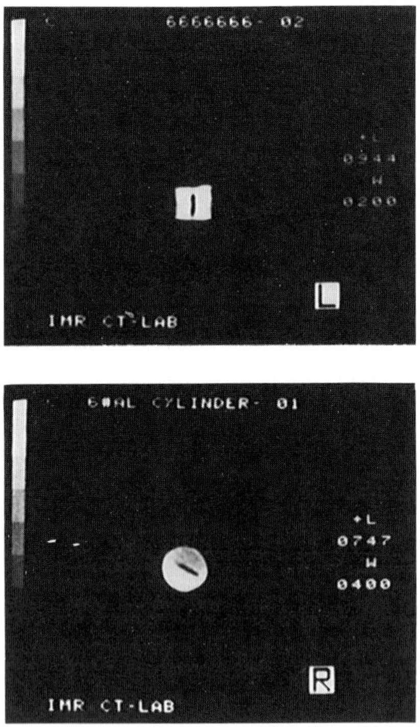

FIGURE 9
CT image along axial and radial directions

that, healing filler and flux were put into the flaw through a small hole drilled toward it. Brazing process was then conducted. X-ray radiographs of the partly and completely healed inner flaw are shown in Figure 10(a) and (b) respectively. Metallography studies showed that the healing of inner flaw was complete.

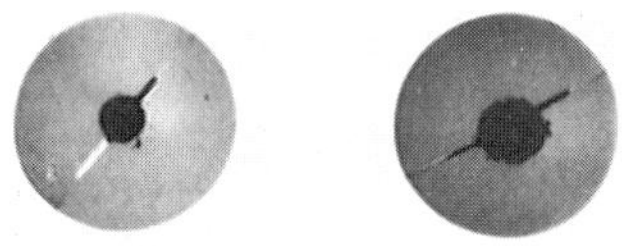

(a) (b)

FIGURE 10
Partly (a) and completely (b) healed inner flaw

The second step of healing studies is the healing of unbounded defects between carbon fibres and aluminum matrix by active surface treatment, and the details of it can be found in another paper[8]. The idea of this kind of inner surface healing comes from bionic simulation of biological functions, and it has seldom been found in literatures. The non-destructive healing technique of inner damage in composite materials is going to be studies in the future.

Sum-up all the points discussed, it can be seen that, what we do is to make the bionic simulation be alike in spirit but not only in appearance, i.e. the simulation of biological regularities. The methodology of it can be summarized as follows:

1. To observe phenomena, receive ideas and get inspirations.

2. To explore regularities, suggest models and calculate them quantitatively.

3. To design the composite structure, improve the properties of materials and make new suggestions.

4. To study further bionic simulation and look for a better solution to the problem.

5. CONCLUDING REMARKS

Owing to the fact that most of bio-materials are found to be in composite structures, thus the bionic simulation of composite materials will be one of the most beneficial among various kinds of materials. This simulation cannot be studied either in the traditional way of biology or only from the view point of engineering science, it must tentatively be studied in a interdisciplinary new way.

REFERENCES

1. Editorial Board, Comment on "Bionic Design", J.Japan Soc. Comp. Mater. 13, 6 (1987) 251.
2. A. Sakakibara, J.Japan Soc. Comp. Mater. 13, 6 (1987) 252.
3. K. One, ibid. 14, 1 (1988) 14.
4. T. Tateishi, ibid, 14, 2 (1988) 49.
5. S. Sato, ibid, 14, 3 (1988) 91.
6. K. Ikegami, ibid. 11, 3 (1985) 108.
7. S.H. Li, Q.Y. Zeng, F.T. Zhang and B.L. Zhou, Symposium B of C-MRS International'90, B3-4 (1990).
8, Y. Guo, B.L. Zhou, Z.J. He, G.H. He, C.F. Xu and Z.Y. Si, Symposium B of C-MRS International'90, B3-3 (1990).
9. H.J. Cao et al, Botany (Publishing House of Chinese Forestry, Beijing, 1978) pp. 98.

Frontiers of Materials Research/
Electronic and Optical Materials
M. Kong, L. Huang (editors)

HYDROTHERMAL CERAMIC VERY FINE POWDERS - OVERVIEW

Shigeyuki SŌMIYA, Kiyoshi ASAGA and Tsuyoshi HAYASHI

The Nishi Tokyo University, Yatsuzawa, Uenohara, Yamanashi 409-01, Japan

ABSTRACT
Very fine powders obtained by hydrothermal techniques. This paper described the results of hydrothermal powders.

Characteristics of hydrothermal powders are:
(1) Very fine particles
(2) Narrow particle size distributions
(3) Little or no macroscopic agglomeration
(4) Homogeneous
(5) Controlled Composition
(6) An hydrons
(7) Well crystallized
(8) Reactive powders - Easy to Sintering
(9) Controlled microstructure
(10) Controlled mechanical properties
(11) Free flowing

1. INTRODUCTION

There are several important points to get advanced technical ceramics. One of the keys for advanced technical ceramics is powders. Requirements of the powders for advanced ceramics are:

1) Fine particle. Average size is less than 1 μm
2) Narrow particle size distribution
3) Little or no macroscopic agglomeration
4) Homogeneous
5) Controlled composition
6) Controlled microstructure
7) Controlled mechanical properties
8) Well crystallized, etc.

Recently, several papers related to hydrothermal powders were appeared in the Powders conferences. For example, two papers of Pb-Zirconate and ZrO_2 were August 3-6, 1986, Boston, USA conference,[1] two papers of ZrO_2 and $PbTiO_3$ were present in November 1-4, 1987, Orlando, Fl., USA conference,[2] four papers of Al_2O_3-TiO_2, ZrO_2, $BaFe_6O_{12}$, and $Bi_{12}GeO_2$ were appeared in October 12-14, 1988 at Berchtesgarden, Germany conference.[3]

Two to four papers related to hydrothermal synthesis for ceramic powders were presented in these conference. And also several hydrothermal papers were appeared in the ceramic journals and publications such as J. Americal Ceramic Society or Bulletin, Zirconia Conferences of I, II, and III,[4,5,6] the first, second, and third Hydrothermal Reactions Symposia, etc.[7]

Hydrothermal process for very fine powders is getting important position in the field of ceramics. Because hydrothermal powders have unique characteristics as follows:

1) Very fine particle less than 1μm
2) Narrow particle size distribution

3) Little or no agglomeration
4) Anhydrous
5) Controlled composition
6) Homogeneous
7) Single crystals to amorphous. It depends on hydrothermal temperature.
   Single crystals have a high degree of crystallinity.
8) Reactive powder in many cases
9) Free flowing
10) Controlled morphology
11) Well-distributed dopant.

Therefore hydrothermal powders is one of the methods to achieve the ideal powders for ceramic use.

What is hydrothermal synthesis? According to R. Roy,[8] the term "hydrothermal synthesis" involves H_2O as a catalyst and occasionally as a component of the solid phase in the synthesis at elevated (>100℃) temperature and pressure (>a few atmospheres).

There are several ways to produce the very fine powders by hydrothermal process. These are the followings:

1) Hydrothermal oxidation
2) Hydrothermal precipitation (Hydrothermal hydrolysis)
3) Hydrothermal crystallization
4) Hydrothermal synthesis
5) Hydrothermal decomposition
6) Hydrothermal dehydration
7) Hydrothermal anodic oxidation
8) Hydrothermal mechano-chemical reactions
9) Reactive electrode submerged arc. (RESA)

Summary papers appeared by S.Sōmiya,[9] L.M. Sheppard[10] and W.J. Dawson[11] for hydrothermal powders and two books for hydrothermal reactions were published.[7,12]

As for hydrothermal processing for powder production, we need autoclaves. It, however, is less polluting, less energy cost, and can use a wide variety of raw materials in comparison with conventional processing.

At present time in Japan, hydrothermal powders are produced several companies for ceramics especially, electronic ceramics, abrasive, etc. R&D in companies are also shown in Table 1.

Table 1. Commercial Development of Japanese Hydrothermal Powders in Japan (Including R&D in Companies)

Name of Companies	Products	Scale	Publication Year
Chichibu Cement Co.	Zirconia	5 ton/month since 1986	
Sakai Kagaku K.K.	Perovskite	10 ton/month since 1987	1986
Showa Denko K.K.	Alumina (Single Crystal)	1 ton/month ~1975~	
Osaka Titanium Manufacturing Co.	Perovskite		1989
Dowa Kogyo K.K.	Ba-Ferrite		
Kunimine Industries	Saponite(Si8-a,Ala) (Mg6-b,Alb)Naa-bO20(OH)4		1985
N.K.K.	Zirconia		1987
Mizusawa Industrial Chemicals Ltd.	$PbTiO_3$		1983
Sony Corp.	$PbTiO_3$		1988
Alumina powders Showa Denko K.K. Nikkei K.K. Mitsui Alumi K.K.		100 Mton/year	
Battlie Columbus		After W.J. Dawson	
Cabot Corp.	Dielectric Ceramic Powders	After W.J. Dawson	
Ferro Corp.	Dielectric Ceramic Powders	After W.J. Dawson	
Degussa Co., Ltd.	ZrO_2	40 ton/year	
	TiO_2	40 ton/year	
	Zeolite Autoclaved Light Weight Concrete		

2. POWDERS

2.1. Al_2O_3

It is well known as the Bayer Process. It is shown in Figure 1. Al_2O_3 is a very popular material and it is not necessary to explain in detail.[13] Single corundum crystal was fine powder was produced by Showa Denko KK.[14]

Recently S. Komarneni et al[15] reported Al_2O_3 from AlOOH gel at 500℃ or 600℃ under 100 MPa for to 24h.

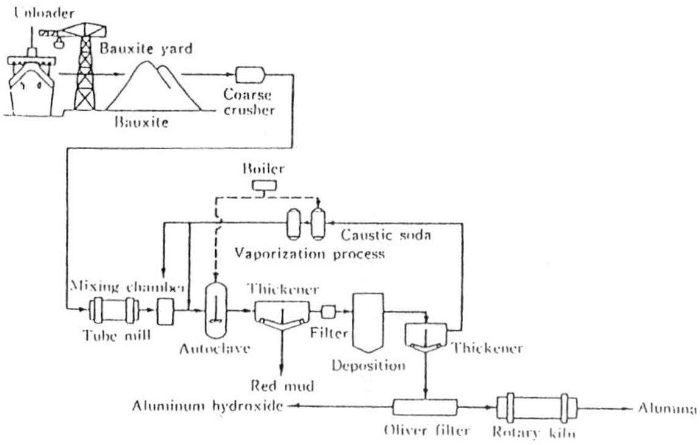

FIGURE 1
Flowchart of the Bayer Process (After Nagai and Kimura)

2.2. Iron Oxide

Iron oxide is produced from ilmenite.[16] Ilmenite is decomposed completely in 10M KOH Solutions 25-35MPa at 500℃. Lattice parameter of magnetite was a=0.8467nm. The Curie point of magnetite formed at 800℃ under 30MPa for 24h in 10M KOH Solution was 573℃.

The hydrothermal oxidation takes place as follows:[17]

$$3Fe + 4H_2O \rightarrow Fe_3O_4 + 4H_2$$

Metallic iron and 10M NaOH solution were used for preparation of magnetite as low as 250 under 980MPa. Well-crystallized magnetite was obtained from 350℃ to 650℃ under 98MPa. As for Wüstite above 650℃ was necessary under 98MPa.

2.3. ZrO_2 and Y-ZrO_2

Flow sheet is given in Figure 2.

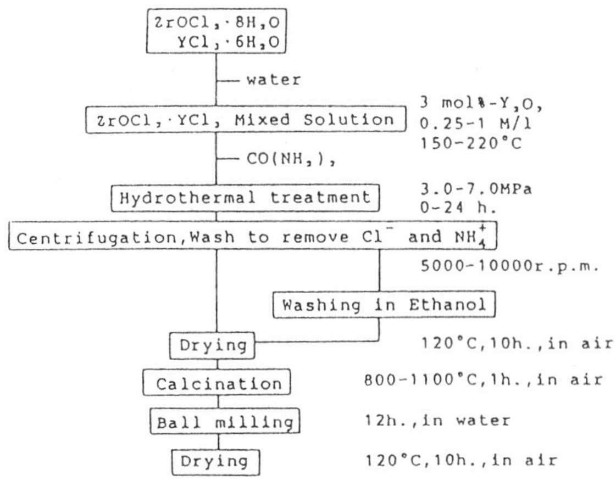

FIGURE 2
Processing flow sheet of hydrothermal homogeneous precipitation method

$ZrOCl_2.8H_2O$ and $YCl_3.6H_2O$ were used as starting materials. KF, NaOH, LiCl, KBr, etc. were used as mineralizer. Crystallite size, hydrothermally crystallized products, etc. were different by operating hydrothermal conditions, especially temperature and mineralizer and concentration of solution.

Sōmiya et al,[18-24] Matsuoka et al,[25] Stamburgh et al,[26-27] Reynen et al,[28] Quon et al,[29] Daimon et al,[30] K. Habenko and W. Pyda,[31] Gunnarsson et al,[32] Adair et al,[32] Komarneni et al,[34] Kriechbaum et al,[35] etc. were published their results.

As for starting materials for Zirconia, $ZrOCl_2.8H_2O$, $ZrCl_4$, $ZrSiO_4$, etc. were used. Processing for preparation of ZrO_2 powder is shown in Figure 2. Products of ZrO_2 were different from starting materials, kind of mineralizers, temperatures, concentration of mineralizers. Effects of mineralizers on products and crystallite size are shown in Table 2.

Mineralizer	Temp (°C)	Average crystallite size (nm)	
		Tetragonal ZrO_2	Monoclinic ZrO_2
KF (8 wt%)	200	Not detected	16 nm
KF (8 wt%)	300	Not detected	20 nm
NaOH (30 wt%)	300	Not detected	40 nm
H_2O	300	15 nm	17 nm
LiCl (15 wt%)	300	15 nm	19 nm
KBr (10 wt%)	300	13 nm	15 nm

Table 2. Phases present and Crystallite Size of Hydrothermally Crystallized Products at 100MPa for 24h.

2.4. Ferrite

S. Komarneni et at[15] reported ultrafine and nearly spherical ferrites such as $NiFe_2O_4$, $ZnFe_2O_4$, $Ni_xZn_{1-x}Fe_2O_4$, $Mn_{0.5}Zn_{0.5}Fe_2O_4$, and $CoFe_2O_4$. These ferrites were prepared under mild hydrothermal conditions by precipitating from metal nitrates with agueous ammonia pH was 10. Temperatures were 135℃, 175℃,or 200℃ for 4 to 20h under saturated steam pressure. Solids were washed with 95% and 100% ethanal. Particle size were 10 to 20nm size and particle was spherical. Sintering temperature was 1000℃.

Dowa Mining Co. was produced Ba-ferrite under hydrothermal condition above 150℃, products were less than 100nm and plate like crystals.[36]

In 1987, Sr-ferrite was published by Lee et al from $SrCl_2.6H_2O$ and $FeCl_3.6H_2O$ at 150-300℃ up to for 48h.[37]

Mechano-chemical reactions are also effective on synthesis of Ba-Ferrite.[38]

2.5. $PbTiO_3$, Perovskite

S. Kaneko and F. Imoto[39] reported $PbTiO_3$ in 1978 from hydrous titanium dioxide and lead monoxide under 49MPa from 200℃ to 500℃ for 5h. Takai et al[40] published their results in 1983. Starting materials were hydrate titanate from precipiting from hot aqueous solutions (60-90℃)of $TiO(NO_3)_2$ with NH_4OH and PbO. Hydrothermal temperatures were from 100-150°C for 3h. Particle size were from 0.2 to 0.4m

Recently many papers were appeared by Lee et al,[41] K. Abe,[42] Watson et al,[43] Suzuki et al,[44] Yamamoto et al,[45] Adair et al,[46] Ichihara et al,[47] etc.

A new hydrothermal processing was appeared by Yoo et al.[48] This process is the hydrothermal anodic oxidation. $BaTiO_3$ was prepared from Ti metal in $Ba(NO_3)_2$ solution with electrochemical reactions. $BaTiO_3$ was produced as low as 250℃.

2.6. Mullite

Synthesis of mullite was studied at Roy's[34] and Sōmiya's laboratory.[49] Gel preparing from Al-iso-propoxde and Si-tetra ethoxide at 300℃ under 50 MPa for 2h and/or hydrolysis at room temperature and then solid was washed by water in order to remove organic materials including carboneous materials. Hydrothermal crystallization of these gel was operated at 600℃, 20MPa for 24h. Mullite peaks by X-ray diffractions were appeared after 2h. Crystallite size by X-ray was 65nm for 2h, 67nm for 24h and 54nm for 72h respectively.

Komarneni et al[34] was used single phase and diphase gel for starting materials preparing by tetraethoxysilane and $Al(NO_3)_3$ in alkohole solution for single-phase and boehmite with ludox for diphase. They did not synthesized mullite. Most stable phase is hydralsite.

2.7. Zircon[34,50,51]

Single phase and diphase ZrO_2-SiO_2 gel were used by Komarneni et al. Zircon was prepared at 450-600℃ under 100MPa for 4 to 12h. Starting materials are single and diphase gel from $ZrO.Cl_2$ and tetraethoxysilane in ethanol.[34,50,51] H. Kido and S. Komarneni[51] were used tetraethoxysilane, tetramethoxysilane, $ZrOCl_2.8H_2O$ were used as starting materials, heated 100-200℃ in a Parr bomb for 3-48h.

2.8. Al_2O_3-TiO_2[52]

G.W. Kriechbaum et al[52] used $TiCl_4$, $Al_2(SO_4)_3 \cdot 16H_2O$ as starting materials for Al_2O_3-TiO_2 powders. They prepared the powders by coprecipitation and hydrothermal aging of the resulting hydroxide gels. Hydrothermal temperatures were 300℃ and 650℃. Crystallite size were from 0.1 to 1μm.

2.9. TiO_2

S. Komarneni et al[34] and Degussa[53] reported TiO_2 crysstals by hydrothermal process. The former used TiO_2 gel. The latter produced 40 ton/year.

2.10. Appatite ($Ca_{10}(PO_4)_6(OH)_2$)[54]

Appatite is one of the important materials as bioceramics. Starting materials were gel preparing from $Ca(NO_3)_2$ and $(NH_4)_2HPO_4$ with pH_{10} NH_4OH solution. Gel was crystallized under hydrothermal condition at 200℃ under 2MPa for 10h. Product was $Ca_{10}(PO_4)_6(OH)_{1.81}CO_3)_{0.05}$. Size was about 25nm x 90nm.

3. SSG DEFINITIONS (R. Roy)

3.1. Single phase gel refers to the single solid phase in the gel which consists then of one finely divided mechanically plastic solid phase with water (or other liquid) in the interstices.

3.2. A di-phasic gel is a material which consists of two chemically or structurally distinct phases (in the Gibbsonian sense), each in the colloidal size range, which form a plastic solid by including some liquid in interstices between the solid particles.

3.3. Xerogels are dried or dehydrated gels, i.e. with the water (or liquid) removed from a diphasic gel. Aerogels are a special subset of xerogels where by using critical point drying the microstructure of the original gel is preserved.

3.4. Di-phasic xerogel is a dried di-phasic gel.

SUMMARY

This paper describes recent activities for hydrothermal powders.

REFERENCES

1. Ceramic Powder Science, in: Advances in Ceramics, Vol. 21, eds. G.L. Messing, K.S. Mazdiyasmi, J.W. McCauley, and R.A. Huber (Am. Ceram. Soc., 1987) p. 825.
2. Ceramic Powder Science, in: Ceramic Transactions, Vol. 1, eds. G.L. Messing, E.R. Fuller, Jr., and H. Hausner (Am. Ceram. Soc., 1988) p. 1107.
3. Ceramic Powder Processing Science, eds. H. Hausner, G.L. Messing, and S. Hirano (Deutsche Keramische Gessellschaft, 1989) p. 960.
4. Zirconia I, in: Advances in Ceramics, Vol. 3, eds. A. Heuer and J.W. Hobbs (Am. Ceram. Soc., 1981) p. 479.
5. Zirconia II, in: Advances in Ceramics, Vol. 12, eds. N. Claussen, M. Rühle, and A.H. Heuer (Am. Ceram. Soc., 1984) p. 842.
6. Zirconia III, in: Advances in Ceramics, Vol. 24, eds. S. Somiya (Am. Ceram. Soc., 1987) p. 1148.
7. Proceedings of the 1st International Symposium on Hydrothermal Reactions, eds. S. Somiya (Gakujuten Bunken Fukyu-kai, 1983) p. 965.
8. R. Roy, personal communication.
9. S. Sōmiya, Hydrothermal Preparation of Fine Powders, in: Advanced Ceramics III (in preparation) ed. S.

Sōmiya (Elsevier Applied Science Publishers, 1990)
10. L.M. Sheppared, Metal Progress Nov. (1986) 47.
11. W.J. Dawson, Am. Ceram. Soc. Bull. 67 (1988) 1673.
12. Hydrothermal Reactions for Materials Science and Engineering: An Overview of Research in Japan, ed. S. Sōmiya (Elsevier Applied Science Publishers, 1990), p. 505.
13. A. Nagai and Y. Kimura, Synthetic Raw Materials for Ceramics, in: Advanced Technical Ceramics, ed. S. Sōmiya (Academic Press, 1989) pp. 27-64.
14. N. Oishi et al, Funtai Kogaku Kaishi 18 (1981) 687.
15. S. Komarneni, E. Fregeau, E. Breval, and R. Roy, J. Am. Ceram. Soc. 71 (1988) C26.
16. MGMD Ismail et al, J. Mat. Sci. Letter 4 (1985) 19.
17. S. Hirano and S. Sōmiya, J. Cryst. Growth 35 (1976) 132.
18. E. Tani et al, J. Am. Ceram. Soc. 64 (1981) 209.
19. E. Tani et al, J. Am. Ceram. Soc. 66 (1983) 216.
20. M. Yoshimura and S. Sōmiya, Rept. R.L.E.M., Tokyo Inst. Tech. 9 (1984) 53.
21. S. Sōmiya et al, Z. Anong Chem. 540/541 (1986) 251.
22. S. Sōmiya et al, Ceramic Powder Science, in: Advances in Ceramics, Vol. 21 (1987) 43.
23. S. Sōmiya et al, Microstructure Development of Hydrothermal Powders and Ceramics, in: Microstructure '86, eds. J.A. Pask and A.G. Evans (Plenum Press, 1987) pp. 465-474.
24. K. Hishinuma et al, Zirconia III, in: Advances in Ceramics, Vol. 24 (1988) p. 201.
25. H. Nishizawa et al, J. Am. Ceram. Soc. 65 (1982) 210.
26. E.P. Stambaugh and J.F. Miller, in: Proceedings of the 1st International Symposium on Hydrothermal Reactions, ed. S. Sōmiya (Gakujiten Bunken Fukyu-kai, 1983) p. 859.
27. E.P. Stambaugh et al, US Patent 4619817, (Oct. 28, 1986).
28. P. Reynen et al, Zirconia I (1981) 464.
29. D.H.H. Quon et al, Interceram 38 (1989) 15.
30. K. Daimon et al, J. Ceram. Soc. Japan 98 (1990) 300.
31. K. Habenko and W. Pyda, in: Zirconia II, eds. N. Claussen, M. Rühle, A.H. Heuer (1986) p. 1774.
32. G. Gunnarson et al, Ceramic Powder Processing Science (1989) 197.
33. J.H. Adair et al, Ceramic Trans., Vol. 1 (1988) 135.
34. S. Komarneni et al, Advanced Ceramic Materials 1 (1986) 87.
35. G.W. Kriechbaum et al, Ceramic Trans., Vol. 1 (Am. Ceram. Soc., 1988) p. 145.
36. Dowa Mining Co., personal communication.
37. K.H. Lee, B.H. Lee, and S.K. Kim, J. Korean Ceram. Soc. 24 (1987) 17.
38. M. Yoshimura et al, J. Ceram. Soc. Japan 97 (1989) 16.
39. S. Kaneko and F. Imoto, Bull. Chem. Soc. Japan 51 (1978) 1739.
40. K. Takai, S. Shogi, H. Naito, and A. Sawaoka, in: Proc. 1st International Symposium on Hydrothermal Reactions (1983) p. 877.
41. K.H. Lee, B.H. Lee, M. Daimon, K. Asaka, W.H. Kang, and H.S. Park, J. Korean Ceram. Soc. 24 (1987) 397.
42. K. Abe et al, Japanese Patents 61-31345, 63-144115, and 63-144116.
43. D.J. Watson, C.A. Randall, R.E. Newnham, and J.H. Adair, Hydrothermal Formation Diagram in the Lead Titanate, in: Ceramic Trans., Vol. 1 (1988) pp. 154-164.
44. M. Suzuki, S. Uedaira, H. Masuya, and H. Tamura, Hydrothermal Synthesis of Lead Titanate Fine Powders, in: Ceramic Trans., Vol. 1 (1988) pp. 163-170.
45. T. Yamamoto, R. Tanaka, K. Okazaki, and T. Ueyama, Japanese J. Appl. Physics 28 (1989) 81.
46. J.H. Adair, T.S. Shrout, and K. Osseo-Asare, Hydrothermal Synthesis of Perovskite Ceramic Powder, in: Proc. MRS International Meeting on Advanced Materials (1989) pp. 141-155.
47. T. Ichihara, T. Tsurumi, K. Asaga, K.H. Lee, and M. Daimon, J. Ceram. Soc. Japan 98 (1990) 155.
48. S.E. Yoo et al, J. Mat. Sci. Letters 8 (1989) 530.
49. S. Sōmiya et al, in: Ceramic Transactions, Vol. 6 (1990) 287.
50. S. Komarneni and R. Roy, Synthesis of Zircon, in: Zircon -- Science and Technology, ed. S. Sōmiya (Uchida Rokakuho Publishing Co., 1989) pp. 289-298.
51. H. Kido and S. Komarneni, in: AMSES Transactions, Vol. 2 (1990, in preparation).
52. G.W. Kriechbaum et al, Hydrothermal Synthesis of Aluminum/Titanium oxide Powders, in: Ceramic Powder Processing Science, eds. H. Hausner, G.L. Messing, and S. Hirano.

53. Y. Matsunaga, personal communication.
54. S. Sōmiya et al, Materials Forum 34-36 (1988) 373.

Frontiers of Materials Research/
Electronic and Optical Materials
M. Kong, L. Huang (editors)

SOME ADVANCES IN X-RAY ANALYSIS OF THIN FILMS

Jinsheng PAN and Kun TAO

Department of Materials Science and Engineering, Tsinghua University, Beijing 100084, CHINA

Abstract

Two modified methods are presented for the analysis of thin films by X-ray diffraction. One is a modified Laue Pattern analysis method used for phase identification and orientation determination of single crystals which have several structure modification and give only 'poor' Laue pattern in the sense that only few or no prominent Laue spots or zones can be discerned from the pattern. An example is given of the analysis of the poor Laue pattern of molybdenum oxide single crystal thin film which has as many as seven structure modifications. The second method is a Seemann-Bohlin diffractometry developed for polycrystalline thin layers analysis . It has been used successfully to characterize some metal and silicide thin films 30~120 Å in thickness by X-ray of 1 KW power in 15 minutes only. The sensitivity of this method is of the same order as that of the parallel beam method published in literature. The developed system is characterized by the ease of interchange between the Seemann-Bohlin method and the conventional Bragg-Brentano method.

I. INTRODUCTION

With the development of materials science thin films play an increasingly important role in modern science and industry and their characterization is of great interest to materials workers in various fields. Nowadays a series of methods and techniques of characterization have been developed, among which X-ray diffraction technique is one of the most powerful tool for the phase analysis of thin films. However, so far as the authors know, up to now no method, either manual or computerized, seems to be available for the phase identification of single crystals from their Laue patterns. For polycrystalline thin films with thickness of tens to thousand Angstrom, the characterization also presents some serious problems.

To overcome these difficulties, two modified X-ray diffraction methods are presented by the authors. One is used for phase identification of single crystals and is based on a Modified Laue Pattern Analysis Method (MLAM), the other is for polycrystalline thin films and based on a Seeman-Bohlin diffractometry. This paper briefly discusses the two methods together with some typical examples.

II. The Modified Laue Pattern Analysis Method

All the Traditional Laue Pattern Analysis Methods(TLAM) are used simply for indexing and orientation determination. Most of them are based on matching the angles between prominent Laue Spots and Zones in the pattern called the control points with those between planes with low Miller indices calculated from given lattice constants [1,2] . Therefore, in TLAM two prerequisites must be satisfied. One is uniquely given lattice constants, the other is enough number of low indices control points. For phase identification, however, the lattice constants can not be uniquely given in advance, and for thin film there may be few or no control points at all in the pattern. To solve these two problems, a Modified Laue Pattern Analysis Method (MLAM) is presented.

In contrast to TLAM, MLAM is based on rotating the reciprocal crystal axes with computer, which greatly reduces the amount of calculation and makes it possible to try different lattice constants corresponding to different structure modifications, and a new criterion for judging true solution, which eliminates the necessity of discerning enough control points. The original MLAM was published in 1982 [3]. In what follows only an outline and recent development of MLAM is given.

1. Principle

The Miller indices of a Laue spot should be proportional to the reciprocal coordinates of its gnomonic projection. To determine the coordinates, it is necessary to know the positions of both the gnomonic projection and the reciprocal axes $\vec{a}^*$, $\vec{b}^*$ and $\vec{c}^*$ of the crystal being studied. The former can be easily deduced from the position of Laue spot , while the latter, of course, can not be known in advance. To determine the real position of the reciprocal coordinate axes, a trial-and-error method is used. First, a set of reciprocal coordinate axes is established at some arbitrarily chosen initial position and with respect to this set of axes the initial reciprocal coordinates of the gnomonic projections are determined (see Fig. 1). Then, this set of coordinate axes is successively rotated with computer until it reaches the real position of the reciprocal axes of the crystal. Referred to the axes at the real position, the reciprocal coordinates of the gnomonic projections will give the true Miller indices, (h k l), of the corresponding Laue spots. Now the question is: How can one know whether the reciprocal axes have already rotated to real position? To answer this question, a proper criterion for judging true solution is needed.

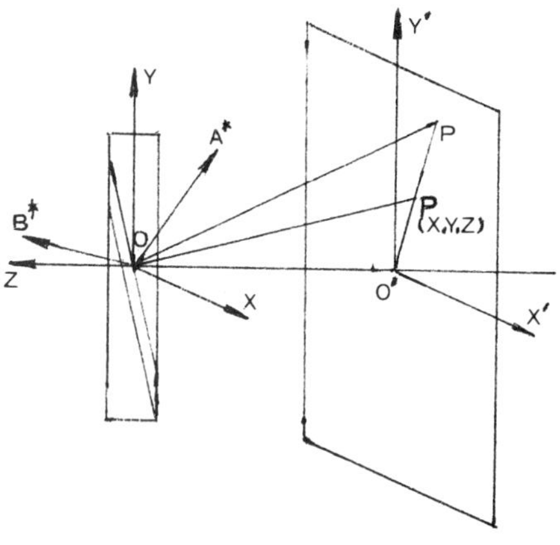

Fig. 1. The relation of Laue spot p and its gnomonic projection P

2. Criterion

Now let us return to the question of how to judge whether the reciprocal axes have rotated to real position. According to TLAM, at real position the control points should have low indices, In practice, however, this low indices criterion often failed to work due to the following experimental facts.

(1) The solution may not be unique. Experiments showed that the less the control points, the more the solutions.If few or even no control points at all, as in some case of thin films, the number of solutions may be infinite.

(2) There is no fixed line of demarcation between 'low' and 'high' indices. It may differ from crystal to crystal. For example, if X-ray beam impinges on a crystal with very large axial ratio, c/a, then most of the spots in the back-reflection Laue pattern will have high indices. This may be called as axial ratio effect.

(3) The Miller indices of spots and the allowable error, E, are interrelated to each other. It is nonsense to talk about one without specifying the other. This may be called as correlation effect.

(4) The same Laue pattern is often compatible to different structure modifications of the crystal, i. e. , for different lattice constants there are still some, though different solutions.

Based on these experimental findings the authors made the following propositions for working out a new criterion:

(1) To judge true solution, as many as possible Laue spots rather than several control points should be taken into consideration.

(2) Considering the axial effect, the apparent Miller indices of the spots, (h k l) , should be replaced by the weighted indices, (h_W k_W l_W), which are defined as:

$h_W=(L_Q/a)h$, $k_W=(L_Q/b)k$ and $l_W=(l_Q/c)l$

(3) Due to the correlation effect, neither the Miller indices nor the error alone can be used as the criterion.

To work out a new criterion for judging true solution, some new parameters need to be introduced on the basis of the above-mentioned propositions. One is the weighted average of the Miller indices, $\overline{M}_W$, defined as

$$\overline{M}_W=\sum_{I=1}^{N}(h_{W_I}+k_{W_I}+l_{W_I})/3N$$

where N is the total number of the spots to be indexed. The other parameter is the average error, E

$$\overline{E}=\sum_{I=1}^{N}E_I/N$$

where E_I is the mismatch in angle between planes (h_I k_I l_I) and (h_I' k_I' l_I'). h_I', k_I' and l_I' are the reciprocal coordinates of the gnomonic projection of spot I , which are proportional to h, k and l respectively, but not necessarily integers. The third parameter is the mean product, $\bar{P}$, which is defined as

$$\bar{P}=\bar{M}_W \times \bar{E}$$

Now the new criterion can be expressed in terms of $\bar{P}$,

$$\bar{P}=\bar{P}_Q$$

for true solution.

This means that for true solution the mean product should be minimum.

Two remarks should be emphasized on the new criterion. First, the true solution gives not only the true Miller indices and orientation but also the true lattice constants. This is the basis for phase identification of single crystal from its Laue Pattern. Second, the criterion imposes no restriction on the number of control points, but if one control point, e. g. the incident direction or the plane perpendicular to it, is available, then the true solution will be reached more quickly. Third, the criterion implies that the less control points one can preselect, the more Laue spots one should taken from the pattern. To test the new criterion, the double exposure method is used, in which two Laue photographs are taken for the same crystal in same oritation, one with white radiation, the other with monochromatic. The first one is used to search out all possible solutions while the second is used to check which solution is in accordance with Bragg equation. The crystals tested include: Ta (bcc), Al, Pt and ZrC (fcc), Sc and Quatz (hexagonal), $La_{2-x}Sr_x CuO_4$ (tetragonal) , $YBa_2 Cu_3 O_{7-x}$ and $La_2 CuO_4$ (orthorhombic) , Sapphire (rhombohedral) and ZrO_2 (monoclinic). Both transmission and back-reflection Laue patterns have been analyzed. All the results fairly confirmed the minimum mean product criterion.

3. Example

As an example, the phase identification of molybdenum oxide single crystal thin film is taken. The film is formed by heating pure molybdenum plate to 900℃ in oxygen. Its chemical formula is MoO_x with x varing from 2.75 to 3, corresponding to several different structures (see table 1) . The transmission Laue pattern is shown in Fig.2, from which only individual spots can be seen. Using MLAM the spots are indexed for each possible structures. The relevant results are shown in table 1. The results clearly show that the oxide is MoO_3 with orthor. structure and the surface of the film is (100).

Table 1 The results of MALM

formula	structure	cell const.	$\bar{P}$	ortn. .
$MoO_{2.8}$	tetr.	a=45.99,c=3.937	2.33	(001)
Mo_4O_{11}	orthor.	a=24.4,b=5.45, c=6.723	1.90	(100)
$Mo_{17}O_{47}$	orthor.	a=21.615,b=19.632 c=3.9515	2.10	(001)
Mo O_3	orthor.	a=3.962,b=13.858 c=3.697	0.96	(010)
Mo_4O_{11}	monocl.	a=24.54,b=5.439 c=6.701, β =94.28	2.33	(100)
Mo_9O_{26}	tricli.	a=8.145,b=11.89 c=19.66, α =95.47 β =90.39, γ =109.97	2.42	(001)
MoO_3	hex.	a=10.531,c=14.876	3.38	(0001)
	(cubic)		3.41	(001)

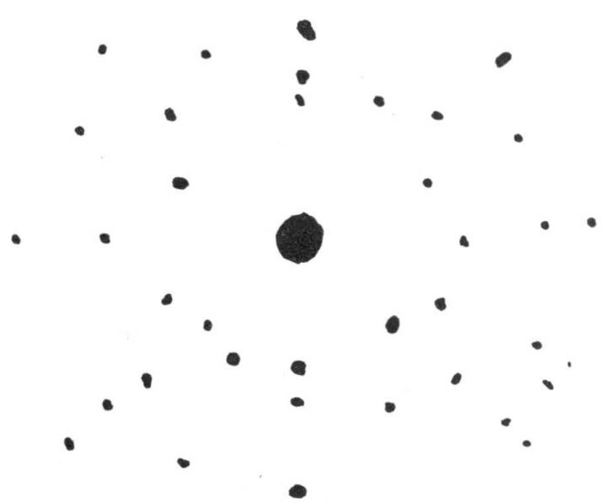

Fig.2. The Transmission Laue pattern

III. The Seemann-Bohlin method

The conventional diffractometer using the Bragg-Brentano parafocusing geometry (B-B method) is not suitable for the ultra-thin layers, because the incident X-ray path in thin layers is not long enough and the diffraction intensity is too weak. The fraction of the total diffraction intensity, from a surface layer of depth x, is given by $I_t/I_\infty = 1-\exp(-2\mu t/\sin\theta)$ for the conventional B-B Method. Thus, for highly absorbing Material such as tungsten with CuK$_\alpha$ radiation the intensity by 500 Å thin films is only 4.5% of that by thick tungsten sample. For Ni and Al films of same thickness, this fraction is 0.6 %and 0.2% respectively.

Investigators have, consequently, made much efforts in developing X-ray procedures to enhance the sensitivity of X-ray diffraction analysis for thin films.

1. Principle

Some procedures have been developed and improved in the past 10 years: Read camera[4], Guinier type method[5], parallel beam (P-B)[6,7]method and Seemann-Bohlin (S-B) diffractometer method[8,9]. The principle of all of these methods is the same: The incident angle, i.e, the angle between the incident X-ray and the sample surface is made smaller to extend the incident path of X-rays in the thin layers. It seems that the Read camera, the P-B method and the S-B method will be used widely in practice.

The Read camera is the simplest method and its cost is much lower than that of the others. However, both its accuracy and resolution are not high enough and it can not be used for quantitative measurement. It is used, therefore, for rough test only.

2. Method

To chatacterize the thin films, in recent years a fully automatic Seemann-Bohlin (S-B) diffractometer is developed in our laboratory. It is modified on the basis of Rigaku D/max-IIIA diffractometer.

In the S-B method the focusing condition remains unchanged over whole data range. For small incident angle, the incident X-ray path in thin films is longer than that of the conventional B-B method. At the same time this distance in the S-B method is practically smaller than that in both the conventional B-B method and P-B method. Therefore, the incident X-ray divergence is larger than that of the P-B method for samples of same size. The above-mentioned factors will greatly increase the diffraction intensity. As a result, the time needed for analizing a sample will be much shorter.

Another advantage of the S-B method is that it can be used to determine the residual stress without varing the incident angle[10], so the glancing incident condition remains unghanged. This is very important for thin films.

There are some disadvantages for the S-B method. The line profile of the S-B method is destorted by the vertical divergence effect in the low 2θ angle range. It is needed to develope a calculation method to correct it. The S-B diffractometer has mechanical limitation when going to very small glancing angle. If the incident angle is very small, the distance from focus of X-ray tube to sample must be very small. However this distance can not be smaller than the radius of the X-ray tube plus the space occupied by the Soller slit and X-ray shutter between them.

In our laboratory the S-B diffractometer has been modified to make the incident angle as small as possible while the focus of X-ray tube is situated on the focusing circle. It has the following features.

A. The incident angle is 10° and measuring range of 2θ is 20°-120° in the following conditions: there is no incident monochromator; there is incident soller slit, the radius of the focusing circle is 185 mm which is same as that of the B-B method.

B. The diffractometer can be used for both S-B and B-B method. It is easy to change from the B-B diffractometer to S-B diffractometer or vice versa in 1 hr (including alingnment).

C.The control system of the B-B diffractometer can be used to the S-B method without any change.

3. Example

Figure 3. A. shows the diffraction patterns recorded from Au films of thicknesses from 35 angstrom to 150 angstrom by the S-B method. Figure 1.B.shows the patterns of the same samples by B-B method.

The data collection time for each pattern shown in Fig. 3. A. and Fig. 3.B. is 20 minutes only in the following condition: 10x10 mm^2 sample size; 1 KW power of X-ray tube. The thicknesses of samples are determined by RBS method. The Au film has no preferred orientation and is suitable for use as standard sample.

A series of η-NiSi films of various thicknesses on Si with slight preferred orientation is checked by S-B method in the same way and in above-mentioned conditions. Four peaks can be observed for 150 Å η-NiSi film.

4. Comparison with the paralled beam method

The P-B method is the most widely investigated

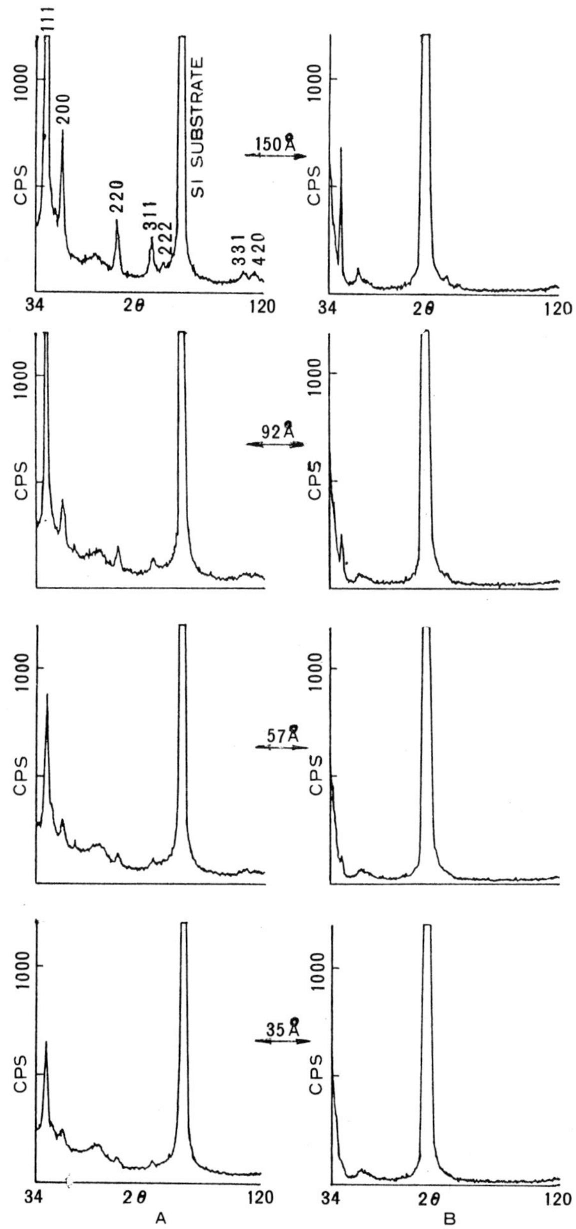

Figure 3: Diffraction patterns of Au films on Si substrate
A: Seemann-Bohlin method
B: conventional B-B method

method. It has the following advantages. First, it is easy to change the incident (glancing) angle from degrees to the total reflections angle (about 0.5°). This provides a possibility of making depth-profile analysis of the structure of surface layer[6,11]. Second, it's line profile distortion is very small.

However, the P-B mehtod has also some disadvantages, because it is, in principle, not a focusing method. When a very small glancing angle, say 1° or less,is used, the effective incident X-ray divergence and its intensity will decrease steeply because of the limited sample size. At larger incident angle, not less than 2° for example, the special parallel slit (of about 0.5° divergence) will limit the increase of the diffraction intensity. Therefore the diffraction intensity is not high enough. Larsen[5] indicated that 5 hrs or more is needed for collecting a pattern of large counts for films of 500 angstrom thickness or less. To shorten the collection time the position sensitive detector and synchrotron radiation[11] may be used but such device is not available for most laboratories.

With same samples and under same conditions, the performance of this S-B method is compared with that of P-B method (1° glancing angle) using " thin films attachment " of Rigaku's D/max diffractometer. The conditions for both S-B and P-B methods are: 1KW X-ray power, 20 minutes data collection time for every recorded time, 0. 6 mm receiving slit. For 42 Å Au film of 1x1 cm² area, 5 peaks (111,200,220,311,331)can be observed by both S-B and P-B methods. The intensity of S-B method is much higher than the P-B methods. for example, the net height of 111 peak (370 cps) is about 14 times that of the P-B method(25 cps). On the other hand, the peak/background ratio of P-B method is about 3 times that of the S-B method. for 10 Å and 20 Å Au film, in the above-mentioned measuring conditions, only the 111 peak can be observed by these two methods; the peak intensity of the S-B method is still much higher than that of the P-B method.

IV. Conclusion

Two modified methods are developed. One is the modified Laue pattern analysis method used for phase identification of single crystals with IBM-PC computer. It is based on the computerized rotating method and a new criterion for judging true solution, the minimum product criterion. The other method is the modified Seemann-Bohlin diffractometry used for polycrystalline thin layers analysis. While keeping the same sensitivity as that of the parallel beam method, the new method has the advantage of time-saving and the ease of interchange between the Seemann-Bohlin method and the conventional Bragg-Brentano method.

References

1. C.Anazia et al., Metall. Trans, 6A(1975) 1751.
2. R.A.Ploc, J.Appl. Crystallogr., 11(1978) 713.
3. J.S.Pan, Acta Metallurgica Sinica, 18(1982) 6,703
4. M.H.Read et al., Thin Solid Fil,s 10 (1972)123
5. R.A.Larsen et al., Advances in X-ray analysis, 32(1988)311
6. G. Lim et al., J. Mater. Res. 2(4)(1987)471
7. S.S. Iyengar et. al., Advances in X-ray analysis, 30(1986)457
8. R. Feder et al., J.Appl. Cryst, 3(1970)372
9. K.Tao, Acta Metallurgica Sinica, 23(1987)B254
10. E.L.Haase, Thin Solid films 124(1985)283
11. F.Pons et al., J. Appl. Cryst., 21(1988)197

Frontiers of Materials Research/
Electronic and Optical Materials
M. Kong, L. Huang (editors)

DEVELOPMENT OF ALUMINUM–LITHIUM ALLOYS

Li Yiyi, Hu Zhuangqi, Zhang Yun, Yao Daping, Jiang Xiaojun

Institute of Metal Research, Academia Sinica, 72 Wenhua Ave., Shenyang 110015, China

The extensive investigation of aluminum–lithium alloys in the last decade has been driven largely by numerous potential structural application in the aerospace industry requiring a low density metallic material having good strength–toughness combination and high stiffness. There has also been increasing interest in the cryogenic properties of these alloys.IMRAS puts great effort focussed on the frontier of development of new materials, and Al–Li alloy is one of them. A brief review of the recent results in the research of Al–Li alloys in IMRAS is compiled, including the role of δ' phase, the PFZ formation, the behaviors of alloying elements, the mechanical properties both at room and cryogenic temperatures, etc.

1. INTRODUCTION

Al–Li alloys are extremely attractive to designers due to their lower density and higher elastic modulus than conventional Al alloys with similar combinations of mechanical properties [1], that is why we have to put emphasis to study this new Al–Li alloy in order to meet the needs of the high tech and economic development.

Al–Li alloys are very attractive candidates of structural materials for aviation and aerospace applications. In early stage, the application was limited because of low ductility. Nowadays, great progress has been achieved on the research of improving their low ductility [2–5]. Recently, Al–Li alloys were also found to have novel properties even at cryogenic temperatures, i.e., the strength, the ductility and the fracture toughness are all improved with decreasing test temperature[6–17]. On the basis of the understanding of Al–Li alloys, a series of Al–Li alloys have been developed and selected to be used in the designing of new aeroplanes. The future prospect of Al–Li alloys is hopeful. Of course, more research works are necessary in order to know deeper and better.

2. ROLE OF δ' PHASE

The metastable $\delta'(Al_3Li)$ phase in the Al–Li system has recently attracted significant interest because of its role as a main hardening precipitate, which has a $L1_2$ type superlattice structure. The non–optimal deformation and fracture behavior of Al–Li alloys has been attributed in part to the strain localization that results from the planar slip of shearable δ' precipitates. Therefore it is necessary to know more about the δ' phase.

The hardness of Al–2.45wt%Li alloy depends on both aging temperature and time(Fig.1), which influence the δ' particle size and amount remarkably, that can be quantitatively expressed with a following equation[18].

$$\alpha = 1 - \exp(-kt^n) \qquad (1)$$

where α is volume fraction of δ' phase, t is aging time, k and n are the constants.

The δ' coarsening related to aging time is obeyed the Lifshitz–Slyozov–Wagner theory, showing that the average radius of δ' particles increase with one third root of time, as shown in Fig.2. From the relationship between the aging temperature and average radius of δ' particles. Following Bulmuth's equation[19], the activation energy for δ' coarsening is 133.7kJ / mol, 131.1kJ / mol, and 123.6kJ / mol for aging time of 5h, 8h and 25h respectively, and they are approximately the same as the activation energy for lithium atom diffusion

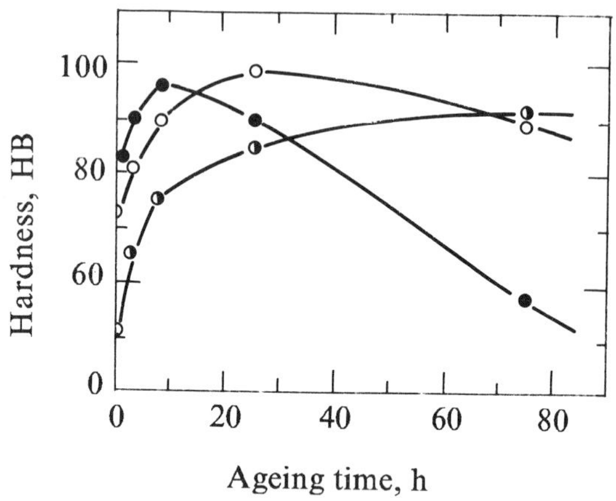

FIGURE 1
The hardness at different ageing temperatures for various times[18]
◑ 160℃, ○ 190℃, ● 220℃

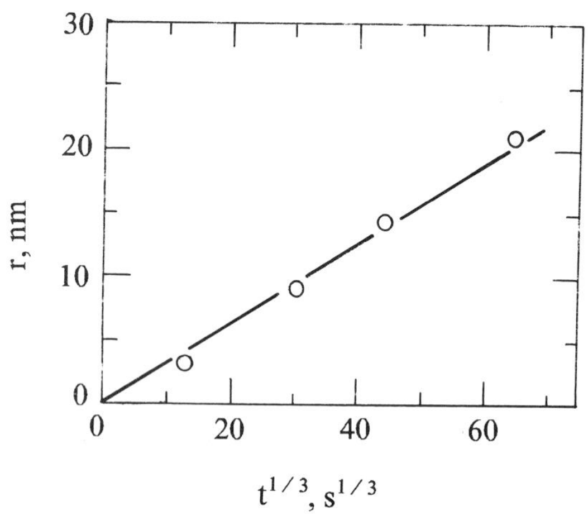

FIGURE 2
The average radius of δ′ particle vs. aging time at 190℃ [18]

as 126.1± 5.2kJ/mol at temperature between 423–513K[20]. It can be clearly concluded that the mechanism of δ′ coarsening is diffusion controlled. Al–2.29Li–3.65Zn–0.08Zr (wt%) alloy also obeys the same rule[21], as shown in Fig.3.

The δ′ precipitation is also influenced by the quenching rate and cold deformation, because both of them change the vacancy concentration and dislocation density. The results prove that the higher

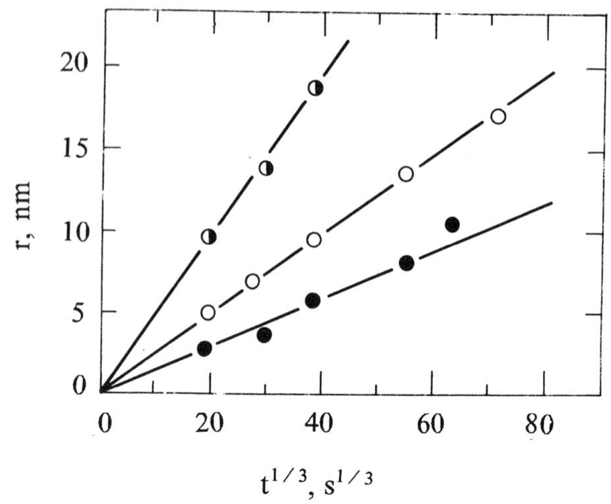

FIGURE 3
Relation between the average radius of δ′ particles and aging time(Al–2.29Li–3.65Zn–0.08Zr alloy)[21]
● 433K, ○ 463K, ◑ 493K

the concentration of vacancy, the faster the δ′ coarsening, the wider the PFZ at grain boundary. The cold deformation prior to aging enhances high density of dislocation in the matrix, which in turn makes the diffusion of lithium atoms easier in Al–2.45wt%Li alloy and accelerates the coarsening rate of δ′ phase, as shown in Fig.4,5,6,7[22]. Because of the large binding energy of lithium atom with vacancy[23], it is easy to form a lithium atom–vacancy complex. During the aging, such complexes move to the grain boundary, that makes the lithium concentration high there. Finally it forms an equilibrium δ phase at grain boundary and a δ′ precipitate free zone near the grain boundary.

Study of PFZ formation and growth in Al–11.9at%Li at 463K indicates that they are closely related to aging time (Fig.8), which depends on the long–range diffusion of lithium atoms[24]. The particle size of δ′ phase becomes smaller near the PFZ interface, and thereafter, it consists of lithium–rich supersaturated solid solution in the PFZ. The half–width of PFZ(w) varies with the

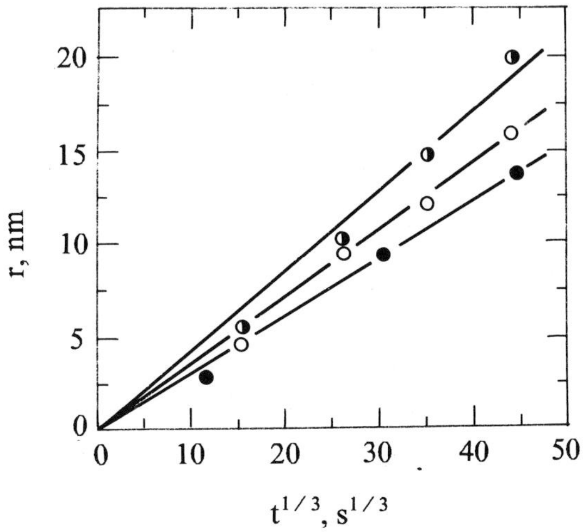

FIGURE 4
The effect of deformation prior to aging on the coarsening of δ' phases[22]

● undeformation
○ 10% deformation
◑ 20% deformation

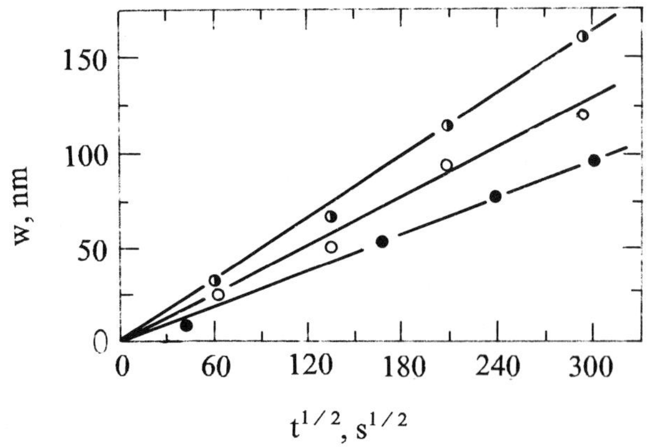

FIGURE 6
The half-width of PFZ at grain boundary vs. aging time at the different amounts of deformation[22]

● undeformation
○ 10% deformation
◑ 20% deformation

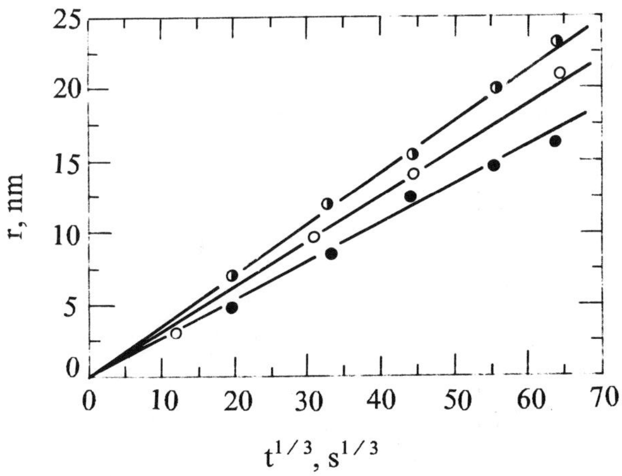

FIGURE 5
The average radii of δ' particles vs. aging times at different quenching rates[22]

◑ Quenched in cold salt water
○ Quenched in cold water.
● Cooled in air

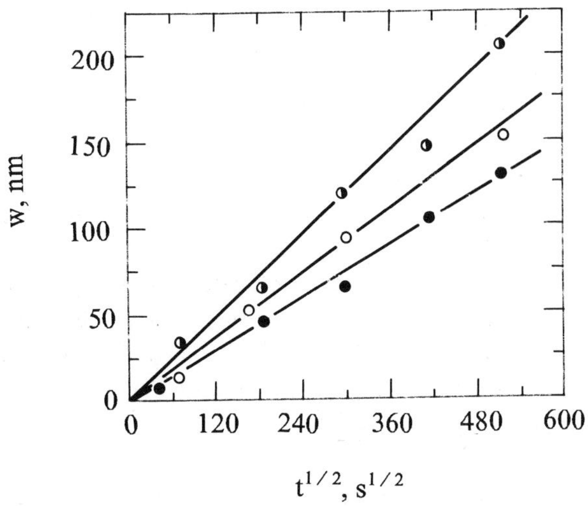

FIGURE 7
Effect of quenching rate on the half-width of PFZ at the grain boundary for various aging times[22]

● Cooled in air
○ Quenched in cold water
◑ Quenched in cold salt water

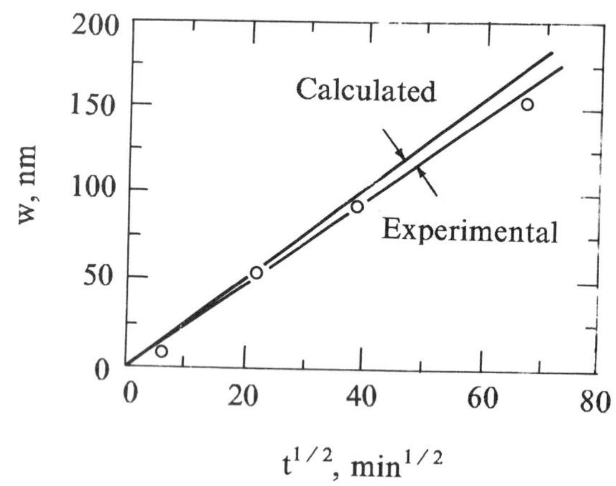

FIGURE 8
The PFZ half–width vs. the square root of aging time in Al–11.9at.%Li[24]

square root of aging time, and it can be calculated by the equation:

$$w = Kt^{\frac{1}{2}} \quad (2)$$

$$K = 2D^{\frac{1}{2}}(C - C_{\alpha}) / (C_X - C_{\alpha}) \quad (3)$$

In Al–11.9at%Li alloy(C), the Li concentration in PFZ(C_X) may be chosen as 26at%(a little larger than 25at% in Al_3Li). The Li concentration in the α phase at the PFZ interface(C_{α}) is 4.5at%Li at 463K according to the phase diagram. The diffusion coefficient of Li in α–Al has been investigated as a function of temperature between 423–513K by Mereau et al as follow[20]:

$$D = 0.37[(-126.4 \pm 5.2\text{kJ}/\text{mol})RT]\ \text{cm}^2/\text{s} \quad (4)$$

Substituting all the above data into eq.2, we get

$$w = 0.33t^{\frac{1}{2}} \quad (5)$$

The calculated half–width of PFZ related to the square root of aging time is also shown in Fig.8, which is in good accordance with the measured half–width. Of course, PFZ in Al–Li alloys is still an important task to understand deeply its formation mechanism and to find some measures to eliminate it.

δ' precipitate dissolution in Al–2.45wt%Li alloy is considered to be a thermally activated process. In the case of age–hardening alloys, formation of precipitate phase is exothermic while their dissolution is an endothermic reaction when tested in a differential scanning calorimeter(Fig.9). The peak A in Fig.9 represents the dissolution of lithium–rich GP zone. The dissolution of δ' phase consists of two stages (B_1 and B_2): In the first stage, it needs much larger thermal activation to break down the superlattice structure so as to form a lithium–rich zone. In the second stage, it is the dissolution of this zone. The calculated apparent activation energies are 278.3kJ / mol and 96.1kJ / mol respectively. The latter activation energy is very similar to the activation energy of GP zone determined to be 88.1kJ / mol. It may be postulated the mechanism of the second stage of δ'

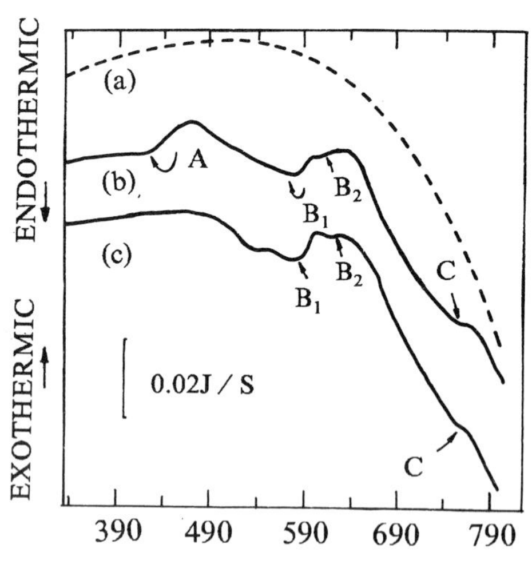

FIGURE 9
DSC results from Al–2.45wt.%Li alloy at 40 K / min[25]
(a)base line
(b)specimen aged at 300K for 2 months(NA)
(c)specimen aged at 463K for 25 h(PA)

dissolution is same as that of GP zone dissolution[25].

The endothermic peak temperature of δ dissolution is shown as C in Fig.9. The δ phase is very stable in the binary Al–Li alloy. Therefore its apparent activation energy for lithium to escape from δ phase is much larger, that was determined to be 565.3kJ / mol(NA) and 536.4kJ / mol(AA)[25]

3. FUNCTIONS OF ALLOYING ELEMENTS

The magnesium acts as an important solid solution strengthening element, which does not change the tendency of age hardening of Al–Li–Mg–Zr alloy. Mg decreases the Li solubility in the matrix, that enhances the coarsening rate of δ' particles, as shown in Fig.10. The alloy plasticity impairs with the increase of Mg in the alloy, and the elongation is the smallest at peak aging temperature[26].

The $\beta'(Al_3Zr)$ phase due to the Zr addition can act as the δ' nucleation site and enhance to precipitate T_1 and S phases. If Zr contents in 2090–T6 alloy and 8090–T6 alloy are intentionally lowered to

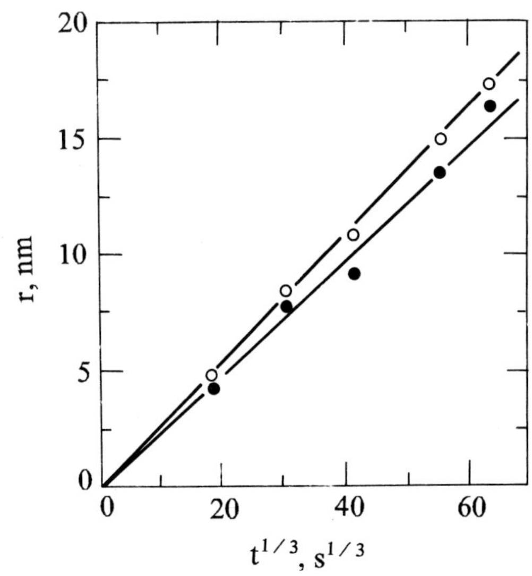

FIGURE 10
The mean radii of δ' particles vs. aging time at 463K[26]
● Al–2.19Li–1.01Mg–0.11Zr (wt%)
○ Al–2.22Li–2.06Mg–0.15Zr (wt%)

0.02% and 0.07% respectively, the ultimate tensile strengths are also impaired from 577MPa to 375MPa in the former alloy and from 553MPa to 469MPa in the latter one[27].

Zn in Al–Li alloy plays the role of solid solution strengthening, lowers the solid solubility of Li in matrix and retards the growth of δ' phase and PFZ[21]. Fig.11 is the PFZ morphology in Al–2.29Li– 3.65Zn–0.08Zr(wt%) alloy when aging at various times at 463K. The longer the aging time, the wider the PFZ. It can be clearly seen that the δ' particles are smaller near the PFZ. Fig.12 shows the relationship of δ' particle size(Fig.12a) and half–width of PFZ(Fig.12b) with aging time.

The microstructure and properties of three kinds of Al–Li–Cu–Zr alloys with Cu / Li ratios of 0.47, 1.17 and 2.39 have been studied. The T_1 platelets grow up during the dissolution of δ' particles, which become smaller and gradually disappear. The sequence of phase precipitation of alloy 1 and alloy 2 is $\delta' \rightarrow \delta' + T_1$, that of alloy 3 is $\delta'+T_1 \rightarrow \delta'+T_1+\theta'$. T_1 phase precipitates both in the grain and on the grain boundary[28]. The tensile strength rises with the increase of Cu / Li ratio and a drop of plasticity(Table 1), accompanying the change of crack mode from transgranular mode to intergranular one, which is caused by the precipitation of equilibrium phase containing Cu at grain boundary[28].

Addition of rare earth element(0.3wt%) to 2090 alloy can refine δ' phase, suppress its growth, enhance strength(Fig.13) and increase the oxidation resistance[29]. After oxidation, the surface of 2090 alloy contains more Si, Mg and Li, while the oxide film of 2090RE alloy contains rare earth elements and less Si, Mg and Li, as shown in Fig.14. That is to say, rare earth elements do not reduce oxidation weight gain, but are substituted for Li and Mg to be oxidized, that induces the reduction of Li and Mg surface depletion of alloys. The widths of surface depletion layer of 2090 and 2090RE alloys after solution treatment at 530℃ for 3h are shown in Fig.15. The width of the latter is nearly two third

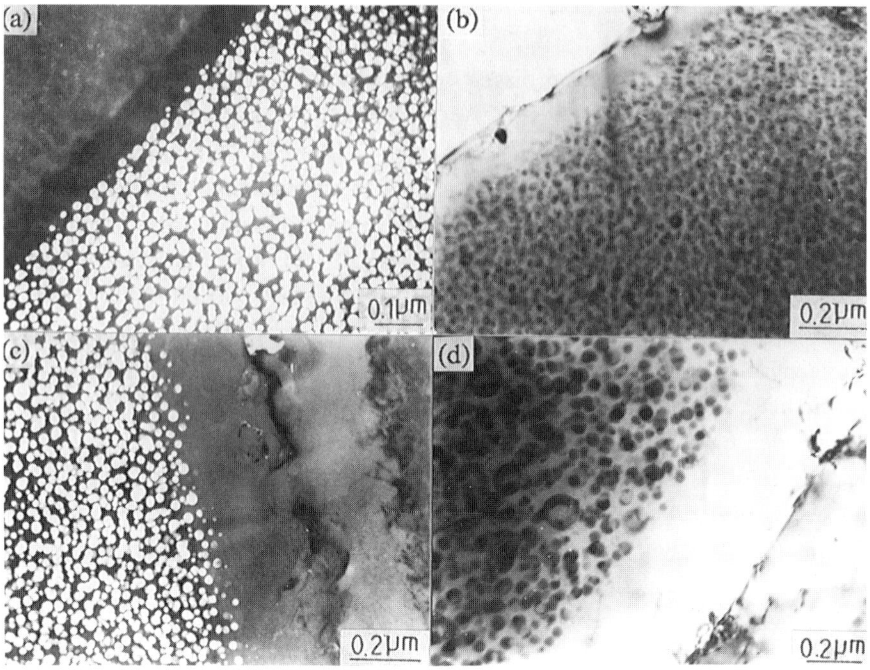

FIGURE 11
The PFZ morphology in Al–2.29Li–3.65Zn–0.08Zr alloy aged at 463K for (a)6h, (b)16h, (c)48h and (d)100h[21]

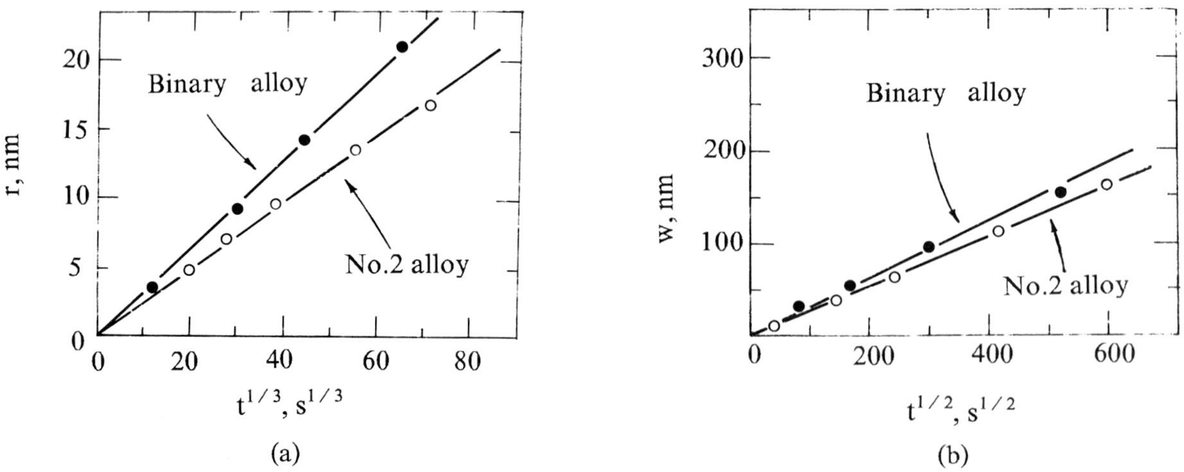

FIGURE 12
Relation between aging time and δ' particle size (a) as well as half-width of PFZ (b) in Al–2.29Li–3.65Zn–0.08Zr alloy aged at 463K[21]

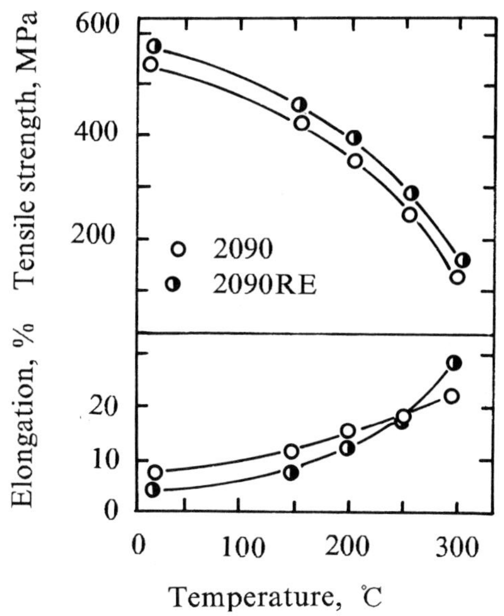

FIGURE 13
Elevated temperature strength and ductility of both alloys[29]

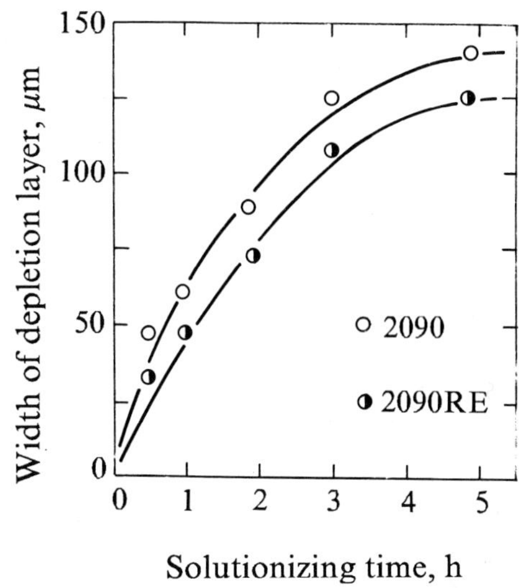

FIGURE 15
Influence of solutionizing time on width of depletion layer at 530℃[29]

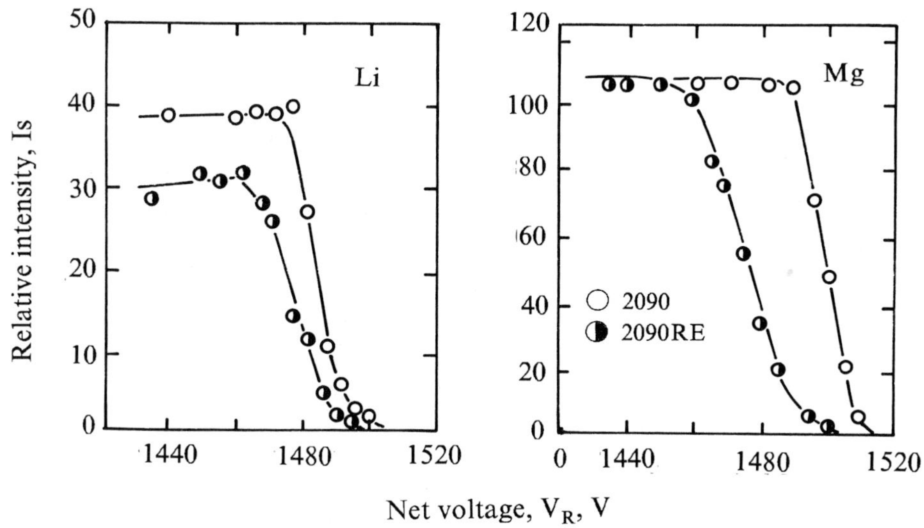

FIGURE 14
Effect of 0.3pct mischmetal addition on relative intensity of secondary Li and Mg ion beam[29]

TABLE 1.
Influence of Cu on tensile properties at room temperature (PA)

Alloy	Li	Cu	Zr	Al	Cu/Li	UTS(MPa)	0.2YS(MPa)	EL(%)
1	2.32	1.09	0.14	bal.	0.47	395	324	8.3
2	2.23	2.62	0.16	bal.	1.17	428	347	7.0
3	2.05	4.90	0.13	bal.	2.39	513	435	5.5

of the former.

4. ENHANCEMENT OF CRYOGENIC MECHANICAL PROPERTIES

Recently Al–Li alloys have been proved to be attractive for their enhanced mechanical properties even at cryogenic temperatures. It is a promising candidate in cryogenic aerospace applications such as hydrogen and oxygen fuel tanks in future aerovehicles.

In Al–Li–Zn–Zr alloys, the tensile strength and elongation are increased with decreasing temperature tested down to 77K as shown in Fig.16[21], but the elongation of OA specimen has a little drop (Fig.17)[30]. The influence of stretching prior aging

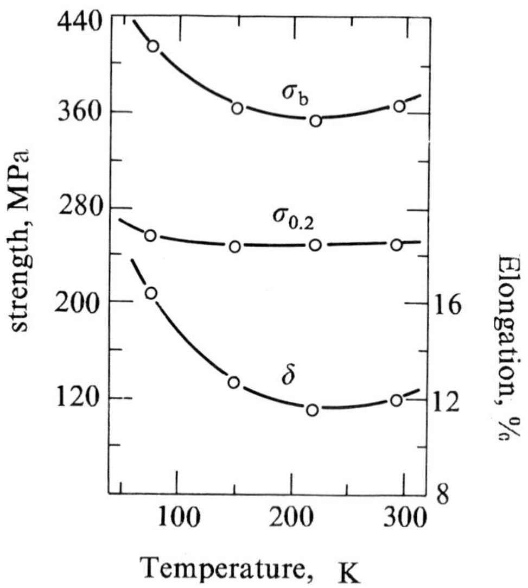

FIGURE 16
The tensile property of Al–2.29Li–3.65Zn–0.08Zr (wt%) alloy at cryogenic temperatures[21]

on tensile properties of Al–Li–Zn–Zr alloys are listed in Table 2. Besides Al–2.75wt%Li, Al–Li–Mg–Zr, Al–Li–Cu–Zr and Al–Li–Mg–Cu–Zr alloys have also been studied. The tensile strength and elongation are also enhanced at liquid nitrogen temperature and listed in Table 3.

The ultimate tensile strength is generally associated with the strain hardening behavior of a metal, and its increase at cryogenic temperature may be attributed to the improving of strain hardening ability. It can also be understood that the strengthening of δ' phase is the most outstanding factor at low temperatures. The ordered nature of δ' phase causes the dislocations to move in pairs during plastic deformation. The antiphase domain energy of the δ' phase is increased so that it is difficult to result in its cutting by dislocation pairs at low temperatures, which can partly account for a more homogeneous deformation mode (Fig.18b), compared with the planar slip deformation mode occurred at room temperature (Fig.18a). A more homogeneous slip distribution and dispersed strain localization at cryogenic temperatures leads to a higer ductility[21].

Fig.19 shows the cracking appears transgranular mainly at room temperature, and changes to an intergranular mode chiefly at 77K. The failure of Al–Li alloy may be contributed in part to the interaction between the grain boundary precipitates and the slip bands. Because of the increase of matrix strength of Al–Li alloys relative to the grain boundary at cryogenic temperatures, large amount of dislocations pile up at grain boundaries, nucleate microcracks there and propagate along the grain boundary, resulting in the

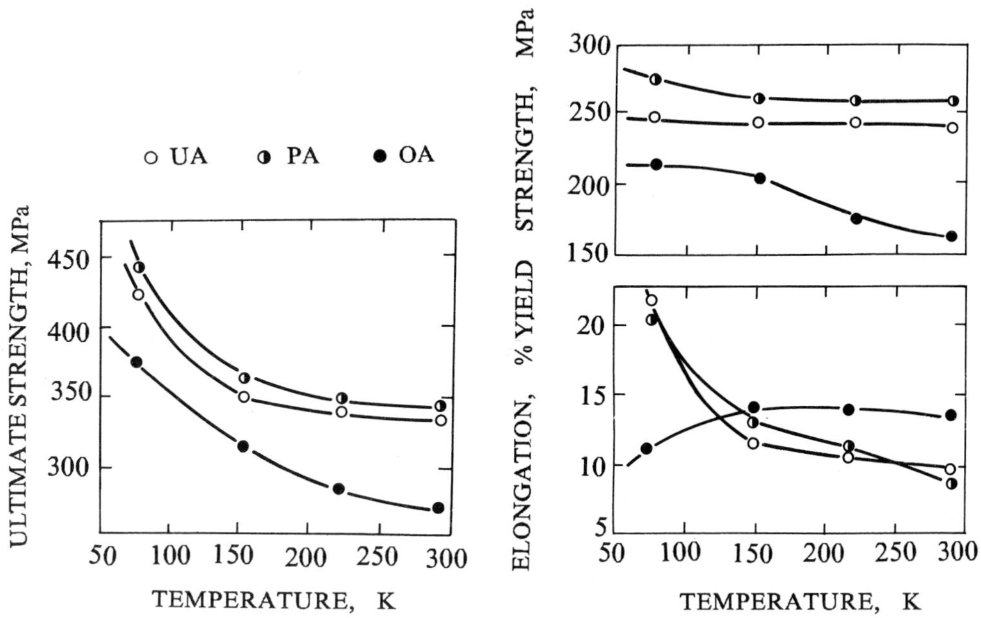

FIGURE 17
Tensile properties of Al–Li–Mg–Zr alloy vs. test temperature[30]

TABLE 2.
Influence of stretching on tensile properties of Al-Li-Zn-Zr alloys[(21)]

Alloy	Stretching(%)	Temp.(K)	0.2YS(MPa)	UTS(MPa)	EL(%)
1	0	300	240	306	9.7
1	2	300	255	315	7.7
1	·5	300	261	315	7.6
2	0	300	251	367	12.0
2	2	300	277	370	8.4
2	5	300	287	381	8.4
1	0	77	244	386	24.1
1	2	77	257	388	20.8
1	5	77	266	390	20.2
2	0	77	254	414	16.0
2	2	77	263	414	11.3
2	5	77	289	437	11.5

Note: Alloy 1 —— Al-1.80Li-1.02Zn-0.12Zr, 798K/1hWQ+403K/48h
Alloy 2 —— Al-2.29Li-3.65Zn-0.08Zr, 798K/1hWQ+433K/16h

TABLE 3
Ratios of mechanical properties of Al-Li alloys tested at cryogenic temperature compared to those tested at room temperature.

Alloy	Treatment Condition	Temp. (K)	UTS	YS	EL	K_{IC}	Ref.
2090	T6	77	1.27	1.17	1.77	1.58	14
2090	T8	77	1.21	1.09	1.45		31.32
2090	T8	4	1.39	1.11	1.88		31.32
2090	T8E41	77	1.27	1.12	1.27	1.42	12.33
2090	T8E41	4	1.45	1.15	1.64	1.78	12.33
2091	T351	77	1.32	1.20	1.60	1.24	31
2091	T8	77	1.18	1.13	1.67	1.33	31
2091	T8	4	1.31	1.25	1.40	1.33	31
2091	T8X	77	1.27	1.20	1.60	1.24	33
8090	T351	77	1.24	1.13	1.41	0.74	13.17.33
8090	T651	77	1.25	1.06	1.67		13.17.33
8091	T351	77	1.37	1.24	1.27	0.74	33
8091	T8X	77	1.20	1.07	2.00	1.90	33
Al-0.93Li	T6	77	1.99	1.13	1.16		34
Al-2.45Li	T6	77	1.21	1.11	1.05		34
Al-2.75Li	T6	77	1.12	1.04	1.50		35
Al-1.80Li-1.02Zn-0.12Zr	T6	77	1.26	1.01	2.48		21
Al-2.29Li-3.65Zn-0.08Zr	T6	77	1.13	1.01	1.33		21
Al-2.32Li-1.09Cu-0.14Zr	T6	77	1.28	1.06	1.83		36
Al-2.23Li-2.62cu-0.16Zr	T6	77	1.21	1.09	2.14		36
Al-2.05Li-4.90Cu-0.13Zr	T6	77	1.01	0.91	2.33		36
Al-2.19Li-1.01Mg-0.11Zr	T6	77	1.25	1.05	3.00		30
Al-2.22Li-2.06Mg-0.15Zr	T6	77	1.22	1.03	1.64		37

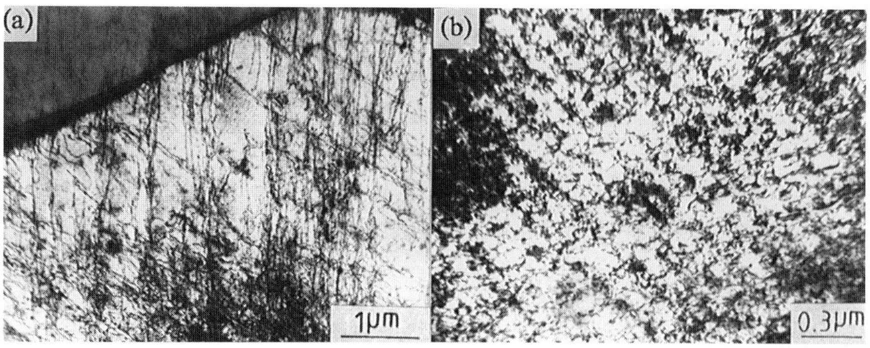

FIGURE 18
The dislocation structure near the tensile fracture of Al–2.29Li–3.65Zn–0.08Zr (wt%) alloy at room temperature (a) and 77K (b)[21]

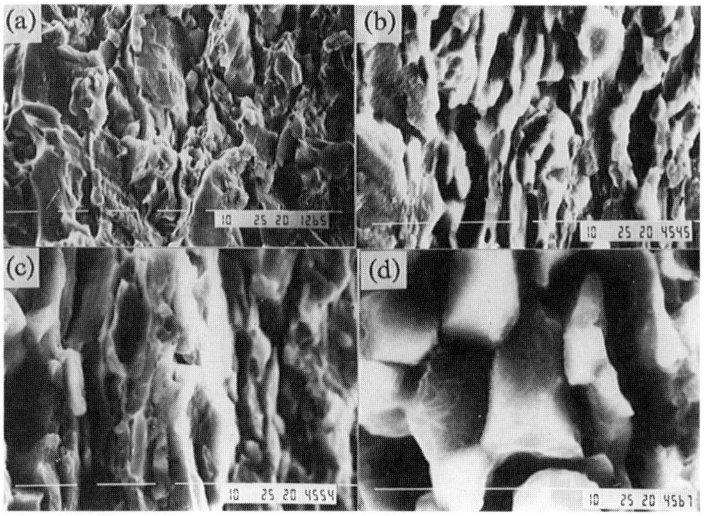

FIGURE 19
Tensile fracture morphology of Al–2.29Li–3.65Zn–0.08Zr (wt%) alloy at 295K(a), 220K(b), 150K(c) and 77K(d)[21]

trend of intergranular fracture.

Optical microscopy observations (Fig.20) of the profile of the fracture surfaces indicate that there is an increasing tendency at low temperature for intergranular delamination (splitting) perpendicular to the short–transverse orientation which is considered as a reason for the improvement of ductility of Al–Li alloys at cryogenic temperature. However it still remains a question of how to explain there is no enhancement of ductility at over–aged condition despite of many intergranular delaminations on fracture surface.

The influence of stretching prior aging on tensile properties of Al–Li–Zn–Zr alloys seems to increase the strength and decrease the ductility both a little bit. The fracture mode at cryogenic temperatures after stretching are same as that tested without stretching, as shown in Fig.21. It means the tensile fracture mode is same and stretching has no pronounced effect on it. Stretching creates a high dislocation density. The higher the deformation, the more the dislocation. Because the δ' precipitation mainly depends on the vacancy concentration and Li diffusion in matrix as already mentioned before, the nucleation and growth of the δ' phase is very rapid. However, the existence of enormous dislocations can accelerate the Li diffusion, but its influence is still not much that makes the strength of Al–Li alloy rise a little. On the other hand, the presence of a large amount of dislocation kinks lowers the strain hardening ability and grain compatibility, that impairs the ductility.

5. CONCLUSIONS

(1). The mechanism of δ' coarsening is diffusion controlled. The activation energy for coarsening is approximately the same as the activation energy for lithium atom diffusion.

(2). δ' dissolution consists of two stages: breaking down the superlattice structure to form a lithium–rich zone and finally dissoluting this zone.

(3). The half–width of PFZ is correlated to the square root of aging time. A equation for the calculation of the half–width of PFZ is proposed.

(4). Zn addition in Al–Li alloy retards the growth of δ' phase and PFZ.

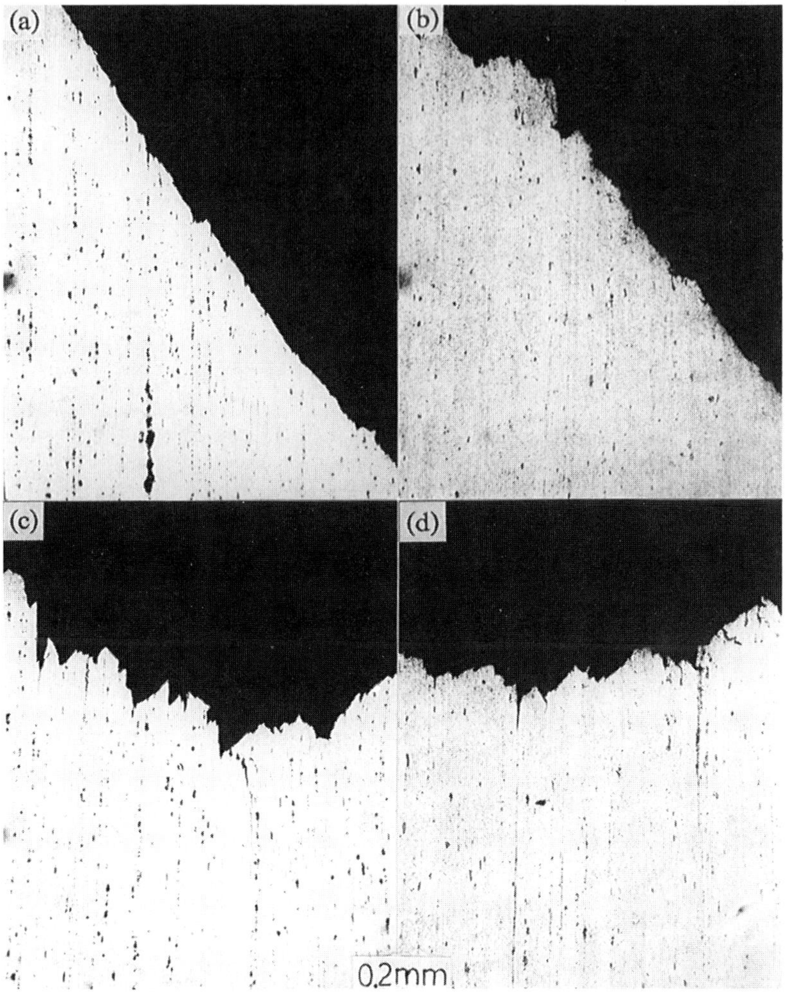

FIGURE 20
Lateral view of tensile fracture of Al–2.29Li–3.65Zn–0.08Zr(wt%) alloy at 295K(a), 220K(b), 150K(c) and 77K(d)[21]

(5). The strength and hardness rise with the increase of Cu / Li ratio and a drop in plasticity in Al–Li–Cu–Zr alloys, accompanying the change of crack mode from transgranular mode to intergranular one.

(6). An appropriate addition of rare earth element can improve strength and ductility as well as increase the resistance to oxidation.

(7). The enhancement of tensile properties can be explained by an increasing tendency of intergranular delamination (splitting) at cryogenic temperatures, that results in the dispersion of strain concentration leading to the shear deformation localization.

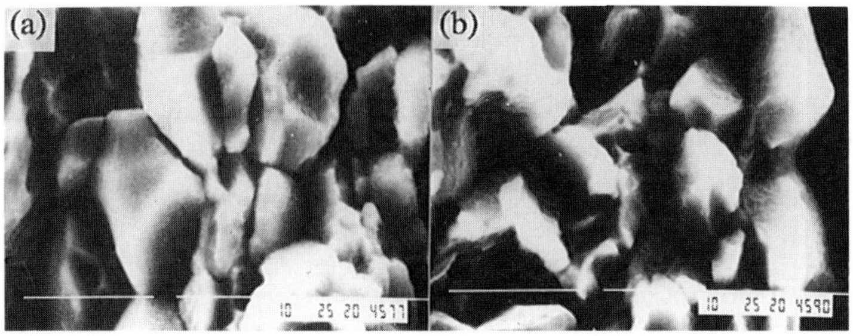

FIGURE 21
Tensile fracture morphology of Al–2.29Li–3.65Zn–0.08Zr (wt%) alloy after stretching tested at 77K[21]
(a) 2% strectching (b) 5% strectching

REFERENCES

1. B. Noble, S.J. Harris and K. Dinsdale, J. Mater. Sci. 17 (1982) 461.

2. K.T. Venkatsewara Rao, W. Yu and R.O. Ritchie, Scripta Metall. 20 (1986) 1459.

3. T.R. Prittchett, Light Metal Age 44 (1986) 10.

4. W.X. Feng, F.S. Lin and E.A. Starke Jr., in: ″Aluminum–Lithium Alloys II ″, eds. T.H. Sanders Jr. and E.A. Starke Jr. (TMS–AIME, Warrendale, PA, 1983) pp. 17.

5. Catalog of Cegedur–Pichiney Co.

6. D. Webster, in: ″Aluminum–Lithium Alloys III ″, eds. C. Baker, P.J. Gregson, S.J. Harris and C.J. Peel (The Institute of Metals, London, 1986) pp. 602.

7. R.C. Dorward, Scripta Metall. 20 (1986) 1379.

8. J. Glazer, S.L. Verzasconi, E.N. Dalder, W. Yu, R.A. Emigh, R.O. Ritchie and J.W. Morris Jr., Adv. Cryo. Eng. 32 (1986) 397.

9. J. Glazer, S.L. Verzasconi, R.R. Sawtell and J.W. Morris Jr., Metall. Trans. A. 18A (1987) 1695.

10. J. Glazer, J.W. Morris Jr. and T.G. Nieh, Adv. Cryo. Eng. 34 (1987) 291.

11. D. Webster, Metall. Trans. A. 18A (1987) 2181.

12. K.T. Venkateswara Rao, H.F. Hayashigatani, W. Yu and R.O. Ritchie, Scripta Metall. 22 (1988) 93.

13. D. Dew Hughes, E. Creed and W.S. Miller, Mater. Sci. Technol. 4 (1988) 106.

14. K.V. Jata, E.A. Starke Jr., Scripta Metall. 22 (1988) 1553.

15. N. Niinomi, K. Degawa and T. Kobayashi, in: ″Aluminum–Lithium Alloys IV″, eds. G. Champier, B. Dubost, D. Minannay and L. Sabetay (Journal de Physique, Paris, 1987) pp. 653.

16. D. Webster, in: ″Aluminum–Lithium Alloys III ″, eds. C.Baker, P.J. Gregson, S.J. Harris and C.J. Peel (The Institute of Metals, London, 1986) pp. 602.

17. K. Welpmann, Y.T. Lee and M. Peters, in: ″Aluminum–Lithium Alloys V″, eds. T.H. Sanders Jr. and E.A. Starke Jr. (Williamsburg, Virginia, 1989) pp. 1513.

18. D.P. Yao, Z.Q. Hu, Y.Y. Li, Y. Zhang and C.X. Shi, Chin. J. Met. Sci. Technol. 6 (1990) 31.

19. E.S. Bulmuth, Scripta Metall. 18 (1984) 301.

20. C. Mereau, A. Allouche and E.J. Knystantas, J. Appl. Phys. 58 (1985) 4582.

21. X.J. Jiang, Master Degree Dissertation, Inst. of Metal Research, Academia Sinica, (1990).

22. D.P. Yao, Z.Q. Hu, Y.Y. Li, Y. Zhang and C.X. Shi, accepted by Acta Metall. Sinica.

23. S. Ceresara, A. Giarada and A. Sanchez, Phil. Mag. 35 (1977) 97.

24. D.P. Yao, Y. Zhang, Z.Q. Hu, Y.Y. Li and C.X. Shi, Scripta Metall. 23 (1989) 537.

25. D.P. Yao, Z.Q. Hu, Y. Zhang and C.X. Shi, Acta Metall. Sinica 26 (1990) A122 (in Chinese)

26.D.P. Yao, Z.Q. Hu, Y.Y. Li, Y. Zhang and C.X. Shi, accepted by Mater. Sci. Prog. (in Chinese).

27.D.P. Yao, Y. Zhang, Z.Q. Hu, Y.Y. Li and C.X. Shi, submitted to Acta Metall. Sinica.

28.D.P. Yao, Z.Q. Hu, Y.Y. Li, Y. Zhang and C.X. Shi, Acta Metall. Sinica 26 (1990) A97 (in Chinese).

29.Y. Zhang, H.E. Zhao, G.Z. Guo, J.Z. Chen, D.P. Yao and Z.Q. Hu, in: ″Aluminum–Lithium Alloys V″, eds. T.H. Sanders Jr. and E.A. Starke Jr.(Williamsburg, Virginia, 1989) pp. 539.

30.D.P. Yao, Y.Y. Li, Z.Q. Hu, Y. Zhang and C.X. Shi, submitted to Adv. Cryo. Eng.

31.J. Glazer and J.W. Morris Jr., in: ″Aluminum–Lithium Alloys V″, eds. T.H. Sanders Jr. and E.A. Starke Jr., (Williamsburg, Virginia, 1989) pp. 1471.

32.K.T.V. Rao, W. Yu and R.O. Ritchie, Metall. Trans. A 20A (1989) 485.

33.K.T.V. Rao and R.O. Ritchie, in: ″Aluminum–Lithium Alloys V″, eds. T.H. Sanders Jr. and E.A. Starke Jr. (Williamsburg, Virginia, 1989) pp. 1501.

34.D.P. Yao, Z.Q. Hu, Y.Y. Li, Y. Zhang and C.X. Shi, accepted by Mater. Sci. Prog. (in Chinese).

35.D.P. Yao, Z.Q. Hu, Y.Y. Li, Y. Zhang and C.X. Shi, unpublished paper.

36.D.P. Yao, Z.Q. Hu, Y.Y. Li, Y. Zhang and C.X. Shi, in: ″Aluminum–Lithium Alloys V″, eds. T.H. Sanders Jr. and E.A. Starke Jr. (Williamsburg, Virginia, 1989) pp. 1533.

37.D.P. Yao, Z.Q. Hu, Y.Y. Li, Y. Zhang and C.X. Shi, unpublished paper.

Frontiers of Materials Research/
Electronic and Optical Materials
M. Kong, L. Huang (editors)

RESEARCH AND APPLICATIONS OF SHAPE MEMORY ALLOYS IN CHINA

D.Z.Yang, M.Zhu, F.T.Wang and F.X.CHen

Research Institute of Shape Memory Alloys, Dalian University of Technology, Dalian 116023, P.R.China

The development of research and application of shape memory alloys in P.R.China, mainly concerning the Cu–base and NiTi alloys, is briefly outlined. The achievements of Chinese reseachers of about 20 Universities and Institutes along with 8 Research Units are vast and have widely received attentions. this paper mainly focused on the grain refinement, phase transformations during heating and cooling, crystal defects of martensite, aging effect and applications of shape memory effect.

1. INTRODUCTION

The research and development of the shape memory alloys (SMAs) began in 1960's since the useful shape memory effect (SME) was discovered in NiTi alloy by Buehler and his colleagues at the U.S. Naval Ordnance Laboratory in 1961[1]. Now, theoretical research and applications of SMAs is growing rapidly.

The study on SMAs in China began in 1977, mainly focusing on TiNi, TiNiX(X=Fe, Co, Cu, Nb etc.) and CuZnAl, CuAlNi, CuZnAlX and CuAlNiX(X=B, V, Ti, Zr, Mn, etc) which are under the scope of two alloy systems known as Nitinol and Cu–base SMAs. Up to now, more than 20 Institutes and Universities along with 8 Research Units have been engaging in projects on the research and applications of SMAs, The present paper trys to outline the achievements of research and applications of SMAs in China, However it should be pointed out that it is quite difficult and impossible to emphasize whole result obtained by Chinese scholars in this field.

2. THE GRAIN REFINING OF SHAPE MEMORY ALLOYS

Current Cu–base SMAs in commercial uses are derived from both Cu–Zn and Cu–Al binary alloy systems. As listed in Fig.1[2]. It is worth to mention here that the Mn additive and its effect in Cu–base SMAs was studied systematically by Chinese scholars and a new alloy system Cu–Al–Zn–Mn–Ni was obtained[3,4]. A problem which is critical to applications of Cu–base SMAs arises from the poor mechanical properties related mainly to the coarse grain of it, and the grain refining is a subject of many works. The effect of alloying elements on the grain refining of Cu–Zn–Al and Cu–Al–Ni alloys has been studied extensively[5], the alloying elements added including Zr, Ti, V, B, Mn, rare earth elements, and so on. It has been proved that Zr is one of the most effective grain refining elements, while Ti can only slightly refine the grains[5]. The different refining effect of Zr and Ti addition can be seen clearly in Figs.2a and b. The grain growth of both refined alloys, especially the Ti–doped one, become faster when the specimen is kept at above 900℃ for a longer time. Scanning electron microscopy and the electron probe analysis indicates that segregation of Ti occurred in the vicinity of grain boundary and the Zr–rich secondary phase particles distributed at grain boundaries and within grains. Neither boundary anomaly nor grain boundary segregation was observed in the Zr–doped alloy. When holding at 950℃ for 2 hrs, no special change of Ti distribution took place, but in Zr–doped alloy, the Zr–riched particles disappeared. It is indicated that "grain boundary anomaly" may be assumed to have no significant effect on suppressing grain growth. The second phase particles retard grain growth to some extent. In refined alloys, the pinning effect of dispersed particles and the solute drag effect due to structure change of grain boundary would effectively suppress the grain growth. The addition of small amount of B and V is also effective to the grain refining and increase the mechanical properties of alloys without significant deteriovation of SME.

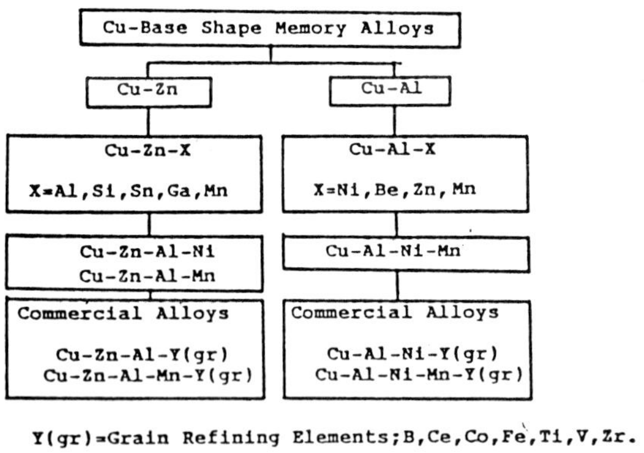

FIGURE 1
The Cu–base shape memory alloy systems developed

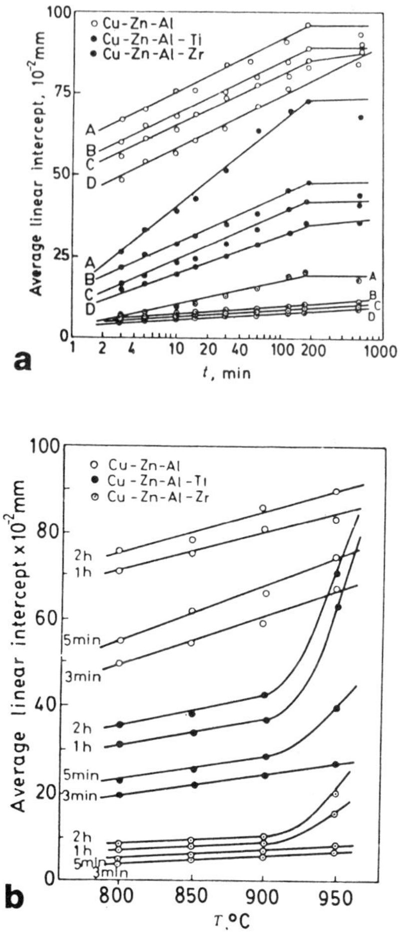

FIGURE 2
The grain refining effect of Ti and Zr, (a). Grain size vs heating time at different temperature, A. 1223K, B. 1173K, C. 1123K, D. 1073K. (b). Grain size vs temperature at different time.

Cu–Al–Ni alloys with practical transformation temperatures are mostly eutectoid or hypereutectoid alloys. The alloys possess little ductility at low temperature because of the existence of γ–phase and the coarse grains with their sizes in mm range. The ductility can be improved by adding Ti and Mn, typical composition of the alloy is Cu–12Al–5Ni–2Mn–1Ti. This kind of alloy not only possess a better workability but also high Ms temperature rising up to 100℃ [6]. Remarkable successes have been achieved by taking rare–earth elements as modifier. The grain size can be effectively controlled at about 30~ 60μm and the formability of Cu–base alloys notably improved. It must be pointed out, however, that the refinement of grains often leads to inferior SME. Thus how to refine the grain size without severe damage of SME is still a topic open for further studies.

In addition, the novel processing techniques, such as powder metallurgy, and rapid solidification, have developed in recent years. Under rapid solidification microcrystal structure can be formed in Cu–base SMAs, of which thin sheets of 0.05–0.4mm in thickness and 30–50mm in width have been made. This kind of strips has pretty good SME, but too thin to be used. Moreover, when strips are heated to 800℃ and over, grain will quickly grow larger. A new method worthy to pay attention is that the TiNi SMAs can be obtained by explosive consolidation of elemental Ni and Ti powders. This method eliminated the sophisticated melting and processing procedure of manufacturing NiTi SMAs[7].

3. PHASE TRANSFORMATIONS IN β–PHASE DURING COOLING AND HEATING

Several kinds of phase transformations happened in Cu–base alloys dealing with thermoelastic martensite, bainite, ordering and decomposition during heating and cooling of β phase of Cu–base SMAs. These transitions are related to SME, reversible SME and aging effect of SMAs.

3.1. Crystallography of Martensitic transformation

When the $\beta_1$phase is cooling below its Ms temperature, β_1–phase parent transforms to the martensite. In general, the martensite has a M18R structure with stacking fault substructure in Cu–Zn–Al alloys, a 2H structure with twin substructure in Cu–Al–Ni alloys[8]. But mixture structure of above two kinds of martensite has also been found to be formed in Cu–Zn–Al–Mn–Ni alloy system[9]. To form whether the 2H or 18R martensite is dependent mainly on the composition of alloys and the heat treatment condition is also an effective factor.

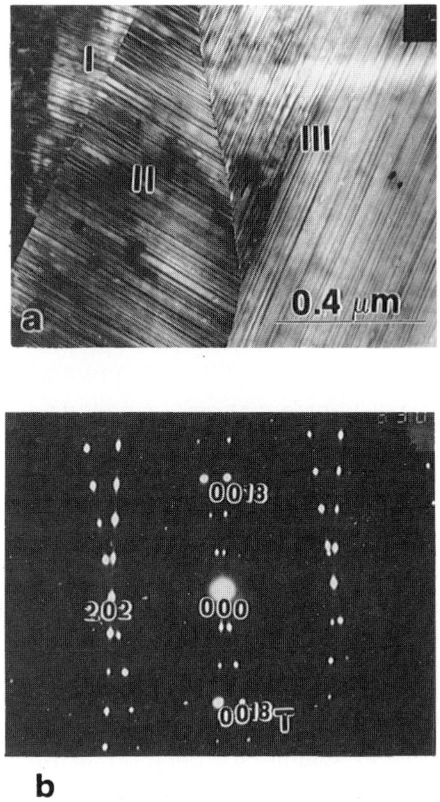

FIGURE 3
(a). Morphology of (101) twin between martensite variants, (b). Diffraction pattern showing the (101) twin relation.

When martensite is formed, several variants exist and there are some definite relationships among them, such as the well known $(128)_M$, $(1010)_M$and $(101)_M$twin relations. Fig.3a and b are morphology and diffration pattern of the $(101)_M$twin. To explain crystallographic relation between variants, a group theory description of the crystallography of thermoelastic martensitic transfromation was proposed[10,11]. The crystallography of martensitic transformation was generally described by a phenomenological crystallographic theory which needs trial mathematical calculation. As pointed out by Guymont[12], however, when a low symmetry phase formed from a high symmetry parent phase in order–disorder transition and paramorphic transition, antiphase domains and orientation domains appear and the domains are related with some symmetry operations which are not the symmetry operations of the phase formed in the parent phase but that of parent phase. By extending this theory to thermoelastic martensitic transformations recently, the crystallographic relations among martensite variants are satisfactorily explained. Fig.4a is the diffraction pattern showing the orientation relationship between martensite and parent phase, and Fig.4b is its stereographic projection. The pole of martensite variant M1(real dot) and the pole of parent phase(square dot) were ploted according to Fig.4a. Since the (101)p is a mirror plane and the symmetry operation belongs to parent phase not martensite, by using the principle mentioned above, another variant M2 can be produced by (101)p mirror reflection, i.e., M1 and M2 are in twin relation. Since the (101)p plane is just coincident with $(128)_M$, the two variants are in $(128)_M$twin relation in the martensite lattice. In this way, all variants and relationships among variants can be obtained simply.

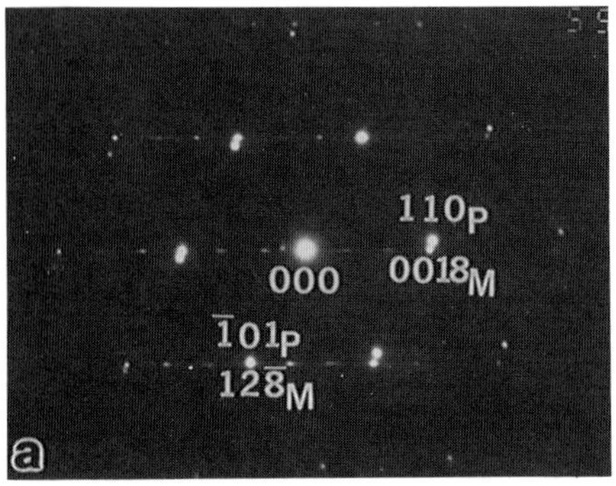

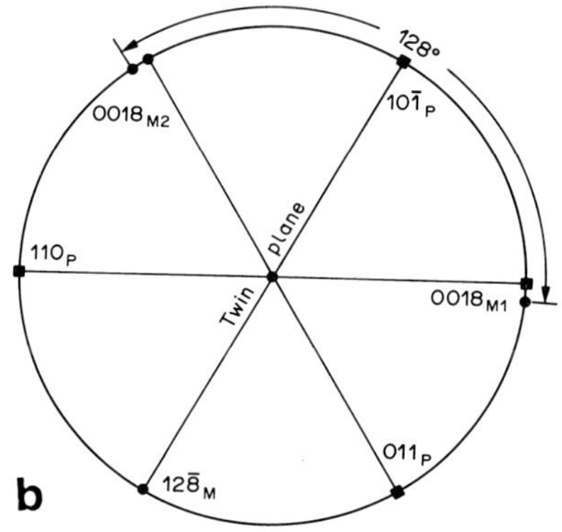

FIGURE 4
(a). Diffraction pattern showing the orientation relationship between martensite and parent phase, (b). Stereographic projection of Fig.4 (a)

3.2. Transformations During Heating

When the $\beta_1$phase is heated, the bainitic transformation takes place at first, which is greatly interested in many recent works[13,14]. The research of bainite is mainly focused on the mechanism of bainitic transformation. Recently a systematic research on prebainitic transformation, which is dealing with the nature of bainitic transformation has been made by professor T.Y.Hsu's group[15,16] Their results prove that there are some structure change within the incubation period, which is corresponding to nucleation process accompanying the local softening and / or lattice instability of the parent phase.

Fig.5 is the morphology of bainite obtained in the $\beta_1$phase of CuZnAl alloy. Its structrue is N9R, which is a little different from that of martensite in the same alloy. However the crystallography of bainitic transformation is identical to that of martensite in this alloy system[17]. Although the crystallography of bainitic transformation is regarded as same as that of

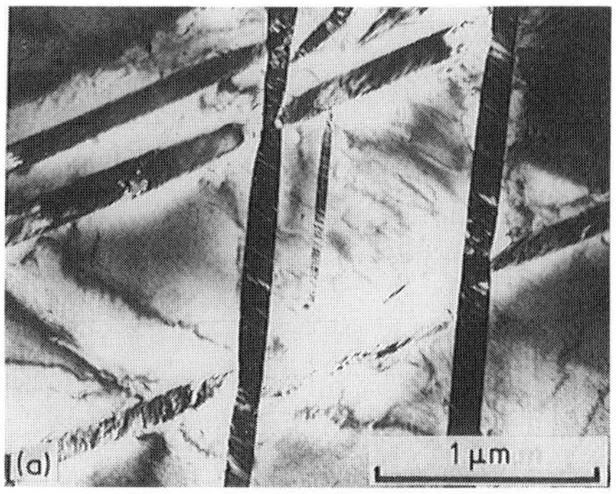

FIGURE 5
The morphology of bainite in CuZnAl alloy

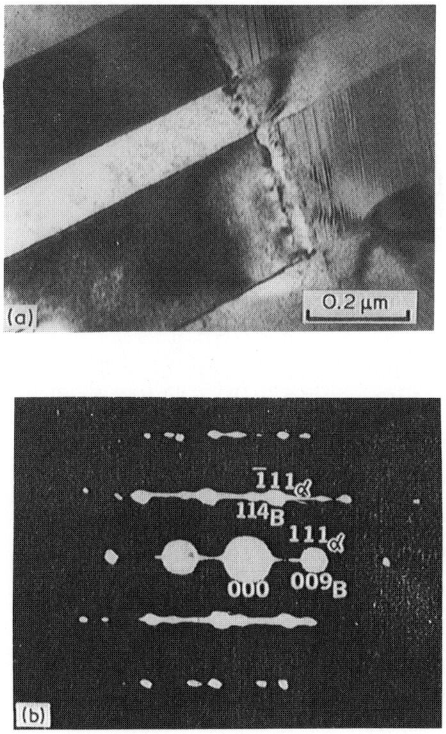

FIGURE 6
(a). the α phase formed inside bainite plate, (b). Diffraction pattern corresponds to (a) indicating the orientation relationship between α and retained 9R.

martensitic transformation, the mechanism of these two transformation is not the same in nature. Thermodynamic calculation of baintic transformation in Cu–Zn alloy has proved that it can only proceed as that of diffusional reaction[18].

During further heating both bainite and $\beta_1$matrix decompose to equilibrium phase[19]. Fig.6a shows that the α phase forms inside former bainite plate and grows along the long axis of plate. Fig.6b is a diffraction pattern showing the orientation relationships between the α phase and retained part of 9R structure. $(111)_\alpha // (001)_{9R}$, $[011]_\alpha // [\bar{1}10]_{9R}$. This relationship is expected, considering the similarity between the structure of f.c.c. and 9R. With the increasing of heating time, the amount of α phase increases and another product, β phase, precipitates in the former bainite plate as shown in Fig.7a. Fig.7b is the diffraction pattern showing the orientation relationship between α and β. It is evident that this relationship is a little deviated from the well known K–S relationship. As having been proved, the orientation relationship between α and β satisfies K–S relation when the α phase is precipitated from β phase during cooling from high temperature[20]. These results emphasized that the crystallography of α phase precipitated from β phase during cooling is different from that of β phase precipitated in the former bainite plate during heating. Accompanying with the decomposition of bainite, the $\beta_1$matrix has also changed. Fig.8a and b is the corresponding morphology and diffraction pattern. It can be observed that a 2H structure has been formed. The modulated contrast was considered to be the multiple diffraction effect, i.e., the moire fringe[21]. For longer time aging, however, the 2H structure vanished and the whole specimen is composed of α and β equilibrium

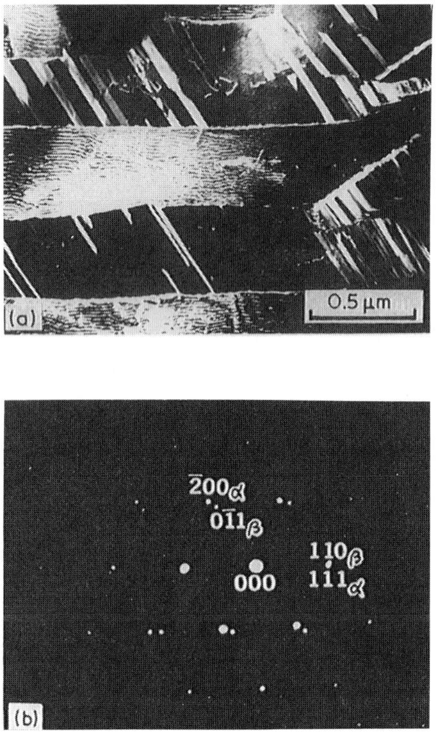

FIGURE 7
(a). The $\beta_1$phase precipitates in the former bainite plate, (b). The composite diffraction pattern obtained inside the plate shown in (a)

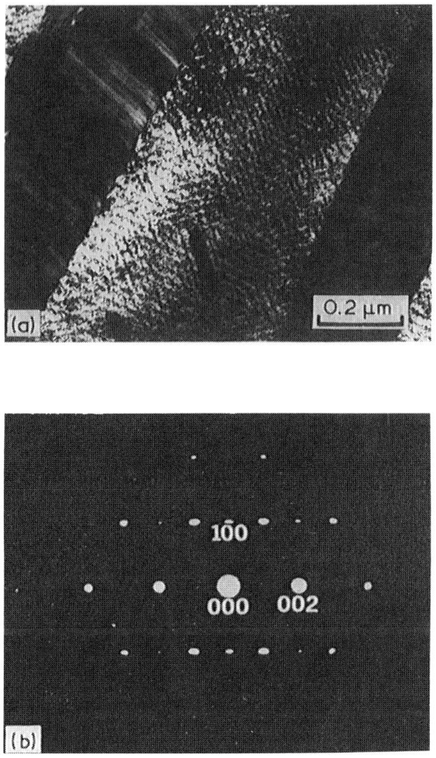

FIGURE 8
(a). Modulated microstructures in $\beta_1$in moderate temperature aging, (b). Diffraction pattern showing the existence of a 2H structure.

phase.

4. CRYSTAL DEFECTS OF MARTENSITE IN Cu–BASE SMAs

The substructure of martensite, in general, can be divided into two types. One is the results of lattice invariant plane strain of martensitic transformation. The other is induced by external influence, such as deformation, thermal cycles etc, and has great influence on behavior of transformation. the latter one is the main interesting present paper.

4.1. Nonbasal Plane Planar Defect in 18R Martensite[22]

When the specimens of CuZnAl alloy are quenched directly into ice water, a kind of defect similar to stacking fault in morphology can be observed as shown in Fig.9a. This kind of stacking fault is not on the basal plane of martensite and is regarded as nonbasal plane planar defect. Fig.9b is the image of it obtained in [210] direction. Two features can be derived from it. One is that it is connected with basal plane stacking fault and the other is that its image becomes a very narrow line of about 2 nm in width, which indicates that the nonbasal plane planar defect is located in the plane parallel to the incident beam direction, [210] direction. By trace analysis, it was determined to be lying on (125) plane. In order to understand the nature

FIGURE 9
The diffraction contrast image of non–basal plane planar defect obtained along [230] direction (a) and [210] direction (b)

of this kind of defect, HREM was used to make a direct observation of it at atomic level. Fig.10a is the HREM image obtained. It can be seen that two α–type basal palne stacking faults break off inside the martensite plate and the atom arrangement between these two α–type basal plane stacking faults has been changed. Fig.10b illustrates the structure characterictics of nonbasal plane defects schematically, which is derived from the result of HREM image. It shows that when an α–type stacking fault is induced in atom plane "P" and interrupted at "0", the positions of atoms in plane "P" are divided into two kinds. One is at position "B" and the other at "C". In this case, the distance between

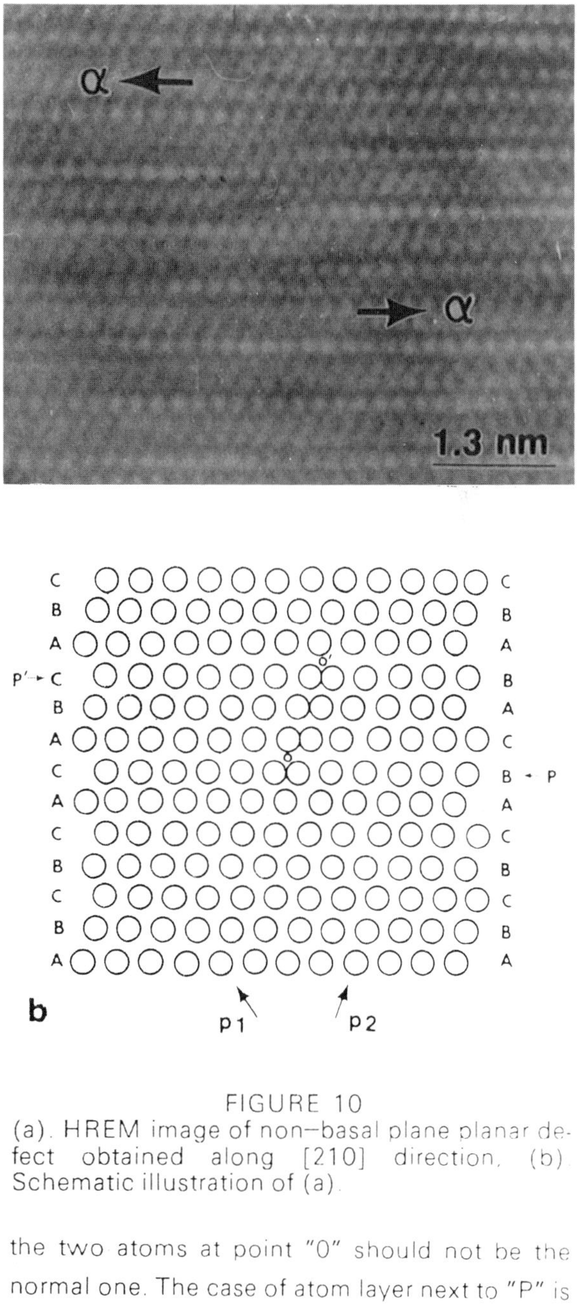

FIGURE 10
(a). HREM image of non–basal plane planar defect obtained along [210] direction. (b). Schematic illustration of (a).

the two atoms at point "0" should not be the normal one. The case of atom layer next to "P" is also the same and should not recover the perfect structure until another incomplete α–type stacking fault appears on the other side of the martensite plate, i.e., the stacking fault appears at "P′" plane and broken at "0′". In this case, a planar defect is formed along "00′" direction

Similar defect has also been found in Cu–Al–Ni alloys[23].

The appearence of nonbasal plane planar defect is considered to be the result of fast quenching rate and results in the incomplete of martensite substructure. So, it is considered to be harmful to thermelasticity and related SME.

4.2. Crystal Defects Formed by Deformation and Thermomechanical Cycles

The martensite undergoes sever deformation during shape memory process and often undergoes thermomechanical cycles. It is important to understand the crystal defects formed and the behaviors of them. By detailed TEM observation of specimen deformed at martensite in different strain[24]. It can be concluded that when deformation is small the strain is contributed by reorientation of variants of martensite. Few crystal defect is induced. As the deformation exceeds the reorientation stage, fine deformation twins formed in martensite. This is the main microstructure formed during the deformation of martensite and the thermoelasticity of martensite lost when it is formed on a large scale. However, the microstructure of martensite induced by thermomechanical cycles, i.e., the two–way shape memory training process, is really different[25] Fig.11 is the general morphology of the microstructure obtained. The main feature of microstructure is that a lot of orientated dislocations are induced by thermomechanical cycles. Trace analysis has proved that the direction of them is all along $[210]_M$direction. It has been found that a comb–like dislocation inside parent phase after thermomechanical cycles and its direction is along $[111]_P$[26]. As have been shown in section 2, the $[111]_P$is parallel to $[210]_M$. Thus, it is believed that the orientated dislocations in martensite correspond to the comb–like dislocations in parent phase. The existence of orientated dislocations in parent phase leads to automatic formation of preferential variants, which results in automatic shape change of specimen when martensite is formed during cooling.

FIGURE 11
Orientated dislocation obtained inside martensite undergone thermomechanical cycles.

Concerning the effects of crystal defects on martensite mentioned above, It is believed that the crystal defects play a very important role in martensitic transformation. The structure of defects and their influence on martensitic transformation is still a challenging topic opening for further study.

5. THE STABILIZATION OF MARTENSITE IN Cu–BASE SMAs

The Cu–base SMAs exhibit severe stabilization of martensite when they are aged both in parent phase and martensite, which is of great importance to the application of Cu–base SMAs. Many works have been made by Chinese researchers in this field[27–30].

As well known, the growth of thermoelastic martensite is dependent on the decrease of temperature. However, It has been found that the martensite grows isothermally in sample quenched from 1073K directly into ice water[31]. The product of isothermal transformation is really thermoelastic martensite which is proved by in–situ observation during heating and cooling. The experiment work revealed that the isothermal growth of thermoealstic martensite is the result of aging. It has been found the Ms temperature increases from 275K to 294K after quenching. So, there is no martensite when observed after quenching immediately at room temperature. But as the aging proceeds, the Ms rises up and the room temperature becomes lower than Ms which results in the formation of thermoelastic martensite. This result proves that the isothermal growth process is an athermal transformation, only that the enviroment temperature is fixed but the Ms increases. So, it is named as pseudo–isothermal transformation.

Further research[32]found that the isothermal transformation is reversible in a two stage aging conditions. When aged at high temperature for short time(20–30 K above A_f), the Ms decreases to a lower value, however, during low temperature aging(near Ms) the Ms recovers its high value, and this process is reversible. The change of Ms mentioned above is one of the phenomena of stabilization of martensite. Since it is a reversible process, the stabilization of martensite of this kind must be related with a reversible transition inside $\beta_1$phase, and is considered to be the ordering process. However the effect of vacancies can not be excluded either, because the existence of vacancies is helpful to ordering process. If the concentration of vacancies is too high, however, the SME of Cu–base SMAs becomes poor[33]. So, no single mechanism is responsible for the stabilization of martensite.

6. THE APPLICATION OF SMAs IN CHINA

Many attempts have been taken in developing the applications of SMAs in China. At present most of the successful application is still achieved by the use of NiTi alloys. In the application of Nitinol in medical field, China has got world–wide reputation. Some clinical applications of NiTi have been made in some departments such as orthopaedic surgery, plastic surgery, orthodontics and so on[34,35]. On the other hand, however, there is still a long way to go for commercial application of Cu–base SMAs. The commercial applications of NiTi alloy on a large scale is also limited due to its very high cost. From the point of view of commercial use of SMAs, the Cu–base alloys are competitive in price. However, the key problem needed to be solved is to improve the mechanical properties, stability and other properties of the Cu–base SMAs.

REFERENCES

1. W.J.Buehler and R.C.Wang, Nickel–Base Alloys. U.S.Patent, 3,174,851, Mar.23, 1965.
2. M.H.Wu, Memory Metals, Inc, Internal Report (1988).
3. L.C.Zhao, T.L.Sun and T.C.Lei, Acta Metall., Sinica, 19A, (1983)271
4. L.C.Zhao and J.H.Yang, Acta Metall., Sinica, 21A, (1985)268.
5. Z.G.Wei and D.Z.Yang, Chin. J. Met. Sci Technol., 4, (1988)305.
6. F.T.Wang, F.X.Chen and D.Z.Yang, Research report, Dalian Research Institute of SMAs.
7. T.C.Li, Y.B.Qiu, J.T.Liu, F.T.Wang, M.Zhu and D.Z.Yang, submmitted to Scr. Metall.
8. L.Delaey, Z.Metallkd, BD 58, (1967)388

9. J.H.Yang, L.C.Zhao, C.S.Zhang and T.C.Lei, Scr. Metall. 21(3), 1987.

10. Zhu Wei–Juang, Chen Wei–Ye and T.Y.Hsu, Acta Metall., 33, (1985)2075.

11. M.Zhu, D.Z.Yang and K.H.Kuo, Acta Metall., 36, (1988)1329.

12. M.Guymont, Phys. Stat. Soli. (a) 38, (1976) 679.

13. E.Hornbogen and H.Warlimont, Acta Metall. 15, (1967)943.

14. G.R.Srinivasan and M.T.Hepworth, Acta Metall., 19, (1971)721.

15. Zhang Jihua, Chen Shuchun and T.Y.Hsu, Materials Science Progress, 3, (1989)105.

16. Chen Shuchun, Zhang Jihua and T.Y.Hsu, Acta Metall. Sinica, 22A, (1986)269.

17. J.Y.An, M.Zhu and D.Z.Yang, Chin. J. Met. Sci. Technol., 5, (1987)44.

18. T.Y.Hsu and Zhou Xiaowang, Materials Science Progress, Sinica, 3, (1989)391.

19. M.Zhu, J.Y.An and D.Z.Yang, J.Materials Science, 25,(1990)2047.

20. D.Z.Yang and M.Zhu, J.Dalian Institute of Technology, 25(2), (1986)80.

21. J.Y.An, M.Zhu and D.Z.Yang, Scr. Metall., 23, (1989)529.

22. M.Zhu, D.Z.Yang and C.L.Jia, Metall. Trans., 20, A, (1989)1631.

23. W.G.Wang, Y.J.Zhange and J.C.Huang, Acta Metall. Sinica, 25, A, (1989)307.

24. M.Zhu, G.B.Li and D.Z.Yang, Acta Metall. Sinica, 26, A, (1990)225.

25. M.Zhu, F.X.Chen and D.Z.Yang, to be published in Acta Metall. Sinica.

26. D.Rios–Jara and G.Guenin, Acta Metall., 35, (1987)109.

27. Cao Mingsheng and Lu Xiaoye, Proc. of Shape Memory Alloy's 86, p286, Guilin, China, 1986.

28. L.C.Zhao, J.H.Yang, C.S.Zhange and T. Q. Lei, ibid, p292

29. Qi Xuan, Jiang Bohong and T.Y.Hsu, ibid, p297.

30. Z.G.Wei and D.Z.Yang, Materials Science and Engineering. A119(1989)145.

31. D.Z.Yang, M.Zhu, M.Qi and G.B.Li, Scr. Metall. 20, (1986)1717.

32. M.Zhu, J.An and D.Z.Yang, Scr. Metall. 21, (1987)1645.

33. L.C.Zhao, J.H.Yang, N.H.Liu, C.S.Zhange and T.Q.Lei, Acta Metall. Sinica, 23, A, (1987)380.

34. Xue Miao and Jia Weitao, Proc. of Shape Memory Alloy's 86, p411, Guilin, China, 1986.

35. K.R.Dai, R.S.Wu, J.X.Yuan and Y.H.Sun, ibid, p430.

Frontiers of Materials Research/
Electronic and Optical Materials
M. Kong, L. Huang (editors)

MATERIALS ASPECTS, ESPECIALLY LATTICE DEFECTS ASPECTS OF COLD NUCLEAR FUSION

Ryukiti R. HASIGUTI

The Science University of Tokyo, Faculty of Engineering, Kagurazaka, Shinjuku-ku, Tokyo 162, Japan

The so-called "cold nuclear fusion" is a phenomenon which can be considered to be the nuclear fusion around the room temperature. This was observed in the electrolysis of heavy water containing some impurities, and was first reported by two groups of researchers almost simultaneously in March 1989. Fusion rates reported heretofore are too high to be explained by the conventional nuclear reaction theory. It is generally accepted that some additional factors, which are not yet clear, must be considered. The present author considers that materials aspects, especially lattice defects aspects of cathode materials, as the above mentioned additional factors, are important, and he proposes to consider interstitial deuterium dynamic crowdions and multi-deuterium atoms trapped at vacancies.

1. INTRODUCTION

In March 1989 two groups of researchers reported almost simultaneously about a phenomenon, which is now called "cold nuclear fusion". These two independent papers were authored by Fleischmann and Pons[1] and by Jones et al.[2]. Since that time many confirmation experiments were performed by many researchers all over the world, but until today the final consensus is not yet obtained.

The purposes of the present paper are, firstly, to point out that some additional factors other than conventional nuclear physics, which are not yet clear enough, must be considered to elucidate the cold nuclear fusion, and, secondly, to describe mechanisms of cold nuclear fusion proposed by the present author from the point of view of materials science, especially from the point of view of lattice defects in cathode materials.

2. THE PRESENT STATUS OF COLD NUCLEAR FUSION

2.1. Overview

There are already many review papers on this subject, so that it would not be necessary to make another comprehensive review here. But it would be necessary to describe the present status of researches in connection with the above mentioned additional factors especially from materials points of view.

As mentioned above, two papers, one by Fleischmann and Pons and the other by Jones et al. opened the field of cold nuclear fusion. One of the most important points in their papers is that the experimental process they used is the electrolysis of heavy water containing small amount of impurities which make heavy water electrolytically conductive. Another important point is that the fusion is considered to occur at the surface or within the bulk of the cathode materials. They and other researchers who made confirmation experiments used in most cases Pd or Ti as the cathode materials. These metals absorb large amount of hydrogen isotopes to such an extent as PdD or TiD_2. The anode materials they used are Pt or Au.

2.2. Reaction Rates

Jones et al. measured[2] the intensity of neutrons accurately which are considered to be produced as a result of nuclear reaction of

$$D+D \rightarrow {}^3He+n+2.5MeV. \quad (1)$$

The reaction rate Λ they obtained is

$$\Lambda = 10^{-23} \text{ fusion/D-D pair/s.} \quad (2)$$

This value of (2) is extraordinarily large value as a nuclear fusion rate at room temperature. According to the conventional nuclear calculation, Λ should be

$$\text{around } 10^{-63} \text{ fusion/D-D pair/s.} \quad (3)$$

This value of (3) can be raised to the value of (2), if various conditions are assumed. For example: (i) if electrons associated with deuterium nuclei have a fictitious effective mass of about $m_e^*/m_e = 5$, the value of (2) is realized. This heavy mass is actually met in the case of the muon catalyzed fusion, but this is not the so-called "cold nuclear fusion".

2.3. Cold Fusion Due to Other Processes

There are several processes other than simple electrolysis method. Klyuev et al[3] . observed neutrons produced when a LiD crystal is fractured by bullets shooting, and this is considered to be due to a cold fusion and is now called "fracto-fusion". It is considered that at the crack tip of a fracture an electrical potensial is induced, and this accelerates nuclear or atomic particles which induces cold fusion [4] . Wada and Nishizawa[5] observed neutron emissions from Pd electrodes, to which 12 kV AC voltage was applied. As the atmosphere is D_2 gas of 1 atm., it is considered that the energy of bombarding particles would be reduced to a rather small value, because of a reduced mean free path. If an experimental facility is constructed, in which the energy of bombarding D ions can be around 200 eV or larger, the results would be very interesting, as will be discussed later. Arata and Zhang[6] observed neutron emissions from Pd cathode of heavy water electrolysis, when they applied temperature cycles (called "on-off effect") to the cathode to change the concentration of D in Pd cyclically and drastically. This rapid concentration change is achieved by rapid absorption and exhaustion of D atoms by the Pd cathode.

3. DYNAMIC PROCESSES

In the preceding section, three processes were described: namely fracto-fusion process, irradiation process and on-off effect (cyclic temperature change) process. By the irradiation process the present author wishes to include the cases where the energy of bombarding particles at the electrode surface is sufficiently high. A common feature in these three processes is that an atomically dynamic process is involved. In other words atomic collisions at the surface and/or in the bulk of electrode materials are considered to be playing an important role.

In the simple electrolysis process the voltages between electrodes used, for example, by Jones et al. are 3 to 25 V. This is too small voltage to induce effective atomic collisions, as will be discussed later. Nevertheless, Jones et al. observed neutron emissions. If this could be explained by assuming some kind of local micro-cracks where atomic particales are accelerated, this could be considered to be a kind of micro-fracto-fusion. There are some indirect evidences for this interpretation. In many electrolysis processes, surface cracks, distortions and swellings of cathode materials are very often reported. These reports are considered to be limited only to the cases where the above mentioned distorsions are remarkable. But there would be many cases where the distorsions are not macroscopically apparent. In such cases micro-cracks which can only be observed by microscopes would be existing. Thus the simple electrolysis process is also considered to be a kind of micro-fracto-fusion. Figure 1 shows examples[6] of cracks, distorsions and swellings of Pd cathode. Although this is the case of on-off effect process and the distorsions are heavy, the general appearance is the same in different cases, even if the distortion is weak.

In summarizing this section, the important common feature in various processes of cold nuclear fusion descrbed above is atomic collision process. The mechanism of cold fusion which is proposed in the present paper is the collisions between two or more deuterium atoms or nuclei with sufficient kinetic energies.

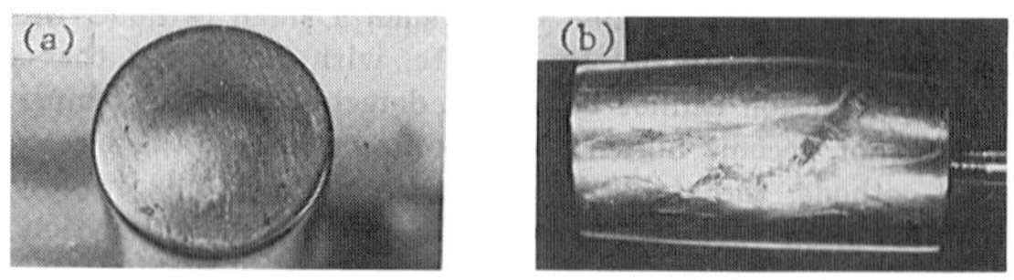

FIGURE 1
Examples of cracks, distortions and swellings of Pd cathodes (Ref.6)

4. PROPOSAL OF NEW MECHANISMS

4.1. Atomic Acceleration Process as a Premise of Lattice Defects Mechanisms

As was mentioned above, an atomic acceleration process and accordingly an atomic collision process are important to increase the fusion reaction rate Λ. This was calculated by several researchers. For example, Oryu[7] calculated approximate values of fusion rates as a function of incident deuterium energies. Here the calculation is based on a two body collision where an incident deuterium atom directly collides with another deuterium atom. A part of his results are shown in Table 1. Incidentally the log(fusion rate) value of -23.1 in Table 1 is almost equal to the value (2) of Jones et al. In section 2.3 it was pointed out that experiments using bombarding D atoms of incident energies of around 200 eV or more would be interesting, and in section 3 it was mentioned that the electrode voltage of 3 to 25 V is too small to induce effective atomic collisions. These statements are based on the above calculations of Table 1.

TABLE 1

D-D fusion rates as a function of colliding D energies and D-D distances (Ref.7)

Incident energy (eV)	Fusion rate (log Λ) (f/D-pair/s)	D-D distance (r_1 Å)
27.2	-63.8	0.51
163.2	-23.1	0.16
190.4	-20.6	0.14

4.2. Importance of Lattice Defects

There are many evidences that lattice defects in cathode materials are important in promoting the cold fusion. For example, it is said that cold worked materials are better than cast materials. This can be considered that dislocations and point defects in cold worked materials play important roles. But the effects of lattice defects in cathode materials on cold fusion have not hitherto been investigated systematically, so that this kind of studies must be performed in the future.

4.3. Multiple D Atoms Trapped by Vacancies

The crystal structure of Pd metal has the face-centered cubic (fcc) lattice. D atoms in solution occupy interstitial octahedral (O) sites as shown in Fig. 2, where the concentration of D atoms is saturated to the extent of PdD (Pd:D=1:1).

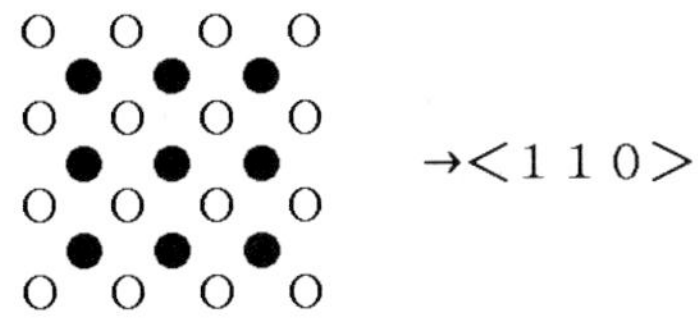

FIGURE 2
(100) atomic plane of Pd crystal containing D to the saturation point of PdD. O : Pd atom, ● : D atom. The solute D atoms occupy interstitial O-sites of Pd cryastal

Now atomic vacancies of the Pd lattice and the trapping of D atoms by Pd vacancies will be considered. We will consider only the case of low D concentration. As the D concentration becomes high, the trapping would become complicated, so that this will be discussed later elsewhere. Besenbacher et al.[8] made detailed investigations of trapping of D atoms by Pd vacancies. According to them, up to six D atoms can be trapped by a single vacancy of Pd crystal. Figure 3 was drawn by the present author following the statements of Besenbacher et al. This shows that four D atoms are trapped by a vacancy. Two more trapped D atoms may be situated along the line perpendicular to the plane of the paper, three dimensionally symmetrically with respect to the center of

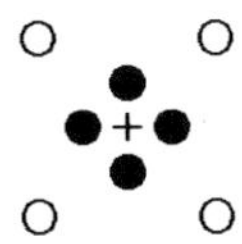

FIGURE 3
(100) face of a unit cell of Pd (O) crystal having a vacant site (+). Four D atoms (●) are trapped in the vacancy.

the vacancy marked +. The distance between two diagonally opposite D atoms was calculated by them to be 1.85Å, which is quite larger than the interatomic distance 0.74Å of D_2 molecule. If these values are compared with the distances r_1 in Table 1, it is obvious that two or more D atoms trapped in a vacancy would not make fusion reaction spontaneously in their static condition at the room temperature.

4.4. Particle Irradiation of Vacancy Trapped D Atoms

As was pointed out in the preceding section vacancy trapped D atoms would not make nuclear fusion spontaneously at the room temperature. Then we need to have the particle irradiation, by which particles, either Pd or D atoms, would collide with vacancy trapped D atoms to transfer enough energy (Table 1) to produce cold fusion reactions.

Now we consider the case where the D concentration, n, is small ($n \ll 1$). The probability that irradiation particles would be D atoms should be smaller by a factor of n compared to the probability that irradiation particles would be Pd atoms. Even if n is small, the vacancy concentration is usually much smaller than n. Then it is considered that almost all the vacancies are trapping two or more D atoms. Then even if the incident particles may be Pd atoms, two or more atoms in each vacancies have a probability to induce fusion reaction upon irradiation. This should be compared with the case where solute D atoms are separately and randomly scattered in regular interstitial O-sites without vacancy trapping. In this case the chance of D-D collisions would be of the order of n^2, which is very small.

It should be added here that the enhancement of cold fusion by vacancy trapped multiple D atoms is not large, because vacancy concentrations are always small and are of the order of magnitude of 10^{-5} to 10^{-4}.

Incidentally, it would be interesting to point out that there would be a possibility that 3D fusion reaction or three body collisions could occur, because three or more D atoms are trapped in single vacancies. Recently Takahashi et al[9]. observed neutron emissions which can be interpreted to be due to 3D fusion reactions.

4.5. Particle Irradiation of Dislocation Trapped D Atoms

The tensile stress side of edge dislocations would trap interstitial solute D atoms. If a dislocation density of 10^{10} /cm^2 is assumed, the number of trapping sites would be around 10^{17}/ cm^3. Thus the contribution of dislocations is almost the same as or a little less than that of vacancies. Therefore, contribution of dislocations should not be neglected.

4.6. Dynamic Crowdions

In the case of high D concentrations, we will propose a dynamic crowdion mechanism. The concept of a crowdion as a configuration of interstitials and of a dynamic crowdion as a one-dimensionally rapidly moving style of a crowdion was first investigated in the case of pure metals by many researchers in 1950's. Here only one literature by Gibson et al.[10] will be cited as a reference, which shows detailed computor calculations and computor modellings. We have done some work on crowdions in ordered Cu-Au alloys [11] which have some similarity to PdD alloys.

As is shown in Fig. 2, the PdD alloy saturated with D has a very regular crystal structure, in which crowdions and dynamic crowdions can be formed along <110> D atomic lines. Figure 4 shows the formation of dynamic crowdion. Particle irradiation consisting of either Pd atoms or D atoms from the left hand side edge of the crystal just formed a crowdion, which is expressed schematically by a compressed three D atoms configuration. This crowdion moves rapidly rightward as a dynamic crowdion. Now let us assume that the energy of about 200 eV or higher (refer to Table 1) was given to the first atom of crowdion, then this D crowdion will run rapidly rightward without loosing much energy for, say, several tens of atomic distances. As the nuclear fu-

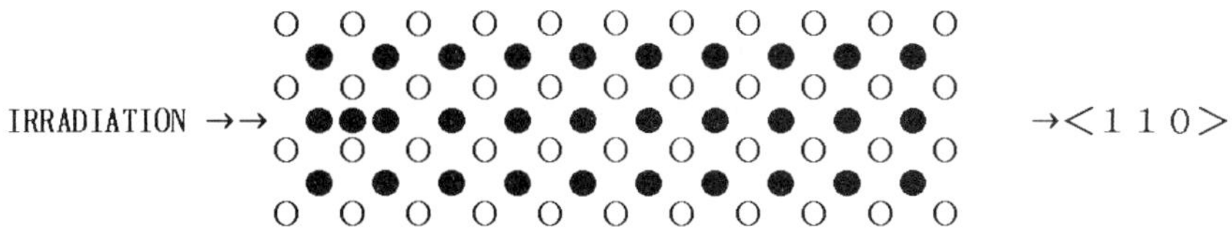

FIGURE 4
(100) atomic plane of PdD crystal. ○ : Pd atom, ● : D atom.
A crowdion of D atomic line is formed at the left hand edge of the crystal, as an atomic particle irradiation is given from the left hand side of the crystal. The D crowdion can move rapidly rightward along <1 1 0> line in this case.

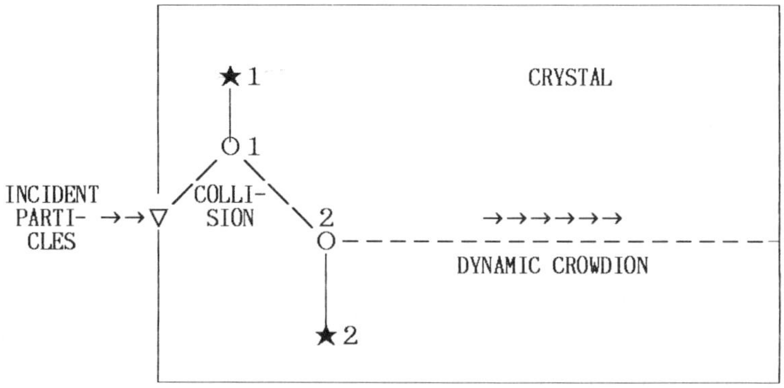

FIGURE 5
A dynamic crowdion is formed after several random displacement collisions. Incident particles bombard the crystal at the point▽, and form vacancies ○1 and ○2 and stable interstitials ★1 and ★2, and if the angle becomes suitable, then a dynamic crowdion is formed, which runs rapidly rightward in this case.

sion cross section, which is considered to be about 10^{-24} cm^2 , is almost ten orders of magnitude smaller than the crowdionic collision cross section, the actual occurence of fusion reaction during the movement of dynamic crowdion for several tens of atomic distances is rather very rare. This is reflected in the extreme smallness of the fusion rate, which is about 10^{-23} fusion/D-D pair/s (Table1).

Therefore, if we assume that one crowdionic collision ends after 30 (the number 30 is just an arbitrary assumption) atomic distance collision chain, the probability shows that it is definitely unlikely that a fusion reaction of such a small reaction rate would occur during this one dynamic crowdionic event. But the interesting point of probability is that this small probability (small reaction rate) might occur by chance in the very first crowdionic collision. Thus we can summarize that if one irradiation particle produces one dynamic crowdion chain of length of m atomic distances we can consider that the fusion rate becomes $m\Lambda$. We arbitrarily assumed above that m is 30, but it would be natural to consider that m is from about 10 to about 100, if the concentration of D is nearly saturation and D atoms are aligned nearly perfectly. If the saturation and the alignment of D atoms are not perfect, the value of m becomes smaller.

Figure 5 shows a case where incident particles do not at first initiate a dynamic crodion. After several random displacement collisions, a dynamic crowdion starts. This kind of events would occur often.

Incidentally, the part of the proposed dynamic crowdion mechanism was preliminarily reported earlier in Japanese[12].

5. CONCLUSIONS

(1) The present status of cold nuclear fusion researches was briefly described in connection with the crystal lattice defects in cathode or electrode materials.

(2) Several lattice defects mechanisms, which enhance the cold fusion rates are proposed. It should be emphasized here that as a premise process, particle irradiations with about 200 eV energy or more are important.

(3) (i) Vacancy trapped multiple D atoms, (ii) dislocation trapped D atoms, and (iii) D atomic line dynamic crowdions are proposed and discussed as new mechanisms.

REFERENCES

1. M. Fleischmann and S. Pons, J. Electroanal. Chem. 261 (1989) 301.

2. S.E. Jones, E.P. Palmer, J.B. Czirr, D.L. Decker, G.L. Jensen, J.M. Thorne, S.F. Taylor and J. Rafelski, Nature, 338 (1989) 737.

3. V.A. Klyuev et al., Sov. Tech. Phys. Lett. 12 (1986) 551.

4. J.S. Cohen and J.D. Davies, Nature, 338 (1989) 705.

5. N. Wada and K. Nishizawa, Japan. J. Appl. Phys. 28 (1989) L-2017.

6. Y. Arata and Y.C. Zhang, Proc. Japan Acad. 66 (1990) 1.

7. S. Oryu, Report of the 133 Committee of the Japan Society for Promotion of Sciences, Sept. 1989.

8. F. Besenbacher, B. Bech Nielsen, J.K. Nφrskov, S.M. Myers and P. Nordlander, Santa Fe Workshop on Cold Fusion Phenomena, New Mexico, May 1989.

9. A. Takahashi, T. Takeuchi, T.Iida and M. Watanabe, submitted, May 1990.

10. J.B. Gibson, A.N. Goland, M. Milgram, G.H. Vineyard, Phys. Rev. 120 (1960) 1229.

11. H. Sakairi, E. Yagi, A. Koyama and R.R. Hasiguti, J. Phys. Soc. Japan, 50 (1981) 3023

12. R.R. Hasiguti, Proc. Topical Meeting on Cold Fusion, Min. of Ed. Sci. and Cult. p.83, March 1990.

Frontiers of Materials Research/
Electronic and Optical Materials
M. Kong, L. Huang (editors)

RESONANCE IONIZATION SPECTROSCOPY AND ITS APPLICATION IN MATERIAL SCIENCE

Keling WEN, Dieyuan CHEN, Wanyun MA, Qing HUI, Wenzheng ZHAO, Liangquan LI and Yuan CHENG

Department of Physics, Tsinghua University, Beijing 100084, China

RIS (Resonance Ionization Spectroscopy) is a photophysics process in which lasers are used to excite and ionize atoms of selected atomic number Z. With commercially available lasers ultrasensitive and ultraselective detection of elemental substances can be realized. As an analysis technique RIS found many applications in various fileds including material science. After a brief description on RIS, the experimental results about RIS analysis of Au and Ru were presented. Detection limits at sub-ppb level have been obtained by autoionizing state approch.

1. INTRODUCTION

The concept of RIS was presented for the first time by Hurst et al. in 1974. An 'exciting potential for ultrasentive detection of elemental substances and other materials' was reported. A demonstrating experiment to show the 'one atom detection ' ability of RIS was done by Hurst et al in 1979, in which single Cs atom was efficiently detected in a buffer gas background of 10^{19} atom/cm^2. So Rutherford's speculation of counting single atoms was realized.

Since then applications of RIS in various field have been preceeded or proposed, such as the measurement of noble gas in deep ocean water, searching for exotic particles and analyzing the products formed in a solar neutrino experiment. Among them, the analysis of ultra-trace elements in materials is certainly an important aspect. More than 40 elements have been now analyzed by RIS technique. Both thermal atomization and ion beam technique have been used.

After a brief review about RIS, the results of RIS analysis on Au and Ru were presented. Detection limits at sub-ppb level were obtained. The autoionizing technique (resonant photoionization via autoionizing state), which had been known for long time, was practically used in these measurements.

2. RIS, RIMS and SIRIS

In RIS tunable dye lasers are used to efficiently and selectively ionize the atoms in certain quantum state of an interested element. Gaseous free atoms were radiated by monocolour or multicolour laser beam as in figure 1.

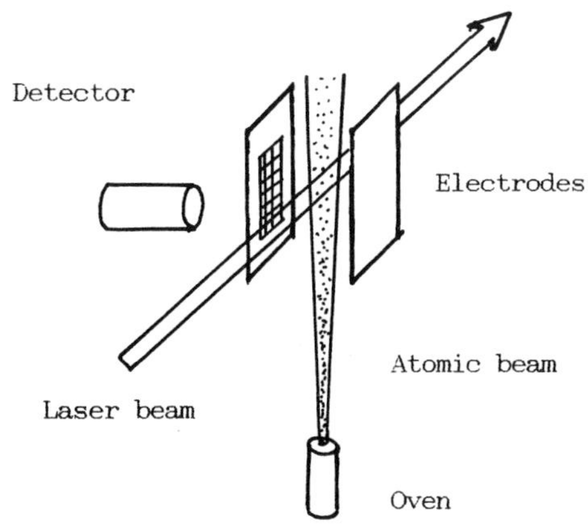

Fig. 1 An example for the explanation of RIS' idea

After absorbing a photon from the laser radiation, the atoms are excited to an intermediate state from its initial state(usually this is the ground state). The energy of this photon must precisely equal to the difference of the atomic energy levels. This can be done by tuning the laser to the correct wavelength. This process is called a resonnant excitation. In order to increase selectivity, it is prefer to repeat this process, by using multicolour laser beams. The excited atoms can be ionized by one of three methods shown in figure 2:

1). Nonresonant photoionization of excited atoms;
2). Electric field ionizatoin of atoms from a Rydberg state;
3). Resonant photoionization via an autoionizing state.

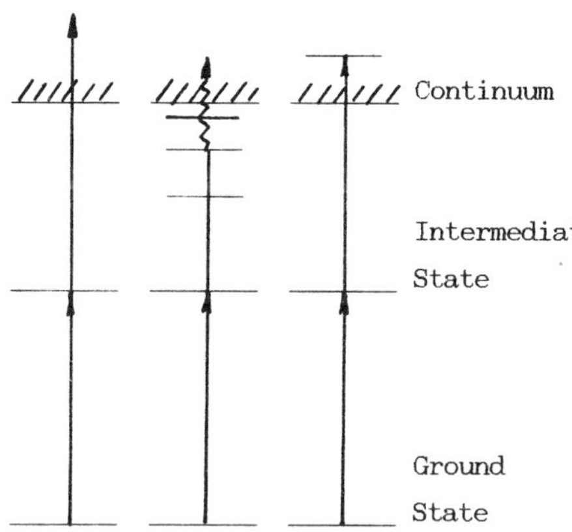

Fig. 2 Schemes of RIS:
(1) Photoionization;
(2) Electric field ionization;
(3) Autoionization.

In method 1, excited atoms absorb a photon with enough energy and go up to the continuum. A laser beam with a power density of 0.01-1 Jule/cm^2 should used to saturate the ionization. In method 2, high voltage pulses (5-20 Kv, 10-20 ns) are used. The necessary electric field strength can be calculated from the effective quantum number n* by the formula:

$$E\ (v/cm) = 3x10^8/n^{*4}.$$

The method 3 is especially interesting. If the atom has a strong autoionizing state, the transition cross section could be as large as about 10^{-15}cm^2, 2-3 order of magnitude greater than that of the nonresonant photoionization. Then the ionization can be saturated with moderate laser power, 1-10 mJ/cm^2 roughly, about the same with the power density for saturated excitation to Rydberg state, but the high voltage pulser is no longer necessary. If a high repetition rate laser (say 20 KHz) was used, extremely high sensitivity could be obtained by this method. Though the existing of autoionizing states has been known for long time, only recently enough information on them has been accumulated and it is possible to use this technique practically.

In this way RIS combines the extreme selectivity with extreme sensitivity. It is selective, by carefully choosing the intermediate state or states, the power density and the bandwidth of the laser beam, principlely only the atoms of a given element are ionized. At the same time, it is sensitive, all the atoms in the laser beam of that element can be ionized with an efficiency close to unity by using commercially available lasers. The technique of detecting ions or eletrons with an efficiency actually equal to unity has been known for long time. In this meaning, RIS has the ablity of single atom detection.

It is possible to integrate lasers into a mass spectrometer as a resonant ion source. Magnet sector, quadrupole or Time-Of-Flight (TOF) massmeter serves as an A selector, and the RIS serves as a Z selector. As a more efficient and selective ion source, Z selectivity of RIS helps to remove interferences from molecules and resolve isobars. A selectivity helps to suppress various stray ion background, improves the detection limit significantly.

In the analysis of solid state sample, some kind of 'atomizer' has to be used to free atoms from the solid. Hot filament and graphite furnace are popular used. This technique is now called Resonance Ionization Mass Spectroscopy, or RIMS. When a graphite or metal crucible is used, the evaporated atoms form an atomic beam, this approach has the advantage of sensitivity.

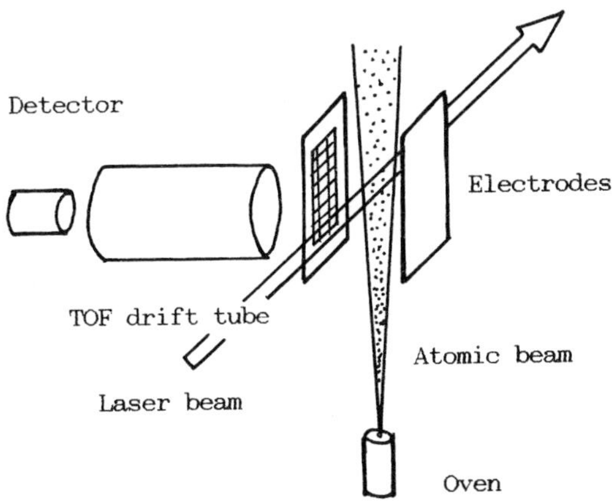

Fig. 3 An example of RIMS: with TOF massmeter

G.L.Bekov et al.[1] reported the analysis of ruthenium trace at the level of 1 part in 10^{-12}. The Ru atoms were evaporated in a thermal atomizer, excited stepwisely by 3 colour laser beams and then ionized from Rydberg state by high voltage pulses. The detection limit of the spectrometer is about 0.5 ppb. Since a fire-assay concentration of Ru into Pb alloy has been used, which gives a concentration factor of 500 for 25g solid state samples or greater then 10,000 for 1 liter seawater, the detection limit of their analysis technique is 1 ppt for solid samples or 0.03 ppt for seawater. Comparing this to the 1 ppb Ru detection limit of conventional analysis method, such as the atomic emission or atomic fluorescence spectrometry, it is a significant improvememt.

In 1986 J.C.Travis et al.[2] gave a survey of elements detected using RIMS, 42 elements were listed in the table, including C, I and metallic elements.

Another technique of atomization in RIS, called Sputtering Initialized Resonance Ionization Spectroscopy, or simply SIRIS, has been developed in many lab for solid sample analysis. An energetic heavy ion beam, such as Ar or Cs, is used to free atoms by sputtering with population characteristic of the solid. This method is suitable to measure the low level impurities in material, as a tool of surface analysis or microprobe, and to do depth profiling.

A well known ion beam technique in material analysis is SIMS. When comparing with it, SIRIS has the virture of higher sensitivity and matrix indepentant nature. By using of certified calibration standards, the measured results of SIRIS may be quantitated. The quantification of analyses for SIMS is difficult, since exact replicate standards do not allways exist. The yield of impurity ions in SIMS is highly dependent on the type of impurity ion, the matrix material and the surface condition.

J.E.Parks et al. first reported the pricniple and results of SIRIS in 1984 [3]. The concentration of Ga and In in silicon, Al and V in stainless steel were measured. With a 45 microamper argon ion source, the detection limit at 2 ppb has been shown for Ga in Si. Since that time more elements more elements have been analyzed by SIRIS. Depth profile of silicon in gallium aresenide was measured too. The results were compared with SIMS' results, the agreement is excellent.

In the 4th international RIS symposium, 1988, K.W.D.Ledingham et al.[4] described the SIRIS machine designed and constructed for trace element analysis in Glaskow University. C.J.McLean et al. presented a SIRIS machine especially designed for profiling semiconductors and the initial results in AlGaAs semiconductors. P.Gelin et al. reported the photoionization schemes, the matrix effects and the actual analysis of trace element (P, Ti, Cr, Fe, Co, Ni, Cu, Zn,....) in Si, GaAs and InP single crystals.

Further more, RIMS or SIRIS is a simple, rapid, and precise tool for isotope ratio measurement. Complicated chemical pretreatment is not necessary since no isobar interference exists. If precisely tuned, narrow bandwidth lasers were used, isotope selectively RIS could be realized. This technique is used to measure the concentration of rare isotope. It is a new isotope seperation method too.

3. EXPERIMENTS

In order to investigate the potential of RIS in trace analysis for geological samples, an electro-thermal atomization resonance ionization time-of-flightspectrometer has been built in Tsinghua University, Laser Single Atom Detection Laboratory.

Graphite furnace with replaceable graphite crucibles was used as the atomizer. The heating time and temperature (as high as 2500 c) can be programed with a microprocessor controler. The evapourateed atoms form a collimated beam, pass through a pair of thermal ion deflecting electrodes and enter the interaction area. The laser system consists of three dye lasers pumped by anexcimer laser, Lambda Physik model EMG 202 MSC. A 1.5 m TOF mass spectrometer has been coupled to the chamber with Eisnel lens and deflect plates. Normally, a 3kv accelerate voltage is used. Fro gold atom (A = 197), the flight time is 20 s, typical mass resolution is 390. The ion signal are recorded by a micro-channel plate detector and fed to a QDC (charge-digit-convertor) or TDC then ADC. The data are storaged and displayed actively by an IBM PC/AT microcomputer.

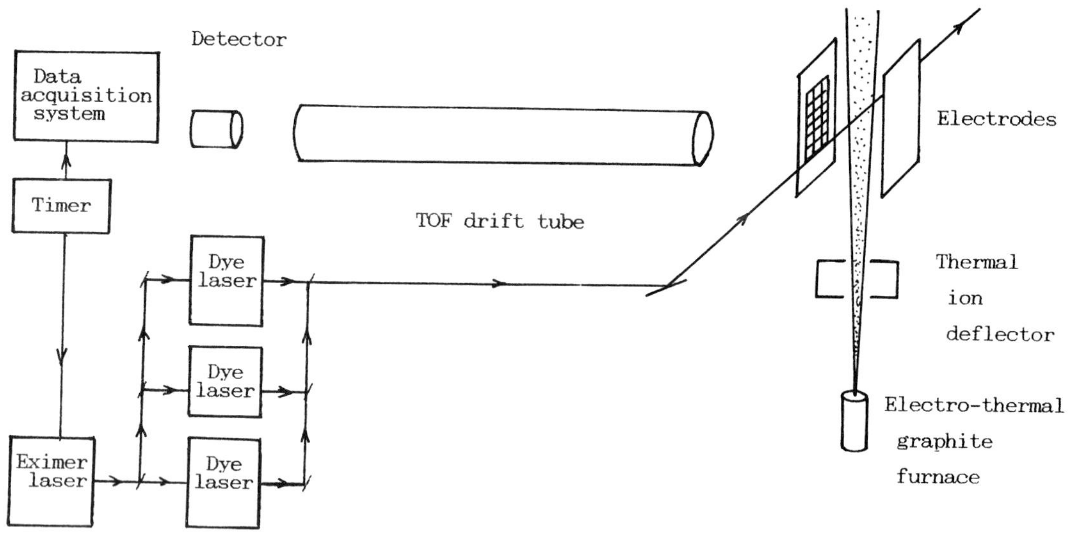

Fig. 4 Scheme of the electro-thermal atomizing resonance ionization time-of-flight spectrometer

The exploration of gold mines attracts great attention for obvious reason. The analysis of gold concentration in geological sample at ppb level is a troublsome problem. So gold was chosen as the first element for our RIS measurement.

Different exciting and ionization schemes have been tried.

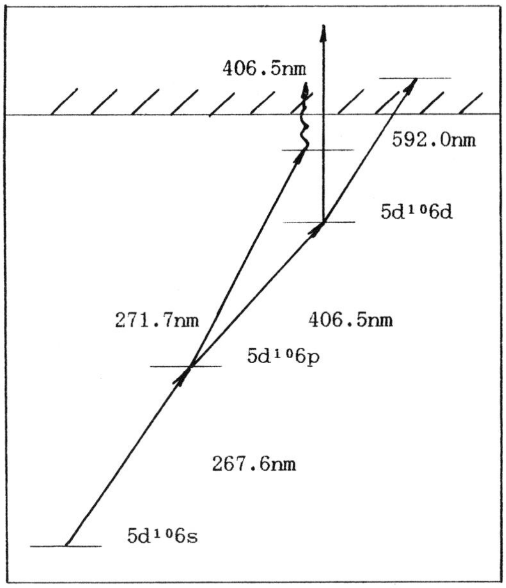

Fig. 5 RIS Schemes of Au

(1) Nonresonant 3 photon ionization:
Au atoms were excited from its ground state to an excited state by the absorption of a ultraviolet photon (2675.95 A), and then to another state by a visible photon (4676.01 A). Ionization occured when a excited gold atom absorbed a photon of either colour. For a 1 ppb, 10 mg gold sample, this approch produced several RIS ions with negligible non-resonant background ions.

(2) Electrical field ionization:
Au atoms in ground state were excited to a Rydberg state by absorption of two ultraviolet photons (2676.95A and 2717 A), and then ionized by a high voltage pulse. Both transition were not saturated, the RIS ion signal was smaller than the photoionization.

(3) Resonant photoionization:
Except the laser beams in (1), a third laser (5920A) was added. Exited gold atoms were prompted to an autoionizing state[5], and then ionized. For the same sample as in (1), this approch produced 100 RIS ions, 10 times of that for nonresonant photoionization. The number of

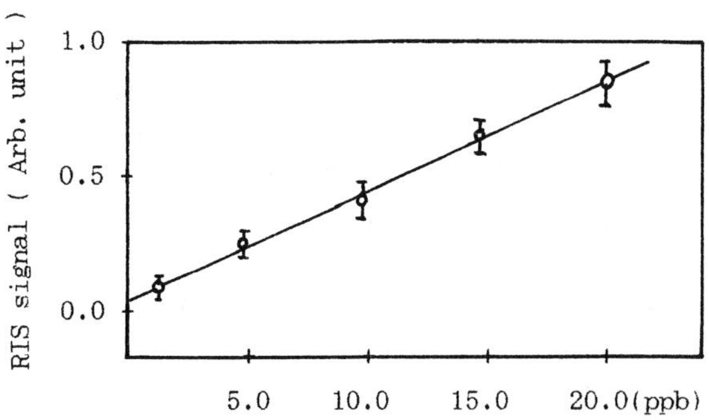

Fig. 5 Calibration curve with 10μl $AuCl_3$ standard solution

background ions was 20, gave a detection limit of 0.2 ppb.

Several standard aqueous solution Au samples were measured, the results were shown in figure The linearity identified the detection limit mentioned above.

Ruthenium is an element of the platinum group. Its distribution in natural objects is interesting for geologists and geochemists. Since its contents in geological samples is in the ppb or even sub-ppb level, the analysis of Ru is a difficult task for normal analysis method.

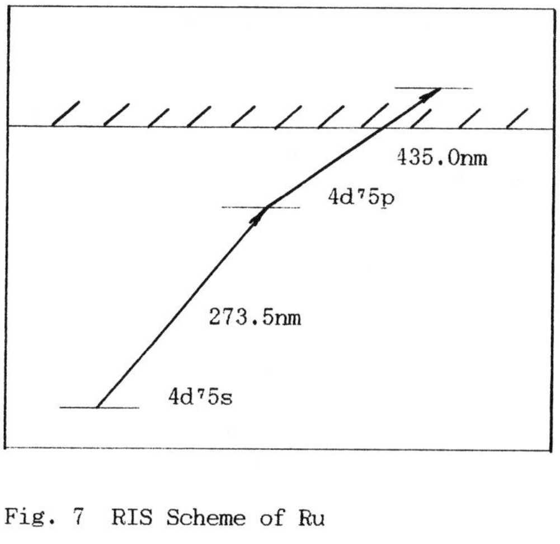

Fig. 7 RIS Scheme of Ru

The RIS measurements of Ru were carried out in our system by both photoionization and autoionization approches. In the former, a 10 ppb, 100 μl aqueous solution standard sample of Ru produced about 400 RIS ions, the nonresonant background ions was 10. This gives a detection limit of 0.25 ppb.

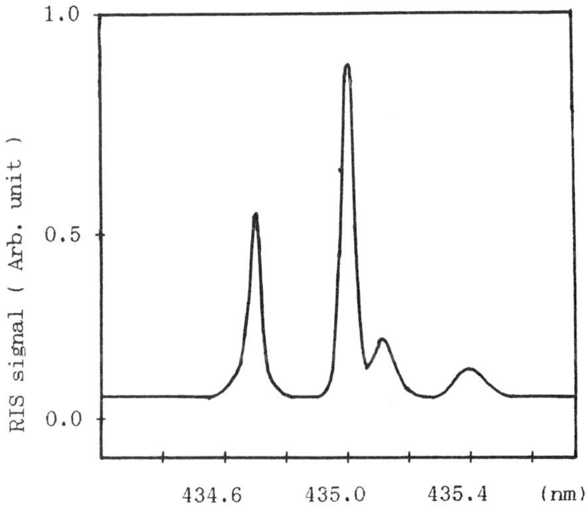

Fig. 8 Autoionizing state scanning of Ru

No autoionizing state of Ru has been reported before in literature, so we tried a RIS scheme shown in figure 7. The second laser was scanned between 4230 A and 4381 A. Four autoionizing state had been found around 4350A. The strongest one is at 4350.74 A, has a width of only 0.3 A. When the laser beam cross section was 2x3 mm , it was found the ionization was saturated at a pulse energy of 0.4 mJ, corresponding to a transition cross section of $0.8 \times 10^{-15} cm^2$. For a 10 ppb, 10μl Ru sample, 3000 RIS ions had been recorded, 80 times greater than that of nonresonant photoionization. The background was 50 ions.

A TOF spectrum of Ru was shown in figure 9, the isotope ratios are coincident with standard values.

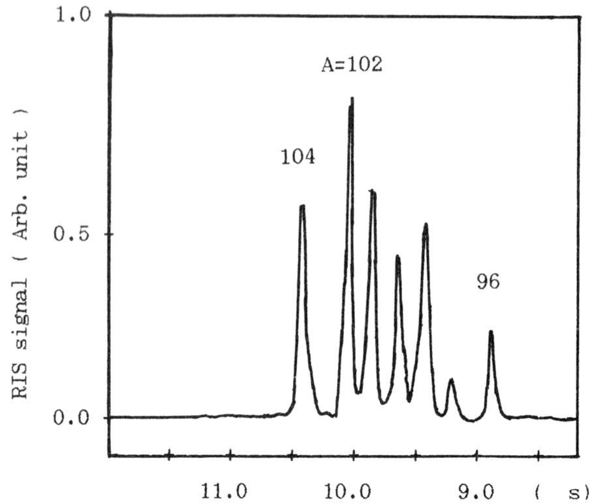

Fig. 9 TOF spectrum of Ru isotopes

Obviously, in RIS analysis the ion signal is proportional to the repetition rate of laser, and so the theoretical maximum sensitivity. It is difficult to raise the rate when nonresonant photoionization approch or electrical field ionization approch was used. In the former, average power of laser would be too large. In the later, there is technical problem to manufact a high rate and high voltage pulser. But no essencial problem exists for autoionization. Even in a high rate, say 20 KHz, the average power is in the 1-10 W range. The progress of laser technique would produce such lasers in a few years.

Many new autoionizing states of gold were found in our experiments. For Ru, new autoionizing states were found in the first run. This may be an incidence. But it looks as if many such states exist, suitable for RIS use. If this is true, then the resonant ionization via auoionizing state will be the best approch in RIS.

4. CONCLUSION

Now some sixteen years after the birth of RIS, once called 'one atom detection', people have a more realistic viewpoint on it. As happens with most new technique, difficulties have risen in practical application. But the general outlook remains optimistic. As an trace analysis technique, for many elements in the period table the detection limit can be now down to sub-ppb level without much difficulty. Since in the next few years economical high repetition rate (10-20 KHz) lasers wuth moderate power may be commercially available, the theoretical detection limit at the sub-ppt level may be a realistic estimation. RIS would find many new applications.

Important exception are H, C, N, O, P, S, At, Se, all halogen elements and noble gas. Either the high ionization potential or the high lying first excited state make it necessary to use two photon transition, and so the very high laser power density. In this case the multiphoton ionization background makes some troubles.

As K.K.Smith et al. indicated, 'each trace analytical technique has its pros and cons. It does not appear that any one method will dominate the analysis field in the future.' 'RIS is hampered by the fact that only one element at a time can be detected and a full analysis of all impurities would be expensive and time comsuming.'

Anyway, RIS is one of the ultrasensitive trace analytical technique which has a potential sensitivity as high as one part in 10^{-13} for most metallic elements. The other one is AMS (Accelelator Mass Spectrometer), which is even more expensive and more complicated than RIS. When extra-high sensitivity is the most important point of choice, RIS may be the best candidate.

REFERENCES

1. G.L. Bekov et al., J. Opt. Soc. Am., 2B (1984) 1554.

2. J.C.Travis et al., Resonance Ionization Spectroscopy 1986, Inst. Phys. Conf. Ser. No. 84, 91.

3. J.E. Parks et al., Resonance Ionization Spectroscopy 1984, Inst. Phys. Conf. Ser. No. 71, 167.

4. K.W.D. Ledlingham et al., Resonance Ionization Spectroscopy 1988, Inst. Phys. Conf. Ser.,

5. U. Kronert et al., Appl. Phys. 44A, (1987) 339.

Frontiers of Materials Research/
Electronic and Optical Materials
M. Kong, L. Huang (editors)

SEARCH, CRYSTAL GROWTH AND APPLICATION OF ACOUSTOOPTIC MATERIALS

Hao-Ran TAN

Shanghai Institute of Ceramics, Academia Sinica, Shanghai 200050, China

Acoustooptic (AO) materials and devices are playing important roles in optical information processing. The present paper decribes the research works in this field carried out by the Shanghai Institute of Ceramics, Academia Sinica. These include the search for new AO crystals, growth and defects of AO crystals, devices fabrication and their applications. Some future trends in AO materials research are also discussed.

1. INTRODUCTION

Acoustooptic (AO) devices including modulators, deflectors and tunable filters operate on the principle of AO interaction in different media. In the past twenty years AO devices have been widely applied in laser editing and typesetting, laser printing, laser facsimile, data recording, precise measurement of vibration and flow velocity, spectral analysis of radar and radio signals, electronic anti-disturbance and fast analysis of optical spectrum. The main advantage of AO technology lies in its ability to process parallelly large quantity of information. There is no doubt that AO technology will play an increasingly important part in laser and optoelectronic technologies.

In view of the importance of AO technology, since the 1970's the Shanghai Institute of Ceramics, Academia Sinica, has been devoting on the research and development of AO crystals and devices. In the following sections, some of the research works in this field including the search for new crystals, growth and defects of AO crystals, devices fabrication and their application are described. Finally some future trends in AO materials research are also discussed.

2. SEARCH FOR NEW CRYSTALS

In reviewing the historical development of AO technology, one would find that the discovery of remarkably high AO figures of merit and the subsequent success in crystal growth of lead molybdate $PbMoO_4$ and tellurium oxide TeO_2 have greatly improved the performance of AO devices and thus have been considered as a milestone in the history of AO technology. Just as in the past, the future development of AO technology will depend to a great extent on the discovery and use of new AO materials.

In order to find out some guidelines for the selection of chemical compounds of AO materials, the relationship between the mean atomic weight and the specific refractive energy of different oxide components was studied.[1] As a result, it was pointed out that the components systems which may be selected for exploration of new AO materials are those consisting of PbO, Bi_2O_3, V_2O_5, TiO_2, CrO_3 and Nb_2O_5 as components. Experimental results about the compounds in PbO-Nb_2O_5 and PbO-CrO_3 systems are in consistent with this conclusion.[2, 3]

By means of X-ray diffraction analysis, the solid state reaction process, crystal structure and AO properties of the six compounds in $PbO-Nb_2O_5$ system[2] and the two compounds in $PbO-CrO_3$ system[3] were studied. The 5:2, 2:1, 1:1, 3:1 and 3:2 compounds in $PbO-Nb_2O_5$ system and the 5:1 and 2:1 compounds in $PbO-CrO_3$ system were found to possess values of AO figures of merit significantly higher than that of lead molybdate and thus are promising new AO materials.

A new method for calculating the molecular refractivity of multi-component metallic oxides by additivity of the ionic refractivity of the component oxides was proposed.[4-6] According to this method the density ρ and mean refractive index n of a certain compound can be calculated simultaneously by solving the combination of the following equations:

$$\text{Lorentz-Lorenz Eq.} \quad \frac{n^2-1}{n^2+2}\cdot\frac{1}{\rho} = \frac{R}{M} \qquad (1)$$

$$\text{Glastone-Dale Eq.} \quad \frac{n-1}{\rho} = \sum_i q_i k_i \qquad (2)$$

where M is the formula weight, R the molecular refractivity, q_i the weight percentage of the i-component and k_i its specific refractive energy. M and q_i can be simply calculated from the molecular formula, while k_i has been published in literature[7]. The molecular refractivity for most simple oxides can be obtained from Eq. (1) as their n and ρ have been known. R for a multi-component oxide compound can be regarded as the sum of molecular refractivities of all the oxide components of which the compound is composed. With this method n and ρ can be calculated with an average error less than 5 and 10% respectively for oxide compounds with mean atomic weight larger than 40.[5]

With n and ρ calculated and acoustic velocity and photoelastic obtained empirically, AO figure of merit can be estimated.

This method has been applied to study systematically the phase relationship and the distribution of AO figures of merit in system $PbO-Bi_2O_3-Nb_2O_5$.[6] A new congruent compound $Pb_3Bi_{0.1}Nb_2O_{8.15}$ ($3PbO\bullet Nb_2O_5:Bi_2O_3$=20:1 in molar ratio) was identified.[8] Its crystallographic parameters were determined as follows. Space group: R3m, R$\bar{3}$m or R32, a=7.457(1) Å and c=48.174(1) Å. Melting Point 1258 °C. Specific gravity measured 7.96 g/cm^3. Transparent single crystals with diameter 1.5 cm and length larger than 2 cm were grown by Bridgman method. These uniaxial crystals are negative birefringence with n_o 2.605, n_e 2.556 and acoustic velocity along <001> 3×10^5 cm/s. The AO figure of merit of this compound is found to be 40, a value about four times as large as that for lead molybdate.

A study on the pseudo-binary system $2PbO\cdot Nb_2O_5-Bi_2O_3$ resulted in finding of another new compound $Pb_2Bi_{0.2}Nb_2O_{7.3}$ ($2PbO\cdot Nb_2O_5:Bi_2O_3$=10:1) with a superlattice structure of pyrochlorite.[9] Its Laue group was determined to be $\bar{3}$m with a=7.467(1) Å and c=66.14(1) Å. Its birefringence is negative. Calculated and measured density values are 7.69 and 7.66 g/cm^3 respectively.

The crystal structure of the compound $PbBi_2Nb_2O_9$ in this ternary system, the existence of which was reported previously,[10] was also determined.[11] It has a space group Fmmm with Z=4 and a=5.4806(7) Å, b=5.4791(7) Å and c=25.416(3) Å.

In conclusion, it is quite interesting

to point out that research works so far performed on $PbO-Bi_2O_3-Nb_2O_5$ system have revealed that Bi_2O_3 addition into the binary compounds of $PbO-Nb_2O_5$ system, though in small quantity, could enhance significantly their AO figures of merit by increasing the density and index of refraction, while at the same time decreasing the acoustic velocity. Therefore, from the view point of search for new AO materials, this ternary systems is worthy for further study.

3. GROWTH AND DEFECTS OF AO CRYSTALS

Single crystals of TeO_2 and $PbMoO_4$ have been found to be the most important AO materials.[12] Two notable features of TeO_2 observed are that a slow shear wave propagating along the [110] direction has an exceptionally low acoustic velocity, which leads to a high figure of merit, and that the crystal has large optical activity. $PbMoO_4$ shows a large value of figure of merit, though not so large as that of TeO_2, as well as a low acoustic loss and hence is very useful for most practical devices. The crystal growth and defects of these two materials have been studied in the Institute.

In the growth of TeO_2 crystal, Bridgman method has been proved to be the most effective technique, which can produce routinely crystals of size up to 40x40x 150 mm with good optical quality provided that the purity of raw-materials is cautiously controlled and a suitable temperature gradient along the growth direction is carefully monitored. The defects which may exist in TeO_2 crystals are comb-like defects, core, growth striation, locally cohered solute layers and solid inclusions, as examined by Tamman interferometer and microscopes.[13] These defects can find their origin in impurity of raw-materials, growth habit, imperfection in seed crystal and fluctuation of furnace temperature and temperature gradient. For TeO_2 crystals grown by Czochralski method under moderate pressure, the density of dislocations is found to be determined by the impurity and stress in the crystal, while the growth striation parallel to the solid-liquid interface depends upon the thermal stability of the pulling system.[14]

The relationship between crystal growth parameters and crystal quality of $PbMoO_4$ has been systematically studied.[15] In this regards, the melt composition was determined to be a decisive factor. Melt composition in excess of MoO_3 leads to the formation of gas bubble inclusions due to supercooling of constituent. On the other hand, crystal growth from melt in excess of PbO always results in formation of scattering platelets with nuclei at their centers. It is interesting to note that these platelets appear on specific faces of (011), $(0\bar{1}1)$ and (101), $(\bar{1}01)$ groups. Further study on these platelets with EPMA reveals that they may have a composition of Pb_2MoO_5[15, 16] or an unknown metastable non-stoichiometric phase[17]. It was also found that these platelets grow and chang their shapes from spots of 1 μm size to typical platelets of 10-30 μm in length and 5-10 μm in width during the process of annealing. This result suggests a diffusion-precipition mechanism.[17] With optimized pulling rate, rotation speed and furnace temperature gradient, large crystals (40 mm in diameter and 60 mm in length) of $PbMoO_4$ can be routinely grown from melt of stoichiometric composition by Czochralski method with a buoyant force balance set-up specially designed for automatic control of diameter of the growing crystal boule.

4. DEVICES FABRICATION AND APPLICATION

On the basis of crystals growth of lead molybdate and tellurium oxide, different types of AO devices have been designed and fabricated using these crystals as AO interaction media and 36° Y-cut lithium niobate crystal plates as transducers. Indium bonding is performed in vacuum under moderate pressure.

AO modulator (AOM) is the most widely used device of the kind. Its design requires a compromise between two performance parameters, i.e. diffraction efficiency and modulation bandwidth, according to the requirement of specific application.[18] For example, AOM for wideband data recording requires a large bandwidth and a moderate diffraction efficiency; while for application in laser switch, single-dot scanning system and Doppler velocimeter, high diffraction efficiency is expected, and a bandwidth of several MHz would be enough. Thus a series of AOM models with $PbMoO_4$ and TeO_2 as interaction media, which operate for laser wavelength 480-1500 nm at a central frequency 40, 80, 100 or 150 MHz with transmission 95% and a diffraction efficiency of 85% at 0.6 watt driven power, have been developed.[19]

AO deflector (AOD), having the advantages of wide instantanious bandwidth, high resolution and low driven power, are the key components of different types of optical signal processors. But in China, the most successful application of AOD in the past years has been spectrograph for radio astronomy, in which the input radio frequency signal is transform into optical deflection pattern via AO interaction. For this purpose, a TeO_2 off-axis AOD was fabricated, and a resolution of 29 kHz with a 1 dB bandwidth of 22 MHz were obtained.[20] This device has been used to construct an AO spectrograph of the 4-m mm-wave radio telescope at Epping, NSW, Australia, for observation of very narrow interstellar molecular lines.

Later, another TeO_2 AOD of the type "on-axis" was designed and fabricated for the mm-wave radio spectrograph of a 13.7-m radio telescope, which was constructed in the U.S. and is now being installed at an observatory in Qinghai Province.[21, 22] High resolution (17.5 kHz) and good stability (with drift only ±0.03 channel/day) were achieved. These distinguishing features make it very suitable for the measurement of fine structure of cosmic molecules and the long term integral of astronomical weak signals. In the design of this AOD, "on-axis" working regime was chosen as it could provid good resolution and stability, which are the two performances required. With proper choice of transducer configuration, the frequency response curve was tuned to smooth out the detrimental dip which usually appears at the middle of frequency response for "on-axis" devices.[21]

To meet the requirement of a radio spectrograph constructed for the observation of solar radio spectrum in the meter-wave band, a wide bandwidth AOD with $PbMoO_4$ crystal as interaction medium was fabricated.[23, 24] A moderate diffraction efficiency has to be selected experimentally so as to avoid the interference between two neighboring diffracted beams while maintain a suitable sensitivity. With this device the AO spectrograph attains a 3 dB bandwidth of 70 MHz and a resolution of 250 kHz when working in 230-300 MHz range. It was set up at Yunnan Observatory, Kunming, and since 1984 has been being used to observe the dynamic spectrum of solar radio burst.[25]

AO tunable filter (AOTF) is a solid-state spectral bandpass filter operating on the principle of AO interaction in an anistropic birefringent crystal. It has come to be of considerable interest and importance in recent years due to its capability to perform real time spectrum analysis and electronic tuning with fast scan rates, wide angular aperture and high resolution. For the implement of fast analysis of spectral response characteristics of color vidicon, a TeO_2 AOTF of non-collinear configuration was designed and fabricated.[26] With optimized design of device structure, optical incident angle and acoustic propagation angle, a diffraction efficiency up to 90% could be obtained when operated in 680-450 nm range and tuned with radio frequencies from 42 to 95 MHz.

For AO devices working at high driven power level, such as $PbMoO_4$ AOD, the thermal effect must be taken into account, because there is only a small percentage of rf power can be converted into useful ultrasonic power, while most of the input rf power is transformed into heat energy. This thermal effect was studied by means of a Mach-Zehnder interferometer, with which the temperature change at any point on the AO crystal in operation can be measured by the variation of number of interference fringes at that point. It was observed that thermal effect could exert harmful influence on device performance by producing an uneven temperature field in the crystal, which results in diffusion of diffracted beam spot and drift of diffraction angles inevitably. An electrode structure with copper heat dissipator was designed and has been proved to be effective by decreasing two thirds of the rise in device temperature at 2-watt driven power level.[27]

5. FUTURE TRENDS

Although research in AO materials has achieved great successes in the past two decades, yet there remains quite a wide field to be explored. Some future trends in this field are discussed as follows.

In the search for new AO oxide crystals, systematic research in systems with PbO, Bi_2O_3, V_2O_5, TiO_2, TeO_2, CrO_3 and Nb_2O_5 as components could lead to discovery of new AO crystals as stated before. Emphasis should be laid upon the anisotropism of properties, especially acoustic velocity and attenuation, and its intrinsic dependence on crystal composition and structure in birefringent materials.

There is an increasing need for IR AO materials to meet the requirements of CO_2 laser technology, optical fiber communication, sensing and information processing in IR range. Several crystals such as Hg_2Cl_2,[28] $ZnGeP_2$,[29] Tl_3AsS_4 and some other chalcogenides[30] have been found to be perspective.

Crystal growth is always a key link in the chain of R&D of new crystals. There are many crystals which have been proved to have good performance but so far can not become available materials. The reason for this is that the feasible methods for their industrial growth are not established as yet. In the field of AO materials research, works on crystal growth have significantly lagged behind. Crystal growth for most of the above mentioned IR crystals remains to be studied and developed. The problems encountered during crystal growth from high PbO content melts, which tend to volatilize intensely, as well as crystal growth of bismuth compounds, which tend to crack along their cleavage faces, have to be solved before these materials become available.

In recent years, most effort to develope bulk wave AO devices has been focused to wideband deflector and tunable filter. The wideband AOD is poised for use in spectrum communication systems, radar warning receivers, adaptive antenna formers, matrix/linear algebra processors, correlators and convolvers,[31] and remarkable results including a 1-GHz bandwidth device with a diffraction efficiency of 100% per rf watt have been demonstrated.[32] AOTF has recently reached technological maturity, moving from the research laboratory to the more demanding commercial and military enviroment.[33] These two kinds of AO devices are expected to find their wide applications in the near future.

Waveguide type AO device, which simultaneously utilizes AO and electrooptic (EO) Bragg diffractions in waveguides on a common lithium niobate substrate, has long been the subject of research, and progress has been made in this attractive field. It is now possible to design and fabricate high-performance planar AOM and AOD with a bandwidth in the order of GHz. Integration of different components on a common lithium niobate substrate to form AO modules has become a reality.[34] These waveguide AO devices should find a number of unique applications, such as real-time spectral analysis of wideband rf signals, optical switching for optical signal processing and optical computing (especially for wavelength-division-multiplex in fiber optic communication),[35] multichannel data processing[36] and tunable frequency shifting for integrated and fiber optic systems[37]. In accordance with the device development, relevant material issues, including growth, processing and non-destructive characterization of large diameter lithium niobate crystals and substrates with high homogeneity and perfection and epitaxial growth of lithium niobate crystal films, have been addressed and should have due attention right now.

REFERENCES

1. Tan Hao-Ran, Ferroelectrics 28 (1980) 411.
2. Xu Yueying, Jin Xiaoyin, Tan Haoran 1 (1983) 143.
3. Xu Yueying, Tan Haoran, J. Inorganic Mater. 3 (1988) 305.
4. Li Deyu, Wang Peiling, Molecular Sci. and Chem. Res. 2 (1984) 231.
5. Li Deyu, J. Chinese Silicate Soc. 12 (1984) 10.
6. Li Deyu, Wang Peiling, Nie Shuyi, Pu Zhifen, Wu Jindi, Prog. Cryst. Growth and Charact. 11 (1985) 305.
7. D. A. Pinnow, J. Quant. Electron. QE-6 (1970) 223.
8. Nie Shuji, Li Deyu, Xu Yueying, Pu Zhifen, J. Inorganic Chem. 3 (1987) 42
9. Wang Peiling, Wu Jindi, Li Deyu, J. Inorganic Mater. 1 (1986) 27.
10. Japan. Pat. 75-05 (1975) 299.

11. Li Deyu, Wang Peiling, Xu Yueying, Acta Physica Sinica 34 (1985) 946.

12. N. Uchida, N. Niizeki, Prod. IEEE 61, 8 (1973) 1073.

13. Zhong Weizhuo, Hua Sukun, Res. Inorg. Mater., Shanghai Inst. Ceram. (1982-1983) 1.

14. Zhou Zheyi, Li Shuyu, Wang Xiaoxian, Xu Shouqi, Wang Zhengying, ibid. (1980) 62.

15. He Chongfan, Lin Youyi, Su Weitang, Shen Binfu, J. Chinese silicate Soc. 9 (1981) 285.

16. Liao Jingying, Chen Jiaping, ibid. 9 (1981) 18.

17. Fan Shiji, ibid. 9 (1981) 279.

18. Tan Haoran, Bian Huian, Hua Wangxian, Res. Inorg. Mater., Shanghai Inst. Ceram. (1982-1983) 159.

19. Catalog of Acoustooptic Materials and Devices, Shanghai Inst. Ceram. (1989).

20. G.C. Huang, J. S. Wang, G. Z. Zhao, Prog. Cryst. Growth and Charact., 11 (1985) 274.

21. Xu Binghuo, Pei Liben, Appl. Acoustics 8 (1989) 10.

22. Xu Binghuo, Zhao Guozhen, Bian Huian, Chu Haiqun, Jin Jian hui, Appl. Laser 8 (1988) 267.

23. Xu Binghuo, Ren Guoxin, Wang Jingsheng, Xia Zhiguo, Chen Jingying, Appl. Acoustics 7 (1988) 5.

24. Min Maolin, Xu Binghuo, Acta Astrophysica Sinica 3 (1983) 250.

25. Wang Jingsheng, Xia Zhiguo, Chen Jingying, Jiang Suyun, Min Maolin, Xu Binghuo, Huang Gengchen, Ren Guoxin, Scientia Sinica, A, No. 11 (1986) 1196.

26. Huang Gengchen, Zhao Guozhen, Dai Yuanchao, Tong Naizhu, Zhou Xingyu, J. Inorg. Mater. 1 (1986) 301.

27. Xu Binghuo, Zhao Guozhen, Appl. Acoustics 8 (1989) 19.

28. N. B. Singh, Mater. Lett. 7 (1989) 397.

29. I. C. Chang, Opt. Eng. 24 (1985) 132.

30. M. Gottlieb, G.W. Roland, Opt. Eng. 19 (1980) 901.

31. I.C. Chang, IEEE Trans. Sonics and Ultrasonics, SU-23 (1976) 1

32. I.C. Chang, 1988 IEEE Ultrason. Sym. Proc. 435.

33. P. Katzka, SPIE Vol.753 (Acousto-Optic, Electro-Optic and Magneto-Optic Devices and Applications) (1987) 22.

34. C.S. Tsai, IEEE Trans. on Circuits and Systems, CAS-26 (1979) 1072.

35. Y. Miyazaki, N. Goto, Y. Kanayama, 1988 IEEE Ultrason. Symp. Proc. 447.

36. D.Y. Zang, P. Le, C.S. Tasi, ibid. 451.

37. C.S. Tsai, Z.Y. Cheng, ibid. 455.

Frontiers of Materials Research/
Electronic and Optical Materials
M. Kong, L. Huang (editors)
Elsevier Science Publishers B.V., 1991

A DISCUSSION OF IDEAL GLASS AND REAL GLASS

Peinan ZHU

Department of Materials Science and Engineering, Tongji University, Shanghai, P.R. China

Litai MA

Shanghai Institute of Ceramics, Academia Sinica, Shanghai, P.R. China

Huizhong WANG

Shanghai Yaohua Glass Factory, Shanghai, P.R. China

A study on the structure of glasses formed from different thermal histories was realized with high resolution electron microscope and selected area electron diffraction. Submicro-crystals or crystallites have been found and the glass structure was still morphologically characterized by one-dimensional preferred orientation model. It is necessary to distinguish the glass in terms of ideal glass and real glass. The authors indicated that the results give a new contents for investigating the nature of glass.

1. INTRODUCTION

The study of glass structure has long been one of the important problems in modern science. A number of hypothesises, theories and models have contributed to explain the structure and properties of glasses. The theory of crystallite and of random network structure were known as two extremely opposite view point in the earlier period. However, they have not been made compromises yet, in the recent decades. As we known, in 1930' the hypothesis of inhomogeneous structure of glass raised, it seemed to give consideration to the statements of both theories mentioned before. But they can not be summarized in one idea which may be generally accepted by the glass scientific workers. In the same time, some researchers used X-ray small angle scattering method to derive the bond length between short-range ordered atoms, according to the results calculated from the function of radial distribution, they obtained a series of significant data. Other scientists based on the theory of polymerization and the scattering theory of super-critical fluctuation to explain the agglomerating states of glass. However, until now, they have not been united in one point of view yet. Recently, Bando[1] used high resolution electron microscope to study the structure of fused silica glass and meassured the size of the crystallites. Goodman[2] suggested the strain mixed cluster random network model to explain the glass structure. Gupta[3] developed a model of vertex sharing regular congruent rigid polytypes to give a quantitative description for the theory of infinite network structure. Pye[4] put forward the continuum model and especially emphasized the importance of structural transition depending on thermal history. Based on the theory of molecular dynamics, Yonezawa[5] used computer to simulate how the unit cluster gathered during the transformation process. Novikov[6] used continuum approximation method to propose a declination model to explain the phenomena revealed in the glasses. Nevertheless, the key problem involving in the glass structure research should concentrate to that whether the theory or hypothesis could appropriately describe the growth and declination between ordered and disordered micro-regions[7]. So, a clear distinction can be devided between crystalline state and vitreous state, just as Zhu[8] indicated before.

From the view point of micro-structure, Zhu and his coworker investigated the morphological aspects of the ordered and disordered micro-regions in glass with high resolution electron microscope and selected area electron diffraction. Then, one dimensional preferred orientation model was proposed. The authors try to use this model for describing the structure of ideal glass or the glass without any crystallite; and hope to reach agreement with the glass scientists who are interested in this topic. However, in practical conditions, one can easily to obtain a glass in which an extremely few crystallites may be present. So, it is necessary to define the real glass, appeared in most common cases, and to distinguish it from the ideal glass[9].

2. EXPERIMENTALS

Just when the author interested in the study of common chemical glass by the electron microscopy, we coincidently found an extremely a few of submicro-crystals or crystallites in situ of the glass. See Fig. 1. The results inspired us to wonder, whether there would be the crystallites in other kinds of glasses or not. Thus, we took the flat glass sample to be ground into fine powder, then separated them by the gravitational method, and inspected them under HREM.

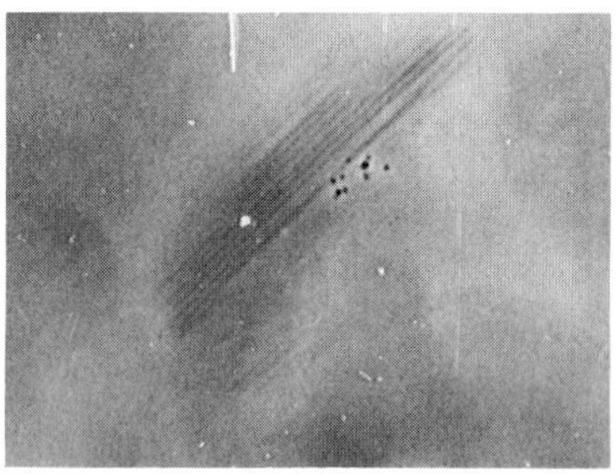

FIGURE 1
Lattice image found in borosilicate glass (under focusing; 1,200,000x)

Electron diffraction was furtherly studied. The results are shown as follows.

Fig. 2 shows the submicro-crystals found in the sample of flat glass separated by gravitational method. The crystallites grew merely in orintated direction. The dark field images corresponding to Fig. 2 were observed in Fig. 3.

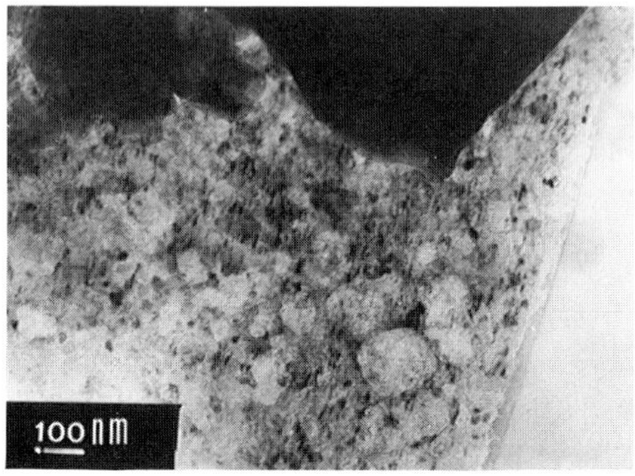

FIGURE 2
Submicro-crystals found in the sample of flat glass sample separated by gravitational method (bright field)

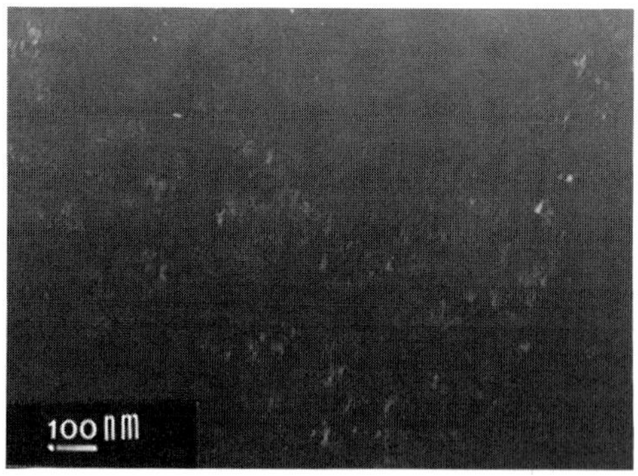

FIGURE 3
Image corresponding to Fig. 2 (dark field)

Fig. 4 is a typical electron diffraction pattern of a single crystal which was found in the flat glass sample. After indexing, we can confirm the name of the crystallite, known as diopside — $CaMgSi_2O_6$. The data calculated by

TABLE 1. Indexing for the electron diffraction pattern shown in Fig. 4

No. of spots	$R_{(mm)}$	$d_{calc.}$	$d_{stand.}$	I	hkl
1	7.25	2.8389	2.991	100	$(22\bar{1})$
2	7.25	2.8389	2.893	30	$(31\bar{1})$
3	10.25	2.008	2.043	14	(041)
4	14.25	1.4194	1.424	14	(531)
5	16.21	1.2697	1.265	6	$(46\bar{1})$

the indexing method were shown in Table 1.

In order to compare the microstructure of vitreous region with that of local devitrified region shown in Fig. 2 and Fig. 3, we can find the characteristic microstructure of ideal glasses shown in Fig. 5 where the HR-micrograph was taken from the sample of alumino-silicate glass and SAED pattern merely appearing dispersive diffraction rings shown in Fig. 6 was observed.

FIGURE 4
Electron diffraction pattern corresponding to the selected area of Fig. 2

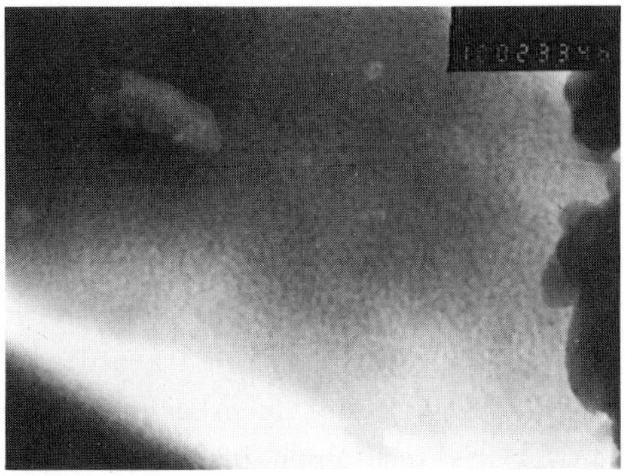

FIGURE 5
HREM image taken from the non-crystalline region of the alumino-silicate glass sample (100,000x)

FIGURE 6
Electron diffraction pattern taken from the selected area of glass sample corresponding to Fig. 5

According to the results of chemical analysis, the composition of glasses are shown in Table 2.

If we study the crystallization of glass in advance, it can be noted that the precipitated crystalline phase should agree consistently with the result inferred from the submicro-crystals appearing in the glass samples. The later are essentially referred to the nuclei which have not grown up yet. So, it will lead us to know

TABLE 2. Chemical composition of different glasses

Kind of glass	Chemical composition (wt%)								
	SiO_2	Al_2O_3	Fe_2O_3	TiO_2	CaO	MgO	K_2O	Na_2O	B_2O_3
Flat glass	72.19	1.56	0.19	0.04	7.18	4.07	0.85	13.97	-
Low-alkali fiber glass	53.92	14.38	0.26	0.54	16.81	4.43	0.10	0.11	8.90
Chemical resistant glass	78.49	2.41	-	-	-	-	0.039	5.41	14.64

an idea that we should not simply regard the glass as an ideal homogeneous matter even to use this idea in discussing the properties of glasses. Actually, it is indicated that submicro-crystal is already grown up to the extent, so much larger than the critical nucleus or embryo existing in the glass as we met in the ordinary conditions. This is the basic experimental verification for distinguishing ideal glass from real glass.

3. DISCUSSION

Normally, we can give some illustrations concerning the high resolution image formation based on the theory of gradient phase contrast. Thus, the distribution of phase contrast intensity created in HREM picture is shown in the following equation

$$I_{(x,y)} = 1-2\sigma\phi_{(x,y)} * F\{\sin x_{(u,v)} \cdot A_{(u,v)}\}$$

where $\sin x_{(u,v)}$, $A_{(u,v)}$ are two functions which will effect the interference contrasts.

If one want to obtain optimum condition for under focusing, after Scherzer condition was adjusted, the highest resolution power will be obtained. That means, under such a condition, one can observe the atoms of an element or the cluster composed by the atoms. However, the materials scientists are not interested in investigating the binding state of atoms or clusters in high resolution electron micrograph. Practically, they are interested to know the morphological informations provided by the ordered and disordered micro-regions; and whether are the submicro-crystals present or not. According to the experimental results, we can elusidate the essential arguments depending on the ideas of ideal glass and real glass. Although we had not give the exact explanation for the high resolution images of multi-component glasses, we could compare the crystalline regions with the non-crystalline or amorphous regions based on the electron micrographs and electron diffraction patterns to clarify the structural features of typical vitreous materials. It should be indicated that one dimensional preferred orientation model is a rational concept for the description of the ideal glass structure thinking in terms of image. Besides, the submicro-crystals found in the real glass, actually, may be regarded as the submicro-inclusion, a defect of glass meassured on structural dimension.

With the view point of the micro-structure of glass, we can make further discussion as follows.

3.1. It is definitely necessary to distinguish glass in terms of two kinds, namely, the ideal glass and real glass. As a matter of fact, oxide systems could easily form glasses. If those glasses were obtained from quenching process, there will be less possibilities to occur submicro-crystals. For this kind of glass, we can call it as ideal glass. In other cases, when a glass may possess some submicro-crystals which probably formed in a definite thermal histories, these extremely tiny crystals will spontanuously grow up to a critical level. Although the appearence of glass shows as same as the ideal glass, but it no longer belongs to the category of ideal glass, other than, we should call them as real glass.

3.2. If one wishes to describe the characteristics of the microstructure of ideal glass, it can be noted that the interweaving network constructed by the one dimensional orientated micro-regions has been recognized before, e.g. in alumino-silicate glass it was already shown in Fig. 5. As for real glass, because the precipitated submicro-crystals are included in the non-crystalline matrix; arounding the crystalline region it is quite possible to appear a structural feature as identical as that of the ideal glass. In similar cases, we will find such a characteristic features for another kind of glass e.g. the fiberglass, without difficulty. Thus, it is believed that the more research works developed, the more evidences will be provided.

3.3. Finally, it is indicated that the presence of submicro-crystals as the submicro-defects in glass and the mismatched properties between crystalline parts and the amorphous matrix would cause the lowering of mechanical strength of the glass. Therefore, it would be expected that to decrease the formation of submicro-crystal in real glass and to improve the homogeneity of ideal glass may have benefit to increase the strength of glass.

4. CONCLUSION

HREM and SAED were used to investigate the structure of glasses. Submicro-crystals were revealed in different kind of glass samples. To distingusih glass in terms of ideal glass and real glass was proposed. One dimensional preferred orientation model is appropriate to explain the structure of ideal glass. The presence of submicro-crystals may be regarded as the submicro-defect in real glass; that would cause the lowering of the strength of glass.

REFERENCES

1. Y. Bando and K. Ishizuka, J. Non-cryst. Solids 33, 3 (1979) 375.
2. C.H.L. Goodmen, Phys. and Chem. of Glass, 26, 1 (1985) 1.
3. P.K. Gupta and A.R. Cooper, Proceed. of 15th Inter. Cong. on Glass, Vol. 1a (1989) 13.
4. L.D. Pye and I. Joseph, ibid. (1989) 19.
5. F. Yonezawa, Nose and Sakamoto, J. Non-cryst. Solids 95-96 (1987) 83.
6. V.N. Novikov, Proceed. of 15th Inter. Cong. on Glass, Vol. 1a (1989) 31.
7. P.N. Zhu and M.X. Wu, J. Non-cryst. Solids, 95-96 (1987) 397.
8. P.N. Zhu, Proceed. of 15th Inter. Cong. on Glass, Vol. 1a (1989) 25.
9. P.N. Zhu, J. Non-cryst. Solids, 112 (1989) 73.

Frontiers of Materials Research/
Electronic and Optical Materials
M. Kong, L. Huang (editors)
Elsevier Science Publishers B.V., 1991

TRENDS IN MEMBRANE SEPARATION MATERIALS RESEARCH

Jiping XU

Changchun Institute of Applied Chemistry, Chinese Academy of Sciences

I. INTRODUCTION

Membrane processes are the most energy-saving among separation processes: here phase transition(where large amount of energy is consumed for the latent heat of vaporization and fusion) is usually absent in comparison with traditional separation processes such as distillation, fractional distillation, crystallization, extraction etc. Separation is achieved by permeation through a separation membrane driven usually by a pressure difference. Application of membrane processes is close related to environmental protection: treatment of waste water and gases. In the present era of shortage of energy resource and serious enviromental problems membrane processes are of high importance and their application is developing at a high speed.

In the beginning stage of membrane separation process polymeric materials for separation membranes were limited to "general purpose" ones such as cellulose acetate, aromatic polyamide, polysulfone and silicone rubber. As a result of developments in polymer science and deepened fundamental researches on the dependence of separation behavior on polymer structure (in different level) more and more new membrane materials with higher flux (or permeation rate) and better rejection (or selectivity) have been developed. Now specific membrane material tailored for certain given separation problem is usually used in practice.

In this paper some new and important from my point of view membrane materials for MF, UF, RO, PD (piezodialysis), gas separation and pervaporation were reviewed.

2. MICROFILTRATION AND ULTRAFILTRATION MEMBRANES

New processes for the preparation of porous membranes and the use of inorganic membrane materials are two of main advances.

2.1. Phase inversion method -- the earliest and most well-studied method for preparing asymmetric membranes. Recently thermally induced phase separation(TIPS) method is used to prepare asymmetric membrane from polymers soluble in some solvent only at higher temperature and for preparation of asymmetric membranes from more concentrated polymer solution.

The "Sunbeam" process of Gelman Sciences used selected solvent in which monomer is soluble but not the polymer. Phase separation takes place as the polymer is formed by photopolymerization of the monomer under a mercury lamp and the solvent is leached out later on. This process has been made continuous and is very productive, up to 350 ft/min at a membrane width of 40 in. The porous MF membrane obtained can be used as vaporpermeable but water-repellent textile materials. Asymmetric membrane with dense skin layer can also be obtained which can be used for nitrogen enrichment from air.

2.2. Sintering method --the oldest method for making porous membranes both inorganic and polymeric. The particles are sieved, and monodispersed particles in micrometer size are pressed and sintered. Such membranes were used for MF but too coarse for UF purposes. In recent years the so-called "sol-gel"process has been developed for making alumina or silica particles of nanometer size. Crosslinked polymeric microgels can also be made into nanometer size. A thin layer of such nanoparticles on MF membrane surface after sintering give asymmmetric UF membrane of controlled surface porosity.

In order to regulate the porosity of support membrane mixed inorganic and polymeric particles are pressed and sintered, then the inorganic component is dissolved out or the polymeric component burned awayduring high temperature sintering to give membranes with higher porosity.

2.3. Stretching method -- homogeneous porous MF membranes were obtained by cold drawing, hot drawing and heat setting films from PP, PE, or PTFE. The stretching process is unidirectional and the pores are needle-like. Recently biaxial stretching of β -PP giving membrane with elliptical pores was succeeded in Shanghai Institute of Organic Chemistry. Porous PET membranes have been obtained by soviet chemists by stretching in an active media like propanol. Porous poly-4-methylpentene-1(PMP) membranes were also obtained by a similar process. It can be expected that more and more membrane materials will be included in porous membrane formation by stretching process.

2.4. Porous alumina membrane by anode oxidation -- All porous membranes formed by above-mentioned methods possess more or less a wide pore distribution. Only nucleopore membranes are known to have nearly monodispersed pores. The Anotec porous alumina membranes were obtained by anode oxidation of Al in acid media. The pores formed are very uniform, the pore diameter is controlled by applied voltage. Asymmetric membranes can also be prepared by stepping down the applied voltage.

The most important area of application of MF and UF porous membranes in the 90's is bio-engineering. Such membranes must stand steam sterilization, solvent or acid wash and likely to be hydrophilic. Inorganic membranes are very suitable for this purpose. Heat stable polymeric materials like polyethersulfone(DUS-40) and polyimides can be used as well. New polyetherketone and polyethersulfone from phenolphthalein (PEK-C and PES-C) developed in our Institute with high Tg are promising materials for this purpose.

Charged membranes are perspective for solving the problem of membrane fouling partially by repelling particles carrying same charge. Sulfonated and tertiary-aminated polysulfones are already in use. Hydrophilicity of the charged membranes also enchanced their flux. The polycationic tertiary-aminated PEK-C UF membrane has been successfully used in the recovery of cationic electrophoretic paints in automobile industry.

Treatment of waste liquor in paper or printing and dyeing industry brings up a serious problem of membrane fouling. For easiness of back washing fluorine containing polymers like polyvinylidene fluoride or membranes modified by grafting silicone or other monomers with low surface tension are generally used.

The chemical balance of catalytic or other reactions can be shifted if one of the reaction product(usually gaseous) is removed partially during the reaction. Increase of yield by several percent can give huge economic effect. Use of membrane supported catalyst formed a new important research field. In Dalian Institute of Chemical Physics encouraging results have been obtained in the dehydrogenation of ethylbenzene. Since the catalytic reactions take place at high temperature inorganic membranes obtained by the "sol gel" process are generally used.

UF in non-aqueous media(for instance, removal of protein and other impurities from edible oil). Oil resistant polyimides are generally recommended.

Cellulose acetate UF membranes can be attacked by moulds and fungi, cyanoethylation of CA (developed by Guanzhou Institute of Chemistry) leads to mould-resisting cyanoethylated CA.

3. REVERSE OSMOSIS

RO membranes for desalination of sea water are working under high pressure in the order of 10MPa, so highly crosslinked polymeric membranes must be used in order to avoid membrane densification. Condensation of furfuryl alcohol and THEIC with some sulfonation gives good RO membrane with high salt rejection and flux, which is now developing in Second Institute of Oceanology jointly with our Institute.

The need in high purity water in electronic industry and for high pressure boilers promoted the development of medium pressure RO process, which use membranes with higher flux. Condensation of polyamine with 1,3,5-benzenetricarbonylchloride gives the most promising materials for MPRO (or nanofiltration), which may found use also for desalination of blackish water in north-west China for drinking purposes.

4. CHARGE MOSAIC MEMBRANE FOR PIEZODIALYSIS

Desalination by electrodialysis using ionexchange membranes has developed into a largest membrane process in China, although the energy cost is much high than RO process, the latter requires the use of high pressure pumps, pipes and fittings and high quality maintenance. If the cationic and anionic resin is made into one membrane in separate domains--charge mosaic membrane, desalination may be realized by ressure driven process avoiding the use of electric field. Simple mixing of cationic and anionic resin powders leads to neutralization of the charges. Other researches are also unsuccessful until Fugimoto synthesized a five-block copolymer with equal block length by anionic living polymerization. After membrane formation the 1, 3 and 5 blocks are crosslinked by vulcanization, and 2,4 blocks are transformed into cationic and anionic segments, the domain size can be regulated to 20-100 nm. Such charge-mosaic membranes have been used successfully for piezodialysis of salt water and also for the separation of inorganic salts from saccharides or aminoacids and also for separation of cations of different valency. Similar charge-mosaic membranes have been obtained recently in East China University of Chemical Technology.

5. GAS SEPARATION MEMBRANES

Main business of membrane processes in the 90's will be in the field of gas separation. The most important processes are: 1)recovery of hydrogen from purge gas in ammonia synthesis and petroleum refinery; 2)oxygen/nitrogen separation from air for oxygen enrichment combustion and nitrogen source for blanketing fuel and inflammable solvent tanks and pipings, for fruit and vegetable preservation under controlled atmosphere during transportation or storage and as laboratory nitrogen source for sweeping modern instruments like FT-IR and NMR or other purposes; 3)dehumidification of industrial gases or air to substitute energy-costing refrigeration or absorption by molecular sieve etc.; 4)recovery of acid gases like carbon dioxide and sulfur dioxide (chief atmospheric pollutants) from flue gas etc.

5.1. Hydrogen recovery -- Prism separator of Monsanto is the first industrialized gas separation set for hydrogen recovery and more than 100 sets are working around the world now. Dalian Institute of Chemical Physics has established similar modules working on polysulfone hollow fibers with silicone coating and put into pilot production. After systematic study of the gas separation behavior of polyimides of different structure our Institute has selected out polyimide membranes with H/N separation behavior appoaching PI hollow fiber modules of Ube Kosan. Modified PES-C and PEK-C show promising as second generation hydrogen recovery membrane material.

5.2. Oxygen/Nitrogen separation from air --most widely used membrane material is composite membrane with active layer from silicone rubber (crosslinked or modified). Spiral wound modules of 4, 6 and 8 in.in diameter and 40 in.in length have been constructed also in Dalian and successfully used in glass-melting furnace with a fuel saving around 20 %.

New membrane materials for O/N separation either with higher permeability or with higher separation factor than silicone are under active research.

Polytrimethylsilylpropyne (PTMSP) has an oxygen permeability tenfold higher than silicone rubber and is intensively developing in Japan, USA and USSR. Air Products reported a vast increase in O/N selectivity from 1.6 to 5 by surface fluorination of PTMSP. In our Institute TMSP monomer and its polymer have been synthesized and surface fluorination by CF4 plasma also enchanced the O/N selectivity to >5.

Other O/N separation material in use include: polymethylpentene(PMP) HF (Dow Chemcal),ethylcellulose(A/G Technology), polyvinyltrimethylsilane(USSR), sulfonated polysulfone(Union Carbide), and Aromatic polyamide and polyimide based on o-substituted diamine (DuPont). We have studied the plasma fluorination of PMP and observed a jump of O/N separation factor to >7 with only minor decrease in permeability.

5.3. Membrane for dehumidification -- water vapor permeates very fast through such hydrophilic membranes as Nafion, polyimide and sulfonated polysulfone or sulfonated PPO, their selectivity of water vapor to common gases usually exceed 1000. The dehumidifying apparatus is simple, compact and requires few maintenance. There is a huge market and Monsanto is now producing hundreds of Prism Cactus modules in month. Ube Kosan has also turned its PI HF modules to the dehumidification market.

5.4. Recovery of acid gases -- carbon dioxide permeates membrane as a fast gas like water vapor. Various kinds of membranes have been used for upgrading the heating value of marsh gas and natural gas and for separation of carbon dioxide from lower hydrocarbons in enchanced secondary recovery of oil by supercritical extraction.

Sulfur dioxide from combustion of S-containing coal is the origin of "acid rains". Its removal from flue gases is an important and cumbersome problem. One of effective methods is by absorption in lime-milk and later oxidized to gypsum which can be used only as building material. China has limited resource of sulfur for producing sulfuric acid, but there are many high sulfur-containing coals(up to 4%). So sulfur dioxide recovery from flue gas to produce sulfuric acid is of vital importance. Polymeric membranes containing sulfolane has been tried but the flue gas must be cooled down before contacting the membrane. A recent report use molten salt (potassium pyrosulfate containing small amount of potassium sulfate and vanadium oxide) membrane supported in porous ceramic matrix between two porous gas permeable electrodes. Hot flue gas flows by the cathode compartment, sulfur dioxide is dissolved in the molten salt,diffused to anode and oxidized there to sulfur trioxide. By the same principle hydrogen sulfide can be recovered as sulfur vapor.

6. PERVAPORATION MEMBRANE

It is estimated that petroleum will be exhausted within one century. Convenient substitute for internal combustion engine fuel would be ethanol from fermentation of biomass and ethanol will become important energy resource in the next century. But existing technology of producing ethanol cosumes more than half of the heat energy obtainable from its combustion. So low energy-wasting process for obtaining pure ethanol from fermentation broth (containing ca. 10% ethanol) must be worked out. Pervaporation is the preferred process for separating water from ethanol. Crosslinked polyvinylalcohol membrane of the GFT is now widely used for obtaining absolute ethanol from ca.90% alcohol. Tens of plants have been built, the largest one produces up to 150 T ethanol daily. But the capital investment is large and both vacuum and cold trap are required and stainless steel frame and support are used. Maybe in China use of dehumidifying modules at temperatures over 100°C (the Vapor Permeation process) will be more realizable than pervaporation. In order to concentrate dilute aqueous ethanol solution prior to dehyration membranes more permeable to ethanol than water must be used. Silicone rubber has been used but the flux is too low, PTMSP shows better, but more effective membrane is still open.

7. MEMBRANE DISTILLATION MATERIALS

Membrane distillation uses microporous membrane from hydrophobic polymeric material with low surface tension like polyvinylidene fluoride. During membrane formation water soluble polymers were added to facilitate pore formation. We have found that the PVP added cannot be disolved out thoroughly due to entanglement of the polymer chains. Addition of inorganic salts during membrane formation has solved the problem successfully. Membranes obtained show higher flux at smaller pore size.

Membrane distillation is found to be applicable also to concentrate solutions up to saturation and crystallization may take place without effecting the membrane distillation process in our experiments.

Frontiers of Materials Research/
Electronic and Optical Materials
M. Kong, L. Huang (editors)

MICROSTRUCTURE DEVELOPMENT OF SILICON NITRIDE CERAMICS AND THE EFFECT ON THEIR MECHANICAL PROPERTIES

T. Y. TIEN

Materials Science and Engineering, The University of Michigan, Ann Arbor, MI 48109
United States of America

ABSTRACT

Sintering and grain growth were studied in Si_3N_4-garnet ($Y_3Al_5O_{12}$) and Si_3N_4-cordierite ($Mg_2Al_4Si_5O_{18}$) systems. Ceramics in both systems have been sintered to full density. β-Si_3N_4 grains were found to have an elongated morphology and to be randomly oriented. The growth rates in both crystallographic directions followed the classic grain growth law with an exponent equal to 1/3. The rate constant in the long direction is about 3 times higher than that in the width direction. The growth rate in both directions in the garnet system was found to be higher than that in the cordierite system. Fracture toughness of the sintered silicon nitride ceramics was found to be dependent on their microstructure.

I. INTRODUCTION

Silicon nitride ceramics is one of the most important engineering materials for high temperature applications. For these applications, it is desirable that these materials have high fracture strength and fracture toughness at room temperature and high creep resistance at moderate temperatures. These properties should depend on the morphology of the β-Si_3N_4 grain and the nature of the grain boundary phase. Lange[1] was the first to point out the anisotropic nature of the hot pressed silicon nitride ceramics. He reported that the microstructure of hot pressed silicon nitride ceramics has an elongated grain morphology orientated preferentially with respect to the hot pressing direction. The direction parallel to the hot pressing direction showed a higher strength and a higher fracture toughness.

In pressureless sintered materials, the β-Si_3N_4 grains are randomly oriented. The toughness of silicon nitride ceramics with random oriented β-Si_3N_4 grains depends on the size and the aspect ratio of the β-Si_3N_4 grains and the nature of the grain boundary phase. Systematic treatment of the property-microstructure relationships in silcon nitride ceramics has never appeared in literature. This paper summarizes our results on the microstructure development of silicon nitride ceramics[2] and their effect on the mechanical properties of these materials.[3,4]

II. MICROSTRUCTURE DEVELOPMENT

Compositions studied in this investigation include two systems: Si_3N_4-garnet ($Y_3Al_5O_{12}$) and Si_3N_4-cordierite ($Mg_2Al_4Si_5O_{18}$). The liquid phase in the garnet system has a comparatively low viscosity and a very high viscosity in the cordierite system. Thermal expansion coefficient of the garnet is much higher than that of silicon nitride and the thermal coefficient of cordierite is much lower than that of silicon nitride. These differences in viscosity will allow us to study the effect of the viscosity of the liquid phase on the sintering behavior and the differences in thermal coefficients of the grain boundary phases will allow us to study the effect of internal stress in the grain boundaries in the silicon nitride ceramics.

Mixtures of starting powders (Si_3N_4, AlN, Al_2O_3 and MgO or Y_2O_3) giving desired compositions containing Si_3N_4 and sintering additives (cordierite or garnet) were weighed and mixtures of the required constituents were dry milled for 6 hours in an agate jar with agate balls using a planetary mill. The milled batches were sieved, cold pressed into pellets and then isostatically pressed at 175 MPa.

Sintering shrinkage of compositions in these two systems were measured with a dilatometer operated at a heating rate of 3°C per minute. These results are given in Fig. 1. These curves showed that the liquid formation occurred at 1100°C and 1300°C for the cordierite and the garnet systems respectively. Shrinkages were observed immediately after the liquid formation.

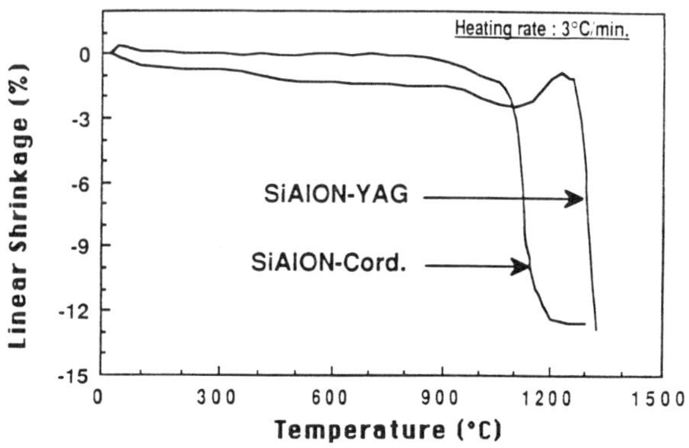

FIGURE 1
The continuous sintering curves of SiAlON-YAG and SiAlON-cordierite ceramics conducted at a heating rate 3°C/min.

The isothermal shrinkage as a function of time (temperature) was plotted in Fig. 2. These curves indicated that a typical liquid phase sintering process occurred according Kingery's model. The initial part of the curve showed the process of particle rearrangement followed by solution-precipitation and then grain growth.

The isostatically pressed pellets were embedded in a powder bed (identical composition with the specimens being sintered) and were sintered in a graphite resistance furnace under one atmosphere of nitrogen. Apparent density, weight loss and volume shrinkage were measured. Phases present were analyzed by x-ray diffraction. For consistency of the data, only specimens having a weight loss less than 3% and containing only β-Si_3N_4 as the crystalline phase were used for further analysis.

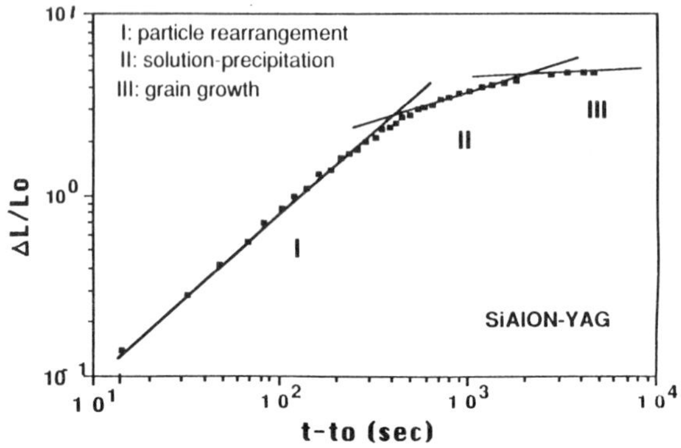

FIGURE 2
The In ($\Delta L/L_0$) vs In(t) plot of the SiAlON-YAG ceramics revealing the three sintering stages -- particle rearrangement, solution precipitation and grain growth.

Near full densities were obtained for all of the specimens sintered above the liquid forming temperatures. A comparison of sintering rates in these two systems are given in Fig. 3. Figure 4 shows the α to β silicon nitride phase transformation rate during sintering. Figure 5 shows the phase transformation always lagging behind the densification.

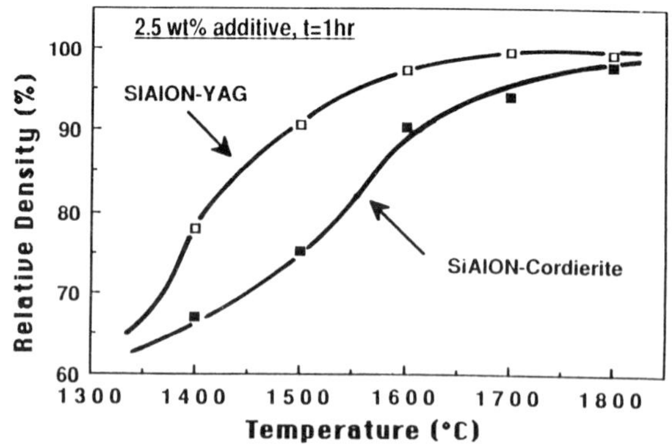

FIGURE 3
The difference of densification results between SiAlON-YAG and SiAlON cordierite ceramics.

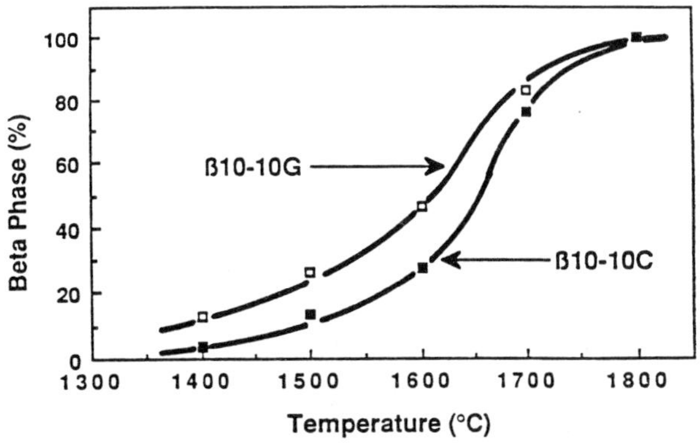

FIGURE 4
The difference of phase transformation in SiAlON-YAG and SiAlON-cordierite systems.

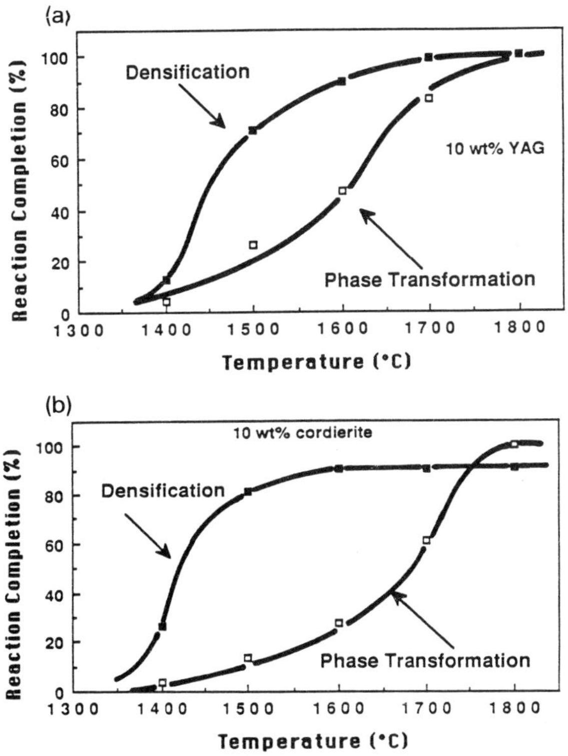

FIGURE 5
The extent of reaction completion of densification and phase transformation as a function of sintering temperature in the systems (a) SiAlON-YAG (b) SiAlON-cordierite.

Specimens were polished and finished with a 0.05 μm alumina slurry. The polished specimens were etched with a molten mixture of 2NaOH•2KOH•LiOH. SEM micrographs were used for measurements of the grain sizes.

A typical grain growth curve[5] is shown in Fig. 6. By analyzing the grain growth data, the growth of β-Si_3N_4 grains in both the length and the width directions follows the empirical equations:

$$G_L^3 = K_L t \quad (1)$$

$$G_W^3 = K_W t \quad (2)$$

Where G is the grain size in length and width directions.

The rate constants K_L and K_W and activation energies for grain growth for both the garnet and cordierite systems are given in Tables 1A and 1B. The grain growth rate should depend on the diffusion rate of the solute in the solvent and interfacial energy between the solution and the solid grains, among other properties. Our results indicated that the growth rate in the garnet system is higher than that in the cordierite

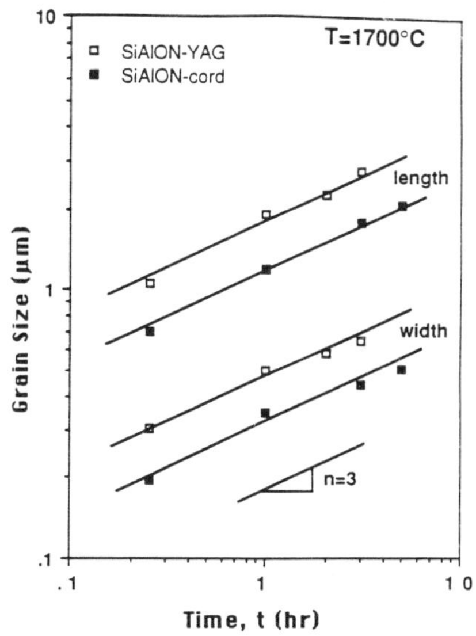

FIGURE 6
Isothermal grain growth data in the two systems: SiAlON-YAG and SiAlON-cordierite.

TABLE 1A
Rate constants, K_L and K_W, determined by the computer analyses based on Eq (3).

Compositions	1800°C		1700°C		1600°C	
	K_L	K_W	K_L	K_W	K_L	K_W
β_{20}-10G*	$4.50 \cdot 10^{+1}$	$4.69 \cdot 10^{-1}$	$6.81 \cdot 10^{0}$	$8.84 \cdot 10^{-2}$	$3.87 \cdot 10^{-1}$	$1.10 \cdot 10^{-2}$
β_{10}-10G	$1.20 \cdot 10^{+1}$	$1.72 \cdot 10^{-1}$	$1.13 \cdot 10^{0}$	$3.12 \cdot 10^{-2}$	$1.03 \cdot 10^{-1}$	$4.40 \cdot 10^{-3}$
β_5 -10G	$6.08 \cdot 10^{0}$	$9.49 \cdot 10^{-2}$	$5.43 \cdot 10^{-1}$	$1.11 \cdot 10^{-2}$	$4.23 \cdot 10^{-2}$	$2.00 \cdot 10^{-3}$
β_0 -10G	$4.78 \cdot 10^{0}$	$5.83 \cdot 10^{-2}$	$6.67 \cdot 10^{-1}$	$9.70 \cdot 10^{-3}$	$2.11 \cdot 10^{-2}$	$8.76 \cdot 10^{-4}$
β_{20}-10C*	$2.84 \cdot 10^{+1}$	$1.37 \cdot 10^{-1}$	$1.76 \cdot 10^{0}$	$2.45 \cdot 10^{-2}$	$5.08 \cdot 10^{-2}$	$2.60 \cdot 10^{-3}$
β_{10}-10C	$8.66 \cdot 10^{0}$	$7.28 \cdot 10^{-2}$	$9.37 \cdot 10^{-1}$	$1.36 \cdot 10^{-2}$	$3.14 \cdot 10^{-2}$	$1.90 \cdot 10^{-3}$
β_5 -10C	$4.40 \cdot 10^{0}$	$3.90 \cdot 10^{-2}$	$4.20 \cdot 10^{-1}$	$7.20 \cdot 10^{-3}$	$1.14 \cdot 10^{-2}$	$8.80 \cdot 10^{-4}$
β_0 -10C	$3.33 \cdot 10^{0}$	$2.79 \cdot 10^{-2}$	$2.53 \cdot 10^{-1}$	$6.50 \cdot 10^{-4}$		

* "G" and "C" stand for YAG and cordierite respectively.

β_0, β_5, β_{10} and β_{20} indicate β-Si_3N_4 solid solution containing 0, 5, 10 and 20 equivalent percent of aluminum.

TABLE 1B
The activation energies ($KJ \cdot mol^{-1}$) of grain growth in the width and length directions in the systems SiAlON-YAG and SiAlON-cordierite.

Direction of growth	YAG system				cordierite system			
	β_{20}-10	β_{10}-10	β_5-10	β_0-10	β_{20}-10	β_{10}-10	β_5-10	β_0-10
width	652	615	602	808	669	707	707	874
length	757	711	728	891	9466	908	954	962

system. As mentioned before, the viscosity of the liquid is lower in the garnet system at sintering temperature than that in the cordierite system. The aspect ratio of the β-grains is obtained by:

$$AR = G_L/ G_W = (K_L/ K_W)^{1/3} \quad (3)$$

Since $$K = K_o \exp[-(E/kT)] \quad (4)$$

Then $$AR = (K_{Lo}/K_{Wo})^{1/3} \exp[-(E_L-E_W)/3kT] \quad (5)$$

Eq (5) indicates that when the rate constants in two growth directions, K_L and K_W, are not equal, anisotropic growth of the β-Si_3N_4 grains will occur. The aspect ratio (AR) of the β-Si_3N_4 grains is dependent on the ratio of the grain growth rate constants, K_L and K_W, and temperature and is independent of time. Figures 7 and 8 show these relationships. Our results showed that the aspect ratio of the β grains is higher in the garnet system than that in the cordierite system.

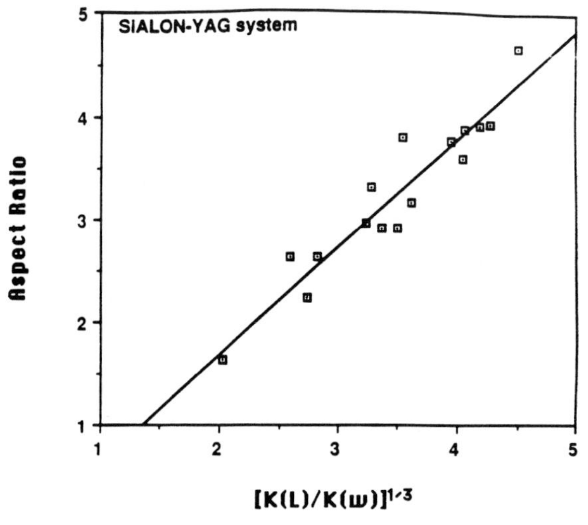

FIGURE 7
Aspect ratio of β-Si_3N_4 solid solution grains in SiAlON-garnet system sintered at 1800°C as a function of the ratio of grain growth rate constants in different crystallographic directions. (After Huang 1988)

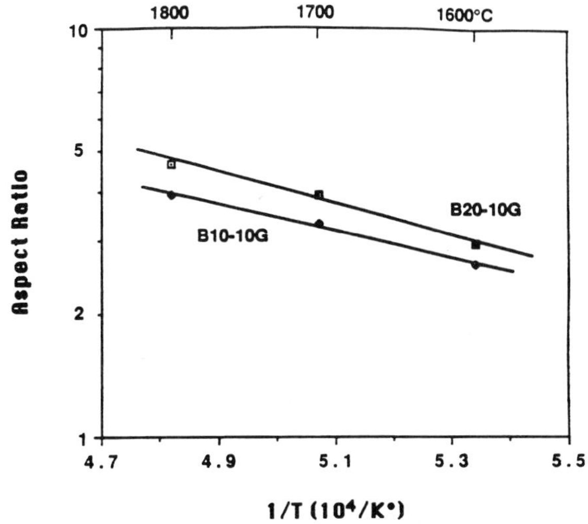

FIGURE 8
Aspect ratio of β-Si_3N_4 solid solution grains in sintered SiAlON-garnet ceramics as a function of sintering temperature. (After Huang 1988)

III. MECHANICAL PROPERTIES

Toughness of the specimens were measured by the indentation method.[6] The results are given in Fig. 9. As shown in the graph, the toughness of silicon nitride ceramics in the SiAlON-$Y_3Al_5O_{12}$ system decreases with increasing size of the Si_3N_4 grains. However, the opposite was observed in the SiAlON-cordierite

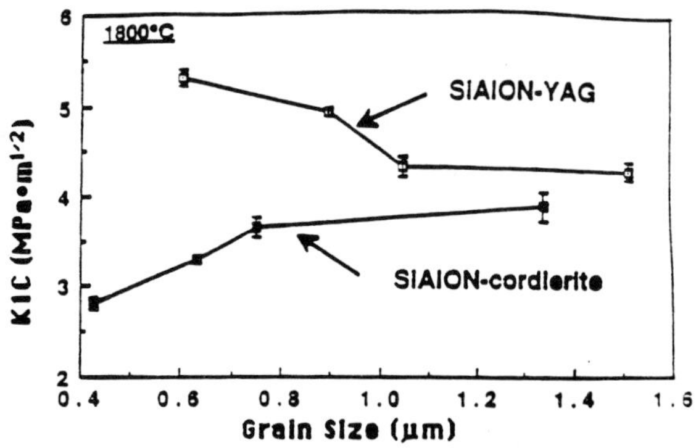

FIGURE 9
Fracture toughness as a function of grain size of sintered SiAlON (20 Eq.% Al) ceramics containing 5 wt.% $Y_3Al_5O_{12}$ and cordierite. (After Hwang and Tien 1988)

system. Figure 10 shows the toughness of specimens in the system SiAlON-$Y_3Al_5O_{12}$ reached a maximum with the increasing aspect ratio of the β-Si_3N_4 grain. With a further increase in the aspect ratio of the β-Si_3N_4 grains, the toughness decreased.[7] Microstructure examinations showed that intergranular fracture dominated in specimens containing grains with a smaller aspect ratio, and a mixed fracture mode was observed in specimens having an aspect ratio larger than two, where the toughness maximum lies.[6]

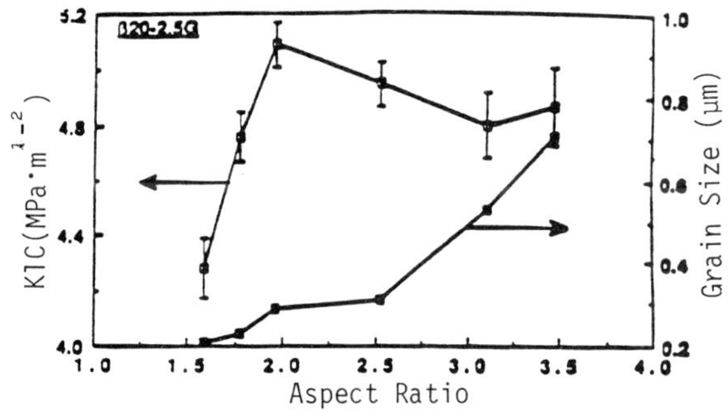

FIGURE 10
Fracture toughness as a function of aspect ratio of β'-Si_3N_4 solid solution grains in sintered SiAlON (20 Eq.% Al) ceramics containing 5 wt.% $Y_3Al_5O_{12}$ and cordierite. (After Hwang and Tien 1988)

The toughness of the β-Si_3N_4 ceramics varies with the composition and amount of grain boundary glass.[8] As shown in Fig. 11, toughness decreases with the amount of grain boundary glass in both the SiAlON-cordierite and the SiAlON-garnet systems. The difference in toughness in these two systems seems to be caused by the differences in properties of the grain boundary glasses. Figure 11 also shows the effect of grain boundary phase crystallization on the toughness of silicon nitride ceramics. Toughness of specimens in the SiAlON-garnet system decreased after crystallization, while the toughness of specimens in the SiAlON-cordierite system increased after the grain boundary glass crystallized. We have attributed this phenomenon to the thermal expansion differences beween the silicon nitride grains and the grain boundary crystalline phase. Garnet has a higher thermal expansion coefficient than that of the silicon nitride grains, while cordierite has a much lower thermal expansion coefficient than that of the silicon nitride grains. Thus, the grain boundary phases are in tension in the garnet system, but under compression in the cordierite system. Since fractures in silicon nitride ceramics are mostly intergranular, the force on the grain boundary phase causes a change in strength and toughness of these ceramics.

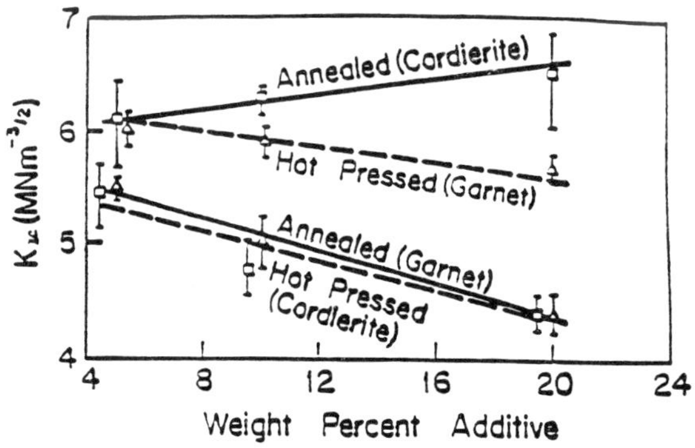

FIGURE 11
Fracture toughness of hot pressed SiAlON ceramics containing garnet and cordierite. (After Bonnell 1986)

Chen[9] reported that the creep rates of hot pressed silicon nitride ceramics varied with the size of the β-Si_3N_4 grain and the nature of grain boundary phase. The results indicated that a better creep resistance can be realized in silicon nitride ceramics with larger grain size and crystallized grain boundary phase. The materials studied were SiAlON containing $Y_3Al_5O_{12}$ and the results are summarized in Fig. 12.

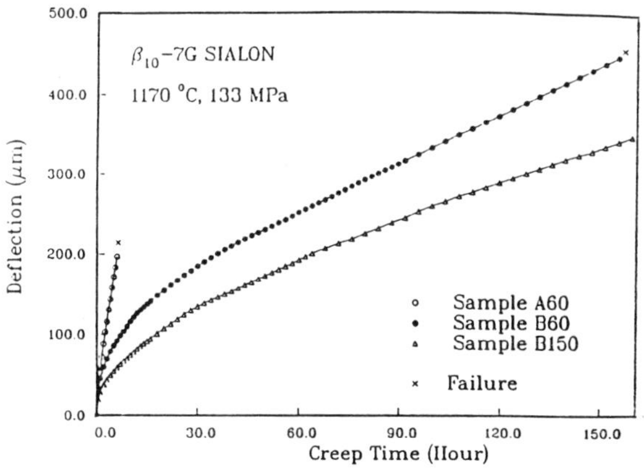

FIGURE 12
Creep curves of SiAlON-garnet ceramics. Creep experiments were performed at 1170°C under a load of 133 MPa. Sample A60 was hot pressed at 1750°C for 60 minutes, Sample B60 was hot pressed and then crystallized at 1250°C for 50 hours. Sample B150 was hot pressed at 1750°C for 150 minutes and annealed at 1250°C for 50 hours. (After Chen 1987)

IV. SUMMARY

Silicon nitride ceramics have been sintered to full density in both garnet and cordierite containing systems. Elongated β-Si_3N_4 grains were found to be randomly oriented. The growth rates in both crystallographic directions followed the classic grain growth law with an exponent equal to 1/3. The rate constant in the long direction is about 3 times higher than that in the width direction. The growth rate in both directions in the garnet system was found to be higher than that in the cordierite system. Fracture toughness of the sintered silicon nitride ceramics was found to be dependent on their microstructure.

REFERENCES

1. F.F. Lange, "Relation Between Strength, Fracture Energy and Microstructure of Hot-Pressed Si_3N_4," *J. Am. Ceram. Soc.* **56** [10] 518-22 (1973).

2. C.M. Hwang, "The Systems SiAlON-$Y_3Al_5O_{12}$ and SiAlON-Cordierite: Sintering and Grain Growth," Doctoral Dissertation, The University of Michigan, 1988.

3. D.A. Bonnell, "Determination of Crystallization Mechanisms and Correlation of Grain Boundary Morphology to Mechanical Properties in Silicon Nitride," Doctoral Dissertation, The University of Michigan, 1986.

4. C.F. Chen, "Creep Behavior of SiAlON and Siliconized Silicon Carbide Ceramics," Doctoral Dissertation, The University of Michigan, 1987.

5. C.M. Hwang and T.Y. Tien, "Microstructure Development in Silicon Nitride Ceramics," Preparation and Properties of Silicon Nitride Based Materials, Materials Science Forum, Vol. 47, pp. 84-109, Eds. D.A. Bonnell and T.Y. Tien, Trans Tech Publications, Switzerland-Germany, 1989.

6. P. Chanitikul, G.R. Anstis, B.R. Lawn and D.B. Marshall, "A Critical Evaluation of Indentation Technique for Measuring Fracture Toughness II: Strength Method," *J. Am. Ceram. Soc.* **64** [9] 534-43 (1981).

7. C.M. Hwang, "Microstructure Dependence of Fracture Toughness of Silicon Nitride Ceramics," To be published.

8. D.A. Bonnell, T.Y. Tien and M. Rühle, "Controlled Crystallization of Amorphous Phase in Silicon Nitride Ceramics," *J. Am. Ceram. Soc.* **70** [7] 460-5 (1987).

9. C.F. Chen and T.Y. Tien, "High Temperature Mechanical Properties of SiAlON Ceramics: Microstructure Effect," *Ceram. Eng. Sci. Proc.* **8** [7-8] 778-95 (1987).

Frontiers of Materials Research/
Electronic and Optical Materials
M. Kong, L. Huang (editors)

DEVELOPMENT OF PROCESSING TECHNIQUES FOR FLEXIBLE COMPOSITES

Kai–Wen Xianyu

Shaanxi Non–Metallic Materials and Technology Institute .P.O. Box 172,Xian 710000. China

The term "flexible composites " is used in the paper to identify composites based upon elastomeric polymers and curved fibers as reinforcements , of which the usable range of deformation is much higher than those of the conventional polymer–based composites.The ability of flexible composites to endure fatigue loading and still provide high load–carrying capacity has been mainly documented in typecord conveyor belt constructions and biological materials etc.
In order to explore and investigate the unique capability of flexible composites, a new molding method is developed here,which is a very practical and useful technique for fabricating flexible composites samples. the method is mainly applicable to a sinusoidal shape carbon fiber with wavelength (λ) and amplitude (α) and ductile matrice which is a silicane elastomer to have up to 100% large elongation . Details of these approaches are discussed.

1. INTRODUCTION

The term " flexible composites " is used here to describe elastomeric polymer based composites . The flexible composites are composed of continuous fibers with sinusoidal waviness in the elastomeric matrix. (Figure 1)

This new ideal material, flexible composites, has been developed and evaluated by Chou and his co–worker. Some researches on nonlinear elastic behavior of flexible composites have made a great deal of progress.

In order to compare theoretical analysis with experiment results, it is necessary to develop a fabrication technique for flexible composites . The purpose of this paper is developing of a processing technique for flexible composites.

Two fabrication techniques, a weaving method and a molding method have been considered.Table 1.,compares their advantage and disadvatages. As a result of the easy in processing, the low cost and the short develop period, the molding method was selected . A key factor used in the molding method is that it can fabricate a curved fiber which can satisfy a described shape . Here , a group of curved fibers , all of the fiber are assumed to be parallel to each other and have a sinusoidal curvature in one spatial direction. This shape is simple, but it can be made into various samples; iso–phase or random phase , longitudinal or transverse,tensile or shear, and single layer or multiple layers.It can satisfy entirely the requirements of mechanical analysis of flexible composites .

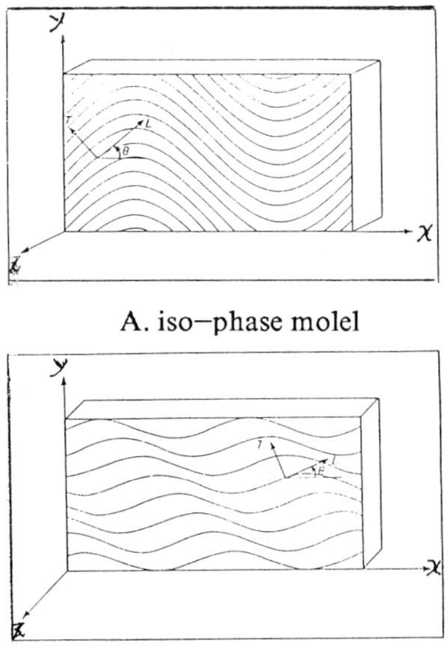

A. iso–phase molel

B. random–phase modle

Figure 1. Flexible Composites

Table 1. Comparison of Two Fabricated Techniques

	weave method	molding method
principle	use computer to control curved shape mechanical operate	use mold to press courved fiber manufacture by hand
cost	expensive	low
develop period sample	long	short
uniformity	excellent	o. k.
labor by hand	less	most

In this paper, flexible composites consist of high modulus fiber (carbon fiber) and ductile matrix (silicone elastomer) ,and are capable of sustaining larger elongation to meet the requirement of mechanical analysis.

2. FIBER and MATRIX MATERIALS

Fiber:

A Thornel–300 and a Celion 3000 carbon fibers are the reinforced material of flexible composites, They are a PANbase fibers. A loose bundle contains 1000 and 3000 filaments, respectively, with a filament diameter of 7μm.

Typical properties of carbon fibers are listed in Table 2.

Matrix:

A Sylgard 184 silicone elastomer was selected as the matrix of flexible composites. The elastomer , made by Dow Corning Co., is transparent and has up to 100 % large elongation.

A Sylgard 184 silicone elastomer is supplied in two parts: a lot–matched base and a curing agent.It can be cured at room temperature with mixed ratio of ten parts base to one part curing agent by weigth. The mechanical properies are shown in Table 3.Curing cycle are shown in Table 4.

The curing time can be appreciably decreased by elevating the curing temperature. Full mechanical strength will be achieved in about 7 days at 25℃.

The majority its physical strength , however , is present after 24 hours. In order to insure adhesion between fiber and matrix , Sylgard prime coat is recommended. It is particularly important that fibers coated with the prime coat can become easily curved form under molding pressure.

3. MOLD DESIGN and MANUFACTURE

The geometric relations between the wavelength (λ), and amplitude (α) of sinusoidal shaped fibers are identified first . The spatial position of a typical fiber in the XYZ coordinated is given by

Table 2. Typical Properties of Carbon Fiber

Fiber	Manufacture	Tensile strength (MPa)	Tensile Moduls (GPa)	Density (g/cm^3)
Thornel –300	Union Carbide Co.	2657	235	1.75
Celion –3000	Celanese Co.	2760	228	1.76

Table 3. Properties of Sylgard 184 Silicon Elastomer

color	transparent
specific gravity (25℃) (g / cm^3)	1.05
tensile strength (MPa)	6.2
elongation (%)	100
tear strength (KN / M)	2.6
special feature	low toxicity, stability over wide temperature range reversion resistant flame resistant, etc

$$y = \alpha \sin\frac{2\pi x}{\lambda}$$

The curved fibers composites with $\alpha / \lambda = 0.09$ can be extended up to 7.6% of its original length just by the straightening of the fibers , if the matrix stiffness is negligible . Wavelength (λ) and amplitude (α) , of 16.93 mm and 1.52 mm respectively, is preferred. The length of the samples is given 305 mm in which the length of all curved part is given as 237 mm.

Table 4. Curing Cycles for sylgard 184 Silicone Elastomer

curing temperature	curing time (hour)
25℃	24
65℃	4
100℃	1
150℃	15 minutes

The mold design is underway as the above listed data. In order to decrease the cost of the molding and the damage to fibers a plastic , Delrin AF was preferred to instead of steel. Delrin plastics properties and the drawing of the mold are shown as Table 5. and Figure 2. ,respectively. The surface of the mold is machined with the help of a computer.

Table 5. Properties of Delrin AF. Polyacetal Filled Homo

specific gravity	1.54
tensile strength (MPa)	52.4
compressive strength 10% (MPa)	89.7
elongation (%)	15–22
hardness Rockwell	R118

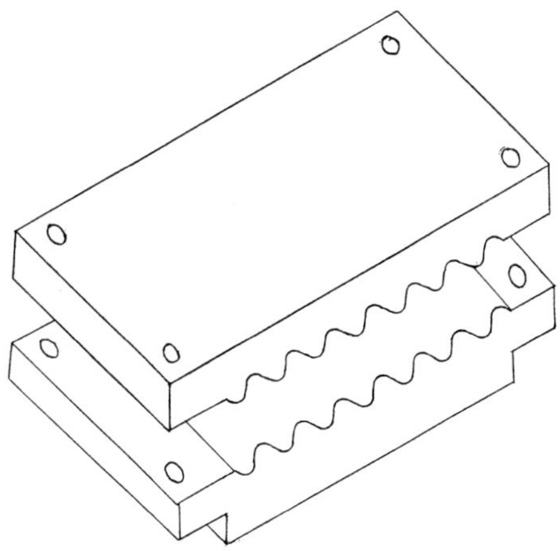

Figure 2. A Pressure Mold Picture

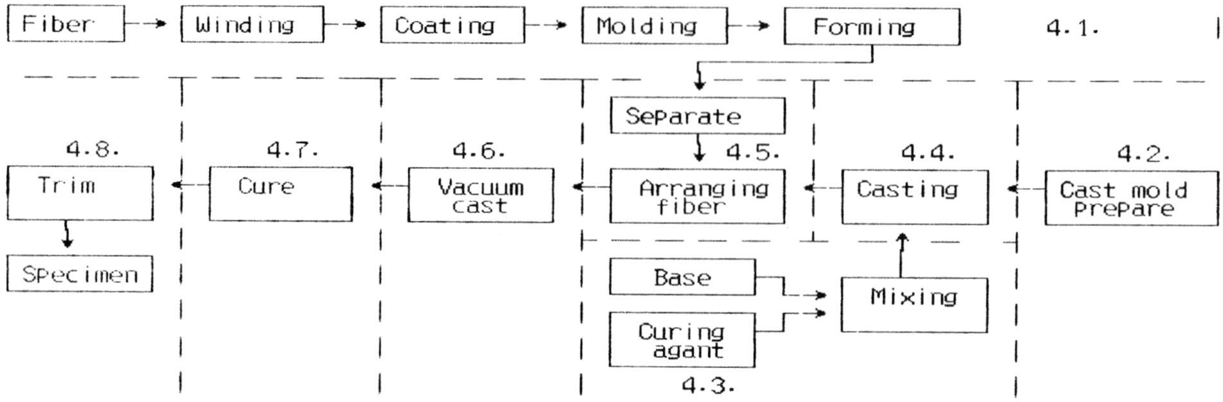

Figure 3. Manufacture Process for Flexible Composites

4. MANUFACTURE PROCESS

The molding method is a practial and useful technique for fabricating flexible composites samples . The method is mainly applicable to a sinusoidal shape fiber with wavelength(λ)and amplitude (α) .Here, carbon fibers were used in the process , however , this method is suitable for Kevlar and Glass fibers , too.

The manufacturing process is shown as Figure 3. and may be divided into eight principal steps :

(1) pressing curved fibers
(2) preparing casting mold
(3) preparing matrix
(4) pouring matrix
(5) arranging for fibers
(6) removing bubbles and repouring
(7) polymerization cycle
(8) finishing samples for test

Here, each step is treated individually and a general chronological sequence is given . All materials and supplies referred to have been carefully selected to produce high–quality samples.

4.1. Pressing Curved Fibers

Pressing curved fibers are shown as Figure 4.

4.1.1. Movable Frame

The structure of a movable frame is shown in Figure 5. A end of the frame is movable. It is made of plexiglas and connected by two cap screws in fixed plate which is made of aluminium . Two cap screws control the distance between the movable plate and a relative fixed plate so that when pressed , stretch fibers on the frame can be adjusted loosely by decreasing the distance between them.

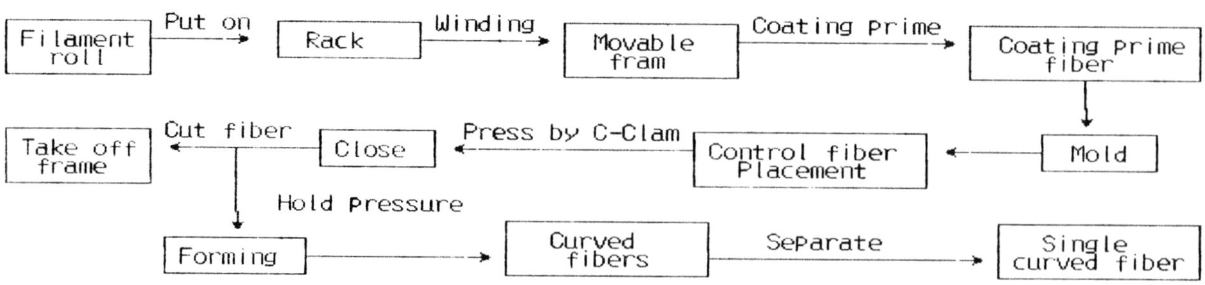

Figure 4. Pressing Curved Fiber Flow Diagram

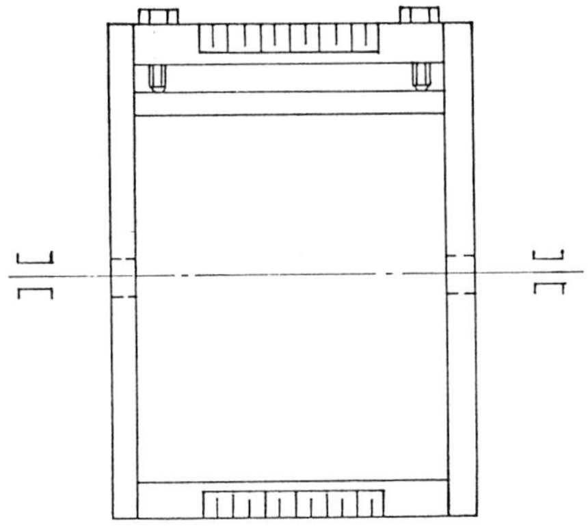

Figure 5. Structure of Movable

4.1.2. Frame Winding

In the frame winding method carbon filaments are carefully wound onto the movable frame with scales on the two opposite ends . When finished winding , fiber ends are attached on the frame by tape. The fibers alignment and uniformity are controlled by the number of fibers per unit length on the frame .(Figure 6)

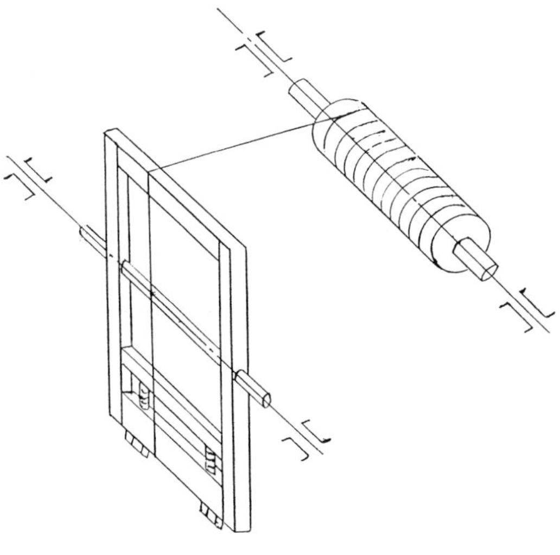

Figure 6. A Frame Winding Method

4.1.3. Coating Prime

The carbon fibers on the frame have to be painted with Sylgard prime coating liquid so that curved fibers can be shaped easily in the pressing mold, and achieved better adhesion with the matrix The Sylgard prime coating liquid is supplied by Dow corning Co. The surface of carbon fibers to be coated should be clean , dry and free of contaminants . All surface of the carbon fibers should be brushed throughly and be air–dried for ten minutes under standard ambient conditions and humidity . Use only with adequate ventilation, avoid prolonged or repeated breathing of vapor,Keep away from heat, sparks and open flame because of the extremely flammable heptane solvent present.

4.1.4. Pressing Curved Fibers As Figure 7. shown the movable frame is put onto a lower mold in which four screws were fixed . Carefully hold the fibers in the same direction with the warp of the lower mold . All frame should be fixed level.

Follow orientation marks, match the upper and the lower molds through four screws , then slowly , carefully and gradually tighten up every screw . The fibers on the frame change its shape and become more tight under molding pressure.

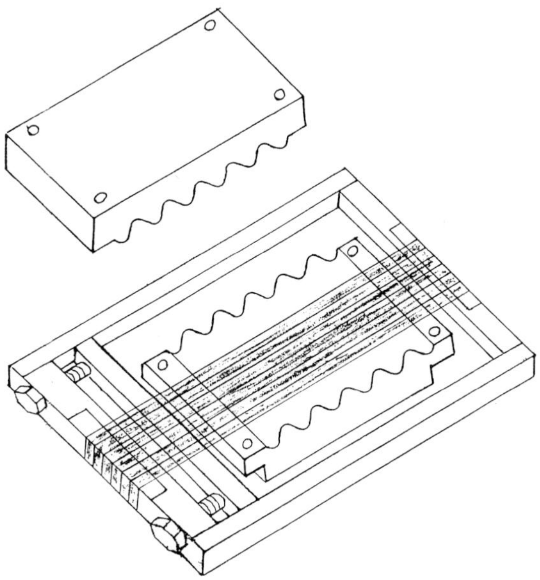

Figure 7. Put Frame onto Mold

At the same time , control two cap screws on the frame so that fibers become flexible , hence , alternatively adjust the fibers loose and tight level, resulting in the upper and the lower molds to be close , which then changed unidirectional fibers into a sinusoidal shape. While the upper and the lower molds are nearing close , clamp whole mold again by three big C–Clam so that the mold can closed entirely . Finally, cut off fibers in the movable frame by a sharp knife. Hold the pressure in the mold for over eight hours. (Figure 8)

4.2. Preparing the Casting Mold

4.2.1. Manufacture of the Casting Mold

A casting mold consists of four parts: (1) an aluminium frame .(2) a sheet of plastic film.(3) a graph

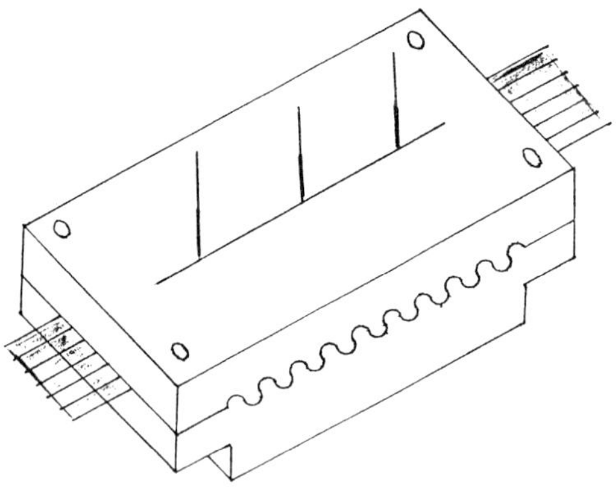

Figure 8.Pressing Curved Fiber

paper and (4) an aluminium plate.

The size of the frame is showm in Figure 9. , the thickness of the frame is also the thickness of the samples. A sheet of plastic film is used to prevent sticking between the sample and the mold after curing . Four slips of double tapes are used between the frame and film , so that there is a reliable seal between them.

A graph paper (marked sinusoidal curve) is now added under the film,and four silps of double tapes run around the inside perimeter of the frame . Again clamp the frame and aluminium plate , then a casting mold is ready.

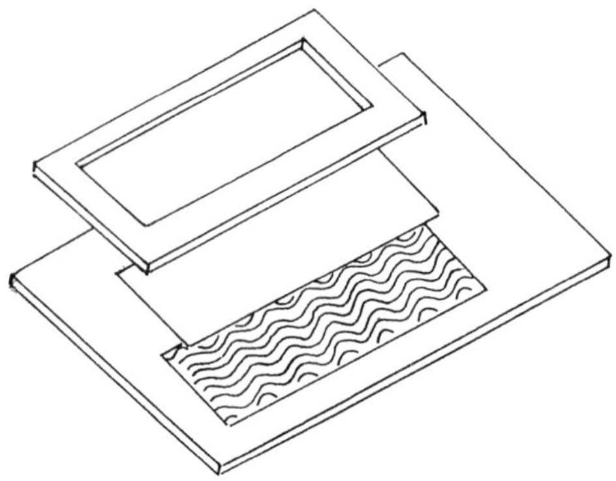

Figure 9. A Costing Mold

4.2.2. Cleaning

Clean the mold surface with a gauze sponge wetted with acetone and wipe dry with a clean , dry gauze sponge.

4.2.3. Leveling the Casting Mold Place the casting mold on a large aluminium plate which should be level on a rigid flat table. Do not move the casting mold from its leveled position while the cast sheet is curing.

4.2.4. Covering

Covering the mold with the plexiglas cover, thus protecting the prepared surface from dust and other contaminants.

4.3. Preparing Matrix

4.3.1. Determining Amount of Base and Curing Agent

Calculate the amount of base and curing agent required . The total amount (as weight) of matrix is determined by

$$W = 1.05\ (g / cm^3) \times A\ (cm^2) \times T\ (cm)$$

Here, W is the total amount needed (g).

A is the area of the sheet to be cast .

T is the desired thickness .

Elastomer density is $1.05 (g / cm^3)$.

Example : For a sheet 305 mm(length) × 38 mm (width) × 2 mm (T), the total amount of matrix required is

$$W = 1.05 \times 30.5 \times 3.8 \times 0.2 = 24.3 (g)$$

Base and curing agent is in a ratio of 10 parts to one part by weight .

Then continu ing with the above example, for a total of 24.3 grams of matrix,the base and curing agent calculation are made as follows :

$$\text{base} \qquad 24.3 \times \frac{10}{10+1} = 22.1(g)$$

$$\text{curingagent} \qquad 24.3 \times \frac{1}{10+1} = 2.2(g)$$

4.3.2. Weighing the Base and Curing Agent

Take an adequate size plastic cup and weigh out the needed amount of curing agent and base on a balance scale in proper order, respectively. Remember,first of all,to be weight out the curing agent because it is a very small amount compared with the base. Remember also to account for the weight of the plastic cup when weighing.

The base can be measured by pouring it directly into the curing agent filled cup while the cup is on the scale. In this case,the scale should be set to the total weight (cup + curing agent +base) , and base added until the scale balances . The base will flow evenly on the top of the curing agent . If excess base is accidently added , it can be conveniently absorbed with paper towel.

4.3.3. Mixing the Base and Curing Agent

In order to obtain a quality sample, it is essential to stir thoroughly but slowly . Do not use a whipping action which will introduce more air bubbles. Stiring to produce a clear, non–streaking matrix is best . After mixing , the mixture should be allow to stand for half of an hour so that the air bubbles disappear .

4.4. Pouring Matrix

4.4.1. Pouring

While pouring , hold the mixing cup close to the casting mold surface and pour gently and continuously (Figure 10). This procedure will mininize bubble formation. On the other hand,while pouring it is advisable to move the cup from an "S" pattern which will improve flow to fill the mold . The final proportion of matrix should be poured along the outside boundary close to the mold. A wooden stem may be used to help spread the matrix over the entire plate surface .

Place a cover over the poured mold to protect it from duct and other foreign matter during polymerization .

Figure 10. Pouring Matrix

4.4.2. Removal of Bubbles

When air bubbles are present in the casting mold,they can be burst by using a little needle or a pointed scribe . When first casting matrix, it is enough to use this method . A vacuum deairing method is also recommended .

4.5. Arranging for Fibers

4.5.1. Define Moment Arrange for Fibers

The best moment to arrange for fiber, is dependent on the matrix curing degree. When the matrix is still liquid, the fibers in the mold cannot fix its shape because the fibers will move in this thin matrix .In the other hand, if the matrix becomes thick as a soild,the fibers cannot be set in the matrix because the cured matrix cannot bind with the fibers .

When the matrix is no longer liquid but adheres to a little needle , and feels sticky to the touch . This moment is considered better to lay up fibers . It occur about 12–16 hours after the base is mixed with the curing agent at 25 ℃ .

4.5.2. Separating Fibers After unload the mold

from C–Clam , carefully part the upper and the lower molds so that pressing curved fibers can be seen . (Figure 11) The curbed fibers to be removed from the mold should be separated one by one . It is noticeable that separation of fibers should begin from two ends of fibers . The separated fibers should be placed on white papers . While separating and placing fibers , it is necessary to be gentle and carful so the fiber will hold curved shape it had in the mold .

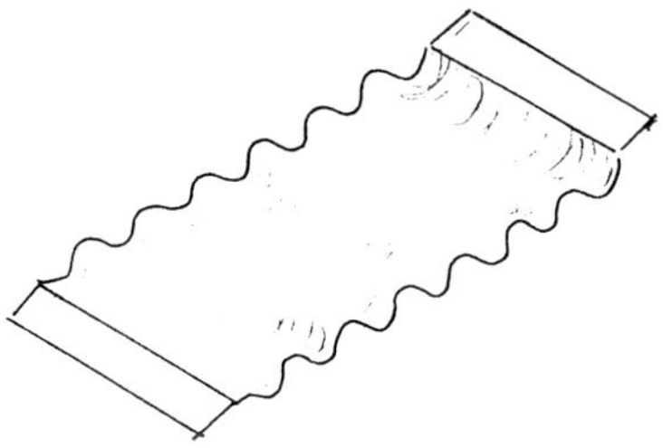

Figure 11. Curved Fiber

4.5.3. Arranging for Fiber As the mark of graph paper in the mold lay up fiber one by one , it may begin from right or left , or middle of the mold , out of every operator habit. With one hand take a needle to trim a fiber and with the other hand hold the other end of the fiber.Every sinusoidal curvation of the fiber should follow the same plane and the correponding curvature on the mark of graph paper . if the fiber gaps from the corresponding curvature of mark on graph paper, eliminate gaps gently with the needle in processing, but operate carefully to avoid air bubbles intruding because this kind of bubbles are difficult to remove . Finally, when all fibers are laid up in the mold,in order to make the fibers to bind tightly in martix , press every fiber tightly with the needle , then do not move the fibers again until casting the matrix secondly .

4.6. Removing Bubbles and Repouring

4.6.1. Removel of Bubble

Generally speaking , the bubbles on the surface can be burst using a needle. But in case of bubbles intruding in between fibers and matrix , it is very difficult to removal these bubbles .

Entrapped air between fibers and matrix can be removed by using a vacuum of 635 to 737 mm of mercury . Large bubbles will form almost immediately and should be collapsed by introducing atmospheric pressure for 30 seconds .Then continue the vacuum until the liquid expands.this may take 15 minutes to 2 hours , depending on the amout of air introduced .

4.6.2. Repouring

Repouring is the same process as 4.4. pouring matrix . Before repouring,there should be no bubbles in the mold. After repouring , the matrix should be full of the mold , and level with four sides of the mold . Surplus matrix can be scraped by a flat cutter

4.7. Polymerization Cycle

4.7.2. Early Stage of Polymerization

Carefully level the mold in a level table for eight hours . The early stage of polymerization of the matrix is in a viscous liquid state . When probed, the matrix still behaves as a liquid , it adheres to the needle and a viscous string of matrix can be pulled up with the needle .

4.7.2.Second Stage of Polymerization During 12–16 hours, the matrix is no longer liquid but adheres to the needle. The surface is easily deformed with a little pressure on the needle and feel very sticky to the touch . It is now suitable to lay up fibers

4.7.3. Third Stage of Polymerization After 24 hours , the surface of the matrix becomes hard . The matrix can be depressed with moderate pressure on the needle . A conner of the sample can be picked up easily from the mold . The majority of samples physical strength is present .

4.7.4. Final Polymerization

If sample is in support of test, get a batter mechanical property,it must be allowed to continue its polymerization cycle to full cure for an additional 6 days at 25℃ or 24 hours at 50℃.

4.8. Finishing Sample for Test

After the sample removel from the oven,it need to remain 24 hours at room temperature for testing.When the sample is removed from the mold, it can be cut the described size for a test. Then,the sample is ready for final preparation ,cleaning , trimming and drawing line .

5. CONCLUSION

Some samples of tensile and shear , iso–phase and random–phase models have completed by the molding method . Non–linear elastic behavior and photoelastic test (PL–3. photoelastic epoxy resin instead of Sylgard 184) of the samples have been examined .

Experimented results show good consistent with the theoretical prediction by Chou etc. , similar to mechanical properties in biomaterial as well.

ACKNOWLEDGMENTS

This work was performed while the author were with Centre Composite Materials of Delaware University. The author wish to acknowledge the assiatance and the guidance given by Prof.T.–W. Chou .

REFERENCES

1. T. –W. Chou and Takahashi, K., ″Non–linear Elastic Bechavier of Flexible Composites″, Composites, Volume 18 No. 1, Jan. 1987, p25–34.

2. S.–Y.Luo and T,–W. Chou. ,″Modeling of Nonlinear Elastic Behavior of Elastomeric Flexible Composites″, Proceeding of The 192nd National Meeting of The American Chemical Society,1986.

3. George Lubin.,″Handbook of Composites″

4. Geroge A.Congdon, ″Information about High Technology Silicone Materials″ Dow Corning Co.

Frontiers of Materials Research/
Electronic and Optical Materials
M. Kong, L. Huang (editors)

FERROELECTRIC MATERIALS FOR APPLICATIONS IN THE 1990S (Abstract)

L Eric Cross

Pennsylvania State University, University Park, PA 16802, USA

It appears probable that through the 1990s major applications for ferroelectric materials will continue to be as dielectrics in capacitors, as piezoelectrics and electrostrictors in transducers, as pyroelectric IR sensors, and as active elements in electro-optic and non-linear optic systems.

In capacitors, the general trend is toward thinner dielectrics and to move complex formulations with the lead based relaxor ferroelectrics taking an increasing share of the market. New interest in very thin (1,000A⁰) ferroelectric films has beer sparked by needs for hysteretic dielectric thin films on silicon and GaAs to use as elements in non-volatile memory. The rapid escalation of the dielectric breakdown strength with decreasing thickness in films, however, permits exceedingly high energy densities and suggests many new possible applications in micro-capacitors, micro-sensors, micro-actuators, and micro-motors.

Piezoelectric composite materials show continuing improvement both for transmitting and for sensing acoustic phased signals. A new generation of agile piezoelectrics permit interesting acoustic phased array systems for ultrasonic electromedical tomography. Composites involving both sensors and responders coupled by active electronics promise a new generation of "smart" materials which could respond automatically to local environmental changes. For actuator systems, there are interesting mew high strain systems and new clues upon the mechanisms of fatigue which limit lifetime. In more conventional materials, mew ways of using actuators give versatility for precise positioning. Clearly with continuing drive to ever smaller feature size in semiconductor devices such ultra precise electrical control of position will be essential in fabrication systems.

Pyroelectric imaging system s appear to be moving toward the use of induced pyroelectric response by field forced ferroelectric phase change, and there is urgent need for materials which retain sharp transitions in samples with very thin (1-10 v meter) sections. Pyrooptic effects offer interesting new alternatives for long wavelength IR imagers.

Many fascinating optical applications are currently materials limited. Ferroelectrics offer the optimum combination of properties for photo refractive systems but there is increasing need for new, larger, and more prefect crystals. Increased activity in thin film ferroelectrics will provide new alternatives for optical systems as techniques for oriented opitaxial single crystal films are improved.

Frontiers of Materials Research/
Electronic and Optical Materials
M. Kong, L. Huang (editors)

PLASMA PROCESSING FOR ADVANCED THIN FILM TECHNOLOGY IN THE 21ST CENTURY

R.P.H. Chang

Materials Research Center and Dept. of Materials Science & Engineering, Northwestern University, Evanston, Illinois, USA

Plasma, a collection of charged species (electrons and ions) is often referred to as the fourth state of matter. Gaseous discharges used in the processing of thin films contain less than a fractional percentage of charged species. The role of charged species is to promote chemical activities in the gas media. Over the past three decades, we have seen the rapid development of plasma processing in advanced thin film technology both for electronic device and protective coating applications. Advantages derived from using plasma processing include: higher quality films, lower processing temperatures, improved adhesion and interfacial properties, superior process and quality control, large area coating, etc.

In this talk, specific examples will be given to illustrate the development of plasma processing for advanced thin film technology in the 21st century. In the area of protective coating, there has been active worldwide research in plasma synthesis of diamond films. These films possess extraordinary properties, including hardness, chemical reactivity, high thermal conductivity, lower coefficient of friction, wide optical bandgap, etc. Plasma processes are being developed in protective coating for metal, ceramic, semiconductor, and polymeric surfaces. In integrated circuit electronics technology, plasma processes are applied to device fabrication, packaging and system integration. Recent developments in these areas will be discussed. Finally, future trends in plasma processing of thin films will be projected.

Frontiers of Materials Research/
Electronic and Optical Materials
M. Kong, L. Huang (editors)

THE PROSPECTS FOR MOLECULAR ENGINEERING IN INORGANIC MATERIALS OF NLO CRYSTALS

Chuangtian CHEN

Fujian Institute of Research on the Structure of Matter, Chinese Academy of Sciences, Fuzhou, Fujian, 350002, China.

Starting from a general quantum-mechanical perturbation theory in the nonlinear optical (NLO) effect in crystals, this talk gives a brief presentation of the basic concepts and calculation methods of the "anionic group theory for the NLO effect of crystals". On this basis, it is possible from theoretical aspects to estimate and foresee the values of the second-order susceptibilities of the anionic group or mulecules; the absorption edge in UV side and the briefringences of the crystals before carrying out the experimental measurements. However, as far as the spacial arrangement of the anionic groups (or molecules) of the crystals in the lattice, the damage threshold as well as the chemical-physical properties of the crystals are concerned, no theoretical scheme is yet available to estimate these parameters before the grown-up of enough large crystals, and therefore a complete set of experimental procedures to determine these properties of NLO materials has been designed, which has effectively prompted our research for new NLO materials. Our discovery of the new ultraviolet NLO crystals, LBO (LiB_3O_5), is just based on the adopting of these theoretical and experimental procedures.

PART 2:
HIGH T_C SUPERCONDUCTORS

Frontiers of Materials Research/
Electronic and Optical Materials
M. Kong, L. Huang (editors)

A NEW METHOD FOR QUANTITATIVE STUDY OF FLUX PINNING IN TYPE II SUPERCONDUCTORS

J.X. Zhang, Z.C. Lin, W.G. Zeng, G.M. Lin, K.F. Liang and Z.L. Du

Department of Physics and Institute of Material Science, Zhongshan University, 510275, Guangzhou, China.

The motion equation with an interaction force between flux and applied H is established and resolved for the moving flux in the process of magnetization of type II superconductors. The expressions of energy dissipation and the dynamic pinning force exerted on the moving flux can therefore be acquired when the low frequency internal friction, modulus and magnetization curve are measured simultaneously in the process of magnetization with a variety of applied magnetization rate $\dot{H}$.

1. INTRODUCTION

It is well known that superconductivity may exist at a very high magnetic field, and the critical current density J_c might also be expected to very high when a type II superconductor is in the mixed state. It is mainly contributed to varied pinning centers which hinder the displacement of flux, and a great high critical current density is obtained in type II superconductor[1].

In a magnetic field, type II superconductors exist in a mixed state of flux bearing vortices in a superconducting matrix. The vortices and their interactions with material defects or pinning centers determine many of properties studied and exploited in superconductors. Therefore, the study of trapping or pinning mechanism are both benificial for appreciating the fundamental properties of type II superconductor, and for the great technological interest.

It is also related that some experiments indicate that the vortices are mobile above and immobile below a reasonably well-defined "irreversibility" temperature $T^*(H)$ [2,3]. The flux motion, flux creep and relaxation effects are pointed out by a number of authors [4-6]. But so far, there is no research report on quantitative calculation of vortex lines motion. One of the authors has proposed a method for investigating the dynamics of magnetic domain walls (MDW_s) in a ferromagnet[7]. In this paper, we will try to suggest a method for investigating the motion of flux in type II superconductor above the "irreversibility" temperature $T^*(H)$.

Literature[8] has reported an internal friction due to flux movement in BSCCO bulk in magnetization process. It has the common IF feature as the IF of transition[9-11], plastic deformation[12-13], and magnetization of ferromagnet[14-15]: a), The characteristic IF appears only with a varying external field, it increases with rate of applied field, and fall down to a back ground value when applied field rate drops to zero suddenly. b), The IF descreases with increasing of measuring frequency. c), The characteristic IF disappears when superconductor-normal state transition takes place. According to the IF results mentioned above, authors suggest a method of quantitative study of flux pinning forces.

2. MOTION EQUATION OF FLUX AND ENERGY DISSIPATION OF MOVING FLUX

Let's consider a flux moved along X-axis in type II superconductor, the vector of applied field may be assumed parallel to the flux averaged. The flux lines penetrate into superconductor from the surface (X=0) by the external field. In addition, an alternating stress was applied for measuring internal friction (IF) simultaneously.

2.1. The forces exerted on flux line

a. Effective driving force F'

Owing to a non-uniform distribution of flux lines in type II superconductor (SC) the driving force F_d can be expressed as

$$F_d = -H\emptyset_0 dn/dx, \qquad (1)$$

where H is external field, $\emptyset_0$ is flux quantum, dn/dx is the gradient of flux density which has an opposite direction with F_d. Because there is a flux pinning force F_p in type II SC, a flux line can move only if $|F_d| > |F_p|$, we define therefore the effective driving force as

$$F' = |F_d| - |F_p| \qquad (2)$$

Because of F_d is direct proportional to field $H=\dot{H}t$, the dependence of F' on H (or t) is therefore less than the linear dependence, so

$$F' = -dn/dx\Phi_0 A_1(H)\dot{H}^m t^m, \ (0<m<1) \qquad (3)$$

where t=0 when $H=H_{c1}$, $A_1(H)$ is a function of H and depended on the nature of dynamic pinning forces.

b. Perturbation force originated from hamonic stress used for IF measurement.

The alternating stress used for IF measurement will conduce to oscillation of flux around their balance position. This kind of force can be written as

$$F_a = \alpha\sigma_A \sin\omega t \qquad (4)$$

Where α is a coupling factor of flux lines and alternating stress and is depended on the measuring frequencies.

c. Main driving force originated from interaction

IF experiment results of type II SC in the process of magnetization have shown[8]: the characteristic IF is associated with flux motion in the process of magnetization; high vaule IF appears only if $\dot{H}\neq 0$ and IF drops to its background vaule if $\dot{H}=0$ although flux lines have penetrated to the body of SC even if H is larger than H_{c1}. This means that the IF in the process of magnetizating depends on the effective driving force F' and alternating stress F_a simutaneously; in other words, the characteristic IF is originated from the interaction between F' and F_a. The expression of interaction force on condition that F' and F_a are not equal to zero simutaneously can be written as

$$F_i = C'F'\alpha\sigma_A \sin\omega t$$
$$= -C'\alpha dn/dx\Phi_0 A_1(H)\dot{H}^m t^m \sigma_A \sin\omega t, \qquad (5)$$

when C' is an interaction coefficient and is independent of measuring frequency, because the frequency effect is contained within the function $\alpha(\omega)$.

2.2. Motion equation of flux line and its resolution

Considering an unit length of flux moves along X direction, the motion equation can be given by

$$\rho\ddot{X} + \gamma\dot{X} + kX = F' + F_a + F_i$$

$$\text{or } \ddot{X} + 2\eta\dot{X} + \omega_0^2 X = C(-dn/dx)\Phi_0 A_1(H)\dot{H}^m t^m$$
$$+ C\alpha\sigma_A \sin\omega t + CC'(-dn/dx)\Phi_0 A_1(H)\dot{H}^m t^m \sigma_A \sin\omega t$$
$$= f(t) \qquad (6)$$

where $\rho = C^{-1}$ is effective mass density of flux lines, η is damping coefficient, K is restored force coefficient and ω_0^2 is the resonant frequency defined by K/ρ.

By using the Laplas transfer, the resolution of EQ.(6) can be given by

$$X = e^{-\eta t/\omega d} \int_0^t e^{\eta t} f(t') \sin\omega_d(t-t')dt', \qquad (7)$$

where $\omega_d^2 = \omega_0^2 - \eta^2$. When coefficients on the left side of eq.(6) are constant, eq.(7) is an accurate resolution for eq.(6). Provided η, ω_0^2 depend weakly on time t, the eq.(7) is also an approximate resolution by segmentation linearizing in each segment. Owing to the distribution of pinning center is fixed in type II SC, η and ω_0^2 are roughly independent of time t, eq.(7) can be considered as a quite accurate resolution.

2.3. Internal friction due to moving flux in magnetization process

The oscillation energy dissipation due to the moving flux per cycle in unit volume of specimen can be calculated by

$$W' = N' \int_0^p \sigma \dot{x} dt.$$

Oscillation energy dissipation in SC bulk is

$$W = \int_0^{x_0} (-dn/dx) \int_0^p \sigma \dot{x} dx dt.$$

Hence IF in the process is defined by

$$Q^{-1} = \Delta W/2\pi W = \int_0^{x_0}\int_0^p (-dn/dx)\sigma\dot{x}dxdt/(\pi\sigma_A/M) \quad (8)$$

where M and P are the modulus and the period of vibration. IF originated from different kind of driving force can be calculated respectively under the assumption that the superposition holds in differentiating equation with constant paraments. When $\omega_0^2 \gg \omega^2$ and $\omega_0^2 > \eta^2$, $e^{-\eta t}$ term can be neglected[7], so the expression of internal friction will be simplified as fellow

a, IF due to effective driving force

$$Q^{-1} = (M/\pi\sigma_A^2)\int_0^{x_0}\int_0^p c(-dn/dx)\Phi_0 A_1(H)\dot{H}^m t^m \dot{x}\sigma_A \cdot \sin\omega t dt dx$$

$$= (c'N(H)\Phi_0 B_0 A_1(H)\alpha M/\sigma_A \omega_0^2)(\dot{H}/\omega)^m, \quad (9)$$

where N(H) is the total number of flux lines under external field H, B_0 is a numberical coefficient, m<2/3 is another parameter involved in pinning points.

b, IF due to the alternating stress

$$Q_s^{-1} = (2cN(H)\Phi_0 \eta\alpha^2 M/\omega_0^4) \quad (10)$$

c, IF due to the interaction force

$$Q_m^{-1} = (cc'N(H)\Phi_0 B_m A_1(H)\alpha^2 M/\omega_0^2)(\dot{H}/\omega)^m \quad (11)$$

Here B_m is a numerical coefficient with order-of-magnitude "1".

Comparing equation (9)-(11), we obtained $Q_m^{-1} \gg Q_0^{-1}$, owing to M/σ_A has its order-of-magnitude 10^5. Because Q_s^{-1} is independent of varying rate of field $\dot{H}$, which is corresponding to flux lines oscillation around its balance position, in the case of $\dot{H}=0$. Q_m^{-1} associated with $\dot{H}\neq 0$, is the characteristic internal friction in magnetization process. It must be pointed out that the comparsion between the Q^{-1} value from eq.(10) and eq.(11) is corrected only if N(H) is the same volue. The IF expression can be rewritten when Q_0^{-1} is neglected.

$$Q^{-1} = A(H)\alpha^2(\omega)\ (\dot{H}/\omega)^m + B(H)\alpha^2(\omega)\omega \quad (12)$$

$$A(H) = cc'N(H)A_1(H)\Phi_0 B_m M/\pi\omega_0^2 \quad (13)$$

$$B(H) = 2cN(H)\Phi_0 M/\omega_0^4 \quad (14)$$

3. CALCULATION OF FLUX PINNING FORCE

The average pinning force can be obtained by

$$F_P = F_D - F'$$

$$= \int_0^{x_0} (-dn/dx\Phi_0)Hdx - A_1(H)\dot{H}^m t^m$$

$$= N(H)\Phi_0 H - A_1(H)(H-H_{c1})^m \quad (15)$$

From eq.(13),(14), we have

$$A_1(H) = (2\pi\eta\ /c'B_m\omega_0^2)A(H)/B(H) = k'A(H)/B(H) \quad (16)$$

$$k' = 2\pi\eta/c'B_m\omega_0^2 \quad (17)$$

Where k' is a parament depended on field H weakly, $N(H)\Phi_0$ is corresponding to magnetic induction strength penetrated in type II superconductor and can be calculated from the difference of DC magnetization curve and graph $I=4\pi M=H$. With a group of internal friction data associated with different appiled field rate $\dot{H}$ and measuring frequencies ω, we can fit Q^{-1} vs H, ω curve, and obtain the function A(H), B(H). Substituting the result into eq.(16), we obtain the flux trapping force associated with appiled field H curve. The treatment was divide into three steps:

3.1. Finding out the relationship of $\alpha(\omega)$ with ω

One of authors have pointed out[9] that Q^{-1} and $\Delta M/M$ obey a relationship

$$Q_m^{-1} / (\Delta M/M) = \alpha(\omega)/2\epsilon_0 \qquad (18)$$

In transition process, ϵ_0 is representative of transition strain and is deformation value for moving dislocation, here, ϵ_0 is corresponding to magnetostriction figure cofficient given by $\mu_0 H_0 (H_c / P)_T$ [16], in superconductivity/normal transition, $\alpha(\omega)$ as a function of ω can be defined by

$$\alpha(\omega) = \alpha_0(\omega'/\omega)^l \qquad 0 < l < 1 \qquad (19)$$

Where l value can be fitted by a group of Q^{-1} and $\Delta M/M$ versus $\dot{H}$, ω data, it is usually at the range from 0.1-0.4 in transformation process.

3.2. Calculation for A(H) and B(H)

Putting l value into eq.(12), letting A(H), B(H) include α_0, we have

$$Q^{-1}/\omega^{l-2l} = A(H)\dot{H}^m/\omega^{l+m} + B(H) \qquad (20)$$

For a group of Q^{-1}, ω, $\dot{H}$ at a certain value of H, linear fitting was carried out for the curve of Q^{-1}/ω^{l-2l} versus $\dot{H}^m/\omega^{l+m}$, when relative coefficient is proximal to number 1, the slop of the line is correspondent to A(H), and interception is correspondent to B(H).

As mention above, A(H) and B(H) are associated with total flux line number N(H) penetrated into the bulk of superconductor, hence the results are only suit to the process which flux line number is the same; flux number is a function of H, the result is correspondent to a certain value of H.

3.3. Calculation for $F_p(H)$ curve

From eq.(15),(16)

$$F_p(H) = N(H)\Phi_0 H - k'(A(H)/B(H))(H-H_{c1})^m \qquad (21)$$

For a certain value of $\dot{H}$, the $N(H)\Phi_0$, A(H) and B(H) is known, but the coefficient k' is still unknown, so eq.(21) can be resolved by utilizing self-consistent method which is useful in data analysis. From the first and the second derivative of $F_p(H)$ with respect to H, the function $F_p(H)$ exists its minimum at field H_v, then, we have

$$K' = N(H)\Phi_0 + N'(H)\Phi_0 H/[(A(H)/B(H)) (H - H_{c1})^m((dL_nA(H)/dH) - (dL_nB(H)/dH) + (m/(H - H_{c1})))] \Big|_{H_v} \qquad (22)$$

We can assume a trial k' value and put into eq.(21), H_v value will be obtained, put the H_v into eq.(22) again, a new k' will given. This method was carried out for 2 or 3 times , we will get k' and H_v accurate vaule, and the graph F_p vs H can be drawn.

4. DISCUSSION

It is well known that flux line in type II superconductor body will jump by a external driving force field, in other word, flux escapes from one pinning centre and jumps to next pinning centre point in a short time. Provided average trapped time is t_0, flux jump takes an average time interval Δt, flux number is much more than $t_0/\Delta t$, it is certainly that there is a statistical rule in the process. It is similar to the DC susceptibility measurement (scaning mode), where magnetization curve changes continually, although flux jumps in point by ponit in the process. It means there is a definite B or M at any instant. For the same reason , it exists statistical average flux moving speed and flux pinning force. Authors aim at determining accurately the average flux pinning force by developing a new method depended on IF measurement. In fact, the jumping flux number is changing with a uniform increasing of magnetic field, for example, initial jumping flux may be trapped, other pinned flux will escape under an effective force, hence, flux number is unconservative. Another article will report detail about its treatment bv statical method. Because eq.(9), (10), (11) give energy dissipation accumulation in a integrated period here, we can describe it as average flux number result in a interval from H to H + $\dot{H}\Delta t$.

Superconductivity/normal transition under external field is a first-order transition, so the treating method is the similar to the phase interface motion in transition. The energy dissipation in flux jumping process can be determined by integrating eq.(21). It is the total energy dissipated in the first-order transition. F_p is the energy dissipation function in the process. By accmulating data of dissipation function, it is benife to both learn the mechanism of flux trapping and know the physical means of index m.

ACKNOWLEDGEMENTS

This work was supperted by The National Center for Research and Development of superconductor and The Science Technology Committee of Guangdong Province of China.

REFERENCES

1. D. Saint-Janes, E.J. Thomas and G. Sarma, Type II Superconductivity, Pergamon Press (1969).

2. K.A. Muller, M. Takashige and T.G. Bednorz, Phys. Rev. Lett, 58, 1143(1987).

3. T.K. Worthington, W.J. Gallagher and T.R. Dinder, Phys. Rev. Lett, 59, 1160(1987). A.P. Malozemoff, T.K. Worthington, Y. Yeshurun, F. Holtzberg and P.H. Kes, Phys. Rev. B38 7203(1988).

4. C.W Hagen and R. Griessen, Phys. Rev. Lett. 62, 2857(1989).

5. E. Zeldov, N.M. Amer, G. Koren, A. Gupta, R. J. Gambino and M.W. Mcelfresh, Phys. Rev. Lett., 62, 3093(1989).

6. C. Rossel and P.Chandhari, Physica, 153c-155 c, 306(1988).

7. W.G. Zeng and J.X Zhang, Acta physica Sinica 36, 37(1987).

8. J.X. Zhang, W.G. Zeng, G.M. Lin, K.F. Liang, Z.C Lin and Z.L. Du, Acta Physica, 39, No.8 (1990).

9. J.X. Zhang and X.J. Li, Acta Physica Sinica, 36, 847(1987).

10. J.X. Zhang, W.G. Zeng, Proceeding of international conf. on Internal Friction and Ultrasomic Attenuation in Solids, Beijing, July, 1989, P133.

11. K.F. Liang, Z.C. Lin and J.X. Zhang, ibid, p365.

12. J.X. Zhang, J.de.Physique, 42, c5-399(1981).

13. J.X. Zhang, Acta Scientiarum Naturalium Universitatis Sunyatseni, 21(5)1(1982).

14. J.X.Zhang, W.G. Zeng and J.M. Wu, Acta Scientiarum Naturalium Universitatis Sunyatseni 23(2)109(1984).

15. W.G. Zeng, J.X. Zhang Z.H. He, Acta Physica Sinica, 35, 1377(1986).

16. L.Y. Zhang, J.L. Zhang and G.J. Cui, Superconductor Physics(in chinese), Electronics Industry Pub., Beijing, 1987, p36.

Frontiers of Materials Research/
Electronic and Optical Materials
M. Kong, L. Huang (editors)

Effect of Oxygen Partial Pressure on the Properties of Superconducting $Nd_{1.85}Ce_{0.15}CuO_x$

Liquan Chen, Yuzhen Huang, Yongqian Liu, Jianqin Bi, T.Nakamura*, Ruixing Liang*, and Zhongxian Zhao

Institute of Physics, Academia Sinica, Beijing 100080, P.R.China

* Research Laboratory of Engineering Materials Tokyo Institute of Technology 4259 Nagatsutacho, Midoriku, Yokohama 227, Japan

Oxygen partial pressure was controlled by mixing O_2 and Ar gas and measured by $ZrO_2-Y_2O_3$ Sensor with successively $Po_2=10^{-1.5}$, 10^{-2}, $10^{-2.5}$, $10^{-3.0}$, $10^{-3.5}$ atm in processing ceramic superconducting $Nd_{1.85}Ce_{0.15}CuO_x$. Four probe DC resistance measurement was employed to measure their resistivity . It was shown that oxygen partial pressure has great influence on superconductivity. Tc^{on} decreases with increasing oxygen partial pressure while the room temperature resistivity increases with decreasing oxygen partial pressure . XPS photoemisson experiments on them clearly shows the existence of both Cu^{+} and Cu^{2+} which support an electron carrier mechanism.

1. INTRODUCTION

Since the discovery of high Tc superconductivity in La-Ba-Cu-O system, many high Tc superconducting cuprates have been found [1-4]. It is generally believed that superconductivity is contributed by 2 dimensional Cu-O plane [5]. In conventional discovered superconducting cuprates , coppers are in the center of octahedral or pyramids with oxygen on the apical site and main charge carriers in normal state are holes. Recently , Y.Tokura et.al. discovered the new n-type superconductor $(Nd_{2-x}Ce_x)CuO_4$ with Tc up to 24K in T'phase of Nd_2CuO_4 structure [6]. In this structure , Cu-O is square planar . After cerium doping and heat treatment in reducing atmosphere, a large number of electron carriers are brought about to contribute to the normal state conduction process [7]. Without oxygen deficiency, superconductivity could not be achieved . Here, the doping and oxygen deficiency act in a symmetry way with respect to p-type superconductor $(La_{2-x}Sr_x)CuO_4$. So, it is oxygen deficiency that makes the carriers participate in the superconducting cooper pairing. To know the role played by oxygen here, we investigate the influence by oxygen partial pressure on both superconductivity and normal state resistivity and some other related properties which determines the charge carrier density.

2. EXPERIMENT AND DISCUSSION

The samples were prepared by mixing high purity compound Nd_2O_3, CeO_2 and CuO in a nominal composition $Nd_{1.85}Ce_{0.15}CuO_x$. Then pressing them into pellets ,the mixtures set in Al_2O_3 crucible were preheated in air at 950°C for 20 hours. After thoroughly ground and again pressed into pellets, we put

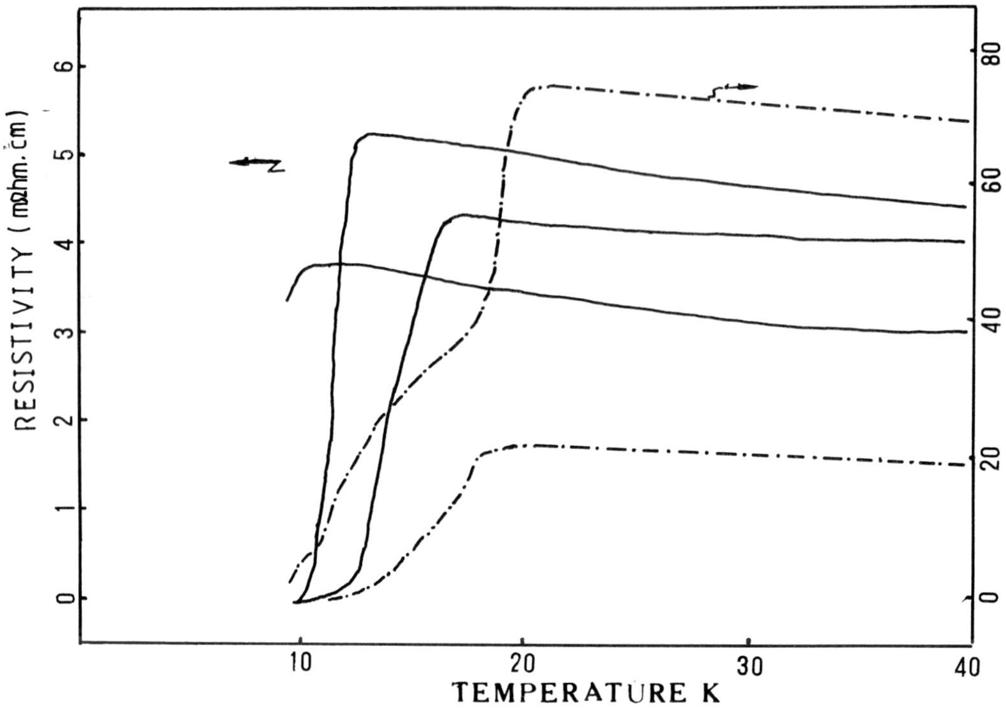

FIGURE 1
Temperature dependence of the resistivity of $Nd_{1.85}Ce_{0.15}CuO_x$ under a) $Po_2=10^{-3}$ atm, b) $Po_2=10^{-1.5}$ atm, c) $Po_2=10^{-2.5}$ atm, d) $Po_2=10^{-2}$ atm, e) $Po_2=10^{-3.5}$ atm

them into furnace in a tube and heated at 1150°C for 24 hours. The oxygen partial pressure was controlled by mixing O_2 and Ar gas which then pass through the tube and measured by $ZnO(Y_2O_3)$ sensor respectively with $Po_2=10^{-1.5}, 10^{-2}, 10^{-2.5}, 10^{-3}$ and $10^{-3.5}$ atm. Measurements of resistivity were performed by four probe method down to 4.2K. The results are shown in Fig. 1. For comparison , Tc^{on} and room temperature resistivities are shown in Figure.2. As can be seen in Fig.2, room temperature resistivities increase with decreasing oxygen partial pressure.

Tc^{on} increases with decreasing oxygen partial pressure. This suggests that oxygen partial pressure adjust the lattice distortion or oxygen vacancies which contribute to the normal state scattering between electrons and phonons. With decreasing oxygen partial pressure , distortion or vacancies becomes greater so that resistivity becomes greater. The strong interaction between electron and phonon enhance the cooper pairing of electrons near Fermi edge . So, Tc increases . From this, we suggest that week electron-phonon interaction still plays some role in cooper pairing in this system which is similar to conventional superconductors.

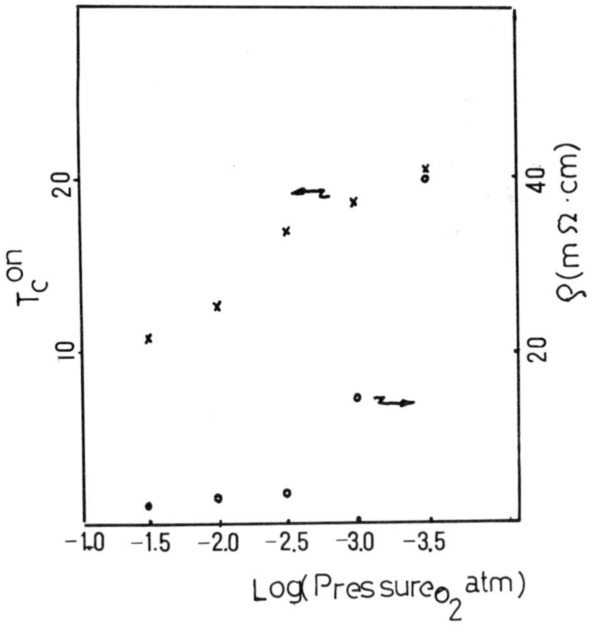

FIGURE 2
Dependence of T_c^{on} and room temperature resistivity on oxygen partial pressure

To investigate the influence played by doping of Ce, we make XPS analysis on our sample as-prepared. The measurement was made using Mg-Kα as source at an accelerating voltage of 12V and a current of 20MA. In Fig.3, we show the result in the energy region of BE=925 eV to 975 eV where Cu 2p appeared.It can be seen that with doping of Ce, the single energy peak corresponding to Cu^{2+} $(2p_{3/2})$ at 933.6 eV moves to a nonsingle relatively broad one with main peak at 932.8 eV and a shoulder at 933.6 eV. The satellite peak of Cu^{2+} $(2p_{3/2})$ around 945 eV is still eminent [8]. This movement give evidence to the existence of mix valence of both Cu^{1+} and Cu^{2+} [9]. The satellite tells us that Cu^{2+} state is still predominant. With the mixing state and balance between electron charge , we could infer that main charge carrier is electron.

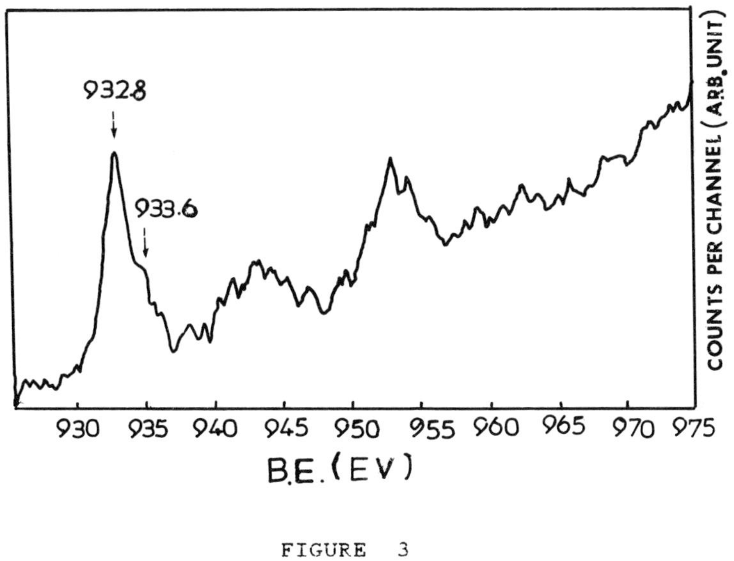

FIGURE 3
The Cu 2p spectra of $Nd_{1.85}Ce_{0.15}CuO_x$

3. CONCLUSIONS

XPS and controlled oxygen partial pressure measurement indicate the following in the new superconductor $Nd_{1.85}Ce_{0.15}CuO_x$

(1) Electron is the main charge carrier in normal state .

(2) Electron-phonon interaction plays some role in cooper pair formation.

REFERENCES

1. J.G.Bednorz and K.A.Müller, Z.Phys.B64 (1986)189
2. Zhongxian Zhao, Liquan Chen et.al. Kexue Tongbao 32 (1987) 1098
3. H.Maeda et.al. Jap.J. Appl. Phys. 27 (1988)L209
4. A.M.Hermann,Z.Z.Sheng et.al. Phys. Rev. B37 (1988)9742
5. Y.Tokura et. al. Phys. Rev.B 38, (1988)7156
6. Y.Tokura, H.Takagi and S.Uchida Nature Jan.26, 1989
7. H.Takagi , S.Uchida and Y.Tokura Phyiscal Review Letters 62 (1989)1197
8. A.Fujimori et. al. Phys. Rev. B35, (1987)8814
9. T.Takahashi et. al. Phys.Rev.B 37 (1988)9788

Frontiers of Materials Research/
Electronic and Optical Materials
M. Kong, L. Huang (editors)

STUDY OF SUPERCONDUCTIVITY AND CRYSTALLOGRAPHY IN $YBa_2(Cu_{1-x}M_x)_3O_{7-d}$ (M=Ni, Co) WITH $d\cong0$ AND $d\cong1$

WEN Tinglian*, Serge VILMINOT, Christian HIS, and R. KUENTZLER

E.H.I.C.S.-D.S.M., I.P.C.M.S., CNRS, 67008 Strasbourg cedex, France

High Tc 123 compound was substituted partially for Cu by Ni or Co in the range of 0.5-6%. Single phase ceramics with nominal chemical compositions of $YBa_2(Cu_{1-x}M_x)_3O_{7-d}$ (M=Co, Ni) with $d\cong0$ or $d\cong1$ were obtained by treatment of the samples under oxygen or nitrogen atmosphere, respectively. The crystalline parameters were determined as a function of substituting concentration. It was found that the materials can remain the superconductivity in lower substitution levels. Ni substitution showed more influence on Tc than Co. Changes of Tc and resisitivity were discussed with regard to the transition between the orthorhombic and tetragonal structure, and the occupation of substitution elements at positions of Cu(1) or Cu(2).

I. INTRODUCTION

Understanding the mechanism of high Tc superconductivity in layer oxides is still an interesting topic for the theoretical and experimental physists. Nevertheless, it has been recognized that the magnetism may play a significant role in the superconducting mechanism. Therefore magnetic 3d transition metal , especially Fe, substitution for Cu in $YBa_2Cu_3O_{7-d}$ superconductors has been investigated by several scientists[1-6]. The influence of 3d transition elements substitution on Tc is related to the position of 3d metal element in Cu sublattice, because the superconductivity originates from the Cu-O arrangement of the crystalline structure in $YBa_2Cu_3O_{7-d}$.

The present work studied the influences of substitution of 3d transition metal Co and Ni for Cu in $YBa_2Cu_3O_{7-d}$ on its crystalline parameters and structure modification as well as Tc for the superconducting phase ($d\cong0$) and for the nonsuperconducting one($d\cong1$).

II. EXPERIMENTAL

1. PREPARATION OF SAMPLES

A. (a) $YBa_2(Cu_{1-x}Co_x)_3O_{7-d}$ with $d\cong0$

Y_2O_3, $BaCO_3$, $CuCO_3.Cu(OH)_2$ and $CoCO_3.Co(OH)_2$ with analytical pure grade were used as the starting materials. The x values were chosen from 0.5% to 6%. After intimate grinding, the starting materials were calcined at 940°C for 15 hours under the air atmosphere. It took 270 minutes to heat from the room temperature up to 940°C, and took 540 minutes to cool the samples from 940°C down to room temperature. The calcined samples were perfect black. After powdering and grinding, pellets were made by a pressure of $2T/cm^2$ in a steel die, and were annealed at 910°C in flowing oxygen for 8 hours. The heating and cooling rate were 300°C/hour and 100°C/hour, respectively. Such annealed samples would not increase their weight when further annealing in oxygen, indicating that d equals nearly to zero.

(b) $YBa_2(Cu_{1-x}Co_x)_3O_{7-d}$ with $d\cong1$

$YBa_2(Cu_{1-x}Co_x)_3O_{7-d}(d\cong1)$ was obtained by annealing $YBa_2(Cu_{1-x}Co_x)_3O_{7-d}(d\cong0)$ in flowing nitrogen at 800°C for 24 hour. When the samples annealed in the nitrogen for another 8

*Shanghai Institute of Ceramics , Chinese Academy of Sciences, Shanghai 200050, P.R.China hours, no weight decrease would be observed, indicating that the first 24 hours annealing had reduced the samples into the formula with d ≅ 1.

B. $YBa_2(Cu_{1-x}Ni_x)_3O_{7-d}$ with d ≅ 0 and d ≅ 1

Except for using $Ni(HCOO)_2.2H_2O$ instead of $CoCO_3.Co(OH)_2$, all the starting materials and the procedures were the same as in the case of $YBa_2(Cu_{1-x}Co_x)_3O_{7-d}$.

2. CRYSTALLOCHEMICAL MEASUREMENTS

The unit cell parameters of samples were calculated depending on the X-ray(CoKa) diffraction data. Powder silicon was used for internal and external standard of X-ray diffraction for the powder and the ceramic samples, respectivly. There was little change in X-ray diagrams before and after the annealing in oxygen at 910°C for 8 hours, indicating that the calcination at 940°C in the air had led to the formation of the compounds with formulas with d≅0. The annealing at 910°C in oxygen only played a role of hardening of the samples, because shrinkage on the samples was hardly observed. The identification of the X-ray diagrams did not show the existence of free metal oxides except for 123 system compounds, which may demonsrate that the products were of homogeneous solid solutions.

3. RESISTANCE MEASUREMENTS IN TEMPERATURE REGION OF LIQUID NITROGEN

The resistance measurment was performed by the standard four-probe technique. Silver paste was used as the probe electric contact. The temperature was measured by using a platinum resistance thermometer with a four probe arrangement. The preciseness of the temperature measurement was ±0.05°C.

III. RESULTS AND DISCUSSIONS

Depending on the data of X-ray diffractions the calculated crystalline lattice parameters of unit cells as a function of x value are listed in Table 1 and 2 for $YBa_2(Cu_{1-x}Co_x)_3O_{7-d}$ with d≅0 and 1, respectively.

From the Table 1 it can be found that a axis of the unit cell of $YBa_2(Cu_{1-x}Co_x)_3O_{7-d}$ with d=0 augments meanwhile b axis tends to reduce as the Co doping concentration increases. When x is equal to or bigger than 5%, a axis equals essentially b axis, i.e. the crystalline structure transforms from the orthorhombical into the tetragonal phase. The c axis shows a little reduction as x increases. However, the volume of unit cell has no obvious change.

Table 1.
Lattice parameters of $YBa_2(Cu_{1-x}Co_x)_3O_{7-d}$ with d≅0

x(%)	a(A)	b(A)	c(A)	Vol. of u.c.(A^3)
0.5	3.830	3.887	11.662	173.577
1.5	3.855	3.882	11.650	174.656
2.0	3.858	3.882	11.640	174.576
2.5	3.858	3.881	11.645	174.402
3.5	3.859	3.878	11.620	173.695
4.0	3.863	3.872	11.621	174.514
5.0	3.870	3.871	11.622	174.148
6.0	3.873	3.873	11.625	174.457

Table 2
Lattice parameters of $YBa_2(Cu_{1-x}Co_x)_3O_{7-d}$ with d≅1

x(%)	a(A)	b(A)	c(A)	Vol. of u.c(A^3)
0.5	3.864	3.864	11.782	175.908
1.5	3.861	3.861	11.813	176.126
2.0	3.866	3.866	11.793	176.233
2.5	3.866	3.866	11.759	175.728
3.5	3.866	3.866	11.793	176.233
4.0	3.867	3.867	11.813	176.689
5.0	3.866	3.866	11.791	176.201
6.0	3.871	3.871	11.754	176.302

From the Table 2 it can be found that $YBa_2(Cu_{1-x}Co_x)O_{7-d}$ with d≅1 exhibits the tetragonal structure regardless of the x value from 0.5% to 6%. None of the lattice parameters(a, b, c axis and volume of unit cell) shows an evident change with x value. Nevertheless, comparing with $YBa_2(Cu_{1-x}Co_x)_3O_{7-d}$ with d≅0, the reduced samples by annealing under nitrogen have a longer c axis and a bigger volume of unit cell.

As the case of $YBa_2(Cu_{1-x}Co_x)_3O_{7-d}$ with d≅0, the X-ray diffraction patterns of $YBa_2(Cu_{1-x}Ni)_3O_{7-d}$ with d≅0 before and after annealing in oxygen show no difference within the preciseness of the instrument. The lattice parameters calculated depending on the X-ray diffraction data are listed in Table 3 and 4 for d≅0 and d≅1.

From Table 3 it can be found that $YBa_2(Cu_{1-x}Ni_x)_3O_{7-d}$ exhibits the orthorhombic structure in x values of 1% to 6% and the lattice parameters nearly do not change with NiO doping concentration investigated. b and c axises remain unchanged but a axis shows a little increase. However, $YBa_2(Cu_{1-x}Ni_x)_3O_{7-d}$(d≅1) behaves as a tetragonal structure(Table 4) in the range of x value from 1% to 6% and shows longer a axis and c axis than $YBa_2(Cu_{1-x}Ni_x)_3O_{7-d}$(d≅0). As a results their volumes of unit cell also enlarge.

Table 3
Lattice parameters of $YBa_2(Cu_{1-x}Ni_x)_3O_{7-d}$ with d≅0

x(%)	a(A)	b(A)	c(A)	Vol. of u.c.(A^3)
1.0	3.803	3.883	11.649	172.083
2.0	3.823	3.884	11.660	173.498
3.0	3.823	3.883	11.650	173.012
6.0	3.831	3.883	11.645	173.212

Table 4
Lattice parameters of $YBa_2(Cu_{1-x}Ni_x)_3O_{7-d}$ with d≅1

x(%)	a(A)	b(A)	c(A)	Vol. of u.c.(A^3)
1.0	3.878	3.878	11.820	177.762
2.0	3.882	3.882	11.834	178.355
3.0	3.882	3.882	11.828	178.273
6.0	3.879	3.879	11.745	176.727

Fig.1 and Fig.2 show the resistance as a function of temperature for $YBa_2(Cu_{0.995}Co_{0.005})_3O_{7-d}$(d≅0) and

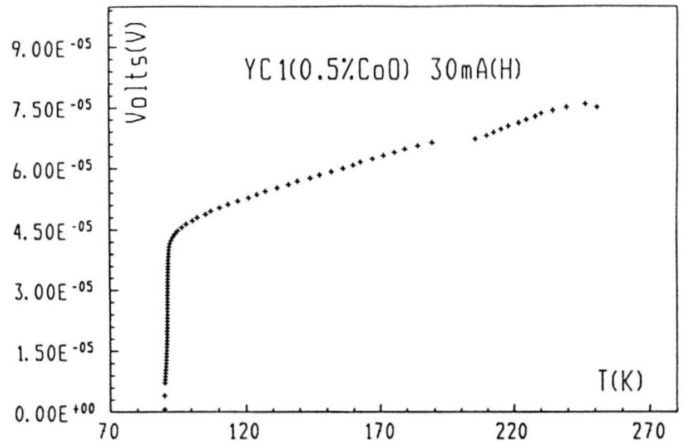

Figure 1
Resistance as a function of temperature for $YBa_2(Cu_{0.995}Co_{0.005})_3O_{7-d}$(d≅0).

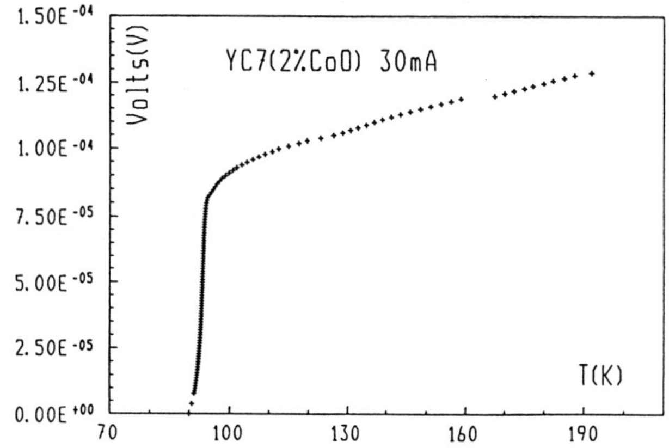

Figure 2
Resistance as a function of temperature for $YBa_2(Cu_{0.98}Co_{0.02})_3O_{7-d}$(d≅0).

$YBa_2(Cu_{0.98}Co_{0.02})_3O_{7-d}$(d≅0), respectively. Both samples exhibit a Tc(R=0) of around 90K. However, the former(with smaller x) has a narrower transition temperature region. When the CoO doping concentration increase beyond 5% the curve of R vs T behaves as a semiconductive characteristics, i.e. the resistance increases as the temperature decreases(Fig. 3), which may be related to the tetragonal structure of $YBa_2(Cu_{1-x}Co_x)_3O_{7-d}$(d≅0) when x is equal to or bigger than 5% as

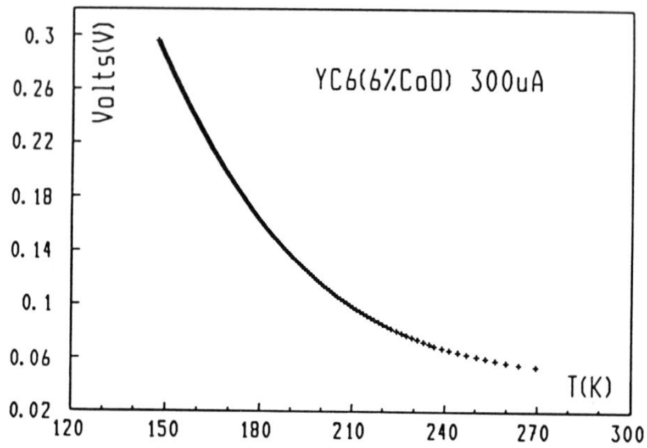

Figure 3
Resistance as a function of temperature for $YBa_2(Cu_{0.94}Co_{0.06})_3O_{7-d}(d\cong 0)$.

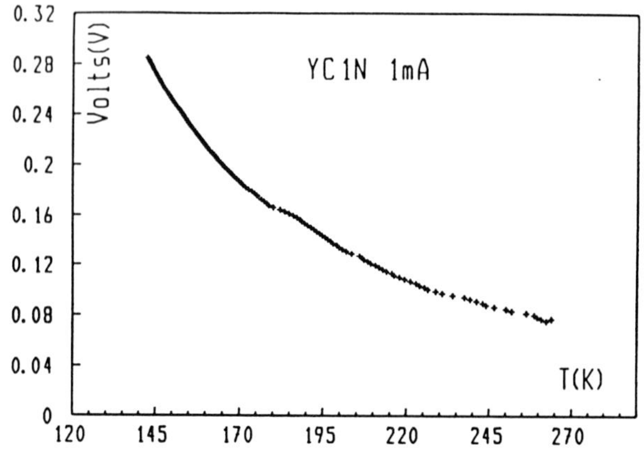

Figure 4
Resistance as a functionn of temperature for $YBa_2(Cu_{0.995}Co_{0.005})_3O_{7-d}(d\cong 1)$.

mentioned above.

Fig. 4 and Fig. 5 shows the curves of R vs T of $YBa_2(Cu_{1-x}Co_x)_3O_{7-d}(d\cong 1)$ for x=0.5% and 2%, respectively. Both curves behaves as a semicondutors, which may be related to their tetragonal structure. The resistivities of the samples with semiconductive charateristics are much larger than that with superconductivity even in the normal state as shown in Fig. 1 through Fig. 5, because all the samples have almost the same dimension.

Fig. 6 shows the temperature dependence of resistance of $YBa_2(Cu_{0.99}Ni_{0.01})_3O_{7-d}(d\cong 0)$. The onset of superconductive transiton takes place at about 90K. It reaches zero resistivity at 87K, which is somewhat lower than that of the sample substituted by CoO with the same concentratin of 1%. When the substitution concentration increases beyond 3% NiO, the superconductive transition is no longer observed in the temperature region of liquid nitrogen, where the resistance either decreases slowly with the temperature decreasing(x=3%, Fig. 7) or keeps unchanged(x=6%, Fig. 8). Maybe, the superconductive transition could happen in much lower temperatures.

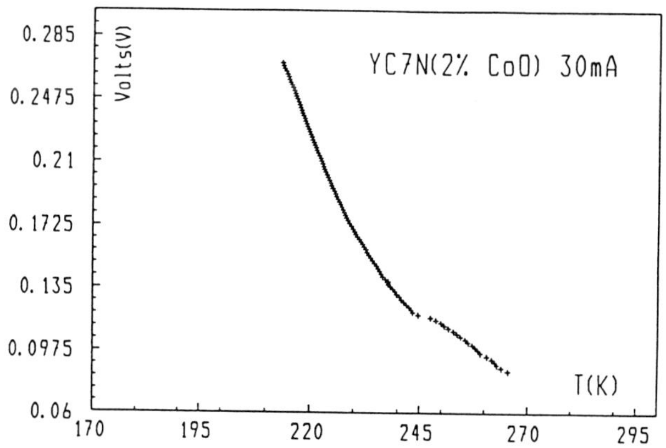

Figure 5
Resistance as a function of temperaure for $YBa_2(Cu_{0.98}Co_{0.02})_3O_{7-d}(d\cong 1)$.

As to the samples of $YBa_2(Cu_{1-x}Ni_x)_3O_{7-d}(d\cong 1)$ for x=1% to 6% with tetragonal sturcture all showed a semiconductive curve of R vs T.

When those samples with d≅1 for both Co and Ni substitution were annealed in oxygen again for several days they could not be recovered to superconductive ones. It may be attributed to very slow recovery of oxygen in matrix.

It has been demonstrated depending on the study of specific heat and magnetism in $YBa_2(Cu_{1-x}Fe_x)_3O_{7-d}$ with d≅0[7] that Fe occupy

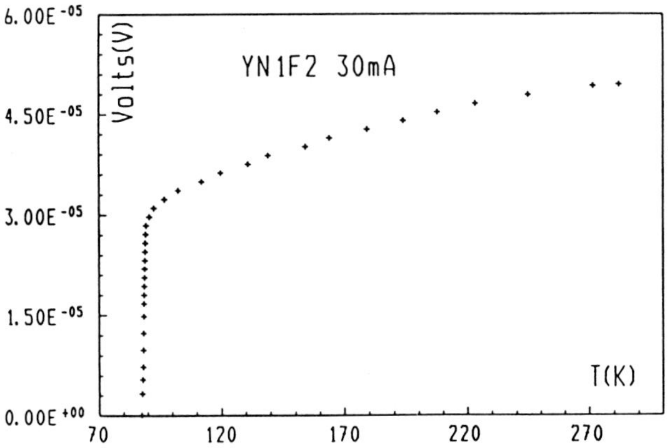

Figure 6
Resistance as a function of temperature for $YBa_2(Cu_{0.99}Ni_{0.01})_3O_{7-d}$ (d≅0).

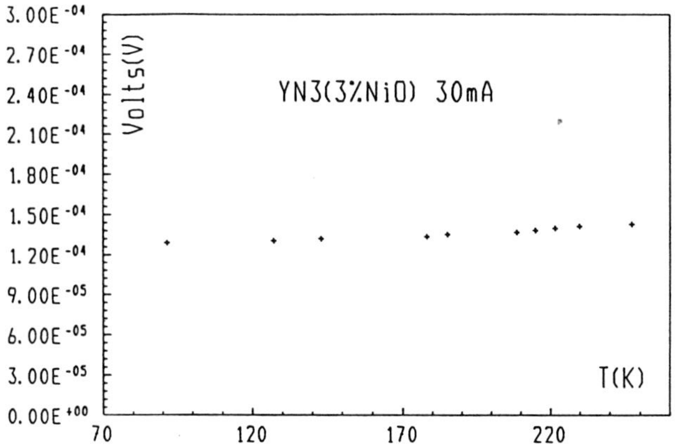

Figure 7
Resistance as a function of temperature for $YBa_2(Cu_{0.97}Ni_{0.03})_3O_{7-d}$ (d≅0).

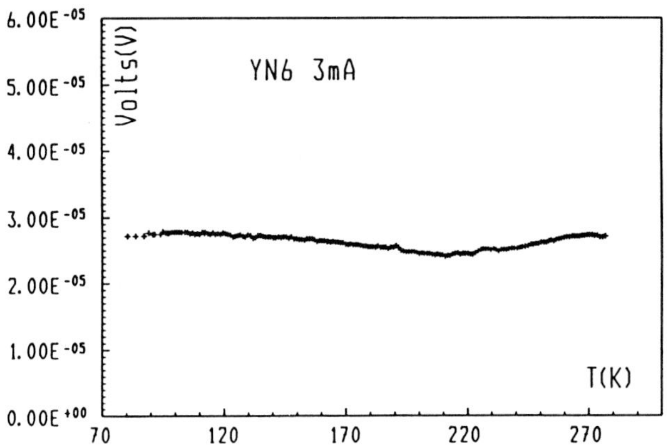

Figure 8
Resistance as a function of temperature for $YBa_2(Cu_{0.94}Ni_{0.06})_3O_{7-d}$ (d≅0).

Cu(1) positions, as a result not affecting the superconductivity and the orthorhombic structure when doping of Fe at lower concentrations. However, more Fe have to occupy Cu(2) positions because 5-6 coordination of Fe can not be accompolished in the Cu(1) chains when doping of Fe at higher concentrations. It might be related to the transition of O-T. Since the superconductivity of 123 compound results from the Cu(2) sheet, higher Fe doping concentration may reduce or lose the superconductivity owing to varying of the original Cu(2) sheet sublattice. Co and Ni as well as Fe are all belong to 3d transition metal, therefore effects of doping of Co and Ni on structure and superconductivity of 123 compound might be similar to that of Fe.

IV. SUMMARY

We have studied the systems of 123 compound which were substituted partially for Cu by Co or Ni, i.e. $YBa_2(Cu_{1-x}Co_x)_3O_{7-d}$ and $YBa_2(Cu_{1-x}Ni_x)_3O_{7-d}$ with d≅0 and d≅1. when annealing at 910°C-950°C in oxygen, the compositions with d≅0 were obtained. However, when annealing again at 800°C in nitrogen, the compositions with d≅1 were obtained. $YBa_2(Cu_{1-x}Co_x)_3O_{7-d}$ with d≅0 with x below 5% appeared to be an orthorhombic structrue and exhibited a superconductivity with Tc equal to around 90K. However, when x value is 5% and 6%, it behaved as a semiconductor and had a tetragonal structure. $YBa_2(Cu_{1-x}Ni_x)_3O_{7-d}$ with d≅0 in the investigated x value of 1%-6% appeared to be an orthorhomic structure. 87K of Tc(R=0) was observed for the composition of x=1%. Others may exhibit the superconductivity at temperatures much lower than the liquid nitrogen. Generally, Tc is lower than systems substituted by Co and drastically decrease with the increase in substitution concentration of Ni. The effects of doping of Co and Ni on the superconductivity of 123 compound may be related to the occupation of Co and Ni on positions of Cu(1) or Cu(2).

$YBa_2(Cu_{1-x}Co_x)_3O_{7-d}$ and $YBa_2(Cu_{1-x}Ni_x)_3O_{7-d}$ with $d \cong 1$ show a tetragonal structure regardless the x value and behave as a semiconductor in the curve of resistance versus temperature, meanwhile they have a relatively high resistivity than the same substituted composition but with $d \cong 0$.

ACKNOWLEDGEMENTS

Wen Ting-Lian would like to thank the C.N.R.S. for financial support

REFERENCES

1. J.M. Tarascon, P. Barboux, L.H. Greene and G.W. Hull, Proceedings of Material research society, Fall Meeting, Nov.30-Dec.4, 1987, Boston, Massech. U.S.A.

2. J. Jing, J. Bieg, H. Engelmann, Y. Hsia, U. Gonser, P. Gultlich and R. Jakobi, Solid state Commun. 66 (1988) 727

3. X. Zhang, S. Labroo, P. Hill, N. Ale, E. Funk and J.R. Gains, Phys. Lett. 130 (1988) 311

4. J.R. Schrieffer, S.G. Wen and S.C. Zhang, Physica C, 153-155 (11988) 21

5. E. Sakagamim Y. Oda, Y. Yamadam, T. Kohara and K. Asayamam J. Phys. Soc. of Japan 56 (11987) 4255

6. T.J. Kistenmacher, Physical Review B, 38 (1988) 8862

7. R. Kuntzler, S. Vilminot, Y. Dossmann, T.L. Wen and C. His, Proceedings of International Conference on Materials and Machanisms of Superconductivity of High-Temperature Superconductors, July 23-28, 1989, Stanford, Califonia, U.S.A.

Frontiers of Materials Research/
Electronic and Optical Materials
M. Kong, L. Huang (editors)

STUDY OF THE GROWTH PROCESS OF LARGE SIZE BSCCO CRYSTALS*

Zhuo WANG, Hong WANG, Zhaohe YANG, Shuxia SHANG and Xiaolin WANG

Institute of Crystal Materials, Shandong University, Jinan 250100, China

The growth process of large size BSCCO crystal has been studied. BSCCO crystals with the size of 15 X 4 X 2 mm^3 have been grown by self-flux method. The a-b plan is spotless and smooth. The results obtained from the measurements of resistivity and a.c. susceptibility show that Tc(zero) is 90K. During the crystal growth process, high temperature gradient and three-step process were used. The single crystals obtained have better cleavage. Differential thermal analysis of BSCCO crystal was also made.

1. INTRODUCTION

After H. Maeda et al.[1] have developed the high Tc superconductor Bi-Sr-Ca-Cu oxide (BSCCO), many researches have been focusing on the structure and physical properties of BSCCO system[2,3,4]. There are some reports on BSCCO single crystal growth[5,6,7]. Sunshine et al[2] have prepared the single crystal of $Bi_{2.2}Sr_2Ca_{0.8}Cu_2O_8$. The crystal structure and physical properties were studied by using a single crystal with dimensions 1 X 0.5 X 0.001 mm^3. S. Nomura et al.[5] have prepared BSCCO crystal of dimensions 7 X 5 X 1 mm^3, whose Tc(zero) is 84K.

In this paper, we report the growth of BSCCO single crystal with the size up to 15 X 4 X 2 mm^3. The Tc(zero) measured is 90K. The crystal grown can be cleaved easily. Differential thermal analysis of BSCCO crystal was made, and the crystal growth process is discussed.

2. EXPERIMENTAL AND RESULTS

BSCCO crystal has been grown by self-flux method with a vertical tubular furnace equipped with an automatic temperature controlling system (see Fig. 1).

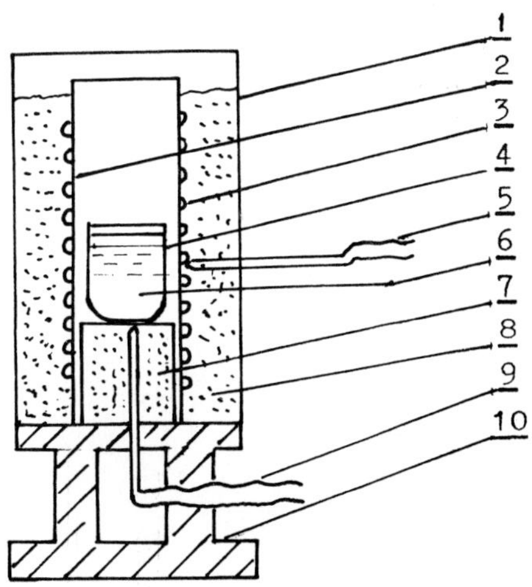

FIGURE 1
Sintering apparatus used for crystal growth: (1) The external Al_2O_3 furnace tube, (2) The internal Al_2O_3 furnace tube, (3) Kanthal heating wire, (4) Pure Al_2O_3 crucible, (5) Pt/Rh thermocouples used for temperature controlling, (6) Mixture of Bi-Sr-Ca-Cu, (7) Refractory block, (8) Insulating material, (9) Pt/Rh thermocouples used for

*This work was supported by the Chinese National Centre for Research and Development on Superconductivity.

measuring temperature, (10) Base.

Powders of Bi_2O_3, $SrCO_3$, $CaCO_3$ and CuO with 99% purity were mixed, and put in a pure Al_2O_3 crucible. The mixture were heated at a rate of 100°C/h to 1050°C and held at this temperature for 8 hours.

The cooling process was divided into three steps: at first, the temperature was rapidly lowered to 900°C and kept for 2-3 hours; then it was lowered to 820°C at a rate of 2°C/h, and held for 8-10 hours. Finally, the temperature was lowered to 100°C. In the whole cooling process, a temperature gradient about 30°C/cm was maintained.

The size of BSCCO crystal obtained has reached to 15 X 4 X 2 mm^3, and the size of the crystal plate cleaved is about 12 X 2.5 X 1 mm^3. Figure 2 shows the shape of a cleaved BSCCO crystal plate.

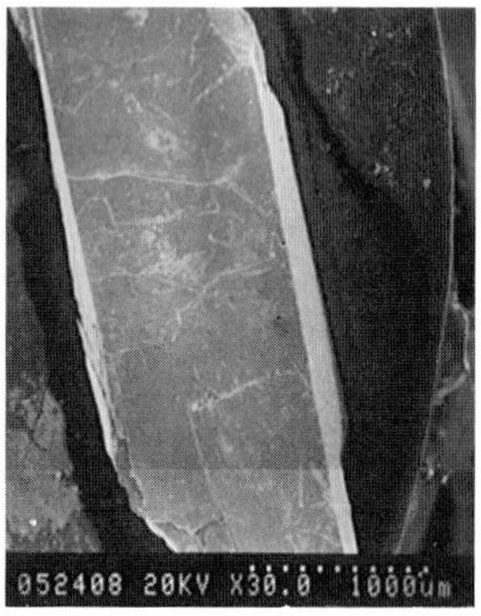

FIGURE 2
Single crystal plate of $Bi_3Sr_3CaCu_{2.5}O_x$

The compositions of BSCCO crystal were determined by using Philips EDAX-9100 fully computerized analysis system equipped with a Hitachi scanning electron microscope. The results show that there is slight changes of compositions in the crystal from surface to bottom in a crucible. The average atomic ratio measured is Bi:Sr:Ca:Cu = 3:3:1:2.5 .

The resistance of BSCCO crystal was measured by using the standard four-probe technique, and a.c. susceptibility was measured by using HP 4192A LF impedance analyzer. The results are shown in Figure 3, the $Bi_3Sr_3CaCu_{2.5}O_x$ crystal reaches zero resistance state at 90K.

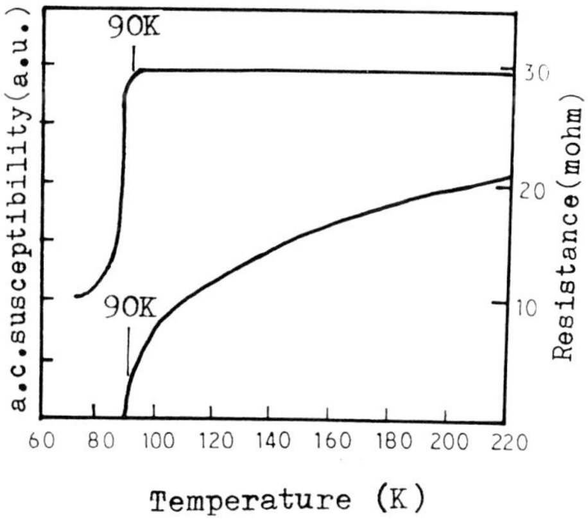

FIGURE 3
Resistance and a.c. susceptibility vs temperature of $Bi_3Sr_3CaCu_{2.5}O_x$ crystal

Thermal analysis of BSCCO crystal was carried out by using a PE DAT1700 high temperature differential thermal analyzer.

3. DISCUSSION

The differential thermal analysis (DTA) figure of $Bi_3Sr_3CaCu_{2.5}O_x$ crystal (Fig. 4) shows that an endothermic peak occurs at 820$\sim$900°C.

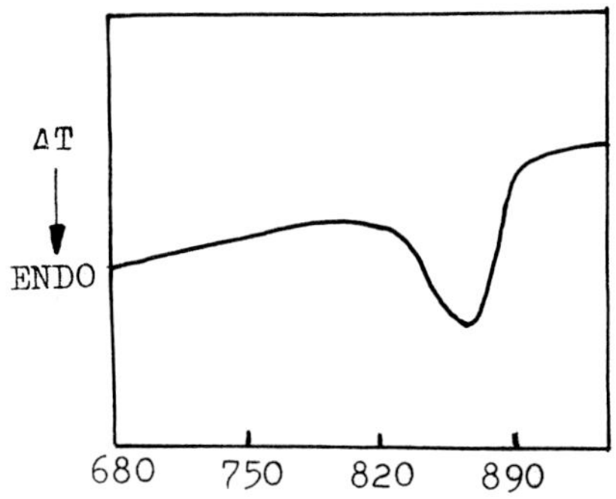

FIGURE 4
DTA curve of $Bi_3Sr_3CaCu_{2.5}O_x$ crystal

From Figure 4, we can see that the initial temperature is 820°C; the extrapolated onset is 836.5°C; the minimum of the peak is 866.5°C and the final temperature is 900°C. Thus, the temperature region from 820°C to 900°C is mainly the crystal growth region.

From our experiments, it is found that in the region of 820~900°C the temperature should be controlled stably and the use of suitable cooling rate is very important for the growing of perfect single crystal with better anti-magnetism property.

During the cooling process, a higher temperature gradient is used (Its maximum is about 30°C/cm.), the BSCCO crystal obtained has larger size and the a-b plan of crystal is more smooth, spotless and perfect. The cleavage of crystal is obviously increased.

REFERENCES

1. H. Maeda, Y. Tanaka et al., J. Appl. Phys. Lett. 27 (1988) 209.
2. S.A. Sunshine, T. Siegrist et al., Phys. Rev. B38 (1988) 893.
3. M. Uehara, Y. Tanaka et al., Japan. J. Appl. Phys. 27 (1988) L665.
4. C. Politis, Appl. Phys. A 45 (1988) 261.
5. S. Nomura, T. Yamasita, H. Yoshino and K. Ando, Japan. J. Appl. Phys. 27 (1988) 1251.
6. T.F. Ciszek, J.P. Goral, C.D. Evans and H.K. Yoshida, J. Crystal Growth 91 (1988) 312.
7. H. Wang, S.X. Shang, Z.H. Yang, Z. Wang, X.N. Shen, H.S. Zhao, M.H. Jiang and H.C. Chen, J. Crystal Growth 99 (1990) 929.

Frontiers of Materials Research/
Electronic and Optical Materials
M. Kong, L. Huang (editors)

THE INFLUENCE OF Sr/Ca RATIO ON THE SUPERCONDUCTIVITY OF Bi-Sr-Ca-Cu-O CRYSTALS

Hong WANG, Shuxia SHANG, Zhuo WANG, Xiaolin WANG

Institute of Crystal Materials, Shandong University, Jinan 250100, China

Crystals grown from Bi-Sr-Ca-Cu-O system may have different contents of Sr and Ca when varying the ratio of Sr/Ca in the starting composition. The Sr/Ca ratio in crystal has important influence on the superconducting property of Bi-Sr-Ca-Cu-O crystal. The lattice parameter c increased from 30.60A to 31.01A and Tc(zero) increased from 84K to 90K with the increasing of Sr/Ca ratio from 2:1 to 3:1 accordingly. This result has shown its importance for understanding high-Tc superconducting phenomenon in Bi-Sr-Ca-Cu-O system further.

1. INTRODUCTION

Since the discovery of high-Tc superconductivity in Bi-Sr-Ca-Cu oxide system by Maeda et al.[1], the crystal growth of Bi-Sr-Ca-Cu-O compounds has been reported by many research groups[2,3,4,5] There are two superconducting phases in Bi-based system. One is low-Tc phase with Tc of 85K, and the other is high-Tc phase with Tc of 110K. The crystals obtained from the melting mixture of Bi-Sr-Ca-Cu-O system all belong to the low Tc phase. The compositions of the crystals are slightly different.(see Table 1.) It is found that a partial substitution of Sr ions for Ca may take place[2,4]. In this paper, we report the influence of Sr/Ca ratio on the superconductivity of Bi-Sr-Ca-Cu-O crystals. The crystals were grown by self-flux method and have different Sr and Ca contents when starting compositions with different Sr/Ca ratio was used. The crystals grown have same orthohombic structures, but lattice parameter c increased from 30.60Å to 31.01Å. The Sr/Ca ratio in the as grown crystal increased from 2:1 to 3:1, and the Tc (zero) measured is also increasing from 84K to 90K. It seems that the superconducting transition temperature is relate to cell parameter c and the ratio of Sr/Ca in crystal composition. This result has shown its importance for understanding high Tc superconducting phenomenon in Bi-Sr-Ca-Cu-O system further.

2. EXPERIMENTAL AND RESULTS

The Bi-Sr-Ca-Cu-O single crystals were grown by the self-flux method. A vertical tubular furnace equipped with a automatic temperature control system was used for crystal growth. Powders of Bi_2O_3, $SrCO_3$, $CaCO_3$ and CuO with 99% purity were weighted and thoroughly mixed, the starting materials is based on $Bi_4Sr_3Ca_{3-x}Cu_4O_y$ (x=0.5,0.7,0.8,1.0) and an excess amount of Bi_2O_3+CuO was added as flux. A total weight of about 70g of mixture were placed in a Al_2O_3 crucible with slight pressure. The charge was heated up to 1100°C at a rate of 200°C/h, hold at this temperature for about 8h, then cooled at a rate of 2-4°C/h to 800°C. Afterwards the crucible was cooled to room temperature.

The bulk crystals with well developed

faces were seperated mechanically. The size of crystals ranged from 3X2X1 mm^3 to 15X4X2 mm^3. The thin plate-like crystals may easily cleaved from bulk crystal. Fig.1 shows the seperated bulk crystal of $Bi_3Sr_3CaCu_{2.5}O_y$ with shining cleaved faces.

The composition of crystals were determined by a philips EDAX-9100 fully computered analysis system on the surface of cleaved crystals. The chemical composition of each crystal was normalized to the three Sr atoms in table 1. The crystal structure was determined by using a Rikaku D/MAX X-ray diffractometer with CuK radiation. The crystals have same orthorhombic structure with nearly same cell parameter a and b, but the lattice parameter c increased apparently from 30.60Å to 31.01Å along with the increasing of Sr/Ca ratio in crystal composition from 2:1 to 3:1 (see Table 1).

The electrical resistance measurements were performed by standard d.c. four-probe method, and the electrical contact were made by using gold electrodes sputtered on the crystal surface The a.c. magnetic susceptibility was measured by using a HP-4192 LF Impendance Analyzer. The Tc(zero) of each

FIGURE 1
A crystal bulk of $Bi_3Sr_3CaCu_{2.5}O_y$.

Table 1 The chemical composition*, lattice constant and Tc(zero) of Bi-Sr-Ca-Cu-O crystal.

Chemical composition Bi:Sr:Ca:Cu				Lattice constant (Å)	Sr/Ca ratio	Tc(zero) (K)
33	30	15	29	30.60	2:1	84
33.5	30	12.5	26.4	30.70	2.4:1	85
32	30	11	26	30.85	2.7:1	87
30	30	10	25	31.01	3:1	90
36.6	30	16.7	32.4		1.8:1	82.5[4]
33	30	12	30		2.5:1	85[3]
30.7	30	11.6	29		2.58:1	84[5]
31.4	30	10	28.6		3:1	~87.5[4]

* The chemical composition of each crystal was normalized to the three Sr atoms in Table 1 by present author.

crystal is listed in Table 1. For comparision, the results obtained by the other researches are also listed.

3. DISCUSSION

Table 1 shows that the crystal grown from Bi-Sr-Ca-Cu-O system have different Sr and Ca contents when the atom ratio of Sr/Ca in starting material for crystal growth was varied. The ratio of Sr/Ca in crystal composition varied from 2:1 to 3:1 in our experiments. The crystal structure is the same but lattice parameter c increased from 30.60Å to 31.01Å with the increasing of Sr/Ca ratio. The superconducting transition temperature Tc(zero) also increased from 84K to 90K accordingly. Fig.2 shows the dependence of Tc(zero) on lattice parameter c, and Fig.3 shows the relation between cell parameter c and Sr/Ca ratio of each crystal. The data of other researchers listed in Table 1 show the similar tendency.

The crystal structure of 2212 phase along the c axis can be described as

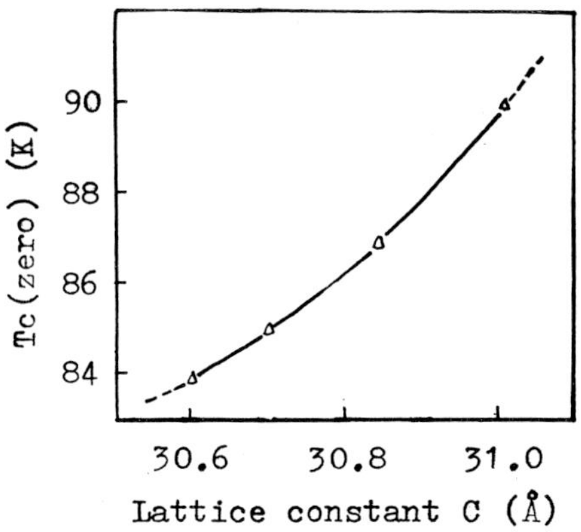

FIGURE 2
Dependence of the Tc(zero) on lattice parametee c of Bi-Sr-Ca-Cu-O crystal.

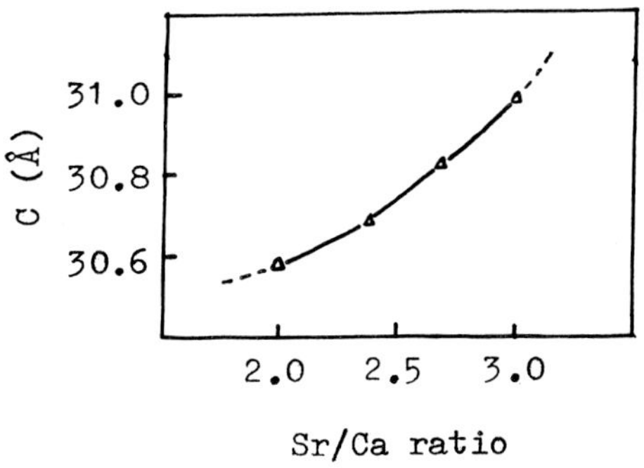

FIGURE 3
The relation between lattice constant c and Sr/Ca ratio of Bi-Sr-Ca-Cu-O crystal.

Ca(Sr) in a cube at the cell center sandwiched between two CuO_5 pyramids which face two Sr polyhedra and that these perovskite-like structures are sepersted by two BiO layers.The two C CuO pyramids seperated by Ca ions are very important for high Tc superconductivity.As a partial substitution of Sr for Ca takes place,some larger ions Sr^{+2} takes the site of Ca^{+2}in a crystal cell, hence the distance between two CuO_5 pyramids increased and the cell parameter c increased too.The interation between the two neighbour Cu-O layers may weaken This probably leads to the increasing of the superconducting transition temperature.

In summary, single crystals grown from Bi-Sr-Ca-Cu-O melt may have different contents of Sr and Ca. The Sr/Ca ratio has an important influence on the superconductivity of Bi-Sr-Ca-Cu-O crystal.

The lattice parameter c and zero resistance temperature Tc(zero) of each crystal increased along with the Sr/Ca ratio increasing and the highest Tc (zero) obtained is 90K with c=31.01Å for $Bi_3Sr_3CaCu_{2.5}O_y$ crystal.

This result is important for further understanding of the high Tc superconductivity in Bi-Sr-Ca-Cu-O system.

REFRENCE

1. H.Maeda, Y. Tanaka et al., J. Appl. Phys. Lett. 27 (1988) 209

2. H. Wang, S.X. Shang et al., J.Crystal Growth 99 (1990) 929

3. S.A. Sunshine, T. Siegrist et al., Phys. Rev. B38 (1988) 893

4. Makota HIKITA, Tsunekazu IWATA, Yukimichi TAJIMA and Akinori KATSUI J. Crystal Growth 91 (1988) 282

5. T.F. CISZEK, J.P. GORAL, C.D. EVANS, and KATAYAMA-YOSHIDA. J. Crystal Growth 91 (1988) 312

Frontiers of Materials Research/
Electronic and Optical Materials
M. Kong, L. Huang (editors)

SYNTHESIS OF HOMOGENEOUS HIGH T_c SUPERCONDUCTORS BY OXIDATION OF RAPIDLY QUENCHED ALLOY PRECURSORS

J. S. LUO, J-P. CHEVALIER and D. MICHEL

Centre d'Etudes de Chimie Métallurgique, C. N. R. S., 15, Rue G. Urbain, 94400 Vitry, France

We have investigated the chemical homogeneity, by means of X-rays diffraction and X-ray micro-analysis, of various high T_c superconductors prepared by oxidation of rapidly quenched alloy precursors. In the case of Eu-Ba-Cu alloy, an essentially amorphous phase is obtained after quenching and the resulting superconductor $EuBa_2Cu_3O_7$ has a more uniform microstructure and a better chemical homogeneity, compared with that of conventional ceramic. This leads to a substantial increase in critical current. For Bi-Sr-Ca-Cu system, the oxidized product contains some impurity phases due to the formation of an intermetallic compound with high melting point (>1200°C) during the alloy preparation stage. The addition of silver in this system is found to be helpful to obtain a homogeneous precursor, by lowering its melting point and through the formation of very fine (~30 nm) microcrystalline structure. A micro-composite Ag/Bi-Sr-Ca-Cu-O superconductor is obtained after oxidation, and its critical current is slightly increased compared with the previous one without Ag addition.

1. INTRODUCTION

High T_c oxide superconductors are usually prepared by solid state reaction of oxides, carbonates, or nitrates. In this case, some undesirable phases are often observed[1]; in addition, the processes are time-consuming and the high calcination temperatures produce coarse particles[2]. Another method for synthesis of superconducting ceramics is high temperature oxidation of a alloy precursor that contains the metallic constituents of the ceramic[3-5]. In this approach, the metallic constituents are first combined to form a homogeneous precursor by rapid solidification processing, and the alloy is then oxidized by heating under oxygen atmosphere, producing the oxide phase. This method has two main advantages. First, if the melt is single phase, then rapid quenching will ensure a better homogeneity of the cast alloy, and it can be further hoped to obtain superconducting ribbons and wires directly by oxidation.

In this paper, we report our results on the chemical homogeneity of the superconductor ceramics prepared by this method, for the systems Eu-Ba-Cu and Bi-Sr-Ca-Cu(-Ag). It has been shown that it is rather simple to form oxides by oxidation of metallic precursors, however, it is somewhat more difficult to ensure that oxidation products have the composition and microstructure necessary to yield good superconducting properties[6, 7]. We also discuss in this paper the relationship between the compositional homogeneity and the resulting microstructure, as well as the electrical properties.

2. EXPERIMENTAL PROCEDURE

The experimental details of the rapid quenching process are described elsewhere[5]. The starting alloys of ~5g with the appropriate metallic constituents are prepared first by radio-frequency melting and then introduced into the quartz glass tube of the planar flow casting system for rapid quenching. The quenched samples are oxidized at temperatures ranging from 800°C to 900°C in the air. Samples are examined by scanning electron microscope (SEM) on a ZEISS DSM950 equipped with X-ray microanalysis. The electrical resistivity is measured using a standard four probe d. c. method with silver paint contacts.

3. RESULTS

3.1 The Eu-Ba-Cu system

After quenching, the metallic precursors obtained are extremely brittle and we are only able to recover fragments. X-ray diffraction (Fig. 1) shows that it is essentially glassy with some traces of microcrystalline phases. These fragments are oxidized at 900°C in air for ~20 hours, ground, compacted and then annealed at 900°C under flowing oxygen. The resulting oxide is single phase with orthorhombic structure, and its transition temperature is 91.5 K. These values are virtually identical to those we have obtained in our own ceramic process $EuBa_2Cu_3O_7$[5]. Moreover, we observe a substantial increase in the critical current for the rapidly quenched sample compared with the ceramic processed one (~500 $A.cm^{-2}$ to ~100 $A.cm^{-2}$ at 77 K).

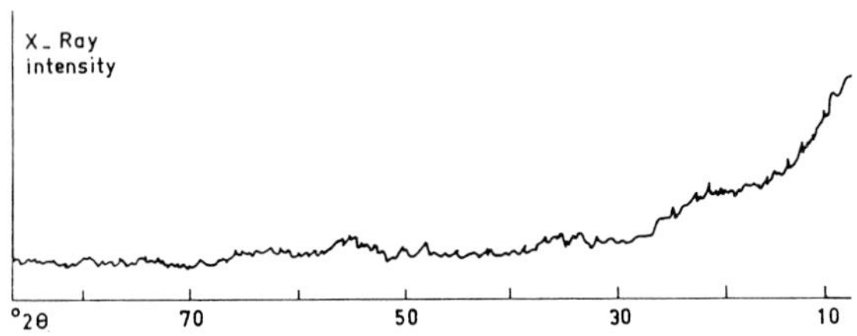

Figure 1: X-ray diffraction (Co K_α radiation) spectrum of quenched $EuBa_2Cu_3$ alloy

X-ray microanalysis maps (Fig. 2) show that the oxidized alloy is more chemically homogeneous in cation distribution than the sample prepared by the ceramic process, notably for copper and barium distribution. Some europium inhomogeneity is however observed in both specimens. In addition, examinations of these specimens by SEM show that the grains of the rapidly quenched sample has a narrow size distribution with an average size of about ~1μm, in contrast to the usual ceramic sample (ranging from 5 to 100 μm) These grains appear also very clean and do not have the barium carbonate contamination which is often observed for the ceramic processed oxide[5].

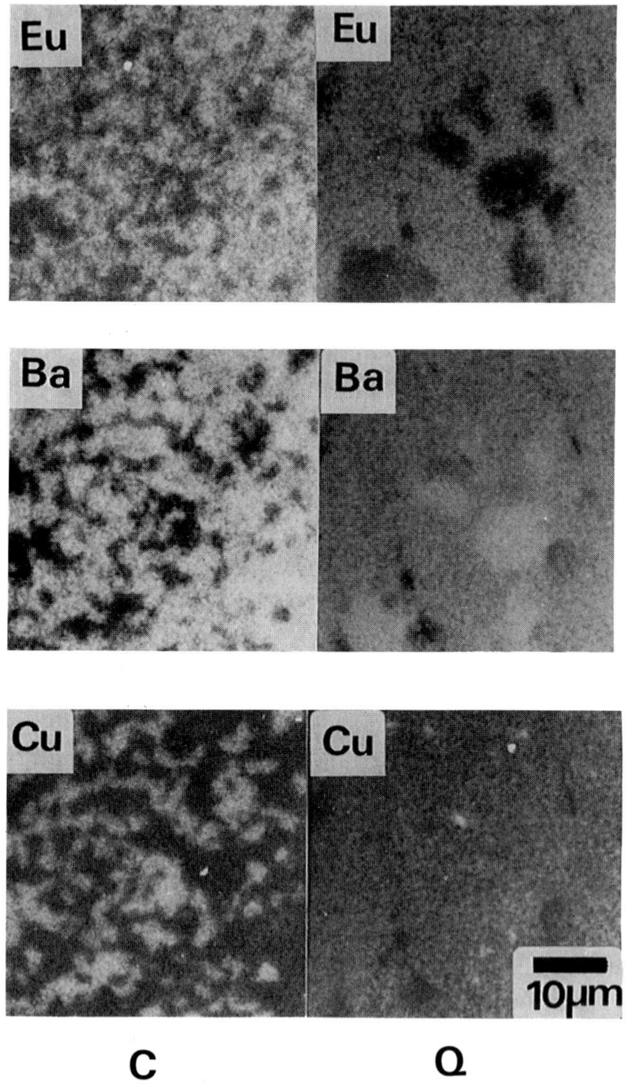

Figure 2: X-ray microanalysis maps for Eu, Ba, Cu of the ceramic processed sample (**C**), and of the sample prepared by rapidly quenching (**Q**).

3.2 The Bi-Sr-Ca-Cu system and the Ag added Bi-Sr-Ca-Cu system

For the Bi-Sr-Ca-Cu system, we also obtain only brittle alloy fragments after rapid quenching. The alloy preparation of $Bi_4Sr_3Ca_3Cu_4$ is difficult due to both the formation of a $Bi_2(Sr_xCa_{1-x})_3$ intermetallic compound with high melting point (> 1200 °C), and a weak electrical coupling. This has two consequences; first, it is necessary to heat the quaternary alloy to a sufficiently high temperature in order to obtain complete dissolution and homogenization in liquid state for the starting ingot. This high melting characteristic makes it also more difficult for glass-forming in the system. Thus, we are only able to obtain a chemical inhomogeneous precursor for Bi-Sr-Ca-Cu alloy system.The quenched foils are not uniform in colour, XRD detects crystalline peaks of f.c.c. copper, indicating that copper solidifies during casting. SEM imaging and microanalysis on surfaces of the quenched alloy confirm the existence of some isolated copper phase of about 20 to 50 μm width, aligned along the solidification direction (Fig. 3a). The distribution of other cations appears to be reasonably homogeneous.

The casting process is made much easier by the addition of the chemically inert element silver. When 25 wt.% silver is added to $Bi_4Sr_3Ca_3Cu_4$, it improves the electrical coupling and lowers the casting temperature to about 1200°C. In addition, the alloy ductibility is significantly enhanced[8]. The as-quenched alloy is now uniform in colour. XRD shows rather broad peaks (Fig. 4) corresponding to bismuth, silver and copper with some modification of lattice constants, indicating formation of metallic solid solutions. The average crystallite dimension is estimated to be about 30 nm from the peak width. X-rays microanalysis of the surfaces of quenched foils by SEM shows that the alloy thus obtained are chemically homogeneous, and no

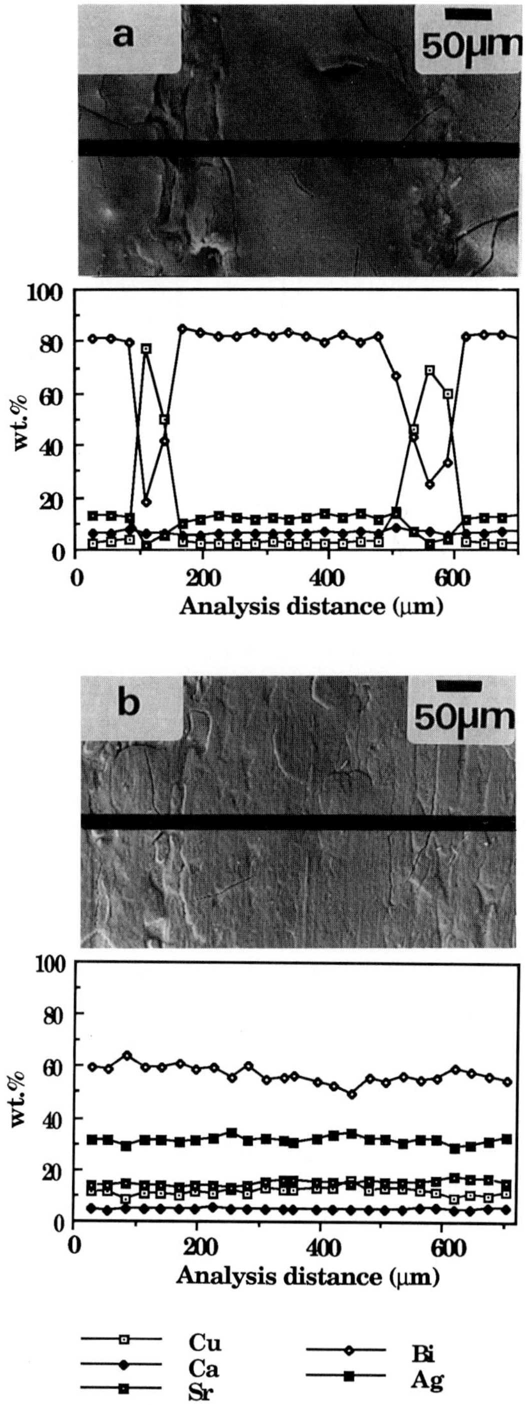

Figure 3:SEM images and EDX microanalysis on the surfaces of the quenched $Bi_4Sr_3Ca_3Cu_4$ ribbon(a) and the Ag added ribbon(b).

phase separation is observed (Fig. 3b). This means that rapid quenching ensures very fine mixing of the components, although not at the atomic scale as is the case for a glassy state.

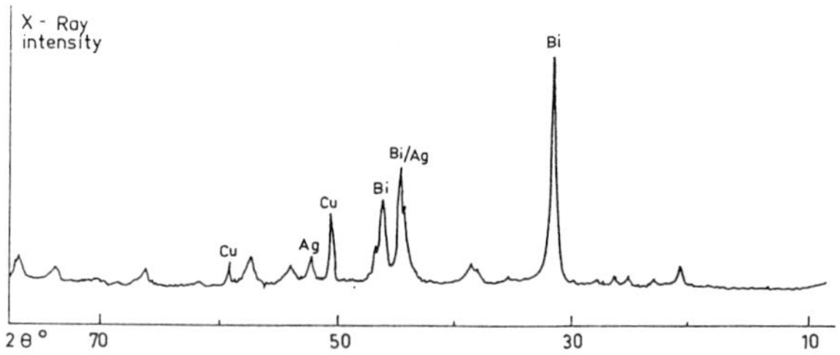

Figure 4: X-ray diffraction (Co K_α radiation) spectrum of quenched $Bi_4Sr_3Ca_3Cu_4$ alloy

The quenched foils for both compositions are heated at 800°C for about 15~20 hours, and then air quenched to room temperature. For the $Bi_4Sr_3Ca_3Cu_4$ alloy precursor, XRD reveals a nearly pure pseudo-tetragonal phase $Bi_4Sr_3Ca_3Cu_4O_{16}$, which is the same as that reported by Tarascon *et al*[9]. Its superconducting transition temperature is ~80 K, with a critical current density of about 25 A.cm^{-2} at 4.2 K. In the case of silver addition, XRD lines of the 4334 phase and of metallic silver, indicate that silver has not been oxidized, moreover EDX in TEM confirms that silver is not incorporated in the superconducting oxide[8]. The sample has the same 80 K transition temperature as that without silver and its critical current density is slightly higher.

SEM imaging on the surfaces of oxidized foils shows some differences for two kinds of samples (Fig.5). The sample without Ag has a wide grain size distribution and contains some particles of ~500 nm, probably a second phase, at surfaces. On the contrary, in the Ag added sample, 4334 plates have a regular size and small spherical silver particles of ~1µm precipitate on the surfaces. The uniform distribution of both Ag particles and the 4334 plates reflects the starting alloy homogeneity.

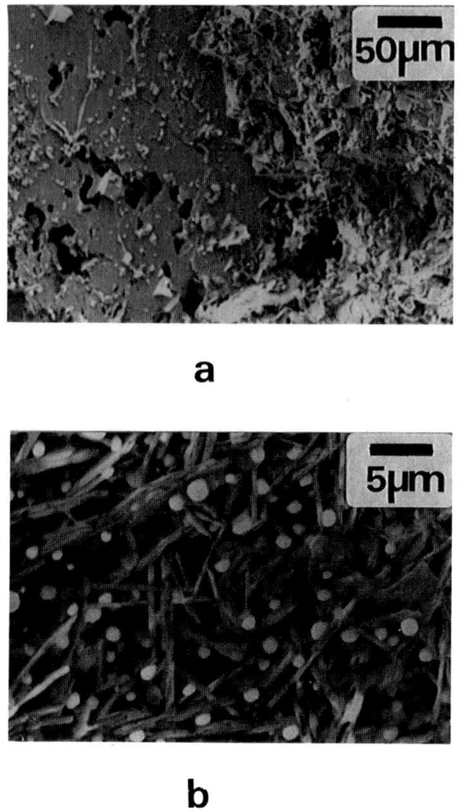

Figure 5: SEM images of $Bi_4Sr_3Ca_3Cu_4O_{16}$ (a) and of Ag/$Bi_4Sr_3Ca_3Cu_4O_{16}$ (b) prepared from rapidly quenched alloy precursors

4. CONCLUSION

We have investigated the chemical homogeneity of various high T_c superconductors prepared by oxidation of rapidly quenched alloy precursor, in correlation with their electrical properties, as well as the processing parameters. It is shown that alloy precursors, either in form of amorphous phase or in form of fine microcrystals, produce a uniform microstructure with a good chemical homogeneity for the resulting oxides. The two latter factors may explain the improved electrical properties, notably for the critical current density, observed in these cases.

ACKNOWLEDGEMENTS

The authors thank A. Dezellus, S. Peynot and J. L. Pastol for essential assistance in alloy preparation and for X-ray microanalysis. We are also grateful to C.N.R.S. (ARC "Microstructure des Nouveaux Supraconducteurs") for partial funding.

REFERENCES

1) D. R. Clarke, T. M. Shaw, and D. Dimos, J. Am. Ceram. Soc., **72**, 1103(1989)

2) R. B. Poeppl, U. Balachandran, J. E. Emerson, S. A. Johnson, M. T. Lanagan, C. A. Youngdahl, and N. G. Eror, To be published in Proceedings of 1st Inter. Ceramic Sci. & Technol. Anaheim, Oct. 31-Nov. 3, 1989

3) K. Matsuzaki, A. Inoue, H. Kimura, K. Moroishi, and T. Masumoto, Jpn. J. Appl. Phys. Lett., **26**, L334 (1987)

4) R. Haldar, Y. Z. Lu, and B. C. Giessen, Appl. Phys. Lett., **51**, 538(1987)

5) J. S. Luo, J-P. Chevalier and D. Michel, Mater. Sci. Eng., **B3**, 325 (1989)

6) D. M. Gruen, W. F. Calaway, V. A. Maroni, B. S. Tani, and A R. Krauss, J. Electrochem. Soc., **134**, 1588 (1987)

7) G. J. Yurek, J. B. Vander Sande, W. X. Wang, D. A. Rudman, Y. Zhang, and M. M. Matthiesen, Met. Trans., **18A**, 1813(1987)

8) J. S. Luo, F. Faudot, J-P. Chevalier, and D. Michel, To be published in Proceedings of E-MRS Spring Meeting, Strasbourg, France, Mai. 29 - June. 2, 1990

9) J.M. Tarascon, Y. Le Page, P. Barboux, B. G. Bagley, L. H. Greene, W. R. Mc Kinnon, G. W. Hull, M. Giroud, and D. M. Huang, Phys. Rev., **B37**, 9382 (1988)

Frontiers of Materials Research/
Electronic and Optical Materials
M. Kong, L. Huang (editors)

STRUCTURAL PROPERTIES OF Bi-BASED SUPERCONDUCTING COMPOUNDS

J.Q. Li and Z.X. Zhao

National Laboratory for Superconductivity,
Institute of Physics, Chinese Academy of Sciences,
Beijing, 100080 China

The main structural properties of $(Bi,Pb)_2Sr_2Ca_{n-1}Cu_nO_y$ with n=1, 2 and 3 and $Bi_2Sr_2YCu_2O_y$ have been summarized. the modulated structures in the Pb-doped and Y-doped Bi-Sr-Ca-Cu-O systems have been studied by electron diffraction and high resolution electron microscopy. Upon in-situ heating, as well as electron beam irradiation, the oxygen/vacancy ordering in all above system have directly been observed by transmission electron microscopes. The defect structures in Bi-based superconducting systems and the relation between T_c and the actual composition of 2212 single crystals have been presented.

1. INTRODUCTION

Three distinct Phases in the homologous series $Bi_2Sr_2Ca_{n-1}Cu_nO_y$ with n=1, 2 and 3 have been identified and isolated. These superconducting phases have the zero resistance at the temperatures of around 10K, 85K and 110K, respectively. The structures of the series appear to be based on alternating double bismuth Bi-O layers and perovskite (Sr, Ca)O-CuO layers sandwiched. It is well known that the structures of Bi and Tl based superconducting phases are almost modulated[1-4], especially in the Bi based superconductors the modulation is very strong, consequently the local structure will deviate significantly from the average structure of these compounds. In this paper the results about the modulated structure, oxygen/vacancy ordering and some substitution experiments in Bi-Sr-Ca-Cu-O system, obtained mainly in the Institute of Physics, the Chinese Academy of Sciences, have been presented.

2. EXPERIMENTAL

Superconducting bulk samples and single crystal of $Bi_2Sr_2Ca_{n-1}Cu_nO_y$ (n=1 and 2) were prepared by conventional methods reported in Refs.3 and 4. Besides the preparation of these compounds a number of preparation experiments were done in order to establish the effect on the superconducting transition temperature, T_c, crystal structure and modulated structure by partial substitution of Bi by Pb, Ca by Y et al..

High resolution electron microscopy was performed with a JEM-200CX electron microscope, equipped with a top entry stage with a ±10° double tilt operating at 200kV. Electron diffraction and EDX analysis were done with the H-9000NA analysis microscope equipped with EDX elemental analysis system operating at 300kV. Samples for electron microscope observations were prepared by grinding as well as ion milling.

3. RESULTS

3.1. The main structural features of $Bi_2Sr_2Ca_{n-1}Cu_nO_y$ (n=1, 2 and 3)

(1). The subcells of these phases are orthorhombic with lattice parameters a=0.541nm, b=0.543nm and c=2.46nm, 3.07nm and 3.68nm, respectively. The structure symmetry is orthorhombic for n=2 and 3, while, due to the different kind of modulation, the structure symmetry of n=1 phase decreases to monoclinic[5, 6].

(2). The modulation is located in the (100) plane for all of the three phases. The components of the modulation wave vectors along the [010] direction are also the same for all of them, and which is given as $q_1=0.217b^*$.

(3). In c direction of the crystal, the perovskite lamellae are separated from each other by double Bi-O layers, weak Bi-O band along c axis makes these compounds as a lay structure with the (001) cleavage plane, and oxygen atoms located at the cleavage planes might be easily removed and disordered during in-situ heating or electron beam irradiation(see below).

(4). The 90° rotation twin domains were frequently observed in the Bi-Sr-Ca-Cu-O system. The typical electron diffraction pattern is shown in figure 1.

(5). The structure of $Bi_2Sr_2CaCu_2O_y$ is dependent on the temperature and oxygen content. Figure 2a and 2b present the diffraction patterns taken at the temperature of 350°C and 550°C, respectively. Detail study showed that there are two phase transitions occurring at about 296°C and 490°C[5].

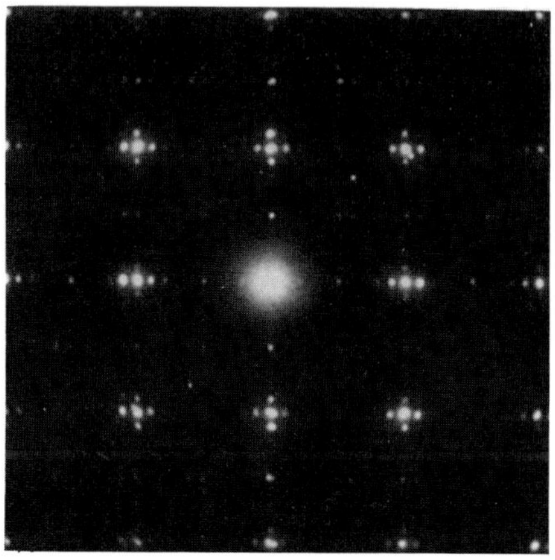

Fig.1. The [001] zone electron diffraction pattern of the 90° rotation twin domain in $Bi_2Sr_2CaCu_2O_y$.

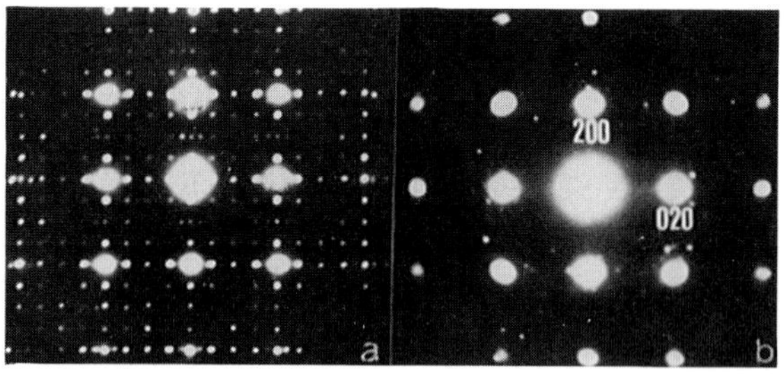

Fig.2 The [001] zone diffraction patterns of $Bi_2Sr_2CaCu_2O_y$ indicating the phase transition. (a) at the interval between 296°C and 490°C and (b) above 490°C.

(6) The actual composition of 2212 single crystal is always found to be poor in Ca and rich in Bi or Sr which

Table 1 The main structural properties of $(Bi,Pb)_2Sr_2$-$(Ca,Y)_{n-1}Cu_nO_y$ at room temperature

	T_c (K)	Subcell	Modulation wave(q)	Super-spacegroup
$Bi_2Sr_2CuO_y$	10	5.41AX5.43AX24.6A	(0, 0.217, 0.62)	$P^{B}\frac{2}{\bar{1}}/\frac{b}{1}$ or $P^{B}\frac{b}{1}$
$Bi_2Sr_2CaCu_2O_y$	85	5.41AX5.43AX30.7A	(0, 0.217, 1)	$N^{B}\frac{b}{1}\frac{m}{\bar{1}}\frac{b}{1}$
$Bi_2Sr_2Ca_2Cu_3O_y$	110	5.41AX5.43AX36.8A	(0, 0.217, 1)	$N^{B}\frac{b}{1}\frac{m}{\bar{1}}\frac{b}{1}$
$Bi_{1.8}Pb_{0.2}Sr_2CuO_y$	10	5.40AX5.44AX24.2A	(0, 0.154, 1)	$N^{B}\frac{b}{1}\frac{m}{\bar{1}}\frac{b}{1}$
$Bi_{1.6}Pb_{0.4}Sr_2CaCu_2O_y$	85	5.40AX5.44AX30.5A	(0, 0.117, 0)	$P^{B}\frac{b}{1}\frac{m}{\bar{1}}\frac{b}{1}$
$Bi_{1.6}Pb_{0.4}Sr_2Ca_2Cu_3O_y$	110	5.40AX5.44AX36.7A	(0, 0.117, 0)	$P^{B}\frac{b}{1}\frac{m}{\bar{1}}\frac{b}{1}$
$Bi_2Sr_2YCu_2O_y$	0	5.39AX5.41AX30.3A	a). (0, 0.25, 1)	$N^{B}\frac{b}{1}\frac{m}{\bar{1}}\frac{b}{1}$
			b). (0, 0.12, 1)	$N^{B}\frac{b}{1}\frac{m}{\bar{1}}\frac{b}{1}$
			c).q_a=(0.28, 0, 0)	
			q_b=(0, 0.22, 0)	

in general causes the variation of T_c. The typical composition with $T_c \approx 85K$ is $Bi_{2.2}Sr_2Ca_{0.8}Cu_2O_y$ obtained by EXD elemental analysis in transmission electron microscope[2,6].

(7) Some main structural properties of $(Bi,Pb)_2Sr_2(Ca,Y)_{n-1}$ Cu_nO_y are summarized in Table 1.

3.2. Experiments of partial substitution aim at influencing the T_c and Modulated structure.

(1). $Bi_{2-x}Pb_xSr_2Ca_{n-1}Cu_nO_y$ with n=2 and 3

A series of experiments was carried out[7-10] whereby Bi was partly substituted by Pb with x ranging from 0 to 0.4. The experimental results show that by such substitution it is easy to prepared superconducting samples with T_c around 110K (2223 phase). Electron diffraction and high resolution electron microscopy indicated that for both n=2 and 3 phases with the increasing of x the modulation vector changed both in direction and value. The typical [100] zone diffraction pattern of $Bi_{1.8}Pb_{0.2}Sr_2Ca_2Cu_3O_y$ is shown in figure 3. In this pattern, in addition to the conventional modulation reflection spots, extra satellite spots due to a new modulation are also observed. Electron diffraction study of $Bi_{1.6}Pb_{0.4}Sr_2Ca_2Cu_3O_y$ shows that in most regions only the new modulated structure exists. High resolution electron microscopy has clearly shown an intensity modulation does occur in the b direction.

(2). $Bi_{2-x}Pb_xSr_2CuO_y$

Several Pb-doped samples with x ranging from 0.1 to 0.5 have been prepared. Most of them showed no superconductivity above 4.2K, only on two samples with nominal composition of $Bi_2Sr_2CuO_y$ and $Bi_{1.8}Pb_{0.2}Sr_2CuO_y$ the superconducting transition temperatures were observed at about 9K.

The structural properties of $Bi_{1.8}Pb_{0.2}Sr_2CuO_y$ single crystal was studied by transmission electron microscopy. It was found that the symmetry of this superconducting phase

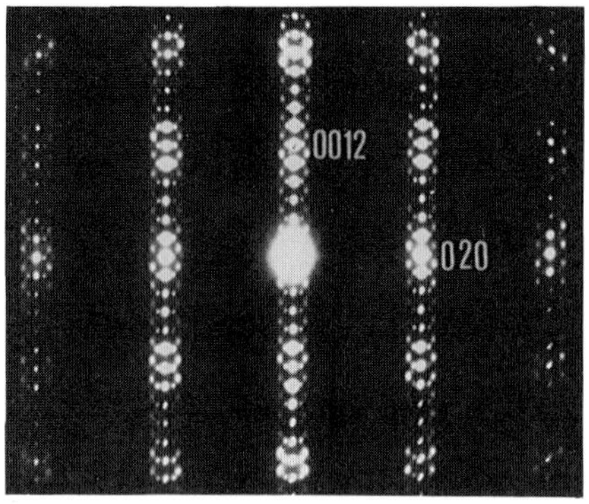

Fig.3 The [100] diffraction pattern from lead-doped sample showing the presence of extra satellite spots.

Fig.4 High resolution images of $Bi_{1.8}Pb_{0.2}Sr_2CuO_y$ along the [101] zone.

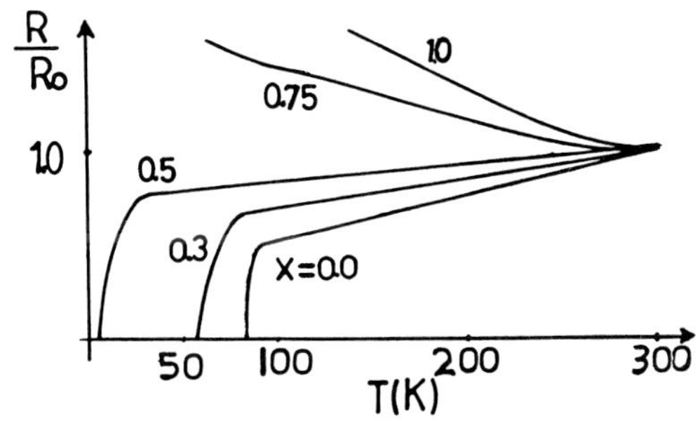

Fig.5 Resistivity plots of Bi_2Sr_2-$Ca_{1-x}Y_xCu_2O_y$ members (R_0: the resistance of the sample at room temperature).

was transformed from monoclinic to orthorhombic. Extinction rules of the of the main reflection and convergent beam electron diffraction pattern indicate that the space group of the average structure is Bbmb[11],which is the same as that of $Bi_2Sr_2CaCu_2O_y$. The modulation wave vector is given as $q_1=0.154b^*+c^*$, the component of this vector along [010] is $0.154b^*$, thus the periodicity of the modulation along the b direction should be 1/q=3.5nm.

The high resolution electron micrograph of $Bi_{1.8}Pb_{0.2}Sr_2CuO_y$, taken with the incident beam parallel to [101] zone, is reproduced in figure 4,. Some dislocation like defects are present in the pattern of the modulation fronts as indicated by arrows. In the b direction the periodicity of the modulation is about 3.5nm, this is consistent with the result obtain from electron diffraction[9-11].

(3) $Bi_2Sr_2Ca_{1-x}Y_xCu_2O_y$

A series of $Bi_2Sr_2Ca_{1-x}Y_xCu_2O_y$ sample were prepared in our Institute. Electrical resistivity measurements on the samples indicated that the samples with x=0, 0.3 and 0.5 compositions are superconductors, while the samples with x=0.75 and 1.0 compositions are both semiconductors as shown in figure 5. the magnetic measurements also show hat the temperature of diamagnetism of $Bi_2Sr_2Ca_{1-x}Y_xCu_2O_y$ decreases progressively from x=0 to 0.5[12].

The structural features of this system has been studied by X-ray diffraction. It was found that the basic plane parameters, a and b , initially decrease with the increasing of x. and the parameter of the c axis decreases progressively from 3.07nm at x=0.0 to 3.02 at x=1.0. Such variation in the lattice parameters as well the unit cell volume was suggested caused by two competing factors, the small size of Y relative to Ca and changes in oxygen stoichiometry with x[12]. Modulated structures in $Bi_2Sr_2YCu_2O_y$ single crystal have been studied by electron diffraction and high resolution microscopy. The results showed that there are three kinds of modulated structures in the compound, two of them are one dimensional modulated structure, and the other is a two dimensional incommensurate modulated structure[13].

3.3. Oxygen/vacancy ordering in Bi based compounds

It is well known that the superconductivity of $YBa_2Cu_3O_y$ are strongly affected by oxygen ordering[14,15]. In $Bi_2Sr_2Ca_{n-1}Cu_nO_y$ superconducting series, most of the latest results seam to support that the origin of the incommensurate modulated structure is related to the extra oxygen in the Bi-O layers[16-18]. Here we will present some results about the oxygen ordering in Bi based compounds produced by in situ heating and by electron beam heating in electron microscopes.

When crystal of $Bi_2Sr_2CaCu_2O_y$ are heated in situ above 290°C, some crystals started to show the presence of an additional superstructure, which have been shown in figure 2a. This superstructure can be described as a unit cell with doubled a and b axes of the basic cell[5], the 2aX2b superstructure did not disappear upon in situ cooling the specimens to room temperature, however on further heating this kind of superstructure and the conventional modulated structure disappear totally as the temperature increases above 490°C, as shown in the figure 2b.

This kind of superstructure is also observed in $Bi_2Sr_2YCu_2O_y$ by in situ heating. Figure 6a and 6b are high resolution electron images directly showing the variation of the structure of $Bi_2Sr_2YCu_2O_y$ produced by electron beam irradiation. In figure 6a only the conventional modulation is visible along the b direction. While in figure 6b two commensurate modulations (superstructure), nearly paralleling to the a and b directions respectively, are superimposed on the conventional modulation. The fronts of the modulations can be clear recognized as black bands. Both of the two commensurate modulations has the period of about 1.08nm[13].

Comparing to the simple cubic perovskite unit cell (a_p) the cell dimensions of the superstructure on the basic plane can be written as $2\sqrt{2}a_p X 2\sqrt{2}a_p$, this superstructure has also been reported in $YBa_2Cu_3O_y$[14]. It was reported that this superstructure only occurs when the sample is heated in situ, heating in a low vacuum(10^{-2} Torr) or in oxygen does not lead to the formation of this kind superstructure[6], this suggest that some oxygen is removed from the structure during in situ heating, an ordering of oxygen vacancies and the corresponding lattice relaxation could lead to this superstructure.

Another kind oxygen ordering excited by electron beam irradiation was observed in $Bi_{1.8}Pb_{0.2}Sr_2CuO_y$ single crystal as the crystal was heated by

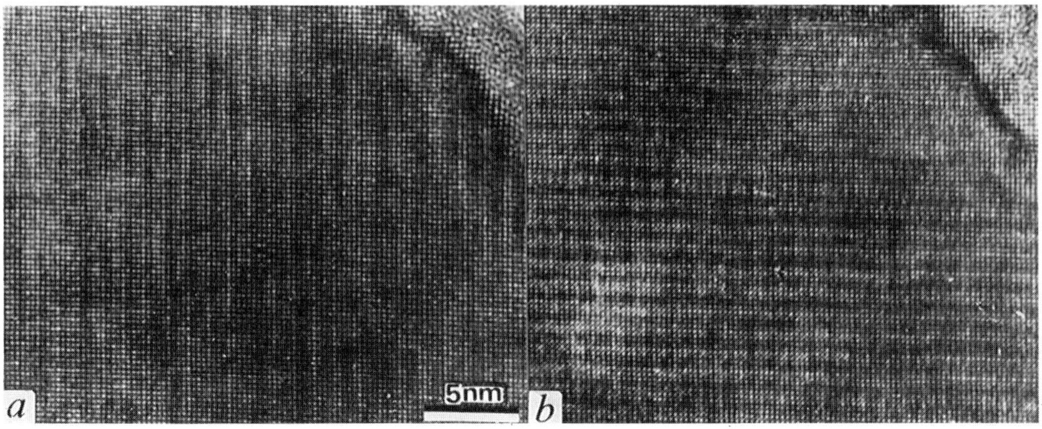

Fig.6 High resolution image of $Bi_2Sr_2YCu_2O_y$ indicating the superstructure introduced by the electron beam irradiation. (a) before irradiation and (b) after beam irradiation.

electron beam irradiation. Figures 7a-7c show a series of selected area electron diffraction pattern of $Bi_{1.8}Pb_{0.2}Sr_2CuO_y$ single crystal irradiated by a low-flux beam of electron at 300kV, First diffraction pattern was taken as soon as the crystal was tilted to the [001] orientation.

Figures 7b and 7c are diffraction patterns of the same crystal under and after electron beam irradiation. It has been observed that the intensities of the satellites due to the conventional modulated structure decreases progressively, and, meanwhile, a new kind of modulated structure gradually appears. Finally only the satellite sequences corresponding the new modulated structure can be seen.

After electron beam irradiation, in some areas the satellites sequences due to the new modulation appear only in one direction, slight by off the direction of a*+b* or a*-b*. Typical [001] zone electron diffraction pattern of such areas is reproduced in Fig. 8a. This pattern contains the linear satellite sequences enclosing an angle of about 42° with the b* axis, and which is represented schematically in figure 8b. These satellite sequences belong to the adjacent rows of the main spots exhibiting spacing and orientation anomalies, no single straight line can passes through the satellite sequences in adjacent bands. The diffraction pattern is thus in general incommensurate. Therefore this new modulated structure is an one dimensional incommensurate modulated structure.

High resolution and electron diffraction study has also indicated that, in general, this new modulation can adopt two orientation in the (001) plane related by mirror symmetry with

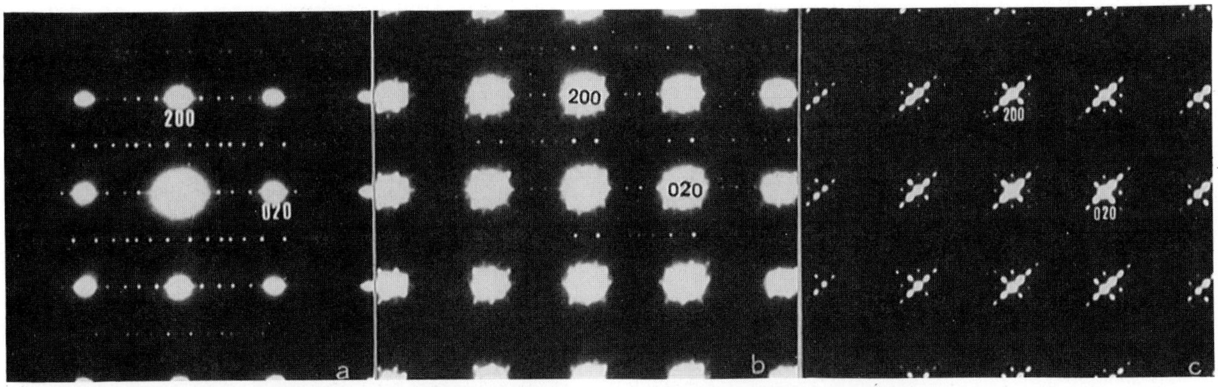

Fig.7 Series electron diffraction patterns from a crystal of $Bi_{1.8}Pb_{0.2}CuO_y$. under beam irradiation.
(a) the [001] zone pattern of the initial $Bi_{1.8}Pb_{0.2}Sr_2CaCu_2O_y$
(b) new satellites appearing after about 10 min.
(c) end product showing only the satellites due to the new modulation.

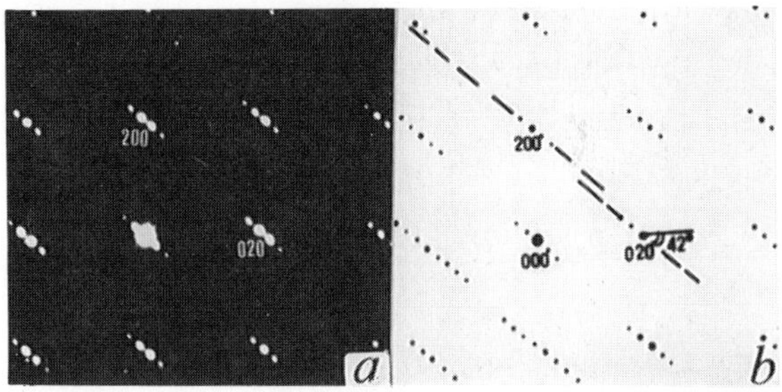

Fig.8 (a) the [001] zone electron diffraction pattern of the new modulated structure; (b) schematic representation of the reciprocal lattice of this new modulated structure projected along [001] zone, spacing and orientation anomalies are present.

respect to (100) or (010) plane. The diffraction pattern showing two orientation variants, with modulation vectors in the same (001) plane, has been reproduced in figure 7c, and the corresponding high resolution image is shown in figure 9. The two variants of modulated structure in twin relationship can clearly be recognized in this image. the spacing of the modulation wave fronts, defined as the period the modulation, is about 2.2nm. This result strongly suggest that the two orientation variants related by (100) or (010) mirror plane are crystallographical equivalent in the parent phase $Pb_{0.2}Bi_{1.8}Sr_2CuO_y$.

3.4. Defect structures in (Bi,Pb)-Sr-Ca-Cu-O system

(1). Intergrowth of lamellae of different phases does frequently occur in these Bi-based system. The typical diffraction pattern of such coherent disorder intergrowth are shown in figures 10. High resolution electron microscopy study showed that the succession of lamellae of different phases along the c direction is irregular[20]. The fact that the frequent appearing of this kind of disorder stacking has some important implications, it would probably to difficulty to grow high quality single

(2). Diffraction contrast microscopy as well as high resolution microscopy have also shown that the $Pb_{0.2}Bi_{1.8}Sr_2CuO_y$ crystallites have some layered morphology in the b direction, through it is not as pronounced as in the c direction[9].

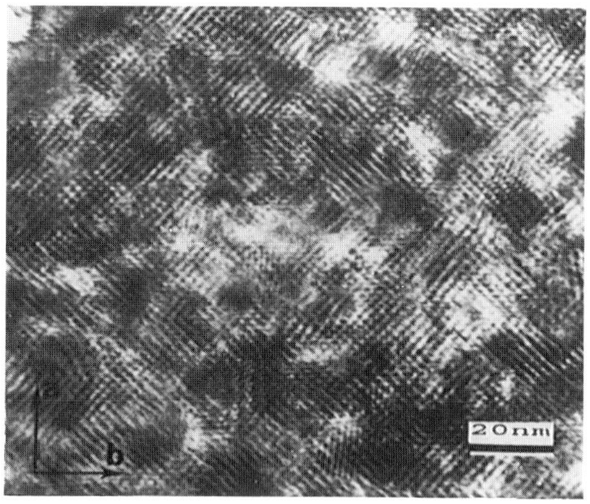

Fig.9 High resolution micrograph with low magnification showing the long period modulations appearing in two twin related variants.

Linear features are often seen within a single grain, the images of such areas are shown in figure 11. These are in fact second phases or regions of stacking disorder[19]

(3) Twining of modulated structure was often observed in this system.

Figure 12 shows a high resolution lattice image taken with the incident

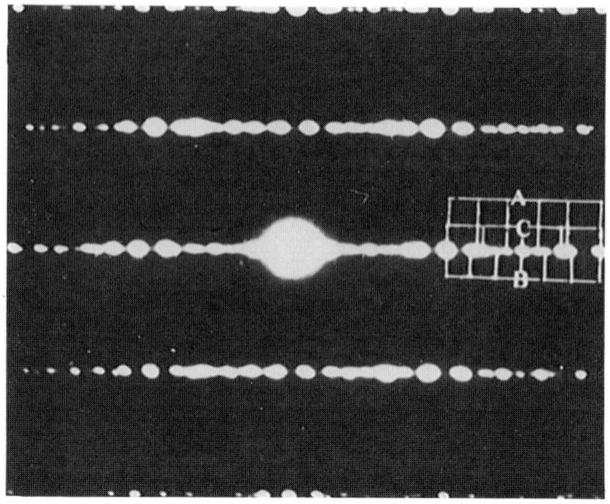

Fig.10 The typical electron diffraction pattern of [110] zone showing intergrowths of lamellae of different phase in Bi-Sr-Ca-Cu-O system.

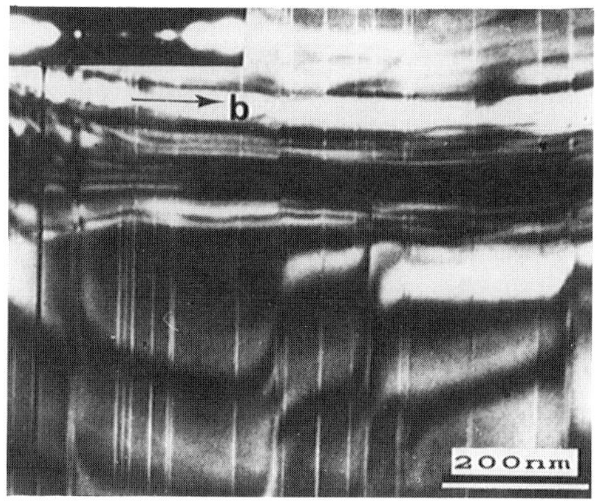

Fig.11 Diffraction contrast image showing the linear contrasts paralleling to (010)plane.

electron beam parallel to c axis. Two sets heavy dark modulation bands can be clearly recognized on left and right part of the image , respectively . They make an angle about 90° and intersect on the boundary nearly parallel to (110) plane as indicated by arrows.

(4) Antiphase domains in $Bi_2Sr_2CaCu_2O_y$ single crystal have directly been observed by transmission electron microscope. A model based on the oxygen vacancies in the Bi-O layer and group theoretical consideration was proposed[21].

3.3. Relation between T_c and compositions of 2212 phase

Several pieces of superconducting single crystal with T_c ranging from 30 to 90K have been selected in the sample of nominal composition $Bi_2Sr_2CaCu_2O_y$. The EDX analysis and x-ray diffraction

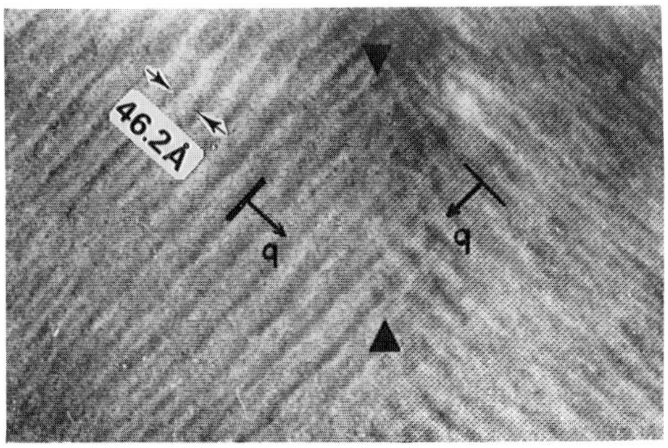

Fig.12 lattice image twin related modulated structures in Pb doped Bi-Sr-Ca-Cu-O system, taken with the incident beam parallel to c axis.

show that the actual composition of these pieces deviated from the ideal composition $Bi_2Sr_2CaCu_2O_y$. Partly substitution of Ca by Bi or Sr et al. were observed in these single crystal pieces. Here only the relation between T_c and the Ca/Bi ratio is plotted in figure 13.

4. SUMMARY

(1) The main structural features of $(Bi,Pb)_2Sr_2Ca_{n-1}Cu_nO_y$ with n=1, 2 and 3 and $Bi_2Sr_2YCu_2O_y$ have been presented in table 1.

(2) In Pb-doped Bi-Sr-Ca-Cu-O system, a new modulation in n=2 as well as n=3 phase has been introduced in the b direction, and the super-spacegroup of them changes from N^{Bbmb}_{111} to $P^{Bbmb}_{1\ 1}$. For n=1 phase, the structure symmetry has been transformed from monoclinic to orthorhombic.

(3) In Y-doped samples $Bi_2Sr_2Ca_{1-x}Y_xCu_2O_y$, Electric measurements show that the samples with x<0.5 were

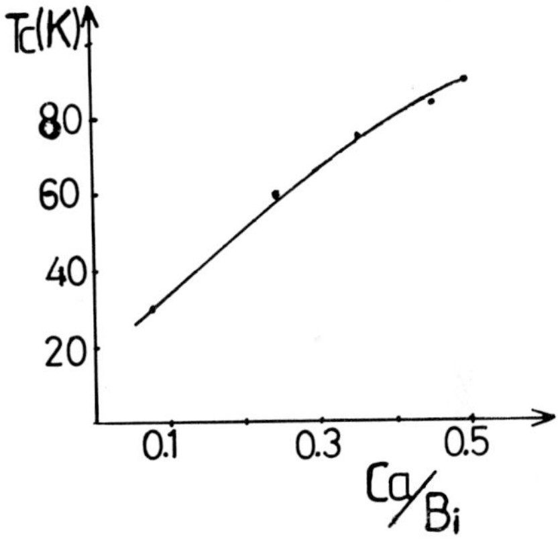

Fig.13 Superconducting transition temperature vs. the Ca/Bi ratio in 2212 phase.

superconductors; while, for x>0.5, all samples were semiconductors. In $Bi_2Sr_2YCu_2O_y$ single crystal three kinds of modulated structure were observed.

(4) Two kinds of oxygen/vacancy ordering have been observed in the Bi-based systems. One is the superstructure in the (001) plane

produced by in situ heating or electron beam irradiation in $Bi_2Sr_2CaCu_2O_y$ as well as $Bi_2Sr_2YCu_2O_y$. The other is an incommensurate modulated structure in the (001) plane which was only observed in the $Bi_{1.8}Pb_{0.2}CuO_y$ as the sample was heated by electron beam irradiation.

(5) The actual composition of 2212 single crystal in general deviates from the ideal composition of $Bi_2Sr_2CaCu_2O_y$, which can cause the variations in the superconductivity and structural features of the single crystal.

ACKNOWLEDGEMENTS

The authors would like to express many thanks to D.Y.Yang, F.H.Li, K.K.Fung, Y.Q.Liu, Z.Y.Ran, Y.M.Ni, S.L.Ja and Q. Lei for their assistance. The work reported here was supported by the National Center for Research and Development on Superconductivity.

REFERENCES

[1] T.M. Shaw, S.A. Shivashankar and D.S. Yee, Phys. Rev. B37(1988)9856.
[2] H.W Zandbergen, W.A. Groen and S. Amelinckx, Physca C156(1988)325.
[3] Z.Y. Ran, C. Chu and Z.X. Zhao, Mod. Phys. Lett. B2(1988)699.
[4] Y.F. Yan and Z.X. Zhao, Mod. Phys. Lett. B2(1988)571.
Y.Q. Liu and Z.X. Zhao, in preparation.
[5] J.Q. Li, C. Chen and Z.X. Zhao, Z.Phys. B68(1989)165.
[6] H.W. Zandbergen and W.A. Groen, Physca C166(1990)282.
[7] C. Chen, Y.H. Yao and W.K. Wang, Sold Stat. Commun. 68(1988)309.
[8] R. Ramesh, G.van Tendeloo and G. Thomas, Appl. Phys. Lett. 53(1988)2220.
[9] N. Fukushima, H. Niu and S. Nakamura, Physica C159(1989)777.
[10] C.H. Chen, D.j. Werder and A.S. Cooper, Phys. Rev. B39(1989)4686.
[11] J.G. Wen, Y.F. Yan and K.K. Fung, Progress in High T_c Superconductivity, Vol.22(1990)423.
[12] A,K. Gangul, R. Nagarajan and C.N.R. Rao, Matter Res. Bull. 24(1989)103.
[13] J.Q. Li and Z.X. Zhao, in preparation
S. Sato and T. Tamegai, Jpn. J. Appl. Phys.27(1988)L209.
[14] R. Beyers, B.T. Ahn and R.A. Huuggins, Nature Vol.30(1989)619.
[15] K. Nakamura and K. Ogawa, Jpn. J. Appl. Phys. 27(1988)577.
[16] C.C. Torrardi, J.B. Parise and A,W. Sleight, Physica C157 (1989)115.
[17] J.M. Torardi et al., Phys. Rev. B39(1989)11587.
[18] E.A. Hewat, J.J. Capponi and M. Marezio, Physica C157(1989)502.
[19] J.Q. Li and Z.X. Zhao, Submitted to Physca C.
[20] J.Q. Li and Z.X. Zhao, Mod. Phys. Lett. B2(1988)1085.
[21] K.K. Fung, R.L. Withers and Z.X. Zhao, J. Phys. Condens. Matter 1 (1989)317.

Frontiers of Materials Research/
Electronic and Optical Materials
M. Kong, L. Huang (editors)

On the Single Crystal of Pb-doped $Bi_2Sr_2CuO_x$: Growth and Effect of Oxygen vacancies

Liquan Chen, Yongqian Liu, Yuzhen Huang, Jianqi Li, Hong Chen, Zhan Chen, Yongming Ni, Shunlian Jia and Zhongxian Zhao

Institute of Physics, Chinese Academy of Sciences P.O.Box 603, Beijing100080, China

Abstract

In this paper , we used flux method in growing large 2201 single crystal in Bi-Sr-Cu oxide . The as-grown crystal was nonsuperconductor and showed semiconductivity in resistivity curve. After reducing it in flowing N_2 at 570°C and appropriate oxygen vacancies were induced, superconductivity with Tc^{on} = 6K was resulted. The great influence of oxygen deficiency to superconductivity is evident . Its influence on structure was characterized by X-ray diffraction and TEM. Laue diffraction pattern was performed to confirm the crystal characterization.

1. INTRODUCTION

Ironically, although first discovered and led to the great achievement of the new high critical temperatures in Bi-Sr-Ca-Cu-O system [1-2], 2201 phase of Bi-Sr-Ca-Cu-O has been proved nonsuperconducting by lots of people in single crystal. This low Tc and nonsuperconducting variants of the layered cuprates allow us to test models of the normal state transport properties , such as electron-phonon scattering in the normal state and residual resistance , at low temperatures , with the aim of gaining a better understanding of the mechanism of high temperature superconductivity [3]. So, the Bi-Sr-Cu-O system which reportedly contains structurally similiar superconducting and nonsuperconducting phases is of particular interest. Recently, Edward Sonder et. al.[4] have successfully obtained superconducting $Bi_2Sr_{1.85}CuO_{6-y}$ single crystal by oxygen obsorption . They believed that oxygen provide mechanism for producing mobile holes in the CuO_2 layers. This suggest that the role of defects introduced by oxygen plays a crucial role in control carrier concentration that are similar to the behavior of 123-phase in YBCO and the new electron type NCCO in leading to superconductivity [5]. In this paper , we outline another approach to grow strontium rich $(Bi_{1-x}Pb_x)_{2.1}Sr_2Cu_{1.02}O_x$ single crystal with large platelets 10x10x0.5 mm^3 and the proheat-treatment in reduced atmosphere to achieve superconductivity in this case. Aspects of structural characterization are also discussed.

2. EXPERIMENT

2.1. Crystal Growth

PbO was used as flux in growing 2201 single crystal . The starting composition was 0.6 Bi_2O_3 , 2 $SrCO_3$, CuO and 0.2 PbO, in which all the compounds were high purity. After thourouly mixed , the mixture was prereacted in flowing oxygen at 840°C in alumine crucible.Then it was heated to 980°C , stay for 4 hours , slowly cooled at 2°C/h to 830°C, further lowered to 400°C at 8°C/h ahd finally the furnace was cooled down to room temperature .

The as-grown crystals were black shining on the upper part of crucible as can be expected by the temprature gradiant arranged with lower temperature on the upper part of the furnace . In the lower part of the crucible, some lead bead could be found . They have partially functioned as flux in low temperature region and formed in the bottom when further cooled due to their heaviness. Crystals were recovered as large , micaeous and very thin platelet. They oriented in the direction with lamina perpendicular to the bottom of the crucible. EDAX 9100 gave the composition: $(Bi_{1x}Pb_x)_{2.01}Sr_{2.00}Cu_{1.02}O_x$. Here, $x\approx0.2$,as an evaluation, for x-ray peak of Bi and Pb of core level is not easily distinguished in energy analyser.

The as-grown crystal was nonsuperconducting. DTA survey showed a peak at 920°C, corresponding to its melting point.We made different proheat treatment to adjust oxygen content in the crystal. Sample A: heating at 570°C in flowing N_2 (340cc/min) for 40 hours then quenching. Sample B: annealing at 600°C in flowing N_2 (200cc/m) for 24 hours then in flowing O_2 (100cc/m) and finally at 700°C in flowing N_2 (300cc/m) for 24 hours also quenching.

2.2. Characterization

We employ four probes DC measurement in resistivity . High quality Au paste was used as electrode and Au filament as lead. The electrodes were arranged in the plane of the plate . The temperature was measured by Pt

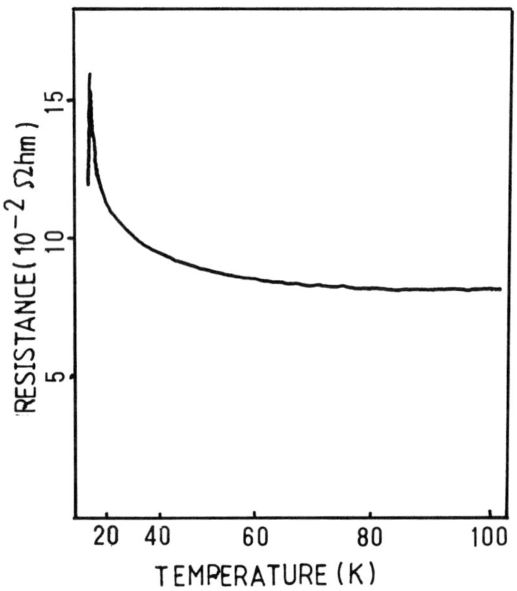

FIGURE 1
Temperature dependence of the resistance of sample A

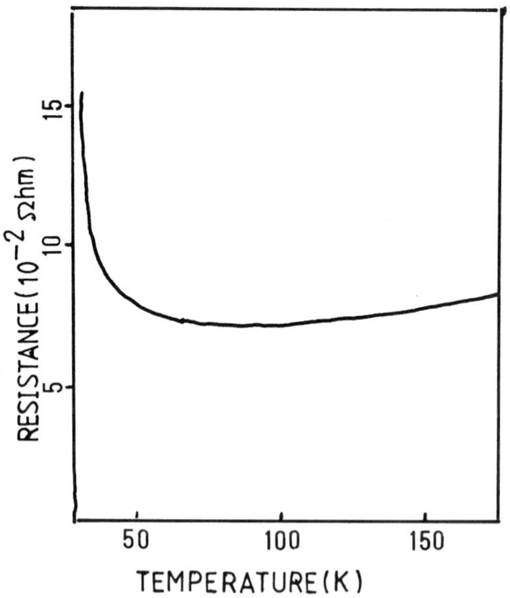

FIGURE 2
Temperature dependence of the resistance of sample B

resistance thermometer.The as-grown crystal is not superconducting down to 4.2K. After proheat treatment , sample A shows superconductivity with T_c^{on}=6K, sample B was not superconducting. Since in reduced atmospheres, the crystal was oxygen deficient , oxygen vacancies were made to controll the carrier density in the crystal . Sample B is more oxygen deficient as indicated by ref.8. So , we conclude that appropriate oxygen vacancies in rich strontium 2201 single crystal contrubute to the superconductivity.

X-ray diffraction analysis were performed on sample A and sample B crystal by RIGAKU PSPC/MDC (Position Sensitive Proportional Counter Microdiffractometer, Cu-kα radiation) [6]. We indexed the lattice parameter to be : sample A crystal is orthombic with cell parameter a=5.386A,b=5.344A and c=24.574A , sample B crystal is orthombic with cell parameter a=5.388A, b=5.62A and c=24.560A. This result is between the commonly poor strontium 2201 phase with c=24.6A and the so called semiconducting collapsed phase with c=23.6A [3]. The oxygen deficiency made c axis smaller. For further studies, we ground the crystals and made powder X-ray diffraction. We found that in sample B crystal , there are two distinct lines different to the sample A crystal corresponding to d=3.956A and 3.039A . We think that it is a minor phase probably corresponding to $Bi_{17}Sr_{16}Cu_7O_x$ with a=5.425A,b=23.254A and c=24.427A [7]. Laue diffraction pattern was also made under voltage 32KV and a current of 12MA, as can be seen in Fig. 3 . As we know , the crystals were stacking in c direction in platelet, it was comprehensible that in Laue pattern several points set around one but with small distances. This is caused by different layer of the crystals which have minor differences in orientation of ab direction. Also we can clearly see a 2 ratational axis which confirm the above index result.

FIGURE 3
Laue diffraction pattern of 2201 single crystal

TEM studies were carried out after peeling off sample A crystals to a thickness of 1 μm and (001) direction diffraction pattern were made. Choosing different small pieces , we found that several different modulation exist in the crystals. The most profound one was a new 2-dim modulation pattern as can be seen in Fig. 4. We can see that the 2-dim modulation has two modulation vectors a*=b*=22.2A and direct in the direction which make an angle a* <110> = b* <110> = 1°. This orientaion anomaly indicates an usually incommensurate structure. The most common b direction one dimensional modulation can be found with a modulation vector b*=29.4A . It is also different from the Pb undoped 2201 structure 9.

3. Discussion

The complexity of the Bi-Sr-Cu-O pseudoternary system may lie in the nonstoichiometric nature. Compositional similar phases near $Bi_2Sr_2CuO_6$ effected greatly by small changes of strontium and oxygen content [7]. The delicate balance between semiconductivity and superconductivity was controlled by oxygen vacancies which may influence the carrier density. In strotium poor single crystals, intercalation of oxygen benefits to superconductivity . In strontium rich case, oxygen vacancies are needed to achieve superconductivity. The effect of oxygen content is to change the carrier concentration to enhance the superconducting exitation. As can be seen from the resistivity curves , semiconductivity was still possesed by the superconducting crystal in our crystals. This may indicate defects in the crystal, or weak locallization or a competetion between superconducting

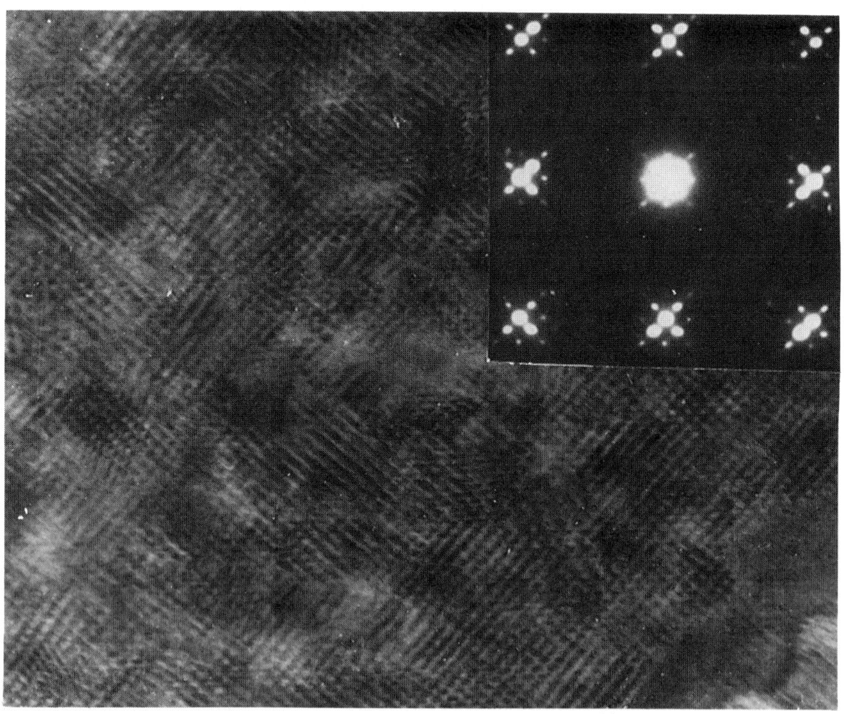

FIGURE 4
High resolution image and diffraction pattern of (001) zone of 2201 single crystal

and semiconducting phases. The X-ray diffraction analysis suggest that minor phase exists in the single crystal probably resulted from an intergrowth mechanism in the growth process. To lessen the minor phase existed in the single crystal is helpful to stablize superconducting phase. Further works are needed to be done on the crystal chemistry and magnetic field effect to demonstrate the cause of superconductivity.

REFERENCES

1. C.Michel et. al Z.Phys. B 68(1987) 421
2. H.Maeda. et. al Jpn.J. Appl.Phys. 27 (1988)L209
3. D.M.Ginzburg Properties of High Temperature Superconductor I, 1988
4. Edward Sonder,.C.Chakoumakos,and B.C.Sales Physical Review B 40 (1989) 6872
5. A.W.Sleight Proceedings of Stanford International Conference on Materials and Mechanism of Superconductivity
6. Klaus Yvon and Michel Francois Z.Phys.B Condensed Matter 76(1989) 413
7. Yasunori Ikeda et.al. Physica C 159 (1989) 93
8. J.M.Tarascon Proceedings of the Stanford Conf. On HiTc, July July 23-29 1989
9. Jianqi Li et. al. preprint

Frontiers of Materials Research/
Electronic and Optical Materials
M. Kong, L. Huang (editors)

ELASTICITY STUDY ON PHASELIKE TRANSITION IN Bi-Sr-Ca-Cu-O MONOCRYSTAL AND POLYCRYSTAL

Yening Wang, Jin Wu, Huimin Shen, Ming Zhu, Linhai Sun, Bin Xu and Zheng Yu

Laboratory of Solid State Microstructures, Nanjing University, Nanjing, P.R.China

*Yifeng Yan and *Zhongxian Zhao

*Institute of Physics, Chinese Academy of Sciences, Beijing, P.R.China

The frequency dependences of heights and positions of the attenuation peaks at 95K and 250K for $Bi_2Sr_2CaCu_2O_8$ single crystal (Tc=84K) show the characters of phase transitions. The pseudo-static ferroelastic $\sigma - \varepsilon$ loops of Pb doped Bi-Sr-Ca-Cu-O polycrystal, in which the superconducting phase with Tc=107K predominates, always appear around 130K and 250K. The energy loss (loop area) is independent of the strain rate and increases with the strain amplitude, which also exhibits the characteristics of static hysteresis loss related to phase transitions instead of Debye relaxation mechanism. Lattice parameter measurements showed a jumpy change at the peak temperatures. A small thermal hysteresis (2-3K) near the C_{66} minima at those peak temperatures was obtained for $Bi_2Sr_2CaCu_2O_8$ crystal, showing a weak first order phase transition. The investigations of the recovery of elastic modulus of coarse grained Pb doped BSCCO as well as YBCO indicate that the kinetics of isothermal transitions has a C-curve relation. Some discrepancies in elasticity measurements including the thermal hysteresis of elastic modulus so far observed for high Tc superconductors have been analyzed.

1. INTRODUCTION

It was reported that phaselike transitions (PLT) exist at tens degrees above Tc and at 210-250K in high Tc oxide superconductors. The anomaly of elastic moduli and internal friction or attenuation peaks in these temperature ranges for $YBa_2Cu_3O_{7-\delta}$ (YBCO) were attributed to a phase transition in some papers [1-12], but not explained in others [13-15]. Quite a few papers reported an anomaly only near 210-250K while not near Tc and in some cases, even no anomalous behavior is observed below room temperature [16,17]. On the other hand, it was reported that an internal friction peak of Debye relaxation type appeared just below and sometimes above Tc with an activation energy of 0.11-0.2ev in YBCO at 1-3KHz [18-20], which might overlap with the internal frictions due to PLT especially when measured at higher frequencies [12]. For Bi-Sr-Ca-Cu-O (BSCCO), similarly, two internal friction and attenuation peaks at 95K and 130K were found to be associated with two superconducting phases with Tc at 84K and 107K respectively [9,21-23] in addition to the peak at 210-250K [9,22,24]. However, someone only measured a broad relaxation peak around 170K [25] and the 250K peak was also attributed a relaxation process instead of phase transition [22]. These discrepancies need to be clarified.

A large thermal hysteresis loop of elastic modulus between 70 K and 260 K in ceramic YBCO has been obtained by ultrasonic velocity measurements [26,27] in the megahertz range and by electrostatic excitation of flexural vibrations in the kilohertz range [28]. Similarly, a hysteresis of ultrasonic velocity in BSCCO was also reported [29]. The authors [26] suggested a hysteretic phase change that occurs between 160 K and 70 K during cooling and between 170 K and 260 K during warming. Others [27,28,30] attributed it to a first order phase transition as well. However, Ewert et al [17] measured the hysteresis of the longitudinal wave velocity only for coarse grained YBCO samples, but neither thermal hysteresis nor softening for fine grained samples. This result has been confirmed by others [15,31]. The thermal hysteresis was ascribed to microstructure [17] or the relaxation of thermoelastic stress caused by the anisotropy of the thermal expansion of grains [15], instead of phase transitions.

In this paper, (1) ultrasonic attenuations were measured for a single crystal of $Bi_2Sr_2CaCuO_8$ and the stress-strain relations by tensile tests were studied for polycrystals of Pb doped Bi-Sr-Ca-Cu-O (BPSCCO), so the characteristics of phaselike transitions in these temperature regions can be confirmed. (2) The elastic constant C_{66} has been measured for the single crystal of $Bi_2Sr_2CaCu_2O_8$ to clarify whether the thermal hysteresis occurs and if exists how large it is. The isothermal recovery of elastic modulus on the hysteretic loop was investigated to explore the origin of the large thermal hysteresis appearing in coarse grained YBCO and BSCCO, and, as a result, the kinetics of isothermal phase transitions has also been studied.

2. EXPERIMENTS

The single crystals of $Bi_2Sr_2CaCu_2O_8$ were prepared by a flux method [32]. The specimen for ultrasonic longitudinal wave measurements had dimensions of 7x4x2 mm^3, with a-b plane parallel to the plane of 7x4 mm^2. X-ray pseudo-Kossel line patterns for the two surfaces parallel to the c-plane show that the whole specimen contains many layers. Each layer is a good single crystal, but has a small tilt or/and rotating angle to the neighboring layers. A $LiNbO_3$ transducer of 5MHz with area 1.5x4 mm^2 was bonded to the sample for the frequency-dependent ultrasonic attenuation measurement, which made the longitudinal waves propagate along the c-plane. An AC-cut quartz transducer of 5MHz with area 1.5x4 mm^2 was used for the shear waves velocity measurements, from which elastic constant C_{66} can be obtained directly. The ultrasonic attenuations and sound velocities were measured by a pulse echo method using a Matec 6600.

The polycrystalline samples of YBCO, BSCCO and Pb doped BSCCO (BPSCCO) were prepared by standard solid state reactions. The specimens used for tensile test made on a Mi-44 micromachine had the dimensions of 19x5.7x2.6 mm^3. An X-cut quartz transducer of 5MHz was used to study the isothermal recovery of sound velocity of BPSCCO. The isothermal recovery of elastic modulus of YBCO was studied by a flexual vibration frequency measurement.

3. RESULTS AND DISCUSSIONS

3.1. Ferroelastic loop due to PLT

Fig.1 shows the stress-strain ($\sigma - \varepsilon$) curves at various temperatures in polycrystal BPSCCO, in which the higher-Tc (107K) phase predominates and coexists with a minor lower-Tc (84K) phase. It is clearly seen that the ferroelastic loops appear near 95K, 130K and 250K, at which the phaselike transition characterized by the abrupt change of lattice parameters, was found to occur [9,21]. The reproducibility of the curves in Fig.1 is verified for about ten times repeated measurements and for several samples. The area

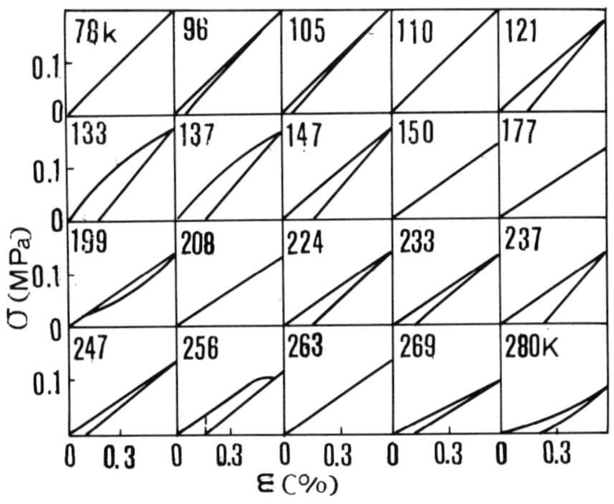

FIGURE 1
Stress(σ)-strain(ε) curves in the range from 280K to 78K for a polycrystalline BPSCCO specimen. The strain amplitude εm=0.6%

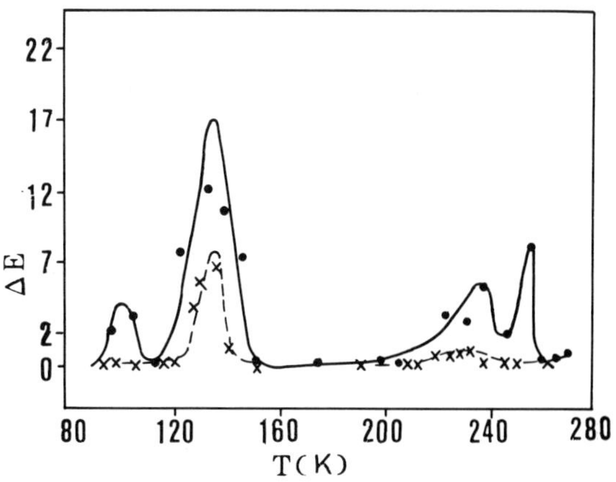

FIGURE 2
Energy loss ΔE (area of $\sigma-\varepsilon$ loop) vs temperature in polycrystalline BPSCCO specimen at two strain amplitudes. (•) εm=0.6% (×) εm=0.4%. ΔE is independent of the strain rate, which corresponds to the frequency independence of the loss.

of the $\sigma-\varepsilon$ loop is proportional to the energy loss per cycle (ΔE). The plots of the area at different temperatures were shown in Fig.2, which correspond to an energy loss curve with a much larger strain amplitude (εm = 0.4%-0.6%) and a much lower frequency than used in internal friction measurements. From the strain rate, the corresponding frequency is estimated at about 10^{-1}-10^{-2} Hz. However, the peaks appear at nearly the same temperatures as measured at 100kHz [10.21] and MHz [9,22] ranges, which indicates the peaks are most likely induced by phase transitions instead of by Debye relaxation mechanism. The fact that the 130K peak is much higher than the 95K one, which appeared only when the strain amplitude is large enough is in accordance with that the higher-Tc phase predominates in the sample for the reason mentioned above[9]. Fig.2 also shows the effect of strain amplitude on the energy loss, which increases with the increase of the strain amplitude εm. The energy loss values at about 95 K, 130 K and 250 K are much smaller for εm=0.4% than those for εm=0.6%. This is another character of static hysteresis instead of relaxation losses in addition to the frequency independence of energy loss values. The latter has been confirmed by the measurements with different strain rates which correspond to the variation of frequency (10^{-1}-10^{-2}Hz). Such hysteresis loss always occurs during first order phase transitions when measured in the low frequency ranges, generally less than MHz. It is well known that the motion of phase interfaces or domain walls induced by the stress is responsible for this internal friction. According to the X-ray diffraction results for BSCCO polycrystal [9,21] as shown in Fig.3, it is considered to be the phase boundary between two phases with slight different lattice parameters.

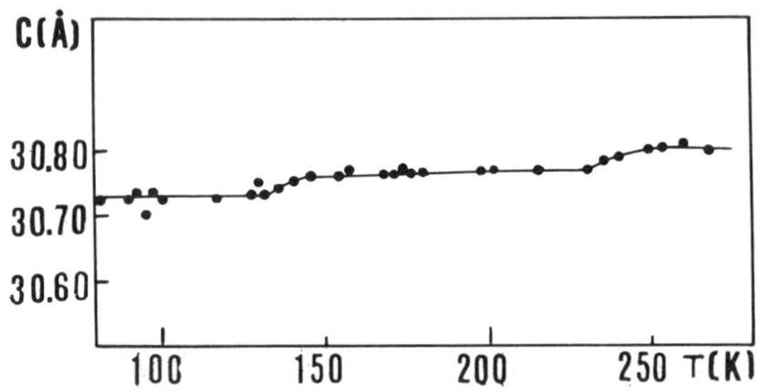

FIGURE 3
Lattice parameter c versus temperature curve for Bi-Sr-Ca-Cu-O ceramic sample.

3.2. Attenuation peaks due to PLT

Two attenuation peaks near 95K and 250K have been measured (Fig.4) for single crystal of $Bi_2Sr_2CaCu_2O_8$ with Tc at 84K. Fig.4(a) and (b) also show the effect of the frequency on the attenuation peaks near 250K (4.7MHz, 8.4MHz, 13.9MHz and 16.7MHz) and 95K (4.7MHz and 8.4MHz) respectively. The peak temperature is found to be independent of measuring frequency , which reveals the contribution of Debye relaxation to the peaks is negligible except the possible contribution on the high temperature side of 250K peak at f=16.7MHz. The attenuation was found to increase with the frequency f. The insets in Fig.4(a) and (b) show the attenuation is nearly proportional

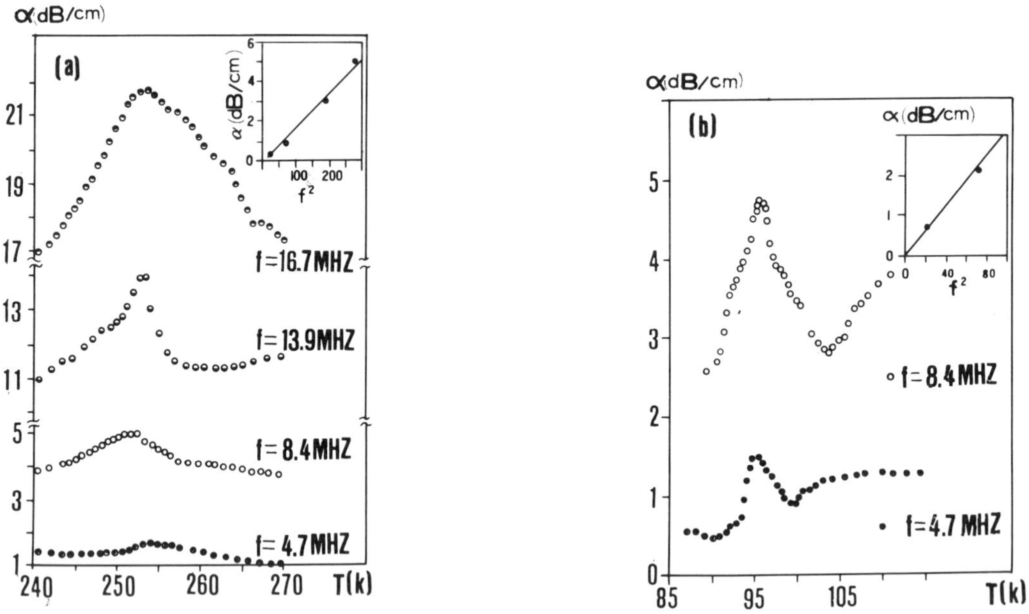

FIGURE 4
Ultrasonic attenuation with longitudinal wave propagated along the c-plane of $Bi_2Sr_2CaCu_2O_8$ single crystal vs temperature, measured at various frequencies (f) near 250K (a) and 95K (b)
Inset: α peak value (with the background substracted) as a function of f^2.

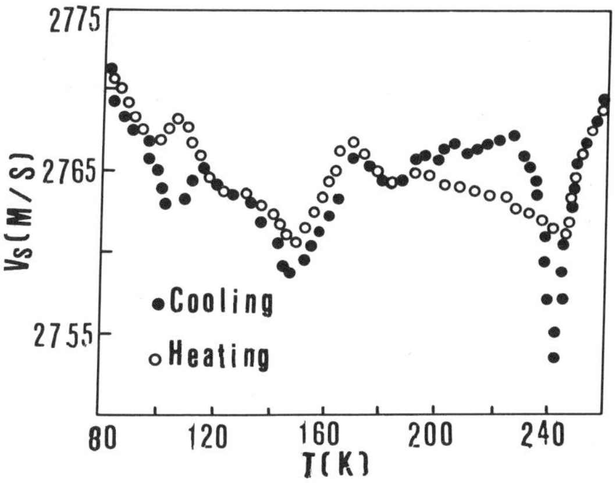

FIGURE 5
Elastic constant C_{66} as a function of temperature for $Bi_2Sr_2CaCu_2O_8$ single crystal on heating and on cooling with $\dot{T}$=0.2K/min.

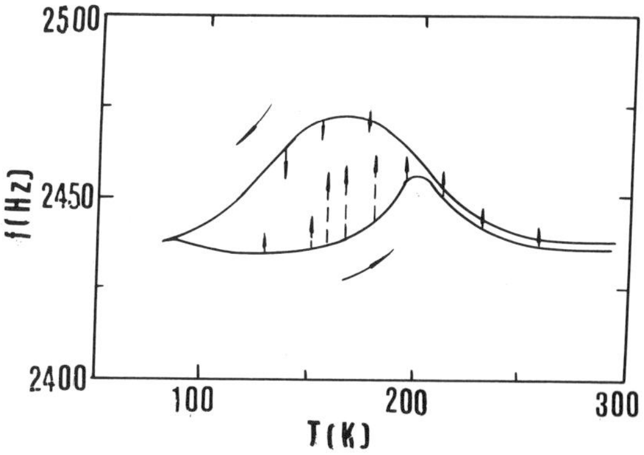

FIGURE 6
Hysteretic loop of resonant frequency f for coarse grained YBCO measured by flexural vibration method. The arrows show the temperatures at which the isothermal recovery of frequency has been measured.

to f^2 for both 250K and 95K peaks, just as predicted by the theory of the ultrasonic attenuation based on the order parameter or phonon relaxation during phase transitions [33], which has quite different mechanism from that of the internal friction peak measured at low frequency. However, the same thing is that the peak always occurs at phase transition temperature no matter what frequency is used. The relation of $\alpha \sim f^2$ for 95K peak is not as good as 250K peak, the stress induced motion of phase interfaces may contribute to the attenuation around 95K, because the elastic softening of C' near 95K is much larger than that of 250K peak [34].

3.3. The kinetics of isothermal phaselike transition and the thermal hysteresis of elastic modulus for coarse grained BSCCO and YBCO

The measurements of elastic constant C_{66} for the single crystal $Bi_2Sr_2CaCu_2O_8$ showed that there are three minima at about 250K, 150K and 95K, giving another verification for a phase transition. As shown in Fig.5, the thermal hysteresis of C_{66} measured on cooling and heating near the minima is found to be about 2-3K, exhibiting the character of a weak first order phase transition, similar to the result of specific heat measurement of the phase transition at 230K for YBCO [35]. However, a large thermal hysteresis (>100K) loop of elastic modulus in coarse grained ceramic YBCO [26-28] and BSCCO [29] has been reported. It needs to explore what the cause of the large hysteresis of elastic modulus in coarse grained YBCO and BPSCCO is. It was speculated that some imperfections were generated near the interconnecting region between grains, which may reduce the elastic modulus, and can not recover until the reverse phase transition occurs or only recover at rather high temperatures such as close to room temperature. We also measured the hysteretic loop of resonant frequency f for coarse grained YBCO (40-50 μm) by a flexural vibration method (Fig.6) and that of longitudinal sound velocity V for BPSCCO (20-30 μm). The isothermal recovery of elastic modulus at various temperatures as shown by the arrows in Fig.6 has been investigated. Fig.7(a) shows the normalized isothermal change of frequency for YBCO as a function of time t at three temperatures. The following creep function is obeyed,

$$f(t)-f(0) = \Delta[\ 1-\exp[-(t/\tau)^{\beta}\ ]]$$

$1/\tau$ reflects the recovery rate. The inset shows the log-log plot of $-\ln[1-(f(t)-f(0))/\Delta\ ]$

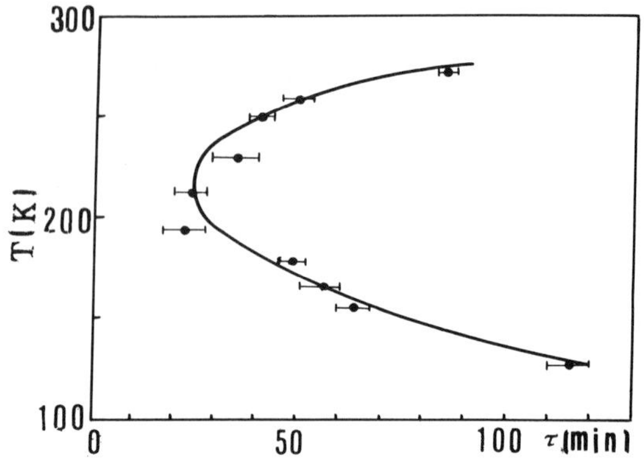

FIGURE 7(b)
Recovery time τ of elastic modulus at different temperatures for coarse grained YBCO.

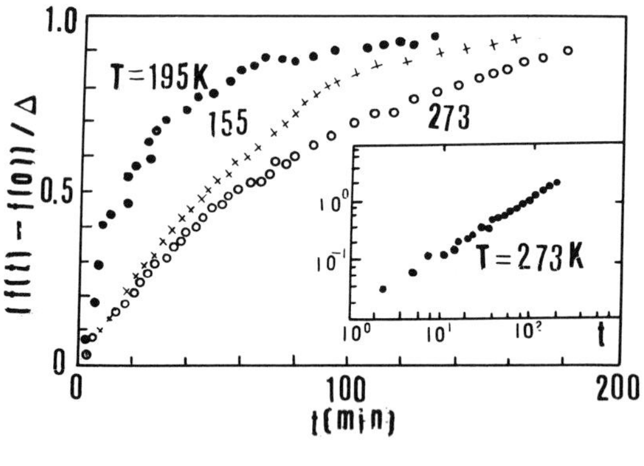

FIGURE 7(a)
Isothermal change of the frequency (normalized) for YBCO as a function of time t at three temperatures.
Inset:The log-log plot of $-\ln[1-(f(t)-f(0)/\Delta\ ]$ vs t at 273K. The slope β=1.

versus t at 273 K, the linear relation is fairly good, from which β was determined to be approximately equal to 1. So, τ can be calculated easily. τ vs. T is ploted in Fig.7 (b). Similarly, the change of the sound velocity for BPSCCO also obeys the creep function. Fig.8(a) shows the isothermal change of the velocity for BPSCCO as a function of time t at several temperatures and β =1 (see the inset). Fig.8(b) shows the τ values at various temperature T for BPSCCO. Obviously, τ (T) disagrees with Arrhenius relation, so the mechanism of oxygen vacancy diffusion [31] should be ruled out. Instead, τ vs. T relation is similar to the C-curve for isothermal martensitic transition. It might be considered

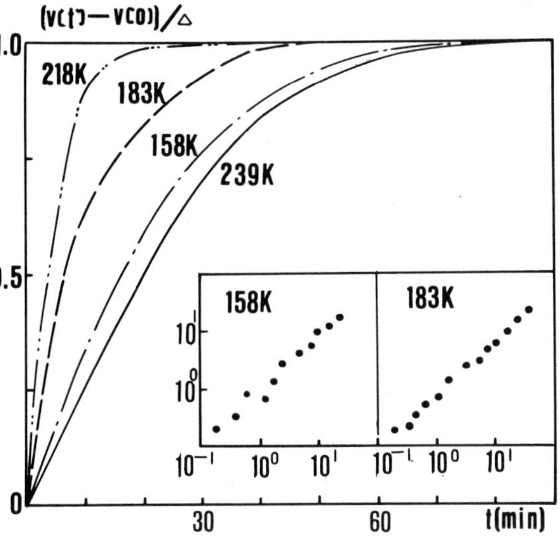

FIGURE 8(a)
Isothermal change of the frequency (normalized) for BPSCCO as a function of time t at four temperatures.
Inset:The log-log plot of $-\ln[1-(f(t)-f(0)/\Delta]$ vs t at 158K and 183K. The slope is approximately equal to 1.

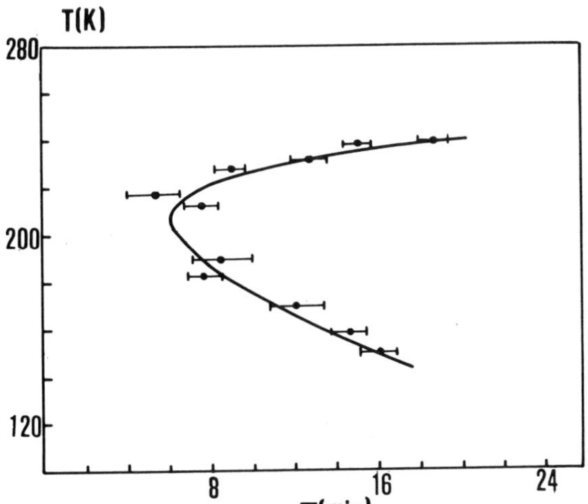

FIGURE 8(b)
Recovery time τ of elastic modulus at different temperatures for coarse grained BPSCCO.

that the defects of pore size leading to the softening and hysteresis decrease only during the reverse phaselike transition. Because the oxygen content in the intergrain region is different from that of intragrain region and it has significant effect on the change of lattice parameters or "apparent thermal expansion" above T_c as shown in Fig.9 [36], the different changes of lattice parameters at various regions result in a high concentration of stress near the interconnecting regions. When the distortion energy increasing with the continuously decreasing lattice parameters on cooling reaches the connecting energy, the defects of pore size may be created e.g. starting at 170 K, thus leading to the decrease

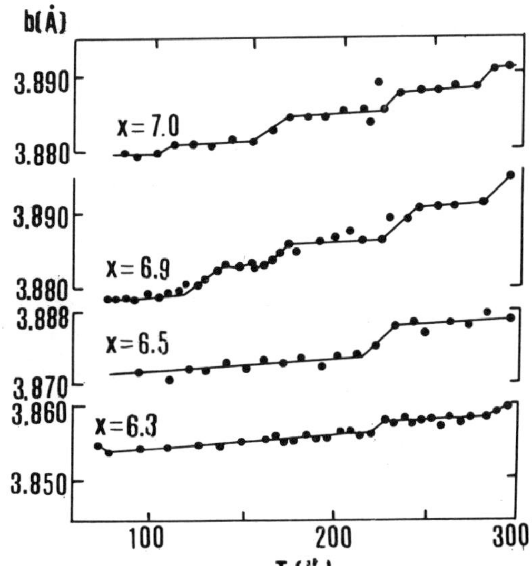

FIGURE 9
Lattice parameter b vs temperature for $YBa_2Cu_3O_x$ with various oxygen content x.

of elastic modulus. The defects decrease only when the reverse change occurs i.e. when the lattice parameters increase to the original values. It can be seen from Fig.9, the rapid change of lattice parameter covers the temperature range of 90-270 K [36,37]. So, the large thermal hysteresis may be explained. This mechanism we proposed is supported by the acoustic emission measurements [38]. The AE counts are concentrated in 90-160K on cooling and in 180-240K on heating, which may be associated with the generation and disppearance of the defects of pore size, respectively. As for the fine grained sample, the distortion energy caused by the apparent thermal expansion which concentrated on the boundaries is smaller for smaller grains so that the imperfections of pore size can not occur.

Because the isothermal recovery rate of elastic modulus is controlled by the isothermal phaselike transition as argued above, the C-curve of Fig. 7(b) and 8(b) must reflect the kinetics of isothermal transition. It is worth noting that the knee temperature of the C-curve for YBCO which represents the temperature of the rapidest recovery rate just coincides with the peak temperature due to PLT near 220K. This is favorable to the view that the imperfections can recover only when the reverse phaselike transition occurs. As for BPSCCO, the knee temperature [Fig.8(b)] is somewhat lower than that of the attenuation peak or velocity minimum temperature in single crystal (Fig.5). It may be attributed to two different superconducting phases. In single crystal of $Bi_2Sr_2CaCu_2O_8$, the 84 K phase predominates, whereas in coarse grained BPSCCO, the 110 K phase predominates.

4. CONCLUSIONS

(1) The 130K peak appears in higher-Tc phase whereas the 95K peak in lower-Tc phase, but 220-250K peak exists in both phases of BSCCO.

(2) The studies of the ultrasonic attenuation and pseudo-static hysteresis losses indicate that these peaks are mainly associated with phaselike transitions characterized by the jumpy change of lattice parameters.

(3) The small thermal hysteresis of C_{66} on cooling and on heating show that the PLTs in BSCCO are of weak first order phase transitions.

(4) The kinetics of the isothermal phaselike transition shows a relation of C-curve.

(5) The ultrasonic attenuation peaks due to PLT appear only when the frequency is high enough. The internal friction peaks at lower frequencies due to PLT may not be measured if the strain amplitude is too low. Some discrepancies in elasticity measurements so far for high Tc superconductors may be understood.

(6) The investigations of isothermal recovery of modulus for coarse grained YBCO and BPSCCO disclose that the large thermal hysteresis may be attributed to the defects of pore size generated and diminished in the intergrain region as a result of PLT and reverse PLT.

ACKNOWLEDGEMENTS

This work was supported by the National Center for Research and Development on Superconductivity of China and the National Foundation of Natural Sciences of China.

REFERENCES

1. Y.N.Wang, H.M.Shen et al, J Phys.C20 (1987) L665

2. X.H.Cheng, L.H.Sun et al, J Phys.C21 (1988) 4603

3. L.H.Sun, Y.N.Wang et al, Phys.Rev.B38 (1988) 5114

4. S.Bhattacharya,M.J.Higgins, Phys.Rev.Lett.60 (1988) 1181

5. G.Cannelli, R.Cantelli et al, Physica C 153-155 (1988) 298

6. H.M.Shen, J.Wu et al, Supercond. Sci.Technol 2 (1989) 52

7. X.D.Shi, R.C.Yu et al, Phys.Rev.B39 (1989) 827

8. M.Saint-Paul, J.L.Tholence et al, Solid State Communications 69 (1989) 1161

9. Y.N.Wang, J.Wu et al, Phys.Rev.B41 (1990) 8981; Physica C 162-164 (1989) 454

10. Y.N.Wang, X.H.Chen et al, Physica C 162-164 (1989) 456

11. Y.S.He, B.W.Zhang et al. J. Phys. F17 (1987) L243

12. K.J.Sun, W.P.Winfree ed al. Phys.Rev.B38 (1988) 11988

13. H.Ledbetter, J.Metals 40 (1988) 24

14. C.Duran, P.Esquinazi et al, Solid State Communications, 68 (1988) 957

15. Yu.B.Burenkov and V.I.Ivanov et al, Sov.Phys.Solid State,30,1837(1988).

16. X.D.Xiang et al,Solid State Communications 69 (1989) 833

17. S.Ewert and S.Guo et al, Solid State Commun., 64, 1153 (1987).

18. H.Mizubayashi, K.Takita and S.Okuda, Phys. Rev. B37 (1988) 9777 and private communication.

19. G.Cannelli and R.Cantelli et al, Phys.Rev. B38, (1988)7200

20. J.J.Du, J.Y.Jiang et al, Modern Phys.Lett. B1 (1988) 425

21. X.H.Cheng, H.M.Shen and Y.N.Wang, Modern Phys. Lett. B3 (1989) 1241

22. J.K.Hu, IC M^2S HTS(Stanford)1989,2B-1; Physica C162-164 (1989) 444

23. M.Saint-Paul, J.L.Tholence et al. Physica C166 (1990) 405

24. Y.S.He, J.Xiang et al. Phys.Rev.B40(1989) 7384

25. F.Zheng, J.Jiang et al. Phys.State.Sol.(a) 115 (1989) K75

26. H.M.Ledbetters and S.A.Kim, Phys.Rev., B38, 11857 (1988).

27. Z.Zhao and S.Adenwalla et al, Phys.Rev., B39, 721 (1989).

28. G.Cannelli and R.Cantelli et al, Europ.Phys.Lett., 15, 271 (1988).

29. P.K.Choi and H.Koizumi et al, Solid State Commun., 70, 1175 (1989).

30. V.Muller and D.Maurer et al, Physica C, 153-155, 280 (1988).

31. P.Lemmens and F.Stellmach et al, Physica C 153-155, 294 (1988).

32. Y.F.Yan et al, Mod. Phys. Lett. B2, 571 (1988).

33. B.Luthi and W.Rehwald, in Structural Phase Transition I, ed. K.A.Muller and H.Thomas, by ISBN 1981, p.131

34. Jin Wu, Yening Wang et al. Phys. Lett. A accepted

35. T.Lagreid and K.Fossheim et al. Nature 330 (1987) 637

36. L.H.Sun and Y.N.Wang et al, Phys.Rev., B38, 5114 (1988).

37. H.You and J.D.Axe et al, Phys.Rev., B38, 9213 (1988).

38. Xu Yuehuang, He Sha et al. J. Wu Han Univ. (CHINA) No.4, 125 (1987).

Frontiers of Materials Research/
Electronic and Optical Materials
M. Kong, L. Huang (editors)
Elsevier Science Publishers B.V., 1991

STUDY OF 130K HIGH-Tc SUPERCONDUCTOR OF Bi-Pb-Sb-Sr-Ca-Cu-O

Qinlun XU, Guangyao MENG, Dingkun PENG, Pinghua YANG, Changyong XIA, Yaozhong RUN, Ling ZHOU, Zhunling WANG, Liezhao CAO, Zhaojia CHEN Jiang LU, Li YANG, Yunlan HUANG, Chengao FAN, Yuheng ZHANG
University of Science and Technology of China, Hefei, Anhui,230026, China

Samples of the ceramic superconductor $Bi_{1.8}Pb_{0.3}Sb_{0.1}Sr_2Ca_2Cu_xO_y$ with a nearly pure 2223 phase were prepared by solid state reaction method. X-ray diffraction revealed that there was neither 2212 nor 2201 phase but some monoclinic phase in these samples. These samples showed a semiconducting behaviour at higher temperature. There was a obvious transition in resistance curves above 125K. Their a.c. magnetic susceptibility had a large drop above 110K. The monoclinic fine particles segregated on the surface of the sample when sintered at higher temperature. The intergrowth of the 2223 and the monoclinic crystals at their boundaries caused the partial distortion of 2223 phase. These distorted 2223 crystals take the responsibility for the 130K superconductivity.

1. INTRODUCTION

Scince the discovery of 130K superconductivity in $Bi_{1.9-x}Pb_xSb_{0.1}Sr_2Ca_2Cu_3O_y$ superconductors [1], a lot of groups attemped to obtain the high-Tc superconductor using different methods. Some groups[2-4] reported a carbonate coprecipitation technique and a solid state reaction method for preparing $Bi_{1.6}Pb_{0.3}Sb_{0.1}Sr_2Ca_2Cu_3O_y$ samples.They all got a nearly single 2223 phase with Tco around 107K. A higher Tc near 125K was obtained from sample $(Bi_{0.85}Pb_{0.1}Sb_{0.05})_2Sr_2Ca_2Cu_3O_x$ prepared by oxidation of liquid-quenched precursor alloys corresponding to a multiphase material [5]. P.V.P.S.S. Sastry[6] observed a sharp drop in electric resistance between 145K and 125K in $Bi_{1.7}Pb_{0.2}Sb_{0.1}Sr_2Ca_2Cu_{2.8}O_x$ sample made by a matrix reaction method. However, the resistance didn't reach zero. A zero resistance at 132K in lead-free $Bi_{1.9}Sb_{0.1}Sr_2Ca_2Cu_3O_y$ was reported by M.R.Chandrachood from India[7].

We had made a series of samples in Bi-Pb-Sb-Sr-Ca-Cu-O systems by controlling Cu nominal compositions, and studied their microstructure and superconductivity. This paper discussed the origin of 130K superconductivity and explained why the 130K superconductivity appeared sometimes and did not at other situation.

2. SUPERCONDUCTING PROPERTIES

The samples were prepared by a solid state reaction method.The nominal composition of samples was $Bi_{1.8}Pb_{0.3}Sb_{0.1}Sr_2Ca_2Cu_xO_y$ with A(x=2.8), B(x=3.0) and C(x=3.4) respectively. The mixed powders of Bi_2O_3, Pb_3O_4, Sb_2O_5,$CaCO_3$,$SrCO_3$ and CuO were calcined at 830°C for 17hrs. They were ground, mixed, pressed into a pellet and finally sintered at 853°C for 60 hrs.in air.

Using CuK_α radiation , We got the x-ray diffraction patterns of these samples (Fig.1). They all show the dominant 2223 phase (It's characteristic peak is located at $2\theta=4.7^o$.) and a little monoclinic phase (2θ between 30°-31°). However, neither 2212 nor 2201 phase is observed in these samples.

the resistances of the samples A and C were meassured by Keithly 181 voltmeter by the standard four lead method. They all show a semiconducting behaviour at higher temperture (Fig.2). For example A, there is a transition in resistance curve at 140K . After that, it shows a metallic behaviour before a sharp drop between 110K and 100K. These transitions were stable upon thermal cycling between 77K and room temperature. As seen from the resistance

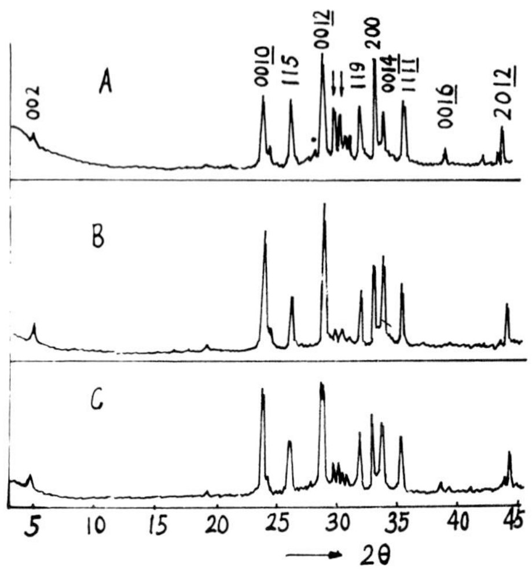

FIGURE1
X-ray diffraction patterns of sample $Bi_{1.8}Pb_{0.3}Sb_{0.1}Sr_2Ca_2Cu_xO_y$
A. x=3.0 ; B. x=2.8; C. x=3.4

plot, the resistance decreases again and reaches zero at 94K. A slowly resisitance decrease from 140K to 125K is observed in sample C. It's zero resistance temperature is 88K.

Employing the mutual inductance bridge with precison 0.1 uv, we observed that the a.c. magnetic susceptilility of samples B only shows one transition at 110K (Fig.3B). A single transition at 125K (Fig.3C) in a.c.magnetic susceptibility curve is observed in sample C. This is corresponding to the drop at 125K in the resistance curve. For sample A (Fig.3A), We can see two superconducting transistions at 135K and 90K respectively , no transition at 110K. These results were repeated three times between 77K and room temperature.

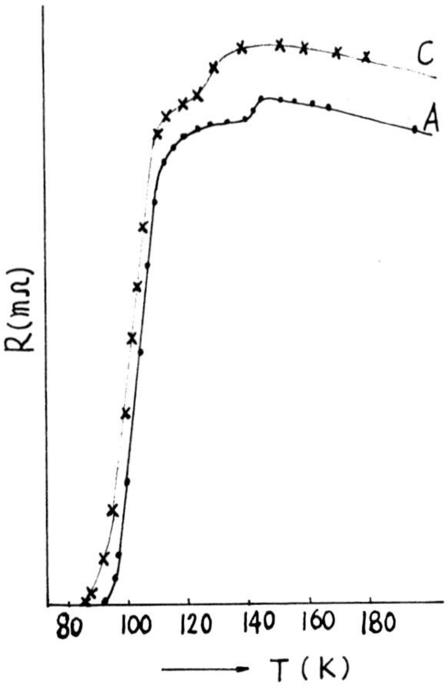

FIGURE2
The curves of resistance versus temperature for samples $Bi_{1.8}Pb_{0.3}Sb_{0.1}Sr_2Ca_2Cu_xO_y$
A. x=3.0 C. x=3.4

After that, we wore off the surface layer of sample A, remeasured again, the transition in a.c. magnetic susceptibility curve started at 125K, it decreased obviously.

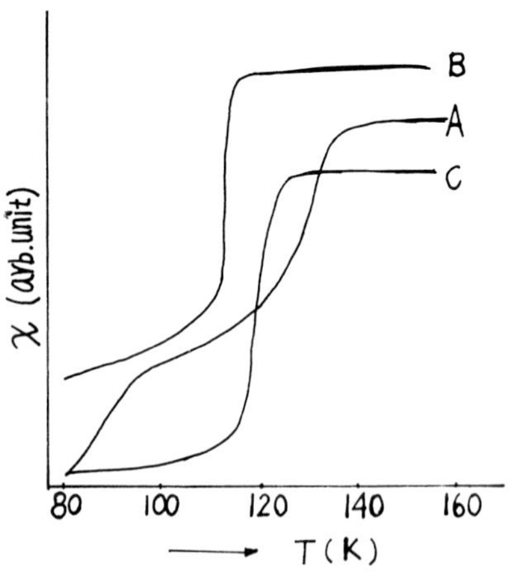

FIGURE3
The temperature dependance of a.c. magnetic susceptibility of sample $Bi_{1.8}Pb_{0.3}Sb_{0.1}Sr_2Ca_2Cu_xO_y$ A. x=3.0, B. x=2.8, C. x=3.4

3. SAMPLES MICROSTRUCTURE

In order to study the mechanism of the 130K supercondictivity formation, we observed their microstructure using scanning electron microscopy (SEM) equiped with wavelength dispersive meter(WDM) and H-800 transmission electron microscopy(TEM). Figure 4 shows the differential thermal analysis results of the sample A. An endothermic peak at 848°C is seen. This indicates a partial melting occurred during sintering time. Figure5 is the SEM photographs of sample A. We can see there are a lot of fine particle clusters and some 2223 thin sheet crystal grains in the surface image of this sample(Fig.5a). Sb line analysis of

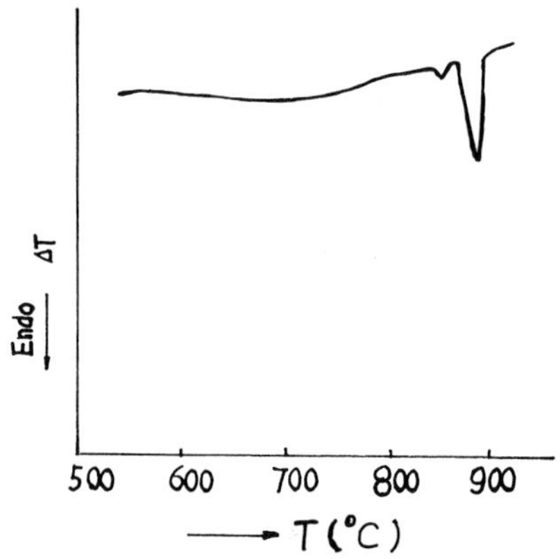

FIGURE4
The DTA results of sample $Bi_{1.8}Pb_{0.3}Sb_{0.1}Sr_2Ca_2Cu_3O_y$

sample's surface by WDM(Fig.5b) indicates these fine particals are Sb rich phase which is the monoclinic phase determined by Tingzhu Cheng[8]. Figure 5c is the section image of this sample . There are a lot of large 2223 thin sheet crystals in different orientation. Little fine particle grew on it. Figure 5d shows there are a few Cu rich grains ($CaCuO_2$) which also can be seen from X-ray (it's characteristic peak is d=3.2). By crushing this sample , we took a series TEM pictures. There

FIGURE5
The SEM images of sample $Bi_{1.8}Pb_{0.3}Sb_{0.1}Sr_2Ca_2Cu_3O_y$ a. the surface picture, b. Sb line analysis of a, c. the section image, d. Cu rich grains

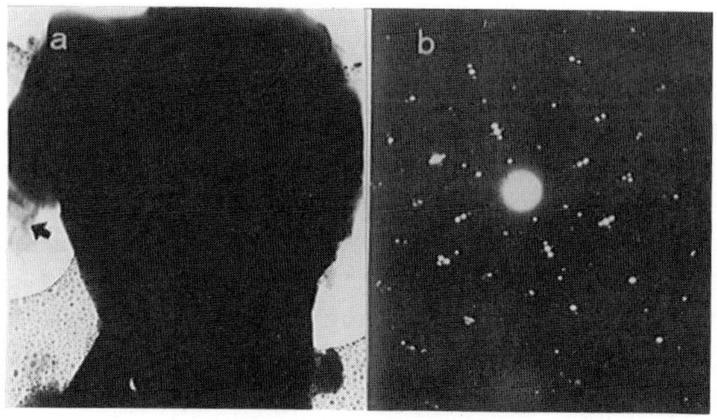

FIGURE6
The TEM images of sample $Bi_{1.8}Pb_{0.3}Sb_{0.1}Sr_2Ca_2Cu_3O_y$ a. thick monoclinic grain with some 2223 sheet growing on it, b. the overlapping diffraction of monoclinic and 2223 crystals at their grain boundaries

are 60% fraction of the large 2223 sheet grains of different orientation, 15% fraction of small 2223 sheet grains, 25%

thick monoclinic crystal grains with some 2223 sheet grains growing on it (Fig.6a). Figure 6b is the picture taken at the boundaries of the 2223 and monoclinic crystals. It shows a overlapping of their electron diffraction patterns. This indicates that there is a intergrowth at the boundaries of the 2223 and monoclinic crystals and suggestes that partial 2223 crystals at their boundaries are distorted. These results are consistent in the report by Chen and Mao[9].

4. DISCUSSION

X-Ray and SEM observation indicates there are only 2223 phase, a little monoclinic phase and trace $CaCuO_2$ phase in these samples. The monoclinic phase is an insulator[10] and the $CaCuO_2$ phase is not superconducting. The 2223 phase with stable structure shows 110K superconductivity. Only those distorted 2223 phase is able to take the responsibility for the 130K superconductivity. SEM observation demonstrated (Fig.2a)that most of monoclinic grains was segregated on the surface of the sample A and the 2223 sheet grains grew on them which were distorted as mentioned above. These distorted 2223 phase may have some proper atom arrangment and hence take the responsibility for the 130K superconductivity .The 130K phase at the surface of the sample produced the shielding effect to the 110K 2223 phase. So,the sample did not show any drop in a.c. magnetic susceptibility curve around 110K which should occur for normal 2223 phase. while the 90K transition could come from the 2212 phase which was too little to be detected by X-ray diffraction.

For resistance versus temperature curve, we can explain that at higher temperature, the samples shows a semiconducting behaviour because of the monoclinic phase which is insulating. When temperature decreases, the disterted 2223 crystals begin to be superconducting state (but does not form a superconducting path) resulting in resistance transition at 140K (A) and 125K(C) respectively. At 110K, large amount of 110K 2223 crystals inside the samples become superconducting state and the resistance drops sharply between 110K and 100K. Because of intergrain weak coupling [11] (Josephson like weak coupling), the zero resistance of these samples are all below 100K.

In order to obtain the information why the sample with similar composition showed the different superconducting behaviour, we observed a series pictures of sample $Bi_{1.92}Pb_{0.38}Sb_{0.1}Sr_2Ca_2Cu_{3.2}O_x$ (D) synthesized from mixture of codecomposition powder of coprecipitated Bi,Pb,Sr,Ca and Cu as oxalats and commercial powder Sb_2O_5[12] . Figure 7a

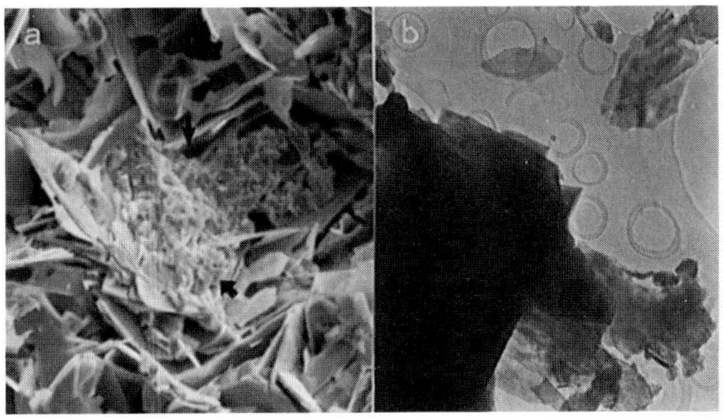

FIGURE7
The SEM photographs of sample $Bi_{1.92}Pb_{0.38}Sb_{0.1}Sr_2Ca_2Cu_{3.2}O_y$
a.a piled up monoclinic particles, b.the 2223 and monoclinic crystals coincided.

is the SEM photograph of the surface of the sample D, we can see the sheet 2223 crystals and a pile monoclinic particles. TEM observation showed no closely connection between the 2223 crystals and the monoclinic particles. They coincided each other in any way (FIg.7b).While the intergrowth of 2223 and monoclinic crystals at their boundaries existed in samples prepared by solid state reaction method. This is because the very fine powders prepared by coprecipitated metallic oxalats and higher chemical activity of the fresh oxides containing Bi,Pb,Sr,Ca and Cu atoms reacted very quickly at lower temperature (843°C) to form 2223 crystals. While commercial Sb_2O_5 powders, which is coase and less activity, reacted slowly. So a pile monoclinic particles tended to appear in the region of Sb rich. For sample A made by a mixture of oxide powders, the 2223 crystals and the monoclinic fine particles formed at the same time at higher temperature(853°C). They grew up alternatively in the case of partial melting ,and this caused the structure distortion of 2223 phase at the grain boundaries of 2223 and monoclinic crystals. This caused the 130K superconductivity of sample A.

5. CONCLUSION

1). Samples of nearly single 2223 phase with a little monoclinic phase have been obtained by solid state reaction method in air.
2).The monoclinic particles tended to segregate on the surface of the samples due to partial melting during sintering time.
3).Intergrowth of the 2223 phase and the monoclinic phase caused the 130K supeucondutivity and led to a shielding effect to 110K 2223 phase.

ACKNOWLEDGEMENTS

This work was supported by National Center for Research and Development on Superconductivity, and the Center of Structure and Element Analysis in the University of Science and Technology of China.

REFERENCES

1. H.B.Liu, L.Z.Cao, L.Zhon, and X.X.Li, Solid State Commun. 69(1989)867.
2. N.D.Spencer, W.R.Grace and Co.-Conn, Jpn.J.Appl. Phys. 28(1989)1564.
3. M.Pissas and D. Niaschos, Physica C in print.
4. B.Dabrowski, S.R.Richards and D.G.Hinks, Physica C 160(1989)281.
5. J.S.Luo, D.Michel, and J-P.chevaller, Appl. Phys. Lett. 55(1989).
6. P.V.P.S.S.Sastry and R.M.Lyer, Solid State Commun. 71(1989)935.
7. M.R.Chandrachood and I.S.Mulla, Appl. Phys. Lett. 55(1989)1472.
8. T.Z.Cheng, C.G.Fan,Z.J.Chen and Y.H.Zhang, Superconducter Science and Technology 3(1990)87.
9. Chen Zhaojia and Z.C.Mao, Beijing Intern.Conf. on high Tc superconductivity,Beijing, 1989.
10. X.D.Wang, G.Y.Meng and D.K.Peng, Solid State Commun. in print.
11. R. Sato, T.Komatsu and K.Matushita, Jpn.J.Appl.Phys. 28(1989)1922.
12. Q.L.Xu, T.Z.Cheng, X.G.Li, C.G.Fan, D.K.Peng, Z.J.Chen and Y.H.Zheng, Physisca C in print.

Frontiers of Materials Research/
Electronic and Optical Materials
M. Kong, L. Huang (editors)

PREPARATION OF HIGH Tc Pb-DOPED Bi-BASED SUPERCONDUCTOR FROM AMORPHOUS SOLID AND FORMATION OF SUPERCONDUCTING PHASE

Chen Meiling, Li Xiaoge, Wang Keyi, Li Suzheng, Xu Shihong, Wu chengjiu, Tong Honggao

Anhui Institute of Optics and Fine Mechanics, Academia Sinica, P.O. Box 25, Hefei Anhui 230031

The Pb-doped Bi-based amorphous solid has been prepared by the anvil quenching method from the melt. The crystallization and phase transition in the amorphous solid was studied by XRD, DSC and SEM. After annealing, we obtained the dense superconductor which onset and zero resistance temperature are 120K and 107K respectively It was observed that there were impure phases in the grain-boundaries by TEM with EDAX.

1. INTRODUCTION

Since it was reported that the partial substitution of Pb for Bi in the Bi-Sr-Ca-Cu-O system was very effective in increasing the volume fraction of high Tc phase[1], many methods have been adopted to prepare this kind of material, such as solid states reaction[1], co-decomposition[2] and casting method[3], etc. In this paper, we report the preparation of Pb-doped Bi-based superconductor from the amorphous solid and the phases transition in the amorphous solid. We hope that this study will be useful to understand the formation of the superconducting phase and improve the superconducting properties.

2. EXPERIMENTAL PROCESS

The pure powders of Bi_2O_3, PbO, $SrCO_3$, $CaCO_3$ and CuO were mixed homogeneously at the atomic ratio of 1.7:0.3:2:2:4. The mixture was heated at 1113K for 4h in an Al_2O_3 crucible, then heated continuously above 1273K till it was completely molten. The melt was cooled rapidly to form the amorphous sheet with thickness less than 0.5 mm by the anvil quenching method[4]. The melt-quenched samples were annealed at temperature ranged from 719 K to 1153 K for various time, then cooled down to room temperature in air. The DSC analysis for the melt-quenched samples were made at a heating rate of 40K/min. The phases and properties annealing samples were determinted by X-ray powder diffraction (XRD) using CuKα radiation, scanning electron microscopy (SEM), and transmission electron microscopy (TEM) with selected area electron diffraction (SAED) and Energy dispersive analysis of X-ray (EDAX). The resistivity was measured by a standard four-probe technique.

3. RESULTS AND DISSCUSSION

Fig.1 shows the XRD pattern of the melt-quenched sample with composition $Bi_{1.7}Pb_{0.3}Sr_2Ca_2Cu_4O_y$.

The DSC curve of the melt-quenched sample was shown in fig.2. From these figures, it can be found that the melt-quenched samples were amorphous solid. In fig.2 we also observe that the glass transition is at 669K and there are four exothermic peaks at the temperature ranged from 700K to 1000K. In order to understand the crystallization process and phase transition in the amorphous solid, the samples of the amorphous solid with the same composition were annealed at 719K, 775K, 847K, 903K, 953K, and 1103K respectively.

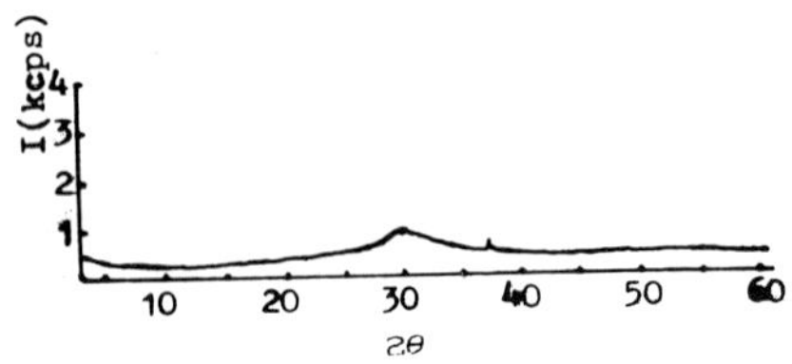

FIGURE 1
XRD pattern of Pb-doped Bi-based amorphous solid

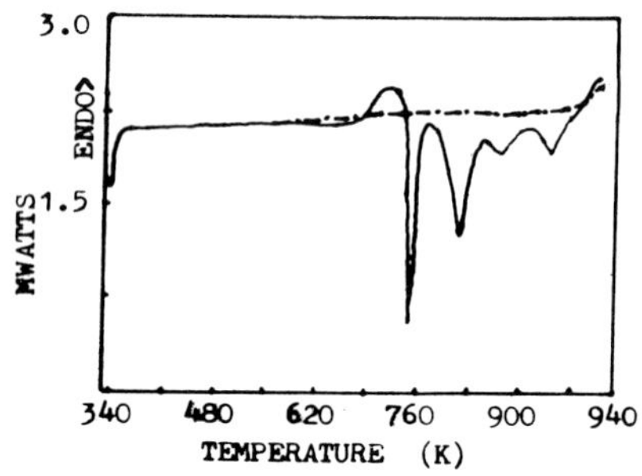

FIGURE 2
DSC curve of Pb-doped Bi-based amorphous solid

Fig.3(a),(b),(c),(d),(e),(f) are the XRD patterns of the samples which were annealed for 24h at 719K, 775K, 847K, 903k, 953K and 1103K respectively. From these patterns, we can observe that the crystallized phase is mainly $Bi_2Sr_2CuO_x$ (2201 phase) phase and a little of CaO phase exists in the sample annealed at 719K.

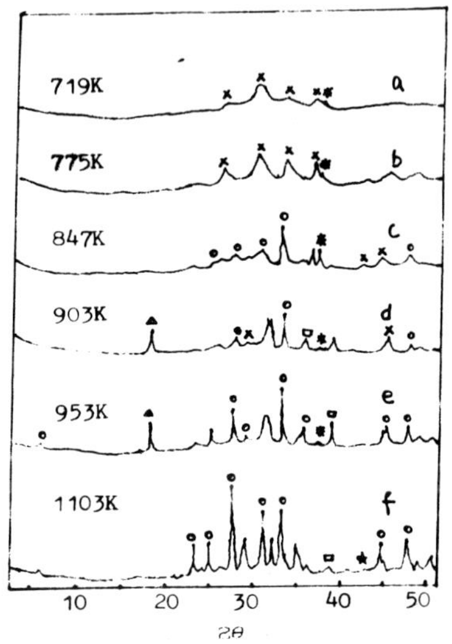

FIGURE 3
XRD patterns of samples annealed at the different temperature for 24h. a: 719K, b:775K, c:847K, d:903K, e:953K, f: 1103K

○ Bi_2Sr_2CuOx phase × $Bi_2Sr_2CaCu_2Oy$ phase
□ CuO ※ CaO ☆ Ca_2CuO_3 △ Ca_2PbO_4

The sample annealed at 847K, not only the $Bi_2Sr_2CaCu_2O_y$ (2212 phase) phase is detected, but also CuO phase can be found. While at 903K and 953K, a lot of 2212 phase are observed in this sample, at the same time, Ca_2PbO_4 phase is also formed. At 1103K, 2212 phase becomes the main phase, Ca_2CuO_3 appears and Ca_2PbO_4 disappears in the sample.

In our experiments, we consider that 2201 phase and CaO phase firstly crystallized in the amorphous solid at low temperature. With increasing the annealing temperature, 2212 phase, CuO and Ca_2PbO_4 phase were formed successively, the exothermic peaks in DSC curve attributed to the crystallization and transition of the phases. At 1103K, although Ca_2PbO_4 and CuO phases disappeared, Ca_2CuO_3 phase appeared, which was characterized by reflection peak at $2\theta=27.5°$ of Fig.3(f). This phenomenon suggests that Ca_2CuO_3 phase was combined by CuO and CaO at about 1103K, meanwhile $CaPbO_4$ decomposed. We can also assume Pb ion did not penetrate lattice completely at least 953K.

Fig.4(a),(b),(c),(d) show SEM photographes of the samples annealed at 1103K for different times, which reveal the growing process of grains in the samples. From these figures we observed that the samples were homogenous nucleation, and with increasing time, the grains in the sample became larger and slender.

Fig.5 is the XRD pattern of the sample annealed at 1103K for 240h. It is shown that there are mainly high T_c phase(2223 phase) in the sample, which formed gradually in it with increasing the annealing time at 1103K. The result may be due to the disproportionation reaction of the 2212 phase, Ca_2CuO_3 and free CuO. We also consider that the temperature of high T_c phase formation is higher than that of low T_c phase formation. The compositon and structure analysis were carried out by SAED and EDAX at the grains and grain-boundaries. There were a lot of plate-like crystals which were superconducting crystals and other polycrystals in the sample by TEM with SAED analysis.EDAX measurement shows that the average composition of the plate-like crystal which abound in the sample is $Bi_2Pb_{0.3}Sr_{1.7}Ca_{1.3}Cu_{2.3}O_y$. We also found

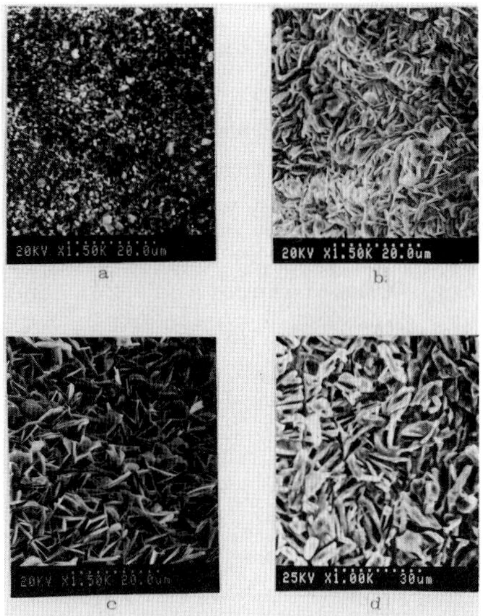

FIGURE 4
SEM photos of different annealing time at 1103K. a:15 min. b:48h. c:96h. d:240h.

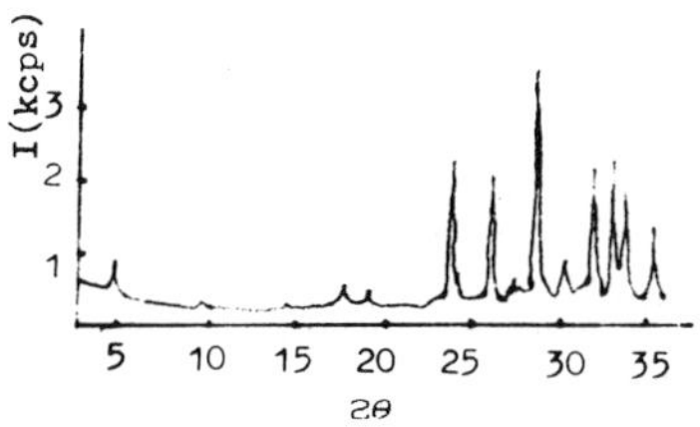

FIGURE 5
XRD pattern of Pb-doped Bi-based superconductor annealing at 1103K for 240h.

some polycrystals, in absence of Cu, and a little of amorphous which did not crystallize during annealing process exists in the grain-boundaries. These impure phases in the grain-boundaries caused in weak-link of the superconducting grains in the sample.

Fig.6 shows the resistivity vs. temperature of the sample annealed at 1103K for 240h, which onset and zero resistance temperature are 120K and 107K respectively. At 77.3K, the J_c was measured about 185 A/cm^2 at zero magnetic field. The further studies of enhancement of J_c is in progress.

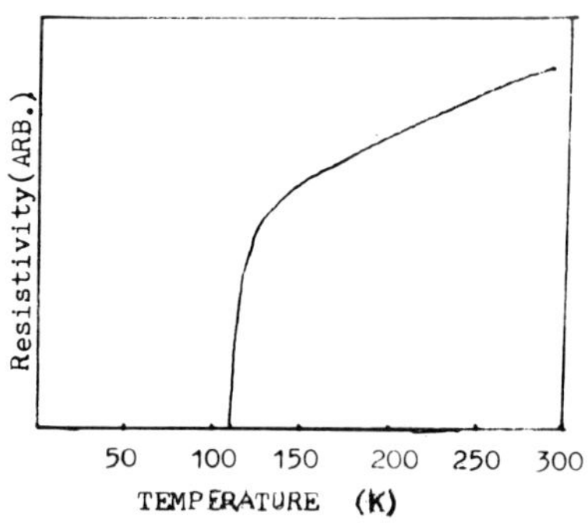

FIGURE 6
Resistivity vs.temperature curve for the sample annealing at 1103K for 240h.

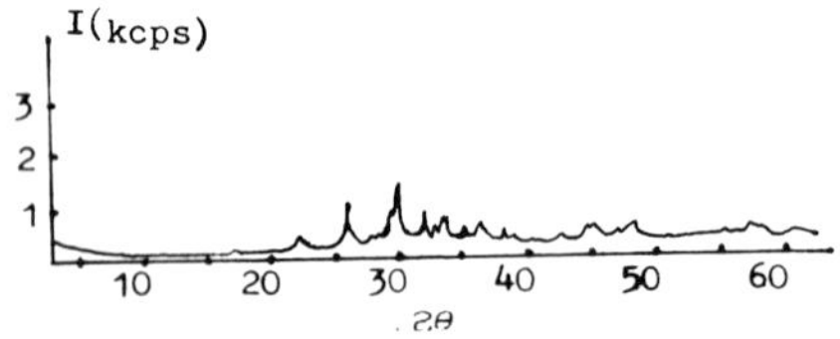

FIGURE 7
XRD pattern of Pb-doped Bi-based superconductor annealing at 1153K for 0.5h.

Fig.7 shows the XRD pattern of the amorphous solid sample annealed at 1153K for 0.5h. It is revealed that there are 2201 phase and 2223 phase in the sample. This result may be due to 2212 phase decomposed.

4. CONCLUSIONS

The superconductor prepared from the amorphous solid is compact and homogenous. During the crystallization process of the amorphous solid, CaO and 2201 phase firstly emerged at 710K, then CuO, Ca_2PbO_4, Ca_2CuO_3 and 2212 phase formed with increasing annealing temperature. The four exothermic peaks in DSC curve are attributed to the crystallization and transition of the phases. At 1103K the 2223 phase in the sample increased with increasing annealing time, and the grains became larger and slender. At 1153K,there existed 2201 and 2223 phases in the sample, which may be due to the decomposition of 2212 phase. The sample annealed at 1103K for 240h enhibited superconductivity with T_c(zero) of 107K and the critical current density (77.3K, zero magnetic field) of 185A/cm^2, due to weak link resulted from the impure phases in the grain-boundaries. The investigation of improving properties is undertaking.

REFERENCES

1. S.A.Sunshine, J.Siegrist, L.F. Schneemeyer, et al., Phys.Rev. B. 38 (1988)893.

2. Satoshi KOYAMA, Ubako ENDO, Tomoji KAWAI, Jpn.J.Apply.Phys.27(1989)1861.

3. CHEN Meiling, WU Chengjiu, WANG Keyi, et al. Chin.J.of Low Temperature Phys. 11(1989)329.

4. Ramanchandrarao, P.and Baunerijee, D. Met.Trans.1(1970)2655.

Frontiers of Materials Research/
Electronic and Optical Materials
M. Kong, L. Huang (editors)

FABRICATION AND PROPERTIES OF HIGH Jc Y—Ba—Cu—O SUPERCONDUCTOR BY (MTG) PROCESS

Ren Hongtao, Xiao Ling, He Qing, Wang Ruikun, Chang Shian, Yu Dingan

General Research Institute for Non—Ferrous Metals, Beijing 100088, China

Cui Changgeng, Li Shanlin

Institute of Physics, Chinese Academy of Sciences, Beijing 100080, China

Wei Chongde, Dai Yuandong

Department of Physics, Beijing University, Beijing 100871, China

Superconducting Y—Ba—Cu—O bulk materials were fabricated by melt—textured growth (MTG) method, where cooling slowly under temperature gradient produces an aligned structure exhibiting high Jc value. The critical current density was measured by continuous dc four—probe method with criterion of 1μV/cm. Jc value has reached 23,800 A/cm^2 at 77K under 2T magnetic field perpendicular to the current. The magnetic field dependences of Jc from 1 to 7 Tesla show the anisotropy for two orientations ($H \perp I$, $H \perp$ a—b plane and $H \perp I$, $H // $ a—b plane). The X—ray diffraction patterns show the crystals preferentially aligned. The SEM and TEM observations of the MTG sample reveal that the sample has dense layered—structure with regular alignment of plate—shaped crystals. Twinning planes of fairly high density exist. A great quantity of dislocations, dislocation loops and stacking faults exist around 211 phase. These crystal defects may be main flux pinning centers. We have measured the zero—field cooled (ZFC) and the field cooled (FC) magnetizations as functions of temperature at various external dc magnetic field. The dependence of irreversibility point T^* on 2/3 power of field H for MTG sample was demonstrated. The flux jumping was found in the hysteresis loop below 10K.

1. INTRODUCTION

Enormous research and tremendous development have been stimulated by the discovery of high Tc superconducting oxides[1,2,3], because of the potentially immense commercial utilization of the superconductors. However, one of the major barriers to the applications of the bulk high Tc superconductors is the "weak—link" problem in the sintered materials which are typified by very low critical current density and their severe degradation in weak magnetic fields[4]. In order to eliminate these weak links and increase the Jc value, a lot of efforts have been paid. Jin et al.[5] synthesized the specimen with Jc up to 17,000 A/cm^2 by the melt—textured growth method in which the samples were heated to 1050～1200℃ and cooled to 900℃ in a temperature gradient of 50℃/cm. Murakami et al.[6] have developed the quench and melt growth (QMG) process where after the Y—Ba—Cu—O powder was melted at 1300℃, and quenched, the melted sample was reheated to 1100℃, cooled down to 1000℃ and followed by slow cooling. Salama et al.[7] applied a liquid phase processing method to the Y—Ba—Cu—O materials with Jc of 18,500 A/cm^2 by continuous dc measurements at 77K in 0T. In this method, no thermal gradient exists in the sample. R. B. Van Dover et al.[8] reported the attainment of a critical current density, Jc, of $\sim 6 \times 10^6$ A/cm^2 at 77K and 0.9T by measuring the magnetic moment in a single crystal of YBaCuO irradiated with fast neutrons to introduce defects. Recently, Ren Hongtao[9] and his coworkers applied melt—textured growth (MTG) method to synthesize the sample with Jc up to 23,800 A/cm^2 at 77K in 2T magnetic field perpendicular to the current. The commen characteristics of all these methods are that the sample was cooled slowly through the peritectic temperature. In this paper, we report the development of a melt—textured growth method for fabricating bulk Y — Ba — Cu — O superconductors with a microstructure favorable to obtain high critical current density. The microstructure and magnetical properties in the MTG sample are also discussed.

2. EXPERIMENTAL

Powder of Y—Ba—Cu—O used in this study was synthesized by solid state reaction. In this process, the powder was well mixed in order to refine and homoginize the distribution of Y_2O_3. The powder was formed into sheet with dimensions of 1.5 × 20 × 60 mm^3 by cold—pressing. The sheet—like samples were sintered at 950℃ for 24 hours in air. Then, the sample was introduced into a temprature gradient (about 30℃/cm) furnace which was preheated to the temperature range of 1050~1150℃. The sample was maintained at the temperature range for 5~30 min. and cooled down to the temperature about 1025℃ at the rate of 10℃/min.. And then the sample was cooled slowly through the peritectic transformation. During the slow cooling, 123 phase was grown continuously. Finally, the sample was annealed in flowing oxygen at the temperature range of 400—600℃. Because the density of the sample approaches the theoritical value, there are absence of porosities and minimum number of grain boundaries in the sample. So a long time heat treatment in oxygen is necessary to replenish the vacancy sites with oxygen which is lost during the melting and make the sample become homogenization.

For the critical current tests, the samples with dimensions of 0.07~0.3 mm by 0.3—3.0 mm by 10 mm were cut along the a—b plane. Silver paste, silver deposition and gold deposition contacts were used as the electric contacts. The silver deposition contacts were found to be the best one because of the low resistivity (about 0.5×10^{-6} Ω/cm^2 at 77K). The silver leads were soldered on the four terminals of the sample with indium. The critical current densities were measured by the standard four—probe method at 77K. The quench criterion of 1μV/cm was taken to define the critical current Ic. Samples tested with continuous direct current were placed in the magnetic fields perpendicular or parallel to a-b plane. A NbTi superconducting solenoid which was worked at the magnetic field range of 0 to 7 Tesla was used as an applied magnetic field. A liquid nitrogen cryostat was inserted in the magnet. The temperature inside the cryostat can be contralled by a pump and a pressure contraller. Magnetization was measured by both vibrating and moving sample magnetometers equipped with a cryostat. The microstructures of the specimens were examined in JEOL JSM 840 scanning electron microscope (SEM) equipped with energy-dispersive spectrometer (EDS). Micro-composition analysis was performed with a EM-400T electron microscope and the microstructure study was also carried out with a JEM-1000 high voltage electron microscope at operation voltage of 1000 KV.

3. RESULTS AND DISCUSSION

Figure 1(a) is X—ray diffraction pattern of cleavage plane in the MTG sample. All the peaks in the pattern were observed to be {00l}. It is obvious that the crystals are preferentially aligned and that the cleavage plane is a—b plane of 123 phase. The extent of C axis alignment was measured by rotating the sample through a Bragg angle as the angle between the X—ray source and detector was kept fixed. From the rocking curves for the (005) peak of the sample (Fig. 1 (b)), we can see that the full—width half—maximum $\Delta\theta$ was 0.6°. The peaks of 211 phase can be seen from the X—ray pattern of the MTG sample which is ground into powder.

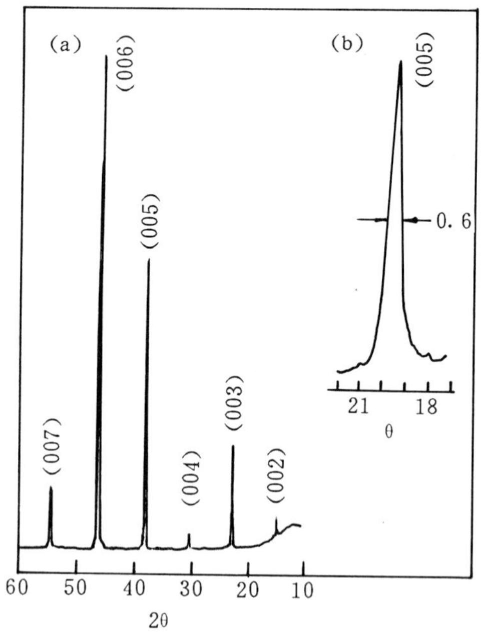

FIGURE 1

(a). X—ray diffraction pattern of cleavage plane in the MTG sample.

(b). The rocking curve for the (005) peak in the MTG sample.

A micro—composition analysis indicates that the superconductor mainly consists of the 123 phase and a few of the 211 phase and CuO dispersed in the 123 phase. The 211 phase is ellipsoidal in shape with a long—axis of 2 to 5 μm or spherical about 0. 5μm in diameter. The CuO particles generally are spherical less than 0. 5 μm in diameter, as shown in Fig. 2.

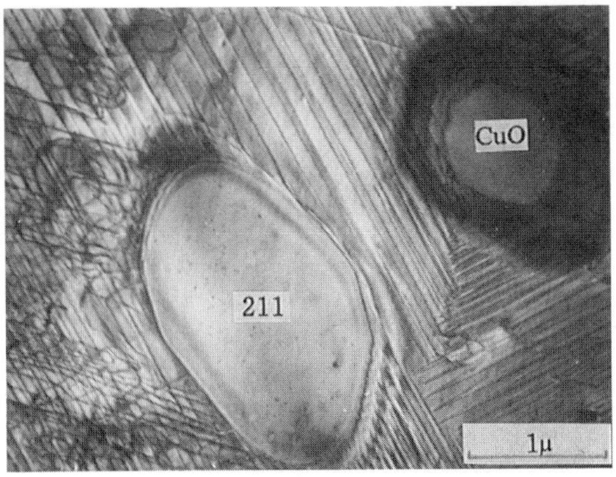

FIGURE 2
211 phase and CuO distributed in 123 phase.

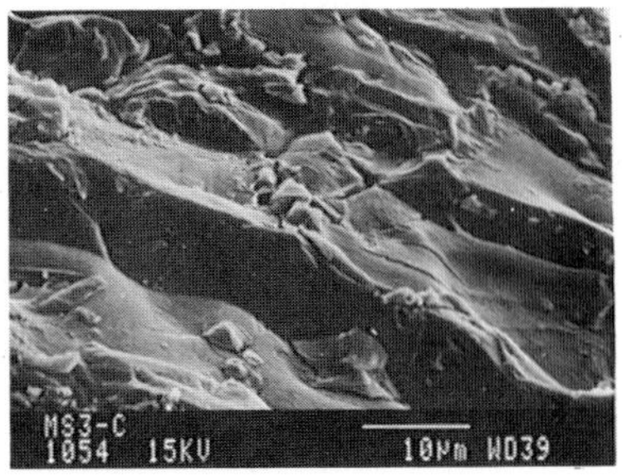

FIGURE 3
The fracture surface of the MTG sample.

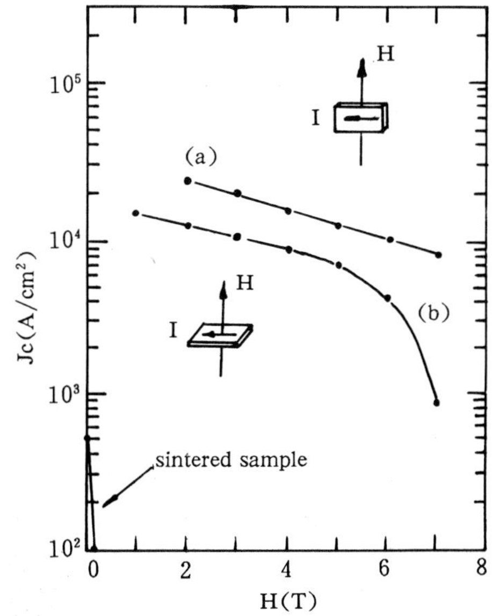

FIGURE 4
The critical current density vs magnetic field for the MTG sample at 77K. Curve (a) is the relation of Jc and H as H perpendicular to the current but parallel to a — b plane. Curve (b) is the relation of Jc and H as H perpendicular to both the current and a—b plane.

The particles of 211 phase and CuO in the 123 phase distribute inhomogeneously. The YBaCuO superconductor prepared by MTG method has an essentially 100% dense structure and there isn' t any microcracks in the a — b basical planes, but the microcracks exist between the a — b basical planes. No impurities was observed at grain boundaries in the superconductor. SEM photographs taken from the fracture surface of the MTG sample were shown in Fig. 3. It can be seen that the sample has dense layer — structure with thickness about 5—10 μm. The grains of 123 phase are 1—2. 5 cm long. Fig. 4 shows the critical current density vs magnetic field for the MTG sample at 77K. Curve (a) is the relation of Jc and H as H perpendicular to the current but parallel to a—b plane. The Jc value has reached 23,800 A/cm² under 2T and reduced to 10,000 A/cm² in 6T. Curve (b) is the relation of Jc and H as H perpendicular to both the current and a—b plane. Because the sample was easily broken due to the overheating of electrical contact, we haven' t obtained the Jc value at 0T.

From the two curves, we can see that the Jc values decrease gradually as the magnetic field increasing. The critical current characteristic in the MTG sample is quite different from that in sintered sample. The weak—link nature of the grain boundaries is greatly improved in the MTG specimen. Compared with the two curves, it can be seen that the critical currents are, as expected, less in the field perpendicular to the a—b plane than in that parallel to the a—b plane. This results mean that the field dependence of Jc is different and there is an anisotropy of Jc(H) for two orientations.

We have measured the zero—field cooled (ZFC) and field cooled (FC) magnetizations as functions of temperature at various external dc magnetic field. Fig. 5 is the typical curves at H = 100 Oe. It exhibit sharp transitions demonstrating the high quality of the sample. The small Meissner signal indicates the strong flux pinning existing in those samples.

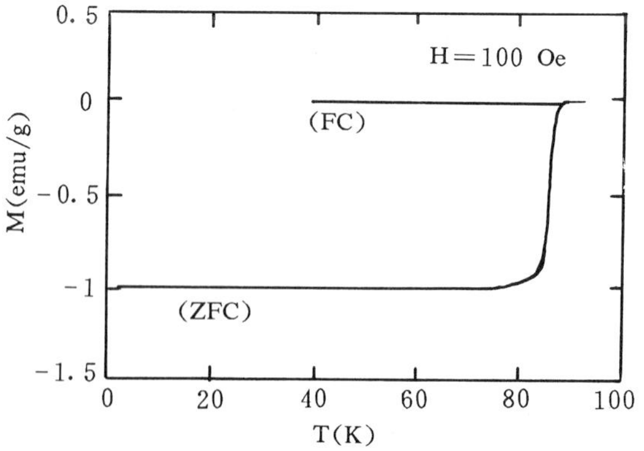

FIGURE 5
The zero—field cooled (ZFC) and field cooled (FC) magnetizations as functions of temperature as H=100 Oe.

From the magnetization curves (ZFC) and (FC), the irreversibility temperature T* can be taken. Fig. 6 shows the dependence of irreversibility temperature T* on the 2/3 power of field H for MTG sample which exhibits zero resistance at 90K. The irreversible points in the different magnetic fields can be well filled by straight line. The obvious flux jumping was observed in the hysteresis loop below 10K (Fig. 7). The flux jumping observed in the hysteresis loop in 1.5K happened more frequently than that in 10K. The flux jumpings at points A and B can be explained by the oblivion of positive and negative flux lines.

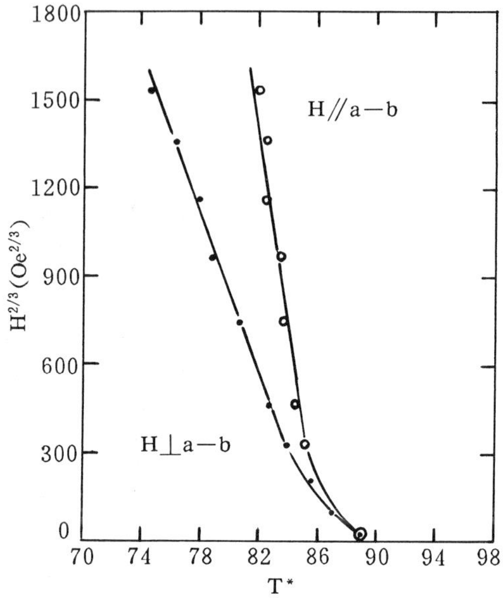

FIGURE 6
The dependence of irreversibility temperature T* on the 2/3 power of field H for the MTG sample.

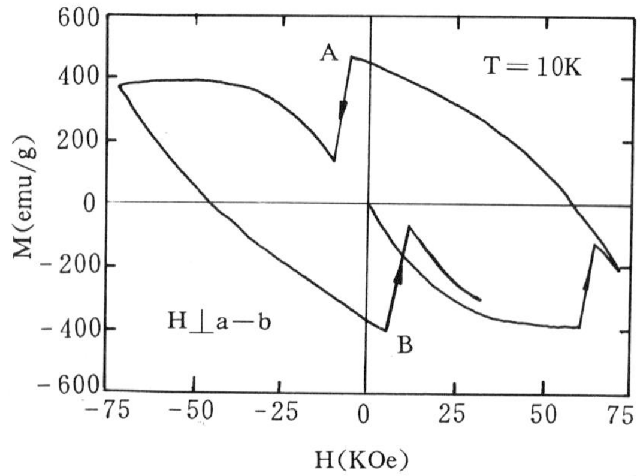

FIGURE 7
The flux jumping was observed in the hysteresis loop at 10K

We have also observed anomalous behavior from the hysteresis loop at 40K (Fig. 8).

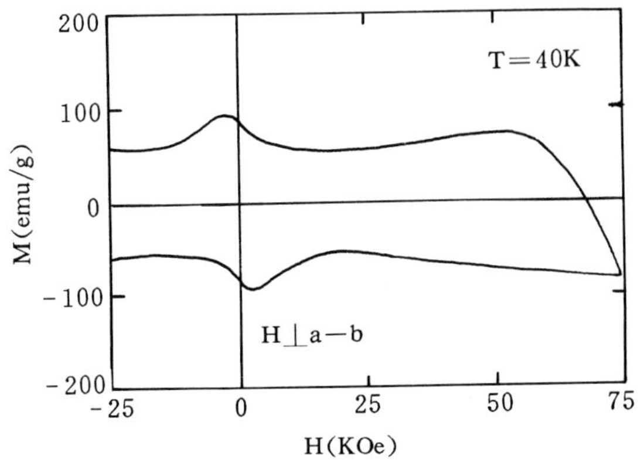

FIGURE 8
The hysteresis loop measured at 40K.

The difference $M^{+}-M^{-}$ is observed to go through a maximum with increasing applied field. The magnetic field Hp corresponding to the maximum Jc(H) values increase with the decrease of the temperature. The phenomena we have observed is similar to that observed by Baker and Sutton[10]. It can be explained that, when the distance between the flux lines matches that between the flux pinning centers, the Jc (H) can reaches the maximum.

From the HVEM photographs of the MTG sample (Fig. 2, Fig. 9~11), we can see that the twinning bands exist in 123 phase. The Twinning bands along (110) and ($1\bar{1}0$) planes cross perpendicularly. The twin spacing is about 0.01—0.1 μm. Twin bands in the MTG sample are thinner than that in the sintered one. Therefore, the area of twin boundaries as pinning center are increased. The high concentrated dislocations, dislocation loops and stacking faults exist around larger 211 phase (Fig. 9). While no concentrated crystal defects has been observed around the particles of 211 and CuO phases less than 0.5 μm in diameter. These crystal defects also exist in an area without the 211 phase and CuO particles in the 123 phase, as shown in Fig. 10, but they are not so concentrated.

Stacking faults are found along the directions of (100) and (010) planes. In two sides of the stacking faults, the partial dislocations can be seen by using a technique of diffraction contrast, as denoted by P in Fig. 11.

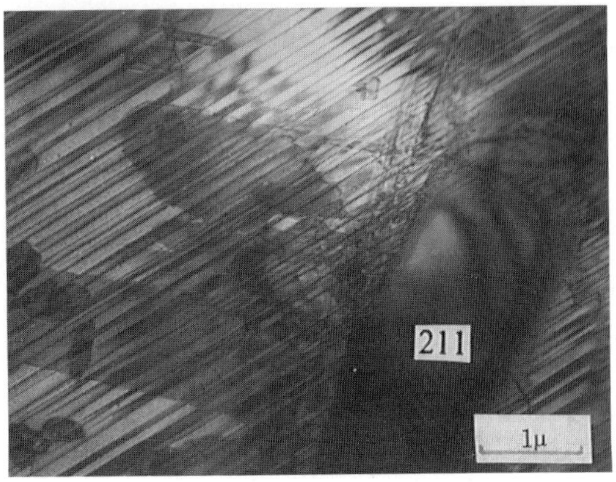

FIGURE 9
The photograph of the MTG sample taken by HVEM. Dislocations and stacking faults exist around 211 phase.

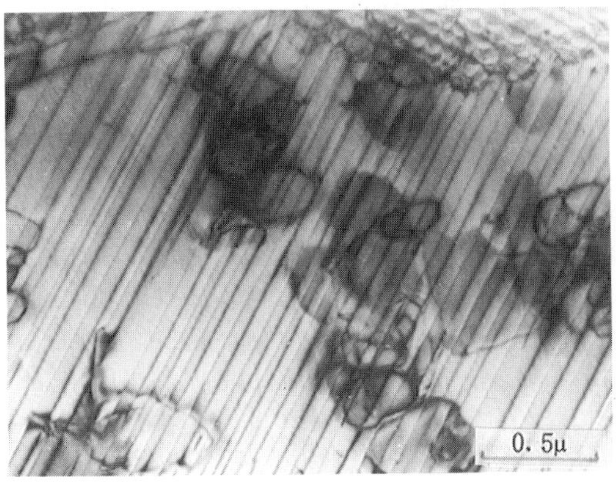

FIGURE 10
The photograph of the MTG sample taken by HVEM. The dislocation loops were also found at the area far away from 211 phase.

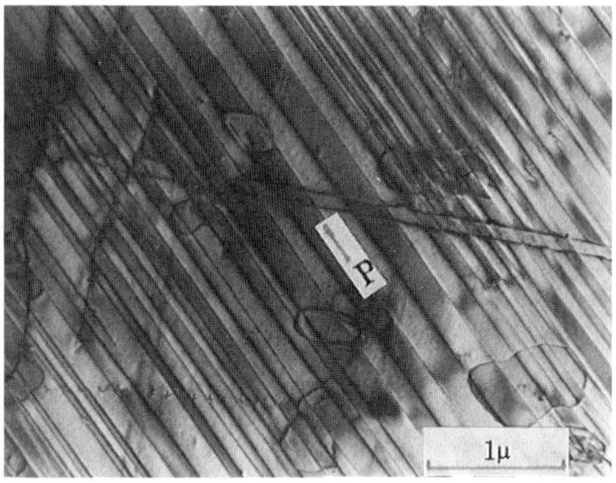

FIGURE 11
The photograph of the MTG sample taken by HVEM. The partical dislocations denoted by P was found.

The high critical current density in high magnetic field may mainly attribute to twinning bands, dislocations, dislocation loops and stacking faults. The bigger 211 and CuO particles can not pin the flux lines because their size is much larger than the coherence length ξ of the superconductor. But they can create a great quantity of crystal defects around them, which are beneficial for increasing Jc value.

ACKNOWLEDGEMENTS
The authors are grateful to Prof. Zhao Zhongxian, Guo Shuquan, Lu Guo, Yan Shousheng and Zhang Jinlong for valuable discussions. We also thank Li Jun, Hua Peiwen, and Liu Ansheng for their technical assistance during the course of this research.
This work was supported by The Chinese National Center for Research and Development on Superconductivity.

REFERENCES
1. J. G. Bednorz and K. A. Mueller, Z. Phys. B64, 189 (1986).

2. M. K. Wu, J. R. Ashburn, C. J. Torng, P. H. Hor, R. L. Meng, L. Gao, Z. J. Huang, Y. Q. Wang, and C. W. Chu, Phys. Rev. Lett. 58, 908 (1987).

3. Zhao Zhongxian, Chen Liquan, Cui Changgeng, Huang Yuhen, Liu Jinxing, Chen Genghua, Li Shanlin, Guo Shuquan, He Yeye, Kexue Tongbao, 32 (1987), 8: 522—525.

4. J. W. Ekin, Adv. Ceram. Mater. 2, 586 (1987).

5. S. Jin, T. H. Tiefel, R. C. Sherwood, M. E. Davis, R. B. Van Dover, G. W. Kammlott, R. A. Fastmacht, and H. D. Keith. Appl. Phys. Lett. 52(24) 13 June 1988 2074—2076.

6. Masato Murakami, Satoshi Gotoh, Hiroyuki Fujimoto, Naoki Koshizuka, Paper presented at AMSAHTS90, Appril 2—6, 1990, Greenbelt, MD..

7. K. Salama, V. Selvamamickam, L. Gao and K. Sun. Appl. Phys. Lett. 54 (23), 5 June 1989 2352—2354.

8. R. B. Van Dover, E. M. Gyorgy, L. F. Schneemeyer, J. W. Mitchell, K. V. Rao, R. Puzniak and J. V. Waszczak, Nature, Vol. 342, November 1989.

9. Ren Hongtao, He Qing, Xiao Ling, Wang Ruikun, Yu Dingan, Cui Changgeng, Li Shanlin, Proceedings of the ICEC—13, Beijing, China, April 24—27, 1990, to be published in Cryogenics.

10. C. Baker, J. Sutton, Phil. Mag. Vol. 19 1223 (1969).

Frontiers of Materials Research/
Electronic and Optical Materials
M. Kong, L. Huang (editors)

SUPERCONDUCTIVITY, STRUCTURE AND OXYGEN DEFICIENCY IN THE $Ba_2YCu_2WO_Y$ SYSTEM

BOKHIMI, M. PORTILLA, And R. PEREZ

Instituto de Física, and Facultad de Química, UNAM, a. p. 20-364, 01000 México D. F., Mexico.

We show that the $Ba_2YCu_xWO_y$ system with $0 \le x \le 2$, is composed of two phases, the first one with the composition Ba_2YWO_6 with a small solution of copper replacing yttrium, it has a cubic ordered perovskite structure with the space group Fm3m, the solution of the copper gives rise to a lattice contraction. The second phase corresponds to CuO, or Cu_2O, depending on the sample preparation. At large copper concentrations the first phase appears as small precipitates embedded in the second one. Reducing the samples at 1150 C in H_2, gives rise to a separation of the phases, segregating the copper in the form of liquid metal. The reduction eliminates from the first phase the dissolved copper, and produces also some oxygen deficiency in the sample, both effects give origin to a lattice distortion in the ordered perovskite structure. Although the samples are black after the reduction, they are not electrical conductors.

1. INTRODUCTION

The announcement last year of a new family of oxide superconductors with a cubic structure[1], with the composition $Ba_2YCu_2MO_{9-y}$, with M = W, Mo, Nb, Ta, Sb, Sn, Bi or Pb, caused expectation, but the fact that the Meissner effect fraction was small, gave rise to suspicion about the results, some people was of the opinion that the presence of Ba, Y and Cu in the sample could give origin to the $YBa_2Cu_3O_{7-x}$ compound, and this to the observed superconductivity[2]. Because there has not been a complete characterization of these new compounds, at a first sight, it was difficult to accept this proposition, although it is plausible. A recent report[3] shows that when copper is substituted by tungsten in the $YBa_2Cu_3O_{7-x}$ superconductor at small W concentrations, the lattice of the system changes from orthorhombic to tetragonal, without changing the transition temperature to the superconductor state.

When W, Y, Ba and O are combined, one of the phases which can be synthesized have a cubic ordered perovskite structure with the formula $Ba_2Y_{0.75}WO_6$[4], by the other side, when W, Cu, Ba and O are combined with the molar relation W:Cu:Ba: = 1:1:2 one obtains a phase with a perovskite related structure and a tetragonal symmetry[5,6]. Recently, it was reported that when 50 % of the Cu is substituted by Y in this compound, a new phase is formed with a cubic ordered perovskite structure with the space group Fm3m[7].

In this work we report the characterization of the $Ba_2YCu_xWO_y$ system for $0 \le x \le 2$, after several heat treatments, with the basic idea to understand the origin of the superconductivity in the $Ba_2YCu_2MO_{9-y}$ system.

2. SAMPLE PREPARATION

The samples were prepared by solid state reaction. We mixed $BaCO_3$, Y_2O_3, CuO, and WO_3 in the molar ratio Ba:Y:Cu:W = 2:1:X:1 with $0 \le X \le 2$, the samples were calcined 2 hours in air at 800 C and sintered also in air at 1150 C for two to twelve hours. They were regrounded several times. After the sintering the samples were heated 12 hours at 450 C in flowing oxygen in order to improve their electrical properties. Some portions of the samples were quenched in liquid nitrogen from temperatures of 1150 C. The reductions were done in flowing hydrogen at 450 C and 1150 C, cooling the samples in the same atmosphere. Some samples were heated at 1150 C also in vacuum, in flowing oxygen, in argon or in nitrogen, cooling the samples in the corresponding atmosphere.

We have prepared some samples in an electrical arc furnace, but the lack of temperature control originated their fusion, which gives rise to the formation of several phases distinct to those obtained at 1150 C by the solid state reaction.

3. SAMPLE CHARACTERIZATION

The structure of the samples was determined from X-ray diffraction at room temperature. The diffractometer works with CuKα radiation, the measurements were done with an step of 2θ = 0.05 degree for the low resolution and of 2θ = 0.005 degree for the high resolution. The transmission electron microscopy has been carried out in a 400 KV high resolution machine. DTA an TG analysis of the samples were done at temperatures between 25 C and 1200 C at heating and cooling rates of 10 C/minute. The magnetization was measured with a sample vibrating magnetometer at temperatures between 77 K and 300 K in magnetic fields between 1 Oe y 500 Oe. The electrical resistance was measured by the four probe technique in a helium closed cycle refrigerator.

4. RESULTS AND DISCUSSION

Figure 1 shows the X-ray diffractograms for the system $Ba_2YCu_xWO_y$ with $0 \leq x \leq 2$. For x = 2 the diffractogram shows the presence of two phases, one of them is CuO and the other one is a cubic phase with the space group Fm3m and a lattice parameter of 0.834 nm. For x = 0.25 the diffractogram shows only this last phase with the same lattice parameter as for x > 0.25, plus small amounts of spurious phases identified as $BaWO_4$ and Ba_3WO_6, which are originated before the formation of the cubic

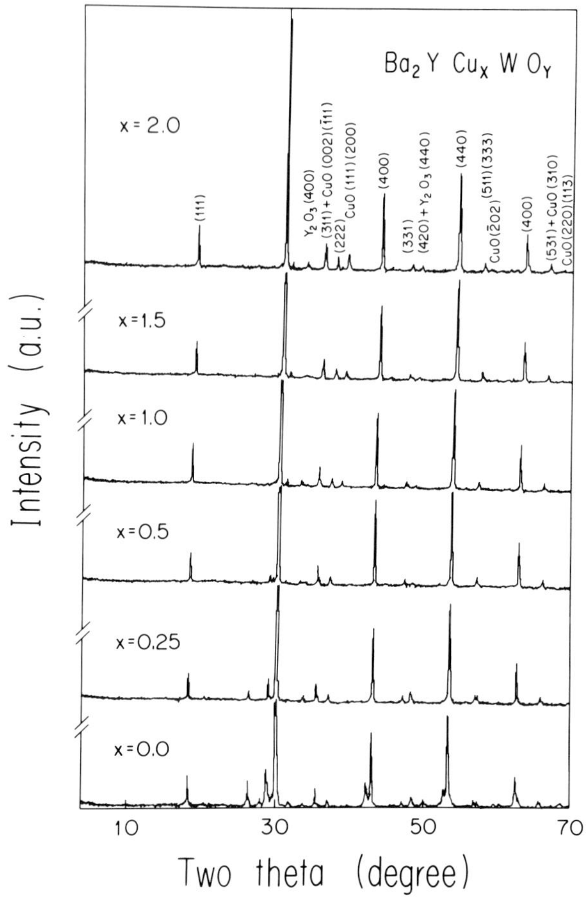

FIGURE 1
X-ray diffractograms of the $Ba_2YCu_xWO_y$ system for different copper concentrations.

phase, they disappear with a longer sintering time. As x grows one observes the presence of CuO. For x = 0 the diffractogram contains only one phase, it is cubic with the space group Fm3m and a lattice parameter of 0.838 nm, the diffractogram shows the same spurious phases observed for x = 0.25. Figure 2 shows a detail of the diffractograms shown in Figure 1. It shows the position of the (422) reflection. From this figure we observe that the lattice parameter of the cubic lattice is shorter for x = 0 than for x > 0 and it does not change with the copper amount. From this observation we conclude that the copper is dissolved in the

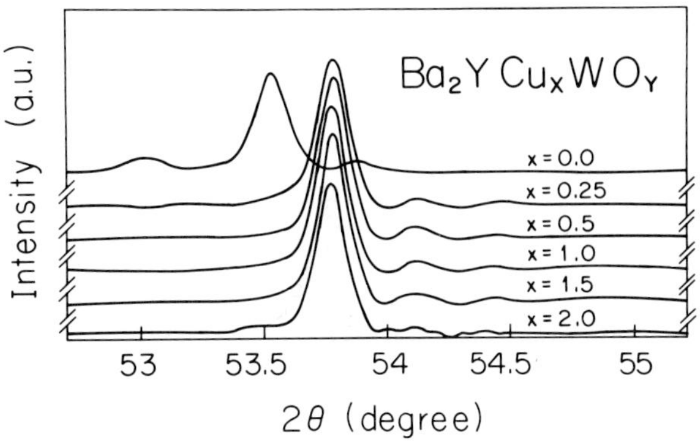

FIGURE 2
(422) X-ray reflection of the $Ba_2YCu_xWO_y$ system for different *x* values.

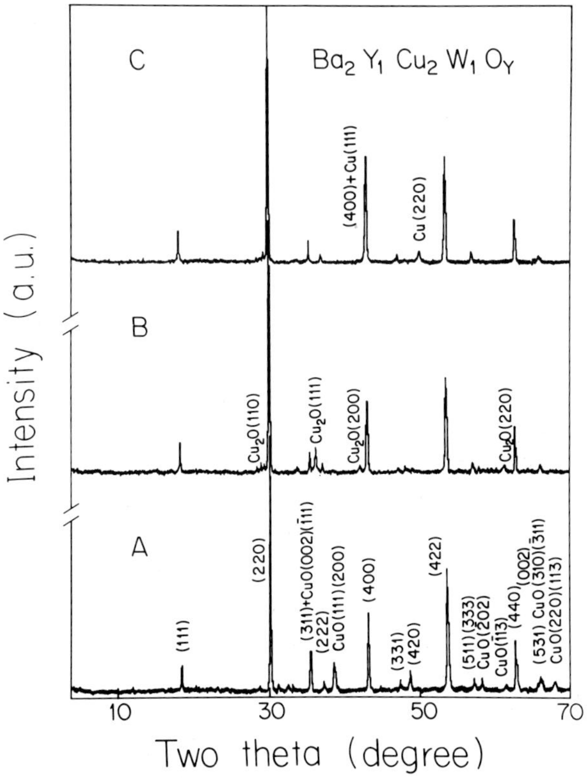

FIGURE 3
X-ray diffractograms of the $Ba_2YCu_2WO_y$ system. A) as prepared. B) After quenching the sample in liquid nitrogen from 1150 C. C) After reducing the sample at 450 C in flowing hydrogen.

basic cubic structure of the compound Ba_2YWO_6, only in a small fraction. Recent results show that the Cu occupies the Y places producing a lattice contraction[7,8].

Figure 3C gives the X-ray diffractogram of the sample with *x* = 2 after its reduction at 450 C in flowing hydrogen. The cubic phase does not change, it has the same lattice parameter of 0.384 nm as before the reduction, that means, that the dissolved copper in this structure stays inside, it is not removed by the reduction at this temperature, in contrast to its behavior when the sample is reduced at 1150 C, as will be seen later. From the diffractogram we observe that the CuO is reduced to metallic copper, and the sample is a good conductor, that means that the copper grains are connected electrically, result which is in agreement with the observation by electron microscopy that the cubic phase grows embedded in a solid CuO matrix (Figure 4). However, after this reduction the sample does not show superconductor behavior.

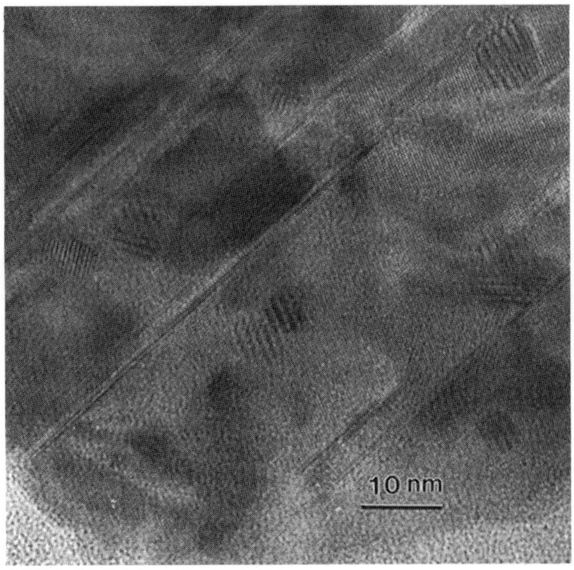

FIGURE 4
High resolution electron micrograph of a superconductor sample of the $Ba_2YCu_2WO_y$ system.

Figure 5 shows the TG analysis of the sample with $x = 2$ in a reducing atmosphere of 80 % N_2 and 20 % H_2, we observe that around 300 C the sample has a large weight loss, that is related to the reduction of the CuO into metallic copper (Figure 3C), the sample does not show any additional change when is heated till 600 C. This weight change corresponds to an oxygen loss of around 1.27 oxygen atoms per formula. TG analysis of the sample in air shows that at room temperature the sample is oxygen deficient with a deficiency of about 0.46 oxygen atoms per formula (Figure 6). That means that at least 1.73 Cu atoms per formula are in the form of CuO. The dynamic conditions involved in the TG analysis and the lack of precise control of the air atmosphere, make impossible to carry out a better determination of the Cu amount in the sample as CuO, however the results give a good idea about the state of the copper in the sample.

Figure 6 shows the TG analysis of the sample in air at temperatures above 900 C, from the figure we observe a loss of weight in the sample which is originated by the transformation of the CuO into Cu_2O, as is observed from the X-ray diffractogram of the samples when they are quenched in liquid

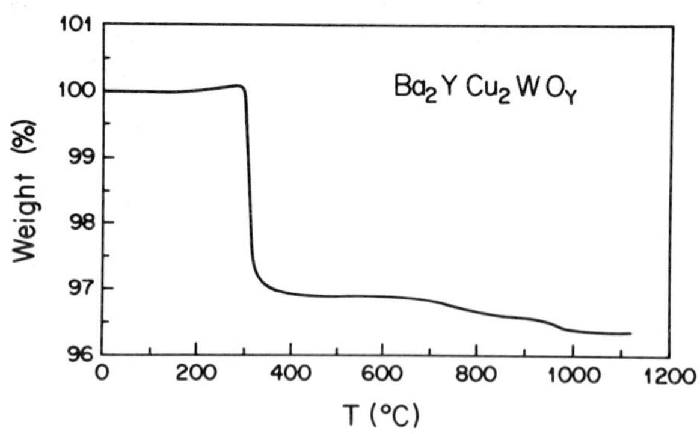

FIGURE 5
TG analysis of the $Ba_2YCu_2WO_y$ system, in an atmosphere of 80 % N_2, 20 % H_2.

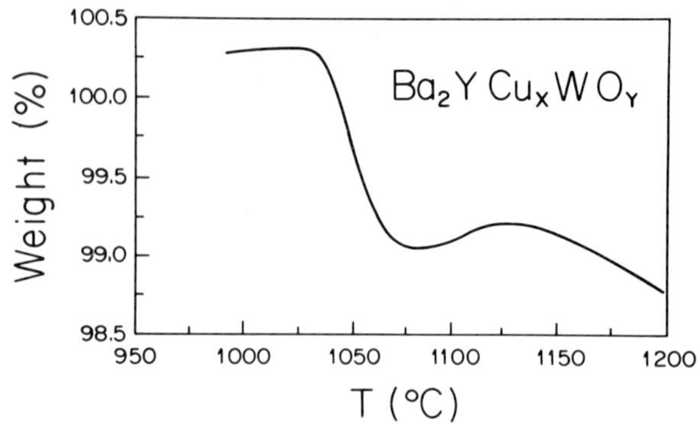

FIGURE 6

TG analysis in air of the $Ba_2YCu_2WO_y$ system.

nitrogen from a temperature above 1050 C (Figure 3B).

From the difference between the calculations of the Cu amount in the form of copper oxide in the sample and the starting nominal composition, together with the observed decrease of the lattice parameter of the cubic structure when the samples contain copper, we conclude that some copper goes into the cubic structure of the Ba_2YWO_6 oxide. Recent results indicate that copper replace yttrium atoms[7,8]. In the next paragraph we give additional experimental evidence that reinforces this proposition.

From the TG analysis of the sample with x=2 in a reducing atmosphere (Figure 5), we observe that the samples loss weight at temperatures higher then 700 C. Starting from this fact we reduced the samples for all x values in flowing hydrogen at a temperature of 1150 C during one hour. Figure 7 gives the X-ray diffractograms of the reduced samples for all x values. The structure is the same for all the samples, and it is almost the same as that obtained for x = 0 without reduction (Figure 1). Figure 8 gives a detail of the diffractograms for the (422) reflection, from the figure we observe that the reflection is broader for the samples with $x > 0$, which is explained by the generation of vacancies in the structure produced by the delivery of the copper from the structure when the sample is reduced at 1150 C. The reduction produces metallic copper, which fuses at 1083 C, and segregates from the sample in the form of metal grains. This result explains also the observed darker black color of the samples with $x > 0$ in comparison to the samples without copper when they are reduced at high temperatures.

We analyzed in air the samples for all x values by TG and DTA analysis, together with samples of $YBa_2Cu_3O_7$ and CuO, in order to try to obtain a correlation between them and to explain the origin of the superconductivity in our system. Figure 9 shows a composition of the DTA curves, from the figure it is difficult to

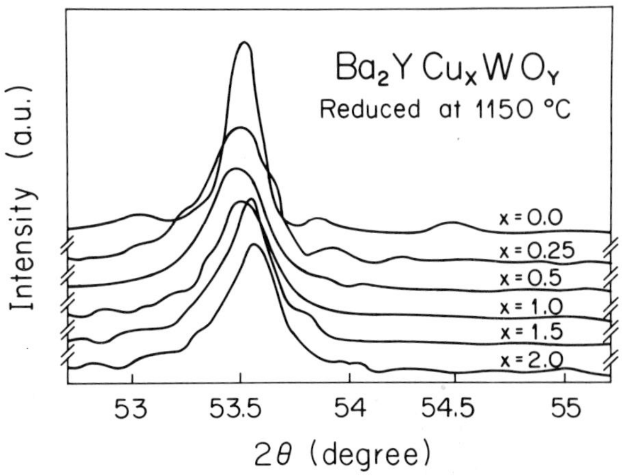

FIGURE 8

(422) X-ray reflection of the $Ba_2YCu_xWO_y$ system for different x values, after reduction of the samples at 1150 C in H_2.

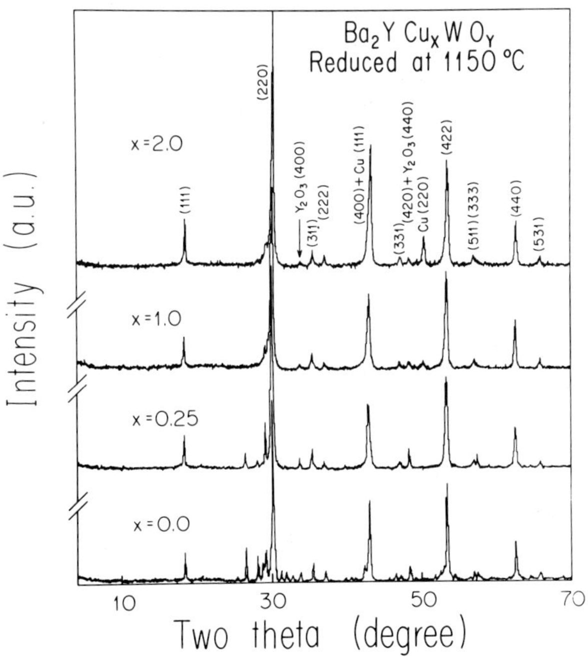

FIGURE 7

X-ray diffractograms after reduction in H_2 at 1150 C of the $Ba_2YCu_xWO_y$ system, for different copper concentrations.

find any correlation between them, however we see that this technique is more sensible to detect small quantities of copper than the X-ray diffraction technique. From the TG analysis it is also not possible to obtain any correlation. Therefore, in order to proof or to discard that the superconductivity in our system is originated by small quantities of $YBa_2Cu_3O_7$ it is necessary to search for new experiments different to those described in this paper.

5. CONCLUSIONS

We have observed that the $Ba_2YCu_xWO_y$ system with $0 \le x \le 2$ is composed of two different phases, one of them is copper oxide and the second one is a phase whose structure is a cubic ordered perovskite, given by the formula Ba_2YWO_6 with some copper dissolved in the structure. The grains of the cubic phase are embedded in the grains of the CuO. TG analysis shows that at least 1.73 copper atoms per formula are in the form of copper oxide. Reduction of the samples in flowing hydrogen at

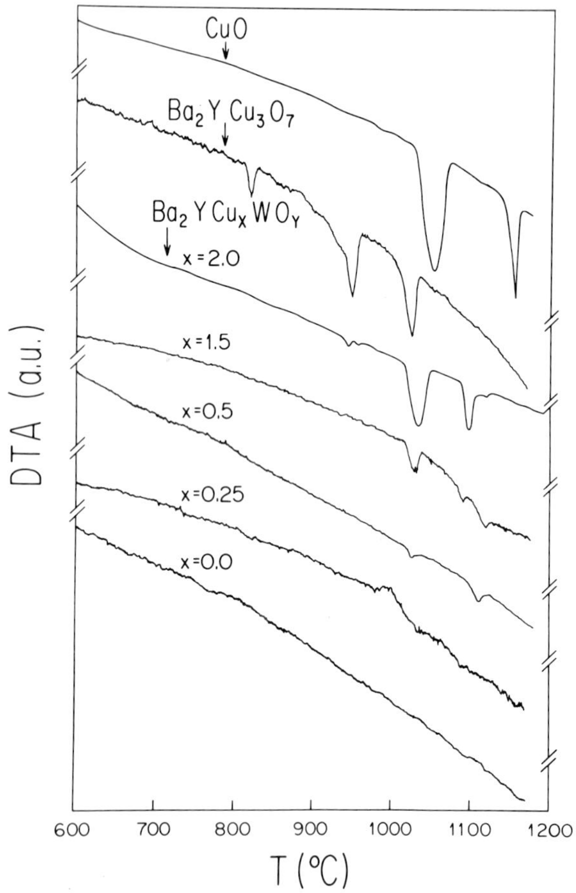

FIGURE 9
DTA analysis in air of the CuO, $YBa_2Cu_3O_7$, and $Ba_2YCu_xWO_y$ systems for $0 \le x \le 2$.

1150 C delivers the copper in the cubic phase, producing lattice distortion. TG and DTA analysis of samples of $YBa_2Cu_3O_7$, CuO, and $Ba_2YCu_xWO_y$ does not show any correlation that could give information about the origin of the superconductivity in our system. That means that new experiments are necessary in order to proof of to discard the possibility that the observed superconductivity in our samples is originated by small amounts of $YBa_2Cu_3O_7$ superconductor.

ACKNOWLEDGMENTS

We thank to A. Morales, M. Aguilar, A. Sánchez for technical support. This work was financial supported by the " Programa Universitario de Superconductores de Alta Temperatura", UNAM, México and by the CONACyT, México under contracts No. P228CCOX891596 and A128CCOE900779-(FI-7).

REFERENCES

1. Bokhimi, and A. García-Ruiz, to be published in Proc. of the 1989 Fall Meeting of the MRS, Boston, Mass. USA.

2. K. Kitazawa, and R. Jarvinen, Superconductor Week 4 (1990) 1.
K. Kitazawa, private communication.
Bokhimi, to be published in Proc. XI winter meeting, Cocoyoc Mor. Mexico celebrated from 14-17 January 1990.

3. Y. Matsumoto, and F. Irie, Jap. J. Appl. Phys., in print.

4. V. L. Balashov, L. N. Lykova, L. M. Kovba, and A. A. Evdokimova, Russ. J. Inorg. Chem. 30 (1985) 1210.

5. A. G. Kapyshev, V. V. Ivanova, and Y. N. Venevtsev, Sov. Phys. Doklady 11 (1966) 195.

6. G. Blasse, J. Inorg. Nucl. Chem. 27 (1965) 993.

7. I. Brynste, submitted to Acta Chem. Scandinavica, 1990.

8. Bokhimi, to be published.

Frontiers of Materials Research/
Electronic and Optical Materials
M. Kong, L. Huang (editors)

PREPARATION AND PROPERTIES OF Ag-SHEATED Bi(Pb)-Sr-Ca-Cu-O COMPOSITE WIRE

Suihua WU, Zhenguo CHENG, Xiaozu WU, Lian ZHOU, Pingxiang ZHANG
Zhengping XI, Huiling ZHENG, Ping JI, Yongchai YE

Northwest Institute for Nonferrous Metal Research
P.O. Box 71, Baoji, Shaanxi 721014, China

We have prepared the flexible Ag-sheated Bi(Pb)-Sr-Ca-Cu-O superconducting composites by the powder-in-tube process. The critical current density Jc has reached 5000 A/cm^2 (77K,0T), and the dependence of Jc on magnetic field is reduced to some extent. The improvement of the critical current density is considered due to the grain alignment and well connection between grains. Using these composites we have succeeded in making gradient coils of magnetic flux tranformers. Some results of bending test are also presented in this paper.

1. INTRODUCTION

The Bi-Sr-Ca-Cu-O superconductor hasn't only higher transition temperature and upper critical field, but also insensitive to oxygen and water in atmosphere[1]. These features are important to practical use. The substitution of Pb for part of Bi in Bi-based system was found very effective on the formation of 110K superconducting phase[1-2]. The powder in tube method offers a simple way to produce high temperature superconducting composite wires. The present paper describes the fabrication and superconducting properties of Ag-sheated Bi(Pb)-based composites.

2. EXPERIMENTAL PROCEDURE

The Bi(Pb)-based powders with nominal composition of $Bi_{1.6}Pb_{0.4}Sr_2Ca_2Cu_3O_y$ were produced by calculating the mixture of Bi_2O_3, PbO, $SrCO_3$, CuO, $CaCO_3$ powders at 800-830°C for 40-90 hours in air. The bulk samples were then reground and packed into a silver tube of 8 mm outer diameter and 7 mm inner diameter. The tube was cold rolled, annealed and cold drawn into wires of 0.7-1.0 mm outer diameter, and finally sintered at 820°C-850°C for 30-90 hours in air. These wires were used to make the gradient coil of flux transformer. The microstructures of the composites were investigated by SEM observation and X-ray diffraction analysis. The critical current densities (Jc) were measured by standard dc four-probe method with 1μv/cm criterion.

3. RESULTS AND DISCUSSION

During the process of the composite fabrication, the grains of Bi-based material in silver tube can be moved and rotated by cold working, and aligned with the working direction. As the sheet plane of Bi-based grain is parallel to the conductive Cu-O plane[4], the alignment of grains made by cold working can improve the transport current density. Fig.1 shows the typical fracture surface of a Bi-based composite tape. The grains in the sample are oriented along the lengthwise of the tape, and the density is higher than that of ordinary sintered bulk material. X-ray diffraction analysis using CuKa radiation was taken on the wide surface of the tape after removing Ag-sheath. The diffraction pattern (Fig.2) indicates that the composite was mainly composed of high Tc (110K) phase. The considerable enhancement of (001) peaks shows that the grains of 110K phase were oriented with the C-axis perpendicular to wide plane of the tape.

The superconducting properties are very sensitive to the heat treatment temperature. A long superconducting composite wire of 0.7mm diameter was cut into short specimens, and rolled into tapes with 0.15mm thickness, then sintered at different temperature in air. Fig.3 shows the dependence of critical current density on the annealing temperature. The temperature 830°C is appropriate to the tape heat treatment. When the temperature is higher than 850°C, the Bi(Pb)-based core is partially melted. EDAX analysis indicates that loss of Pb is about 86-96 percent, and the contents of Cu and Ca are also decreased. The composites are, therefore, nonsuperconducting.

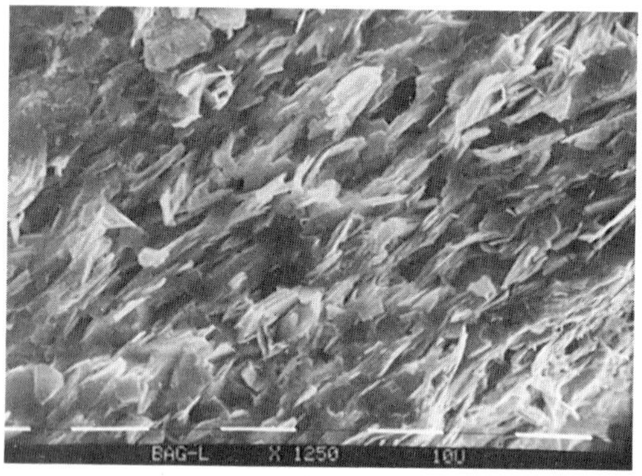

FIGURE 1
Fracture surface of Ag-sheathed Bi(Pb)-based tape

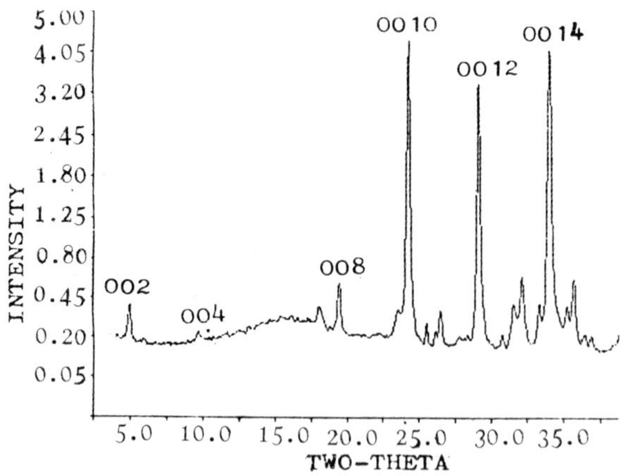

FIGURE 2
X-ray diffraction pattern of Bi(Pb) based composite tape after removing Ag-sheath

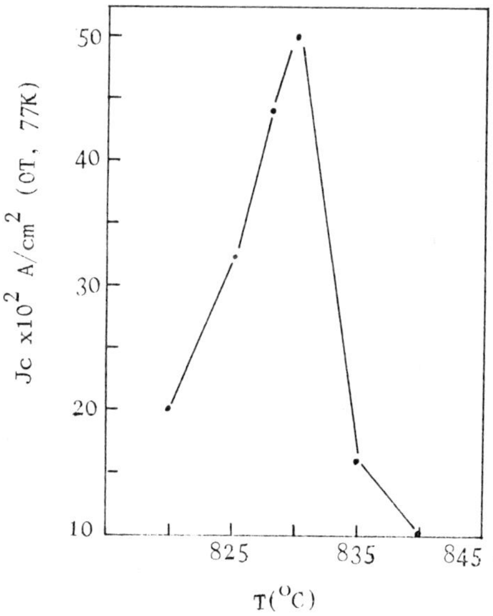

FIGURE 3
The dependence of Jc (77K,0T) on sintering temperature

Cold working results in high density and well-oriented grains. Moreover, proper heat treatment improves the contacts of grains. A wire of 0.7mm outer diameter was rolled into a tape of 0.15mm thick, and annealed in proper condition, the Jc values increased from 960 A/cm^2 (77K,0T) to 5000 A/cm^2 (77K,0T). Fig.4 shows the dependence of Jc on magnetic field. The decrease of the Jc in magnetic field is reduced to some extent compared with the Ag-based YBCO tape[3]. The improvements of current transport capacity in the Ag-sheathed Bi(Pb)-based composite are mainly due to the alignment of the grains. Since the weak links at grain boundaries are not overcome by solid state diffusion reaction in the Bi-based composite, the dependence of Jc on magnetic field is stronger than that of MTG YBCO.

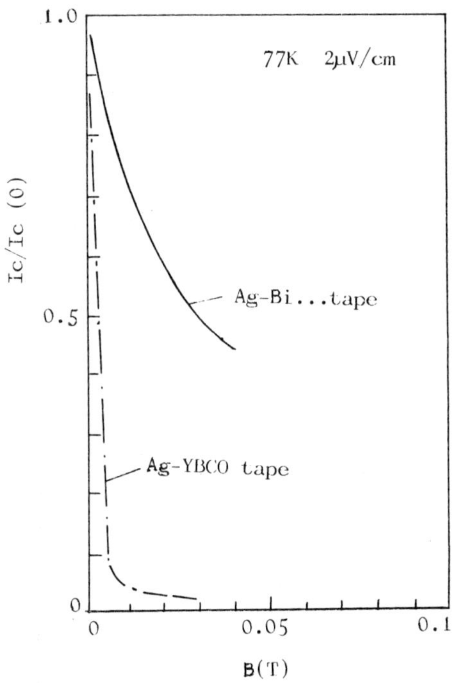

FIGURE 4
The dependence of Jc on magnetic field

The bending character of superconductor is important to the practical use. The alignment of Bi(Pb)-based sheet grain made the composite wires easy to bend. When a wire of 0.7mm diameter was bent to zigzag shape with the angle of 90 degree, the Jc was decreased from 960 A/cm^2 (77K) to 300 A/cm^2 (77K,0T). This kind of wire with above properties can be used to prepare some superconducting devices which do not demand high critical current density. For example, we have succeed in producing magnetic flux transformer using the wire of 0.7mm-1.0mm diameter and 1.5 m length (Fig.5)

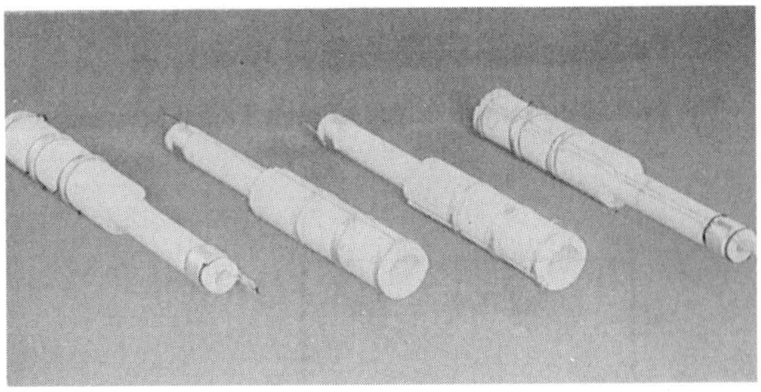

FIGURE 5
High Tc S.C. magnetic flux transformer

It is found that the device has a good receiving and transfering properties for magnetic flux.

4. CONCLUSION

The Ag-sheathed Bi (Pb)-Sr-Ca-Cu-O superconducting composite can be fabricated by means of cold working plus correlated annealing. The superconducting properties are sensitive to the annealing condition. Both Jc value and bending property have improved by grain alignments which were formed in the process of high degree of cold working. Using these composite wires, the magnetic flux tranformers have been successfully fabricated.

REFERENCES

1. M.Takano, J.Takada, K,Oda, H.Kitaguchi, Y.Ikeua, Y.Tomii and H.Mazaki, Jpn. J.Appl. Phys. (1988) 27 L1041-L1043

2. U.Balachandran, D.Shi, D.I.Dos Sanlus, S.W.Graham, M.A.Patel, B.Tani, B.Tani, H.Claus and R.B.Poeppel, Phys. C (1988) 156 649-651

3. M.Okada, A.Okayama, T.Morimot, T.Matsumot, K.Aihara and S.Matsuda, Jpn. J.Appl. Phys. 27 (1988) L185

4. J.M.tarascon, W.R.Mckinnon, P.Barbox, D.M.Hwang, B.G.Bagley, L.H.Greene, G.Hull Y.LePage, M.Sotaffel and M.Giroud, "Preparation, Structure and Properties of the Superconducting Compound Series $Bi_2Sr_2Ca_{n-1}Cu_nO_y$ with n=1,2 and 3." Preprint.

Frontiers of Materials Research/
Electronic and Optical Materials
M. Kong, L. Huang (editors)

MICROSTRUCTURE OF YBaCuO FILMS GROWN ON MgO BY MAGNETRON SPUTTERING.

M.Sidorov, S.Oktyabrsky

P.N. Lebedev Physical Institute, Department of Solid State Physics, Leninsky Prospect 53, Moscow 117924, USSR.

A microstructure of the YBaCuO films grown on (001)MgO substrates by magnetron sputtering has been examined by plan and cross-sectional TEM. The films appear to be crystalline consisting of three main types of epitaxial domains 90° misoriented from each other with their c-axes parallel to three <100>MgO directions. This domain structure results from a growth rate anisotropy which depends on the growth conditions. The domain boundaries contain neither separate boundary phases nor amorphous layers. A large density of (001) planar stacking defects is seen in domains with c-axis parallel to the substrate surface (a,b-type). These very domains exhibit difficiency of yttrium and abundance of copper and barium as established by X-ray microanalisis. The film with the higher density of the a,b-domains of a smaller size has shown the lower T_c .

1. INTRODUCTION

A real structure of the high T_c superconductor (HTSC) films influences greatly the most important physical properties such as critical temperature (T_c), critical current density (J_c), transition width etc. Moreover, the absence of clear understanding of the HTSC mechanism makes it necessary to distinguish the effects due to fundamental properties of the material from the phenomena owing to the microstructure. We report here the results of $YBa_2Cu_3O_{7-x}$ (hereafter referred to as YBaCuO) film microstructure investigation performed by a transmission electron microscopy (TEM).

2. EXPERIMENT

The films are prepared by dc magnetron sputtering in the oxygen-argon mixture from YBaCuO ceramic pellets on the mechanically polished (001)MgO substrates at a substrate-holder temperature of $T_{substr} \approx 680^oC$ without postgrowth annealing at a higher temperature. The deposited films thickness was about 200 nm. As-grown films exhibit zero resistivity at about 86K and a critical current density $J_c \approx 10^5$ A/cm^2 at 77K. A detailed preparation techniques are described elsewhere[1].

The microstructure of the films were investigated by a plan and cross-sectional TEM and a high resolution TEM at 300 keV on PHILIPS CM30/(S)TEM equipped with a X-ray energy dispersive spectrometer (LINK AN10000) for the compositional analysis. The samples for the TEM observation were prepared by the ordinary method without using a water. Ion-milling was performed using 5keV Ar^+ beam at an incident angle of 15 degrees.

3. RESULTS AND DISCUSSION

A typical low-magnification bright-field plan-view image of the YBaCuO film is shown in Fig.1a. It is easy to dis-

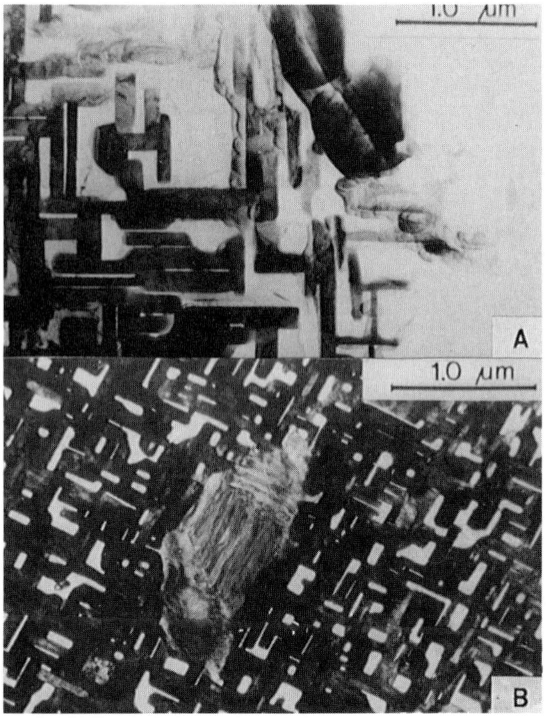

FIGURE 1
A typical TEM plan-view images of YBaCuO films grown on MgO substrates at different substrate-holder temperatures: (a) T_{substr} 690 C, (b) T_{substr} 680 C

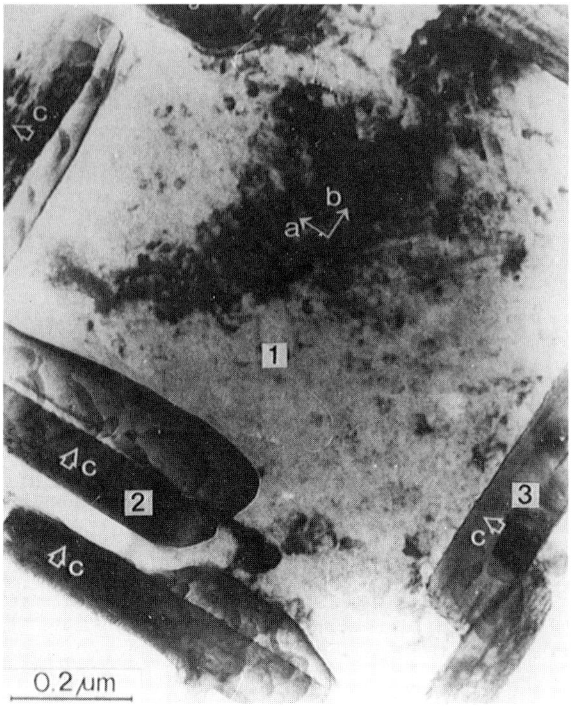

FIGURE 2
Three main types of epitaxial domains in the YBaCuO films (1- c-type, 2- a-type, 3- b-type).

tinguish three main types of single crystalline epitaxial domains 90° misoriented from each other with the c-axis parallel to three equivalent <100>MgO axes. The film domain structure observed by TEM occurs as bar-shaped blocks with c-axis of YBaCuO parallel to the substrate surface (a,b-domain type) of approximately 100x500 nm size immersed in a media with c-axis perpendicular to the substrate surface (c-domain type) (Fig.2). Electron microscopy cannot allow to distinguish [100] from [010] direction in YBaCuO without special methods[2,3] due to small differences in their lattice parameters and further we shall not draw a line between a- and b-types.

It is necessary to note that the MgO substrates have a rather large lattice mismatch (~ 9%) with YBaCuO, MgO has a cubic unit cell with a= 0.42nm and YBaCuO is orthorombic with a≈ b≈ 0.39nm, c≈ 1.17nm. In spite of this fact the epitaxial growth of YBaCuO on MgO is surprisingly successful. Figure 3 shows a typical cross-sectional image of this very film, a,b-oriented domains appear as a pillar and these domains spread from the bottom to the top of the film. Selected area diffraction pattern shows distinct diffraction spots corresponding to two misoriented domains and the substrate. The mutual position of these spots confirms orienting effect of the

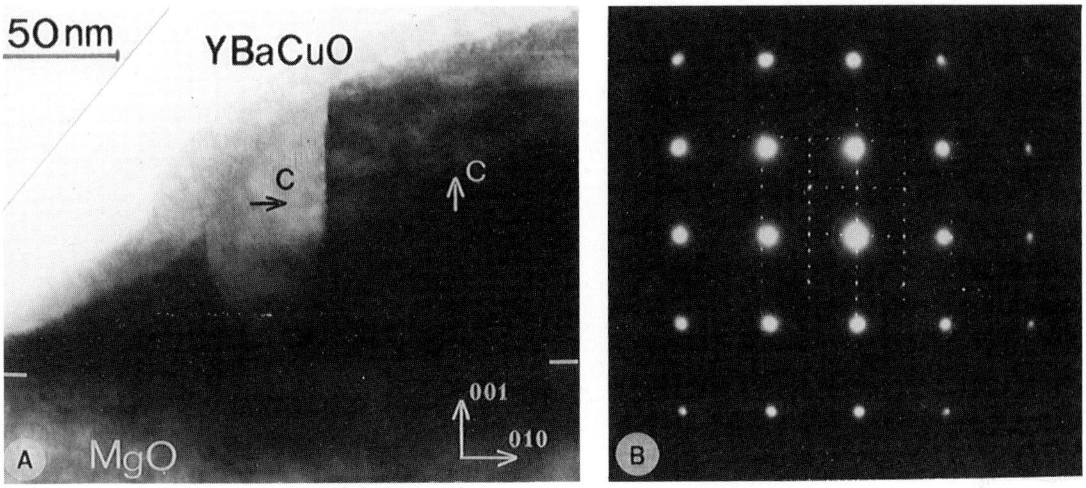

FIGURE 3
(a) Cross-sectional TEM image of YBaCuO film on MgO substrate. The a,b-oriented domain immersed in the c-domain is visible. (b) Selected area diffraction pattern of two domains and the substrate.

MgO substrate (Fig.3).

X-ray diffraction studies of similar films[1] indicate that they have a perfect single crystalline structure with (001) orientation, but it is worth mentioning that analysis of the thin layers of YBaCuO by X-ray diffraction is weakly sensitive to the existence of the a,b-oriented domains[2,4,5]. For example the film shown in Fig.1a appears to be fully c-oriented as established by X-ray diffraction while a,b-oriented domains have been detected only in the films with some higher volume contents of such domains (Fig.1b).

The morphology of the grown films discussed above corresponds to the anisotropy of the growth rate of the YBaCuO[5]. From the observed aspect ratio of the a,b-domains it follows that the growth rate along c-axis is about 5-10 times slower than the growth rate along the perpendicular directions to the c-axis. The aspect ratio of the domains in Fig.1a is 8-10 while in Fig.1b is 5-6. This difference is due to difference in the substrate temperatures- 690°C in the first case and 680°C in the second case. Thus the increase of T_{substr} leads to the reduction of the critical nuclei concentration and acceleration of the surface diffusion of the components to the growing steps. Consequently these stimulate the tangential growth of the (001) planes of YBaCuO.

Hence, the anisotropy of the growth rate is in fact responsible for the increase of the volume portion of the a,b-domains in the thicker films. It is important to notice that the disorientation of the YBaCuO films far away from the interface[6,14] and the domination of the a,b-orientation in the relatively thick films (>250-400nm) are repeatedly

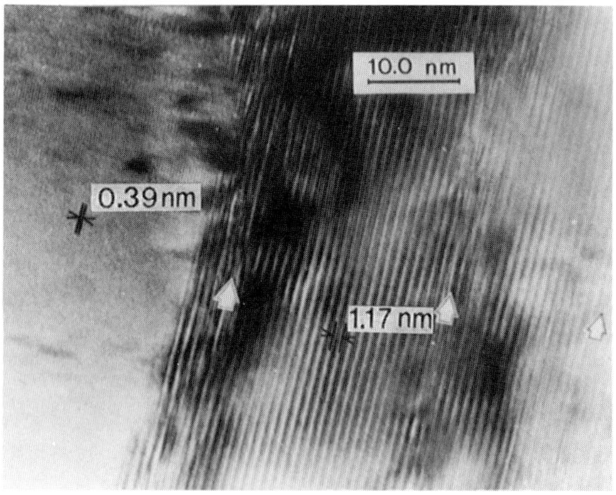

FIGURE 4
High resolution image of the grain boundary between c- and a,b-oriented domain. The arrows indicate planar stacking defects in a,b-domain.

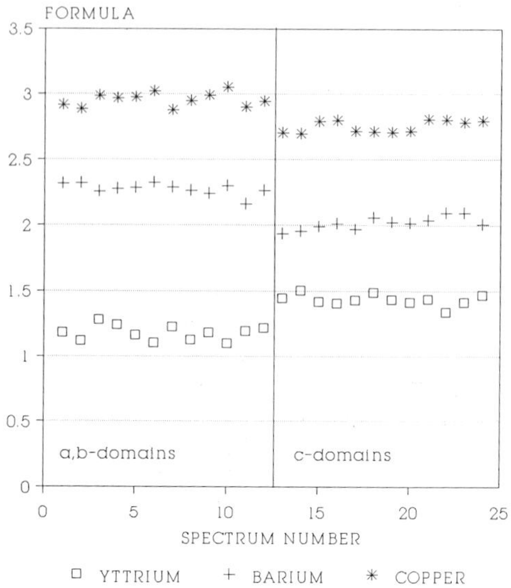

FIGURE 5
The results of X-ray microanalisis of 24 different domains of different orientations.

observed in the X-ray diffraction measurements[7,14].

To our opinion these results may be qualitatively explained by the anisotropy of the growth rate and it is important to notice that a,b-oriented nuclei appear on all other substrates, including $SrTiO_3$ and lead to the degradation of the c-orientation. On the other hand the anisotropy of the growth rate decreases with diminishing of the growth temperature and disappears when $T_{substr}<650^oC$[8].

This growth mechanism, with fast growth along [hk0] directions, promotes the incorporation of planar defects parallel to the (001) plane[4] which are clearly seen in Fig.4. The existence of these planar defects is explained by insertion of an extra CuO layer thus forming $(CuO)_2$ double layer between the two BaO layers[3,9]. In the case of our films the X-ray compositional microanalisis confirms the abundance of the copper in the a,b-domains. The results of the microanalisis of the domains with the different orientations are summarized in Fig.5. Here the relative cation concentrations are shown. Calculations are performed by RTS (Ratio Thin Section) method from characteristical X-ray spectra. The measurements were produced in 24 different domains with different thicknesses. There is obviously the abundance of Cu and Ba in a,b-domains in comparison with c-domains. We believe that the higher concentration of Cu-atoms results from the extra CuO planes-planar defects. As far as extra Ba is concerned a,b-domains have a rather worse cristallinity (Fig.6) which is responsible for a high density of Ba-rich defects, for example substitutional defects observed in Ref.10. Furthermore, changing of growth conditions (temperature, components ratio etc.) one can

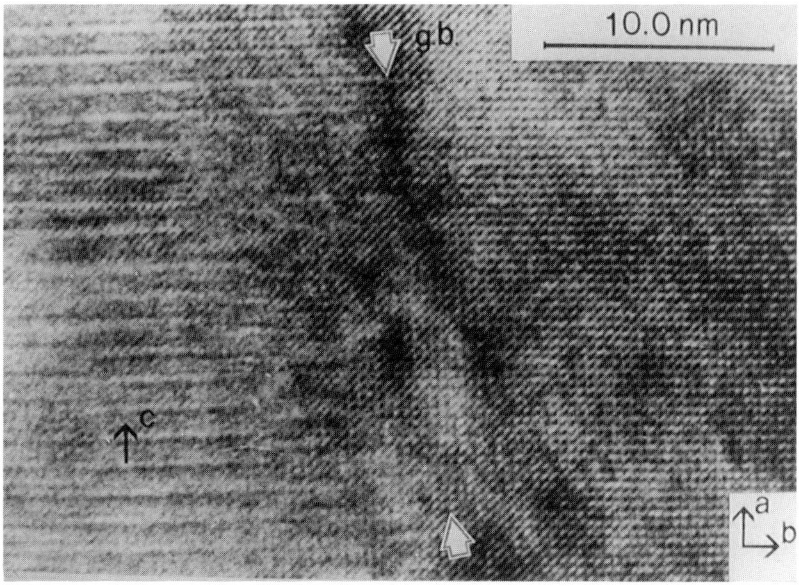

FIGURE 6
High resolution image of the grain boundary between c- and a,b-oriented domains. The grain boundary indicated by arrows.

move the phase diagram of the YBaCuO towards the predominant condensation of either one or another orientation. Hence, the underestimation of peculiarities of the microstructure may result in the errors when studying the dependence of the physical properties upon the technology parameters.

Some of the well-known (110) mirror twins of 30-50 nm in width occur in the c-oriented domains. Such twins are formed to accommodate the strain energy of the tetragonal-to-orthorhombic transition.

The main feature of the films investigated is a lot of boundaries formed by 90° misoriented domains. These boundaries are coherent and free of any second phase and other decorations (Fig.4,6). It is worth noting that the boundary in the image is not sharp because the domains are not confined by high-symmetry crystalline planes. As it is has been already shown in cross-sectional view (Fig.3) the domain boundaries are smoothly inclined owing to the anisotropy of the growth rate.There are no other boundaries in these films which contains impurity phases while even in the best of the bulk samples their fraction is about ~10-25% [9,12]. Also, there is no precipitation like Y_2BaCuO_5, Y_2O_3, $BaCuO_3$ etc in the film volume, which we have observed in the films grown by ion-beam and laser deposition like other authors have [13,14].

4. CONCLUSION

In summary, we have investigated the microstructure of the YBaCuO films grown on (001)MgO substrates by magnetron sputtering. The films appear to be crystalline consisting of three types of epitaxial domains 90^0 misoriented from each other with their c-axes parallel to three equivalent <100> directions of MgO. Such a morphology results from the anisotropy of the growth rate which depends on growth conditions such as temperature, stoichiometry etc. The main differences of the films grown at different substrate-holder temperature are shown in table 1. A number of 90^0 mis-

TABLE 1
Main differences of the films grown at different substrate-holder temperature.
* As established by X-ray diffraction.

T_{substr}	680^0	690^0C
T_a	85K	86.5K
Volume portion of the a,b-domains	>	<
Orientation*	a,b,c	c
Aspect ratio of the a,b-domains	5-6	8-10
Mean size of the domains	100x500 [nm]	100x800 [nm]

oriented domains creates a number of boundaries, however these boundaries are free of any second phase and other decorations. There are no any extraneous phases segregation elsewhere in the films but the available crystalline defects and chemical fluctuations are distributed over the whole film volume, especially in the a,b-domains, which contains many of Cu-rich planar defects as well as abundance of barium.

ACKNOWLEDGEMENTS

The authors thank Dr. E.V.Pechen and his group for film preparation and electrical measurements.

REFERENCES

1. A.I.Golovashkin, E.V.Ekimov, S.I. Krasnosvobodtsev, V.P.Martovitsky, E.V. Pechen, in the Proceedings of the M2S-HTSC,Stanford, (1989).

2 H. Shibahara, J.P.Zhang, L.D.Marks, J.Electron Microscopy, 38, No.4, (1989), p.314.

3. H.W.Zandbergen, R.Gronsky, K.Wang, G.Thomas, preprint of Lawrence Berkeley Laboratory, LBL-24671, (1987).

4. D.M.Hwang, T.Venkatesan, C.C.Chang, L.Nazar, Appl.Phys.Lett. 54, 17, (1989), p.1704.

5 Siu-Wai Chan, D.M.Hwang, R.Ramesh, S.M.Sampere, L.Nazar, AIP Conference Proceedings, No.200, (1990), p.172.

6 Escribe-Fillipini, P.L.Reydet, J.Markus, M.Brunel, J. Less-Common Metals, 151, (1989), p.263.

7. A.F.Marshall, R.W.Barton, K.Char, A.Kapitulnik, et. al.Physical Review B, 37, No.16, (1988), p.9353.

8. J.Chang, M.Nakajima, K.Yamamoto, A.Sayama, Appl. Phys. Lett., 54, No.23, (1989), p.2349.

9. H.W.Zandbergen, G.Thomas, Phys. Stat. Sol.(a), 107, (1988), p.825.

10.G.Van Tendeloo, H.W.Zandbergen, T. Okabe, S.Amelinckx, Sol.Stat.Comm., 63, (1987), p.969.

11.M.Inoue, E.Takase, Jap.J. of Appl. Phys., 28, No.9, (1989), L1578.

12.S.E.Babcock, T.F.Kelly, P.J.Lee, J.M.Seuntjiens, et. al. Physica C, 152, (1988), p.25.

13.L.T.Romano,P.R.Wilshaw, N.J.Long, C.R.M.Grovenor, Supercond.Sci.Technol. 1, (1989), p.258.

14.F.E.Luborsky, R.F.Kwasnik, et. al. J.Appl.Phys., 64, No.11, (1988), p.6388.

Frontiers of Materials Research/
Electronic and Optical Materials
M. Kong, L. Huang (editors)

THE PERCOLATION PROCESS IN HIGH Tc SUPERCONDUCTOR

Maosheng Zheng, Xiaoran Liu, Enze Luo, Tianshe Zhu

Phys.Dept. Xidian University, Xi'an, 710071, China

The percolation process in high Tc superconductor $Ba_xY_{1-x}Cu_{3-y}$ system is fully studied by site–bond percolation model in the present paper. It is shown that, there are two superconductivity phases in this system. The first one is the well–known $Ba_2YCu_3O_{9-\delta}$ with the initial turn temperature 100K, and the second is the phase containing more YCuO and less Ba with the initial turn temperature 60K. Furthermore, It is found that the effective methods to increase the critical current density and the zero resistance temperature are as follows, 1, decreasing the average junction resistance of normal state; 2, increasing the fraction of superconductor phase; 3, increasing the initial turn temperature Tco.

1. INTRODUCTION

Although the statement that the tial phenomenon of R–T curve in high Tc superconductor implies a percolation was pointed by J.G. Bednorz and K.A. Müller early in 1986[1], this phenomenon has not been fully discussed up to now. In the present paper, we try to conduct this problem by using site–bond percolation model.

As is known, the high Tc superconductor system consists of " superconductor grains" (S.C.G. for the sake of simplicity) and " non superconductor grains" (N.S.C.G. for the sake of simplicity) randomly, therefore, it's superconductivity has the property of " random site–bond percolation" . The S.C.G. and the N.S.C.G.take the part of sites randomly, while the coupling between two S.C.G.behaves as bond. With the decreasing of temperature, the "S.C.G." enters "S.C." state first, further, the "S.C.G." couples through the Josephson junction between them, while the coupling energy of a junction is greater than that of the destruction of heat. When the total coupling probability in the system is greater than the percolation threshold, an infinite S.C.G. cluster forms, and the bulk resistance becomes zero, this is the rule.

Because the theory of random site–bond percolation has not been established, the ordered site–bond percolation theory developed lately[2] is used to deal with this problem approximately. In the ordered site–bond percolation process, in order to form an infinite percolation cluster in the system, the relationship of phase boundaries between site probability P_S and critical bond probability P_{bc} may hold,

$$P_{bc} \cdot (A \cdot P_S + B) = C \quad (1)$$

where, A and B are constants depending on the microstructure of system, and C is a constant depending on the topological dimension of the system[2]. In general, A and C are positive.[2]

From eq.(1), it can be seen, that when P_S decreases, P_{bc} will increase to keep the infinite percolation cluster exist. In this paper, We substitute the face centre cube for the random dense packing of S.C.G. and N.S.C.G.. For F.C.C. structure, A = 13.385, B = –1.2, C = 1.45.[2]

If the $Ba_2YCu_3O_{9-\delta}$ is the only superconductor phase in $Ba_xY_{1-x}Cu_{3-y}$ system, then, for $X < \frac{2}{3}$, we have, the site probability of S.C.G. phase,

$$P_S = \frac{x}{[2(1-x)]} \quad (2)$$

and for $X > \frac{2}{3}$,

$$P_S = \frac{(1-x)}{(2x-1)} \quad (3)$$

By assuming the distribution of the junction resistance between S.C.G. in the normal state is a mormal distribution, the problem is solved as follows.

2. PERCOLATION PROCESS

Generally, the normal distribution of a quantity x with the average x_0, the square difference (S.D.)δ, is written as

$$f(x) = \frac{1}{\sqrt{2\pi}\,\delta} \exp\left[-\frac{(x-x_0)^2}{2\delta^2} \right] \tag{4}$$

and it is normalized in $(-\infty, +\infty)$. But, if $\delta / x_0 < 1$, then [3],

$$\int_{-\infty}^{0} f(x) \cdot dx \ll 1 \tag{5}$$

therefore , for the first order approximate, if $\delta < x_0$ there is,

$$\int_{0}^{\infty} \mathbf{f(x) \cdot dx} \approx 1 \tag{6}$$

Therefore, the normal distribution function of resistance is

$$F(R) = \frac{1}{\sqrt{2\pi}\, R_{S.D.}} \exp\left[-\frac{(R-\overline{R})^2}{2 \cdot R_{S.D.}^2} \right] \tag{7}$$

where, $R_{S.D.}$ and $\overline{R}$ are the square difference and the average value of junction resistance in $Ba_xY_{1-x}Cu_{3-y}$ system, respectively.

If $R_{S.D.} < \overline{R}$, then,

$$\overline{R} \approx \int_{0}^{\infty} \mathbf{R}\cdot F(R) \cdot dR \tag{8}$$

The total probability for R from 0 to R_x is,

$$P_b = \int_{0}^{R_x} F(R) \cdot dR \tag{9}$$

For a system with different fraction of S.C.G. or site probability P_s, the critical bond probability P_{bc} (p_s) can be derived from eq.(1). Let p_b in eq.(9) equal to P_{bc} (P_s), then the relationship between R_x and P_s can be derived, i.e.,

$$R_x = R_x(\overline{R},\ R_{S.D.},\ P_S) \tag{10}$$

If the junction with resistance R_x is conducting, then, the junctions with the resistance less than R_x are all conducting.

In order to keep the junction with resistance R_x conducting, the coupling energy of this junction E_J may be greater than the distruction energy of heat, $3/2K_b \cdot T$, i.e.,

$$E_J \geqslant \frac{3}{2} K_b \cdot T \tag{11}$$

In which, E_J is the Josephson coupling energy, and,

$$E_J = \frac{\pi}{4} \cdot \frac{K}{e^2 \cdot R} \Delta(T) \tanh \frac{\Delta(T)}{2K_b \cdot T} \tag{12}$$

where $\Delta(T)$ is the energy gap function. For $Ba_xY_{1-x}CUO_{3-y}$ high Tc S.C., the values of $\Delta(T)$ in Ref.4 is used.

Substituting eq.(10) into eq.(11) and eq.(12), the zero resistance temperature can be obtained for definite $\overline{R}$, $R_{S.D.}$, and P_S.

According to A.Barotoff[5], the current density of a single junction is,

$$j_{co} = \frac{\pi}{2} \cdot \frac{\Delta(T)}{e \cdot R \cdot \delta} \tanh\left(\frac{\Delta(T)}{2K_b \cdot T} \right) \tag{13}$$

For high Tc S.C., the area δ of the junction is about $2.5 \times 10^{-9} cm^{-2}$.[4]

Furthermore, the total current density for system with bond probability P is [6]

$$j_c = j_{co} \cdot \left[\frac{(P-P_{bc})}{(1-P_{bc})} \right]^{1.65} \tag{14}$$

This is the rule.

3. CALCULATING RESULTS

Based on G.Deutscher[7], Let $R_{S.D.} = 0.6\overline{R}$, and Z.Dong[8], $\overline{R} = 1 \times 10^{-3} \Omega \cdot cm$, $T_{CO} = 100K$, the calculating results of middle point turn temperature $Tc_{1/2}$ is shown in Fig.1 by Solid line, meanwhile, the experimental result [8] is shown in Fig.1 by small circle too. It can be seen, that in the region of $X < 0.286$, the theory deviates the experiments obviously. Thinking of many experiments[8, 9, 10], we assume that the phase with more YCuO and less Ba is another superconductor phase, with initial turn temperature 60K, thus, the problem can be solved easily, as is shown in Fig.1 by dotted line.

By changing $\overline{R}$, the curves of $Tc_{1/2} \sim \overline{R}$, $J_c \sim T$, etc., are obtained, as in Fig.2, Fig.3, and Fig.4.

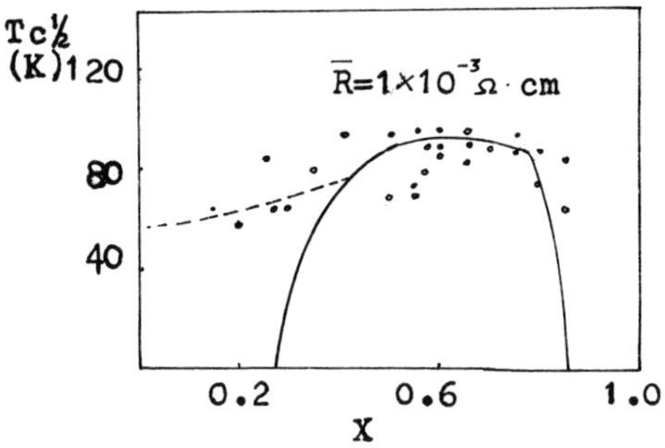

Fig.1 $Tc_{1/2} \sim X$, $\overline{R} = 1 \times 10^{-3}\Omega cm$

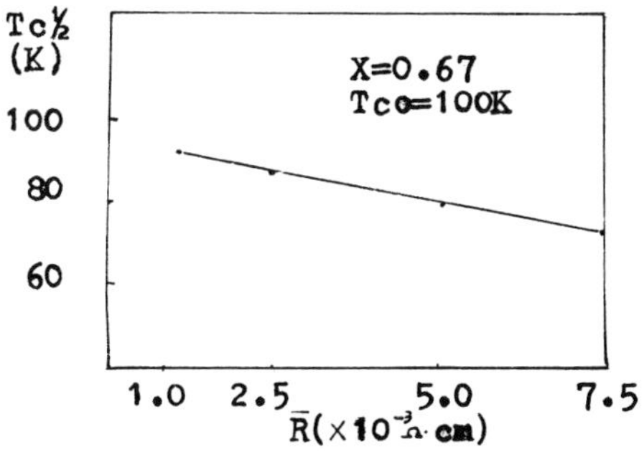

Fig.2 $Tc_{1/2} \sim \overline{R}$, X=0.67

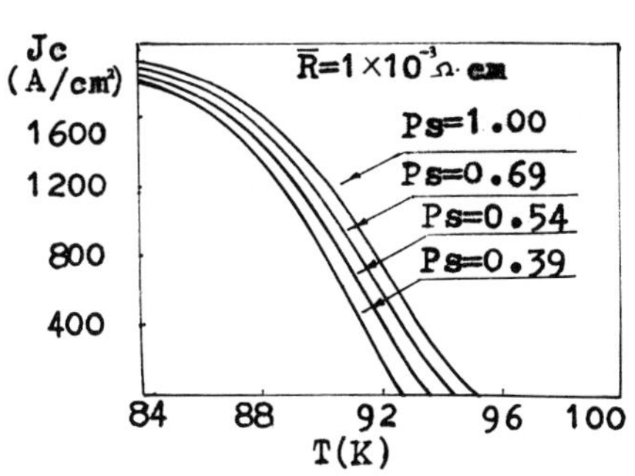

Fig.3 $J_c \sim T$, $\overline{R} = 1 \times 10^{-3}\Omega cm$

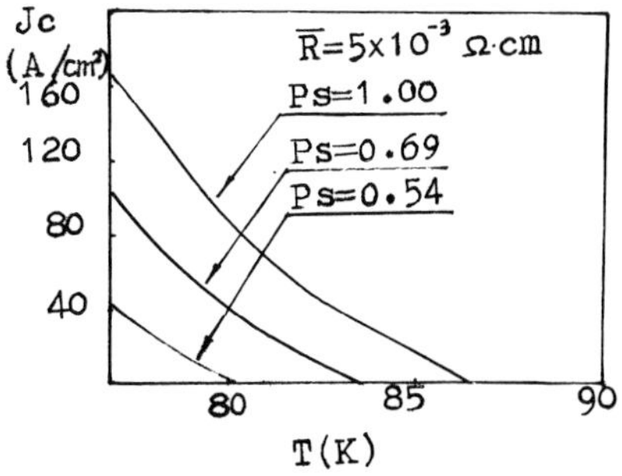

Fig.4 $J_c \sim T$, $\overline{R} = 5 \times 10^{-3}\Omega cm$

4. CONCLUSION

It can be seen, from the above analyses that the approaches in this paper is aceptable; there are two superconeluctor phases in $Ba_xY_{1-x}CuO_{3-y}$ system; the effective methods to increase J_c, and $Tc_{1/2}$ or Tc are, 1, increasing the fraction of S.C.G.; 2, decreasing $\overline{R}$; 3, increasing the initial turn temperature T_{CO}.

REFERENCE:

1. J.G. Bednorz, K.A.Müller, Z.Phys.B64, (1986), 189.

2. Y.Liu, M.Zheng, Chin.Phys.Lett., 7, (1990), 234.

3. K.Fang,J.Xu,Stastic Distribution,(Sci.Press, Beijing, China, 1987. in Chinese).

4. Y.He, Proceding ofHigh Tc Symposium, V2, (Baoji, China, 1988), p.308, (in Chinese).

5. A.Barottoff, Phys.Rev.11, (1963), 486.

6. R.Zallen, The Phys.of Amorphous Solids. (Awiley Inter.Sci.Publication, U.S.A.1983).

7. G.Deuscher, et.al., Phys Rev., B21, (1980), 5041.

8. Z.Dong. et.al., Chin.Low Temp.Phys., 9, (1987), 216 (in Chinese).

9. Z. Hua et.al., Fudan Journal,26, (1987), 241 (in Chinese).

10. C. Jia, Chin.Low Temp. Phys., 9, (1987), 260, (in Chinese).

Frontiers of Materials Research/
Electronic and Optical Materials
M. Kong, L. Huang (editors)

ULTRASONIC INVESTIGATIONS ON THE LATTICE INSTABILITY OF $La_{2-x}Sr_xCuO_{4-y}$ AND $Nd_{2-x}Ce_xCuO_{4-y}$ SUPERCONDUCTORS

HE Yusheng, XIANG Jiong, JIN Sheng and SUN Xiangzhong

Physics Department, Tsinghua University, Beijing 100084, P.R.China

Ultrasonic measurements were carried out on a series samples of $La_{2-x}Sr_xCuO_{4-y}$ and $Nd_{2-x}Ce_xCuO_{4-y}$ by 10 MHz longitudinal and transverse waves in the temperature range between 10 K to 300 K. A striking velocity softening was observed and this softening temperature T_0 decreases almost linearly with the increasing of x. A restiffening of velocity takes place at even lower temperature, implying another possible structural phase transition occurred. The anomalous changes in both attenuation and velocity around 170 and 200 K were found for the superconducting samples of $Nd_{2-x}Ce_xCuO_{4-y}$ (x=0.15) during cooling, whereas no anomalies can be found in this region for non-superconducting samples. The implication of these phenomenon is discussed.

1. INTRODUCTION

The discovery of high-T_c superconductivity in $La_{2-x}M_xCuO_{4-y}$ (M=Ba,Sr)[1,2] has led to numerous studies on the properties of these kind of perovskite superconductors. In early 1989, a new family of superconductors $Nd_{2-x}Ce_xCuO_{4-y}$ with T_c above 20 K was discovered by Tokura et al[3,4]. Measurements of Hall and Seebeck coefficients seemed to indicate that the majority charge carriers in the new superconductors are electrons, rather than holes as in the $La_{2-x}Sr_xCuO_{4-y}$ as well as other previously known high T_c superconductors.

Both the $La_{2-x}Sr_xCuO_{4-y}$ and $Nd_{2-x}Ce_xCuO_{4-y}$ superconductor are of multilayered tetragonal K_2NiF_4 structures. $La_{2-x}Sr_xCuO_{4-y}$ has the T-phase structure, which is composed of Cu-O octahedra, while $Nd_{2-x}Ce_xCuO_4$ has the T'-phase structure, which is composed of sheets of Cu-O squares. The structural instability of these materials, on approaching superconducting transition from room temperature, has drawn much attention. In this paper, we will report the results of recent ultrasonic investigations on these two kinds of superconductors and give the comparisons of the lattice instabilities of La-based and Nd-based superconductors.

2. EXPERIMENTAL

$La_{2-x}Sr_xCuO_{4-y}$ and $Nd_{2-x}Ce_xCuO_{4-y}$ samples were synthesized from La_2O_3, $SrCO_3$, Nd_2O_3, CeO_2 and CuO by standard solid-state reaction method[2,3,4]. Superconducting ceramics of single phase were achieved.

The ultrasonic measurements were carried out with a MATEC 7700 series equipment. 10 MHz longitudinal and transverse $LiNbO_3$ transducers and Nonaq grease were used. The ultrasonic attenuation and velocity are measured by means of pulse-echo method in the temperature ranged from room temperature to 10 K.

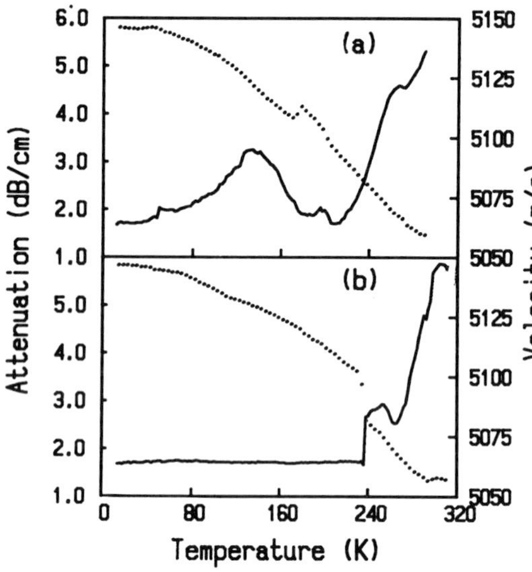

FIGURE 1
Temperature dependence of the attenuation (solid curve) and velocity (dotted curve) at 10 MHz for superconducting $Nd_{1.85}Ce_{0.15}CuO_{4-y}$ in (a) cooling run and (b) heating run

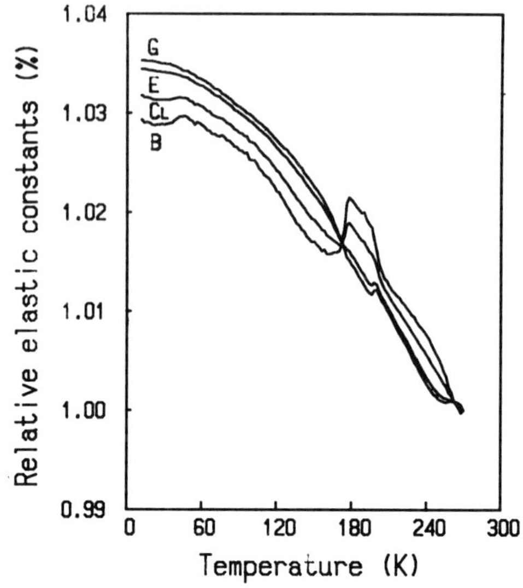

FIGURE 2
Temperature dependence of shear modulus (G), Young's modulus (E), longitudinal modulus (C_L) and bulk modulus (B) for superconducting $Nd_{1.85}Ce_{0.15}CuO_{4-y}$

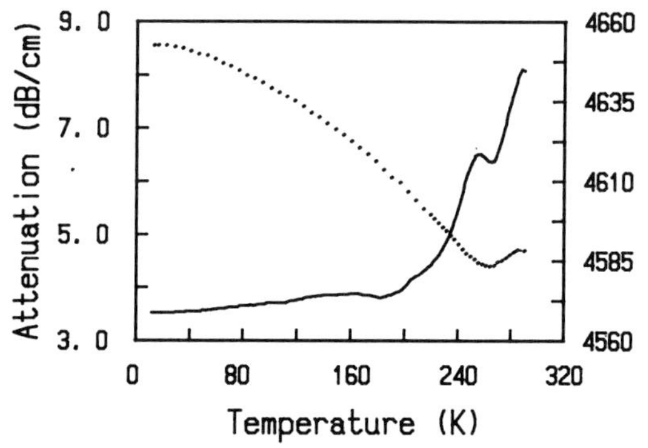

FIGURE 3
Temperature dependence of the attenuation (solid curve) and velocity (dotted curve) at 10 MHz for non-superconducting $Nd_{1.85}Ce_{0.15}CuO_{4-y}$ in cooling run

3. RESULTS

Typical results of the temperature dependent longitudinal ultrasonic attenuation and velocity of the superconducting $Nd_{2-x}Ce_xCuO_{4-y}$ ceramics (x=0.15) are shown in figure 1. The temperature dependent elastic constants, which calculated from the velocities of longitudinal and transverse waves in the cooling run, are shown in figure 2. Figure 3 shows the temperature dependent longitudinal ultrasonic attenuation and velocity for a non-superconducting sample in cooling run, while those in the heating run are similar with figure 1(b).

The temperature dependence of longitudinal velocities for samples of various compositions are shown in figure 4. A drastic drop in Velocity with decreasing temperature was observed for x=0.075, 0.15, and 0.20. The velocity softening temperature T_0 depends strongly on the strontium content x. It is important to point out that in addition to the remarkable mode softening, there are also a mode hardening at even lower temperatures in

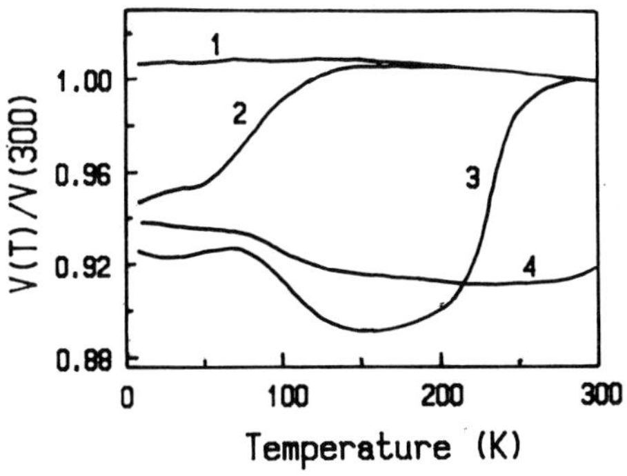

FIGURE 4
Temperature dependence of V_L for $La_{2-x}Sr_xCuO_{4-y}$ with different x. 1.x=0; 2.x=0.20; 3.x=0.15; 4.x=0.075 (compare to the V_L before softening)

Table 1. The characteristic temperatures for $La_{2-x}Sr_xCuO_{4-y}$ with different x

x	T_c (K)	T_o (K)	T_o' (K)	T_t (K)	T_t' (K)
0	/	530	/	/	/
0.075	28	320	291	100	106
0.15	38	231	215	106	111
0.20	34	88	70	/	/

figure 4 for x=0.15 and x=0.075. Such noticeable changes must have profound physical origin, hence, we denote these characteristic hardening temperatures as T_t. Table 1 summarized several parameters obtained from our experiments for $La_{2-x}Sr_xCuO_{4-y}$, where T_o' and T_t' denote the temperatures of the corresponding attenuation anomalies which are about 10-30K away from T_o or T_t. It can be seen that T_o decreases almost linearly with increasing of x.

Figure 5 shows the temperature dependence of the ultrasonic attenuation coefficient $\alpha(T)$, the longitudinal velocity V_L and its derivative with respect to temperature dV/dT for $La_{2-x}Sr_xCuO_{4-y}$ with x=0.15. It is

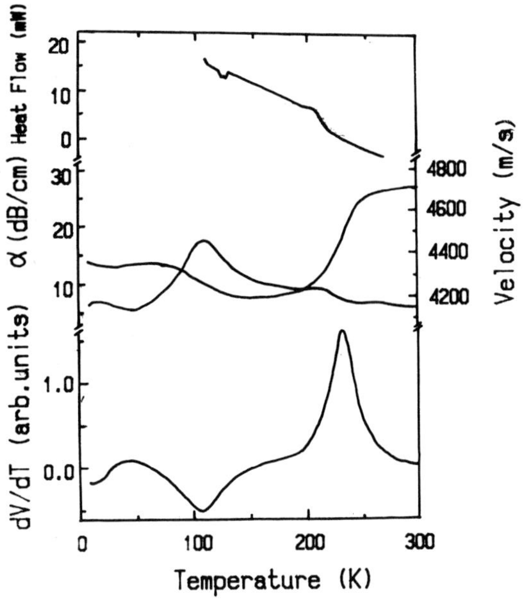

FIGURE 5
Temperature dependence of the result of DSC analysis, $\alpha(T)$, V_L and dV/dT for $La_{1.85}Sr_{0.15}CuO_{4-y}$

interesting to notice that the attenuation peak at T_t' is even much higher than that at T_o'. A DSC curve is also shown at the top of this figure. A heat flow dip and a specific heat anomaly appeared near T_t and T_o respectively, implying possible structural phase transitions occurred.

4. DISCUSSIONS

(1) It is known that the pure La_2CuO_4 has the tetragonal K_2NiF_4 structure at elevated temperatures but undergoes a tetragonal to orthorhombic (T-O) structure phase transition at 530 K, and the T-O phase transition is considered to be a continuous (second-order) one. The partial substitution of La by Sr or Ba stabilizes the tetragonal phase at room temperature and with the increasing of the Sr

composition x, the oxidation on CuO_2 plane and the valence of copper would be enhanced, so the T-O transition temperature T_0 decreases with the increasing of x.

The experimental results show that in addition to the already known large fall in sound velocity at higher temperature, $La_{2-x}Sr_xCuO_{4-y}$ also exhibit a modulus stiffening at lower temperature near 100K, accompanied by a large attenuation peak around this temperature range. And DSC result confirmed that in this temperature range the structure undergoes some changes or instabilities, which has a characteristics of first order phase transition. It is reported[5] that in addition to the already known high temperature tetragonal (HTT) to low temperature orthorhombic (LTO) phase transformation, a further transformation to a second tetragonal (low temperature tetragonal, hereafter referred to LTT) phase occurs at lower temperatures in $La_{2-x}Ba_xCuO_{4-y}$. An examination of the reported phase diagram shows that the substantial softening near 230K measured in $La_{1.85}Sr_{0.15}CuO_{4-y}$ is associated with the HTT-LTO transformation, while the restiffening occurs at lower temperatures where the LTT phase develops. For $La_{1.8}Sr_{0.2}CuO_{4-y}$, we didn't found the restiffening at lower temperatures, this may be because that the two phase transitions (HTT-LTO and LTO-LTT) temperatures coincide with each other and the drastic softening conceal the restiffening.

The softening and restiffening of sound velocity reveals the complex behavior of the lattice instabilities of $La_{2-x}Ba_xCuO_{4-y}$. It is interesting to learn whether the LTO to LTT transition is closely related to its superconductivity or if the LTT phase is responsible for the high T_c. At present stage, the question seems to be still open. The fact, however, that for La_2CuO_{4+y} only tetragonal phase is superconducting and its T-O phase change is also a first-order transition[6] might give us some clue. Moreover, it has been reported that an increase in magnetic susceptibility occurs at the LTO-LTT transformation.[7] As a consequence, rearrangements in the structural buckling at LTO to LTT transformation may profoundly influence the interplanar magnetic coupling and thus superconductivity.

(2) For $Nd_{2-x}Ce_xCuO_{4-y}$ system, no such mode softening and restiffening was observed, but comparing the results of superconducting and non-superconducting $Nd_{2-x}Ce_xCuO_{4-y}$ samples, some ultrasonic anomalies related to its superconductivity could be found. An attenuation peak and a reflection point in longitudinal velocity curve near 200 K could be found in figure 1(a). Furthermore, a mode softening near 170 K in velocity while cooling can be clearly seen, corresponding with a broad ultrasonic attenuation peak around 140 K, in which a shoulder could be found near 170 K. Anomalies can also be clearly seen in the charts of the elastic constants (figure 2). So it can be deduced that some structural instability or phase transition near 170 K and 200 K does exist and might relate to its superconductivity for

$Nd_{2-x}Ce_xCuO_{4-y}$ compound. It should be pointed out that the velocity result reported by A. Al-Kheffaji et al.[8] showed similar slope changes near the temperature mentioned above. Extraordinarily, the anomalies in both attenuation and velocity in cooling run are not found in heating run (figure 1(b)), manifesting a possible hysteresis between cooling and heating, which is at least 40 K, and this gives a characteristics of first order phase transition, whereas no such hysteresis exists in La-based superconductors.

The above anomalous behavior is very much like that found in Bi-Sr-Ca-Cu-O systems.[9] X-ray diffraction experiments proved that the framework of the lattice of both Nd and Bi based systems are not changed down to very low temperature. However, fine structure rearrangements are still possible. Dmowksi et al.[10] proposed a model of short-range ordering due to displacements of Tl (or Bi), O atoms. The ordering remains very much short range and does not alter the average symmetry, resulting in no distinct change of diffraction pattern. For the $Nd_{2-x}Ce_xCuO_{4-y}$, because the Nd(Ce)-O bond is much softer than the Cu-O bond, the slight rearrangements of Nd(Ce), O atoms within the framework of the lattice could be possible and might lead to the anomalous ultrasound changes.

(3) It is well known that the crystal structures of $La_{2-x}Sr_xCuO_4$ and $Nd_{2-x}Ce_xCuO_4$ are very similar. Their parent materials are all of the form Ln_2CuO_4, where Ln is one of the light trivalent lanthanides. The only difference, however, between the two structures is that, unlike La_2CuO_4, the tetragonal unit cell for Nd_2CuO_4 has no apical oxygen atoms.

In the La_2CuO_4 structure, the superconducting active layer consists of a single CuO_2 sheet whereas the inactive layer of variable oxygen concentration is $(LnO)_2$ rocksalt layer. From structural-chemical point of view, bond-length mismatch at the interface between the two layers plays an important role in the lattice instability. Goodenough[11] suggested a tolerance factor $t=(Ln\text{-}O)/\sqrt{2}(Cu\text{-}O)$ as a measure of the bond length matching, where Ln-O and Cu-O are the respective bond lengths. Perfect matching occurs for the 180° Cu-O-Cu bonds if $t=1$. Considering, however, thermal expansion of the softer La-O bond is significantly greater than that of the harder Cu-O bond, t decreases with decreasing temperature. A $t<1$ places the CuO_2 sheets under compression and the $(LaO)_2$ layers under tension. The internal stress finally lead to a cooperative tilting of the CuO_6 octahedra about [110] axis which bends the Cu-O-Cu bond from 180° and reduces the crystal symmetry from tetragonal to orthorhomic. It is worth pointing out that such a bend of Cu-O-Cu bond from 180° appears in some other p-type high T_c copper oxide (e.g. $YBa_2Cu_3O_{7-y}$) and the changes of the bond angle may also have some relations with the observed lattice instabilities.

For the Nd_2CuO_4 structure, the situation is completely different. The displacement of the c-axis apical oxygen atom to the interstitial site transforms the rocksalt $(LnO)_2$ layer to a fluorite $Ln-O_2-Ln$ layer. The $Ln-O_2-Ln$ layer has a larger a axis and the expansion tends to place the CuO_2 planes under tension and the Cu-O-Cu bond angle remains 180° to the lowest temperature. Hence, no structural changes in the framework of the lattice can be found in this system. However, as discussed above, some fine structural changes of short range ordering due to rearrangement of $Ln-O_2-Ln$ layer to meet the t=1 requirement are still possible, which would also lead to a lattice instability, resulting in some anomalous behaviors as observed in our ultrasonic measurements near 200 K.

(4) It is a common feature of conventional A15 superconductors that the transition temperature is raised by the proximity to a structural instability. Equally, there has also been a long suspected link between lattice instabilities and high temperature superconductivity since the discovery of the high T_c copper oxides. Many theoretical models[12,13] have been proposed to explain the mechanism of such instability and superconductivity. But, clearly, it seems that there is still a gap between the theoretical models and the experimental results. Nevertheless, the experimental evidence for p-type La-, Y-[14] and Bi-[9] based systems and n-type Nd-based system strongly suggest that the existing perovskite multilayer superconductors do have some lattice instabilities on approaching their superconducting transitions and such lattice instabilities do correlate with their superconductivity in some way.

ACKNOWLEDGEMENTS

This work was supported by the National Commission of Education of China and the National Center for R&D on Superconductivity of China.

REFERENCES

1. J.G. Bednorz and K.A. Muller, Z. Phys. B 64 (1986) 189.

2. Zhao Zhongxian et al., Kexue Tongbao (Science Bulletin) 32 (1987) 177 (in Chinese); 32(1987) 522 (in English).

3. Y. Tokura, H. Takagi and S. Uchida, Nature 337 (1989) 345.

4. H. Takagi, S. Uchida and Y. Tokura, Phys. Rev. Lett. 62 (1989) 1197.

5. J.D. Axe et al. Phys. Rev. Lett. 62 (1989) 2751.

6. J. Zhou, S. Sinha and J.B. Goodenough, Phys. Rev. B 39 (1989) 12331.

7. M. Sera et al., Solid State Commun. 69 (1989) 851.

8. A. Al-Kheffagi, J. Freestone, D.P. Almond, G.A. Saunders and Jing Wang, J. Phys. Condens. Matter 1 (1989) 5993.

9. He Yusheng et al., Phys. Rev. B 40 (1989) 7384.

10. W. Dmowksi, B.H. Toby and T. Egami, Phys. Rev. Lett. 61 (1989) 2608.

11. J.B. Goodenough, Supercond. Sci. Technol. 3 (1990) 26.

12. B.S. Lee, Physica C 160 (1989) 141.

13. P.B. Litterwood, C.M. Varma and E. Abrahams, Phys. Rev. Lett. 63 (1989) 2602.

14. Y. He et al., J. Phys. F 17 (1987) L243; Int. J. Mod. Phys. 3 (1989) 1581.

Frontiers of Materials Research/
Electronic and Optical Materials
M. Kong, L. Huang (editors)

Bi-BASED SUPERCONDUCTING FIBERS WITH HIGH CRITICAL PARAMETERS

Huo Yujing, He Yusheng+, Liu Menglin, Mao Sining, Cai Liying, Wang Ying, Zhang Jincang*, He Aisheng# and Wang Jinsong

Tsinghua University, Beijing 100084, P.R.China

Superconducting fibers of Bi(Pb)-Sr-Ca-Cu-O high Tc superconducting materials have been prepared by means of the laser-heated pedestal growth (LHPG) method. The highest zero resistance temperature T_{c0} reached is 114 K, and the highest critical current density J_c (77 K, 0 T) is greater than 5000 A/cm^2. As-grown superconducting fibers were successfully fabricated without post-growth heat treatment. Amorphous materials were used for the first time to make high quality fibers. The influence of growth conditions, thermal treatment and the composition of the fibers were discussed.

1. INTRODUCTION

Following the discovery of high temperature superconducting transitions in the copper oxide perovskite systems by Bednorz and Muller[1] in 1986, great efforts have been made to fabricate high quality superconducting fibers for measurements of properties and potential applications. The laser-heated pedestal growth (LHPG) method has been shown to be a powerful method for rapidly growing small diameter fibers, particularly oxides[2,3].

In this paper, we will report our recent results of Bi-based superconducting fibers made by LHPG method. The structure, morphology, growth conditions, thermal treatment and composition of the fibers will be discussed.

2. PREPARATIONS

The source materials in these experiments were prepared from raw Bi_2O_3, PbO, $SrCO_3$, $CaCO_3$ and CuO powders with analytic purity. The appropriate amounts were mixed, finely ground, and calcined at 800-820°C for 24 h. For preparing polycrystalline source materials used in fiber growth, the calcined materials were then ground again, pressed into pellets, and sintered in air at 840-860°C for more than 60 hours and furnace cooled. While for making amorphous source materials, the calcined materials were rapidly heated to 1150-1250°C and then quenched to room temperature at the cooling rate of more than 300°C/min. Rectangular rods 0.5x0.5x20 mm in size were cut out and used both as the source materials and the seeds.

Fiber growth was accomplished by two steps. Firstly, the source rod was pulled at the rate of about 2 mm/min to become round, smooth and dense. Then the fibers were grown at the rate of 0.04-1 mm/min. The diameters of the resulting fibers were 35-340 μm.

+ Mailing address: Physics Department, Tsinghua University, Beijing 100084, China
* Visiting scholar from Henan Normal University, Xinxiang 453002, China
# Visiting scholar from North University of Technology, Beijing 100041, China

Table 1 Main results of Bi-based superconducting fibers prepared by the LHPG method

sample	Source material	Nominal composition Bi(Pb)-Sr-Ca-Cu	Post-growth heat treatment	Phase	T_{co} (K)	J_c(77K,0T) (A/cm^2)
45Q-1	polycrystal	2(0.4)-2-2.8-7.5	--------	2212	76	----
FB152	polycrystal	2-2-2-4	840°C/46h	2212	80	5150
11-3	polycrystal	1.84(0.36)-1.91 -2.03-3.06	850°C/48h	2223	110	2000
45Q-3	polycrystal	2(0.4)-2-2.8-7.5	820°C/4h+842°C/12h	2223	114	300
M67-2	amorphous	2(0.4)-2-2-4	840°C/6h	2212 (main)+2223	84	>5000

Amorphous materials, polycrystals with strong textured structure and single crystals were used as the seed rods.

3. MAIN RESULTS

Table 1 summarized some main results of our Bi-based superconducting fibers prepared by LHPG method. It can be seen that the as-grown fiber of sample 45Q-1 was superconducting with T_{co}=76 K without post-growth heat treatment. Unfortunately, because of large contact resistance of the electrodes, measurements of the critical current density could not be done. According to our present results, highest J_c (77 K, 0 T) is 5150 A/cm^2. It is interesting to notice that, although the transition temperature of the 2223 phase superconducting fibers is much higher than that of the 2212 phase fibers, the critical current density of the former is much lower. Amorphous materials have been used for the first time to fabricate superconducting fibers. The highest critical current density is also greater than 5000 A/cm^2.

Structure analysis shows that the fibers have strong textured structure and highly aligned morphology. SEM photograph of the as-grown fiber together with its source rod is given in figure 1. The crystal grains in the fiber aligned in good order, comparing to the chaos state of source rod, which shows that strong textured structure was produced during fiber growth. As shown in figure 2, XRD spectrum for the side surface of the fiber prepared from amorphous materials indicates that only the (00L) peaks appear. Especially in the XRD spectrum for the cross section of the same fiber, the (200) peak takes dominant precedence, all of the (00L) peaks, even the (220) peak almost disappear. It indicates that the a-axis is parallel to the growth direction of the fibers. The highly aligned morphology with a-axis parallel to the growth direction will provide optimized path for superconducting current along the fibers, which is closely related to the high critical current density.

FIGURE 1
SEM photograph of the as-grown fiber (upper part) at its source material

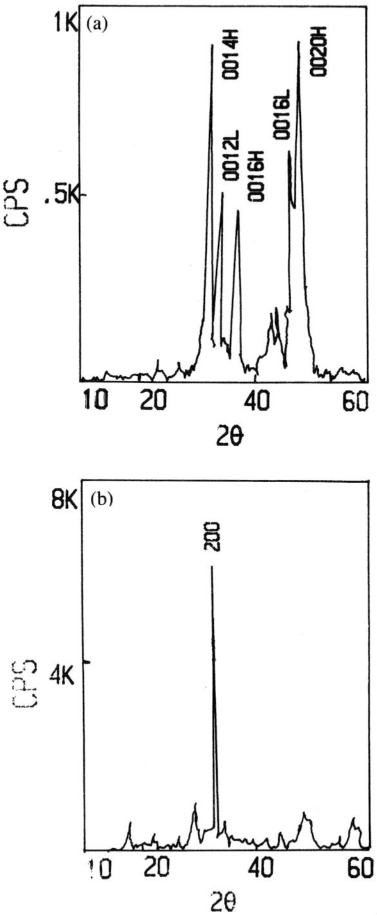

FIGURE 2
XRD spectrum for the side surface (a) and the cross section (b) of the fibers prepared from amorphous materials

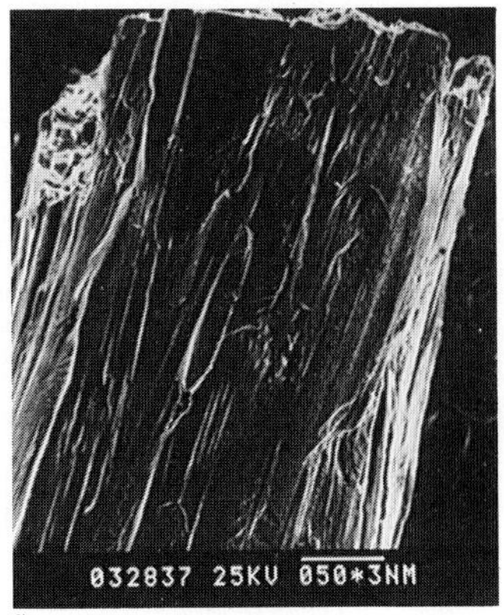

FIGURE 3
SEM photograph of the morphology of the side surface of the fiber made from amorphous materials

Table 2 Main results of the fibers prepared from amorphous materials

No.	nominal composition Bi(Pb)-Sr-Ca-Cu	T_{co} (K)	J_c(77K,0T) (A/cm^2)
M40	2-2-2.8-7.5	84	1920
M45	2(0.4)-2-2.8-7.5	81	2520
M67	2(0.4)-2-2-4	84	>5000

4. DISCUSSIONS

For the practical applications of superconducting fibers, it is essential to raise their critical current density. As we know, J_c is influenced by many factors. In this paper, the effects of source material, growth conditions, post-growth heat treatment and homogeneity of composition on superconducting properties of the fibers will be discussed.

4.1 Source materials

For growing superconducting fibers with high J_c, it is vital to ultimately increase superconducting portion and eliminate non-superconducting defects in the fiber. Hence, high quality source materials with proper composition are needed and the knowledge of the composition changes after growth will be important. EDAX composition analysis showed that the compositions of the fibers were obviously deviated from those of the source materials. It seems that only the element Sr had the nearly same portion in the fibers as in the source rods, and could be regarded as a stable element. To avoid composition loss during growth, one of

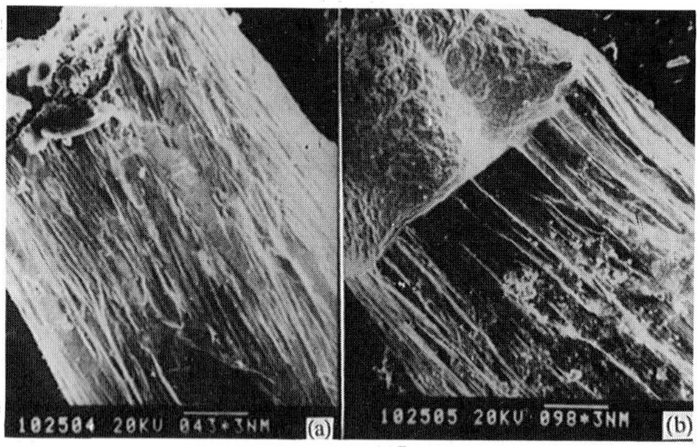

FIGURE 4
SEM photographs of the side surface of 110 K superconducting fibers grown at the rate of (a) 1 mm/min and (b) 0.04 mm/min

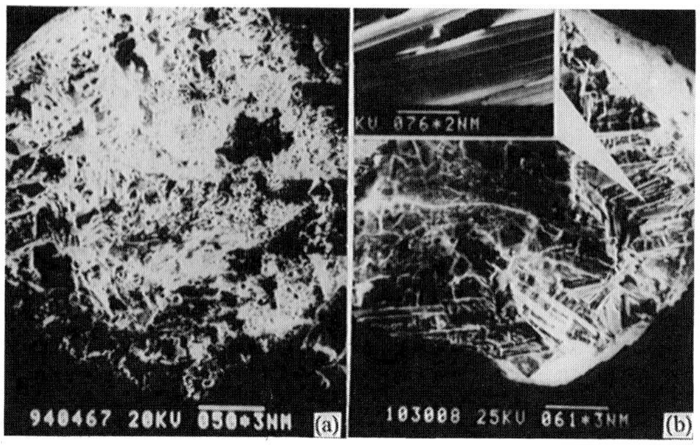

FIGURE 5
SEM photographs of the cross section of 110 K superconducting fibers grown at the rate of (a) 1 mm/min and (b) 0.04 mm/min

the best techniques is to reduce the melting temperature of the materials by introducing proper flux. In fact,for sample 45Q-1, excess CuO was added in the source material as self-flux. As-grown superconducting fiber was successfully obtained (see table 1), which demonstrates that proper flux is profitable for fabricating high quality fibers.

It is worth emphasizing that comparing to the ordinary ceramic polycrystalline source materials, amorphous material with its comparatively less composition loss and much higher density, showed great potential in making high quality superconducting fibers. Table 2 showed some of our primary results of superconducting fibers made from amorphous materials. It has been proven that by using amorphous materials, it is easier to achieve their superconductivity and obtain high J_c. It also shows that the best nominal composition of amorphous materials may be different from that of polycrystalline materials, for relatively short period high temperature treatment in the melt and quench method gives rise to small composition deviation. The density of the amorphous materials ranges from 6.0 to 6.2 g/cm^3, which is much greater than that of the sintered polycrystalline materials (4.8-5.5 g/cm^3) and close to the theoretical density of Bi-based superconducting materials. This is obviously beneficial for eliminating voids and other defects and fabricating high quality superconducting fibers with larger crystal grains, as it can be seen from figure 3 that the textured structure is very strong and the crystal grains are quite large.

4.2 Growth rate

The growth rate of the fibers also strongly affects their micro-structure and hence their superconducting properties. SEM observations were performed for the 110 K superconducting fibers grown at different rates prepared from same source rod. As shown in figure 4a,

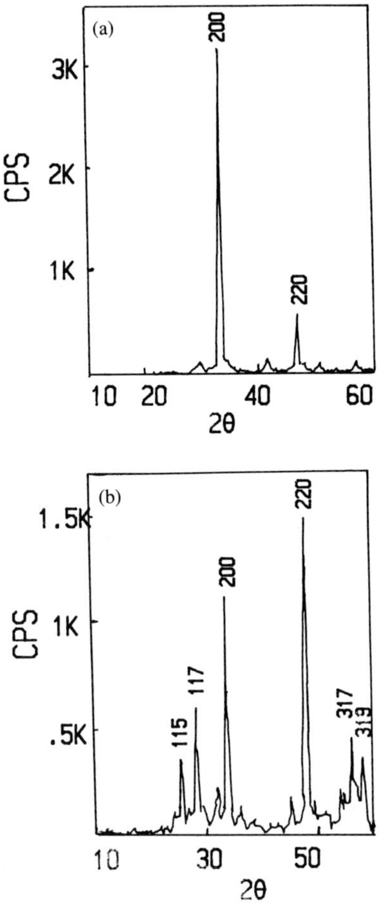

FIGURE 6
XRD spectrum for the cross section of (a) sample 45Q-1 and (b) sample 45Q-3

for the fiber grown at the rate of 1 mm/min, the side surface had a number of growth wrinkles aligned parallel to the growth direction. When the growth rate reduced to 0.04 mm/min, the side surface of the fiber became flat and smooth, and facets of larger size could be seen (figure 4b), which obviously resulted from regular arrangement of single crystals with their c-axis vertical to the growth direction. This phenomenon can also be seen from SEM

Table 3 Results of thermal treatment of the as-grown superconducting fiber

No.	Heat treatment	Phase	T_{co} (K)	J_c(77K, 0T) (A/cm^2)
1	--------	2212	76	----
2	820°C/4h	2212	85.5	4200
3	820°C/4h +842°C/12h	2223	114	300

photographs of the cross section of the fibers. For rapidly grown fibers, as shown in figure 5a, crystal grains are quite small, and there are many "holes" in the fiber. With the slowing down of the growth rate, the number of the "holes" decreases remarkably, and the grains become much larger. As shown in the inset of figure 5b, many plane-like single crystals are aligned with their a-b plane parallel to the growth direction. It indicates that slower growth rates yield fibers with fewer, larger grains. Finally, it can also be seen from figure 5b that random arrangements of the crystal grains still exist in some extent, which may be attributed to the influence of post-growth heat treatment.

4.3 Post-growth heat treatment

Post-growth heat treatment strongly influences the critical current density of the fibers. Table 3 gives the results of thermal treatment of the as-grown fiber of sample 45Q-1. It can be seen that after a long period heat treatment, zero resistance temperature increased from 76 K to 114 K, but the critical current density decreased to 300 A/cm^2 (sample 45Q-3). The decrease of the critical current density J_c may be attributed to the comparatively long period thermal treatment, as it is believed that post-growth heat treat-

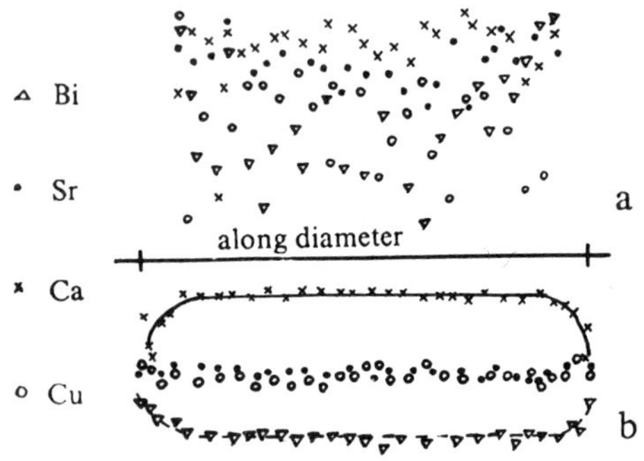

FIGURE 7
X-ray energy dispersion analysis results of two superconducting fibers with Jc equal to (a) 100 A/cm2 and (b) 1150 A/cm2

ment will destroy the strong textured structure in some extent, hence the best path for superconducting current will be cut off, leading to a dramatic decrease of J_c. The above idea is supported by our X-ray diffraction results. XRD profiles for the fractured cross section of sample 45Q-1 and 45Q-3 are shown in figure 6. It is clear that for sample 45Q-1, the (200) peak is much higher than the others, and the intensity of the (220) peak is only 19% as high as that of the (200) peak, while all of the (00L) peaks disappear completely. It indicates that the as-grown fibers have highly aligned structure with the a-b plane parallel to the growth direction. Whereas for sample 45Q-3, the strong alignment of a-b plane is weakened in some extent due to long time thermal treatment. It can be seen that (117), (115), (317) and (319) peaks are considerably high, implying that the c-axis is not completely perpendicular to the growth direction. Such a change will surely influence the critical current density of the superconducting fibers.

4.4 Homogeneity of composition

Figure 7 shows the X-ray energy dispersion analysis results of two superconducting fibers with J_c equal to 100 A/cm^2 (fiber A) and 1150 A/cm^2 (fiber B). They were obtained by scanning the electron beam along the diameter of polished sections. It can be seen that in most regions of fiber B the composition is homogeneous and close to that of the 2212 superconductor while the composition of fiber A is rather inhomogeneous. This implies that the homogeneity of composition is closely related to the value of J_c.

ACKNOWLEDGEMENTS

This work was supported by the National Commission of Education and the National Centre for R&D on Superconductivity of China.

REFERENCES

1. J. Bednorz and K.A. Muller, Z Phys B64 (1986) 189.

2. D. Gazit and R.S. Feigelson, Journal of Crystal Growth 91 (1988) 318.

3. R.S. Feigelson, D. Gazit, D.K. Fork, and T.H. Geballe, Science 240 (1988) 1642.

4. Y.J. Huo et al, Proc. Int. Conf. High T_c Supercond. (BHTSC'89) Beijing, 1989, Z.X. Zhao et al eds., World Scientific, P162.

5. Zhang Jincang et al, MRS Fall Meeting-symposium 7.242 (1989) Boston.

Frontiers of Materials Research/
Electronic and Optical Materials
M. Kong, L. Huang (editors)

THE SYNTHETIC PROCESS OF $Bi_2Sr_2Ca_2Cu_3O_y$ BULK SUPERCONDUCTORS*

Xiangdong REN and Zutao YAO

Department of Chemistry, Tsinghua University, Beijing 100084, CHINA.

We have obtained bulk superconductors with zero resistance near 110K. From the results of Differential Thermal Analyze (DTA) and X-ray Diffraction (XRD) we assumed that the synthesis of superconductors proceed on step by step, and CuO play an important part in the growth and connection of 110K phase. To enhance the ability for CuO to permeate in the crystal of 110K phase and to promote the growth of 110K phase, some samples were prepared by means of two step synthesis. It has been shown that two step process on adding Bi_2O_3 lately may impel CuO to penetrate into the lattice of the superconductive phase and accelerate the growth of the high T_c phase. The results of electronic property were also reported in this paper.

1. INTRODUCTION

Soon the oxide superconductors in the Bi-Sr-Ca-Cu-O system without any rare earth element were found by Maeda et al.[1], some researches on this materials had been carried out. The synthetic processes, properties, compositions and structures of the superconductors in the Bi-Sr-Ca-Cu-O system were studied widely[2-7]. Efforts have been poured to isolate two different superconductive phases with T_c at 110K and 80K respectively. Recently some results with zero resistance temperature near 110K were reported by several research groups[8-9]. Takano et al[10]. also succeeded to synthesize bulk samples with T_c=108K by the Pb-doping in the Bi-Sr-Ca-Cu-O system. Even so, their samples were still heterogeneous, containing both high T_c phase and low T_c phase. Undoubtedly, it is very useful to study further on the synthetic process of the superconductors and the behavior of each component in the Bi-Sr-Ca-Cu-O system. Some results of TG-DTA, XRD experiments and two step synthesis were reported in this paper. The chemical reactions in the process and the effect of sintering conditions on forming 110K phase were discussed.

2. EXPERIMENTAL PROCEDURE

The samples were prepared by solid state reaction method. Appropriate amounts of analytical purity Bi_2O_3, $SrCO_3$, $CaCO_3$ and CuO powders were weighed stoichiometrically, ball-milled, dried and ground in an agate mortar.

For one step method the mixture with nominal composition of $Bi_2Sr_2Ca_2Cu_4O_y$ were used as initial composition. The well-mixed mixtures were pressed into pellets, calcined at 780-880°C for 20-40 h and regrounded. The final mixtures were pressed into a cylindrical pellets under 2.5-3 T/cm^2, 3 mm in thickness

*This project is supported by the found of the National Research Center of the Superconductive Technique.

and 13 mm in diameter, and sintered at 860-880°C for 20-60 h in air.

For two step method the triple mixture $Bi_1Sr_1Ca_1$ and $Sr_1Ca_1Cu_2$ were prepared separately. The well-mixed mixtures were calcined at 800-1000°C for 20 h in air. We chose the calcining temperature range at 850-880°C for the Bi-Sr-Ca mixture and 950-1000°C for Sr-Ca-Cu. Then the triple powders were mixed with appropriate amount of CuO or Bi_2O_3 to yield both mixtures with nominal compositions of $Bi_2Sr_2Ca_2Cu_4O_y$. The mixtures were ground carefully and pressed into pellets, finally sintered at 850-880°C for 4-40 h. All samples were cooled in the furnace or quenched rapidly in air. The standard four-probe technique was used for resistance measurement with measuring current 1-7 mA and voltage resolution $1*10^{-8}$ V.

TG-DTA were adopted to study the solid state reaction for the one-step mixtures and the two-step precursors. XRD was used to investigate the phases formed after calcination and sintering. The samples for the XRD experiments were pellets with different chemical compositions, which were prepared from powder reagents of Bi_2O_3, $SrCO_3$, $CaCO_3$ and CuO. The pellets were sintered at different temperatures for 24 hours and were quenched in air.

3. RESULTS AND DISCUSSION

The TG-DTA curves of the mixtures with different compositions are shown in Figure 1, in which the compositions of DTA curve (a), (b) and (c) are $Bi_1Sr_1Ca_1$, $Sr_1Ca_1Cu_2$ and $Bi_2Sr_2Ca_2Cu_4$ respectively. As we know, For pure calcium carbonate there is a strong endothermal peak at 950°C. Pure Bi_2O_3 has two endothermic peaks at 720°C and 817°C, which relate to phase transition and melting-point of Bi_2O_3 respectively. The phase transition peak at 720°C of Bi_2O_3 in the mixtures became wide by compared with pure Bi_2O_3, and there was a peak at about 750°C. The TG curves of $Bi_1Sr_1Ca_1$ and $Sr_1Ca_1Cu_2$ showed that the mixture of $Bi_1Sr_1Ca_1$ began slightly to lose weight at 710°C, but the mixture of $Sr_1Ca_1Cu_2$ was no weight loss until 850°C.

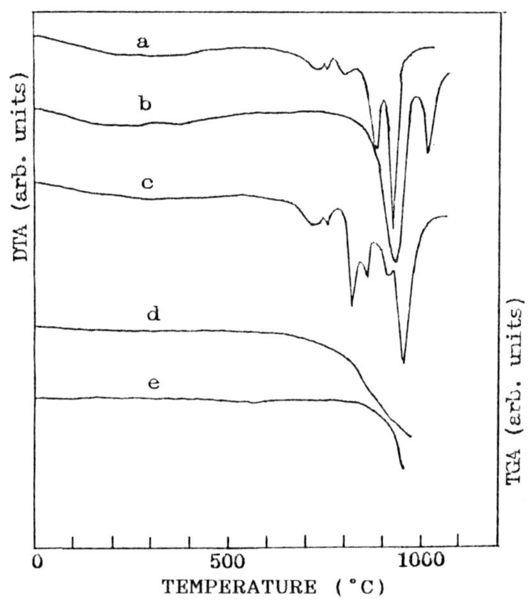

FIGURE 1
The TG-DTA curves of the mixtures with different compositions. (a)--$Bi_1Sr_1Ca_1$(DTA); (b)--$Sr_1Ca_1Cu_2$(DTA); (c)--$Bi_2Sr_2Ca_2Cu_4$(DTA); (d)--$Bi_1Sr_1Ca_1$(TG); (e)--$Sr_1Ca_1Cu_2$(TG).

The TG-DTA curves proved that Bi_2O_3 is very different from CuO in the ability of reacting with carbonates. The reactions between the original compounds in the $Bi_1Sr_1Ca_1$ mixture started at 720°C, but no reaction in the

$Sr_1Ca_1Cu_2$ mixture could be identified until 850°C.

The synthesized Bi-Sr-Ca precursor was multiphase material. The results of XRD experiments showed that there were $(Sr,Ca)_7Bi_{10}O_{22}$ and $(Sr,Ca)_7Bi_6O_{16}$ phases in the Bi-Sr-Ca mixture. The XRD patterns for the prereacted Sr-Ca-Cu precursor indicated that the prereacted products at high temperature were different from those calcinated at lower temperature. After the mixture of Sr-Ca-Cu was calcined at 850°C the diffraction peaks of the original compounds decreased considerably and the new pattern produced. Except the peaks of $CaCuO_2$ phase, there were three strong diffraction peaks at 2θ=31.04°, 31.35° and 33.43° showing no different from those of $SrCu_2O_3$ reported by Li Youmo[11]. The products heated at 1000°C was most probably a single phase, which was $(Sr,Ca)CuO_2$.

There were many complex solid reactions in the $Bi_2Sr_2Ca_2Cu_4$ system according to the results of XRD. The XRD patterns of $Bi_2Sr_2Ca_2Cu_4$ are shown in Figure 2. The results showed that all the diffraction peaks of Bi_2O_3, $SrCO_3$ and $CaCO_3$ disappeared after calcined at 800°C, simultaneously $(Sr,Ca)_7Bi_{10}O_{22}$, $(Sr,Ca)_7Bi_6O_{16}$ were identified in the sample. But the peaks of CuO remained mostly. It is suggested that no or only a little CuO participate in reaction at this temperature. Calcined at 860°C the 80K phase turned up, but no peaks of 110K phase were identified in the sample. And a certain amount of CuO still remained at 860°C. The peaks of CuO vanished in the sample calcined at 880°C. Three more peaks, which were closely related to high-T_c phase, emerged at 2θ=4.86°, 32.04° and 33.80° by comparison with the chart of 80K phase.

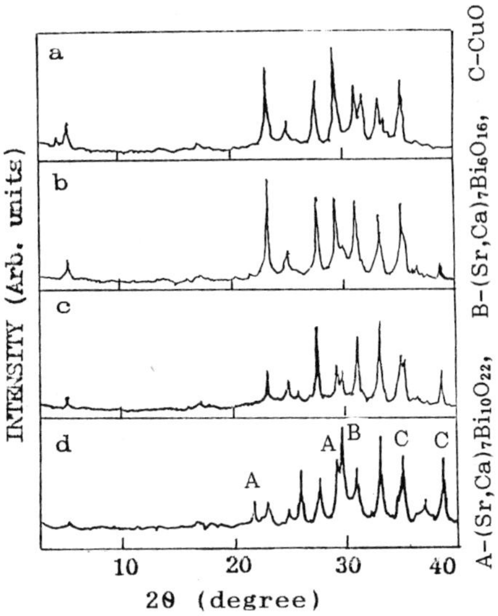

FIGURE 2
The X-ray Powder diffraction patterns of $Bi_2Sr_2Ca_2Cu_4O_y$ samples calcined for 24 h. (a) calcined at 880°C; (b) calcined at 860°C; (c) calcined at 840°C; (d) calcined at 800°C.

The typical curves of the resistance and the ac magnetic susceptibility vs. temperature were shown in Figure 3. The different sintering conditions result in different superconductivity of the samples. For one step method, the sample sintered at lower temperature was not superconductive. When temperature increased to 860°C for 60 h, sample 1-a exhibited superconductivity at 91K, but did not get zero resistance above 77K. The result of XRD showed that there was a certain amount of CuO in the sample. The electric resistance of sample 1-b sintered at 900°C for 5 min plus at 880°C for 60 h was beginning to fall near 123K and drop to zero at 111K. The superconductor was multiphase.

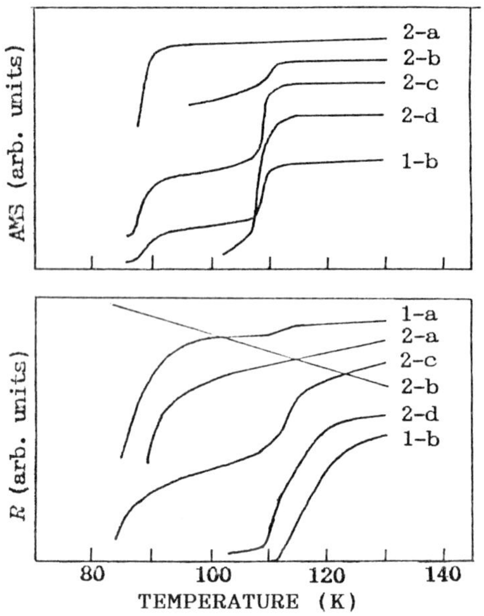

FIGURE 3
Resistance vs. temperature (R-t) and AC magnetic susceptibility vs. temperature (AMS-t) for typical samples.

The different synthetic methods affected obviously the properties of resistance and magnetic susceptibility. The sample 2-a and 2-b were synthesized by two step method, i.e. (Bi-Sr-Ca)+CuO. After the sample 2-a was sintered at 860°C for 40 h, its transition temperature was 90K, and no drop of resistance and no change of magnetic susceptibility could be observed near 110K. The sample 2-b was sintered at 880°C for 5 h. there was weak Meissner's effect near 110K, but the material was semiconductor in the region of liquid nitrogen temperature. The sample 2-c and 2-d were synthesized by (Sr-Ca-Cu)+Bi_2O_3. After sintered at 860°C for 40 h, the resistance curve fell at 110K and 80K respectively. It differs from the curve of the sample 2-a, in which only low Tc phase could be identified, that the sample 2-c included both high T_c and low T_c phases. In the pattern of XRD there were two peaks at 2θ=4.86° and 5.75° corresponding to C axis=36.8Å and 30.8Å. The sample 2-d sintered at 880°C for 5 h. The main phase of the sample 2-d was 110K phase. Its resistance curve dropped rapidly at 121K, but with a resistance tail.

To compare the different synthetic pathways, i.e. one step method, (Bi-Sr-Ca)+CuO and (Sr-Ca-Cu)+Bi_2O_3 in same sintering conditions, we found that one step method and two step method on (Bi-Sr-Ca)+CuO were similar in the properties of the samples. And the results of XRD and DTA also showed that the phase change processes during sintering were same roughly for both synthetic pathways. It suggests that the synthesis of nominal composition of $Bi_1Sr_1Ca_1Cu_2$ could proceed on steps. Sintering at lower temperature the reaction firstly went on in the triple Bi-Sr-Ca, and then a part of CuO reacted with the Bi-Sr-Ca triple compounds and produced 80K phase. To raise sintering temperature the more CuO penetrated into the lattice to form high T_c phase. Because the melting-point of Bi_2O_3 was low, while the samples were sintered, there was the potential for significant liquid phase formation prior to the complete decomposition of the carbonate components, leading to erratic product characteristics. This was due in part to the macroscopic segregation of transient carbonate-oxide fluids with composition different from the composition of the samples. Once the carbonate decomposition was complete, it was difficult to rehomogenize the sample by solid state diffusion.

For two step method, i.e. (Bi-Sr-Ca)+CuO and (Sr-Ca-Cu)+Bi_2O_3, the superconductivity of the latter were better than that of the former. The latter was easy to form 110K phase. It indicated that the two step process using precalcined precursor $Sr_1Ca_1Cu_2$ and Bi_2O_3 is favorable to the growth and the connection of high T_c phase. Although sample 2-b, sintered at 880°C for 5 h, is semiconductor, but the curve of the magnetic susceptibility presented weak Meissner's effect at 110K. It shows that the structure of 110K phase formed in the material, but its quantity is not enough to connect between them. It seems possible that the dynamics of the growth of 110K phase is poor and the chemical ability of CuO is low and the diffuse of copper ion is slow.

4. CONCLUSIONS

In the synthetic process of $Bi_2Sr_2Ca_2Cu_4O_y$, Bi_2O_3 is very different from CuO in the ability of reacting with carbonates. And CuO plays an important part in the growth and connection of 110K phase. The two step process using precalcined precursor $Sr_1Ca_1Cu_2$ and Bi_2O_3 may help CuO to penetrate into the lattice of the superconductive phase and to accelerate the growth of high T_c phase.

ACKNOWLEDGEMENTS

We are grateful to B.L. Gu, L.Y. Cai and S.Y. Chen for technical help.

REFERENCES

1. H. Maeda, Y. Tanaka, M. Fukutomi and T.Asano Jpn. J. Appl. Phys. Lett. 27, L209(1988).

2. J.M. Tarascon, Y. Lepage, P. Barboux, B.G. Bagley, L.H.Greene, W.R. Mckinnon, G.W.Hull, M. Giroud, and D.M. Hwang, Phys. Rev. B37, 9382(1988).

3. R. Ramesh and G. Thomas, Appl. Phys. Lett. 53(6) 520-522(1988).

4. I. Iguchi and A. Sugishita, Phys. C 152, 228 230(1988).

5. Ram Srinivasan, Maribeth Saum, Robert J. DeAngelis, J.O. Deasy, and J.W. Brill, J. mater. Sci. Lett. 8, 383- 386(1988)

6. T.W. Huang, N.C. Wu, and M.P. Hung, J.W.liou W.N. Wang, S.E. Hsu, P.C. Yao, M.F. Tai,H.C. Ku, and T.S. Chin, J. Mater. Sci. 24, 2319-2323(1989).

7. R.H. Arendt, M.F. Garbauskas, and L.L. Schilling, J. Mater. Res. 5(1), 33-38(1990).

8. K. Kuwahara, S. Yaegashi, K. Kishio, T. Hasegawa and K.Kitazawa, Proc. Lain-American Conf. on "High Temperature Superconductivity", Rio de Janeiro, May 5-6(1988).

9. S.N. Mao, H.S. Huang, M.L. Lui, Z.Y. Wu, Z. Gao, X. Zhao, and Z.L. Huang, Chinese Phys. Lett., 5(9), 385(1988).

10. M. Takana, J. Takeda, K. Oda, H. Kitaguchi, Y. Miura, Y. Ikeda, Y. Tomii and H. Mazaki, Jpn. J. Appl. Phys., 27(1988).

11. Youmo Li, The Second Conf. on Inorganic Solid Chem. and Synthesis Chem. Hefei, China (1988).

Frontiers of Materials Research/
Electronic and Optical Materials
M. Kong, L. Huang (editors)

OBSERVATION OF PEROXIDE KIND OF OXYGEN IN $YBa_2Cu_3O_{7-x}$ SUPERCONDUCTOR

Liang zhong ZHAO and Yu xiu HU

Institute of Chemistry, Academia Sinica, Beijing, 100080, CHINA

High-Tc superconductor YBCO and the related compounds have been studied by means of iodometric titration, the wet experiments and XPS analysis. The results have shown the presence of peroxide kind of oxygen in YBCO superconductors and the related Ba-containing compounds annealed at 600°C. Under the same heat-treatment conditions, the content of peroxide kind of oxygen in YBCO superconductors is higher than that in non-superconducting Y-Ba-Cu-O compounds. XPS determinations have confirmed the result of absence of trivalent copper ions in YBCO superconductor. However, the nature of peroxide kind of oxygen in YBCO superconductor may be different from that in barium peroxide. The possible ways of formation of peroxide kind of oxygen in YBCO are discussed.

1. INTRODUCTION

Research on $YBa_2Cu_3O_{7-x}$(YBCO) superconductor has been focused on identifying the mechanism responsible for high Tc. Despite the intensive efforts, the mechanism has remained unclear. It is sometimes assumed that a mixed Cu^{3+}/Cu^{2+} valence state is responsible for the high-Tc superconductivity. The iodometric titration technique has been used to determine the Cu^{3+}/Cu^{2+} ratio in superconducting YBCO[1,2]. However, arguments can be found in the literature where more and more experimental evidences have been found for the absence of Cu^{3+} ions. In this work we present the results of the iodometric titration, the wet experiments and XPS analysis on superconducting YBCO and the related compounds. The results show that the "two-electron oxidant" in iodometric titration of superconducting YBCO may be not Cu^{3+} ions, but peroxide -like species (O_2^{2-} or O^{1-}). We suggest that the latter may play a key role in the high-Tc superconductivity.

2. EXPERIMENTAL

The samples listed in table 1 were prepared by mixing appropriate amount of oxides (Y_2O_3, BaO and CuO). The heat-treatment conditions of the mixtures are also shown in table 1. All the samples were analyzed by iodometric titration method[2]. In the determination of the sum of "one-electron oxidant" (Cu^{2+}) and "two-electron oxidant"(Cu^{3+}? or O_2^{2-}?), the sample was dissolved in 1.2mol KI/0.6mol HCl, then neutralized with ammonium hydroxide. Acetic acid was added and the liberated iodine was titrated with standard $Na_2S_2O_3$ solution. $NaCuO_2$ was prepared according to method described in literature[3]. XPS analysis were carried out on an KRATOS ES300 electron spectrometer.

3. RESULTS AND DISCUSSION

Table 1 shows the analytical results by iodometric titration. It is clear that all the high-Tc superconducting YBCO and the related Ba-containing compounds annealed at 600°C contain the "two-electron oxidant". As a comparison,

Table 1. The content of "two-electron oxidant" (expressed by BaO_2) in YBCO superconductors and the related compounds

Approximate formula	Zero-R Tc (K)	Theoretical total Ba(%)	$\frac{\text{Ba in BaO}_2}{\text{total Ba}}$ (%)	The heat-treatment conditions
$YBa_2Cu_3O_{6.9}$	78	41.3	15.6	preheated at 900°C for 12h, sintered at 950°C for 4h, slow cooled, annealed at 600°C for 8h in air
$YBa_2Cu_3O_{6.9}$	88	41.3	19.1	
$YBa_2Cu_3O_{6.9}$	90	41.3	17.2	
*$YBa_2Cu_2O_{5.5}$		47.5	10.4	
*$YBa_2Cu_1O_{4.5}$		55.0	9.0	
*$Ba_2Cu_3O_5$		50.4	8.9	
*$YCu_3O_{4.5}$		0	0	
*BaO		89.6	70.7	
*BaO		89.6	3.1	air quenched from 950°C
*BaO		89.6	0.4	Liq N quenched from 950°C
*$YBa_2Cu_3O_{6.9}$		41.3	12.6	air quenched from 950°C
*$YBa_2Cu_3O_{6.9}$		41.3	0	Liq N quenched from 950°C

* Initial composition

no "two-electron oxidant" can be found in the Ba-free compound and the liquid nitrogen quenched YBCO. The wet experiments show a rapid evolution of oxygen when superconducting YBCO powder or the Ba-containing compounds annealed under 600°C is dissolved into a mixed acid (5% HCl - 2% H_3PO_4), but no evolution of oxygen can be observed for the Ba-free compound and the liquid nitrogen quenched YBCO. These results reflect that the "two-electron oxidant" is related to the presence of Ba in the compounds, and the annealing condition at low temperature (600°C or lower) is essential for the formation of the "two-electron oxidant". In fact, it is well-known that barium oxide strongly favour the incorporation of peroxide units, and barium peroxide is formed easily from BaO by exposure to air at atmospheric pressure at low temperature (400°C - 600°C, for example). Besides, the evolution of oxygen from peroxide solution is also a well-known phenomenon, and the decomposition of peroxide is catalysed by transition metals. Shafer et al[4] have reported that the oxygen evolved by the acid dissolution of YBCO superconductor comes from the solid and not the solvent. Therefore, we suggest that the "two-electron oxidant" is not Cu^{3+} ions, but a kind of peroxide ions.

By comparison of the Cu2p spectra of $NaCuO_2$ and YBCO, the present XPS determinations have confirmed the absence of Cu^{3+} ions in YBCO superconductor. The $Cu2p_{3/2}$ binding energy of YBCO is 1.2eV lower than that of $NaCuO_2$ (see spectra a and b in Fig.1). However, no shoulder can be observed on the higher binding energy side of $Cu2p_{3/2}$ peak in spectrum a. Our XPS analysis also show that, $NaCuO_2$ must be prepared in O_2 atmosphere. If it is prepared by annealing in air at 450°C, the product is mainly a Cu^{2+}-containing compound (spectrum c). Furthermore, Cu^{3+} ion is very unstable, for example, Cu^{3+} ions in $NaCuO_2$ powder have converted to Cu^{2+} ions after exposed to moist air at room temperature for only 2 days (spectrum d); and no Cu^{3+} ions can be detected after heating $NaCuO_2$ at 920°C and subsequent annealing at 600°C in air (spectrum e). In contrast, YBCO superconductor can be prepared in air by heating above 920°C and subsequent annealing at 600°C, and it can be kept in air for more than several months without loss in superconductivity. These differences also imply that the presence of Cu^{3+} ions in YBCO superconductor is highly unlikely.

Since there are only two possibilities which can compensate the charge mismatch in YBCO compound, i.e. some of the Cu^{2+} ions are transformed to Cu^{3+} ions or oxygen holes are created, absence of Cu^{3+} ions implies that the higher oxidation state of oxygen (O_2^{2-} or O^{1-}) are present. Table 1 also shows that the content of peroxide kind of oxygen (expressed in terms of BaO_2) is more than 15% of the total Ba for superconducting YBCO, and in general is less than 10% of the total Ba for non-superconducting Y-Ba-Cu-O compounds.

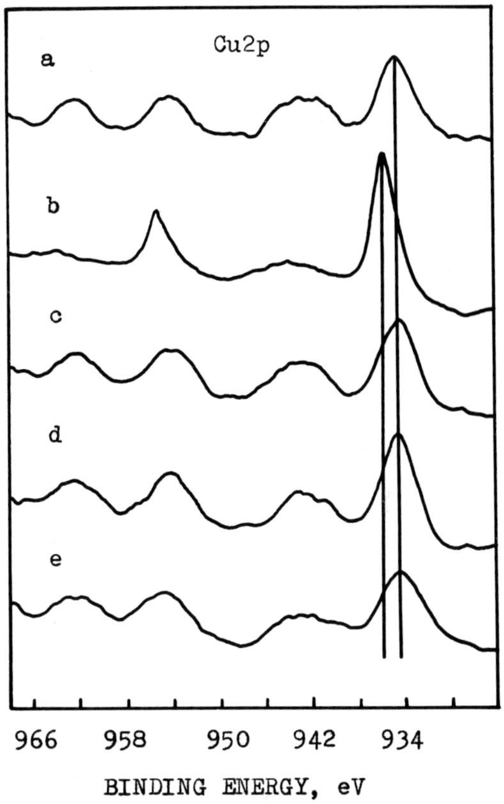

FIGURE 1
Cu2p XPS spectra of a) YBCO superconductor (Tc=90K); b) freshly prepared $NaCuO_2$ by annealing at 450°C in oxygen; c)the sodium cuprate sample prepared by annealing at 450°C in air; d) the sample in b) after exposing to moist air for 48h; e) the sample in b) after heating in air to 920°C for 2h and subsequent annealing at 600°C for 4h.

We suggest that the peroxide kind of oxygen in YBCO may play a key role in high-Tc superconductivity. However, so far we have failed to detect peroxide ions in acidic YBCO solution by reaction with $KMnO_4$. Further, upon the addition of a few drops of dilute $K_2Cr_2O_7$ solution, we were unable to observe the blue color in ether layer. The failure to chemically detect peroxide in YBCO has been attributed to rapid decomposition of peroxide ions catalyzed by Cu^{2+} at surface of the dissolving solid before reaction with $KMnO_4$ or $K_2Cr_2O_7$ can occur. It is also possible that the nature of peroxide kind of oxygen in superconducting YBCO is different from that in BaO_2 because the chemical bonds in YBCO are more covalent in character. At present time, however, the nature of peroxide kind of oxygen in YBCO is still far from clear. We speculate that the possible ways of formation of the peroxide kind of oxygen may be as follows: 1. It may come from trace unreacted BaO which can form peroxide during annealing; 2. it may originate from the superconducting phase itself, for example, since the O(1) - O(4) distance is quite short (ca. 2.6Å), the O(1) may couple with O(4) near Ba to form peroxide-like species. Our XPS analysis has revealed that O1s and Cu2p spectra changed their intensity or line shape concurrently with Ba4d spectrum after heating YBCO in vacuum. This implies that the released oxygen may have bonding with Ba and Cu ions in the crystal; 3. it may originate from chemisorbed oxygen which exists in the form of peroxide-like species. If the cases of 2) and 3) are true, we would expect that the peroxide kind of oxygen in YBCO may be not in a strict sense peroxide with usual -O-O- linkage, but a higher oxidation state of oxygen O^{1-} or O_2^{2-}. The O^{1-} can be thought of as a hole in the oxygen band, but to our knowledge there are no solids, other than peroxides, where the occurrence of an O^{1-} entity has been identified. In order to clarify the nature of peroxide kind of oxygen in YBCO superconductor, extensive research is required.

REFERENCES

1. D. C. Harris and T. A. Hewston, J. Solid State Chem., 69 (1987) 182.
2. X. S. Yuan and Y. M. Li, Anal. Chem. (in Chinese), 16 (1988) 708.
3. P. Steiner, V. Kinsinger et al., Z. Phys. B, 67 (1987) 497.
4. M. W. Shafer, R. A. de Groot et al., Physica C, 153-155 (1988) 836.

Frontiers of Materials Research/
Electronic and Optical Materials
M. Kong, L. Huang (editors)
Elsevier Science Publishers B.V., 1991

MICROSTRUCTURAL FEATURES, Cu-O PLANES BEHAVIOUR AND Jc IN $YBa_2Cu_3O_{7-x}$ SUPERCONDUCTING MATERIALS

SONG XiangYun, WEN ShuLin, ZHAO ZhongXian, ZHAO Meiyu, LI Chenen and JIN Deling

Open Research Lab. of High Performance Ceramics and Superfine Microstructural Studies
Shanghai Institute of Ceramics, Academia Sinica, Shanghai 200050, China

The purpose of this investigation is to reveal the correlation among the microstructural features, curitical current density (Jc) and Cu-O plane behaviour in YBCO. The Tc value of three samples was approximately the same (90K) but the Jc values were quite different: $30A/cm^2$, $300\ A/cm^2$ and $2000\ A/cm^2$. The results have indicated that the grains of $c/a > 3$, the grains with twin domains and the grain boundaries of parallel-straight play a major role in materials. Our analyses have shown that these microstructural features are dependent on Cu-O planes behaviour of two dimensions and the key to explain the mechanism of the relation between microstructure and Jc.

I. INTRODUCTION

It is well known that the microstructural feature is crucial to the superconducting characteristics of $YBa_2Cu_3O_{7-x}$. Electron microscopy is obviously a very important tool for studying microstructure of ceramic superconductors. Since the twin domains and grain boundaries have been characterized[1-3] several microstructural features have been suggested to explain the superconducting phenomena[4-11]. However, the relationships between microstructure and superconductivity are so far not established in spite of their careful electron microcope studies. This is due to the fact that, in some case, the superconducting properties of the corresponding samples have not been investigated because only a few crystals in most of samples have been investigated, i.e. these which are perfectly ordered and cannot be mostly considered as a representative of the ceramic samples. In this paper the corrlation between microstructure and curitical current density (Jc) with three different Jc values in YBCO ceramic superconductors were examined by transmission electron microscopy (TEM). The analysis have indicated that these microstructural features are dependent of Cu-O plane behaviour and the key to explain the mechanism between microstructure and Jc.

II. EXPERIMENTAL

The superconductivities of three YBCO samples, which have the same composition and different processings, were measured and compared. The Tc values of three samples were approximately the same (90K) but the Jc values were quite different: $30A/cm^2$ for sample S_1, $300A/cm^2$ for sample S_2 and $2000A/cm^2$ for sample S_3.

In order to pursue the different of magnitude in Jc value, hundreds of pictures for three samples were randomly taken in low magnification by TEM. The microstructural information, including the distribution of grain size, the ratio of length c to width a of grains, the content of grains with twin domain and grain boundaries between parallel-straight grains etc. were analysed by statistics from TEM pictures. The distribution of grain size with the range from ≤ 1, 1-3, 3-5 um, etc., to ≥ 9 um was divided. 577 grains of sample S_1, 589 grains of sample S_2 and 601 grains of sample S_1 were measured. The relationships between microstructure and Jc are shown in Table 1.

Table 1.
Relation Between Microstructures and Jc

Sample		S_1		S_2		S_3
Critical current density Jc(A/cm^2)		30		300		2000
Total grains		577		589		601
≤1 um grains	72	12.4%	26	4.41%	58	9.65%
1-3 um grains	231	40.00%	222	37.69%	233	38.77%
3-5 um grains	137	23.74%	207	35.14%	198	32.95%
5-7 um grains	74	12.82%	95	16.13%	66	10.98%
7-9 um grains	49	8.49%	34	5.77%	20	3.33%
≥9 um grains	14	2.43%	4	0.68%	27	4.49%
c/a ≥3 grains	50	8.67%	97	16.47%	182	30.28%
Grains with twin domain	43	7.45%	163	27.67%	316	52.58%
Grain boundaries with parallel-straight	55	9.35%	123	20.88%	221	36.77%

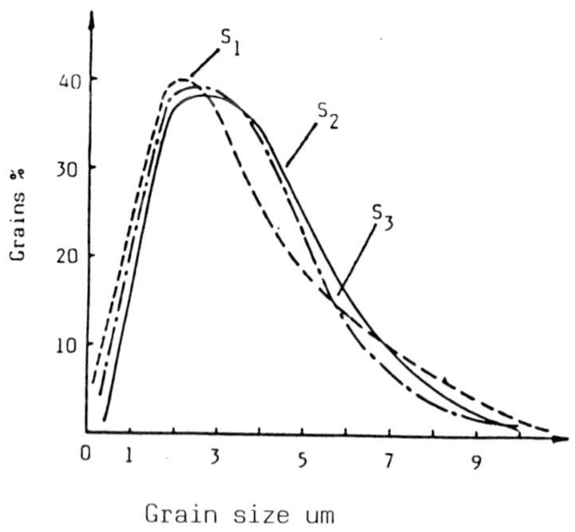

FIGURE 1
Distribution of grain size in three samples

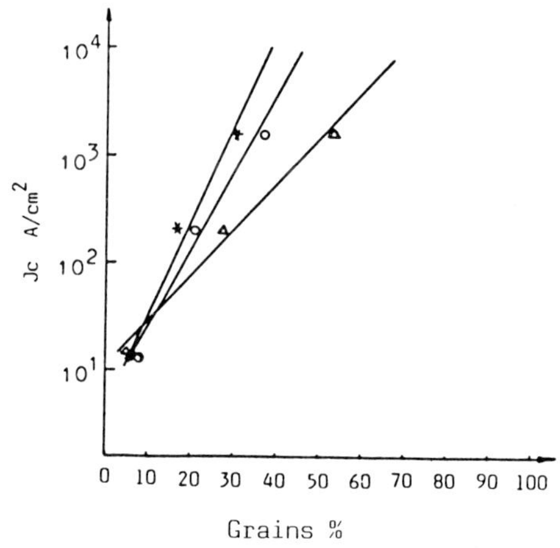

FIGURE 2
The correlation Between Microstructure and Jc in three samples.
* c/a 3 grains
Δ grains with twin domain
o parallel-straight grain boundaries

III. RESULTS AND DISCUSSION

Table 1. illustrates the correlation between the distribution of grain size and Jc showing in Fig. 1. It has been indicated that sample S_1 has a rather narrow distribution range in the grain size, sample S_2 and S_3 have wider distribution.It may be considered that the more grains with well-distribution in grain size will get the higher Jc.

The grains with the radio of length c to width a ≥ 3 often existing in YBCO ceramic superconductors have been found from a lot of observation. However,whether this kind of grains is the orthorhombic superconducting phase and what is the correlation between c/a ≥3 grains and Jc, it still hasn't been verified. The result from Table 1 has indicated that the c/a ≥

3 grains as a function of Jc are approximately linear shown in Fig. 2. The TEM observation have revealed that the c/a ⩾ 3 grains are of the orthorhombic superconducting phase and verified the certain orientation in crystallography. As shown in Fig. 3, the length c of grains corresponding to (001) orientation, it means that the length side of grains is parallel to ab planes which are connected to sheets parallel to the Cu-O plane (also called CuO_2 plane) and the width side of grain is perpendicular to this Cu-O plane. The result has also indicated that the grains with c/a ⩾ 3 exhibit the Cu-O planes in two dimensions and their conducting direction of the superconducting electrons, the more grains of c/a ⩾ 3 the higher Jc in YBCO ceramics.

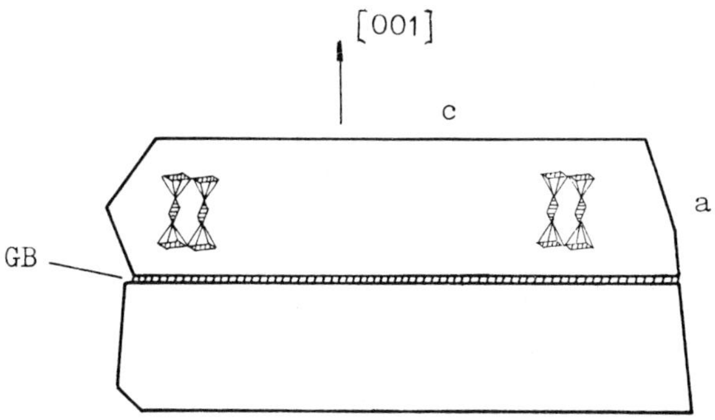

FIGURE 3
Schematic illustration of the c/a ⩾ 3 grain and parallel-straight grain boundary (GB).

The grain boundary is the basic characteristics of any ceramics. There are many kinds of grain boundaries such as the twin boundary, coherent grain boundary connected two grains without any grain boundary phase and the grain boundary with second amorphous phase etc..[2,3] A number of observations by TEM have found that the grain boundary between parallel-straight grains are very narrow and there are hardly any boundary phase in between. Therefore, it has been paid special attention to this kind of grain boundary during studying superconducting materials. It has been shown in Table 1 that the content of the grain boundary between parallel-straight grains increases as Jc value is increased and as a function of Jc as shown in Fig. 2. TEM study has proved that this grain boundary is usually parallel to the (001) plane, i.e. Cu-O planes in the grains on the two sides of this grain boundary are of parallel to the parallel-straight grain boundary as shown in Fig. 3. The structural features of this grain boundary has indicated that they are the expression of optimizing orientation in grains on two sides of this grain boundary. The grain boundary itself might be a very weak link or unlink for Cu-O planes. Therefore, the more grain boundaries between parallel-straight grains in YBCO ceramics, there will be the more conducting plies for superconducting electrons and will get the higher Jc.

A number of investigators have reported that the crystal grains of YBCO have invariably contained at least one set of parallel (110) twin domain. The grain with twin domain has been confirmed to be a superconducting phase of orthorhombic structure and the relation between twin domains and their textures in YBCO reported[4]. However, no further examination has indicated whether and why the grains with twin domains to enhance the superconducting performance. The results from Table 1 have shown that there is certain correlation between the content of grains with twin domains and Jc in YBCO. The Jc variation as a function of grains content with twin domain has been shown in Fig. 2. According to the orthorhombic superconducting structure, their angles are shared in Cu-O planes with square-planar when Cu-O planes link along (100) or (010) plane as shown in Fig. 4a. Howherver, when Cu-O planes link along (110) plane, the Cu-O square-planar are shared between their edges as shown in Fig. 4b. The Cu-O planes

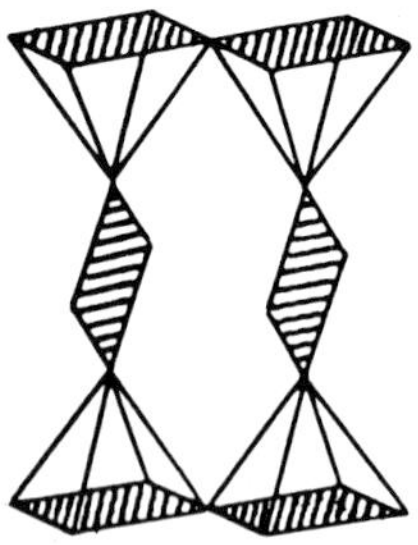

FIGURE 4a
Schematic illustration of Cu-O planes link along (100) or (010) plane.

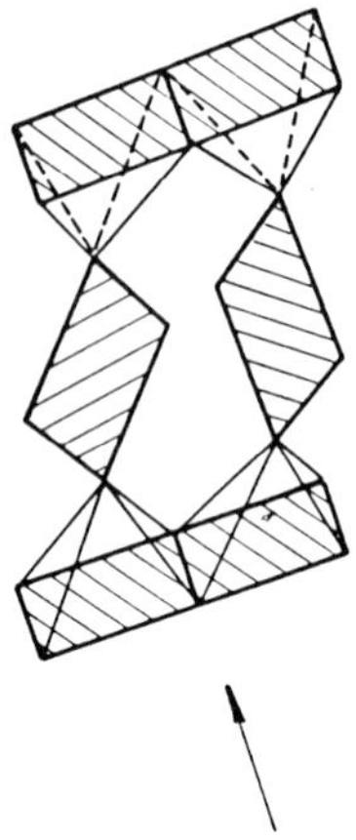

FIGURE 4b
Schematic illustration of Cu-O planes link along (110) plane

presented a good link and the conducting might be an origin of the twin domain for enhancing Jc. It has been also expressed that the width of twin domains and their textures are dependent on the distribution of oxygen vacancy and the degree of order of oxygen vacancies[12].

On the other hand, it can be seen from Table 1 when the Jc value increas an order of magnitude i.e. its, range from 10^1 to $10^3 A/cm^2$ and the grain content of twin domanis, and ratio $c/a \geqslant 3$ as well as parallel-straight grain boundaries all of them increas about a time. If the grains with twin domains are risen to 100 percent the Jc values could be reached $10^4 A/cm^2$. The Jc values might be reached above $10^5 A/cm^2$ if the content of the $c/a \geqslant 3$ grains and grain boundaries with parallel-straight are further risen.

IV. CONCLUSION

The correlation among the microstructural features, Jc and Cu-O plane behaviour have been established and the following experimental results obtained:

(1) The values of curitical current density are dependent on three kinds of microstructural features made up of three parts: the $c/a \geqslant 3$ grains, the grain boundaries with parallel-straight and the grains with twin domains.

(2) The Cu-O planes in two dimensions presented a good link and the conducting are the origin of these microstructures for enhancing Jc values.

(3) According to the relation between microstructures and Jc, the Jc values might expect tendency to rise to 10^4 and 10^5 A/cm^2 and hardly reach 10^6 A/cm^2 in YBCO ceramic superconductors.

REFERENCES

(1) M. Hervieu, B. Domenqes, C. Michel, G. Heqer, J. Provost, and B. Raveau, Physical Review B, 36(7)(1987)3920.

(2) SONG Xiangyun, Feng Jingwei, LI Chengen and WEN Shulin, KEXUE TONGBAO (Science Bulletin) 33(22)(1988)1905.

(3) WEN Shulin, SONG Xiangyun and FENG Jingwei, Materials Letters, 6(1988)385.

(4) Y. Kawamata, H.U. nissen, C. Beeli, F. Hulliger and H.R. Ott, Z. Phys. B--Condenseed Matter, 72(1988)345.

(5) G. Roth, D. Ewert and G. Heger, et al., Z. Phys. B--Condensed Matter, 69(1987)21.

(6) M. Hervieu, B. Domenges, B. Raveau, and Post, Materials Letters, 8(1989)73.

(7) Ichihashi, T., Iijima, S., kubo, Y., Tabuchi J. Jpn. J. Appl. Phys. Lett., 27, L1187 (1988).

(8) Zandbergen, H.W., Thomas, G., Phys. Status Solid(a), 107 (1988)825.

(9) H. Krishnan, V. Sankaranarayanan, et al., Phys. Rev. B, 40(4)(1989)2639.

(10) M. Ausloos, Ch. Laurent, M. Laguesse, S.R. Patapis, et al., Mod. Phys. Lett. B3, (1989)167.

(11) SONG Xiangyun, LI Chengen, ZHAO Meiyu and WEN Shulin, Chinese Science Bulletin, 35(5) (1990)375.

(12) SONG Xiangyun, ZHAO Meiyun, LI Chengen, and WEN Shulin, Chinese Science Bulletin, 35(13)(1990).

Frontiers of Materials Research/
Electronic and Optical Materials
M. Kong, L. Huang (editors)

INTERNAL FRICTION PEAK DUE TO MOVING FLUX IN BSCCO BULK SUPERCONDUCTORS

J.X. Zhang, W.G. Zeng, G.M. Lin, Z.L. Du, K.F. Liang and Z.C. Lin

Department of Physics and Institute of Material Science, Zhongshan University, Guangzhou 510275, P.R. China

The low-frequency internal friction in the magnetization process is measured as a function of applied field H for $(Pb_{0.2}Bi_{0.8})SrCaCu_2O_x$ at a temperature below T_c. An internal friction peak is found at H 45x10^{-4}T and is concluded to associate with the moving flux. A new method of investigating the dynamics and the pinning mechanism of flux in high-T_c superconductor is discussed.

1. INTRODUCTION

The importance of the study of flux dynamics in high T_c superconductor has theoretical and practical reasons. Because the character of quantum magnetic flux not only is an elementary property of a high -T_c superconductor but also associates with the flux pinning and the problem of high critical current density J_c. The study of flux noise and flux creep had been reported by M.J. Ferrari et al for YBCO high-T_c superconductor [1], Y. Yeshurun et al [2] reported a giant flux creep in BSCCO crystal, C.W. Hagen and R. Griessen [3] reported the distribution of activation energies for thermally activated flux motion, but the dynamics behaviour of flux has not yet reported.

Two of the authors had reported a low frequency internal friction (IF) peak originated from the moving magnetic domain walls (MDWS) in the magnetization process for the ferromagnetic materials [4,5]. The dynamic property of MDWS can be obtained from the dependence of IF peak on the variation rate of applied field $\dot{H}$ and the frequency of IF measurement [6]. The relationship between average velocity of MDWS, the effective driving force in the magnetization process and the resistance exerted on moving MDW can also be calculated as a function of applied field H [7]. Consequently, it is interesting to see whether an analogous IF peak associated with the moving flux exists or not in the magnetization process for high T_c superconductor. We report such results for the BSCCO bulk superconductor.

2. EXPERIMENTS AND SPECIMEN

A muti-function inverted torsion pendulum system (MFIFA-1, made by the institute of Solid state Physics, Academia Sinica) modified with a helix is used for measuring internal friction (IF) in the process of magnetization. The specimen is situated at the center of helix and parallels to the axis of helix. The specimen of size 1x5x50 mm^5 is enclosed with the torsion system, and immersed in a liquid-Nitrogen Dewar in order to obtain a wide range of temperature variation. The IF is measured at two or four selected frequencies in the range of 10^{-2} to 5HZ simultaneously with the forced vibration mode. The strain amplitude IF measurment is about 5x10^{-8}. The measurements prints, plots and the temperature controll are carried out automatically by an IBM computer. Sample was prepared by a powder method, using BiO_2, $SrCO_3$, $CaCO_3$. CuO and PbO powders as starting material. The nominal composition of the sample is $(Bi_{0.8}Pb_{0.2})SrCaCu_2O_x$(1112). The mixed powder is calcined at 820 C for 3 hrs, grinded and pressed into a rectangular piece of size 1x5x 60 mm and then sintered at 840 C for 45 hrs.

To ensure the sample has a same initial state (when H=0, M=0), after each measurement with magnetic field, the sample is heated to normal state (110K) and measured with field, then cooled in zero field to superconducting state (95K).

3. EXPERIMENTAL RESULTS

The graphs of resistance R versus T and a.c. susceptibility vs. T of (1112) specimen are shown in Fig.1. It is shown that the (1112) specimen has two superconducting (SC) phases. The T_c (onset) and T_c (finished) are found to be 105K and 97K for the high-T_c phase and 90K, 80K for low-T_c SC phase respectively. The T_c (R=0) is found to be 97K in the graph of R vs T. Because of the coexistence of two superconducting phases, the results of IF measurement in the process of magnetization at 95K are associated merely with the moving flux in high-T_c SC phase.

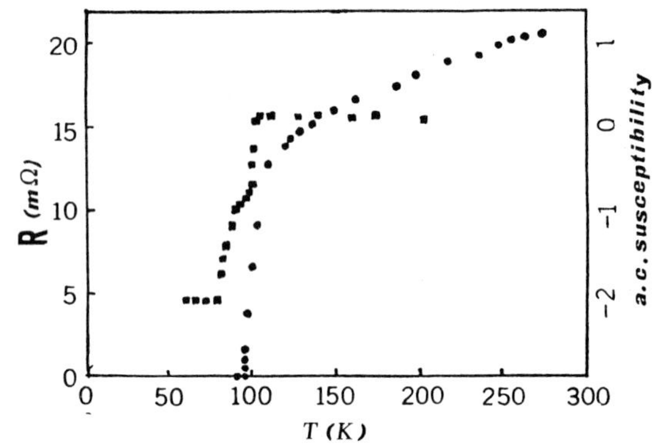

FIGURE 1
The graphs of R vs T and a.c. susceptibilty vs T of (1112) specimen.

Fig.2 shows the temperature-dependence of IF with applied field of 15×10^{-4}T for 1112 specimen. Two frequencies are used for this diagram: f_1= 0.1Hz, f_2=5Hz. It is shown that there is no relaxation IF peak to be found but the frequency-dependence of internal friction is inverted at about 109K.

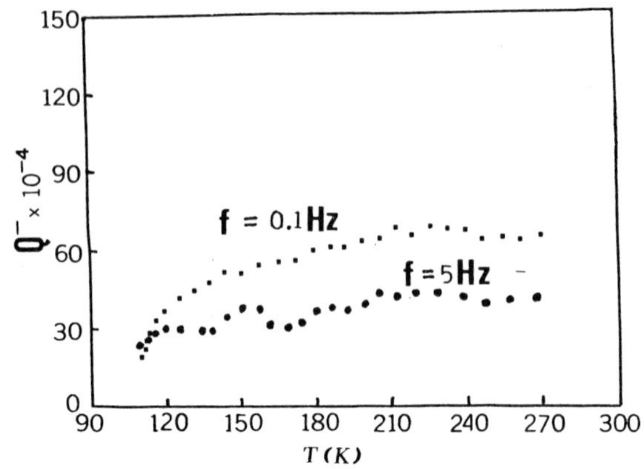

FIGURE 2
The temperature-dependence of Q^{-1} f_1=0.1Hz, f_2= 5Hz, H=15×10^{-4}T.

Fig.3 shows the effect of varying rate of applied field $\dot{H}$ on the IF in the process of magnetization. The graphs of Q^{-1} vs H with f= 1Hz for superconducting state specimen (95K±0.5 K) are shown in Fig.3(a). An IF peak appears at the Q^{-1}-H curves when $\dot{H}\neq0$, and the peak

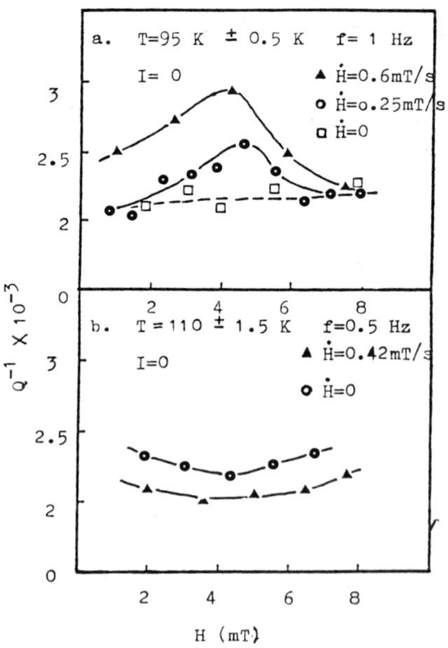

FIGURE 3
The effect of $\dot{H}$ on Q^{-1} curves. a, SC state; b, normal state.

height increases with an increasing of $\dot{H}$ but the IF peak disappears when $\dot{H}$=0. The intensity of magnetic field H_p at which the IF peak situated is about $44x10^{-4}$T. The graphs of Q^{-1} vs H with f=0.5 Hz for normal state are shown in Fig.3(b). It is clear that there is no IF peak to be found for both $\dot{H}$=0 and $\dot{H}\neq0$. Contrary to the SC state, the value of IF for $\dot{H}$=0 in this case is even higher than that for $\dot{H}\neq0$.

Fig.4 shows the influence of measuring frequency on IF in the process magnetization. The graphs of Q^{-1} vs H with $\dot{H}=6x10^{-4}$T/S for SC state (95$\pm$0.5K) are shown in Fig.4(a). It is shown that IF peak with f=1Hz is a bit higher than that with f=5Hz. The graphs of Q^{-1} vs H with $\dot{H}$ =$4.2x10^{-4}$T/S for normal state are shown in Fig. 4(b). Contrary to the SC state again, the Q^{-1}-H curve for f=0.5Hz in this case is lower than that for f=5Hz in the case of $\dot{H}\neq0$.

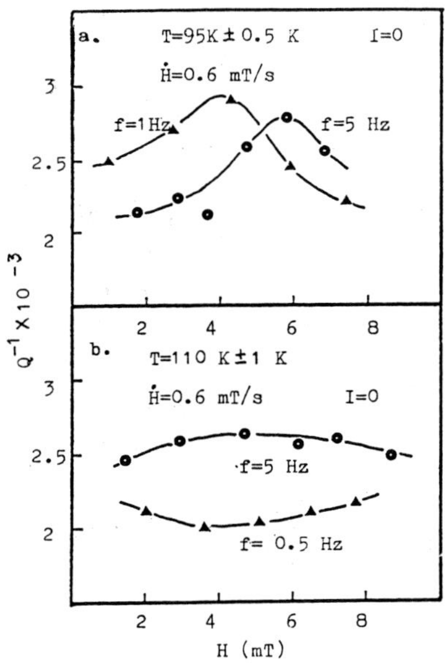

FIGURE 4
The effect of mearsuring frequency of Q^{-1}-H curves.
a, SC state; b, normal state.

Fig.5(a) shows the effect of current, i.e. the crosswiss magnetic field, on the Q^{-1} curve for SC state (95K$\pm$0.5K) when f=5Hz and $\dot{H}=6x10^{-4}$ T/S. It is shown that the existence of current (2mA) causes an increasing of IF peak and a shift of H_p to lower value. Contrary to SC state, the Q^{-1}-H curve is independent of the current in the case of normal state (Fig.5b 112.5K$\pm$0.5K).

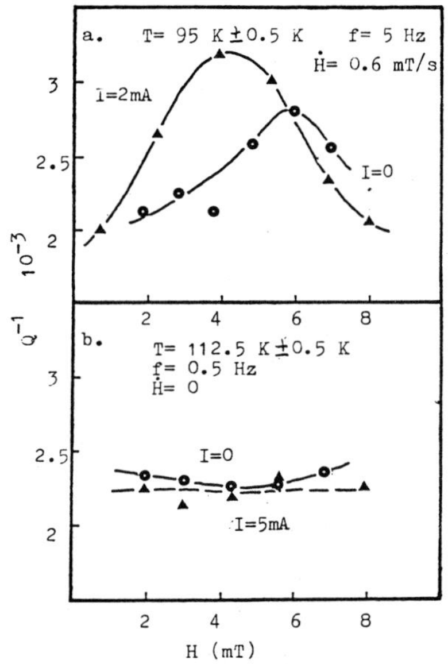

FIGURE 5
The effect of current on Q^{-1}-H curves.
a, SC state; b, normal state.

Fig.6 shows the influence of $\dot{H}$ on Q^{-1}-H curve when current exists. The graphs of Q^{-1} vs. H with I=2mA and f=5Hz for SC state (95K$\pm$0.5K) are shown in Fig.6(a). It is shown obviously that the peak height with $\dot{H}=6x10^{-4}$T/S is much higher than that with $\dot{H}=2.5x10^{-4}$T/S, and the increase of $\dot{H}$ causes the H_p shifts to a lower field, but the Q^{-1}-H curve is indenpendent of the existence of current and the variation of $\dot{H}$ for the normal state (110K$\pm$1.5K, Fig.6(b)).

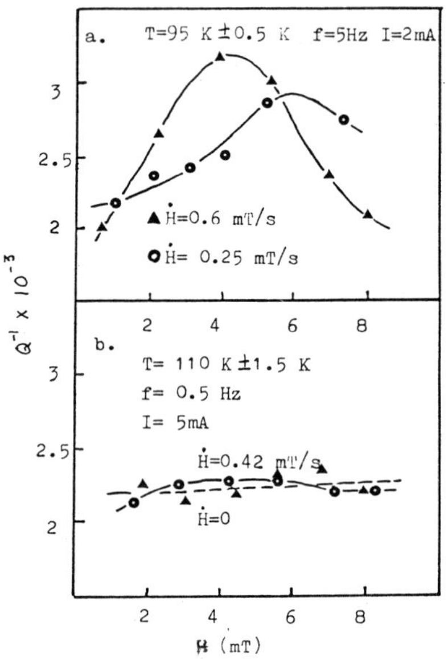

FIGURE 6
The effect of Ḣ on Q^{-1}-H curves when current existing.
a, SC state; b, normal state.

The experimental results described in Fig.3-Fig.6 can be summarized as: 1, An IF peak can be found in the process of magnetization when specimen is in superconducting state, the peak height of IF increases with an increasing of Ḣ but decreases with an increasing of measuring frequency; 2, The existence of crosswiss magnetic field causes an increase of peak height and makes H_p shift to a lower field. 3, When specimen is in normal state, the characteristic IF peak disappears.

4. DISCUSSION

Experimental results described above are analogous to the IF peak in the process of magnetization in ferromagnetic materials (Fe, Ni and permalloy etc)[5], where the moving magnetic domain walls cause the IF peak with the same character: a, peak height Q_p^{-1} increses with an increasing of Ḣ; b, Q_p^{-1} decreases with an increasing of f; c, when Ḣ=0, the peak disappears.

H. Kumakura et al [8] have reported that the lower critical field H_{c1} of 1112 BSCCO specimen is found to be 45×10^{-4}T at 77K and it is about the same value of H_p at which the IF peak appeared. In the process of magnetization for the magnetic field range used in the present work, the quantum flux can thus enter into the SC phase of the two phases 1112 specimen at 95K and makes the SC phase to be a critical state. Therefore the IF peak appeared in the process of magnetization can be explained to associate with the entering of flux. Since the peak height Q_p^{-1} increases with an increasing of Ḣ and the peak disappears when Ḣ=0, the IF peak is consequently involved with a rate process and associated inevitably with the moving flux.

If the magnetization curves are determined with variable sweep rate of applied field Ḣ, the average resistance exerted on the moving flux can therefore be calculated [9]. Hence, the results of present paper can be used to develope a new method for investigating the flux dynamics.

ACKOWLEDGEMENTS

This work was supported by The National center for Research and Develpment on Superconductor and The Science Technology Committee of Guangdong Province of China.

REFERENCE

1. M.J. Ferrai, M. Johnson, F.C. Wellatood, J. Clarke, P.A. Rosenthal, R.H. Hammond and M.R. Beasley. Appl. Phys. Lett. 53(8), 22, 695 (1988).

2. Y. Yeshurun and A.P. Malozemoff, Phys. Rev. Lett., 60 2022 (1988).

3. C.W. Hagen and R. Griessen, Phys. Rev. Lett., 62 2857 (1989).

4. J.X. Zhang, W.G. Zeng and J.M. Wu, Acta Scien. Natur. Vni. Sunyatseni. 24 109 (1984).

5. W.G. Zeng, J.X. Zhang and H.Q. Lin, Acta

Physica Sincia 35 1377 (1986).

6. W.G. Zeng and J.X. Zhang, Acta Physica Sinica. 36 37 (1987).

7. J.X. Zhang and X.J. Li, Acta Physica Sinica 36 847 (1987); Cinese Physics 8 308 (1988).

8. H. Kumakura, K. Takahashi, D.R. Dietderich, K. Togano and H. Maeda, Appl. Phys. Lett. 52(24) 2064 (1988).

9. J.X. Zhang and W.G. Zeng, Acta Scientiarum Natur. Universitatis Sunyatseni, 29 No.2, 1 (1990).

Frontiers of Materials Research/
Electronic and Optical Materials
M. Kong, L. Huang (editors)

EFFECT OF Sb OR Cd ADDITION ON THE FORMATION OF SUPERCONDUCTING PHASES IN THE Bi-BASED SYSTEM

Xiaolin WANG, Hong WANG, Shuxia SHANG, Zhuo WANG, Zhikuan LU and Minhua JIANG

Institute of Crystal Materials, Shandong University, Jinan 250100, P.R. China

The effect of Sb or Cd addition on the formation of superconducting phases in the ceramic samples of $Bi_{1.84}Pb_{0.34}Sb_{0.1}Sr_2Ca_2Cu_3O_y$ and $Bi_{1.84}Pb_{0.34}Sr_2Ca_2Cd_xCu_3O_y$ (x=0.1 to 1 in step of 0.1) has been studied by using a direct sintering and air quenching technique. The measurements of resistivity and a.c. susceptibility show that there are two main phases in the Sb-doped sample, one is 2223 phase, the other is 2234 phase with Tc(zero) and cell parameter c of 95K and 43.35Å, respectively. For Cd-doped samples, Tc(zero) decreased slightly with the increased doping of Cd, and the X-ray diffraction patterns reveal that 2212 phase is the main phase in these samples.

1. INTRODUCTION

It has been well established that there are three superconducting phases i.e. 110K (2223) phase, 85K (2212) phase and 7K (2201) phase with three, two and one Cu-O plane in each half unit cell. Hongbao et al.[1] proposed a new superconducting phase with Tc of 132K would probably be formed in Sb-doped Bi-Pb-Sr-Ca-Cu-O sample prepared by conventional sintering method, but its structure is still unknown. Hideaki et al.[2] synthesized thin film with four Cu-O planes in a unit cell by sputting method. They found that the Tc is lower than that of 2223 phase, but the Tc(zero) was not determined. Several researchers[3,4,5] has focused on the Sb-doped Bi-Pb-Sr-Ca-Cu-O superconductor by different heat treatments. Although their results were different from one another, they all observed some unknown peaks in XRD pattern at $2\theta\sim30^\circ$ which might be related to to a new phase. The preparetion of Cd-doped Bi-Sr-Ca-Cu-O superconductor was also reported.[6] Yoshimura[7] observed that only 2212 phase was inside the thin film of Pb-free $Bi_2Sr_2Ca_{2-x}Cd_xCu_3O_y$ prepared by the rapid-quenching-after-rapid-melting technique. In this paper, we report the superconductivity of $Bi_{1.84}Pb_{0.34}Sb_{0.1}Sr_2Ca_2Cu_3O_y$ and $Bi_{1.84}Pb_{0.34}Sr_2Ca_2Cd_xCu_3O_y$ (x=0.1 to 1 in step of 0.1) prepared by the direct sintering and air quenching technique. The 2223 phase and 2234 phase with Tc(zero) of 95K and cell parameter c of 43.35A are main phases inside the Sb-doped sample. For Cd-doped samples, the content of 2223 phase lowered and the 2212 phase become main phase with the increase of Cd doping. The role of Sb or Cd for the formation of superconducting phases is also discussed.

2. EXPERIMENTAL

Samples with nominal composition of $Bi_{1.84}Pb_{0.34}Sb_{0.1}Sr_2Ca_2Cu_3O_y$ and $Bi_{1.84}Pb_{0.34}Sr_2Ca_2Cd_xCu_3O_y$ (x=0.1 to 1 in step of 0.1) were prepared by a direct sintering and air quenching method. The starting materials were high-purity powders. They were well mixed and pressed into disk like pallets, directly sintered at 840°C for 24h or 76h for Sb-doped sample, and at 830°C for 120h for Cd-doped samples, respectivly, then rapidly quenched in air. Resistance measurements were performed by a standard four-probe method. The a.c. susceptibility were measured by using HP-4192-ALF Impendance Analyser. X-ray powder diffraction patterns were obtained with a Japan Rikaku D/Max RA X-ray diffractometer using CuK_α radiation for these samples. The composition of crystal grains in these samples were analyzed by using Philips EDAX, PV-9100 energy dispersive X-ray spectroscopy.

3. RESULTS AND DISCUSSION

3.1. Superconductivity of $Bi_{1.84}Pb_{0.34}Sb_{0.1}Sr_2Ca_2Cu_3O_y$

Figure 1 shows the temperature dependence on the resistivity and a.c. susceptibility of the Sb-doped sample after sintered at at 840°C for 24h (sample A) or 76h (sample B). The curve of sample A in Figure 1 (a) reveals that the sample is multiphased with three superconducting transition in the temperature ranges between 77K and 130K. The frist is 107K corresponding to 2223 phase, the second is 95K to 2234 phase, and the third is 83K to 2212 phase. The XRD pattern of sample A is shown in Figure 2 (a). At low angle region (2θ<8°), there are four peaks with d value of 21.86Å, 18.1Å, 15.7Å and 12.32Å, respectively. These four peaks indicate that there are four phases, 2234 phase, 2223 phase, 2212 phase and 2201 phase in this sample, and the main phases are 2223, 2212 and 2234 phase. After sintered for 76h, it is found that there are only two superconducting transition steps and the zero resistance reaches 95K from a.c. susceptibility and resistance measurements as shown in Figure 1 (b) (curve B) and (curve A). The peaks of 2212 phase is nearly disappeared, while the peaks of 2234 phase and 2223 phase are still clear. (See Figure 2 (b)). These results show that the sample sintered for 76h contains mainly 2223 phase and 2234 phase. The XRD analysis was carried out to indentify the structure of the 95K (2234) phase. About 20 lines can be indexed by using a=5.41Å, b=5.44Å, c=43.35Å. Some unknown peaks at 2θ~30° cited in literature[3,4,5] are also indexed.

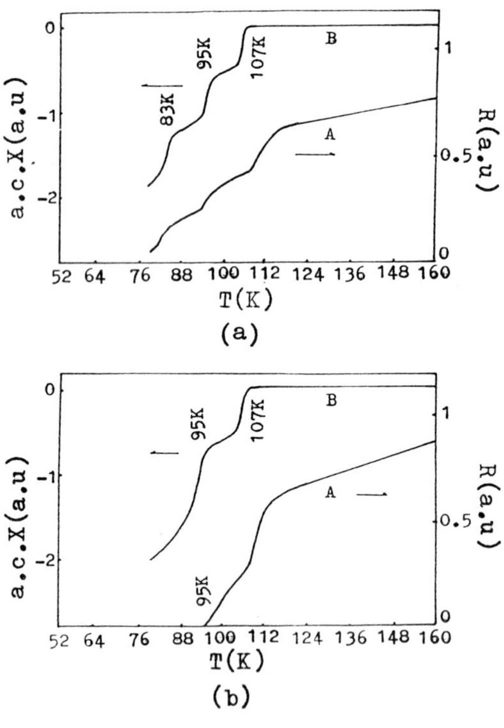

FIGURE 1
Temperature dependence of the resistivity and a.c.X. of Sb-doped sample after sintered at 840°C for 24h (a) or for 76h (b).

The EDAX measurements were made on the flat surface of the crystal grains in the sample. It is found that there are mainly two types of grains, whose compositions are close to (BiPb):Sr:Ca:Cu of 2:2:2:3 and 2:2:3:4 respectivly. The Sb element has not been detected in these grains. The role of Sb addition is that it may change the free energy for the formation of 2223 phase and 2234 phase as we discussed in our another paper.[9]

3.2 Superconductivity of $Bi_{1.84}Pb_{0.34}Sr_2Ca_2Cd_xCu_3O_y$

The temperature dependence of the resistivity of Cd-doped samples sintered at 830°C is shown in Figure 3. It shows that for sample of x=0.1, there are two superconducting transition around 110K and 85K, which means that at least two phases i.e. 2223 phase and 2212 phase exsist in the sample, while for other samples (x=0.2 to 1), only one superconducting transition around 80K with Tc(zero) decreased from

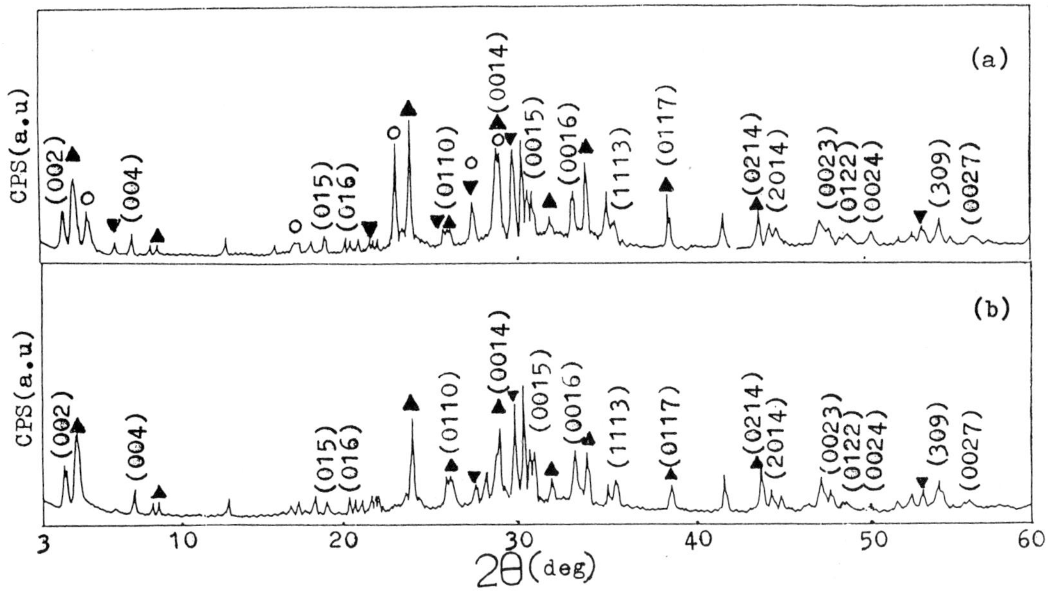

FIGURE 2
X-ray powder diffraction patterns of $Bi_{1.84}Pb_{0.34}Sb_{0.1}Sr_2Ca_2Cu_3O_y$ sintered at 840°C for 24h (a) or 76h (b). Peaks marked ▲, ○, ▼, are from 2223 phase, 2212 phase and 2201 phases, respectivly. Indexed lines are from 2234 phase.

80K to lower than 77K gradually. The results of XRD measurements of Cd-doped samples show that the 2223 phase disappeared and the 2212 phase become dominant with the increase of Cd doping. For samples of x=0.3 to 1,XRD patterns indicate that only 2212 phase existed. Figure 4 (a) and (b) show the patterns of the sample

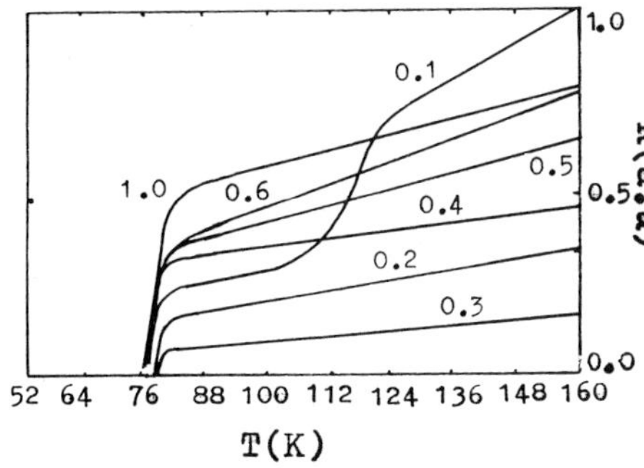

FIGURE 3
Temperature dependence of the resistivity on various Cd concentration. Samples were sintered at 830°C for 120h.

of x=0.4 and x=1, respectivly. The EDAX measurement was carried out to determine the actual doping level. The composition of a grain inside the sample of x=1 is $(BiPb)_{3.04}Sr_{3.35}Ca_{0.98}Cd_{0.64}Cu_{1.87}$ close to (BiPb):Sr:(CaCd):Cu=2:2:1:2, which means that the actual doping level is close to Cd:Ca=1:2,and is in good agreement with the nomonal doping level. Because the ionic radius of Cd^{2+} is smaller than that of Ca^{2+}, the Cd^{2+} substitution for Ca^{2+} may cause the decrease of lattice parameter of c, which results in the lowering of Tc (zero). (See Figure 3). In order to investigate the role of the Cd element for the formation of 2212 phase together with the disappearence of the 2223 phase, DTA measurement was carried out.[10] It is found that the melting point of Cd doped samples decreased with the increase of the doping level. This is due to the substitution of Cd for Ca, which bring about the lowering of the optimum sintering temperature.

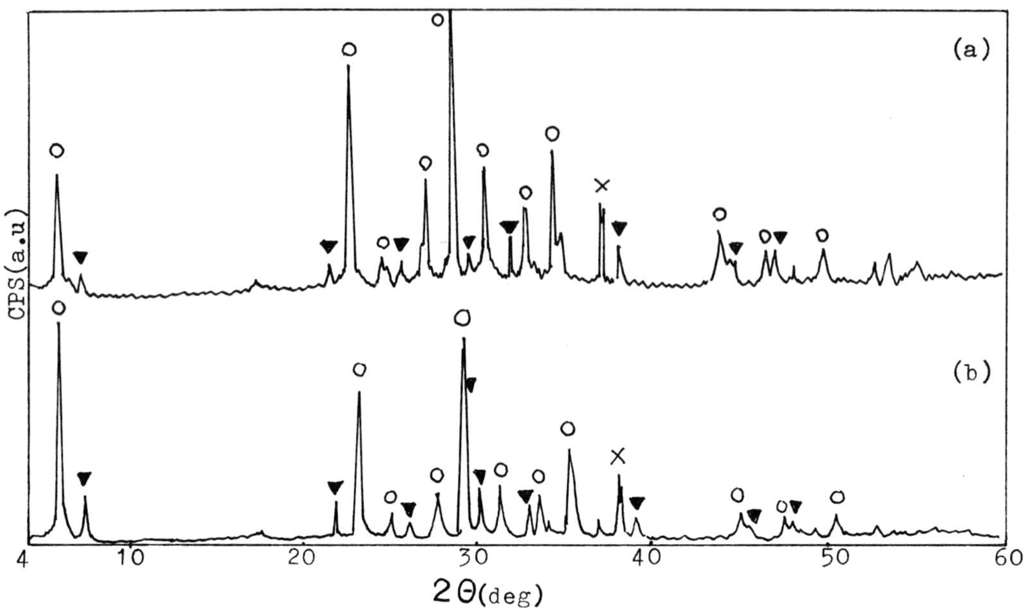

FIGURE 4
X-ray powder diffraction patterns of $Bi_{1.84}Pb_{0.34}Sr_2Ca_2Cd_xCu_3O_y$ sintered at 830°C for 120h. (a):x=0.1 (b):x=1. Peaks marked o, ▼, x, are from 2212 phase, 2201 phase and unknown phases.

The 2223 phase can easily be descomposed at high temperature of 830°C. The disappearence of 2223 phase by the addition of Cd element to the Bi-system also means that the crystal structure of 2212 phase become more stable than than that of 2223 phase by the substitution of Cd atom into Ca sites.

4. CONCLUSIONS

The effect of Sb or Cd addition on the formation of superconducting phases in the ceramic samples of $Bi_{1.84}Pb_{0.34}Sb_{0.1}Sr_2Ca_2Cu_3O_y$ and $Bi_{1.84}Pb_{0.34}Sr_2Ca_2Cd_xCu_3O_y$ (x=0.1 to 1 in step of 0.1) has been studied by using a direct sintering and air quenching technique. It is found that there are two main phases in the Sb-doped sample, one is 110K (2223) phase, the other is 95K (2234) phase with cell parameters of a=5.14Å, b=5.44Å, c=43.35Å. The role of Sb element addition is that it can change the free energy for the formation of superconducting phases, which results in the increase of the growth of 2223 phase and 2234 phases. For Cd-doped samples, because of the Cd substitution for Ca, the zero resistance temperature decreases with increase of Cd-doping, and the XRD patterns reveal that 2212 phase is dominant inside the Cd-doped samples.

ACKNOWLEDGEMENTS

The authors would like to thank Mr.S.M. Dong and Mr. W.T. Yu for their asistance in XRD measurements.

REFRENCES

1. H.B. Liu, L.Z. Cao et al., Solid State Commun. 69 (1989) 867

2. H. Adachi, S. Kohiki et al., Jpn. J. Appl. Phys. (1988) L1883

3. T. Komtsu, R. Sato et al., Jpn. J. Appl. Phys. 28 (1989) L1159

4. J.S. Luo, D. Michel et al., Appl. Phys.Lett.

14 (1989) 1449

5. P.V.P.S.S. Sastry, J.V. Yakhimi et al., Solid State Commun., 5 (1989) 935

6. T. Kanai, T. Kamo et al., Jpn. J. Appl. Phys. 28 (1989) L551

7. M. Yoshimura, T.H. Sung et al., Jpn. J. Appl. Phys., 28 (1989) L424

8. H. Wang, X.L. Wang, S.X. Shang, Z. Wang, Z.K. Lu and M.H. Jiang, Appl. Phys. Lett., to be published

9. X.L. Wang, H. Wang, S.X. Shang, Z. Wang and M.H. Jiang, to be published.

10. X.L. Wang, H. Wang, S.X. Shang, Z. Wang, Z.H. Yang, Z.K. Lu and M.H. Jiang, to be published.

Frontiers of Materials Research/
Electronic and Optical Materials
M. Kong, L. Huang (editors)

THE OXIDE SUPERCONDUCTING COATING USED FOR MAGNETIC SHIELDING

Shengwei QIAN, Chengfu LIN, Kan GUAN, Lishi WEN AND Yuzhi ZHUANG,

Institute of Metal Research, Academia Sinica, 2-6 Wenhua Road, Shenyang 110015, China

The plasma spraying techniqe was used to prepare oxide superconducting coating. By this method, it is easy to make magnetic shielding devices with large area and complex shapes. The preparation , microstructure and diamagnetism of the coating materials were disscused in detail and a shielding device made by this method shows a good shielding effects.

1. INTRODUCTION

A fundamental requirement for the utilization of superconducting electronics is a well magnetic shielding enviroment. The oxide superconductor can perform it successfully. In the earlier reports[1,2], the shielding devices are made of bulk superconducting materials by sintering process. By that way, it is very difficult to make devices with large area and complex shapes due to the high brittleness of these materials, The plasma spraying, which is very flexible in forming superconductor with a great variety of geometries and complex shapes[3,4], is used in this work to prepare superconducting coating on manifold substrates to form a shielding devices. The microstructure, phase transformation and superconductivity of the coating materials are disscussed in detail, and its diamagnetic behavior in low magnetic field are emphasized.

2. THE PREPARATION AND THE MICROSTRUCTURE OF THE COATING

Y_2O_3, $BaCO_3$ and CuO were mixed with appropriate ratio, sintering in air at 950℃ for 10H and then grinding into feedstock powder.Fig.1a shows its X-ray diffraction pattern. The plasma spraying was carried out in air using GDP- 80KW equipment and the thickness of the coating is about 0.3mm.

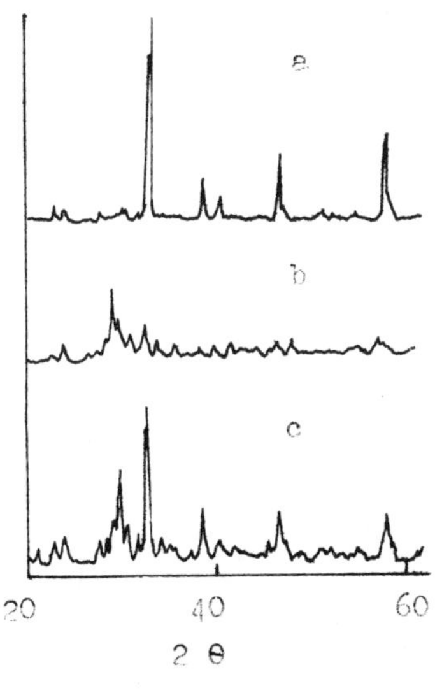

FIGURE 1
X-ray diffraction pattern of the amples: (a) feedstock powder; (b) as-sprayed coating; (c) annealed coating.

The as-sprayed coating is black in colour and poorly crystalized(Fig.1.b). The micrograph of its fracture surface shows a lamella structure usually typical for plasma sprayed coating with microcracks and amorphous states(Fig.2). The as-sprayed coating usually have no superconductivity in LN temperature.

The project was supported by the National Natural Science Foundation of China

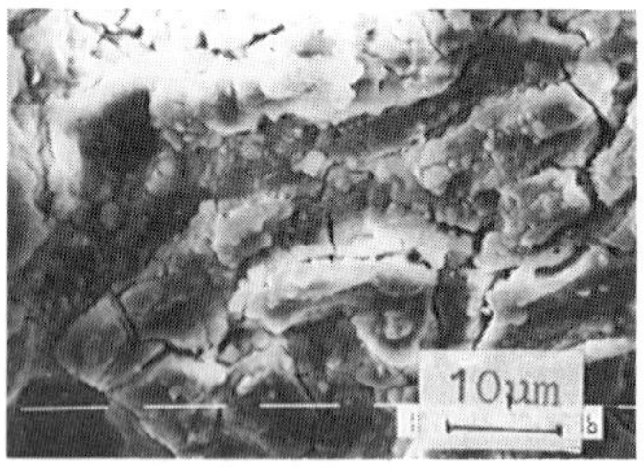

FIGURE 2
The SEM micrograph of the as-sprayed coating.

After the coating was properly annealed in about 900℃, Its superconductivity can be restored with zero-resistance temperature higher than 90K, The properties of the coating vary a lot depend on the annealing conditions. Fig. 3 shows a typical microstructure of the coating which is similiar to that of the bulk materials, Fig.1c is its X-ray diffraction pattern.

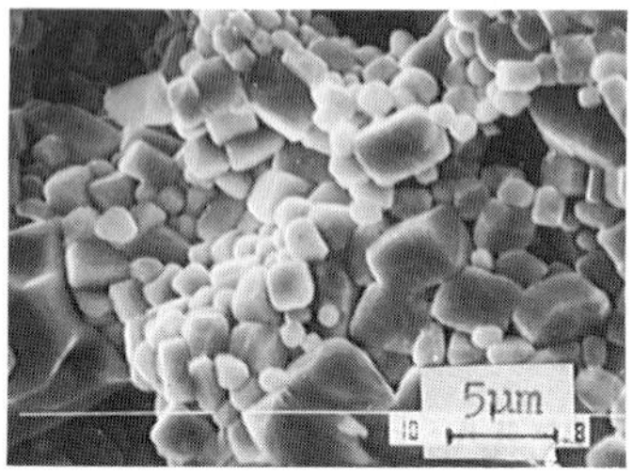

FIGURE 3
The SEM microgagh of the annealed coating

3. THE DIAMAGNETIC SHIELDING PROPERTIES OF THE COATING MATERIALS

It is gernerally recognized that the H_{c1} of the Y-Ba-Cu-O material is below 500G at LN temperature[5] , but the magnetic flux can enter these material at only serveral gauses , that was because the oxide superconductors are composed of large amount of single crystal particals with random orientations. Between the particals are non-superconducting phases. The superconducting current passes these phases by Josephson effects. That not only reduces the J_c value, but also causes magnetic flux line entering the superconductor through weak-linking between the particals.

We have carefully measured the low field behavior of the coating material. From fig. 4, We can see the magnetization curve departure from straight line at 5G which is nearly the same as the bulk materials. So we can expect that the maxinum external field the material can shield is 5G, but it is enough for practical uses.

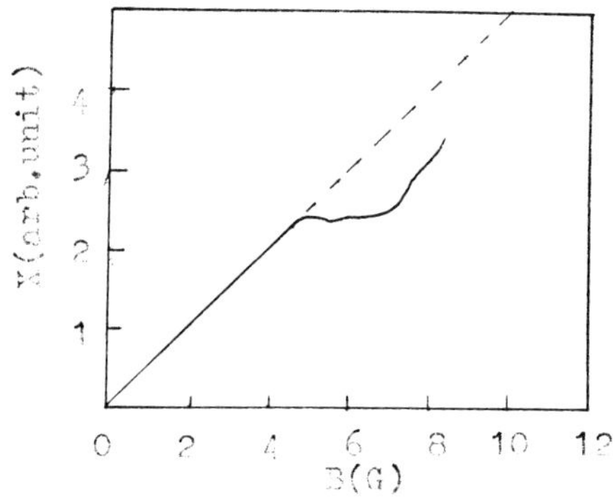

FIGURE 4
The magnetization curve of the coating at low field

4. THE MAGNETIC SHIELDING COMPONENTS MADE BY PLASMA SPRAYING

The surface current producing on material surface will exclude the external magnetic field and thus shield the internal space encircled by superconductor. The shape of the shielding components is depended on the usage and is usually cylinder with one or two open ends. When the cylinder is long enough, the magnetic fields attenuates after entering the open end and will be zero at the center of the cylinder.

The shielding components were prepared by spraying superconducting coating on the outer wall of a stainless steel tube with the length of 100mm and diameter of 20mm. the coating thinkness is about 0.3mm.

A variable magnetic fiels was produced by a electrmagnet. the shielding cylinder was put in liquid nitrogen at zero field, and the magnetic field in it is detected by ballistic galvanometer.

The measuring results of the shielding effects are shown in Fig. 5. We can see that below 3G, no megnetic field can be detected within the devices. And if we encircle the device with high susceptibility meterial, the results will be much better.

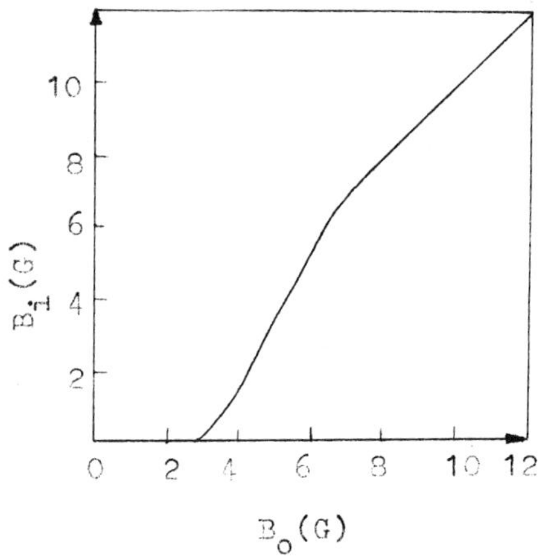

FIGURE 5
The shielding effects of the shielding cylinder

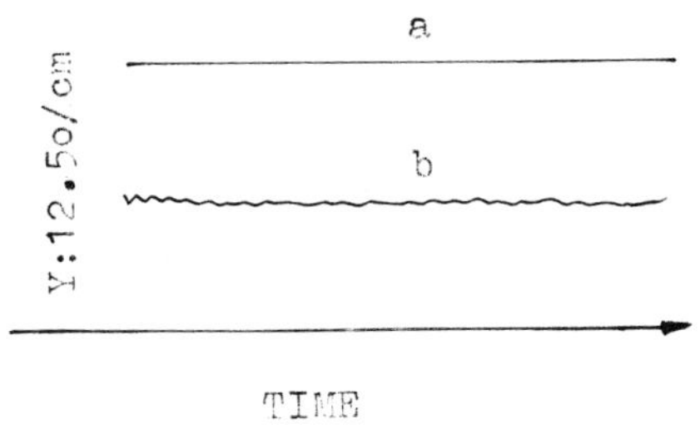

FIGURE 6
The shielding effect of the shielding device to a RF-SQUID electronic. (a) RF-SQUID in the shielding devices, (b) RF-SQUID outside the shielding devices.

To examine its practical shielding effects, a RF-SQUID was placed in the shielding device to measure its output singnal, then compared it with the singnal that is not shielded. The results in fig. 6 indicates that the shielding effects are very remarkable.

In summary, the plasma spraying can be used to prepare magnetic shielding devices with similiar shielding propertis of bulk material, but with large range of geometries and complex shapes. So it will be easily applied to practical usage.

ACKNOWLEDGEMENTS

We would like to thank Prof. J. W. Qiu for his kindful helping in the RF-SQUID measurment.

REFERENCE:

1. E. Tjukanov, R.W.Cline et al., Physical Review B, 36(1987), 7244.
2. J. W. Qiu, J.R.Wang et al., The magnetic shielding properties of high-T_c superconductor., In The Proceeding of The National Conference on Superconductivity, Baoji, CHINA, 1988,No.2, P.420.(in Chineses).
3. Qian Shengwei, Hu Qingyu et al., J. Inorg. Mater., 2(1987), 364(in Chineses).
4. L.S.Wen, S.W.Qian et al, Thin Solid Films, 152(1987), L143.
5. R.J.Cave, B.Batlogg et al., Phys Rev. Lett. 58(1987), 1676.

Frontiers of Materials Research/
Electronic and Optical Materials
M. Kong, L. Huang (editors)

COMPLEX SUSCEPTIBILITY OF $(Bi_{1-x}Pb_x)_2Sr_2Ca_2Cu_3O_y$ SUPERCONDUCTORS WITH DIFFERENT PHASES

Huaqing YIN, Yuping SUN, Jianyi JIANG, Fanchun ZEN, Jiaju DU

Institute of Solid State Physics, Academia Sinica, Hefei 230031, P.R. China

The ac susceptibility of both 2223 single phase and 2212, 2223 multi-phase $(Bi_{1-x}Pb_x)_2Sr_2Ca_2Cu_3O_y$ superconductors in small ac magnetic field has been performed in terms of complex susceptibility X=X'-iX". The real part X' of the susceptibility spectra shows two types of characteristics for two kinds of samples: One contributed by intra-grain is insensitive to the amplitude h_0(6mOe-12.0Oe) of the external ac field, and the other contributed by the inter-grain is sensitive to h_0. In the latter case, the transitive width is remarkably broadened and there exists a corresponding X" peak, the temperature of which shifts to lower temperature as h_0 increases. Differently, for the multi-phase sample, there exist two steps of transition and two corresponding X" peaks which are sensitive to h_0. The frequency dependence of the complex susceptibility has also been investigated, but there is no obvious frequency dependence.

INTRODUCTION

The ac susceptibility of the superconducting transition without dc field was originally made for the detection of the filamentary structure in superconductors in terms of a peak appearance in the imaginary part of the susceptibility, but this kind of investigations intended to reveal the nature of the superconducting transition has so far been very limited. In recent years, considerable interests have been focused on the coupling properties of granular superconductors.

The high transitive temperature of the oxide ceramic superconductors attracts many researchers to work toward the practical application. The granular superconductivity of high-T_c superconductors limits the critical current density. The coupling properties of polycrystalline materials have been investigated by means of complex susceptibility measurement recently[1,2,3], but there is no general understanding of granular superconductors.

Recently, Mazaki, Takeda et al[1,2,3] studied the fundamental susceptibility X'-iX" of La-system, Y-system, Bi-system high-T_c superconductors. They found that both X' and X" were quite sensitive to the amplitude of ac magnetic field. The weak-link loop was established to explain the results of complex susceptibility. Several other models were also established [4,5,6] to explain the results.In present work, we study the Complex Susceptibility of $(Bi_{1-x}Pb_x)_2Sr_2Ca_2Cu_3O_y$ superconductors with different structures in a small ac magnetic field. The samples were prepared with sintering method.

EXPERIMENTAL

The samples of $(Bi_{1-x}Pbx)_2Sr_2Ca_2Cu_3O_y$ were prepared by co-decomposition of metal nitrate[7] Bi_2O_3, PbO, $SrCo_3$, CaO and CuO were dissolved in nitric acid with the cation ratio of Bi:Pb:Sr:Ca:Cu for 1.8:0.4:1.9:2.1:3.1. The solution was stirred and heated until it was dried and became light blue color. The nitrates were then co-decomposed at 800 °C for 12 hr. The powder prepared was ground and pressed into pellets (Φ10x3mm). The pellets were sintered at 850 °C for 100-370 hr in air, and then cooled slowly.

The resistance measurement results measured with standard four-probe method showed that the T_{co} is 107.0 K for sample A and 109.0 K for sample B. The X-ray patterns show that sample A is of single 2223 phase with orthorhombic cell (a= 5.366 Å, b=5.396 Å , c=36.574 Å) and sample B consists of two kinds of phases (2212 phase and 2223 phase).

The measuring system of ac susceptibility X'-iX" consists of mutual inductance bridge and lock-in amplifier. The mutual inductance bridge contains primary coil and secondary coil. The secondary coil consists of two coaxial cylindrical coils wound in counter direction, each one contains 7200 turns of Cu wire (0.025 mm in diameter) coated by polyvinyl formal. The sample was placed on one secondary coil and held by a ruby cylinder which has good thermal conductivity. Displex CSA-202E closed cycle refrigeration system produces 10 K to 300 K temperature. The temperature was detected with Au (0.07% Fe) vs. Cr (APC) thermocouple within uncertainty of 0.1 K. Curves of X' and X" were drawn by an X-Y recorder.

RESULTS

The ac susceptibility X'-iX" of $(Bi_{1-x}Pb_x)_2Sr_2Ca_2Cu_3O_y$ sample A(with 2223 single phase) and B (with 2212 and 2223 phase) was measured with respect to temperature as well as amplitude of ac magnetic field (h_0 6mOe-12.0Oe, f: 200 Hz). In sample A two transitions appear in X' curves, an X" peak appears on second transitive temperature region of X'.

The effects of ac field on the transition of X' and the X" peak are shown in Fig.1(a) and Fig.1(b). The following characteristics can be observed:

1. Two types of transition appear on X' curves. Corresponding to low temperature transition, there is an X" peak in the middle point of the low temperature transition.
2. Two types of transition have different characteristics. The high temperature transition do not depend on the amplitude of ac field(h_0). The low temperature transition width becomes broad with increase of the amplitude of ac field, but the onset temperature do not change with the amplitude of ac field.

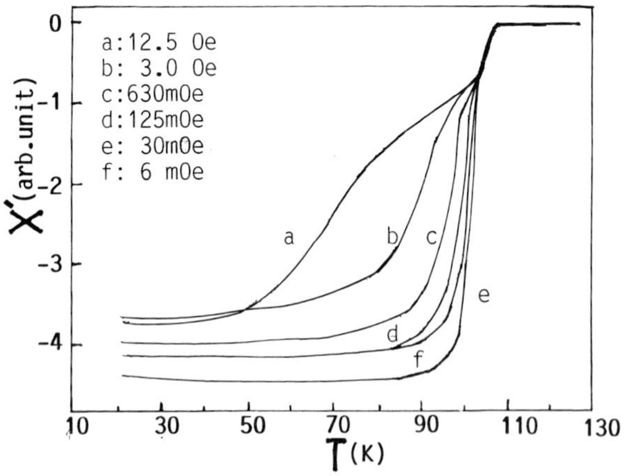

FIGURE 1 (a)
The real component of ac susceptibility in different ac field for sample with 2223 phase (h_0: 6mOe - 12.0Oe)

3. The X" peak changes with ac field

amplitude. The peak temperature shifts to low temperature and the peak width broaden with the increase of ac field amplitude, but the onset temperature remains unchange.

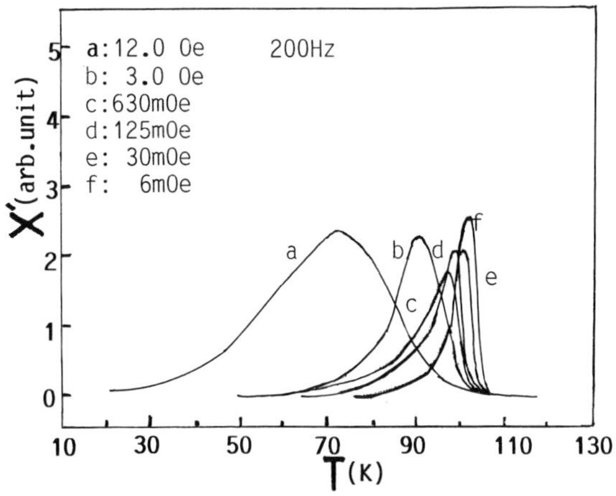

FIGURE 1 (b)
The imaginary component of ac susceptibility in different ac field for sample with 2223 phase (h_0: 6mOe - 12.0Oe)

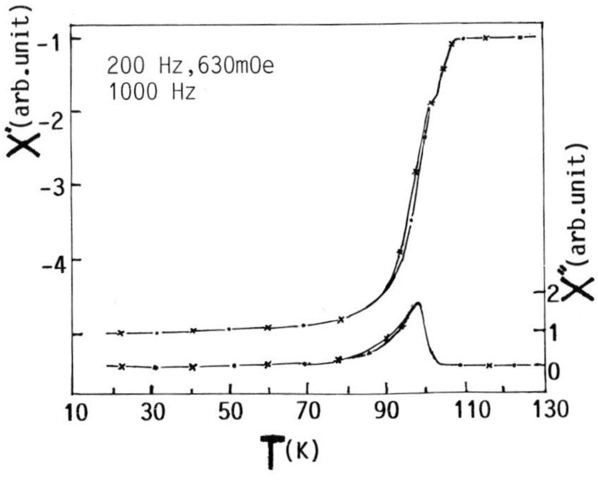

FIGURE 2
The frequency dependance of complex susceptibility (200 Hz and 1000 Hz) for sample with 2223 phase

The frequency effects on X' and X" were performed in a constant amplitude field (Fig.2). 200 Hz and 1000 Hz were applied. For both X' curves and X" curves, no remarkable difference was observed for difference frequency.

The similar phenomena were observed in sample B(Fig.3). Differently, The complex X' curves of sample B exhibit three transition regions, and two X" peaks. This two peaks locates on middle temperature of later transition separately. The characteristics of the effect of applied ac field amplitude and frequency on complex X' are the same as that in sample A.

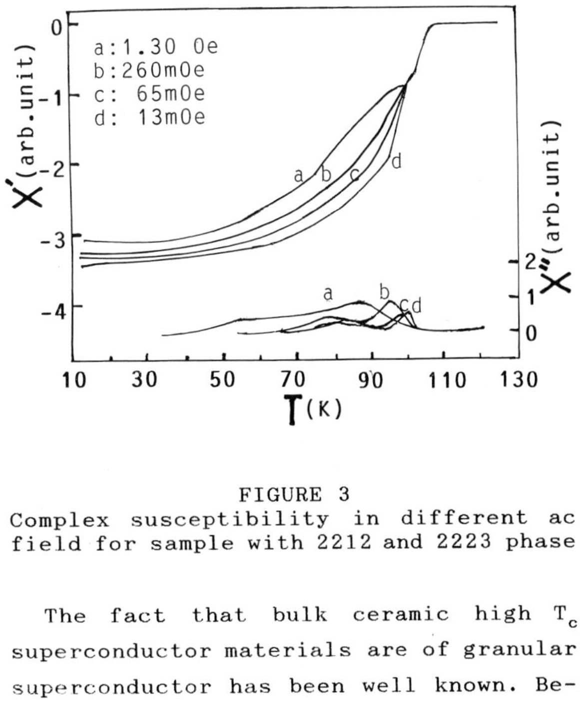

FIGURE 3
Complex susceptibility in different ac field for sample with 2212 and 2223 phase

The fact that bulk ceramic high T_c superconductor materials are of granular superconductor has been well known. Between two grains, there exists a week-link junction. Therefore the bulk superconductor is a kind of three dimensions weak-link junction network, and consequently the I-V characteristics strongly depend on the junction network involved. To explain the complex susceptibility, a weak-link loop model is established[8]. In this model, in stead of considering the complicated three dimensions junction net-work, a weak-link loop which is closed through a weak-link junction is analyzed. The low temperature transition of X' is related to the junction network. Assuming j is the shield current of weak-link loop which is necessary to shield the loop against the external ac magnetic field h(t), and $J_0(T)$ the critical current, when j exceeds $J_0(T)$, a magnetic flux penetrates into the loop, and $4\pi M$ changes with $n4\pi\Phi_0$. So, the relation between the low temperature transition of X' and h(t) depends on the superconducting junction characteristics. The X" peak is related to magnetic hysteresis on superconducting junctions. The frequency independence of X" peak conforms this kind of consideration. The high temperature transition is contributed by Meissner effect of intergrain.

Based on above understanding, the fact that the curves of multi-phase sample is more complicated than that of single phase sample is understandable. So far, the understanding of microstructure of bulk superconductors is very limited, and superconducting transition width is quite broad compared to the conventional superconductors. therefore, there exists a spread of superconducting junction, and the junction type is very difficult to be determined.

REFERENCE

1. Hiromasa Mazaki, Mikio Takano, Ryoji Kanno and Yasuo Takeda, Japanese J. Appl. Phys. Vol.26, No.5, (1987)L780

2. Hiromasa Mazaki, Mikio Takano, Yasunori Ikeda, Yoshichika Bando, Japanese J. Appl. Phys. Vol.26, No.10, (1987)L1749

3. Yasuo Takeda and Hiromasa Mazaki, Japanese J. Appl. Phys. Vol.27, No.4, (1988)L531

4. J.W. Ekin, A.I. Braginski, A.J. Panson, M.A. Janocko, D.W. Capone,II, N.J. Zaluzec, B. Flandermeyer, O.F. de lima, M. Hong, J. Kwo, and S.H. Liou, J. Appl. Phys. Vol.62, (1987)4821

5. N. Garcia, S. Vieira, A.M. Baro, L. Vazquez, J. Gomez, A. Aguilo, S. Bourgeal, A. Buendia, J. Tornero, M. Hortal, M.A. Lopez de La Torre, M.A. Ramos, R. Villar, K.V. Rao, D.X. chen, and J. Nogues, In proceedings of the European Workshop on High T_c Superconductors and Potential Applications, 1-3 July 1987, Genoa, Italy, pp.97-98

6. C.H. Chen, D.J. Werder, S.H. Liou, J.R. Kao, and M. Hong, Phys. Rev. B, Vol.35, (1987)8767

7. Jiang Jianyi, Sun Yuping, Zen Fanchun, Yin Huaqing, and Du Jiaju, Z. Phys. B, Vol.77, (1989)193

8. T.Ishida and H.Mazaki, Japaness J Appl. Phys. Vol.52, No.11, (1981)6796

Frontiers of Materials Research/
Electronic and Optical Materials
M. Kong, L. Huang (editors)

GROWTH OF SUPERCONDUCTING WHISKERS

Yunlong GE, Qingmin LIU, Hua JIN and Zhuangqi HU

Institute of Metal Research, Academia Sinica, Shenyang 110015, P.R.China

The (1112) and (2212) amorphous strips of Bi(Pb)-Sr-Ca-Cu-O were prepared by non-equilibrium solidification of melt-quenching method. A lot of superconducting whiskers have been grown up on the strip at 840-860ºC in a stream of oxygen for 50-150 hours, which are usually several to 10 odd millimeters in length, several μm in thickness and more than 10 μm in width. These whiskers have very excellent flexibility. The Tc(onset) is 106K measured by AC susceptibility.

1. INTRODUCTION

In polygrain oxide superconducting materials, owing to the weak link resulted from grain boundary and grain orientation, the increase of critical current density (Jc) is limited. There are some unfavourable factors such as the micro-crack at grain boundary, the unsound connection of grain boundary caused by voids, the deviations of composition and structure due to the segregation of impurities, and the anisotropic conduction induced by various grain orientations. In order to overcome these defects, remarkably enhance the critical current density (J_c) and deeply study the crystal structure, superconductivity and other physical properties of oxide superconductors, it is important to produce large single crystal with good property, especially with excellent flexibility to meet the need of practical application. Some new methods have been used in succession such as laser-heated pedestal growth[1-3], floating zone melting method[4] and flux method[5]. In Y-system, the melting texture growth[6] and quench-melting growth[7] have been the important methods to increase the critical current density.

A novel growth phenomenon of superconducting whiskers on the (1112) and (2212) melt-quenched strips of Bi(Pb)-Sr-Ca-Cu-O have been studied. The growth process, feature, composition and structure of the whiskers were examined in detail. The superconducting characteristic was measured by the method of AC susceptibility.

2. EXPERIMENTAL

The chemical purity powders of Bi_2O_3, $SrCO_3$, $CaCO_3$, CuO and PbO were mixed in molar ratio of (1112)+Pb and (2212)+Pb, and ground in an agate mortar. The mixed powders were tamped into Al_2O_3 crucible and were then melted in an electric furnace at 1100°C for 30 minutes. Like the method of making amorphous materials, the molten oxide was poured onto a steel plate and pressed quickly with a steel roller into 0.5-1.5 mm thickness strips. These strips were then heated at 840-860°C in a stream of oxygen for 50-150 hours and cooled to room temperature. On the surface of strips many whiskers were grown up which have long length, even diameter and excellent flexibility. The feature of strip with whiskers or single whisker taken down from the substrate was observed by SEM. The composition and in depth composition of whiskers were measured by electron probe and Auger spectroscope. The structure and lattice constant of whiskers were determined by XRD. The layer structure, crystal lattice and modulated structure of the whiskers grown on the melt-quenched strip were also analyzed by electron diffraction and high resolution electron microscope.

3. RESULTS AND DISCUSSION

The non-equilibrium rapid solidification method of preparing oxide superconductor has overcome not only the defects of low density and unsoundness in sintering method, but also the compositional deviation due to segregation. By using this method, the uniform and highly dense material close to theoretical density can be obtained. Fig.1 shows the compositions of melt-quenched strip at various locations analysed by EDAX.

Owing to the Bi-system having low melting point and containing Bi_2O_3, which is easy to be amorphous, it can be transformed into glassy state even at a lower cooling rate of 10^3K/s. Fig.2 shows the X-ray diffraction pattern for the melt-quenched strip, which shows a typical amorphous structure.

Whiskers can be found on the strips of (1112)+Pb and (2212)+Pb, which have about 10 mm in length, several μm in thickness and several ten μm in width. Some whiskers are long and straight (see Fig.3), and some are curved.

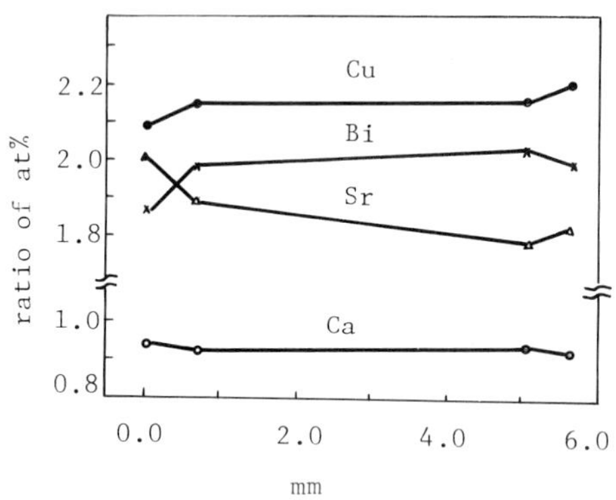

FIGURE 1
Compositions of melt-quenched strip at various locations analysed by EDAX.

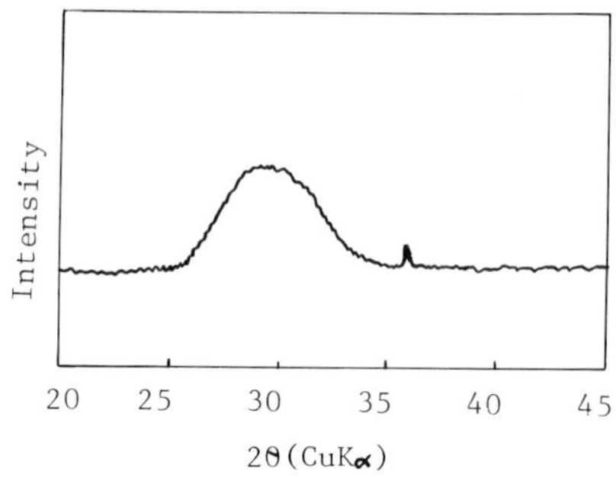

FIGURE 2
XRD pattern for melt-quenched strip.

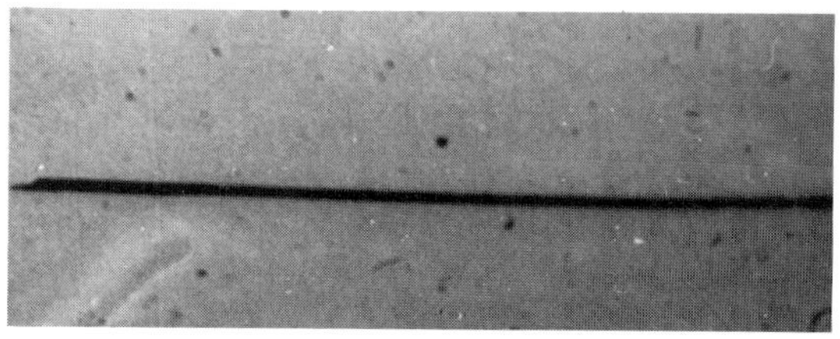

FIGURE 3
Macroscopic morphology of whisker (3mmx50μmx 2μm)

It is observed by SEM that most whiskers have smooth surface and are thin and long (see Fig.4), and some whiskers have striations of dark and bright colors on their surfaces (see Fig.5). The compositions in dark and bright regions are same and the different colors are caused by the difference of height analyzed by modulated scanning method in SEM. The whisker exhibits poly-prism.

The whiskers were not uniformly distributed on the strip, but grown somewhere. Most whiskers were grown at the grainy substrates where usually have high content of Cu (see Fig.6).

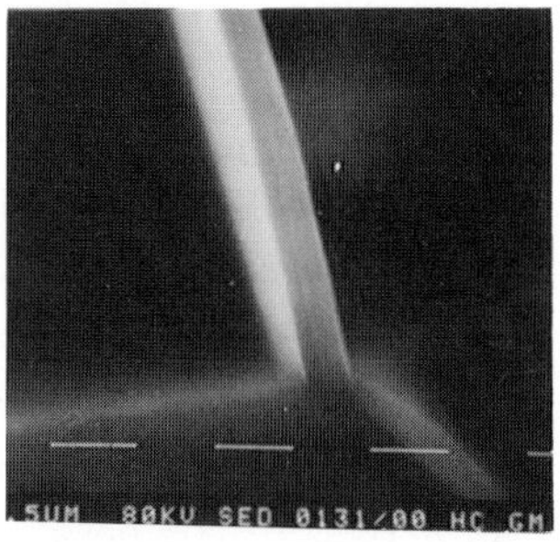

FIGURE 4
The whisker with smooth surface

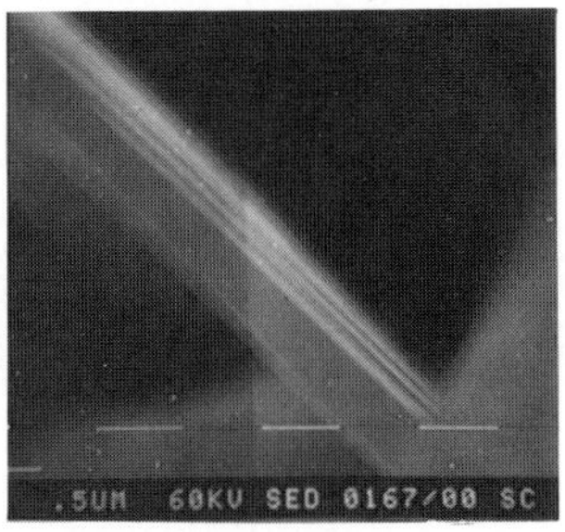

FIGURE 5
Poly-prismatic whisker

The composition of whiskers taken down from the substrate were analyzed by several methods for more than thirty pieces of whiskers, which are listed in Table 1. It is clear that the compositions of the superconducting whiskers are (2212) in molar ratio.

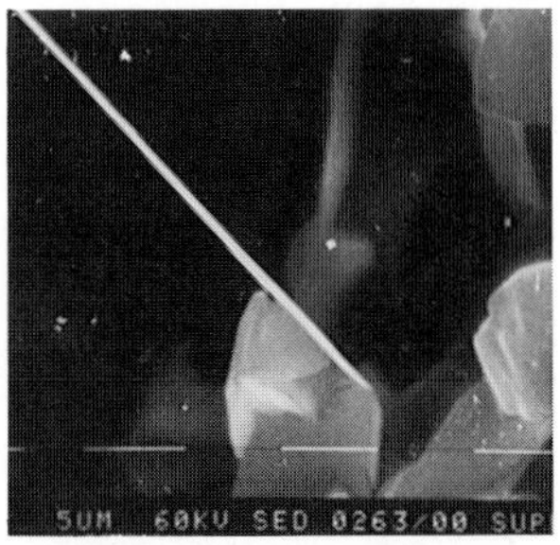

FIGURE 6
Whiskers grown from grainy substrate where has high content of Cu.

It is shown that the whiskers are orthorhombic structure (a=0.54 nm) or pseudo-tetragonal structure (a=0.534 nm, b=0.518 nm, c=3.065 nm) analyzed by XRD for single whiskers (see Fig.7). The diffraction pattern of (00L) is the characteristic of single crystal growth along c axis. The length of c was not easily determined by electron diffraction due to the cleavage of a-b planes. Fig.8 shows the electron diffraction pattern of [001]. The whiskers have a structure of (2212) superconducting crystal according to structure analysis.

The result of AC susceptibility measurement is shown in Fig.9. The whiskers are superconductor with Tc(onset)=106K which is higher than that of (2212) single crystal reported by I. Matsubara[8] (Tc(onset)=75K) and J. Jung[9] (Tc(onset)=76K). The result can be compared with

TABLE 1
Analysis of whisker composition

Whiskers	Method	Num of whisker	Composition (at%) Bi	Sr	Ca	Cu	Pb	phase
Free whiskers on (2212)	SEM EDAX	3	32	22	12	27	0.	(2212)
	EPMA	3	31	24	9	26	2.	
Whiskers on 1112+Pb bulk	SEM EDAX	8	30	24	14	31	trace	(2212)
	TEM EDAX	9	29	24	13	30	2.6	
Whiskers on 2212+Pb bulk	TEM EDAX	9	30	22	13	31	1.3	(2212)

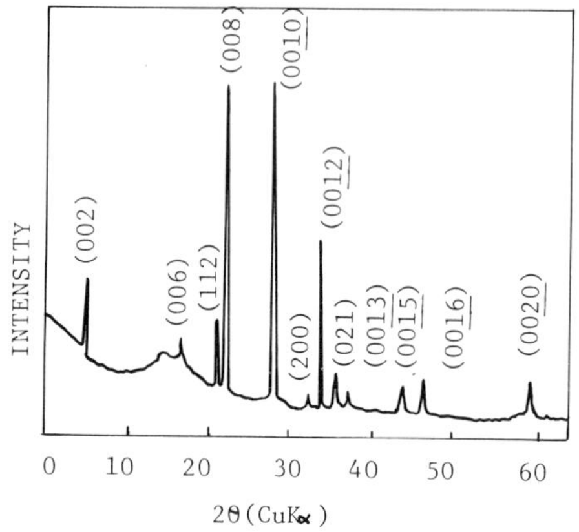

FIGURE 7
XRD pattern of single crystal.

FIGURE 8
Electron diffraction pattern for whisker powders.

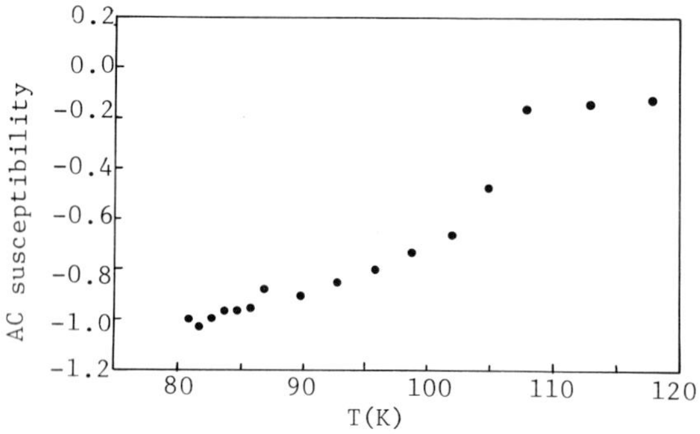

FIGURE 9
AC susceptibility of 2212 whisker

not only that of polygrain (2212) materials (Tc(onset)=102K[10], Tc(onset)=104.5K[11]), but also that of (2212) single crystals (Tc(onset)= 105K[1,4], Tc(onset)=110K[11]). Owing to its perfect structure and accurate composition, the whisker will have good superconducting properties. These high quality superconducting whiskers also promote the deep understanding of their crystal structure and physical properties.

4. CONCLUSION

The (2212) superconducting whiskers have been grown up successfully on the surface of melt-quenched strips of (1112)+Pb and (2212)+ Pb. The whiskers are superconductor with Tc(onset)=106K. Owing to their perfect structure and accurate composition, these whiskers are of good superconductivity and excellent flexibility.

ACKNOWLEDGEMENTS

The authors wish to gratefully acknowledge valuable suggestions with professor Zhao Zhongxian. This work is supported by key project of National Natural Science Foundation.

REFERENCES

1. R.S.Feigelson, et al Science 240(1988)1642.
2. D.Gazit,et al J. Crystal Growth 91(1988)318.
3. H.D.Brody,et al J.Crystal Growth 96(1988)225.
4. Y.Kobo, et al Jpn. J. Appl. Phys.28(1989)606.
5. K.W.Goeking, et al Mater. Lett. 8(1989)432.
6. S.Jin, et al Appl. Phys. Lett. 52(1987)2047; Phys. Rev. B.37(1988)7850.
7. Y.Ibara, et al Jpn.J.Appl. Phys.28(1989)37.
8. I.Matsubara,et al Jpn.J.Appl.Phys.28(1989) 1121.
9. J.Jung, et al Phys. C 156(1988)494.
10. T.Komatsu,et al Jpn.J.Appl.Phys.27(1988)533.
11. T.Shishido,et al Jpn.J.Appl.Phys.28(1989)791.

PART 3:
OPTOELECTRIC MATERIALS AND FUNCTIONAL CRYSTALS

Frontiers of Materials Research/
Electronic and Optical Materials
M. Kong, L. Huang (editors)

A NEW ORGANIC MOLECULAR NLO CRYSTAL---3-METHOXY-4-HYDROXY-BENZALDEHYDE (MHBA)

Minhua JIANG, Xutang TAO, Duorong YUAN, Nan ZHANG and Zongshu SHAO

Institute of Crystal Materials, Shandong University, Jinan 250100, P.R. China

This paper reports a prospect new organic materials MHBA whose powder SHG intensity is 30 times higher than that of Urea and cut-off wavelength is 370nm. The crystal structure, single crystal growth, linear and nonlinear optical properties of MHBA are also reported in detail. MHBA belongs to monoclinic system; space group is $P2_1$, and the unit cell parameters are: a=14.057Å, b=7.875Å, c=15.037Å and β=115.45°. The nonlinear susceptibilities are: $d_{11}=25.0d_{36}^{KDP}$, $d_{12}=10d_{36}^{KDP}$, $d_{13}=33.3d_{36}^{KDP}$ and $d_{14}=8.3d_{36}^{KDP}$.

1. INTRODUCTION

Some organic molecular crystals have quadratic NLO efficiency at least one order of magnitude higher than that of classical inorganic materials such as KDP and $LiNbO_3$[1,2]. The other advantages of organic materials are their relatively high damage threshold and short response time to optical excitation. The organic materials have a lot of practical applications in integrated optics and the most attractive one is as diode laser frequency doublers. A device based on frequency doubled GaAs laser will find a variety of practical applications in optical chemistry, optical storage, underwater communication and other important technologies[3]. But, the fundamental wavelengths of GaAs series diode laser are shorter than 1.0μm, and their power is low and easy to disperse. As we know, for a useful organic material for such applications, the most important properties it should has are SHG efficiency, transparency, phase-matching characteristics and crystal growth parameters. Yet, there appears no such material to satisfy these requirements up to now. In this paper we present a new organic material----3-methoxy-4-hydroxy-benzaldehyde (MHBA). Its powder SHG intensity is 30 times higher than that of area, and its cut-off wavelength is 370nm.

2. CRYSTAL STRUCTURE AND SINGLE CRYSTAL GROWTH

The structure of MHBA crystal was measured by using single crystal X-ray diffraction. The results show that MHBA belongs to monoclinic system; space group is $P2_1$; point group is 2 and the lattice constants and other structural data are summarized in table 1.

TABLE 1
Crystal structural data of MHBA

Formula: $C_8H_8O_3$
Formula weight: 152.14
Lattice parameters:
a=14.057(3), b=7.875(1), c=15.037(3) Å
β=115.45(1)°
V=1503.30 A^3, Z=8, d_c=1.34 g/cm^3
M_o K_a radiation: Lamd=0.71069 Å

MHBA is a donor-acceptor(D-A) substituted benzene derivative. It has a charge transfer interaction between the para donor hydroxy (-OH) and carbonyl ($-C\lessgtr^{O}_{H}$) acceptor groups. The methoxy provides a minor contribution to the nonlinearity, but plays an important role in promoting the formation of acentric crystal structures. Because the prototype compound of MHBA is 4-hydroxybenzaldehyde[4], it is centrosymmetric and null for macroscopic SHG. Figure 1 is the projection of half of an unit along the crystallographic b axis. It should be noted that there are 8 molecules in one unit cell. The intermolecular hydrogen bonds are situated between carbonyl and hydroxy groups. The hydrogen bonds play an important role in the arrangement of molecules in the crystal, which resulted in the increase of the susceptibility of MHBA.

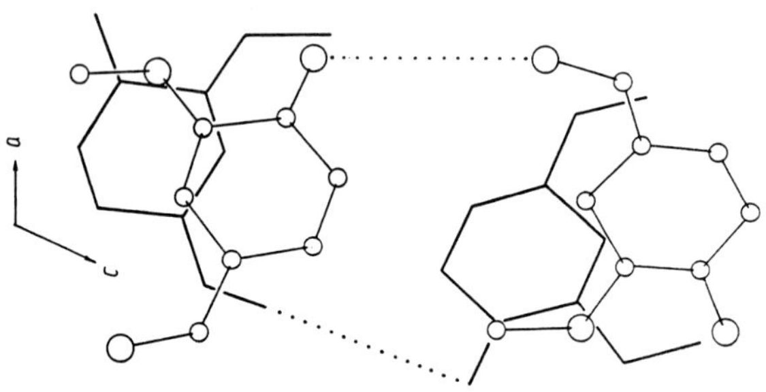

Figure 1
Projection of an asymmetrical unit of MHBA along the crystallographic b axis

MHBA is soluble in several common organic solvents and can be grown easily from solution by temperature lowering method. In a 20-day period, high quality MHBA single crystals with dimensions of 40 X 30 X 6 mm^3 can be obtained.

3. LINEAR OPTICAL PROPERTIES

By using V prism method, we measured the refractive indices of MHBA in the visible regions (See Table 2). The refractive indices at wavelengths of 1064, 532, 830 and 415 nm were obtained by the simulation of Sellmeier equation of refractive index dispersion:

$$n^2=A+B/(Lamd^2-C)-D/(F-Lamd)^2$$

where Lamd is the wavelength in microns and the constants A, B, C, D and F are given in Table 3.

Table 2
The refractive indices of MHBA

Lamd	n_x	n_y	n_z
404.7	1.63352	- -	- -
435.8	1.60345	- -	- -
447.1	1.59644	- -	- -
546.1	1.55840	1.70018	1.89349
587.5	1.55143	1.69045	1.80896
589.3	1.55127	1.69039	1.79235
656.3	- -	1.68352	- -
667.8	1.53996	1.67963	1.77105
705.7	1.53673	1.67668	1.76812
532.0	1.56246	1.70331	1.81376
106.4	1.51438	1.66261	1.74390
415.0	1.62183	1.76697	1.96996
830.0	1.52717	1.66944	1.75470

Based on the refractive indices at the wavelengths of 1064, 532, 830 and 415 nm, the phase-matching loci of MHBA at fundamental wavelengths 1064 and 830

Table 3

Sellmeier coefficients for MHBA

n	n_x	n_y	n_z
A	2.3226	2.7339	2.9925
B	0.02514	0.26352	0.27386
C	0.09263	0.064986	0.11651
D	0.04720	-0.23227	-0.22437
F	-------	0.05997	0.11651

nm, harmonic wavelengths 532 and 415 nm have been obtained. The results are shown in Figure 2 and Figure 3. They

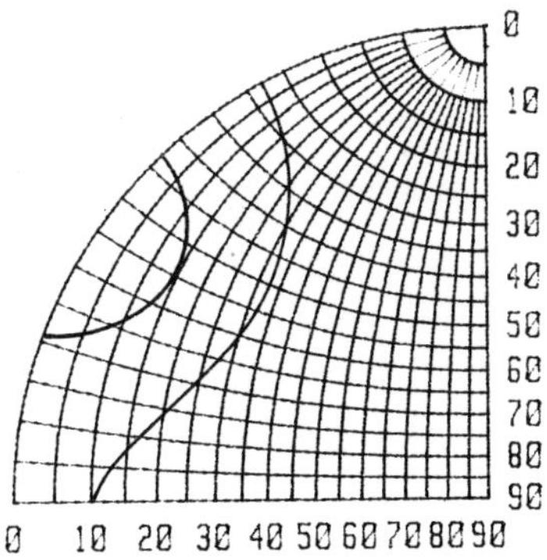

Fig.2
PM loci of SHG in MHBA from 1064 to 532

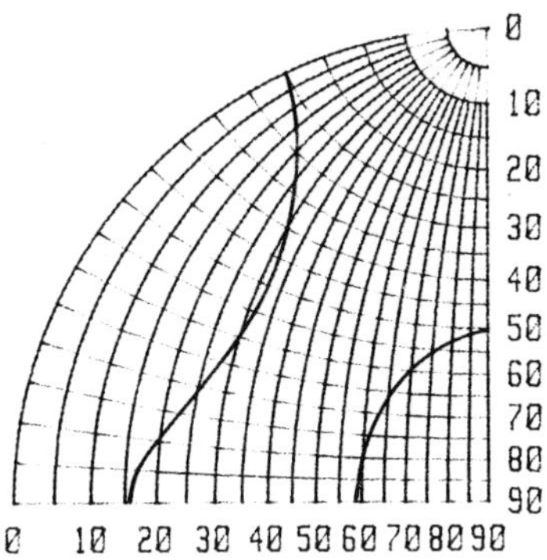

Fig. 3
PM loci of SHG in MHBA from 830 nm to 415 nm

belong to the third type in the 13 possible PM loci of biaxial crystals described by Hobden[5].

The transparency curve of MHBA was measured by using Hitachi Model 340 UV-V-NIR recording spectrophotometer. The thickness of the sample slice used was 2.7 mm of [010] direction. The transmission curve is shown in Figure 4. The transparent regions are from 375 to 1700 nm and the cut-off wavelength is 370 nm.

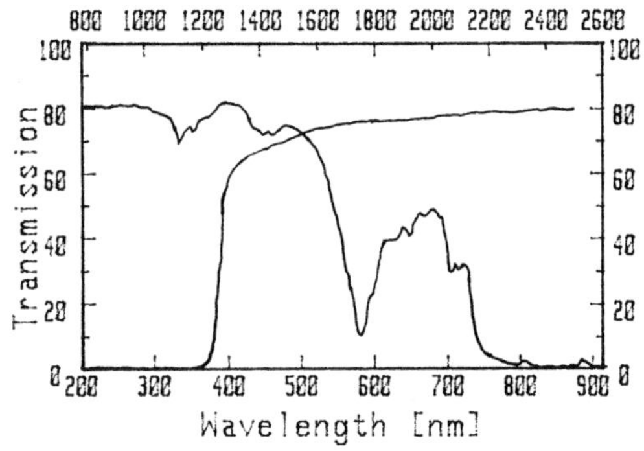

Figure 4
The optical transmission spectrum of MHBA

4. NONLINEAR OPTICAL PROPERTIES

MHBA belongs to the monoclinic system and point group is 2, so there are 4 independent nonzero second-order NLO coefficients. Marker Fringe method were used to measure these coefficients. A mode-locked Nd:YAG laser operated at 1064 nm was used in the experiment. The results are:

$d_{11}=25.0d_{36}^{KDP}$

$d_{12}=10.0d_{36}^{KDP}$

$d_{13}=33.3d_{36}^{KDP}$

$d_{14}= 8.3d_{36}^{KDP}$

5. CONCLUSION

MHBA is a new organic nonlinear optical material, whose single crystal can easily grown from solution by temperature lowering method. MHBA has high optical nonlinearity and short cut-off wavelength. All these indicate that aromatic molecular NLO crystals with weaker donors (such as CH_3O-, $-OH$) and acceptors are promising candidates in considering the trade-off between efficiency and transparency. Therefore, MHBA is a promising new organic NLO material with potential applications.

REFERENCES

1. D.J.William, Editor, Nonlinear Optical Properties of Organic and Polymeric Materials (ACS symposium Series 233, 1983).
2. D.S.Chemla and J.Zyss. eds., Nonlinear Optical properties of Organic Molecules and Crystals, Vol. 1 and Vol. 2 (Academic, New York, 1986).
3. G.T.Boyd, J. Opt. Soc. Am. B, Vol. 6 (1989) 685.
4. F.Iwasaki, Acta Cryst. Sect. B, B33 (5) 1646.
5. M.V. Hobden, J. Appl. Phys., Vol.38 (1967) 4365.

Frontiers of Materials Research/
Electronic and Optical Materials
M. Kong, L. Huang (editors)

GROWTH AND PHOTOREFRACTIVE PROPERTIES OF KNSBN:Co CRYSTAL

Jiang QUANZHONG, Chen HUANCHU, Song YONGYUAN and Sun DALIANG

Institute of Crystal Materials, Shandong University, Jinan, P.R.C.

Xu KEBIN, Xu HAIYING, Yu YOULONG and Yuan YIANG

Department of Applied Physics, Herbin Institute of Technology, Herbin, P.R.C.

Growth and photorefractive properties of KNSBN:Co crystal are reported for the the first time.Crystals were grown by CZ technique and the growth habits of them are severly affected by the amount of Co ions.New energy levels are formed in the energy band gap. Comparing to undoped KNSBN, KNSBN:Co has larger two-wave coupling coefficient, shorter response time and larger photo-induced free electron concentration in the experiment of two-wave coupling. Phase conjugation reflectivity of 56% and response time of 150ms were determined in the experiment of degenerate four-wave mixing.

1. INTRODUCTION

Potassium sodium strontium barium niobate(KNSBN) has been developed in our country for the first time[1]. It is a new ferroelectric with A-sites thoroughly occupied by K^+,Na^+,Sr^{2+} and Ba^{2+} ions.Comparing to $BaTiO_3$, KNSBN crystal has no phase transition at room temperature, large hardness coefficient and 180° ferroelectric domains easy to be polarized.KNSBN will be one of the most prospective photorefractive(PR) materials that can be used in holographic storage, image amplification and optical phase conjugation[2]. The purpose of this paper is to describe the effect of Co ions on the PR properties of KNSBN.

2. CRYSTAL GROWTH

$Co_z(K_xNa_{1-x})_{0.2+z}(Sr_{1-y}Ba_y)_{0.9-2z}Nb_2O_6$ (KNSBN:Co,x=o.5, y=0.39) crystals were grown by CZ technique using MCGS-3 system. The starting materials are 99.99%$BaCO_3$, $SrCO_3$, K_2CO_3, $NaCO_3$, Nb_2O_5 and Co_2O_3. They were mixed according to the crystal compositions as z=o.oo3, 0.006 and 0.02. After balling in pure alcohol for 4hours, the mixtures were dried and pressed into disks of 3omm in diameter and 25mm in thickness. The disks were placed in a platium crucible and heated to melt by induction heating in the air.After the melt was superheated for 15 minutes,it was cooled down to close the melting point and the crystal was pulled out from the melt. The growth conditions are approximately the same as that reported in reference 1. Striation-free unsplit as-grown samples with size of 15mmx15mmx15mm were obtained.

From the appearances of the as-grown crystals, it can be seen that the amounts of Co ions severely affects the growth habits of them. For crystal of z=o.oo3, it appears eight facets with good symmetry, which are (100), (110) their symmetrical facets, and is very similar to the undoped KNSBN(z=0). For crystal of z=o.oo6, the cross section perpendicular to C axis is square, and at the places equal to its edges there are (210), (110), (120) and their symmetrical facets. Only two facets of (100) appear for crystal of z=0.02.

After annealing at 1050°C for 24h, the crystals were cut along a, b and c axes into cuboids with size of 5.5mmx 5mmx5mm. Before experiment they were also poled and polished.

3. TRANSMISSION SPECTRA

Measuring the transmission spectra is one of the most effective and reliable means for the determination of Co ion induced energy levels in the energy band gap. According to the data obtained, effects of Co ions on the PR properties of KNSBN can be predicted. For comparision, the transmission spectra of doped and undoped KNSBN are given in Figure 1.

From Figure 1, it can be seen that the transmission spectra are seriously affected by Co ions. Firstly, the energy gap is narrowed,for the undoped crystal the gap is 375nm(represented by wavelength) and for the crystals of z= 0.003, 0.006 and 0.02, the gaps are 420nm, 475nm and 490nm respectively. Secondly, althrough the value of z is changed,increasing absorption to the light with wavelength of 675nm appears for all of the three doped samples;this is corresponding to the characteristic energy level of Co ions in KNSBN. These two aspects have proved that new energy levels have formed in the band gap for the dopants of Co ions, and new centers for free charge carrier stimulation and recombination can greatly improve the PR properties of KNSBN.

4. TWO-WAVE COUPLING IN KNSBN:Co

Two-wave coupling[3] usually means the energy transformation between two coherent beams because of PR effect. Owing to the inhomogeneous distribution of energy of the interference pattern, an optical induced free charge carrier concentration gradient is formed, and through the photovoltaic effect, diffusion and drift, an electric field E(x) is formed in the PR crystal. For small diffusion length, the electric field can be represented by:

$$E(x)=-(mK_BT/q)(K/(1+(K/K_0)^2))\sin(Kx)$$

where $m=I_s/I_r$, I_s and I_r are energy densities of the singal beam and the pump beam respectively, K_BT is the thermal energy of crystal lattice, K is the period of the interference pattern, q is the charge of the mobile carrier, and K_0 is a constant which can be represented by:

$$k_0 = (Nq^2/\epsilon\epsilon_0 k_B T)^{1/2}$$

where N is the effective charge carri-

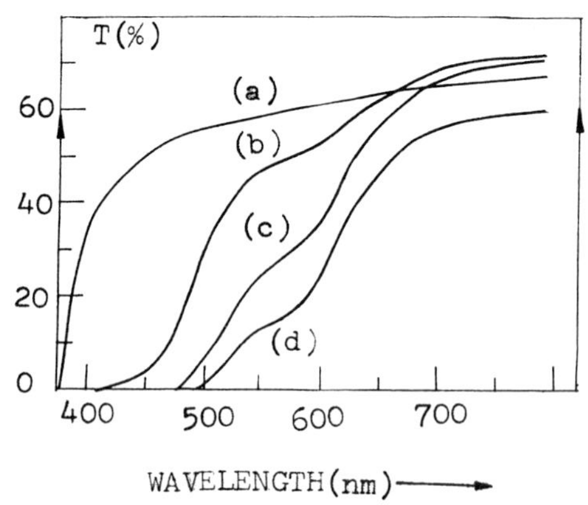

FIGURE 1
Transmission spectra of four sample, a) undoped, d=4.3mm; b)z=0.003, d=4.4mm; c)z=0.006, d=4.6mm; d) z=0.02, d = 4.8mm.

er concentration. Through electro-optic effect, an index grating is formed. It is because of the phase difference between the interference pattern and the grating that energy can be transformed from one beam to the other.The change of energy density of the signal beam can be represented by:

$$I_s(d)/I_r(0)=(1+m)e^{\Gamma d}/1+me^{\Gamma d}$$

where Γ is the two-wave coupling gain coefficient and d is the thickness of a sample. In our experiments, $m=500^{-1}$ and d=3.8mm, then:

$$\Gamma=d^{-1}\ln(500R/501-R)$$

where $R=I_s(d)/I_s(0)$.

Figure 2 shows the arrangment for two-wave coupling. A beam of light from Ar^+-laser(λ =488nm) with its polarization in the input plane becomes two beams (I_s and I_r), as passing through a beam splitter and a reflector, to impenetrate into the sample. A photoelectric detector connected to a X-Y recorder was used to record the time response property.

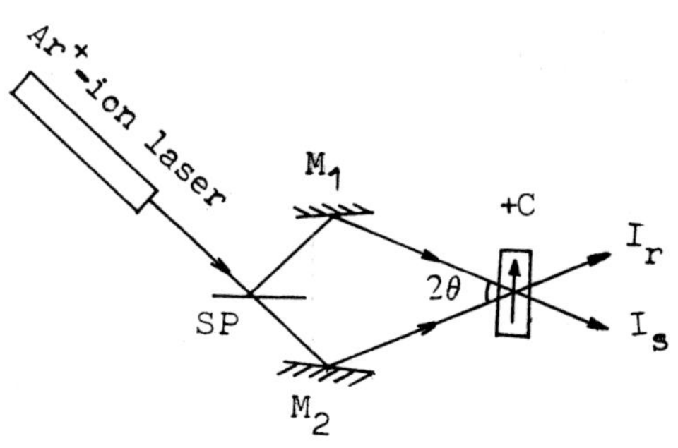

FIGURE 2
Experimental set-up for two-wave coupling.SP,beam splitter; M, mirror.

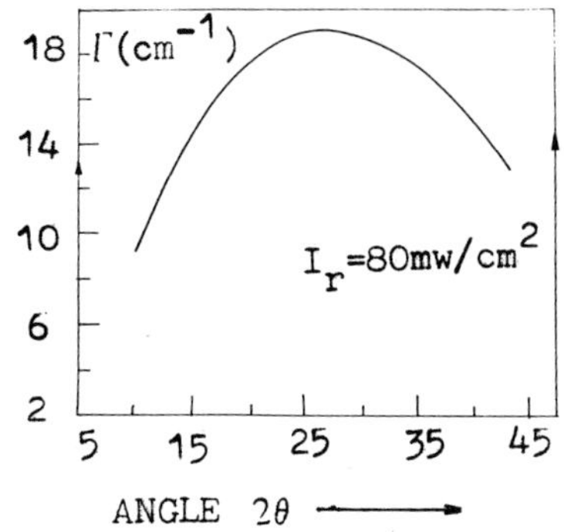

FIGURE 3
The relation between the gain Γ and the angle 2θ .

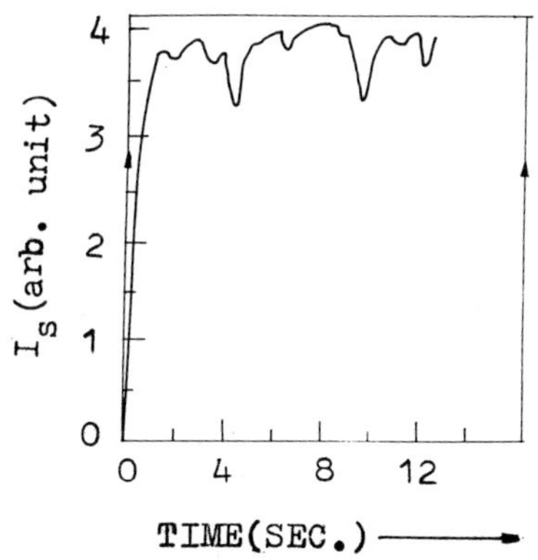

FIGURE 4
The time response property of KNSBN:Co.

Figure 3 shows the relation between the gain coefficient and the angle 2θ . for crystal of z=0.003. Figure 4 is the time response property. The maxmium Γ_{max} corresponding to the angle $2\theta=25°$ is $18cm^{-1}$,approximately two times of that of undoped KNSBN. The optical induced free charge carrier density N is estimated to be in the order of 10^{17} cm^{-3},more than one order of magnitude higher than that of undoped KNSBN.

Therefore, the optical conductivity of KNSBN increases and the response time is shortened.

5. HOLOGRAPHIC STORAGE PROPERTIES OF KNSBN:Co

Because of the uses of KNSBN for read-write applications(because they are optically "erasable") and for read-only applications(because they are fixable), it can play an important role in volume optical holography, real-time associative memory,[4] etc.. The relation between the diffraction efficiency η of KNSBN:Co and the angle 2θ is shown in Figure 5. The efficiency obtained is 90% at $2\theta=20°$ and in the range from 15° to 60° it is over 80%. Figure 6 shows the time response property of storage and the response time is 250ms. The PR sensitivity is estimated to be 10^{-3} J/cm^2, two orders of magnitude higher than that of $LiNbO_3$:Fe.

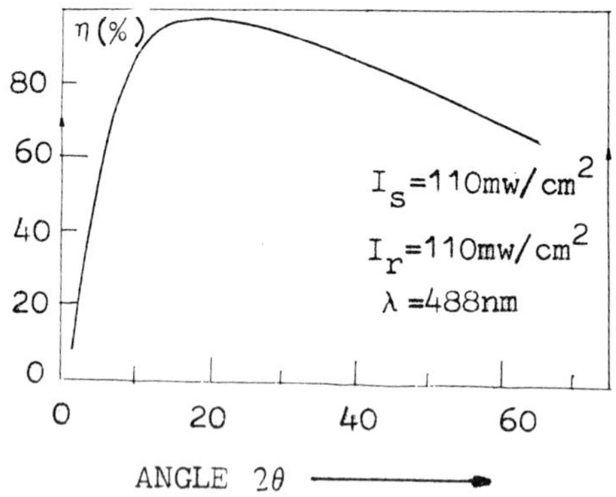

FIGURE 5
The relation between the diffraction efficiency and the angle 2θ by two coherent beams.

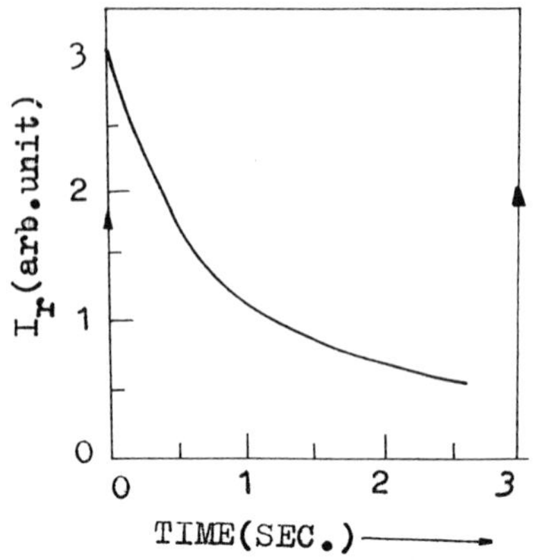

FIGURE 6
The time response property of storage in KNSBN:Co.

6. DEGENERATE FOUR-WAVE MIXING IN KNSBN:Co

One distinguishing feature of PR materials to be used in degenerate four-wave mixing[5](DFWM, Figure 7) is that, according to dynamic self-diffraction theory, the phase conjugation beam I_4 has large dependence on pump beam I_3. It is assumed that the DFWM process is recording a grating by the signal beam I_2 and a pump beam I_1 and

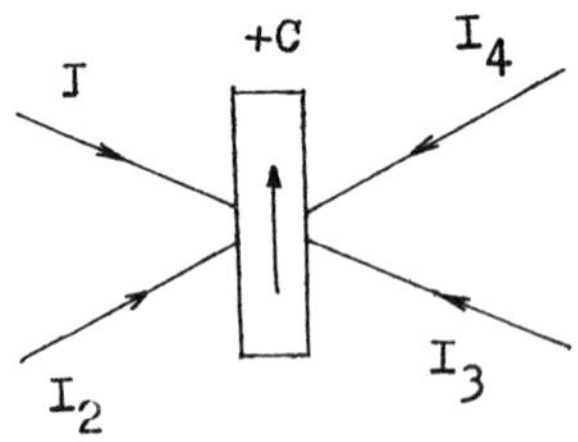

Figure 7
Diagram for degenerate four-wave mixing

reading by the other pump beam I_3 to produce a phase conjugation beam I_4 antiparallel to the signal beam. Under the condition of a weak signal beam, the reflectivity I_4/I_2 can be greater than unit, that is to say, the signal can be amplified using DFWM in PR materials. Besides, image edge enhancement (image differention) can be broght out in PR materials based on DFWM under two conditions: large ratio of I_2/I_1 or transformation, using Fourier lenses, of an input image. So, DFWM properties can be some more important to KNSBN:Co.

Experimental set-up of DFWM is shown in Figure 8. The intensity of the beams I_1=670mw/cm^2, I_2=11omw/cm^2, I_3=510mw/cm^2 and the angle 2θ =18°. The stable phase conjugation reflectivity is 56% and the transient value is 120%. The time reponse properties is shown in Figure 9. The transient peak is at the time of 50ms.

7. CONCLUSION

(1)Striation-free KNSBN:Co crystals with the size of 15mmx15mmx15mm can be grown in the air by CZ technique. The color and the growth habits of the crystals are seriously affected by the amount of Co ions.

(2)New energy levels in the gap are caused by Co ions and new centers for free charge carrier stimulation and recombination have been increased to improve the PR properties of KNSBN.

(3)From our experiments it can be seen that large two-wave coupling coefficient, high diffraction efficiency, as well as short response time enable KNSBN:Co crystals to be used in image amplification, optical storage and phase conjugation.

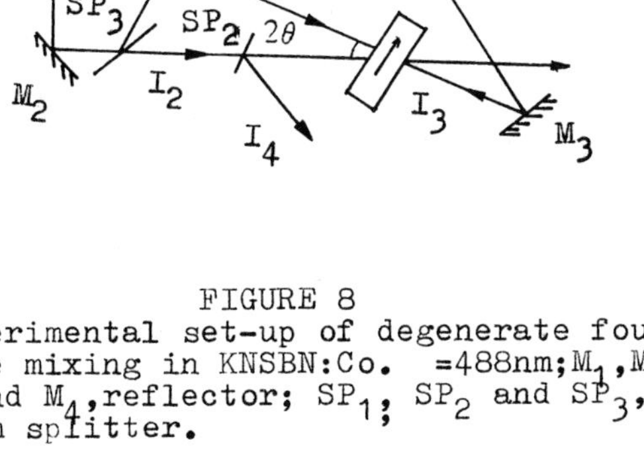

FIGURE 8
Experimental set-up of degenerate four-wave mixing in KNSBN:Co. =488nm; M_1, M_2, M_3 and M_4, reflector; SP_1, SP_2 and SP_3, beam splitter.

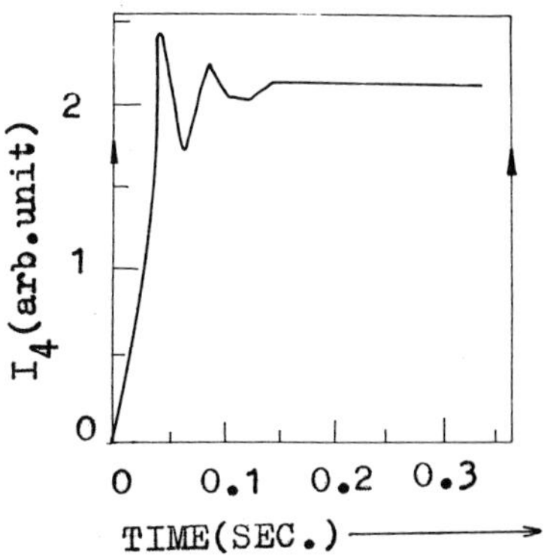

FIGURE 9
The time response properties of DFWM in KNSBN:Co.

REFERENCES

1. Chen Huan-chu et al; Physics 10(1981)729(in Chinese)
2. R.R.Neurgaonker et al, Optical Engneering 5(1987)392
3. D.W.Vally, J.Appl.Phys. 46(1975)3510
4. P.Gunter and J.-P. Huignard, Topics in in Appl. Phys. Vol. 61 and 62.
5. R.A.Rupp et al, Appl. Phys. B29(1986)223.
6. J. Feinberg, Opt. Lett., 5(1980)33 .

Frontiers of Materials Research/
Electronic and Optical Materials
M. Kong, L. Huang (editors)
Elsevier Science Publishers B.V., 1991

GROWTH AND CHARACTERISTICS OF HOMOGENEOUS KTN CRYSTALS

Jiyang WANG, Yaogang LIU, Jingqian WEI; Qingcai Guan and Min WANG

Institute of Crystal Materials, Shandong University, Jinan 250100, P. R. China

The cubic phase $KTa_{1-x}Nb_xO_3$ (KTN) has been grown by flux pulling method. The contents of Ta and Nb distributed homogeneously in the crystal. The size of well-faceted and transparent crystal obtained is up to 35 X 33 X 18 mm. Some properties, such as dielectric constants, dielectric losses, pyroelectric coefficients and others, have been measured. The abrupt variation behaviors of KTN crystal exhibited during the phase transitions clearly demonstrate the homogeneous distribution of Ta and Nb contents in the crystal. The observation of photorefractive phenomenon has been carried out preliminary.

1. INTRODUCTION

Potassium tantalate niobate (KTN, $KTa_{1-x}Nb_xO_3$) is a photorefractive crystal attracting people's great attention during recent years. Possessing strong electric-optic, photorefractive nonlinear optical and pyroelectric effects[1-6], KTN crystal expects prospect applications in the fields of electro-optic switchers, light modulating[7], light holographic memories[8] and parametric osciallation[5]. As we know, KTN is a continuous solid solution of $KTaO_3$ (x=0) and $KNbO_3$ (x=1), and its crystal structure belongs to perofskite one. The ferroelectric phase transition temperature of KTN single crystal can be altered within the range of 0-708K by adjusting the Ta and Nb contents in the crystal[9,10]. When $0<x\leqslant 0.008$, the ferroelectric phase transition of KTN has quantum effect, so it is called the quantum ferroelectrics[11]. When $0.05\leqslant x\leqslant 1$, the structural phase transition of cubic(m3m) $\rightleftharpoons$ tetragonal(4mm) $\rightleftharpoons$ orthohombic(mm2) $\rightleftharpoons$ trigonal(3m) take place. The cubic phase is paraelectric phase and the other are ferroelectric phase. Therefore, KTN single crystal is an object for fundamental theory studies.

In this paper, we mainly report the growth of homogeneous cubic KTN crystal by flux pulling method, and some of its properties.

2. EXPERIMENTAL

2.1. Crystal growth

(1) Apparatus

Crystal growth was carried out in the improved flux system. The flux pulling method was used and the heating elements are JM-1700 Si-Mo rods. The maximum temperature of the furnace is 1700°C. The temperature controlling accuracy is ±0.5°C. The furnace wall is made up of alumina tube, and there is an observing window providing in-situ observations for the real growing conditions of the growth procession.

(2) Procedures

According to the corresponding phase diagram, KTN crystal is a continuous solid solution of $KTaO_3$ and $KNbO_3$. Any slight fluctuation of the growth temperature and the small change of the solution component will cause great variation of Ta and Nb distribution in the crystal. For the purpose of improving homogeneity, some measure e.g. the using of large crucible, narrow growth temperature range, low growth velocity and uniform pulling rate have been taken.

Following the solid and liquid lines of KTN phase diagram, the contents of Ta and Nb in the raw materials were calculated. Excess K_2CO_3 were used as flux. Some other metal oxides or carbonates were added as stablizer and dopant. The raw materials were mixed molten and soaked for more than 24 hours at the temperature about 200°C higher than the melting temperature. Tentative seed method was used to determine the saturation temperature of the flux system. After seeding, well-designed temperature lowering procedure was carried out. By careful observations, when the seed crystal formed shouder to some extent, suitable pulling rate was used according to the real growth conditions. At the desired temperature range (e.g. 10°C or 15°C), the crystal grew to a certain size (about 15-20 mm long). At that time, the crystal was pulled out of the flux. Then, another well-designed temperature lowering procedure was carried out. Finally, colorless and transparent crystal up to the size of 35 X 33 X 18 mm can be obtained. No inclusion and other macrodefects were observed in the crystal except at the shouder-bounderies.

(3) Relation between seed orientation and crystal habit

The quality and habits of the as-grown KTN crystal are closely related to the quality and orientation of the seed crystal. The orientations of seed crystal we used are [100],[110] and [111]. The habits of KTN single crystal are shown in Fig. 1. Because the existing of growth section bounderies, the transparent and inclusion-free part of the [100] orientated as-grown KTN single crystal is larger than those of the crystals grown taking the other orientations.

2.2. Determination of crystal content and the homogeneity

The properties of KTN crystal grown are closely related to the Ta:Nb proportion in the crystal, which depends on the original proportion of Ta and Nb as well as the growth conditions. The determination of crystal content and homogeneity is very important for

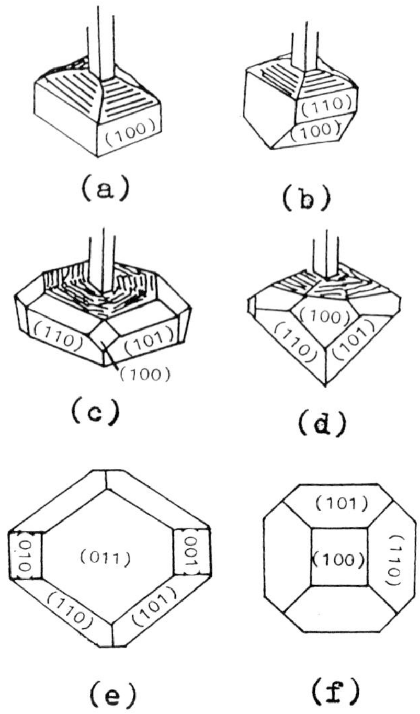

FIGURE 1
The different shapes of as-grown KTN crystals grown by seed orientations
(a) [100] seed; (b), (c) [110] seed; (d) [111] seed; (e) schematic diagram for (c); (f) schematic diagram for (d).

the characteristics of the crystal. The component content and homogeneity of the crystal can be determined by using specific weight, electronic probe and back scattering methods.

The specific weight of KTN crystal is linearly proportional to Ta content. It can be expressed approximately as

$$d = 0.0233x + 4.64 \qquad (x = Ta\%)$$

The Ta content can be calculated by using the formula after determining the specific weight of the crystal with floating method. The value obtained is the mean value of the bulk crystal.

The component contentes at different parts of the crystal can be determined by using electronic probe, and the homogeneity can be determined by using the back scattering analysis method.

The Nb content of a KTN crystal grown from the melt containing 71% Nb is 37%; that is to say, the as grown KTN crystal can be written as $KTa_{0.63}Nb_{0.37}O_3$. The inhomogeneity of the crystal in growth orientation is: $\Delta x = 8.5 \times 10^{-3}$/mm, which is in good agreement with the releationship between Tc and the content we will describe later on in this paper.

2.3. Transmitting curve and refractive index

The transmitting curve of $KTa0.63Nb0.37O3$ was measured by using Hitachi 340 UV-V-NIR spectrophotometer. According to the results obtained, the crystal is almost transparent in all these regions, and the absorption band limit is about 390 nm.

The refractive index of $KTa_{0.65}Nb_{0.35}O_3$ (cubic) crysatl is n = 2.266 (λ=6328 Å) measured by using least-bias angle method.

2.4. Dielectric property and phase transition

For unpolarized and polarized KTN crystal of x = 0.37, the typical result of the temperature dependence of dielectric constants is shown in Fig. 2.

From Fig. 2, we can see that the phase transitions are all first order and similar to those of $BaTiO_3$ and $KNbO_3$ crystals. The abrupt variation of dielectric constant occured at the temperature close to phase transition temperature has shown the homogeneous Ta and Nb distribution in the crystal, and also reflects the quality of the crystal.

2.5. The thermal property

The DSC specific heat curves were measured. The phase transition temperaures can be observed clearly. They are 282 K (8.8 ℃), 221 K (-52.2 ℃) and 183 K (-90.2 ℃), which correspond to Tc, T_I and T_{II}, respectively. The phase transitions are all considered to be first order The average spcific heats of the four different phases are 0.0911 (trigonal), 0.0967 (ortho.), 0.1042 (tetragonal) and 0.1077 (cubic) cal/g.deg.

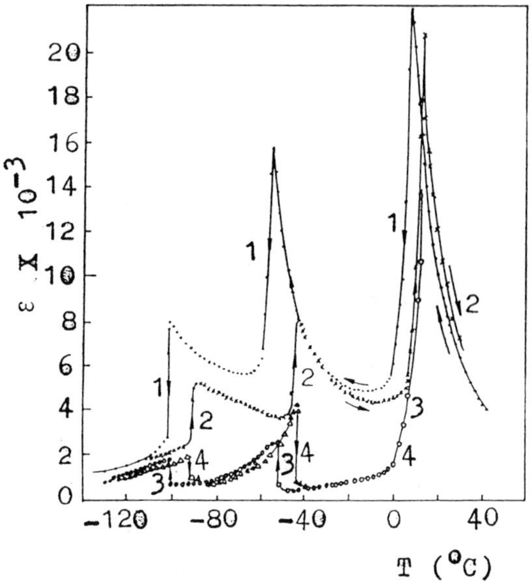

FIGURE 2
The relationship between dielectrie constant and temperature of KTN crystal
Curves 1 and 2: unpolarized; Curves 3 and 4: polarized.

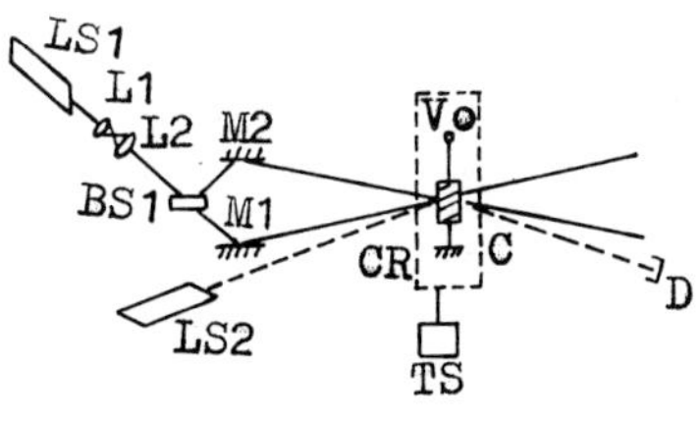

FIGURE 3
The setup for photorefractive effect measurement
LS1: Ar laser; LS2: He-Ne laser; L1, L2: Beam expander lens; BS1: Beam splitter; M1, M2: Mirrors; CR: Cryostat; C: Crystal; D: Detector.

2.6. Photorefractive effect

By applying electric field, the photorefractive effects of KTN were measured. The experimental setup is shown in Fig. 3. The refractive efficiency has close relations with the phase, quality of the crystal and the voltage applied.

3. DISCUSSION

In many literatures related, emphasis has often been laid on the information that at the boundery of cubic and tetragonal phase at room temperature, $KTa_{0.65}Nb_{0.35}O_3$ crystal has the largest second order electro-optic coefficient and light storage sensibility, which make it become an ideal holographic bulk storage material. From the results of photorefractive property measurement of KTN after the applying of electric field, it can be seen that the tetragonal phase is superior than cubic phase. But, the crystal growth, polarization and processing of tetragonal phase are all more difficult than that of cubic phase. Therefore, it is worthy for us to consider the problem of how to find a suitable component material to use after taking the factors of high quality performance and convenient applications into consideration.

REFERENCES

1. W. Haas and R. Johannes, Appl. Opt. 6 (1967) 2007.
2. A. J. Fox and P. W. Whipps, Electrinics Lett. 7 (1971) 139.
3. F. S. Fang, J. Appl. Phys. 38 (1967) 418.
4. R. Orlowski, L. A. Boatner and E. Kratzig, Optic. Comm. 35 (1980) 45.
5. S. K. Kurtz and T. T. Perry, J. Appl. Phys. 39 (1968) 166.
6. K. W. Goeking et al. Ferroelectrics 92 (1989) 89.
7. A. J. Fox, Appl. Opt. 14 (1975) 343.
8. L. A. Boatner, E. Kratzig and R. Orlowski, Ferroelectrics 27 (1980) 247.
9. D. Rytz and H. J. Scheel, J. Crystal Growth 59 (1982) 59.
10. S. Triebwessen, Phys. Rev. 114 (1959) 62.
11. D. Rytz, Phys. Rev. Lett. 54 (1985) 2038.

Frontiers of Materials Research/
Electronic and Optical Materials
M. Kong, L. Huang (editors)

GROWTH, SPECTRA AND SELF–FREQUENCY–DOUBLING PROPERTIES OF $Nd:Mg:LiNbO_3$ SINGLE CRYSTAL

Xianlin MENG, Zongshu SHAO, Youxuan SUN, Enquan LIU, Yuming SUN

Institute of Cyastal Materiais and Department of Optics, Shandong University, Jinan, 250100, China

In this paper, growth and polarization conditions, absorption spectra, laser and self–frequency–doubling properties of $Nd:Mg:LiNbO_3$ single crystal have been reported. The high quality crystal with diameter larger than 30 mm has been grown. Laser oscillation has been demonstrated in the $Nd:Mg:LiNbO_3$ crystal and green self–frequency–doubling laser of 542nm wavelength has also been observed.

1. INTRODUCTION

Lithium niobate ($LiNbO_3$ or LN) single crystal has good electro–optical and nonlinear optical properties. However, its low photorefractive thresholds have limited the applications. Doping 5 mol.% or more magnesium oxide (MgO) into LN can remarkably increase the photorefractive thresholds of the crystal[1].

LN is also a laser host matarial. Flash–lamp–pumped and laser–pumped Nd:LN lasers were made[2, 3]. However, the photorefractive effect in the crystal caused the efficiency of the devices greatly reduced.

Nd–Mg–doped $LiNbO_3$(Nd:Mg:LN) single crystal combines the excellent laser properties of Nd^{3+}ion, the ideal anti–photorefractive effect of Mg^{2+}ion and the good electro–optical, nonlinear optical properties of LN crystal together. So, Nd:Mg:LN ctystal is a promising multi–functional material. Many investigations have been carried out on the crystal growth and devices of this material[4,5].

We have investigated the crystal growth and polarization conditions of Nd:Mg:LN and have grown high quality single crystal. The absorption spectra and laser properties of the Nd:Mg:LN crystal have been determined. Laser oscillations and self–frequency–doubling operation have been realized.

2. CRYSTAL GROWTH AND POLARIZATION

High purity Nb_2O_5, Li_2CO_3, MgO and $Nd_2O_3$are used for growing Nd:Mg:LN crystal. The equipment is a radio frequency heating single crystal growing furnace with computer control. The crystal was pulled along the y or z axis of the crystal from platinum crucible by Czochralski method in atmosphere.

The Nd:Mg:LN single crystals grown by using our growth conditions have diameter larger than 30 mm and exhibit purple color. There are no macroscopic defects observed in the crystals (Fig.1).

Fig.1

$Nd:Mg:LiNbO_3$ single crystals

Transmissivity of the crystal is up to 95–99% in 950–2100nm wave band along the directions parallel and perpendicular to the growth axis. Extinction ratio of the crystal is above 30 db for He–Ne laser with 632.8nm wavelength. Laser oscillation and self–frequency–doubling operation have been realized.

The Nd:Mg:LN crystal was polarized at temperature above its Curie point T_C. Polarizing the crystal can change its polydomain structure into single domain structure. An important character of the single domain crystal is that the etching properties of its +Z and −Z sections are greatly different.

After +Z and −Z sections of the single domain Nd:Mg:LN crystal were polished, the sample was etched in HF and HNO_3mixed etchant for half an hour at 100℃ . The etching figures of the crystal are shown in Fig.2.

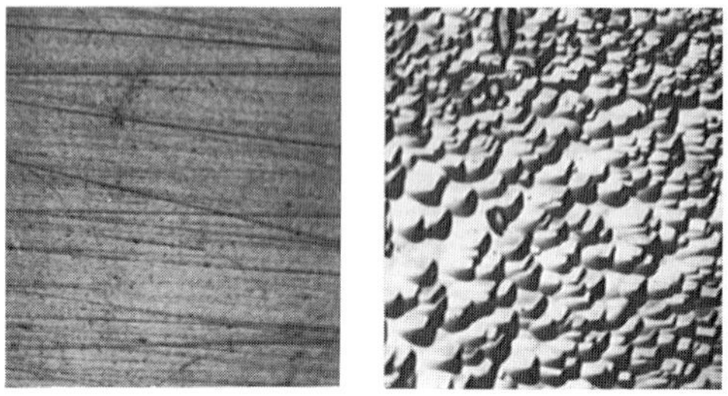

a 120X b 120X

Fig.2
Etching figures on +Z(a) and −Z(b) sections of Nd:Mg:$LiNbO_3$ crystal

From Fig.2, it can be seen that the etching figure on −Z section is triangle pyramid. The etching figure on +Z section is more difficult to carry out than that on −Z section. The etching properties of single domain Nd:Mg:LN crystal is similar to that of pure single domain LN crystal.

Etching figures on both Z sections of unpolarized LN crystal are about the same. However, the etching figures on both Z surfaces of unpolarized Nd:Mg:LN crystal are very different from eath other. One of them is similar to the etching figure on +Z section of single domain crystal and the other is similar to that on −Z section. This means that the Nd:Mg:LN crystal has been polarized to a certain extent in the growth process, in spite of that the polarization is not quite complete.

3. ABSORPTION SPECTRA

Absorption spectra along X or Y direction of the Nd:Mg:LN crystal samples of 7mm thick have been measured by HITACHI−340 UV−V−NIR recording spectrophotometer. There are seven high absorption peaks in the range of 420−2600nm. The corresponding wavelengths and light absorption ratio are listed in Table 1.

Table 1
Wavelength and absorption ratio of the peaks of Nd:Mg:$LiNbO_3$ crystal

Wavelength (nm)	533	598	753	813	880	2390	2550
Absorption ratio (%)	70	91	75	70	30	31	39

It should be noted that the self−frequency−doubling wavelength of Nd:Mg:LN crystal is 542nm, which is deviated from the absorption peak of 533nm. This fact opens up a possibility of making self−frequency−doubling laser devices with Nd:Mg:LN crystal.

Table 1 provides important experimental criteria for the selection of pumping light source and device design.

4. LASER AND SELF−FREQUENCY−DOUBLING PROPERTIES

4.1 Sample preparation and experimental arrangement

Crysal used for sample preparation was grown from $LiNbO_3$ melt doped 1 mol % Nd_2O_3 and 5 mol.% MgO. The crystal samples of 5mm × 7mm × 5mm (X.Y.Z) were not so strictly directed and polished. Then they were coated on the Y end faces. Pumping source used is DATACHROM−5000 dye laser. The diameter of the pulsed dye laser beam is 4mm. Its repetition rate is 10 times per second and the pulse width is 10 ns. The dye laser pumps the crystal sample along Z axis. Experimental arrangement is shown in Fig.3 which indicates that the arrangement is very simple.

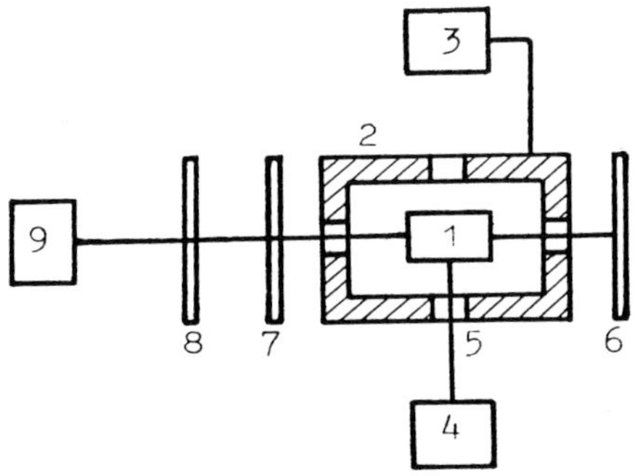

Fig 3.
Schematic diagram of the Nd:Mg:LN laser experimental arrangeunment
1–Nd:Mg:LN crystal sample
2–thermostat
3–auto–control temperature system
4–dye laser
5–pumping light beam
6–1084nm and 542nm reflective mirror
7–1084nm reflective mirror (98% or 100%)
8–color filter
9–detector

4.2. Experimental and results

4.2.1. Selection of the pumping light wavelength

According to the absorption spectra of Nd:Mg:LN crystal, the sample has a maximum absorbability to the light whose wavelength is close to 598nm.

At constant temperature and pumping energy, we have determined the second harmonic energies under the condition of making the pumping light wavelength change from 595nm to 603nm. Experimental results show that better pump effect can be obtained at the pumping light wavelengths ranged from 597nm to 601nm. The second–harmonic energy is greatly reduced when the wavelength is above 601nm or below 597nm. The dye pumping light of 598.6nm wavelength was selected and used in our experiment.

4.2.2. Determination of fundamental and self–frequency–doubling laser wavelengths.

The laser oscillation and self–frequency–doubling can be realized by using dye laser with 46mJ energy and 598.6nm wavelength to pump the Nd:Mg:LN sample at room–temperature. The 1084nm reflective mirror and monochromator were used for the determination of self–frequency–doubing light wavelength and 1084nm 98% reflective mirror and color filter were used for fundamental wavelength.

The fundamental wavelength obtained is 1084nm and the self–frequency–doubling laser wavelength is 542nm.

4.2.3. Determination of pumping thresholds

The pumping energy and second–harmonic energy can bc measured by using dye laser with higher energy to excite the crystal sample at a constant temperature. We reduced the pumping energy gradually and observed the variation of the green self–frequency–doubling laser. The pumping threshold is exactly the pumpimg energy when the green laser is about to disappear at the given temperature.

The results obtained show that the thresholds are 9mJ at 80℃ and 7.7mJ at 86℃ respectively.

4.2.4. Relation between temperature and energies of frequency–doubling or self–frequency–doubling light.

The fundamental light used for frequency–doubling experiment is pulse laser generated by Nd:YAG crystal. Laser wavelength is 1064nm. Its repetition rate is 10 times per second and pulse width is 10ns. The fundamental light beam with 44mJ energy was passing through the Nd:Mg:LN sample along Y axis. We have detected the frequency–doubling light energies through the filter at temperature range from 20–165℃ The results are shown in Table 2.

From Table 2, we can see that the crystal has frequency–doubling property at temperature range from 20–165℃ and its phase–matching temperature (T_{pm}) is 100.4℃.

Table 2

Relation between frequency–doubling energy and temperature (fundamental wavelength:1064nm, energy:44mJ)

Temperature (℃)	20	60	95	98	99	100	100.4	100.9	101.5	102	110	140	165
Frequency–doubling energy (μJ)	18	25	102	250	1268	4821	5536	3036	1250	714	54	36	18

Table 3

Relation between self–frequency–doubling energy and temperature (pumping wavelength:598.6nm, energy:46mJ)

Temperature (℃)	20	40	60	80	90	94	97	100	104	110	120	140	165
Self–frequency–doubling energy(μJ)	4	10	18	10	8	13	12	13	12	12	13	7	4

The dye laser beam with 46mJ energy was used to pump Nd:Mg:LN crystal sample along Z axis. We have measured the energies of self–frequency–doubling laser projected along Y axis at temperatures range from 20–165℃ . Experimental results are shown in Table 3.

The self–frequency–doubling determination indicates that the crystal has self–frequency–doubling ability in a quite wide range of temperature. The differences of second–harmonic outputs at different temperatures are not so large. No abrupt change of the self–frequency–doubling energy appears at the temperature close to T_{pm}.

5. DISCUSSION

5.1. Crystal growth and domain structure

Nd:Mg:LN crystal is more difficult to grow than pure LN crystal. Macroscopic defects such as cracking, cellular structure and growth striation are easy to form in Nd:Mg:LN crystal. These problems should be solved.

Cracking problem can be solved by using special–designed platinum auxiliary heater. The heater may create a temperature field in which temperature gradient near the phase interface is quite large whereas that above the interface is very small. In this temperature field, the cellular structure can be completely overcome by adopting appropriate pulling rate and rotating rate. The pulling rate and rotating rate are correlated with crystal diameter and the doping concentration. Using crucible of suitable shape, the growth stration can be minimized. For the purpose of improving the optical properties of the crystal, it should be annealed at 1150℃ for 24 hours.

The Nd:Mg:LN crystl has been polarized to a certain extent in the growth process. This can be attributed to the entering of Nd^{3+} ion into the crystal. The radius of Nd^{3+} ion is 1.00Å, but these of Li^{+}ion and Nd^{5+} ion are 0.68Å and 0.70Å, respectively. The Nd^{3+} ion in the crystal occupies the lattice sites of Li^{+} ion or Nb^{5+} ion, which not only results in lattice distortion but also leads to charge compensation and point defect removal in the crystal. As a result, electric field may be created near the phase interface and it can partialy polarize the crystalline ions.

5.2. Crystal properties

At phase–matching temperature (100.4℃) the frequency–doubling output is more powerful than those at other temperatures, but the self–fre–

quency—doubling energy exhibits no such phenomenon at this temperature. This question will be discussed in other paper.

This paper only reports our initial research work and better experimental results will certainly be obtained after the crystal processing, laser cavity and experimental equipment are improved.

ACKNOWLEDGEMENTS

The authors would like to thank Zhaohe YANG, Xunmin LIU, Airong ZHANG, Reiping CHENG, Jiguo ZHANG, Li ZHU and Reihua WANG for their helps in the spectrum measurements, crystal growth, coating and processing.

REFERENCES

1. G. Zhong et al., Proceeding of 11—th Internatinal Quantum Electronics Conference; IEEE Catalog No. 80 CH 1561—0 (New York, 1980) 631.

2. L.F. Johnson et al., J. Appl. Phys. 40, (1969) 297.

3. I.P. Kaminow et al., IEEE J. Quantum Electron. QE—11, (1975) 306.

4. Xu Guanfeng et al., J. Synthetic Crystal, Vol. 17, No. 3—4, (1988) 198.

5. T.Y.Fan et al., J. Opt. Soc. Am. B/ Vol.3, No.1, (1986) 140.

Frontiers of Materials Research/
Electronic and Optical Materials
M. Kong, L. Huang (editors)

NONLINEAR OPTICAL PROPERTY OF SOME ORGANO-TITANIUM AND -ZIRCONIUM COMPOUNDS*

Jingui QIN, Xiaoping GONG, Xiangdong WU, Jinlin LIAO, Caoyang DAI, Jianxin Gu, Shenggang MA, Daoyu LIU

Department of Chemistry, Wuhan University, Wuhan 430072, China

Chuangtian CHEN, Baichang WU

Fujian Institute of Research on the Structure of Matter, Fuzhou 350002, China

This paper describes the design, synthesis, characterization and powder second harmonic generation (SHG) effect of some organo-titanium and -zirconium compounds of the type of $Cp_2M(Cl)$-EAr (Cp = cyclopentadienyl, M = Ti or Zr, E = O, S).

1. INTRODUCTION

The intensive investigation and development of nonlinear optical (NLO) materials in the last twenty years or so has been related with their applications in optical information processing, telecommunication and integrated optics. The classical and practically useful NLO crystals are inorganic salts such as KH_2PO_4 (KDP), $KTiOPO_4$ (KTP) and β-BaB_2O_4 (BBO). Unfortunately, their NLO coefficiency is not very large. Many organic compounds have been found to possess the second order NLO coefficiency which is much greater than that of inorganic materials. However, it is very difficult to grow these organic crystals to a large size with good optical homogeneity [1].

Recently, the interest in investigation of NLO property of organometallic compounds is growing so as to search for new type of NLO materials in the hope that they may combine advantages of both the inorganic and the organic compounds [2].

The chemistry of Cp_2MCl_2 (M = Ti or Zr, Cp = cyclopentadienyl) and their various derivatives has been studied extensively [3]. The nonlinear optical property of these compounds, however, has not been explored yet. These compounds can be easily synthesized and many of them are lightly coloured or white and stable in air. In addition, $[Cp_2Ti]$ and $[Cp_2Zr]$ are both 14 electron species which have a very strong tendency to achieve 16 electron configuration. Therefore, these compounds

* Supported by the National Foundation for Natural Science of China

may be considered as the candidates for a new type of NLO materials.

This paper presents the preliminary study on the synthesis, characterization and powder second harmonic generation (SHG) effect of such compounds.

2. RESULTS AND DISCUSSION

The reaction of Cp_2MCl_2 (M=Ti or Zr) with substituted phenol or thiophenol (HEAr, E=O or S) in the absence or presence of a base (Et_3N or $NaNH_2$) gave corresponding monosubstituted, $Cp_2M(Cl)$-EAr, or disubstituted product, $Cp_2M(EAr)_2$, depending on the molar ratio of the starting materials and reaction conditions. In many cases, the mixture of both compounds were obtained. The expected products were isolated and purified normally by recrystallization.

The reaction were as follows:

$$Cp_2MCl_2 + HEAr \longrightarrow Cp_2M(Cl)\text{-}EAr$$

$$Cp_2MCL_2 + 2HEAr \longrightarrow Cp_2M(EAr)_2$$

(M= Ti or Zr , E= O or S)

The products were characterized by microelemental analysis and infrared spectroscopy. SHG effect of these compounds was measured by a modified version of Kurtz powder technique using Nd:YAG laser emitting 1.06 μm as a source and KDP as a reference.

Both mono- and di-substituted titanocene derivatives were prepared. The SHG measurement indicated that disubstituted compounds did not show SHG effect at all, while many of mono-substituted compounds did exhibit SHG (see Table 1). The reason for this may be that the symmetry of the disubstistuted compounds is higher than that of monosubstituted one, leading to more chance of the formation of centrosymmetric crystal structure.

Table 1. Structure and SHG signals of the Titanocene Compounds

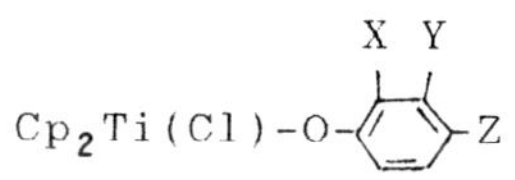

No.	X	Y	Z	SHG (× KDP)
1	H	H	Me	0.2
2	F	H	H	0.3
3	H	Me	Cl	0.3
4	H	H	Br	1.0
5	H	H	CO_2Me	1.7
6	NO_2	H	NO_2	0

Then eleven monosubstituted zirconocene derivatives were synthesized, most being white. Suprisingly, it was found that ten of them (91%) exhibited SHG effect with a maximum value of 4.5 × KDP (see Table 2).

It is known that there are only about twenty percent of the crystals which belong to noncentrosymmetric space groups according to several extensive investigations[4], the compounds of the

Table 2 Structure, λ_{max} , and SHG effect of Zirconocene Compounds

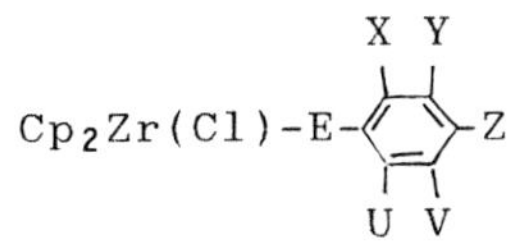

NO.	E	X	Y	Z	V	U	λ max (nm)	SHG (×KDP)
1	O	H	H	NO_2	H	H	328	2.0
2	O	NO_2	H	H	H	H		1.4
3	O	NO_2	H	NO_2	H	H		0.0
4	O	H	H	F	H	H	315	0.15
5	O	F	H	H	H	H	338	2.0
6	O	H	H	Br	H	H	334	0.3
7	O	Cl	H	H	H	H	344	2.0
8	O	H	H	CO_2Pr	H	H	330	1.2
9	S	H	H	F	H	H	316	1.2
10	S	F	F	H	F	F	336	4.5
11	S	F	F	F	F	F	363	0.15

type of $Cp_2Zr(Cl)$-EAr seem quite promising for SHG effect with such a high percentage of the compounds showing SHG phenomenon.

It was well known that an asymmetric stretching absorption of a nitro group, $\nu_{as(NO_2)}$, of nitrobenzene derivatives on their infrared spectra had shifted to a lower frequency if the para- or ortho- position to the nitro group was occupied by an electron donor (e.g. $-NH_2$, -OH, Table 3). The infrared spectra of the compounds $Cp_2Ti(Cl)$-O-C_6H_4-NO_2 and $Cp_2Zr(Cl)$-O-C_6H_4-NO_2 were recorded, and their $\nu_{as(NO_2)}$ have both shifted towards lower frequency (see Table 3), indicating that [$Cp_2M(Cl)$-O] (M = Ti, Zr) fragments are both electron donors in this situation. Further study is in progress.

Table 3. Asymmetric stretching absorption of the nitro group in nitrobenzene derivatives

compound	$\nu_{as(NO_2)}$ (cm^{-1})
C_6H_5-NO_2	1525
H_2N-C_6H_4-NO_2	1475
HO-C_6H_4-NO_2	1496
$Cp_2Ti(Cl)$-C_6H_4-NO_2	1488
$Cp_2Zr(Cl)$-C_6H_4-NO_2	1490

3. EXPERIMENTAL

All manipulations were carried out under an inert atmosphere of nitrogen using standard vacuum-line and Schlenk tube techniques. All solvents used were predried over sodium wires and then distilled from sodium-potassium alloy.

Microelemental analysis was carried out on a Carlo Erba 1106 Analyser. Infrared spectra were recorded on either Shimadzu IR 450 or Nicolet FTIR 170SK Spectrophotometer. SHG signals were measured using a modified version of Kurtz powder technique with Nd-YAG laser emitting 1.64 μm as a source and KDP as a reference.

Cp_2TiCl_2 [5] and Cp_2ZrCl_2 [6] were prepared

by the literature method. Disubstituted titanocene derivatives, i.e. $Cp_2Ti(OAr)_2$, were prepared following the literature [7].

Synthesis of $Cp_2Ti(Cl)$-OAr:

To a refluxing solution of Cp_2TiCl_2 (2 mmol) and Et_3N (2 mmol) in benzene (20 ml) was added dropwise a solution of substituted phenol (3 mmol) in benzene over a period of 30 min. After the addition was completed, the mixture was refluxed for further half an hour. The white precipitates were filtered off. The filtrate was concentrated to ca.20 ml, to which was added a suitable amount of petroleum ether (30-60°C). The precipitates (mainly being unreacted phenols) were filtered off. The filtrate was concentrated to ca. 15 ml. Petroleum ether was added to make hot saturated solution which was allowed to cool down slowly, resulting in the formation of crystals which were then recrystallized from benzene.

Zirconocene derivatives, $Cp_2Zr(Cl)OAr$, were prepared in a similar way except using equimolar starting materials. In the cases that thiophenols were used as the starting materials, the base (Et_3N) was not used.

4. CONCLUSIONS

Mono- and di-substituted titanocene and zirconocene derivatives with the formula of $Cp_2M(Cl)$-EAr and Cp_2M-$(OAr)_2$ (M = Ti or Zr ; E = O or S) were prepared, and their SHG effect was measured by the Kurtz powder teshnique. Zirconocene derivatives showed surprisingly high percentage (91%) of compounds which exhibited SHG effect.

5. REFERENCES

1. C. T.Chen, G.Z.Liu, Ann. Rev. Mater. Sci., 16 (1986) 203.

2. M.L.H.Green, J.Qin, D.O'Hare, J.Organomet.Chem., 358 (1989) 375.

3. V.Thewalt, Organotitanium Compounds, Part 4, in Gmelin Handbook of Inorganic Chemistry, 8th Ed., No.41,(Springer-Verlag,Berlin,1984).

4. D.S.Chemla and J.Zyss, eds., Nonlinear Optical Properties of Organic molecules and Crystals, (Academic Press, Orlando, 1987).

5. T.G.A.Stone, et.al., Advances in Organometallic Chemistry, Academic Press, New York, 2 (1964) 373.

6. E.Samual, R.setton, C.R.Acad.Sci., Ser.C, 254 (1962)308.

7. S.Chen,Y.Liu, Science Sinica, 10 (1981) 1225.(in Chinese).

Frontiers of Materials Research/
Electronic and Optical Materials
M. Kong, L. Huang (editors)

THE RELATIONSHIP BETWEEN THE SECOND-HARMONIC EFFECT AND CRYSTAL STRUCTURE IN N-(2,4-DINITROPHENYL) AMINOPROPIONATE PYRIDINIUM SALT

Li Songxian, Liu Shixiong, Lin Lifeng, Lin Fan, Chen Jianzhong

(Center for Research in Solid Materials, Fuzhou University)

A title compound (PAP) crystal with excellent nonlinear optical properties has been synthesized. The crystal is monoclinic with space group $P2_1$ and composed of N-(2, 4-dinitrophenyl) aminopropionate anion (AP^-) and pyridinium cation (PyH^+). There is a strong hydrogen bond N–H···O between N–H of PyH^+) and oxygen atom of carboxyl group. G and $\overline{P^2(2\omega)}$ of PAP, MAP and NaAP·HAP [N-(2,4-dinitrophenyl) aminopropionate sodium salt] are 565, 130, 2.10 and 0.111, 0.0166, 9.44×10^4, respectively. Their second-harmonic intensity $I_{2\omega}/I_{2\omega(KDP)}$ are 100, 50, 5.0 times as large as that of KDP, respectively. The close agreement between the calculated G and $\overline{P^2(2\omega)}$ and measured $I_{2\omega}/I_{2\omega(KDP)}$ can predict the frequency doubling properties.

1. INTRODUCTION

In recent ten years, the exploration of the nonlinear optical organic molecular crystals is an active research area with the advance of molecular designing engineering.[1,2,3] In this kind of crystal materials, N-(2,4-dinitrophenyl) aminopro pionate methyl ester (MAP) with excellent nonlinear optical property was advanced more than ten years ago. But owing to its low melting point, weaker mechanical strength and harsh growing condition, up to now it has not seen any report concerning its practical application. In order to overcome its defect, the authors attempt that its salts are used instead of MAP. The crystals of N-(2,4-dinitrophenyl) aminopropionate lithium salt, sodium salt, potassium salt, ammonium salt, barium salt, thallium salt, tetramethyl amino salt and pyridinium salt has been synthesized, respectively. Their relative intensites of the second order harmonic generation $I_{2\omega}/I_{2\omega(KDP)}$ are determined and shown that the property of N-(2,4-dinitrophenyl) aminopropionate pyridinium salt (PAP) was the best of these eight crystals.[5]

This paper reports the synthesis, single crystal structure, nuclear magnetic resonant hydrogen spectrogram and second harmonic effect of PAP. Based on the crystal structural parameters, we calculated the space factor G and the mean square value of the unit second order polarizability $\overline{P_0^2(2\omega)}$ of PAP,MAP and NaAP·HAP. The regularities of the theoretically calculated G and $\overline{P_0^2(2\omega)}$ were in fair agreement with the measured $I_{2\omega}/I_{2\omega(KDP)}$ values.

2. SYNTHESIS AND ELEMENTAL ANALYSIS

Equimolar 2,4-dinitrofluorobenzene and amino-propionic acid reacted under stirring for 5 hours at 40°C in acetone-alcohol solvent. After the extraction, separation and purification, N-(2,4-dini-trophenyl) aminopropionate (HAP) was obtained from the mixture. HAP was soluted to the alcohol, and then pass through into HC1 gas and pyridine was dropped slowly into the alcohol

The project supported by National Natural Csience Foundation of China.

solution, PAP crystals was preciepitated gradually. The melting point of PAP crystal is 109±1°C.

Chemical analysis: found (calculated for $C_{14}H_{14}N_4O_6$)%: C,50.28(50.29); H, 4.24(4.23); N, 16.60(16.76). It explains that the close agreement between the synthetical compound and predictable molecular formula.

The polycrystalline diffraction figures of PAP, NaHP·HAP, MAP and HAP were obtained on a D/max-r diffractometer and shown that PAP and NaAP·HAP are new compounds.

3. CRYSTAL STRUCTURE AND NMR SPECTRA

3.1. Diffraction experimental

A yellow prismatic crystal with dimensions of 0.52×0.41×0.08 mm was mounted on a AFC5R diffractometer with graphite-monochromatized MoK_α radiation. The title compound crystal is monoclinic with lattice paprmeters a=4.521(1), b=9.601(6), c=17.892(2)Å, β=96.12(1)° and v=772.8Å^3, Z=2, Dc.=1.44g.cm^{-3}. A total of 2036 unique reflections was collected by $\omega-2\theta$ scan technique in the range of $\theta \leq 27°$, in which 1094 reflections with $I \geq 3\sigma(I)$ were considered to be observed and used in structure calculation. Intensity data were corrected for L_p factors and DIFABS empirical absorption. The systematic absences shown by the X-ray diffraction indicate that the possible space groups are $P2_1$ and $P2_1/m$. Patterson function analysis shows that there is not m(010) in this crystal. Therefore the space group should be $P2_1$. This has also been confirmed by subsequent structure determination. The structure was solved by the direct method and refined by full-matrix least squares with anisotropic thermal factors for all non-hydrogen atoms and isotropic thermal for all hydrogen atoms obtained from difference Fourier map and HYDROGEN program. The final discrepancy factors R and Rw were 0.038 and 0.049 $(W=(\sigma^2|F_o|^{-1}))$ espectively. All calculation were carried out on a Micro VAX II computer with TEXSAN software package.

3.2. NMR spectra

^{1}H NMR spectra for the title compound were recorded in DMSO-d_6 (dimethyl sulfoxide) on a FT-80A NMR Spectrometer at 80 MHz using tetramethyl silane as the internal reference at room temperature. Another NMR spectra for the title compound were recorded after the solution exchanged with D_2O. The ^{1}H NMR spectra as shown in Fig. 1.

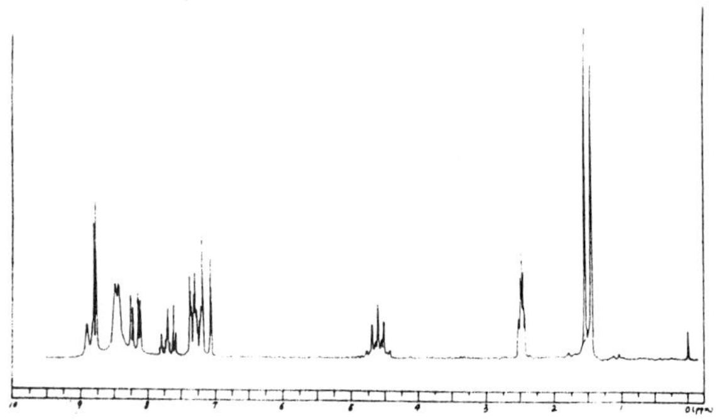

FIGURE 1

^{1}H NMR spectra

A multiple peak at δ=4.60 ppm may be assigned to the proton in -ĊH-, which could be explained by that there is a coupling of the proton in -NH- with the proton in -ĊH- when ammonium salt formed by -NH- group decreases exchange rate. The multiplet peak become a simple quadruplet peak in the spectra exchanged by D_2O, because active hy drogen in -NH- is exchanged by D_2O.

A doublet peak at δ=8.47 ppm could be attributed to the overlap of proton absorptions in α-H of pyridinium and in hydrogen of -NH-. This due to disappearance of the -NH- absorption peak.

The integrated area ratio of all absorption peaks in NMP spectra was in good agreement with the number radio in all kinds of protons.

3.3. Discussion

Atomic coordinates with thermal parameters, is

given in Table 1. Fig. 2 and 3 show molecular configuration of PAP and molecular packing in a unit cell. The structure of title compound consists of anion O_2N-  $-NHCH(CH_3)COO^-(AP^-)$ NO_2 and pyridinium cation PyH^+. The value of bond lengths and angles in the AP^- anion is similar to that in MAP [6]. It is worth to note that N(1)-C(1) bond length is 1.334 Å and N(1)-C(7) 1.445 Å. The shortening of N(1)-C(1) bond was also found in MAP molecule [6]. This fact indicates that N(1) atom takes part in conjugation effect of benzene ring.

As shown in Fig. 3, there is a strong hydrogen bond N(4)-H(01)-O(6) between O(6) atom of carboxyl group and N-H of pyridinium cation, N···O distance being 2.541 Å.

4. THE CALCULATION OF THE MEAN SQUARE OF THE UNIT SECOND ORDER POLARIZABILITY $\overline{P_o^2(2\omega)}$ OF A CRYSTAL

4.1. The definition of $P_o^2(2\omega)$

In an applied electrical field of a unit intensity $\vec{E}=\vec{E}_o \exp(\omega_o - wt)$, the second order polarizability of a crystal is

$$P_o(2\omega) = \chi(2\omega)\vec{E}^o\vec{E}^o \exp[2(\omega_o - \omega t)] \qquad (1)$$

E^o is the unit vector of the applied electrical field. $\chi(2\omega)$ is the three scale tensor of the second order susceptibility

$$\chi(2\omega) = NL(2\omega)B(2\omega). \qquad (2)$$

Among formula the N is the number of frequency doubling function groups per unit volume of a crystal, N=n/v, the n is the equivalent point of a unit cell, the v is the volume of a crystal cell, $L(2\omega)$ is the local field factor, $B(2\omega)$ is a tensor of the second order susceptibility without inclosing the volume of the crystal, its term is b_{hkl}.

According to the model of the second order susceptibility vector for nonlinear optical frequency doubling effect [7]. In the applied electrical fields

Table 1. atomic coordinates and thermal parameters

atom	x	y	z	B(eq)
O1	0.334(1)	0.3994	-0.0054(2)	8.3(2)
O2	0.683(1)	0.4694(8)	0.0764(2)	8.3(2)
O3	0.9415(8)	0.2138(6)	0.2970(2)	5.9(2)
O4	0.6885(8)	0.0382(7)	0.3289(2)	6.2(2)
O5	0.1656(9)	-0.2434(7)	0.3372(2)	6.1(2)
O6	-0.168(1)	-0.3661(7)	0.2649(2)	6.4(2)
N1	0.2948(8)	-0.0786(7)	0.2291(2)	4.0(2)
N2	0.487(1)	0.3871(8)	0.0550(2)	5.5(2)
N3	0.739(1)	0.1298(8)	0.2840(2)	4.3(2)
N4	-0.298(1)	-0.4864(7)	0.3836(2)	4.5(2)
C1	0.345(1)	0.0346(7)	0.1892(2)	3.5(2)
C2	0.184(1)	0.0542(8)	0.1169(2)	4.1(2)
C3	0.230(1)	0.1678(8)	0.0747(2)	4.3(2)
C4	0.437(1)	0.2683(8)	0.1010(3)	4.1(2)
C5	0.599(1)	0.2541(8)	0.1696(3)	4.1(2)
C6	0.555(1)	0.1383(7)	0.2134(2)	3.8(2)
C7	0.090(1)	-0.1896(7)	0.2055(2)	3.8(2)
C8	0.207(1)	-0.2889(8)	0.1482(3)	5.2(2)
C9	0.033(1)	-0.2700(8)	0.2766(3)	4.5(2)
C10	-0.164(1)	-0.4594(7)	0.4511(3)	5.2(3)
C11	-0.228(1)	-0.5341(9)	0.5124(3)	5.3(3)
C12	-0.438(1)	-0.6353(8)	0.5046(3)	5.2(3)
C13	-0.578(1)	-0.6617(8)	0.4336(3)	5.3(2)
C14	-0.501(1)	-0.5850(8)	0.3750(2)	4.8(2)
H1	0.4029	-0.0870	0.2775	4.8
H01	-0.2487	-0.4350	0.3414	5.5
H2	0.0424	-0.0135	0.0981	5.0
H3	0.1194	0.1789	0.0269	5.2
H5	0.7395	0.3233	0.1872	4.9
H7	-0.0918	-0.1500	0.1839	4.6
H10	-0.0203	-0.3869	0.4571	6.2
H11	-0.1259	-0.5150	0.5605	6.4
H12	-0.4877	-0.6868	0.5468	6.3
H13	-0.7245	-0.7325	0.4259	6.4
H14	-0.5972	-0.6030	0.3261	5.7
H81	0.0646	-0.3602	0.1355	6.2
H82	0.2401	-0.2385	0.1041	6.2
H83	0.3885	-0.3296	0.1693	6.2

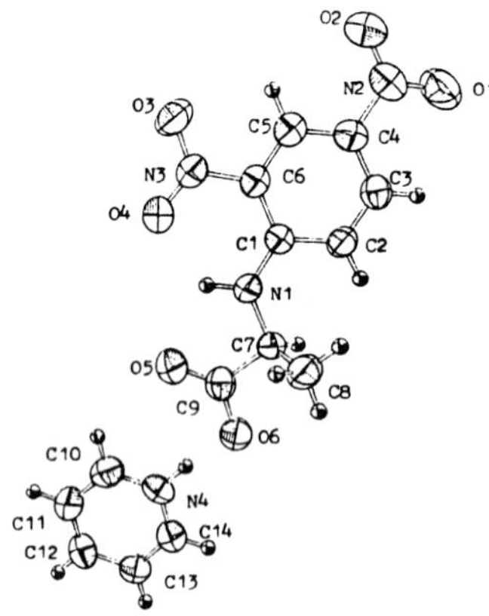

FIGURE 2
Configuration of $C_{14}H_{14}N_4O_6$ molecule

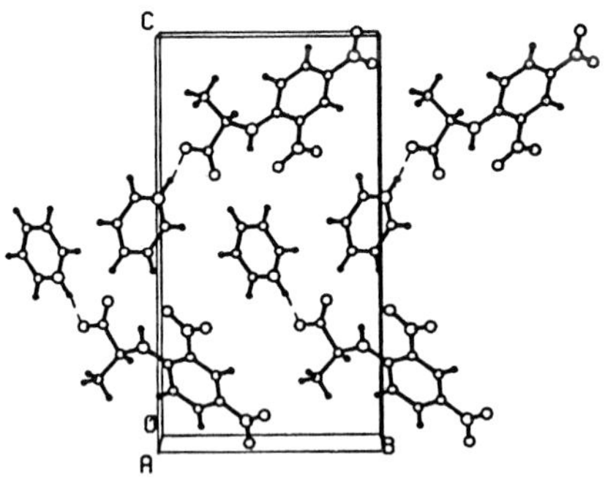

FIGURE 3
Packing diagram of $C_{14}H_{14}N_4O_6$ molecules in a unit cell along a axis

with different directions, the mean square of the unit second order polarizability of a crystal is

$$\overline{P_0^2(2\omega)} = 1/4\pi \oint\!\!\oint_S [\chi(2\omega)\vec{E^0}\vec{E^0}]^2 ds. \tag{3}$$

The intensity of the second order harmonic generation of the crystal powder

$$I_{2\omega} \propto \overline{P_0^2(2\omega)} \tag{4}$$

We suppose there is only one function group in symmetric independent region of a unit cell, and the function group can be approximated to be as one dimensional function group, then

$$\chi(2\omega) = NL(2\omega)b_{hkl} \tag{5}$$

$$b_{hkl} = (1/n) \sum_{i}^{n} \beta_i \delta_i r_{ih}^{o} r_{ik}^{o} r_{il}^{o} \ . \tag{6}$$

The summation is performed over all in function group (indexed by) in the unit cell, the r_{ih}^{o} coefficients are the scalar products $\vec{h}$ a of unit vectors along the crystal axis h and the axis a of the one dimensional function group, the β_i is the magnitude of the second order susceptibility of the i function group, the δ_i is the direction-changing factor between vector $\vec{\beta_i}$ and vector $\vec{r_i^o}$, $\delta_i = \pm 1$.

With the application of Kleiman symmetry and of the symmetry of a crystal, we can simplity the calculation of $\overline{P_0^2(2\omega)}$. For a crystal which point group is 222

$$\overline{P^2(2\omega)} = N^2L^2(2\omega)\beta_i^2G \tag{7}$$

$$G = (4/5)(r_{ix}^{o} r_{iy}^{o} r_{iz}^{o})^2 \tag{8}$$

G is called as a space factor of a crystal, for a crystal which point group is 2

$$\overline{P_0^2(2\omega)} = N^2L^2(2\omega)\beta_i^2G \tag{9}$$

$$G = (1/15)(8r_{iy}^{o6} - 12r_{iy}^{o4} + 7r_{iy}^{o2}) \tag{10}$$

4.2. The calculation of G and $\overline{P_0^2(2\omega)}$

According to the crystallographic parameters of PAP, MAP, NaAP·HAP, the atom coordinates of a crystal cell are changed to the cartesion coordinates. The main frequency doubling function group of the three moleciles PAP, MAP and NaAP·HAP is O_2N–  –NH, which is approximated to be a one dimensional function group
NO_2
in this paper, the axis a of the function group points is from the N atom of the NO_2 group to the N atom of NH group. The calculated G and $\overline{P_0^2(2\omega)}$ are showed in table 2.

4.3. The measurement of the second order harmonic generation

The intensities ($I_{2\omega}$) of the second order harmonic generation of the crystal poders (PAP, MAP, NaAP·HAP) are 100, 50 5.0 times of that of the KDP crystal powder, repectively (table 2). We can see that the relative magnitude of the space factor G and the mean square of the unit second order polarizability $\overline{P_0^2(2\omega)}$ which are calculated with one dimensional approxi,ation of the function group, agree with measure relative intensities $I_{2\omega}/I_{2\omega(KDP)}$ of the second order harmonic generation of the crystal.

Table 2. G, $\overline{P_0^2(2\omega)}$ and $I_{2\omega}/I_{2\omega(KDP)}$ value of crystals

Crystals	G	$\overline{P_0^2(2\omega)}(L^2(2\omega))$	$I_{2\omega}/I_{2\omega(KDP)}$
PAP	0.11	5.6×10^{-16}	100
MAP	0.017	1.3×10^{-16}	50
NaAP·HAP	9.4×10^{-4}	2.1×10^{-18}	5.0

5. CONCLUSION

5.1. The determination of single crystal structure of PAP indicates that it be composed of N-(2.4-dinitrophenyl) aminopropionate anion (AP^-) and pyridinium cation (PyH^+). There are strong hydrogen bond N-H···O betweem in N-H of PyH and oxygen atom of the carboxyl group.

5.2. We have calculated the space factors G and the mean square of the unit second order polarizability of the crystals (PAP, MAP, NaAP·HAP), their relative magnitudes are in accord with their relative intensities of the second order harmonic generation of the crystal powders. Among the larger is space factor G, the higher is intensity of the second order harmonic generation of the crystal power. The space factor G of the crystal can indicates the relations between the intensity ($I_{2\omega}$) of the SHG of the crystal powder and the symmetry of the crystal, while the $\overline{P_0^2(2\omega)}$ which combines the space factor G of the crystal (crystal level) and the SHG of the function group (molecular level) can show the relative magnitudes of the SHG of the crystals.

5.3. According to PAP with excellent frequency doubling property, it remains further to progress.

ACKNOWLEDGEMENT

We would like to thank Mr. Yang Wenhuo, Mrs. Lu Getan and Mr. Wu Baichang for their help in NMR measurement and in SHG relative itensity measurement.

REFERENCES

1. B.F. Levine, J. Chem. Phys., 63(1), 115 (1975).
2. B.F. Levine, et al., J. Appl. Phys., 50(4), 2523 (1979).
3. J. Zyss and D.S. Chemla, J. Chem. Phys.,74(9), 4800 (1981).
4. J.L. Oudar and R. Hierle, J. Appl. Phys.,48(7), 2699(1977).
5. Li Songxian, et al., J. Fuzhou University, No.4 90(1983).
6. M. Knossow, et al., Cryst. Struct. Comm.,54(4), 719(1976).
7. Lin Fan and Li Songxian, It remains to be published.

Frontiers of Materials Research/
Electronic and Optical Materials
M. Kong, L. Huang (editors)

INVESTIGATION OF IMPROVEMENT OF LASER QUALITY OF TUNABLE Ti:Al_2O_3 CRYSTALS

Deng Peizhen, Chai Yue, Yan Shenhui, Zhang Guanming, Qiao Jinwen and Hu Bing

Shanghai Institute of Optics and Fine Mechanics, Academia Sinica, P.O. Box 800-216, Shanghai, P.R.China

The process of growing large diameter and lower residual infrared absorption single crystals of Ti:Al_2O_3 by a new growth technique named induction field up-shift method (IFSM) is described, and some of the problems in producing high laser quality single crystal are discussed using different technique have been given.

1. INTRODUCTION

Ti:Al_2O_3 single crystal is one of the most attractive broadly tunable solid state laser material[1]. Both cw and pulsed lasers have been demonstrated with very high efficiency over tuning range in excess of 300 nm centered at 800 nm. In order to develop this material to meet the needs of current commercial laser systems a varety of growth techniques have been used to grow large diameter, high purity and laser quality Ti:Al_2O_3 single crystals. However, laser performance has generally been limited by the infrared absorption band present in the laser output wavelength region which is primarily due to the presence of Ti^{3+}-Ti^{4+} ion pairs[2].

In this paper we report a description of a new crystal growth technique capable of producing high purity, large diameter and laser quality Ti:Ai_2O_3 single crystal with much lower residual infrared absorption. In principle, this technique is the modified Stockbarger method, but the new technique differs considerably from the original one, not only in heating method and utilizing much higher temperature because of the higher melting point of Ti:Al_2O_3 (about 2000°C), but also in several other important respects, in particular, the crucible material molybdenum rather than graphite or other metal material. The starting material is whole-melting rather than zone-melting before crystal growth, and the growth is seeded rather than unseeded. In addition, the flowing ambient of puritified hydrogen gas is used during crystal growth.

Ti:Al_2O_3 single crystals have been grown by the induction field up-shift method (IFSM) and the figure of merit (FOM) as high as 100 have been obtained without post-growth annealing. In primary laser experiments on samples cut from as-grown crystals, room temperature pulsed and tunable operation with lower threshold has been obtained from the high optical quality rods containing between 0.034 and 0.065 Wt.% Ti_2O_3.

2. CRYSTAL GROWTH

The HEM, VSOM and VGF techniques developed by Khattak[3], Powell[4] and Strauss[5] have been successful in producing large diameter and laser quality Ti:Al_2O_3. Based on the original work[6] a new technique (IFSM)[7] developed in our Lab also proved successful in growing large diameter and high laser quality Ti:Al_2O_3 crystals.

In ordinary Stockbarger growth technique, some factors limited this technique in production of large diameter and high quality crystal with higher melting point, such as Al_2O_3 and Ti:Al_2O_3. First the furnace temperature is lower, this problem can be solved by using induction heating method and molybdenum crucible, which allow growing thus crystals that have both higher melting point and high specific gravity. Next the propagation of the growth interface during crystal growth is achieved by moving the crucible downwards, it can be easily result in vibration and thermal flucturation, and produce thermal stress in the cooling crystal, such stress not only decreases the optical quality of crystal but generally also results in microcracks, particularly for $Ti:Al_2O_3$ crystal growth the larger vibration and thermal flucturation can cause severe constitution supercooling and other growing defects due to smaller distribution coefficient of Ti^{3+} ion in grown crystals, which are detrimental to the production of laser quality crystals. So the crystal should be grown in a vibration-free environment and with a planar or slightly convex growth interface. This problem can be solved by keeping the crucible in stationary state and only raising the heating coil to realize the moving of thermal field up, although the problem discussed above cannot be elimilated entirely, by using the IFSM technique this effect can be singnificantly minimized. In additin, for the purpose of protecting the crucible from oxidation, and specically minimizing the IR absorpion the flowing ambient of purified hydrogen gas is used during the melting before crystal growth.

A Schematic diagram of IFSM crystal growth apparatus is shown in figure 1. The starting material consisting of highly purified α-Al_2O_3 and TiO_2 powders are loaded into a molybdenum crucible with a oriented sapphire seed placed in the center of crucible. A radio frequency (R. F.) induction coil at 2.5 Mc is used to heat the crucible and is raised or lowered by means of a precise mechanical device. The temperature at seed is measured by a W/W-Re thermal couple.

The system operations as follows: The furnace is evacuated with a mechanical pumper to a pressure of about 10^{-3} Torr, backfilled the flowing growth ambient of hydrogen gas at the pressure about 0.1-0.4 kg/cm^2. and heated sufficiently to melt the charge. Then adjust the temperature by observing the image of melt through a prism, keep the temperature constant for several hours and stop flowing H_2, subsequently raise the heating coil at a definite rate. After the melt is condensed totally stop moving the coil and let the grown crystal cool gradually.

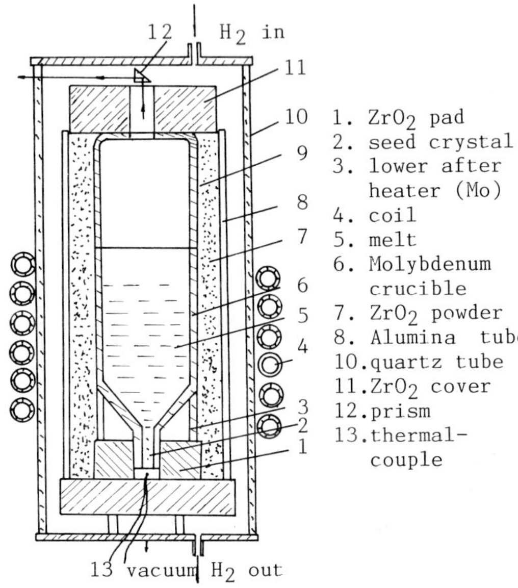

FIGURE 1
Schematic diagram of apparatus for IFSM method

3. CRYSTAL QUALITY EVALUATION

$Ti:Al_2O_3$ crystals, each weighting about 500 g, have been grown in 5.5 cm-diameter crucible. Figure 2 is a photograph of as-grown crystals which exhibit less strain and can be cut and polished without cracking. The crystal sides are partially covered with a thin, strongly adhering dark coating which has been examined by X-ray power diffraction. Figure 3 is the X-ray diffraction spectra, the peaks represented by arrows correspond to the characteristic peaks of Ti_2O_3 and others to the peaks of Al_2O_3. It shows that the coating are formed as a result of the rejection of Ti^{3+} ion from the solidifying $Ti:Al_2O_3$ crystal, and differs completely from that observed by Strauss[5]. In their crystals the dark coating was determined to be the mixed-valence phase of Ti_3O_5. This difference may result from the different ambient used. In strong reduction hydrogen ambient almost all of Ti^{4+} in TiO_2 to be converted to Ti^{3+} in IFSM grown crystals. Figure 4 is a typical absorption spectra of as-grown crystal sample in π polarization, near flat 800 nm absorption band shows that IR residual absorption is very weak.

FIGURE 2
IFSM grown $Ti:Al_2O_3$ single crystals. left: polished crystal, right: crystal boule

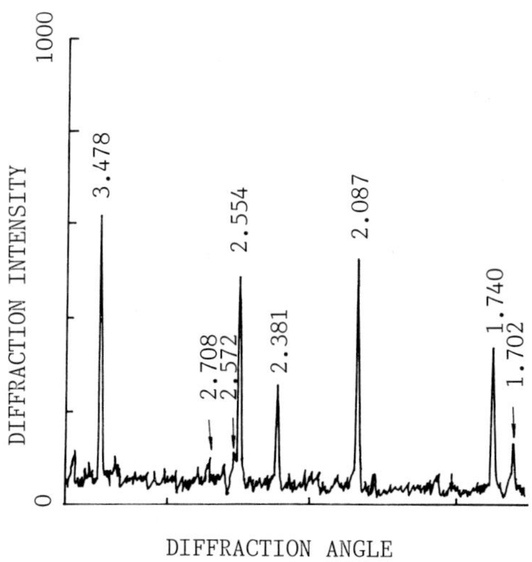

FIGURE 3
X-ray diffraction spectra, arrows represent the characteristic peaks of Ti_2O_3.

The principal internal defects observed in as-grown crystals are small bubbles, inclusions and scattering particles.

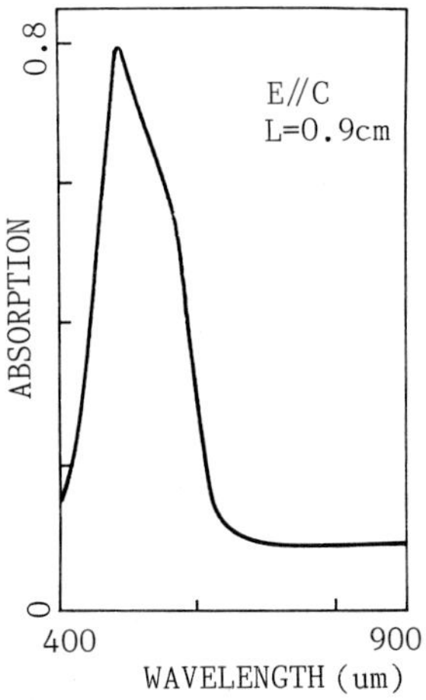

FIGURE 4
Absorption spectra of $Ti:Al_2O_3$ showing lower IR absorption.

The inclusions are generally associated with bubbles and located at the regions near crucible bottom, which appeared as spherical or filiform configurations with the characteristic of supercooling constitutional defects[8]. It has been found that the occurrence of these defects can be gradually reduced by changing the crucible shape from taper bottom to near flat one. Therefore, the formation of these defects can be attributed to the unsuitable temperature field resulted from the poor induction coupling between the coil and crucible bottom. In fact, with the increasing distance from the crucible bottom the induction coupling improved so that these defects mentioned above disappeared gradually.

Scattering particles have been revealed by the laser light scattering tomography (LLST)[9]. They are metal particles of molybdenum identified by EDS. Figure 5 is the tomograph taken in the middle part of crystal boule showing less scattering centers. Experimentally, the concentration of scattering particles apparently varied with the purity of hydrogen ambient and the growth temperature, and decreased with the increasing distance from the seed crystal. It is desirable to keep the oxygen level in gas ambient as low as possible. For molybdenum crucible the low concentration of oxygen can cause oxidation of the crucible which leads to entrapment within the crystal of metal platelets. So in technology, their entrapment can be reduced by using flowing growth ambient of puritified hydrogen gas in entire growth run and choicing the molybdenum-tungsten or tungsten crucible which have much higher melting point.

FIGURE 5
LLST tomograph of improved $Ti:Al_2O_3$ crystal showing less scattering particles (x150).

After the gradual modification in technology and growth parameters the quality and perfection of crystals have been improved greatly, in which the main macrodefect and submicrodefect, such as mosaic structure and grain boundary etc. have been eliminated in most of as-grown crystals. Figure 6 is the interference pattern of a crystal shown in the left of figure 2 with Zygo Mark III interferometer, which shows that except for the region A the crystal has high optical homogeneity with $\Delta n=1.5\times10^{-5}$. A number of high quality of laser rods have been cut and fabricated from thus crystal boules.

The arrangement , property and density of dislocations in crystal boules have been revealed by X-ray transmission topograph and chemical etching method[10], which are similar to that observed in TGT grown sapphire crystals[11]. The results show that the IFSM grown $Ti:Al_2O_3$ crystal have high optical quality and perfection.

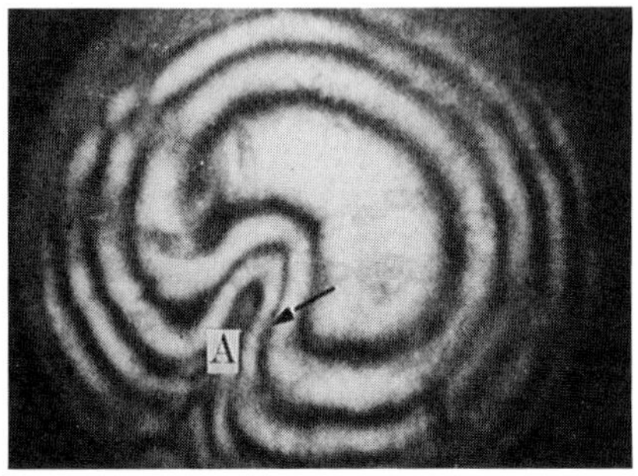

FIGURE 6
The intererence pattern of 55x25 mm $Ti:Al_2O_3$ by Zygo Mark III interferometer (double beam path), The optical homogeeity of crystal is high except for a region A.

4. DISTRIBUTION OF Ti^{3+} IN CRYSTALS

The distribution of Ti^{3+} concentration in as-grown $Ti:Al_2O_3$ crystals were monitored by measuring the optical absorption band peaking at 490 nm, the absorption coefficients, α_{max} along the growth axis are plotted in figure 7. There exists a relatively large concentration gradient of Ti^{3+} in growth axis.

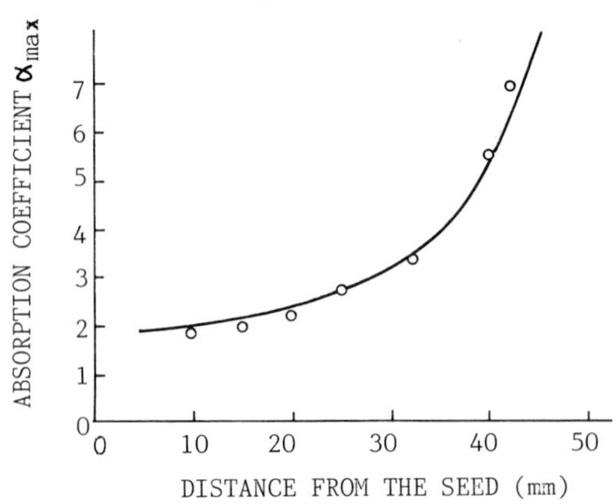

FIGURE 7
Absorption coefficient α_{max} vs distance from seed

According to the theory for normal freezing, the Ti^{3+} ion distribution along the path parallel to growth axis of crystal growth from the melt is given by the equation:

$$C = K C_o (1 - g)^{k - 1} \quad (1)$$

where C is the concentration of doping ion in the crystal at the point, where a fraction of the liquid from the data shown in figure 7, the distribution coefficient K for the Ti^{3+} ion in Al_2O_3 can be evaculated from the equation (1) Figure 8 is a log-log of α_{max} vs (1-g) based on the data of figure 7, from which the effective distribution coefficient K_e and the equilibrium distribution coefficient K_o can be obtained to be 0.19 and 0.084 respectively. The highest measured value of α_{max} in as-grown crystal reaches 7 cm^{-1}, from the relationship of Wt.% Ti^{3+}=0.032 α_{max}^{5}, this value corresponds 0.22 Wt.% which is believed to be the highest one so far obtained for Ti:Al_2O_3 crystal, it shows that IFSM technique is a very effective and practical growth method for obtaining high Ti^{3+}doped Ti:Al_2O_3 crystals.

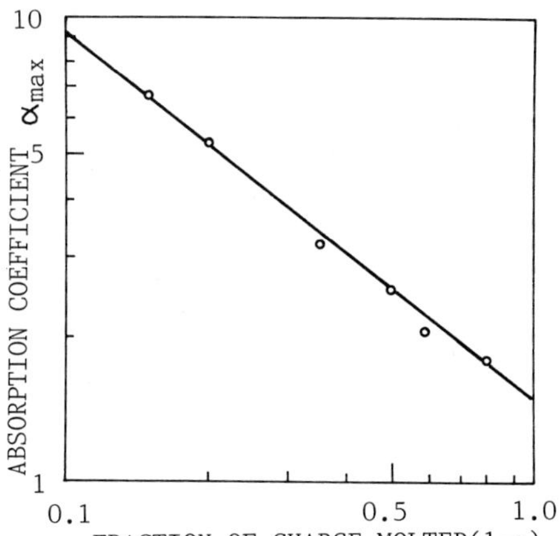

FIGURE 8
Peak of Ti^{3+} absorption coefficient α_{max} vs fraction of growth charge remaing melting.

5. LASER OPERATION

Samples from as-grown crystals have been fabricated into laser rods with end faces that are either plane parallel with antireflection coating or cut at the Brewster angle. Room temperature pulsed and tunable laser operation have been obtained for all the rods so far tested, which have Ti_2O_3 concentrations ranging from 0.033 to 0.068 Wt.%, as determined from the measured values of α_{max}. The samples had the FOM> 100 and better optical quality.

The laser setup in this experiment is the typical four-plate tuner which is very similar to the one reported by Rapoport[12]. The 532-nm pump was generated by a Q-switched doubled frequency Nd:YAG laser using a telescopic resonator producing a near-diffraction-limited TEM_{oo} mode beam, this was then amplified in a single pass saturated amplifier producing 100 mJ/pulse at 5 Hz, the output was doubled to 532 nm using KDP and generating 45 mJ of high quality green energy. In figure 9 the pulsed output energy at 770 nm is plotted against the incident pump output for the rod from the crystal with absorption coefficient of 1.9 cm^{-1}, the slope efficiency and the slope quentum efficiency varied with the mirror reflectivity T.

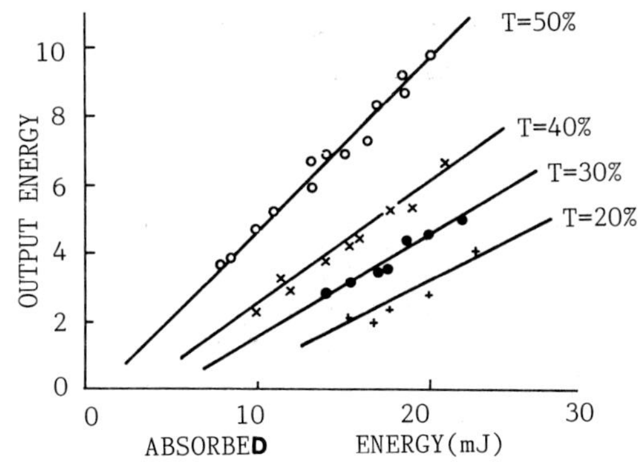

FIGURE 9
Output of room-temperature pluse Ti:Al_2O_3 laser at 770 nm vs incident pump energy.

When the mirror reflectivity T=50% the laser

threshold is 1 mJ, the highest output energy is 10 mJ which was obtained for a incident energy of 20mJ, the slope efficiency is 50% and the corresponding slope quantum efficiency is 75%. In these samples the tunable laser ranging from 660 nm to 938 nm have been obtained with the output linewidth less than 6 nm.

The above results are only the primary work, a number of laser experiments including cw laser have been doing, and the results are to be published further.

6. CONCLUSION

A new growth technique named induction field up-shift method (IFSM) has been used to grow the large diameter and laser quality $Ti:Al_2O_3$ crystals with much lower residual infrared absorption and high Ti^{3+} doping concentration from charges containing between 0.2 and 1.0 wt% TiO_2. In this technique the furnace is relatively simple in structure and can be operated easily, by using flowing growth ambient of purified hydrogen gas the crystal with the FOM value as high as 100 can be obtained without growth-post annealing, and the samples with α_{max} as high as 7 cm^{-1} can be choiced. The main defects present in as-grown crystals are the molybdenum scattering particles which affect the laser performance, with the improvement in technology the larger diameter and higher quality $Ti:Al_2O_3$ single crystals can be grown for more wide laser applications.

REFERENCES

1. R. F. Moulton et al, J. Opt. Soc. Am. B3. (1986)125.
2. Antonio Sanchz et al, IEEE J. Quantum Electron, Vol.24, No.6, (1900)995.
3. C. P. Khattak et al, Proc. SPIE. Vol.831, (1987)56-61.
4. Richard C. Powell et al, J. appl. Phys. Vol.58, No.6, (1985) 2331-2337.
5. A. J. Strauss et al, Proc. SPIE. Vol.831, (1978) 62-69.
6. A. Ross., Mat. Res. Bull. Vol.18, (1983) 405-409.
7. Chai Yue et al, Chinese Cera. Soc.'in press!
8. Zhang Qiang and Deng Peizhen, Cryst. Res. Technol. 'in press'.
9. Deng Peizhen and Qiao Jiwen, J. Crystal Growth, Vol.82 (1987) 579-583.
10. Zhang Qiang and Deng Peizhen. Materials Letters, Vol.4 (1990)128-132.
11. Zhang Qiang and Deng Peizhen, Materials Letters Vol.3/4, (1989) 105-109.
12. W.R. Ropoport et al, Applied Optics Vol.27, No.13 (1986) 2677-2684

Frontiers of Materials Research/
Electronic and Optical Materials
M. Kong, L. Huang (editors)

Mg_2SiO_4:Cr —— A NEW PHONON-TERMINATED LASER CRYSTAL

Peicong PAN, Shenhui YAN, Peizhen DENG, Bing HU, Hongbin ZHU and Fuxi GAN

Shanghai Institute of Optics and Fine Mechanics, Academia Sinica, P.O.Box 800-216, Shanghai 201800, P.R.China

Large single crystals of Mg_2SiO_4:Cr have been grown by Cz method for the first time in china and the growth technique is also introduced. Vibronic side bands of the fluorescence spectra in our crystal range from 700nm to 1400nm. The fluorescence covering 1000nm to 1400nm may be attributed to a new radiative "center" other than Cr^{3+}. The distribution coefficient for Cr in Cz grown crystals is reported as 0.15.

1. INTRODUCTION

Cr-doped forsterite (Cr:Mg_2SiO_4), a new phonon-terminated laser crystal, has been introduced in recent years. Many scientists [1-3] have reported the room-temperature pulsed laser action centered at 1235nm with a bandwidth of 25nm and CW laser action at 1244nm with a bandwidth of 12nm. The effective emission cross section in this crystal is estimated to be $1.1x10^{-19}cm^2$. The tunable stimulate emission of Mg_2SiO_4:Cr in the 1.2um is of technological importance for eye safe rangeing, remote sensing, especially for optical communication. It has a high potential to be a very useful and practical laser system for relative large emission cross section as well as it is composed of that low-price raw material.

2. CRYSTAL GROWTH OF Mg_2SiO_4:Cr

Before 1988, a number of efforts to obtain single crystals of forsterite that is of considerable geoscientific interest, have reported by many investigators. Such as by flux technique [4], the Bridgman method using a carbon or a Mo crucible [5,6], Verneuil method [7] and by Cz method [8]. Scientists found that preparation of large and strain-free crystal was not so easy.

Mg_2SiO_4 is stable in air at a congruent melting point of 1890±20°C [9]. We grew Cr-doped crystals of Mg_2SiO_4 by the Cz method for the purpose of studying spectroscopy and laser action. In the crystal growth the Ir crucible of Ø50x40mm and RF generator are employed. The starting materials were reagent grade MgO and SiO_2, of stated 99.99%. In order to ensure stoichiometric melting of the row material, the oxides were blended and prefired by a resistance furnace at about 1100°C for 20 hr in air. In the first experiment the Ir rod of 2mm in diameter was used as a seed, nucleating onto an Ir rod. By the necking technique the resulting crystals were large enough to provide seeds of different orientation. The large transparent single crystals with different growth directions, such as [100], [010], and [001] have been grown successfully. As shown in Fig.1. The typical growth conditions are: pulling rate 4-mm/hr, rotation

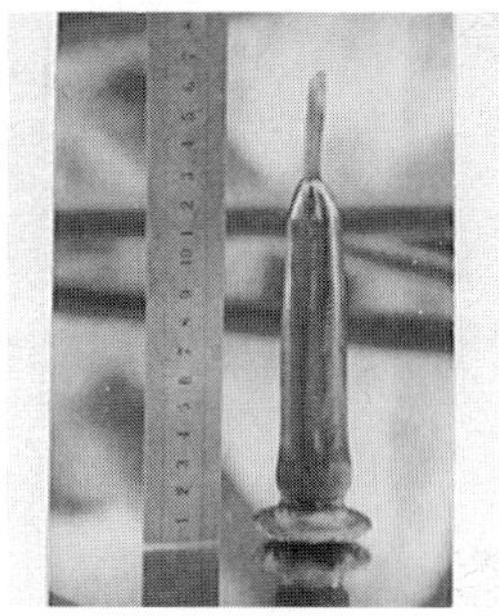

FIGURE 1
As-grown crystal of Mg_2SiO_4:Cr

rate 35 rpm, atmosphere N_2.

The crystal we have grown contains 0.02 atomic percentage of Cr ions. The crystal appears different colors when looked at from different directions. Along the [010] axis it appears to have a bluish hue, along [001] it appears violet, while along the [100] axis it looks greenish.

Typical Mg_2SiO_4:Cr crystals exhibit two main types of imperfections, inclusions around core and cracking. In our experiment, the core was correlated with a convex, facet forming crystal interface. Around the core, there are a lot of inclusions. These inclusions form a boundary of the core, as shown in Fig.2. The result of electron probe analysis show that the most of the inclusions are rich in SiO_2, as shown in table.1 By increasing MgO in burden, the inclusions that can be seen by naked eye are disappeared, as shown in Fig.3.

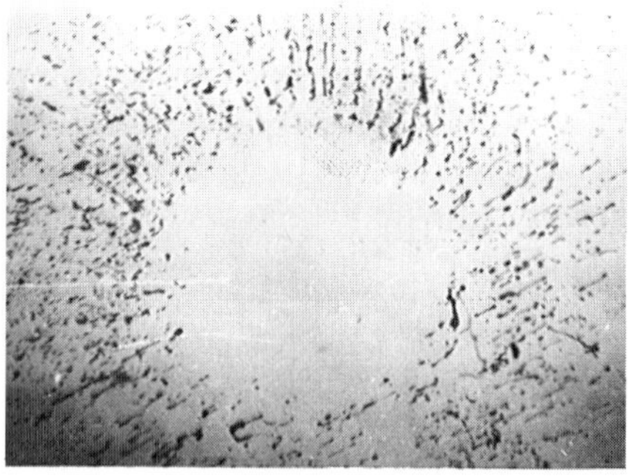

FIGURE 2
A cross section of heavily cored crystal revealing clear inclusions around the core x10

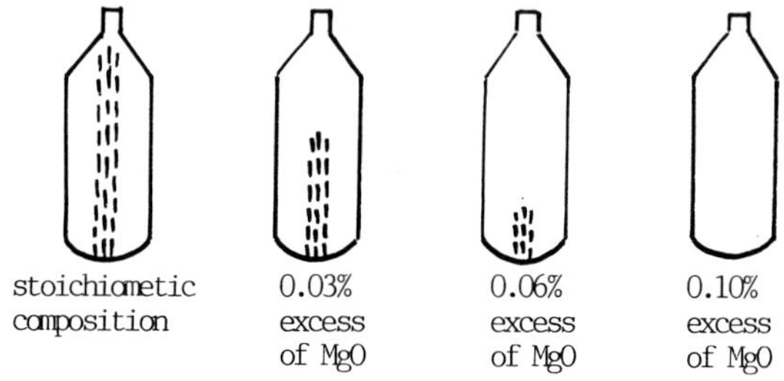

FIGURE 3
Relationship between crystal quality and "excess" of MgO

This fact lead to the consideration that the congruent melting position of Mg_2SiO_4 in the $MgO-SiO_2$ pseudo-binary system is between 0.03 and o.10 weight% excess of MgO over stoichimetry, as was already discussed by shankland [10]. When the burden calculation based on the formula of Mg_2SiO_4 was done, the surplus SiO_2 may form some molecular clusters in the melt such as SiO_2 and $MgSiO_3$, and these clusters are captured in

Table 1
Analytical data of inclusions in forsterite crystal by electron probe

probe site	probe parameters 10^{-8}A 15kv	concentration of element %									
		SiO_2	Al_2O_3	MgO	Cr_2O_3	CuO	FeO	CaO	PtO_2	ZrO_2	Au_2O_3
matrix crystal	1.0	42.4618	0.0000	57.0053	0.0074	0.0000	0.0000	0.0000	0.0000	0.0000	0.1453
matrix crystal	1.0	42.1981	0.0305	57.3453	0.0419	0.0000	0.0000	0.0000	0.0738	0.0000	0.0442
inclusion	3.0	57.0227	2.0654	38.6890	0.0100	0.0044	0.1082	1.7993	0.0000	0.0000	0.0000
inclusion	3.0	58.0510	1.8326	38.7776	0.0687	0.0000	0.4402	0.0304	0.0000	0.0000	0.0000
inclusion	3.0	54.3210	3.1239	36.7979	0.0000	0.0229	0.4390	0.9595	0.0000	0.0000	0.0360
inclusion	3.0	84.0234	0.8063	8.1761	0.0000	0.0000	0.2981	2.5534	0.0000	0.0000	0.0094
inclusion	3.0	83.5588	1.9943	10.6622	0.0147	0.0111	0.0609	0.0622	0.0000	0.0000	0.0000
inclusion	3.0	81.1253	2.3265	10.548	0.0123	0.0154	0.1820	0.0000	0.0210	0.0000	0.0132

growth process.

According to the Fig.2 the fact is found that there are less inclusions in the core area than in the other. We suggest that it is much easier to capture inclusions in nonsingular interface (out of facet area) than in singular interface (facet area).

The crystal tend to crack into some pieces if you do not lift the crystal from the melt after finishing the growth, and the boules pulled in the [010] direction are more liable to crack along the (010) face. The cracking may be caused by the thermal strain due to the anisotropy of thermal expansion. Once the cracking happen, a rapid cooling of the cracked crystal did not cause further cracking.

3. CRYSTAL STRUCTURE AND GROWTH HABIT OF Mg_2SiO_4

Crystal structure Mg_2SiO_4 has been determined. It is an orthorhombic structure with space group Pbnm and contains four formula units per unit cell. The structure can be described as being based on a nearly hexagonal close-packed array of oxygens with silicon in the fourfold and magnesium in the sixfold coordinated sites. The cell parameter are: a_o=10.1932Å; b_o=5.97875Å; c_o=5.97875Å. the results are consistent with the literature [11,12].

The miller indices of all developing facets in the crystals grown along a,b,c axis are determined by means of x-ray and optical methods. In the crystals grown along a axis, the developing facets are (010), (021), (001); In the crystals grown along b axis, the developing facets are (101), (001); and along c axis are (010) (110) facets were revealed. In these experiments the morphological importance of developing facet is in order of (010),(110),(101),(021),(001).

According to the Hartman theory, the relations between morphology of forsterite and its structure are explained. We have found out seven strong PBC vectors. They are A[100],B[110], C[001],D[010],E[111],F[021],G[011], From these PBC vectors seven F faces can be defined by using a structural model. Depending on values of average attachment energy, the morphological importance of these facets is (010),(110),(221), (101),(021),(001),(011). The results calculated by PBC theory agree with the results of experiments approximately.

4. DISTRIBUTION COEFFICIENT OF Cr IN Mg_2SiO_4

The distribution coefficient for Cr in Cz grown Mg_2SiO_4 is reported in this paper in order to predict specific dopant levels in the crystal.

The Cr contents of the crystals were determined by X-ray fluorescence spectrum. Representative samples for X-ray analysis were chosen from high quality, bubble-free, strain-free material so that errors arising from segregated second phases, precipitates, etc. could be eliminated.

The values of distribution coefficient K were obtained by solution of the equation [13]:

$$C_g=C_oK(1-g)^{K-1} \quad (1)$$

Where g is the fraction of the melt crystallized, C_g and C_o are the crystal and initial melt Cr concentrations.

Cutting two thin slices approximately 1mm thick on the top and bottom place respectively for each of the boules. As shown in Fig.4.

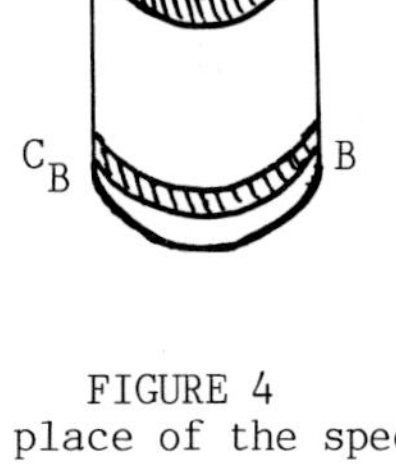

FIGURE 4
Section place of the specimen

From the equation (1), we obtain

$$C_{gA}=C_oK(1-g_A)^{K-1} \quad (2)$$

$$C_{gB}=C_oK(1-g_B)^{K-1} \quad (3)$$

Then we get

Table 2
Concentration and distribution data

No.	Pulling rate (mm/h)	C_o(ppm)	g	C_g(ppm)	K	C_o^*(ppm)
1	4	1234	A:0.0126	159.702	0.1436	1100.18
			B:0.6286	368.984		
2	6	1974	A:0.0487	317.035	0.1608	1890.73
			B:0.7142	868.739		
3	4	1974	A:0.1778	318.148	0.1429	1882.48
			B:0.3164	372.695		
4	4	2469	A:0.0020	324.642	0.1426	2280.04
			B:0.2289	404.978		
5	3	2469	A:0.2945	433.550	0.1357	2315.29
			B:0.5000	578.823		
6	6	3086	A:0.0111	426.685	0.1552	2723.81
			B:0.3472	606.096		
average					0.1471	

* calculated by K

$$K = \frac{\lg[C_{gA}/C_{gB}]}{\lg[(1-g_A)/(1-g_B)]} + 1 \qquad (4)$$

The data calculated from Eq.(4) for six crystals are presented in table 2.

It is interesting to note that there is slight variation in the values of K given in table 2 as a function of pullingrate.Thus we think that the values of K presented in this paper may be considered effective distribution coefficients.

From the table 2, it is be noted that the initial value C_o^* calculated by K is slight smaller than the real initial value C_o. This is because a loss of Cr_2O_3 due to vaporization during growth caused the K calculated by Eq.4 become more large, thus the initial volue C_o^* calculated by Eq.2 appear a little decline.

5.SPECTROSCOPIC STUDY OF Mg_2SiO_4:Cr

At room temperature, we used a Perkin-Elmer lambda-9 spectrophotometer to measure the transmission spectrum of Mg_2SiO_4:Cr, shown in Fig.5. There are two main broad absorption peaks, the wavelenth positions of the peak value are 460.0-nm and 650.0nm respectively, corresponding to the $^4A_2$ to $^4T_2$ and $^4A_1$ to $^4T_2$ absorption transitions. There are also a broad and weak absorption band in 850nm to 1120nm region. The origin of this absorption is not transitions in the Cr^{3+} ions for the reason that the band is not observed in the excitation spectrum [14]. We suggest that the band may be arised from a new radiative "center".

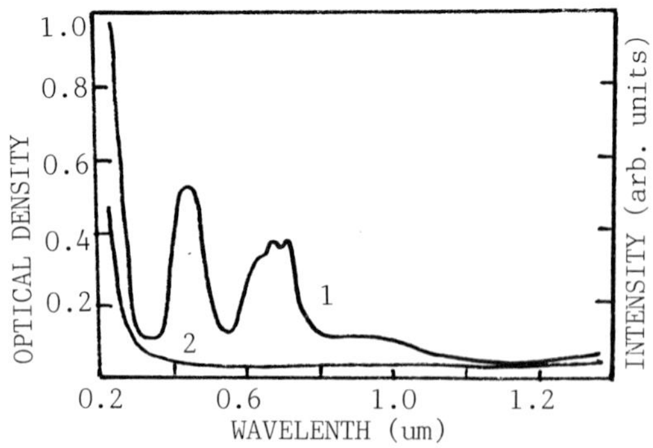

FIGURE 5
Absorption spectrum of forsterite
1. Cr-doped crystals
2. Undoped crystals

The fluorescence spectrum of Mg_2SiO_4:Cr was excited by 1.06um and 0.53um radiation from a Q-switched Nd:YAG/DYE laser system operating at a 10 Hz repetition rate and 10-ns full width at half maximum. Shown in Fig.6. The curve 1 excited by 0.53um radiation is characterized by a broad band covering 700-1400nm. When excited by 1.06um radiation, the fluorescence spectrum span quite a diffirent spectral range of 1000-1400nm.

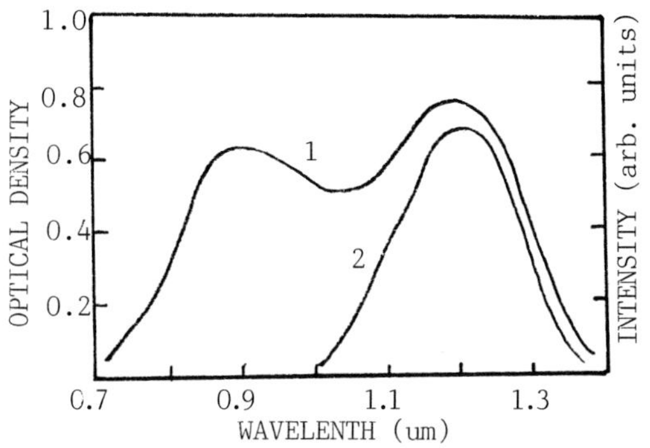

FIGURE 6
Fluorescence spectrum of Cr-doped forsterite
1. by 532nm radiation from an frequency-doubled YAG:Nd^{3+}
2. by 1064nm radiation from a YAG:Nd^{3+}

Becauce 1.06um radiation is located in the weak absorption band we atribute the fluorescence covering 1000-1400nm in curve 2 to the new radiative "center".

The key question that remains to answered is what is the origin of this new radiative "center".

Fe^{3+} is a commonly occurring impurity in olivines and exhibits a similar absorption band in the near infraed. By analysis on absorption spectra of Cr-doped and undoped samples, shown in fig.5. The possibility of fluorescence "center" come from prasitic impurities is excluded for no infrared absorption was found in curve 2

According to the literature [15] and [16], the Cr^{4+} substituting for Si in tetrahedral site and Cr^{2+} for Mg^{2+} in the inversion (M1) site may occur and be stable in the forsterite. These two "centres" may account for the observed infraed absorption based on the prediction of the crystal field theory.

The exact identification of the radiative "centre" in Cr-activated forsterite still remains an interesting problem for further investigation.

The works on laser action of Mg_2SiO_4:Cr are being made arrangements.

ACKNOWLEDGEMENTS

This work is supported by Shanghai science and technology committee for encouragement of young scientist.

REFERENCES

1. V.Petricevic, S.K.Gayen, and R.R.Alfano, APPLIED OPTICS Vol.27, No.20 (1988) 4162.

2. V.Petricevic, S.K.Gayen, and R.R.Alfano, APPLIED OPTICS VOL.28, No.9 (1989) 1609.

3. V.Petricevic, S.K.Gayen, and R.R.Alfano, OPTICS LETTERS Vol.14, No.12 (1989) 612.

4. L.Vu Tien, et al., J. Crystal Growth 13/14 (1972) 601.

5. W.Jordan and J.J.Naughton, Am. Miner. 48 (1963) 200.

6. J. Wilson, unpublished work; briefly reviewed in the report of a Geophysical Discussion held on July 22, 1964.

7. T.J.Shangland and K.N.Hemmenway, Am. Miner. 48 (1963) 208.

8. C.B.Finch and G.W.Clark, J. crystal Growth 8 (1971) 307.

9. N.L.Bowen and O.Anderson, Am.J.Sci.37(1914) 487.

10. T.J.Shankland, Gordon Mackay lab. Technical Report No.HP-16 (1966).

11. J.D.Birle, G.V.Gibbs, et al., Am.Miner. 53 (1968) 807.

12. S.Ghose, Solid State Communication Vol.63, No.11 (1987) 1045.

13. Min Nai-ben Physical Fundamentals of Crystal

Growth, (China, Shanghai, 1982) P.55.

14.V.Petricevic, S.K.Gayen, and R.R.Alfano, PHOTONICS SPECTRA, MARCH (1988) 95.

15.R.G.Burns, Contrib. Mineral. Petrol., 51, 213-221 (1975).

16.B.E.Scheetz and William B.White, Contr. Mineral. und Petrol., 37,221-227 (1972).

Frontiers of Materials Research/
Electronic and Optical Materials
M. Kong, L. Huang (editors)

RESEARCH AND MORPHOLOGY OF LASER CRYSTAL LMA:Nd^{3+}

Jun XU, Xiaoshan MA, Guangzhao WU, Yafang SHEN, Xinmin ZHANG, Xiorong ZHANG

Shanghai Institute of Optics and Fine Mechanics, Chinese Academy of Sciences,
P.O.Box 800-216, Shanghai 201800, P.R.China

The large sized, not cleaved and quality LMA:Nd^{3+} crystals are grown by Czochralski method. In order to diminish cleavage, we choose different seed directions, charging and suitable temperature fields. The phenomena of interface inversion and inclusions are also studied. The relationship between crystal morphology and its structure is investigated firstly with PBC theory in our paper. A easy method to decide the impotance of F-faces preliminarily is achieved.

1. INTRODUCTION

At the begin of eighties, $LaMgAl_{11}O_{19}:Nd^{3+}$ (LMA) was proved to be a good laser crystal.[1] It has stable physic-chemical properties, high hardness and appropriate melting point (1890°C). It can be doped with Nd concentration 6 times higher than YAG:Nd . The activator Nd is the lattice component and its distribution coeff. is 1. The material is very cheap. It is attracting the attention of more and more specialists in the world.

2. CRYSTAL GROWTH OF LMA:Nd^{3+}

The furnace used for the growth is designed by ourselves similar in construction to those used for the growth of $BeAl_2O_4:Cr^{3+}$.[2,3]

It is heated by middle-frequency inducing generator under nitrogen or argon atmospheres. The pull rate is 1-2 mm/hr. The rotation rate is 16 rpm and the interface shape is convex. The first growth started from an iridium wire with 2 mm in diameter. It is out for seed along [100] direction.

Charging is very necessary because La_2O_3 power in air will soon become $La(OH)_3$ power. X-ray crystal powder diffraction pattern, electronic probe and chemical analysis all proved this idea. If La_2O_3 powder is chargedup to 12.2 wt%, the quality of crystal is excellent.

A typical sample is approximately 23 mm in width by 15 mm in breath and 90 mm in length.

The inversion of the solid-liquid interface and the formation of various defects are described firstly. The cause of defects formation during interface is analysed and the formation mechanism of fringers due to stress is investigated. Two methods of controlling interface inversion is growing quality crystals are available: a) variable rotation rates; b) slow rotation rate.

The usual defect of this crystal is the noticeable cleavage (cleavage face (001)). The cleavage property is determined by the structure with the lamellar character. This cleavage strongly influence the crystal quality. But if we choose a-axis as our seed direction and suitable temperature distribution, we can diminish its cleavage and cracking phenomena. The suitable temperature gradients around interface: T_{max}=30°C/mm, T_{min}=5°C/mm, T_{normal}=10°C/mm.

In grown LMA crystals, a novel inclusions has been found. The inclusions are quantitatively analysed by electronic probe. The mechanism for forming these inclusions is connected with the growing habit.

3. MORPHOLOGY OF LMA CRYSTAL

The [100], [010] crystals have been grown by CZ. technique. We have determined the orientations of all developing facets in the crystals by means of X-ray and optical methods, and hence determined their Miller or indices. The facets in crystals grown along different orientations are shown in figure 1.

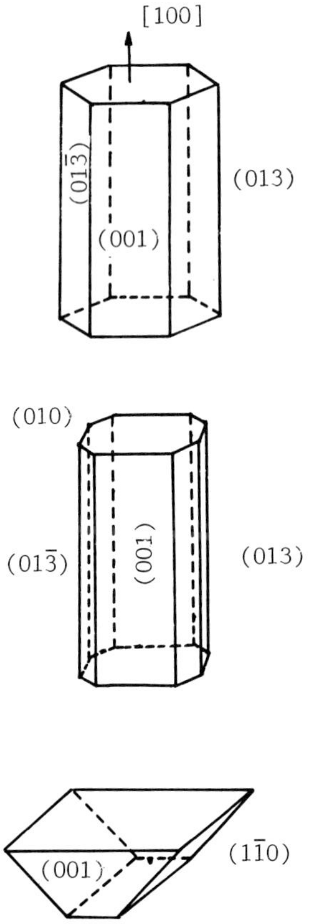

Fig.1 The diagrams of morphology of LMA crystal grown along [100], [010] directions.

In the crystals grown along [100] the developing facets are (001), (013), $(01\bar{3})$ and (010); in the crystals grown along [010], the developing facets are (001), (103), $(10\bar{3})$ and (100); in sometimes, $(1\bar{1}0)$ facets were revealed. In these experiments, the morphological importance of developing facets is in the order of (001); (013), $(0\bar{1}3)$, (103), $(10\bar{3})$; (010), (100); $(1\bar{1}0)$.

According to Bravais rule and Donnary-Harker theory, no satisfactory explation to the above experimental results can be given. Therefore PBC theory was suggested by Hartman-Perdok during the 1950's.[4] Hartman-Perdok suggested that the strongest periodic bond chain (PBC) of crystal is the orientation with the greatest growth rate. Then the PBC vectors defined the facets as F face(flat face), S face(step face), K face (kink face). And the F face must contain two or more PBC vectors. Morever, the importance of developing F faces can be determined qualitatively by using average attachment energy. The smaller the average attachment energy, the more easily the F face developes.

Depending on the bond energy and atomic site, four groups of stronger PBC vectors included with 25 kinds of orientations can be calculated:

A [100] [010] [110] [11/20] [1/210] $[1/2\bar{1}/20]$

B [03/21/2] $[0\bar{3}/21/2]$ [3/201/2] $[\bar{3}/201/2]$ [3/23/21/2] $[\bar{3}/2\bar{3}/21/2]$

C [001/2]

D $[1\bar{1}/21/2]$ [1/23/21/2] $[\bar{3}/2\bar{1}1/2]$ [13/21/2] $[\bar{3}/2\bar{1}/21/2]$ $[\bar{1}1/21/2]$ $[\bar{1}/2\bar{3}/21/2]$ $[3/2\frac{1}{2}\frac{1}{2}]$ $[\bar{1}\bar{3}/2\frac{1}{2}]$ $[3/21\frac{1}{2}]$ $[\frac{1}{2}\bar{1}\frac{1}{2}]$ $[\frac{\bar{1}}{2}1\frac{1}{2}]$.

The strongest PBC vectors are [100], [010], [110], $[1\frac{1}{2}0]$, $[\frac{1}{2}10]$ and $[\frac{\bar{1}}{2}\frac{1}{2}0]$ directions. Therefore, the crystal should get the highest growth rate along these orientations. This agrees with the experimental data.

From above stronger PBC vectors, many F-faces (about 134) are defined and compound with computer. Now, we almost cann't calculate the average attachment energies of so many F faces one by one. So that we think that PBC theory has its disadvantage for so complicated structural crystal. We give out all PBC vectors on all F faces, make a comparesion among their number, strength and arrangement, and achieve an easy method to decide the importance of F faces. According to this idea, we have successfully explained the relationship between morphology of LMA crystal and its structure.

We calculate the attachment energies of the main developing facets: (001), 3.7080; (103),

$(10\bar{3})$, (013), $(01\bar{3})$, 3.9802; (100), (010), 4.7552; $(1\bar{1}0)$, 4.7835(e^2A^{-1}).

Depending on the values of the average attachment energies, the morphological importance of these faces is in the order of (001); (103), $(10\bar{3})$, (013), $(01\bar{3})$; (100), (010), $(1\bar{1}0)$. The theoretic diagram of the morphology of LMA crystal shows in figure 2.

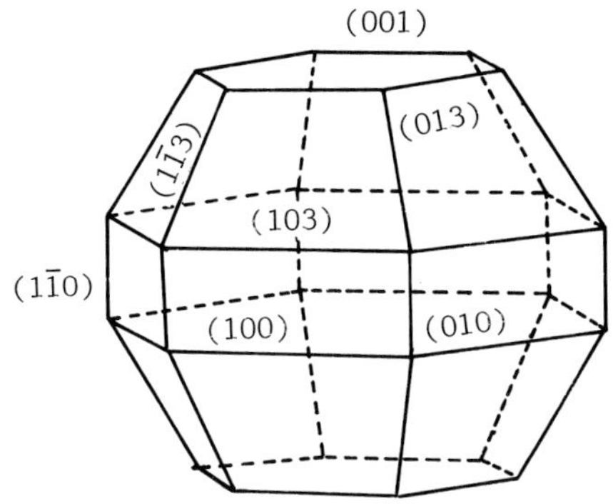

Fig.2 The theoretic diagram of the morphology of LMA crystal

During Czochralski method growth of the crystals, the developing facets can be related to the crystal structure, the shape of isothermal surface, the temperature gradient, the growth direction, the impurity composition and concentration. Therefore, it might appear that figure 1 is different from figure 2.

ACKNOWLEDGEMENTS

I would like to thank my colleagues Zhu Rude and Bao Wenhu who helped me in the growth of LMA crystals.

REFERENCES

1. D.Saber and A.M.Lejus, Mater.Res.Bull.16 (1981) 1325.

2. Ma Xiaoshan, et al., Chinese Journal of Lasers, 15,7 (1988) 403.

3. Xu Jun, et al.,Chinese Phys. Lett., 6,12 (1989) 557.

4. P.Hartman, et al., Acta Cryst., 8 (1955) 49; 521.

Frontiers of Materials Research/
Electronic and Optical Materials
M. Kong, L. Huang (editors)
Elsevier Science Publishers B.V., 1991

SINGLE CRYSTAL GROWTH OF MERCURIC IODIDE BY THE VAPOR PHASE METHOD

Guanxiong CHEN Shifu ZHU Zhenghui LI Yujun RAN Beijun ZHAO Shujun YIN and Bin HAN

Department of Materials Science, Sichuan University, Chengdu, 610014, China

Modified Temperature Oscillation Method (TOM) was applied in this paper to grow HgI_2 single crystal. Actual process was described in detail and perfection of the grown crystal was studied. The used apparatus is rather simple and easier to control. The growing temperature is relatively lower and fewer defects are created in crystal. Reported results suggest that TOM be an excellent method for qualified HgI_2 single crystal

1. INTRODUCTION

Mercuric iodide (HgI_2) is a new type photoelectric material which started to develop from the 1970s[1] and its constituents have high atomic numbers (80, 53). It has a relatively large bandgap (2.14eV) and a high density (6.4 g/cm^3) as well as a good performance of the electron transport. Therefore HgI_2 is a pretty well behaviour material for new choice of room temperature radiation detector. Low-noise room-temperature X-ray and γ-ray detectors might be constructed from HgI_2[2]. Furthermore a novel photodetector consists of HgI_2 coupled with scintillation crystals[3]. It opens vast vistas for the development of utilization[4].

Intensive investigations have been carried out for many years for the preparation and application of HgI_2 crystal and much progress has been made. Growing in the solution, the HgI_2 crystal is smaller in size and in the crystals solvent inclusion occurs easily[5,6]. At present, HgI_2 crystal grows mainly by means of vapor phase. In particular the crystal weight could reach hundreds even one thousand grams prepared by the temperature oscillation method[7]. The mercuric iodide single crystals mentioned in this paper were created by the modified TOM, in which the crystal weight achieved about 200g, and the density of dislocation was low. The preliminary study on crystal perfection was performed as well.

2. CRYSTAL GROWTH

2.1. Purification of raw material

Before growing crystals raw material must be purified. The purification methods have several different types. Usually raw HgI_2 material is purified by repeated sublimations and distillations[8]. Two steps purification used is as follows: (1) repeated sublimation in a closed tube. Each closed sublimation cycle is designated by the letter S; (2) distillation in an open system under continuous evacuation. The distillation process is designated by the letter M. All glassware is cleaned prior to each use in a glass etch solution of ACS reagent-grade nitric acid and hydrofluoric acid. After etching, all apparatus is rinsed with 4 MΩ deionized H_2O.

Here is the detailed procedure for purification. Approximately 50g of reagent grade HgI_2 powder (99.5% pure) was put into a straight glass tube 80 cm long, evacuated and sealed under 10^{-5} Torr vacuum and placed in a two-zone horizontal furnace. The temperature was about 140℃ in the sublimation zone and 60℃ in the cold crystallization zone. After each sublimation the tube was shifted along the furnace. The residue remained at the end of the tube. This procedure is repeated 14 times and the product is called 14S material.

The distillation was carried out in vertical two heating zones quartz distiller with continuous pumping. The sublimation temperature was increased to 255℃ and the HgI_2 vapor superheated to 300℃. 255℃ distillation proved to be more efficient that low temperature sublimation due to more advanced dissociation of the impurity-HgI_2-complexes and enhanced decomposition of the heavy organic impurities. Impurity analyses by Inductively Coupled Plasma/Atomic Emission Spectrometry indicated that the content of inorganic impurities was about tens of ppm in high-purity HgI_2 materials(14SM).

2. 2 Crystal growth

The mercuric iodide single crystal were grown by the modified TOM. The apparatus of single growth and the typical temperature profiles are shown as fig . 1 . The purified material (200g 14SM) was

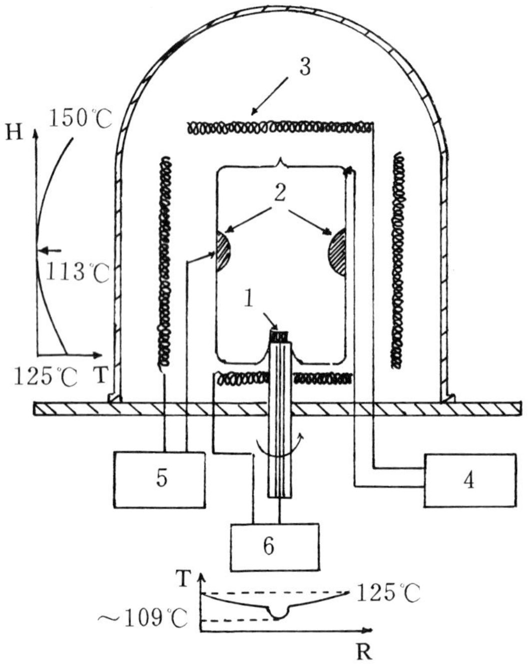

FIGURE 1

Schematic representation of the furnace showing temperature distribution at the outside wall of the ampule. 1. Crystal 2. Source Material 3. Heating Coil 4. Top Temperature Control 5. Axial Temperature Control 6. Bottom temperature control

enclosed in a glass growth ampule(20cm in diameter and 25cm high) under a vacuum $< 10^{-5}$ Torr and placed in the vertical furnace. The center of the bottom of the ampule was raised so that a pedestal was formed. The vertical heater coil provided an axial temperature gradient over the height of the ampule. A circumferential heating coil under the ampule controled the temperature of the bottom. A bell jar covered the system to minimize heat losses. The ampule is supported by the copper axis. Crystal growth could be achieved by cooling the pedestal, which made a coolest area in the ampule and created radial temperature gradient. The axial heater also introduced an azimuthal temperature gradient so that the source material concentrated on two different spots of the ampule wall. The azimuthal thermal asymmetry caused in the ampule is determined by the geometry of the heating coils, the power input, and the rotation speed of the ampule. Higher temperature settings for the bottom coil, for example, will increase the axial height at which the source material collects.

The actual growth process is conducted in the following way: The ampule revolved 4 rotations per minute during growing crystal. When the source material had reached a stable configuration on 1/2 height of the ampule, the pedestal temperature was adjusted slowly to the growth temperature Tc (~110℃). After a few seeds (red α-phase) had nucleated, a good quality seed was selected by periodical oscillations of the crystal temperature (Tc oscillates 109—115℃, the source temperature Ts is kept at~113℃). The seed[001]direction paralleled to the pedestal surface. once the seed had been positioned correctly, growth was continued by judicious adjustment of the temperature of the ampule bottom and the source material, and the temperature of the center of the pedestal. By our modified TOM, a 200g HgI_2 single crystal was grown in about three weeks and sometimes even a large size single crystal could be grown.

3. OBSERVATION OF THE CRYSTAL PERFECTION

3. 1. Crystal morphology and crystal structure

The surface form of grown crystal relates to the growth condition in general. Usually the single crystal have the smooth natual crystal planes under the typical growth condition. Fig. 2 shows 200g crystal grown by our modified TOM. The extrapolation of our x-ray powder data leads to a tetragonal structure with unit cell parameters of about a = 4. 367 Å and c = 12. 441 Å fitting with ASTM. The fourfold symmetry of the HgI_2 single

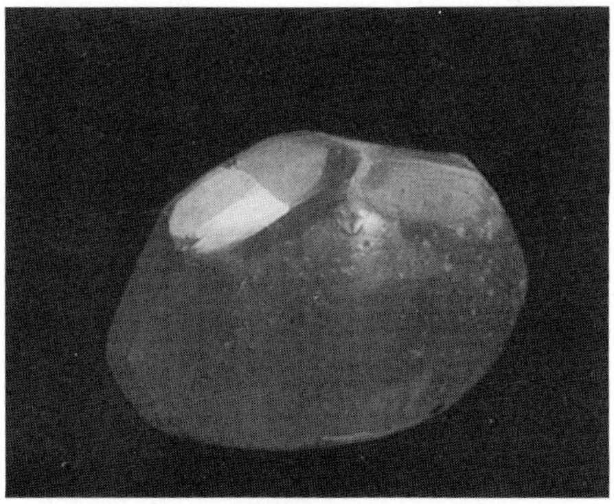

FIGURE 2

Photograph of HgI_2 crystal weighing about 200g growing by the TOM

crystal on the [001] direction appears in Laue photographs clearly. The aforesaid results indicate that the HgI_2 crystal belongs to the tetragonal system, and its growth interfaces are fairly stable, the growth is better and perfect in structure on the whole.

3. 2. Stoichiometry and observation of the etch pits The stoichiometry of HgI_2 crystal was determined by scanning electron microscopy (SEM) and ZAF corrected. The results had shown that most routinely growth crystal had a stoichiometry ratio I/Hg = 1. 99 : 1 close to the ideal stoichiometry ratio 2. The iodinelacks because it volatilizes more easily than the mercury in the purification process.

HgI_2 does not solve in H_2O, but it can solve in some organic solvents and in KI water solution. In this paper, the lattice defects of the single crystal are investigated by means of chemical etching and SEM. The samples, slices of ~ 1mm in thickness, are prepared by sawing crystal with a wet line or cleaving directly from a single crystal, and then polished in 10% KI water solution. The density of dislocation emerging on the (001) face is of $10^3 cm^{-2}$ order (fig. 3) in agreement with those reported[9].

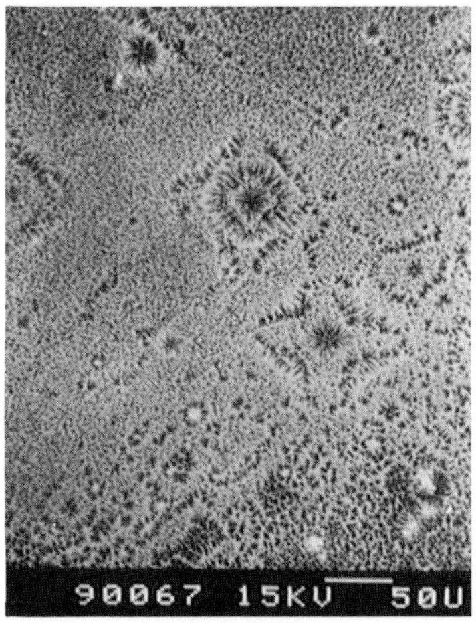

FIGURE 3

Dislocation etch pits on a (001) face

4. CONCLUSIONS

The experiment results indicate that the sublimation distillation is a simple and effective way. The modified TOM is a very hopeful method of growing HgI_2 single crystal. It has the merit of simple equipment, easier to control, and lower growth temperature. Furthermore, it also has the advantages of large size, high purity, lower dislocation density, and better crystal.

REFERENCES

1. H. L. Malm, IEEE Trans. Nucl. Sci, Ns — 19 (1972)263

2. Proc. 5th Intern, workshop on HgI_2 Nucler Radiation Detectors, Ed. M. schieber/Nucl. Instr. Methods, 213(1983)1—154

3. J. S. Iwanczyk, et al. , IEEE Trans Nucl. sci, Ns—30(1983)363

4. Zhenghui Li, et al. , physics, Vol. 17, No. 5, (1988)291

5. I. F. Nicolan and J. P. Joly, J. Crystal Growth, 48 (1980)61

6. P. Bohae, J. crystal Growth, 61(1983)163

7. N. L. skinner, et al. , J. crystal Growth, 89(1988) 86

8. Y. F. Nicolan, J. chem. Tech. Biotechnol, 33A (1983)350

9. P. T. Randtke and C. ortale, IEEE Trans. Nucl. sci, NS—24(1977)129

Frontiers of Materials Research/
Electronic and Optical Materials
M. Kong, L. Huang (editors)

STRUCTURAL FEATURES AFFECTING THERMOMAGNETIC RECORDING CHARACTERISTICS IN Tb-TM THIN FILMS

Soon Gwang KIM, Seh Kwang LEE and Jong Chul PARK

Materials Design Laboratory, Korea Institute of Science and Technology, P.O.Box 131, Cheongryang, Seoul 130-650, Korea

Dependence of the shape, size and substructure of thermomagnetically written magnetic domains(bits) on the composition and sputtering condition of amorphous Tb-Transition Metal(TM; Fe,Co) thin films has been systematically investigated and the influence of structural features on the domain formation process has been elucidated through microscopic observation of dynamic domain propagation process and transmission electron microscopy of the film structure. Thermomagnetic recording property of TbFe films varies with the composition roughly symmetrically about the *critical composition* of 24 at.% Tb where compensation temperature(T_{comp}) is close to Curie temperature(T_c). In structurally homogeneous films produced at low sputtering pressure, the recording process is controlled by thermomagnetic properties. Introduction of columnar structure in the film which is promoted at higher sputtering pressure changes the recording into a structure-controlled process. Recording process controlled by thermomagnetic properties renders a single-domain bit of irregular shapes near 24 at.% Tb due to the dominance of domain wall energy. As the composition deviates away towards opposite ends, the bit shape changes gradually circular, consisting of mutiple subdomains at low applied magnetic field. Columnar structure, however, promotes circular bits of larger size by facilitating easier domain nucleation and inhibiting domain wall migration. Addition of Co changes the recording properties indirectly through modification of thermomagnetic properties.

1. INTRODUCTION

Magneto-Optic(MO) recording has received great attention for years as a new mass data storage technology. The first generation MO disk which has recently been commercialized consists of magnetron-sputtered amorphous Tb-TM thin film as the recording layer. Current research and development are directed largely towards production scale-up and direct overwriting technology for Tb-TM MO disk. The performance of MO disk is essentially determined by the thermomagnetic properties of the recording layer, but significant discrepancies among the published data suggest that the thermomagnetic properties of Tb-TM films are controlled not only by the composition but also by the sputtering condition. Since there are a number of sputtering parameters which can alter the properties, it may be difficult or even meaningless to evaluate and control the effect of individual sputtering parameters on the properties as they are often interrelated in complex manner. The design concept of Tb-TM alloy should therefore be based on the fact that it is the composition and structure of the material which determines the properties of the film directly. The sputtering process controls the properties only indirectly through control of film structure.

However, little work has been reported so far on the relationship between microstructure and thermomagnetic recording properties of Tb-TM film, probably because of the limited characterization techniques available for amorphous films and of the high reactivity of Tb to oxygen. It may therefore be worthwhile to examine the magnetic domain structure as a tool for probing the material microstructure since it is one of the most structure-sensitive properties of magnetic materials. In particular, strong perpendicular magnetic anisotropy and magneto-optic effect of Tb-TM films facilitate observation of clear domain images simply by using a polarizing microscope. In this paper, the compositional and structural dependence of thermomagnetic recording characteristics was investigated mainly for TbFe films because of the simplicity of alloy system, by thermomagnetic recording test, transmission electron microscopy (TEM) and magnetic domain observation, and the effect of sputtering argon pressure on the film structure was discussed. The effect of Co addition was also investigated for limited compositions.

2. EXPERIMENTAL

TbFe films about 1000 Å thick were deposited on slide glasses by DC magnetron sputtering. To prevent oxidation of the films Si_3N_4 layers of about 700 Å-thickness were applied on both sides of MO films with RF magnetron sputtering. For TbFe films, a composite target made up of 6"-diameter Fe plate with 10 mm x 10 mm Tb chips arrayed on it was used. The base pressure before sputtering was below $7x10^{-7}$ Torr and the MO films were prepared at different Ar pressures between 1 mTorr and 10 mTorr. The target-to-substrate distance was 142 mm. To make TbFeCo films, Co chips were placed additionally. Temperature dependencies of the saturation magnetization (M_s) and coercivity (H_c) were measured respectively with the vibrating sample magnetometer and polar Kerr hysteresis loop tracer. Perpendicular magnetic anisotropy constant was measured by torque magnetometer. Film composition was determined by inductively coupled plasma spectroscopy. Both the magnetic domain dynamics during magnetization reversal process and thermomagnetic recording characteristics were investigated with a polarizing microscope. A diode laser of 830 nm wavelength was used in recording with the applied field strength between +300 Oe and -300 Oe. Positive sign denotes the direction which helps the bit formation. The laser power and pulse duration for recording were set at 9 mW and 14 μsec. The microstructure was observed for 400 A TbFe film prepared on carbon-coated Cu grid with TEM at the acceleration voltage of 160 kV.

3. RESULT AND DISCUSSION

3.1. Compositional dependence of recording characteristics

1) TbFe

The influence of applied magnetic field(H_a) upon thermomagnetically written domain structure has been studied and appropriate models were proposed to account for the observed recording characteristics[1-3]. Two types of domains have so far attracted interests because of their radically different response to H_a depending on whether the composition belongs to Fe-rich or Tb-rich regime, namely, type A and B as shown in figure 1. Recording on type A films with a minimum recordable H_a produces a circular bit consisting of sparsely dispersed, tiny subdomains, whereas it produces a small single-domain bit of irregular shape in case of type B. As H_a is increased, subdomain in type A bit increases in density and eventually forms a single-domain bit while maintaining more or less constant bit size. Type B bit, in contrast, grows in size proportionately with H_a assuming increasingly circular shape and finally becomes a circular bit.

In this work, both types of recording properties were confirmed again in the films with compositions deviated over 4~5 at. % from 24

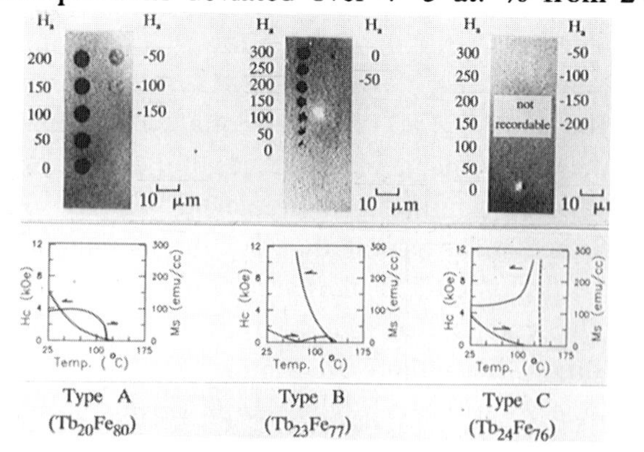

FIGURE 1
Compositional and structural dependence of thermomagnetic recording characteristics of TbFe films

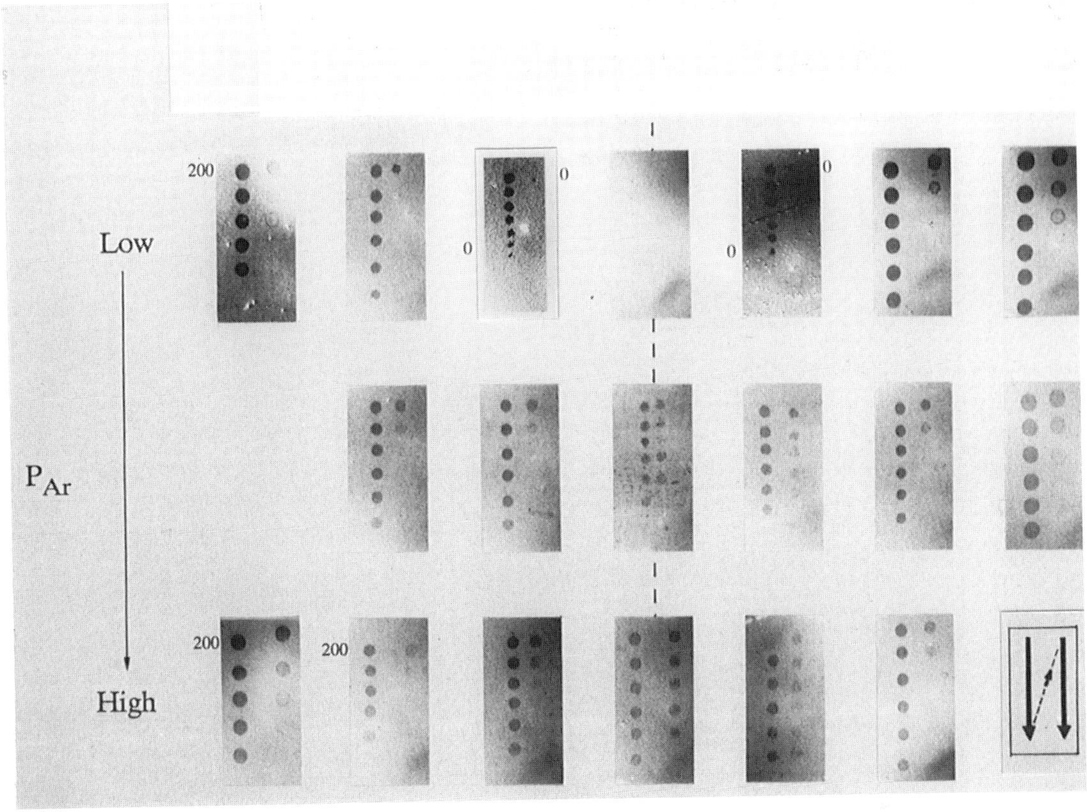

FIGURE 2
Compositional and structural dependence of thermomagnetic recording characteristics of TbFe films
*Order of bits with decreasing H_a by 50 Oe from $H_{a,max}$(300 Oe) is denoted at lower right corner of the figure.

at. % Tb. However, thermomagnetic recording tests for a range of composition between 19 and 29 at. % Tb revealed a variety of recorded domain structure depending on the composition as well as the structure of the film. Figure 2 shows the photomicrographs of recorded bits for the range of composition and sputtering pressure. It can be seen that the domain structure changes continuously in both directions along the composition axis from the center at around 24 at. % Tb towards Fe-rich and Tb-rich so that the tendency apppears to be roughly symmetric with respect to 24 at. % Tb, i.e., changing to type B and finally to type A accompanied by gradual increase of bit size. Hence, 24 at. % Tb can be considered as a *critical composition* in thermomagnetic recording characteristics of TbFe films. The significance of 24 at. % Tb may be found in that T_{comp} is close to T_c at this composition and the film is virtually unrecordable when sputtered at 1 mTorr.

Figure 3 summarizes the Kerr hysteresis loops, coercivity(H_c), magnetization(M_s), recording characteristics, typical dynamic domain structure during magnetization reversal process and H_c(T) and M_s(T) for the range of composition.

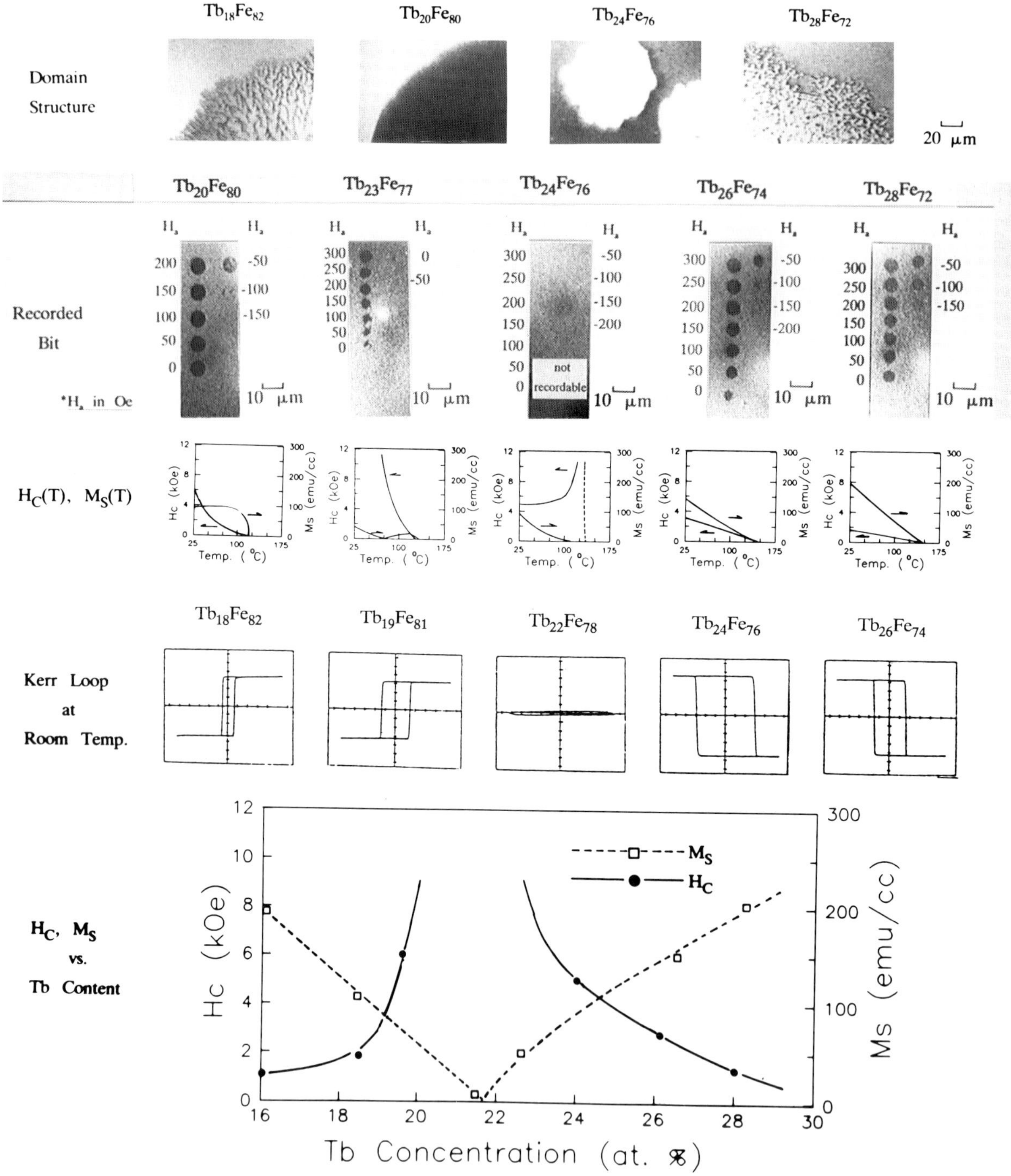

FIGURE 3
Compositional dependence of thermomagnetic properties and recording characteristics of TbFe films (P_{Ar}= 1 mTorr)

Perpendicular magnetic anisotropy can be confirmed by the squareness of the Kerr hysteresis loops for all films. Although dynamic domain structure during magnetization cycle is varied widely from film to film even for the same sputtering condition, films sputtered at 1 mTorr share a common characteristic feature of domain dynamics regardless of the composition; i.e., uniform radial growth of domain nuclei into roughly globular shape. High mobility of the domain boundaries provides an evidence of structural homogeneity of the film which is also supported by TEM observation, which will be shown in later section, although irregular boundaries observed in some films may possibly result from the pinning by unresolved structural defects.

Close coincidence between the symmetrical change of recording characteristics with composition and the symmetry of $M_s(T)_{T\sim T_c}$ and $H_c(T)_{T\sim T_c}$ change implies that the compositional dependence of recording characteristics for 1 mTorr-films is dominated by $M_s(T)_{T\sim T_c}$ and $H_c(T)_{T\sim T_c}$ and therefore can be explained satisfactorily by extending the appropriate models of Suits et. al.[1] for type A and B, since other structures can be regarded as transients.

(a) Compositions of high $M_s(T)_{T\sim T_c}$ and low $H_c(T)_{T\sim T_c}$

If the recording laser pulse induce a Gaussian thermal profile[4] in a film with the maximum temperature above T_c, creation of a paramagnetic circle causes a geometrically-induced concentration of demagnetizing field (figure 4), nucleating a ring domain just outside the paramagnetic circle due to the high $M_s(T)_{T\sim T_c}$ during cooling. Width of the ring can be defined by the isothermal radii of T_c and of a threshold temperature at which magnetization reversal is balanced by H_c, domain wall tension, etc., and therefore be determined by the slope of the thermal profile as well as H_a. Since the concentration of demagnetizing field disappears by the nucleation of a ring domain at peak temperature, domain reversal of the remaining region during subsequent cooling may be triggered by relatively uniform H_a but high $M_s(T)_{T\sim T_c}$ and low $H_c(T)_{T\sim T_c}$ will prefer energetically more stable multidomain substructure unless the field is sufficiently high. Tendency for multi-domain structure of this composition can be demonstrated in figure 5 by the stripe dynamic domain structure at elevated temperatures, which is well contrasted to the blocky structure of a composition with small $M_s(T)_{T\sim T_c}$ and large $H_c(T)_{T\sim T_c}$. Relative insensitiveness of domain size on the intensity of H_a is mainly due to the large slope of $M_s(T)_{T\sim T_c}$ curve and steep temperature gradient.

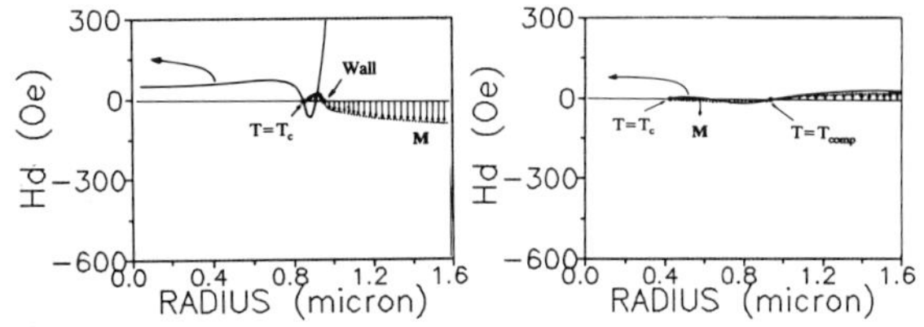

(a) Type A (b) Type B

FIGURE 4
Demagnetizing field distribution in laser-heated region for (a) Type A and (b) Type B films during thermomagnetic recording process. (reproduced from ref.1)

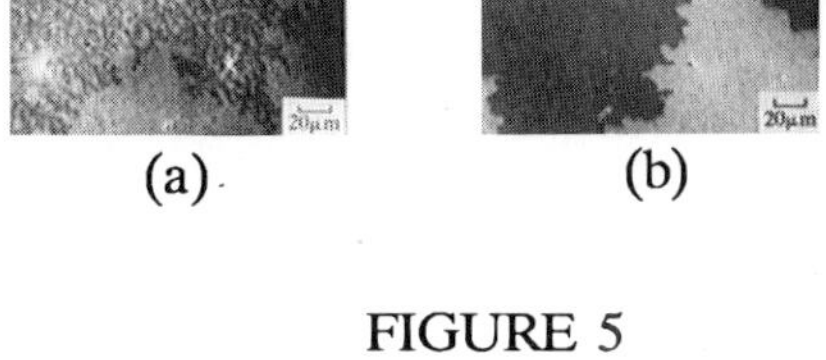

(a) (b)

FIGURE 5
Domain structures at elevated temperatures
(a) Type A ($Tb_{20}Fe_{80}$, 107°C)
(b) Type B ($Tb_{23}Fe_{77}$, 108°C)

(b) Compositions of low $M_s(T)_{T\sim T_c}$ and high $H_c(T)_{T\sim T_c}$

The same line of argument can still be applied for type B but much smaller $M_s(T)_{T\sim T_c}$ of type B films may not be effective enough to produce circumferential ring domains (figure 4). Instead, the formation of a single domain may be energetically more favourable because of small $M_s(T)_{T\sim T_c}$, but in this case domain wall energy controls the process. Since the major driving force for domain reversal is therefore H_a, the domain size must increase proportionally with increase of H_a.

The domain boundary irregularity, on the contrary to our previous paper[3], is found to arise from submicronic compositional or structural inhomogeniety as suggested by Takahashi et. al.[2] and Nagato et. al.[5]. The domain shape did change by succesive pulses but, careful observation of some selected spikes revealed that they did not appear at random but on specific sites with specific shapes. This observation suggests that although the microstructure of 1 mTorr-film appears clean and the recording process is evidently governed by the magnetic properties, even the unresolvable defects could significantly influence the recording properties particularly in the films of low $M_s(T)_{T\sim T_c}$ which receive great attention in the research for single layer direct-overwriting[6,7].

For the compositions with intermediate $M_s(T)_{T\sim T_c}$ and $H_c(T)_{T\sim T_c}$, a compromise between the demagnetizing field and wall energy may produce a variety of domain structures. The reason why $Tb_{24}Fe_{76}$ film is unrecordable can also be easily understood from the infinitely large $H_c(T)_{T\sim T_c}$ and smallest $M_s(T)_{T\sim T_c}$

2) TbFeCo

Thermomagnetic recording test on TbFeCo films has shown a variety of domain configurations depending on the composition as shown in figure 6. It was not possible to categorize them into a few distinct types as in binary TbFe film. However, the dependence of recording characteristics on the $H_c(T)$ and $M_s(T)$ found in

	θ_k(°)	H_c(KOe)	T_c(°C)	T_{comp}(°C)
(1)	0.24	10	210	25
(2)	0.26	4	290	120
(3)	0.28	>13	370	15
(4)	0.30	5	420	<-20
(5)	0.20	1.7	360	220
(6)	0.28	5	?	<-20

	θ_k(°)	H_c(KOe)	T_c(°C)	T_{comp}(°C)
(7)	0.23	4	400	80
(8)	0.20	4	400	120
(9)	0.20	2	380	165
(10)	0.16	0.8	130	-
(11)	0.32	1.5	?	-
(12)	0.33	2.5	290	-

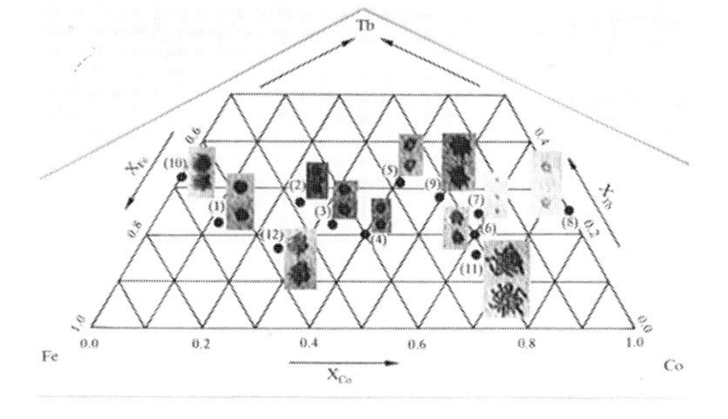

FIGURE 6
Compositional dependence of thermomagnetic recording characteristics of amorphous Tb-Fe-Co films (P_{Ar}= 1 mTorr)

TbFe film is essentially not affected by the addition of Co provided that the composition of TbFeCo is adjusted so as to produce similar $H_c(T)_{T\sim T_c}$ and $M_s(T)_{T\sim T_c}$ relationship. Figure 7 shows temperature-dependence of H_c and M_s for three different compositions of TbFeCo films, which are chosen in such a way that the thermomagnetic properties are equivalent to those of the respective type of TbFe films. It can be seen that the recording characteristics are quite similar to their corresponding types of TbFe, suggesting that it is the thermomagnetic properties which control the recording characteristics. The domain size, however, is markedly reduced by Co addition due to the increase of T_c.

3.2. Structural dependence of recording characteristics

1) Domain dynamics

Having noticed the faithful reproducibility of domain dynamics for any experimental films through repeated magnetization, the effect of some practically important sputtering parameters on the domain dynamics of Tb-TM films has been investigated. The essential features of domain dynamics as shown typically in figure 8 were little affected by the variation of deposition rate and target-substrate spacing, but changed most

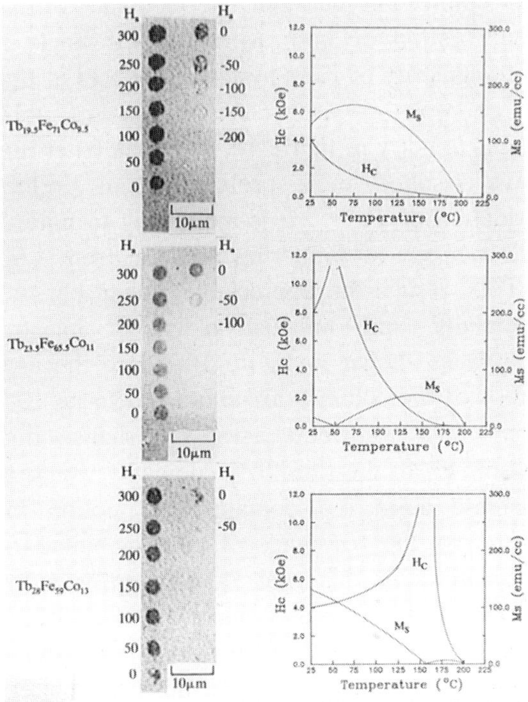

FIGURE 7
Recording characteristics and $H_c(T)$, $M_s(T)$ of TbFeCo films (P_{Ar}= 1 mTorr)
*H_a in Oe

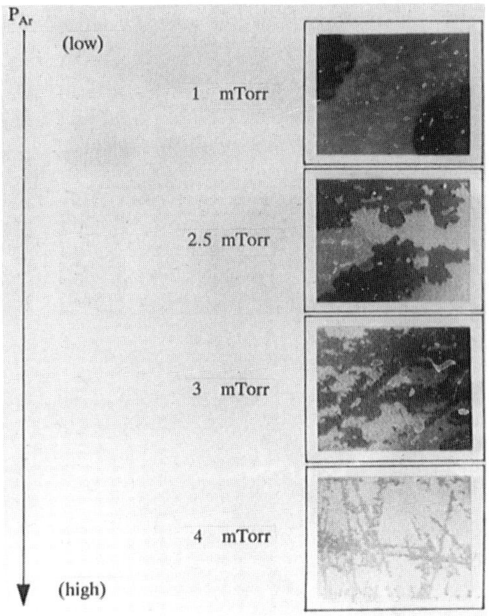

FIGURE 8
Domain structures at different sputtering pressures

sensitively by the sputtering pressure and also by the deposition angle to a lesser degree. Figure 8 shows typical domain structures at different sputtering pressures. The globular domains were found to be developed only at 1 mTorr-films. With increase of pressure, domain dynamics changes to a dendritic growth mode. As the sputtering pressure becomes higher, the thickness of branch-like domains tends to be thinner and almost unresolvable above ~ 5 mTorr.

The most informative domain dynamics is the growth mode during the magnetization reversal process. Growth of globular domains occurs rapidly and continuously and the speed of domain wall migration does not appear to slow down until complete saturation through colliding and merging with neighbouring domains. This is reflected by the squareness of the Kerr hysteresis loops as shown in figure 3. Dendritic domains, on the other hand, grow discontinuously with irregular speeds and steps and often stop advancing either by collision with other domains or defects. Saturation occurs gradually with the increase of magnetic field by the sporadic reversal of small left-out regions trapped by surrounding dendrites. This results in the curved tails of Kerr hysteresis loops, the curvature being greater with the increase of sputtering pressure as in figure 9. Over 5 mTorr, the structure is too fine to be resolved and the rectangularity of the Kerr hysteresis curve is disrupted with deteriorated Kerr rotation, presumably due to oxidation.

2) Microstructure

Figure 9 shows that no structural details are resolved in 1 mTorr-films by TEM, but sputtering at higher than about 3 mTorr produces cellular structure. The cell structure developed in 3 mTorr-film is barely resolved due to obscurely defined boundaries with cell size of 50~100 Å.

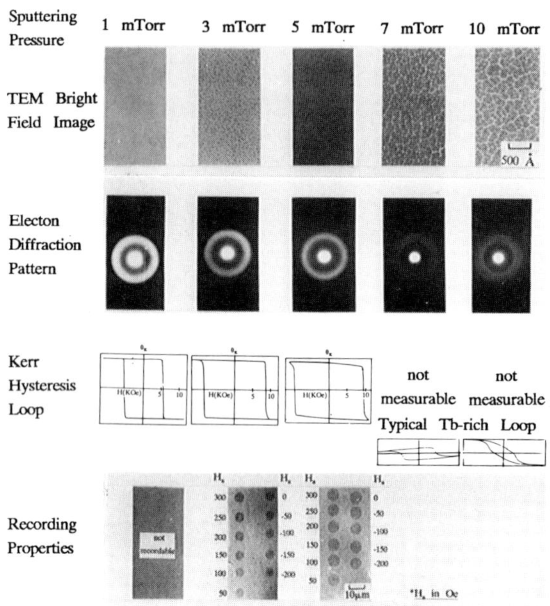

FIGURE 9
Microstructures and recording properties of $Tb_{24}Fe_{76}$ films at different sputtering pressures

Considering the film thickness of 1000 Å, the cells can actually be regarded as columns. As the pressure is increased, the structural details are resolved more clearly revealing well-defined boundaries while maintaining comparable cell size. Over 5 mTorr, duplex structure is developed by creating new, thick boundaries which embrace several cells. The selected area diffraction pattern for up to 5 mTorr-film is typified by a broad halo due to the amorphous structure, corresponding to crystalline lattice spacing of 3.2 - 1.9 Å range. But the ring halo at higher pressure splits into sharper doublets which are equivalent to the lattice spacing of 3.2 - 2.7 Å and 2.3 - 1.9 Å, respectively. Considering the (111) d-spacing of crystalline Tb_2O_3 and the (110) d-spacing of Fe are 3.1 and 2.03 Å, respectively, the split ring may possibly be due to the preferential oxidation of Tb probably at the cell boundaries since such doublets have also been obtained by annealing of similar alloys[8-10]. This may also be supported by the connection between the doublets with the deterioration of both Kerr rotation angle and rectangularity of Kerr hysteresis loop as in figure 9 as well as from the results of torque magnetometry in figure 10, since the latter might have resulted most probably from preferred oxidation of Tb[11,12]. It is interesting to note that the increase of sputtering pressure from 1 to 3 mTorr results in appreciable enhancement of coercivity and uniaxial magnetic anisotropy energy as shown in figure 10. Since the aspect ratios of the columns are estimated to be 10-20, this would contribute additional shape-induced anisotropy to the film. The coercivity would also increase since the domain wall motion must overcome the resistance of column boundaries. Discontinuous dendritic domain growth in 3 mTorr-films is most likely caused by the interaction of the domain boundaries with the columnar structure. For higher sputtering pressure, such effects could even be greater due to the improved isolation of the individual columns, but the results suggest that they are more than offset by adverse effect of oxidation. Therefore, films sputtered at higher than 5 mTorr were excluded in further investigation.

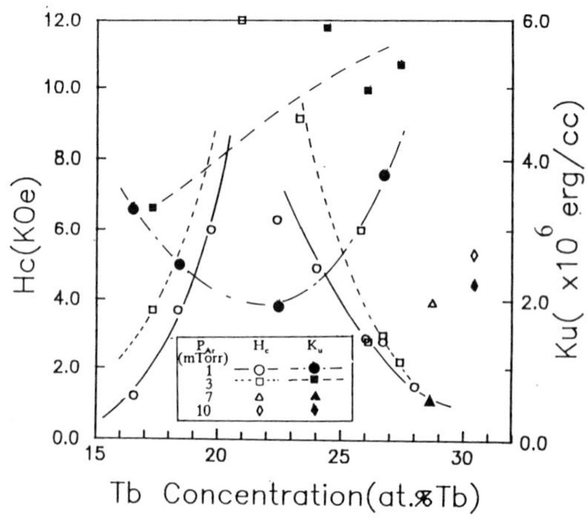

FIGURE 10
Dependence of coercivity (H_c) and uniaxial magnetic anisotropy constant (K_u) of TbFe films on composition and sputtering Ar pressure (P_{Ar})

3) Recording characteristics

Figure 2 shows that the symmetry itself of the recording characteristics across the *critical composition* of 24 at. % Tb is still maintained at higher sputtering pressure, but the recording characteristics are gradually modified with increasing pressure, which are most pronounced around the *critical composition*.

Detailed results may be summarized as follows; (a) The composition range where the effect of sputtering pressure is most pronounced coincides with that for $M_s(T)_{T\sim T_c}$ valley and thus domain wall contraction plays a dominant role while cooling in the recording process in structurally homogeneous films. Type A behaviour which is governed by demagnetizing field, however, is affected little. (b) As the sputtering pressure increases, films with around 24 at. % Tb gain recordability and the recording characteristics of around 22-27 at. % Tb change to type A. Also, the minimum recordable H_a decreases so that recording for a given H_a produces larger and less irregular bits, depending on the composition as shown in figure 11. (c) The sensitivity of recorded domain structure on H_a becomes weaker as the composition approaches the *critical composition* and as the sputtering pressure becomes higher. Eventually, it is possible to produce a series of domains of identical contrast which are completely independent of H_a, that is, thermomagnetically unerasable (see figure 9). (d) Film with just around the *critical composition* produces a bit of grey level contrast when sputtered at relatively high pressure regime but, each domain contains a small extra domain at the center. If the composition of Tb is less than 24 at. %, the bit contains a small, white spot at the center at positive H_a, but it turns dark at negative H_a. With Tb greater than 24 at. %, on the contrary, the response of contrast to H_a is completely opposite as shown in figure 12. Opposite behavior across the *critical composition* of 24 at. % Tb implies that the extra inner domain

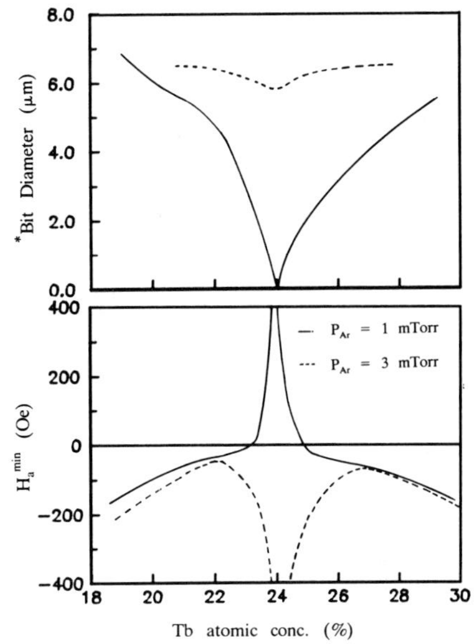

FIGURE 11
Compositional/structural dependence of bit diameter and minimum recordable H_a. Bit diameters were measured at H_a= 0 Oe.

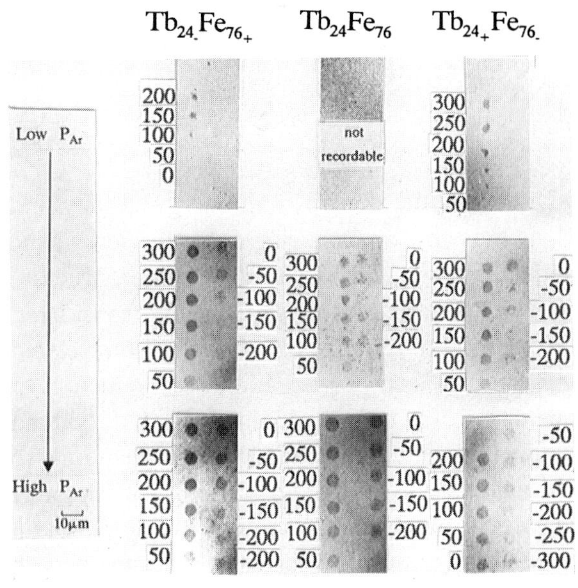

FIGURE 12
Recording characteristics of TbFe films around 24 at. % Tb
*Numbers denote applied field in Oe.

must closely be associated with the reversal of magnetization, because 24_- at % Tb-film encounters magnetization reversal during cooling cycle, but 24_+ at. % Tb-film does not. Note that the polarity of applied magnetic field for recording must also be changed across the *critical composition* due to the reversal of magnetization. Variations of the domain size and minimum recordable H_a with respect to the composition are compared in figure 11 for different sputtering pressures. Remarkable effect of the structure on the recording characteristics can be readily noticed particularly around the *critical composition*. Since $M_s(T)_{T \sim T_c}$ is not a structure-sensitive property unless the film contains excessive defects or has been oxidized, such transition can be attributed entirely to the structure of the film.

4) Mechanism of structure-controlled recording process

As suggested by Huth[13], the force balance on the wall of a cylindrical domain of radius r in a homogeneous film with perpendicular magnetization can be expressed with the following equation:

$$-\frac{\sigma_w}{2r} - \frac{\partial \sigma_w}{2\partial r} + MH_d + MH_b \leqq MH_c$$

where σ_w is the domain wall energy density and H_d is the self-demagnetizing field. In view of the above equation, the effect of columnar structure on the recording process may be considered through 3 different but mutually interrelated factors; provision of nucleation site, reduction of demagnetizing factor and inhibition of domain boundary migration. Since the field required to nucleate a reverse domain is approximately inversely proportional to $M_s(T)_{T \sim T_c}$, domain nucleation must be increasingly difficult and dominated by the domain wall energy, as T_{comp} approaches T_c, that is, around 24 at. % Tb. This may be true in the structurally homogeneous film like those sputtered at low pressures. On the other hand, if the film has columnar structure like the case sputtered at higher pressures, the domain wall energy will no longer control nucleation since a single domain can be nucleated in each column, if the column diameter is sufficiently small.

The columns having aspect ratio of 10-20 will also promote nucleation greatly since the demagnetizing factor can be reduced by 50-150. On the other hand, inhibition of domain boundary migration will not allow the change of the size and shape of a bit during cooling cycle, once nucleated at the maximum heating, and therefore result in more circular domain. Effect of columnar structure on the recording characteristics and domain wall mobility is dramatically demonstrated in figure 13, where dragging of a stripe domain wall by laser scanning observed in 1 mTorr-film is completely inhibited in 4 mTorr sample.

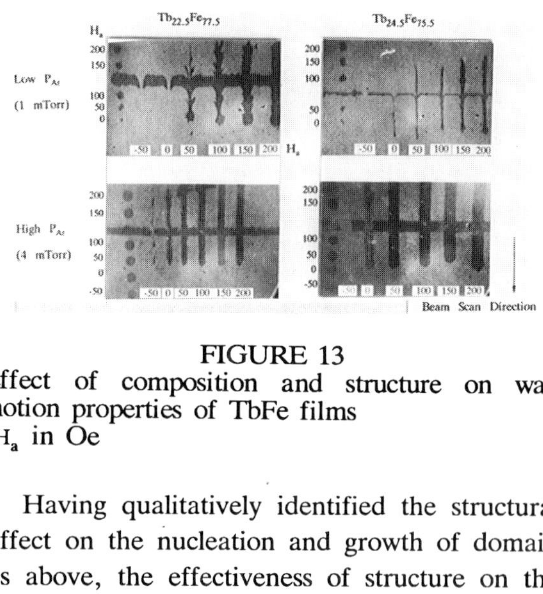

FIGURE 13
Effect of composition and structure on wall motion properties of TbFe films
*H_a in Oe

Having qualitatively identified the structural effect on the nucleation and growth of domain as above, the effectiveness of structure on the modification of recording process must therefore depend on the size, uniformity, aspect ratio of the columns and also on the structure of the column boundaries. Since the columnar structure appears to be developed as the sputtering pressure is higher, the recording sensitivity and circularity would be improved with increasing pressure as observed. On the other hand, decrease of bit contrast with increasing sputtering pressure near the *critical composition*, which may be characterized by the grey level region surrounding the dark center spot in the duplex domain structure, implies strongly that the grey level domains consist of small subdomains with the polarity anti-parallel to each other since the Kerr rotation angles of two films remain unaffected up to 5 mTorr-films.

Although a film of the *critical composition* with $T_{comp}=T_c$ is non-magnetic and can not be thermomagnetically recorded, compositional segregations which are common in practical films can produce recordability because of the local segregation of Tb-depleted and Tb-enriched regions which are both magnetic, though macroscopically still non-magnetic. If the wavelength of segregation is large enough to be resolved optically by the Kerr effect polarizing microscope, thermomagnetic recording may produce bit contrast but of a grey level. However, the subdomain structure and hence the contrast of the bit should be completely independent of the external magnetic field because the magnetic polarity of Tb-depleted regions reverses by cooling down through their T_{comp} while Tb-enriched regions maintain their original polarity which is parallel to the external field. Such a mechanism can also operate even though the composition deviates slightly from the critical point, but the substructure and domain contrast must be dependent on the polarity and intensity of the external field, which in turn is determined by the amplitude of segregation. The segregation-induced recording will eventually diminish as the composition deviates further. It is obvious that introduction of columnar structure by high pressure sputtering greatly promotes such a subdomain structure for any given amplitude of compositional modulations, which would otherwise not possible in column-free films, and therefore extends the composition range which actually develops substructure. Effectiveness of the columnar structure itself on such a segregation-induced recording should of course increase by increasing the sputtering pressure, but preferential oxidation of Tb at the column boundaries would contribute more. However, the reason how an extra domain can be formed at the center of such a grey level domain is by no means clear at the moment and further study is in progrss.

4. CONCLUSION

Thermomagnetic recording response of TbFe films was investigated for the composition range between 19-29 at. % Tb and the effect of microstructure on the recording properties was studied by using transmission electron microscopy. In structurally homogeneous films, the recording properties were found to be controlled by $M_s(T)_{T\sim T_c}$ and $H_c(T)_{T\sim T_c}$. Columnar structure developed by high pressure sputtering up to 5 mTorr modified recording properties markedly, which was particularly pronounced near *critical composition* of 24 at. % Tb, at which $T_{comp}\sim T_c$. Columnar structure was found to increase the recording sensitivity, resulting in circular bits through the reduction of demagnetizing factor, promotion of heterogeneous domain nucleation and inhibition of domain boundary migration. Around the *critical composition*, columnar structure enhanced segregation-induced subdomain structure in the bits of grey level contrast.

ACKNOWLEDGEMENT

The authors wish to thank the Central Research Laboratory of Gold Star Co. Ltd. for torque magnetometry.

REFERENCES

1. J.C. Suits, D. Rugar and C.-J. Lin, J. Appl. Phys., 64 (1988) 252.

2. M. Takahashi, T. Niihara and N. Ohta, J. Appl. Phys., 64 (1988) 262.

3. S.K. Lee and S.G. Kim, IEEE Trans. Magn. Vol.25 (1989) 4039.

4. P. Hansen, J. Appl. Phys., 63 (1988) 2364.

5. K. Nagato, A. Kawamoto, T. Sato and T. Yorozu, J. Appl. Phys., 63 (1988) 3856.

6. H-P.D. Shieh and M.H. Kryder, Appl. Phys. Lett., 49 (1986) 473.

7. M.H. Kryder and M.D. Schultz, Proc. Int. Symp. on Optical Memory, Tokyo, 1989, Jpn. J. Appl. Phys., Suppl. 28-3 (1989) pp.3-9.

8. H.J. Leamy and A.G. Dirks, J. Appl. Phys. 50 (1979) 2871.

9. A.G. Dirks and J.R.M. Gijsbers, Thin Solid Films, 58 (1979) 333.

10. H.S. Gill and J.H. Judy, J. Appl. Phys., 50 (1979) 1648.

11. R.B. van Dover, E.M. Gyorgy, R.P. Frankenthal, M. Hong and D.J. Siconolfi, J. Appl. Phys., 59 (1986) 1291.

12. H. Heitmann, A.M.J. Spruijt, P. Willich and H. Wilting, J. Appl. Phys., 61 (1987) 3343.

13. B.G. Huth, IBM J. Res. Develop., 18 (1974) 100.

Frontiers of Materials Research/
Electronic and Optical Materials
M. Kong, L. Huang (editors)

RESEARCHES OF ORGANIC PHOTOCONDUCTIVE MATERIAL TPD TO BE USED IN LASER PRINTERS

Jianhua FANG, Daocheng PAN and Hekang ZHANG

Department of Applied Chemistry, Shanghai Jiao–Tong University, Shanghai200030, China *

A novel synthetic method of an excellent charge transport material N, N′–diphenyl–bis (3–methyl phenyl)–(1,1′–biphenyl)–4.4′–diamine (TPD) is proposed. The dual–layered photoreceptors with TPD dispersed in bisphenol–A–polycarbonate as the charge transport layer are prepared using a–Se and Se–Te alloy as charge generation layer. The results of xerographic discharge measurements show that TPD produced in such a way is highly satisfactory with all the requirements fulfilled.

1. INTRODUCTION

Photoreceptors are crucial to laser printers which usually work at wavelength 780nm and even longer. The two–layered structure of the photoreceptors consisting of charge generation layer and charge transport layer is generally accepted to optimize the photoconductive properties, which is followed by several new requirements such as the compatibility of different materials within such a system[1–3]. Among these materials, the charge transport compound TPD is considered to be optimizied with respect to the different requirements to be fulfilled for satisfactory operation of the photoreceptors[2,4]. A novel synthetic method of TPD is suggested and sample–photoreceptors are prepared to be tested for the satisfactory operation of wavelength about 780 nm.

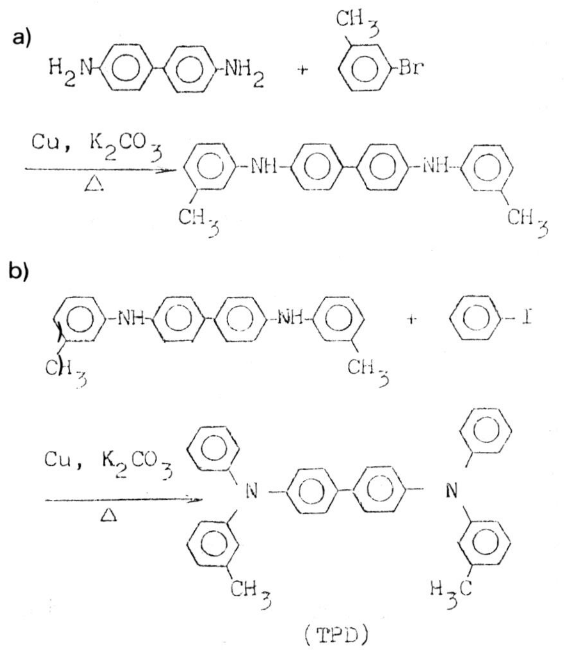

2. EXPERIMENTAL

2.1 Synthesis of TPD

On the basis of Ullman reaction, a novel synthetic method of TPD is attempted using benzidine and aromatic halide as reagents:

In a 3–necked flask fitted with a mechanical stirrer and blanketed with nitrogen, are placed the reagents and catalysts in two steps. when the reactions are finished, the mixture is allowed to cool down and benzene is added. The dark slurry

* Supported by Chinese National Science Foundation

obtained is then filtered and extracted 4 times with water. The black product is column chromatographed using neutral alumina as absorbent with benzene and cyclohexane (2:8 by volume) as washing agent. Colorless crystals of the compound are obtained by recrystallization from n–octane.

2.2 Preparation of Photoreceptors

Two dual–layered photoreceptors are prepared with aluminium substrate, the charge–generation layer being a–Se and Se–Te alloy correspondingly, and the outside coating consisting of 25% TPD and 75% bisphenol–A–polycarbonate to be used as charge–transport layer. The intermediate inorganic layer is vacuum deposited on the substrates while the organic materials are coated by dip–draw technique and dried at 40℃ for several hours under vacuum conditions.

2.3 Measurements of Photoconductive Properties

Xerographic discharge method is used to measure the photoconductive properties of the prepared photoreceptors. The radiation source is a 500W wolfram–halide lamp, with a suitable monochromator placed in front of the lamp source. The photoreceptors are charged negatively with corona discharge at −5.6 KV. The photo–induced discharge curve of the first sample, i.e., with a–Se as the charge–generation layer is shown in Fig 1, where white light is used as source and the light intensity, at 10 lux.

The second sample photoreceptor is characterized by Se – Te charge generation layer, which is sensitive to long–wavelength such as 769 nm. Its photo–induced discharge curve is shown in Fig 2.

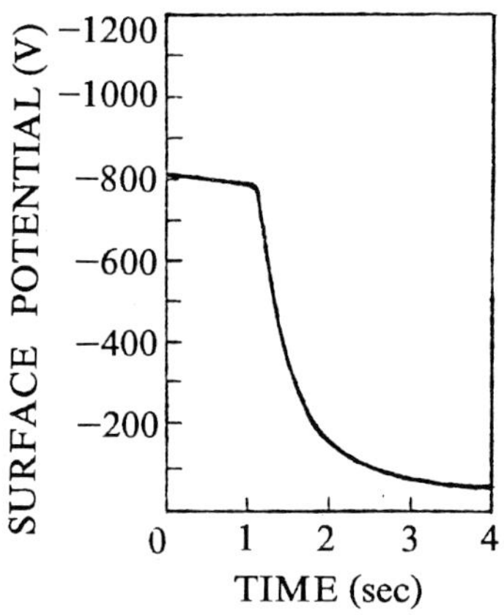

Figure 1.
Photo–induced discharge curve of photoreceptor a–Se / TPD obtained under white light exposure.

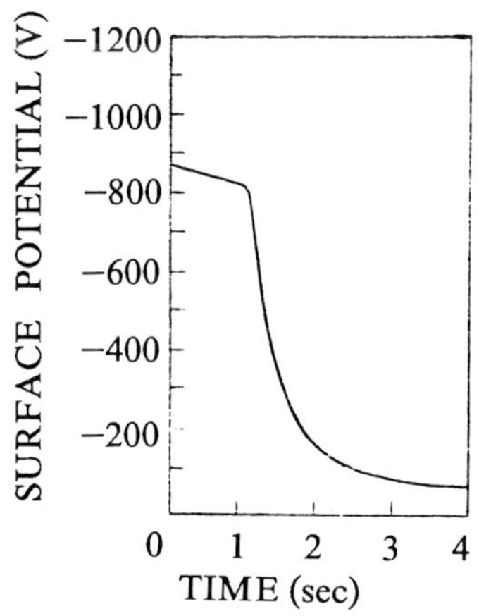

Figure 2.
Photo–induced discharge curve of photoreceptor Se–Te / TPD obtained under 769 nm exposure.

3. DISCUSSIONS AND CONCLUSIONS

In the synthetic route mentioned above, the reaction is of Ullman recction type[5], in which halides of aromatics are used as reagents. The activation series in these cases is

$$ArCl < ArBr < ArI$$

For the first step of reaction, when primary amine is converted into secondary amine, the activities of both ArBr and ArI are high enough to allow the reaction to be completed. However, for the accomplishment of the second step of the reaction, ArI is required because of its high activity. Accordingly the proposed route is well based and seems reasonable.

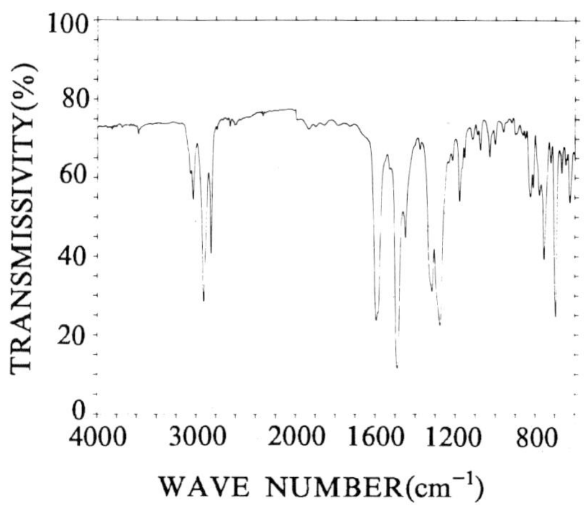

Figure 3.
IR spectrum of the synthesized product.

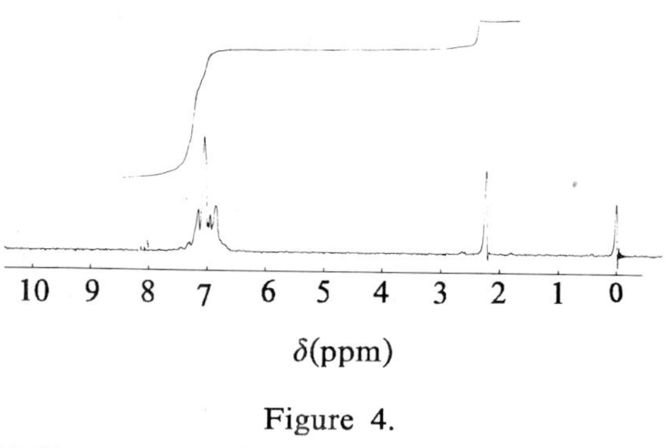

Figure 4.
NMR spectrum of the synthesized product.
Solution: $CDCL_3$

From the IR spectrum (Fig. 3), it follows that there exist groups: $-CH_3$(2924cm^{-1}, 1449cm^{-1}), C–N (1314cm^{-1}, 1276cm^{-1}), benzene ring (3031cm^{-1}, 1593cm^{-1}, 1490cm^{-1}), 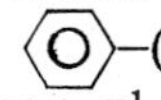–(721cm^{-1}, 696cm^{-1}), 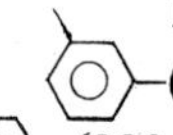–(896cm^{-1}, 754cm^{-1}, 721cm^{-1}), and 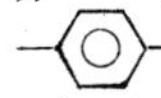–(820cm^{-1}), in close agreement with the structure of TPD.

In NMR spectrum (Fig.4) of the synthetic product, there appear a peak at 2.23ppm, indicating the existance of 6 hydrogen atoms in two methyl groups; Besides, there is a peak group distributed between 6.60 ~ 7.47ppm, which are representative of the existing 26 hydrogen atoms attached to the six benzene rings. All these are shown to agree the expected formula of TPD.

From the photo–induced discharge curve of the first sample photoreceptor (figl), the photosensitivity $E_{1/2}$is equal to 3.8 lux • s with the light intensity at 10 lux. At the same time, from Fig. 2, with monochromatic light of 769nm, and light intensity of 2.12μW / cm^2 the photosensitivity is found to be 0.84μJ / cm^2.

By the measurements of photoconductivities of the photoreceptor samples, it is shown that the double layered systems with TPD may be used in visible light region and may be extended to the near infra–red region, required by the laser–printers.

REFERENCES

1. Atsushi Kakuta, Yasuki Mori, Shigemase Takano, Manabu Sawada, and Lkutoshi Shibuya, J. Imaging Technol. 11(1985)7.
2. R. O. Loutfy, A. M. Hor, and A. Rucklldge, J. Imaging Sci. 31(1987)31–33.
3. M. Stolka, J. F. Yanus, D. M. Pai, J. Phys. Chem. 88(1984)4707.
4. Miian Stoika, Damodar M. Pai, and john F. Yanus, U. S. Pat. 4,265,990(1981).
5. Sylvie Gauthier, Jean M. J. Frechet, Synthesis, 4(1987)383–385.

Frontiers of Materials Research/
Electronic and Optical Materials
M. Kong, L. Huang (editors)

OPTICAL AND RECORDING PROPERTIES OF MAGNETO-OPTICAL MEDIA FOR ERASABLE OPTICAL DISKS

Gan Fuxi, Jiang Qi*, and Shen Defang**

Shanghai Institute of Optics and Fine Mechanics, Academia Sinica, P.O.Box 800-211, Shanghai, China
Department of Physics , suzhou University, China*
Shanghai Institute of Metallurgy, Academia Sinica, China**

The main problems in optical recording of magneto-optic media for erasable optical disks have been analysed. The effects of the temperature profiles on magneto-optical polar Kerr rotation angle and recording domain size have been discussed in detail.

1. INTRODUCTION

Recently laser induced optical and magneto-optical property change of amorphous thin films has been paid more attention due to their application in optical data storage[1]. Since the computer was invented, storage and retrieval of digital information has been a major challenge. Engineers have continually sought to develop more convenient storage methods to hold more data and make the data easier to access. The emergence of optical disk recording technology was a great event in the storage and retrieval field. For rewritable optical storage a great number of studies were performed concentrating mainly on laser induced magnetic micro-domain formation of rare-earth-transition metal (RE-TM) alloy films. A lot of problems have to be solved concerning the write, read and erase laser power, signal-to-noise ratio, number of erase cycles, material fatigue and stability. Therefore, it is worth investigating the magneto-optical and recording properties of these materials in detail.

For a suitable magneto-optical material, it needs: (a) high Kerr rotation angle, (b) high stability of recorded bits, (c) no grain boundary and film uniformity, (d) thermal and chemical stability. Among the optical and recording properties, the signal-to-noise ratio S/N is the most important factor in optical recording system. S/N ratio can be simply expressed as

$$S/N = A \cdot R \cdot P_o \mathrm{Sin}^2 2\theta_k \qquad (1)$$

where R is the reflectivity of multilayer film, P_o is the laser power, A is the constant of the sensing system. We propose an evaluation factor F. When the Kerr angle is less than 5°, we have[2]

$$F = (P_o \cdot R)^{\frac{1}{2}} \cdot \theta_k \qquad (2)$$

2. OPTICAL AND RECORDING PROPERTIES OF MULTI-LAYER RECORDING FILMS

For increasing the Kerr angle θ_k the multilayer interference enhancement effect can be used. By interference enhancement of dielectric layers the Kerr angle increases up to 1°. Therefore, for getting high recording performances it is important to design a multi-layer structure for getting the optimized S/N, P_o, R, and θ_k values. A new computational method was proposed by us, which can easily evaluate the relationship between the values mentioned above[3-5]. As a practical example, we used a four-layer structure film, the optimized parameters of four-layer structure are listed in Table 1. According to equation 2, the S/N ratio is proportional to reading laser power, but raising the laser power causes the increment of temperature of reading spot, that the thermal effect decrease the θ_k value. We have obtained the numerical solution for the relationship between the Kerr rotation angle θ_k and temperauture T. The calculated results are shown in fig.1, which is

Table 1 Thickness and refractive index of quadrilayer films

Material	Thickness (nm) l_i	Refractive index n_i
Antireflective film (SiO)	90	1.5
Magneto-optical film (GdFe)	20	$\sqrt{\varepsilon_o}=3\sim6$ g=0.44+0.012i (20°C)
Intermediate film (SiO)	150	1.5
Reflective film (Al)	300	1.2-6.9i
Substrate (PMMA)	∞	1.46

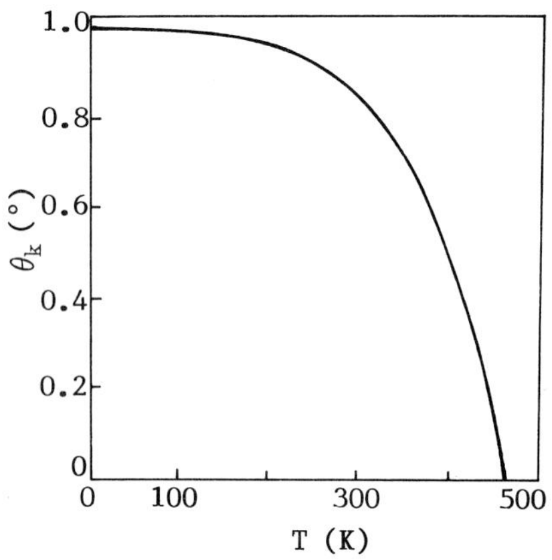

FIGURE 1
Dependence of Kerr rotation angle θ_k of the quadrilayer structure of magneto-optical film $Gd_{24}Fe_{76}$ at temperature T.

in good accordance with the experiments. The θ_k value decreases quickly at T=200K and at Curie temperature Tc, $\theta_k = 0$. Fig.2 shows the central reading temperature T vs. the reading laser power P_o, in which curve (1) and (2) corresponding to the reading time $\Delta\tau$ =50ns and 100 ns respectively, it is obtained by the theory model presented by us. It clearly shows that

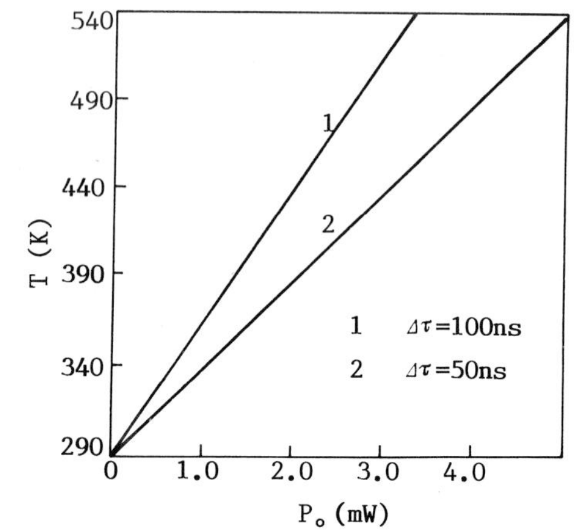

FIGURE 2
Relation between the central recording temperature T and the reading laser power P_o

the central reading temperature T is in direct proportional to the reading laser power P_o. By the data fig.2, the figure of merit F_T vs. reading laser power P_o for GdFe media can be obtained in Fig.3, It is observed that optimum

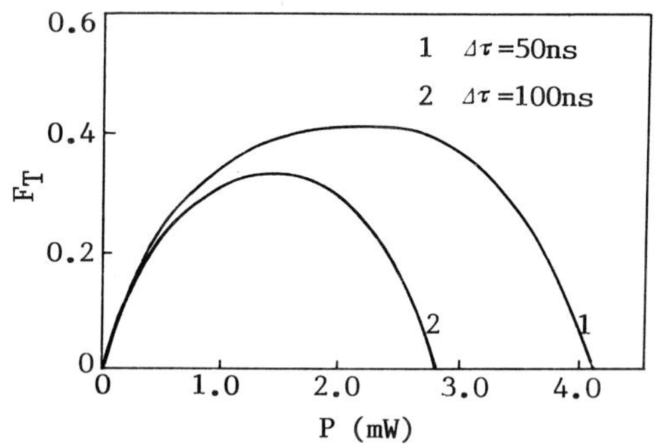

FIGURE 3
The figure of merit F vs the reading laser power P for the GdFe media

reading laser power P_r is 1.4mW and 2.1mW at the different reading time 50ns and 100ns. The figure of merit F_T for 50ns is absolutely higher than that of 100ns at the same reading laser power.

For developing erasable magneto-optical disk we put more emphysis on multilayer TbFeCo films. Table 2 lists the measured values θ_k, R, F of double layer and quadrilayer TbFeCo films. The

Table 2 Measured values , R, F of double-layer and quadrilayer films

film	θ_k [deg]	R%	$F^{*}=R^{\frac{1}{2}}\theta_k$ (min)
SiO/TbFeCo	1.0	2.0	26.8
SiO/TbFeCo/SiO/Al	1.2	24	35.3
Si_3N_4/TbFeCo (AlN) /Si_3N_4/Al (AlN)	0.85	20	25

recording characteristics of TbFeCo multilayer films are also studied. With the quadrilayer structure mentioned above, the recording domain sizes vs recording laser power and time are shown in fig.4, it is coincident with theoretical calculation.

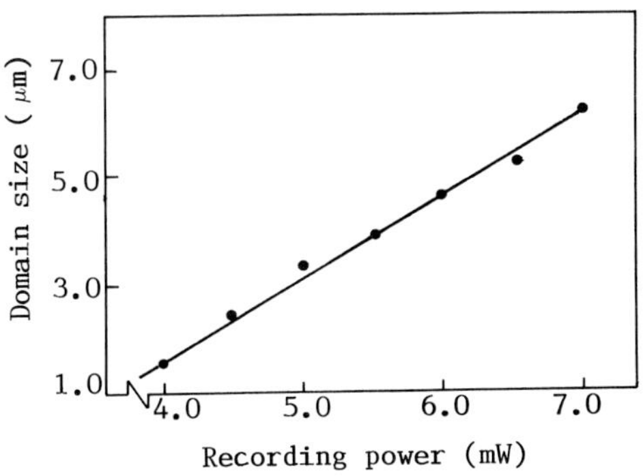

FIGURE 4
Experimental diagram of recording domain sizes vs recording laser power

3. RECORDING PERFORMANCES OF MAGNETO-OPTICAL MEDIA MEASURED BY STATIC AND DYNAMIC TESTINGS

The multilayer AlN/TbFeCo/AlN/Al films are used to make erasable optical disk, we measured the recording performances on the home-made static testing machine. The experimental results are shown in fig.5. Process value represents the relative value of the evaluation factor of the film. Five curves in the figure

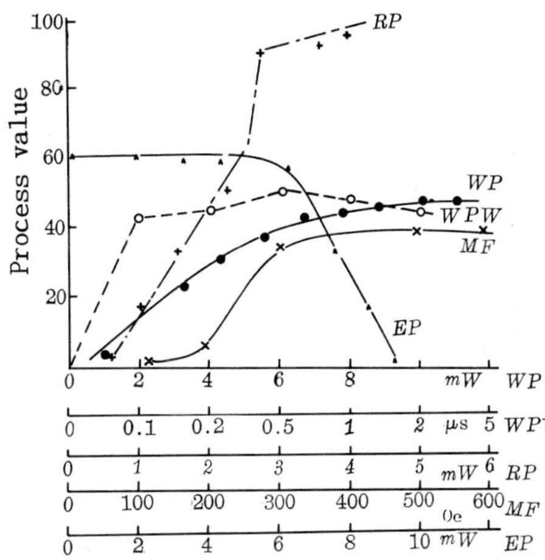

FIGURE 5
Optical recording performances of multilayer AlN/TbFeCo/AlN/Al film by static testing
WP-- Writing laser power test at 400 Oe magnatic field, 500ns writing pulse width and 100ns reading pulse width.
WPW--Writing pulse width test at 400 Oe magnatic field, 11.2 mW writing power and 100 ns reading pulse width.
RP-- Reading laser power test at 400 Oe magnetic field, 11 mW writing power and 100ns writing and reading pulse width.
MF-- Magnetic field test at 9.27 writing power, 100ns reading pulse width and 500ns writing pulse width.

demonstrate the relationship of the process value with writing power (WP), reading power (RP), erasing power (EP), pulse width of writing beam (WPW) and bias magnetic field (MF). From these changing curves the optimal recording parameters can be chosen . The writing laser power and pulse width is 8mW and 0.1 μs respectively. The reading and erasing laser power is 2.0 mW and 6mW respectively. The external bias magnetic field is about 300 Oe.

The dynamic testing has been carried out on a special apparatus using 5.25 in optical disk at rotation speed 1800 rpm. The writing and erasing laser power is 8 mW , and the reading power is 1.5 mW. These are in agreement with static test results. Fig.6 shows the signal to noise ratio of the disk with the working frequency. The S/N ratio is about 50db at 1 MHz. The bit error rate (BER) is uniformly distributed and at 10^{-5} range.

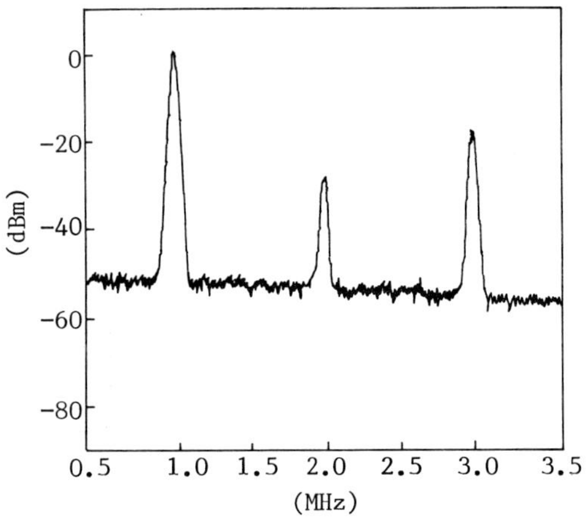

FIGURE 6
S/N ratio curve of 5.25 in magneto-optical disk at different frequency.

REFERENCES

1. Gan Fuxi, Chinese J. Lasers, 12(1985)705.

2. K. Egashira and T. Yamada, J. Appl. Phys., 45(1974)3634.

3. Jiang Qi, Shen Defang and Gan Fuxi, Acta Physica Sinica, 37(1988)1689.

4. Jiang Qi and Gan Fuxi, Acta Optica Sinica, 9(1989)925.

5. Jiang Qi, Shen Defang and Gan Fuxi, Acta Electronica Sinica, 18(1990)104.

Frontiers of Materials Research/
Electronic and Optical Materials
M. Kong, L. Huang (editors)
Elsevier Science Publishers B.V., 1991

MULTI–ION–BEAM REACTIVE CO–SPUTTERING (MIBRECS) TECHNIQUE AND ITS APPLICATIONS IN DEPOSITION OF MULTI–COMPONENT FERROELECTRIC (Pb, La) TIO_3 THIN FILMS*

Xiao Dingquan[a], Xiao Zhili[a], Zhu Jumu[a], Miao Jingsong[a], Cheng Honglie[a], Wan Derui[b], Guo Huachong[c] and Xie Bizheng[c]

a) Department of Materials Science, Sichuan University, Chengdu 610064, China
b) Institute of Materials Science and Technology, Sichuan University, Chengdu 610064, China
c) Institute of Nuclear Science and Technology, Sichuan University, Chengdu 610064, China

Considering the advantages of ion–beam sputtering deposition over other sputtering processes and the needs for fine controlling the composition and microstructure in preparing multi–component oxide thin films, multi–ion–beam reactive co–sputtering deposition technique was developed. In this paper, MIBRECS apparatus and related techniques which allow the fine control of stoichiometry and microstructure of multi–component oxide thin film are described, and epitaxial thin films of lead titanate (PT) and lanthanium modified lead titanate (PLT) thin films deposited on (0001) sapphire substrates for the first time by MIBRECS are reported.

1. INTRODUCTION

A variety of thin films was widely used in everyday life and high technical applications. They are also invaluable materials for investigating new physical phenomena.

At present, many deposition processes for various thin films have been developed. However, for preparing multi–component metal oxide thin films such as high temperature ceramic superconductor films and lanthanium modified lead zirconate titanate (PLZT) films, it is important to adjust finely the composition and microstructure of the films. This is imperative particularly for preparing thin films whose properties are dependent strongly on the stoichiometry. For example,the pyroelectric,piezoelectric and electro–optic properties of (Pb, La) (Zr, Ti)O_3 thin films are very sensitive to their composition. Therefore a technique which could finely control the stoichiometry and microstructure is desired. In order to control the composition of multi–component thin films, the best way is to finely control the deposition rate of each component during deposition processes. Based on this consideration some co–deposition techniques such as multi–source co–evaporation and molecular beam epitaxy (MBE) were developed. Because the sputtering deposition is an atom–by–atom momentum transfer process, a multiple targets magnetron co–sputtering system was established in 1985[1]. This technique has been used for preparing multi–component PLZT[1] and high temperature superconductor YBCO thin films[2].

It is well known that ion–beam sputtering technique has many attractive characteristics[3~5]. The deposition pressure is lower than that in other sputtering processes. There is no plasma in main deposition chamber. Especially, an ion source for radiating substrates can be used to adjust the micro–structure and to improve the crystallo–

* The project was supported by National Advanced Materials Committee of China (NAMCC)

graphic properties of thin films[6]. Thus ,the single and double ion–beam sputtering techniques have been developed in recent years. In 1986, we proposed the idea of constructing a multi–ion–beam reactive co–sputtering (MIBRECS) system and developing the related techniques for preparing multi–component metal oxide thin films. And the MIBRECS apparatus was successfully established in 1988[7]. Now this system is used for preparing PLZT, high Tc superconductors and other thin films. Here we describle the novel system and report the results for preparing PLT thin films by using MIBRECS technique. The thin films were analysed by electron probe X–ray microanalyzer (EPMA), reflection high energy electron diffraction (RHEED) and scanning electron microscopy (SEM). From the results of crystallographic analyses of the thin films, one can see that the epitaxial thin films with prospected stoichiometry and smooth surface have been obtained. The experiments show the MIBRECS technique is a really novel method for preparing the multicomponent thin films.

2. MIBRECS APPARATUS

The schematic illustration of MIBRECS system is shown in Figure 1. There are four ion sourecs in the system. Three focused ion sources which are distributed around the substrate stage center are mouted on the vacuum chamber for sputtering three targets respectively. A group of rotary target stages are mounted in front of the ion sources. The targets for preparing complex oxide thin films may be metals and metal oxides with high purity. The prospected composition of deposited thin films could be got by controlling strictly and individually the parameters of each sputtering ion source. The fourth ion source without filament is mounted in front of the substrate in order to clean substrate before sputtering and to make dynamic mixing during depoiting. For in–site growth of homogeneous thin films, the substrate could be heated and rotated. The main parameters of the system are listed in table I.

The advantages of MIBRECS apparatus are:

1. Because of the lower operating pressure (about $10^{-1} \sim 10^{-2}$ Pa), less gas incorporates into

TABLE Ⅰ Some Parameters of MIBRECS System

Vacuum chamber	600mm in diameter × 650mm in deep, cylindrical, stainless steel	
Ion source	Three Kauffmann ion sources with 90mm in diameter for sputtering. ion beam density 25mA/cm^2 and maximum energy 1.5–2.0KeV. One microwave ion source for precleaning substrate and dynamic mixing	
Vacuum system	Back ground pressure	$\sim 10^{-4}$ Pa
	Operating pressure	$\sim 10^{-2}$ Pa
Substrate stage	Rotating with rates from 10–120 turns per minute.	
	Heated from room temperature up to 700℃	
Target holdings	Multi–sided and water–cooled, 60mm in diameter, stainless steel.	

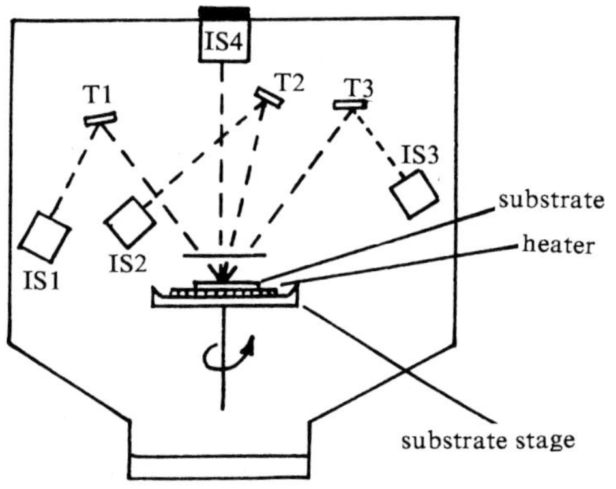

FIGURE 1
The schematic illustration of multi–ion–beam reactive co–sputting system.

the the films.

2. There is no plasma in main deposition chamber, thus there are no plasma bombardment of substrate, target and films, no plasma bombardment chamber wall or substrate holder, and no plasma backscattering and cross–contamination. This is a favorable condition for preparing good thin films.

3. The stoichiometry of deposited thin films could be controlled precisely by adjusting independently the parameter of each ion source.

4. The MIBRECS system is favorable to prepare complicated multicomponent thin films and multi–layer thin film with more than one component in a layer.

5. Because of the radiation of substrates by reactive gas ion during film depoition, the reaction probability will increase and the properties of thin film will be improved.

6. The system and related techniques we developed can be used for deposition of various thin films, among which multicomponent ferroelectric thin films, high Tc superconductor thin films, semiconductor thin films, buffers and dielectric superlattice thin films are important ones.

3. PREPARATION AND PROPERTIES OF PLT THIN FILMS

For deposition of PLT thin films, high purity metal targests of Pb(99. 999%). La(99. 9%) and Ti(99. 9%) were used and the locations of the targets were finely adjusted for effective deposition. Considering lattice matching, thermal expansion, crystalline structure and the applications of the films for optoelectronic devices[8], the (0001) sapphire wafers which were optically polished were used as substrates. The conditions for the deposition of PLT thin films are given in Table II.

The as–growth $PbTiO_3$ and PLT thin films were pale yellowish in colour and almost transparent . Their crystallographic properties were examined by X–ray diffraction (XRD) and reflection high energy electron diffrcation (RHEED).

Figure 2 gives the typical XRD pattern of the prepared PT and PLT thin films. It indicates that the as–growth thin films are (111) completely oriented. The results of RHEED pattern of the thin film is shown in Figure 3. It indicates the epitaxial characteristics of these films. Therefore it is concluded that the epitaxial PT and PLT thin films are obtained on (0001) sapphire substrates by MIBRECS technique and the epitaxial relation is:

(111) PLT / / (0001) sapphire

which is the same as that of the literature[9]

The cross–section and surface morphologies of the as–sputtered films have been observed by scanning electron microscope (SEM). The micrographs are shown in Fig. 4. It can be seen from Fig. 4 that the surface of thin films is very smooth and the boundary between thin film and substrate is very clear. The composition of PLT thin film is examined by electron probe X–ray micro–analyzer (EPMA) measurments. The results show that the stoichiometry of the thin films can be finely controlled by adjusting the parameters of each ion source. In our experiments we obtained the prospected stoichiometry thin films $Pb_{0.72}La_{0.28}Ti_{0.93}O_3$ (known as PLZT 28 / 0 / 100).

These results indicates that MIBRECS technique is very convenient to deposit multi–component thin films with different stoichiometry. The homogenety of the thin films was examined by X–ray diffraction and EPMA point by point acrossing the substrate surface with diameter of 25 mm through different direction. The results are almost the same for different point of the surface.

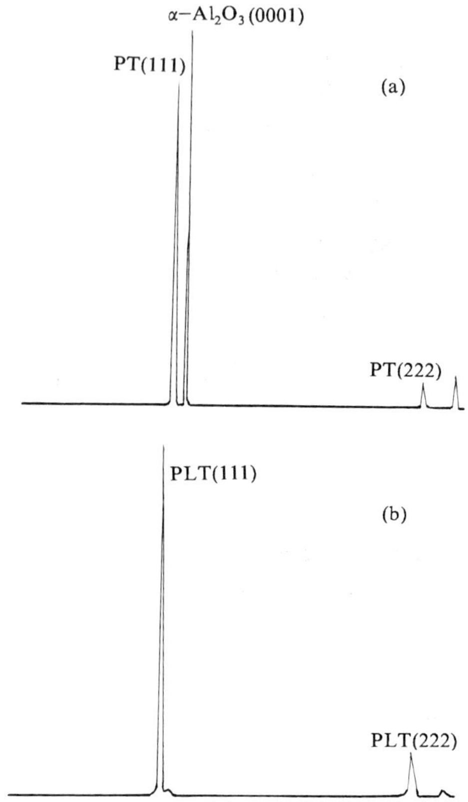

FIGURE 2

X–ray diffraction patterns of as–sputtered $PbTiO_3$ (a) and PLT ($Pb_{0.72}La_{0.28}Ti_{0.93}O_3$) (b) respective.

4. SUMMARY

A new systen named as MIBRECS and related

TABLE Ⅱ Summary of Sputtering Conditions

Gas pressure	$3\sim 8\times 10^{-2}$ Pa
Sputtering gas	$Ar / O_2 = 1/0.2\sim 1/1$
Substrate temperature	600~650℃
Substrate rotation rates	80~90 / min
Sputtering ion source parameners	
for Pb target energy	0.3~0.6 keV
current	2~6 mA / cm^2
for Ti target energy	0.8~1.2keV
current	5~12 mA / cm^2
for La target energy	0.4~0.7keV
current	5~8mA / cm^2
Deposition rates	3~10 nm / min

FIGURE 3

Typical reflection high energy electron diffraction (RHEED) pattern of $Pb_{0.72}La_{0.28}Ti_{0,93}O_3$ films with thickness of about 250 nm sputtered on (0001) sapphires.

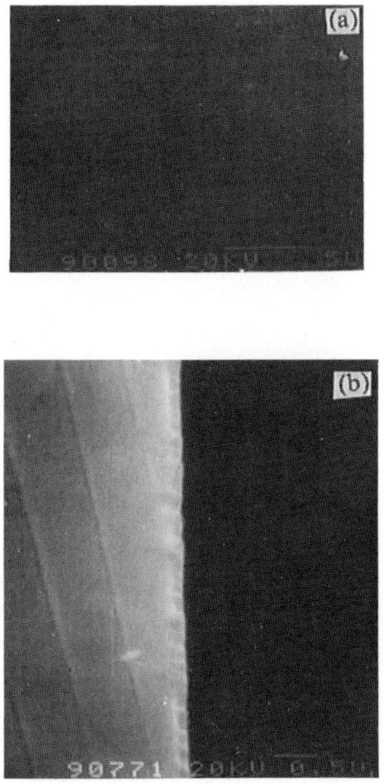

FIGURE 4

Scanning electron micrographs of (a) the surface and (b) cross section of epitaxial PLT thin films deposited on (0001) sapphire

techniques for depositing multicomponent thin films have been developed, and the epitaxial (Pb. La)TiO_3(PLT)thin films have been prepared by using this technique. The crystallographic analyses by XRD, RHEED, EPMA, SEM show that the deposited thin films have good quality. The composition of the thin films could be precisely controlled during the deposition. Complex oxide thin films with good quality and ideal stoichiometry could be obtained without much difficulty by using puring metal targets. The present works have shown that the MIBRECS technique is a useful method for depositing complex multicomponent thin films . Now the preparation of PT, PLT and PZT thin films on α–Al_2O_3, Si and GaAs substrates, high Tc superconductors thin films, PT / PLT superlattice thin films on different substrates are in hand.

REFERENCE

1. H. Adachi, T. Mitsugu et al., Jpn. J. Appl. Phys., Suppl. 24–3, (1985)13

2. Y. Terashima, M. Sagoi et al. , Jpn. J. Appl. Phys., 28(1989) L653

3. D. Bouchier et al., J. Appl. Phys. ,49 (1978)5896

4. P. Madakon et al., J. Appl. Phys., 63 (1985)2046

5. M. Naoe et al., Jpn. J. Appl. Phys., 55 (1986)11

6. D. J. Martin, J. Mate. Sci., 21(1986) 25

7. Guo Huachong, Xiao Dingquan et al. , Presenten at the Seventh International Meeting on Ferroelectrcity. Saarbrucken, F. R. Germany, Aug, 28–Sept. 3(1989); to be published.

8. H. Adachi et al., J. Appl. Phys,60 (1986)736

9. H. Adachi et al., Appl. Phys. Lett., 42 (1983)867

Frontiers of Materials Research/
Electronic and Optical Materials
M. Kong, L. Huang (editors)

Zr-EXAFS AND MD STUDIES OF FLUOROZIRCONATE GLASSES

Yoji KAWAMOTO, Ryoji KANNO, Yukihiro UMETANI and Kazuyuki HIRAO*

Faculty of Science, Kobe University, Kobe 657, Japan
*Faculty of Engineering, Kyoto University, Kyoto 606, Japan

The variation of local structures around Zr with glass constituents and glass compositions in the ZrF_4-BaF_2, ZrF_4-BaF_2-CsF, ZrF_4-LaF_3-MF (M: alkali metals) and ZrF_4-LaF_3-LiF glass systems was examined using Zr-EXAFS and MD simulation. The $(100-x)ZrF_4 \cdot xBaF_2$ (x=30-50) and $(95.0-x)ZrF_4 \cdot 5.0LaF_3 \cdot xLiF$ (x=20.0-47.5) glasses showed no variation of the F coordination environment of Zr with composition. In the $55.0ZrF_4 \cdot (45.0-x)BaF_2 \cdot xCsF$ (x=0-40.0) and $52.5ZrF_4 \cdot 5.0LaF_3 \cdot 42.5MF$ (M: alkali metals) glasses, the local structures around Zr remarkably changed with the alkali-fluoride addition. That is, a splitting of the Zr-F interatomic distances into two groups occurred, accompanying a decrease in average Zr-F interatomic distances. The reliability of structural parameters determined by the present EXAFS experiment is also discussed from the EXAFS result obtained for crystalline Li_2ZrF_6, Cs_2ZrF_6, α-$BaZrF_6$, β-$BaZrF_6$ and β-ZrF_4.

1. INTRODUCTION

Fluorozirconate glasses, i.e. ZrF_4-based glasses, have potential applications in fields where traditional oxide glasses can not be used. For example, applications to mid infrared-transmitting materials, ultra-low-loss optical fibers, high-power lasers, infrared-pumped visible lasers, high fluoride-ion conductors and so on are anticipated.

On the other hand, the fluorozirconate glasses have a structural feature that Zachariasen's criteria on glass formation are probably inapplicable to these glasses. For instance, it is presumed that the F coordination numbers of Zr may be from 6 to 8 and these coordination polyhedra may be linked by sharing not only the corners but also the edges, depending on glass compositions and/or glass-constituents.

Such a structural peculiarity seems to be closely related to the appearance of peculiar properties in these glasses. However, structural studies of the fluorozirconate glasses have not been so much advanced. At present, even the local environment around Zr is still a matter of controversy. Therefore the relation between structure and properties in fluorozirconate glasses is little well-defined yet.

The present EXAFS and MD (Molecular Dynamics) studies have been undertaken in order to examine the variation of local structures around Zr (the F coordination environment of Zr and the linkage manner of ZrF_n polyhedra) with glass constituents or glass compositions in the glass systems ZrF_4-BaF_2, ZrF_4-BaF_2-CsF, ZrF_4-LaF_3-MF (M: alkali metals) and ZrF_4-LaF_3-LiF.

2. EXPERIMENTAL PROCEDURE

2.1. Sample preparation

Ultra-pure chemicals of fluorides were employed as the starting raw materials.

Glasses $(100-x)ZrF_4 \cdot xBaF_2$ (x=30-50), $55ZrF_4 \cdot (45-x)BaF_2 \cdot xCsF$ (x=0-40), $52.5ZrF_4 \cdot 5.0LaF_3 \cdot 42.5MF$ (M=alkali metals) and $(95.0-x)ZrF_4 \cdot 5.0LaF_3 \cdot xLiF$ (x=20.0-47.5) were prepared. About 6g batches of the raw materials with the addition of small amounts of $NH_4F \cdot HF$ fluorinating agent were melted in platinum crucibles in an Ar gas atmosphere. The melts were kept at 800-900°C for 20min and then quenched in flat-plate forms by pressing them with two preheated brass plates. No inclusion of crystalline phases was ascertained by observing the quenched samples under a polarized optical microscope. The prepared glasses were annealed at the respective glass transition temperatures.

As reference crystals, Li_2ZrF_6, Cs_2ZrF_6, α-$BaZrF_6$, β-$BaZrF_6$ and β-ZrF_4 were synthesized. About 6g batches of the raw materials with the addition of about 1-2g $NH_4F \cdot HF$ were melted at about 850°C for 10min in platinum crucibles under an Ar gas atmosphere. Subsequently the molten Li_2ZrF_6, Cs_2ZrF_6 and α-$BaZrF_6$ batches were kept for about 10min at the respective crystallization temperatures and then slowly cooled to room temperature in the furnace. On the other hand, the molten β-$BaZrF_6$ batch was rapidly quenched from the crystallization temperature to room temperature by dipping it into liquid nitrogen. The synthesized compounds were confirmed to be the respective single phases by X-ray powder diffraction. The raw material ZrF_4 itself was used for the β-ZrF_4 since it was ascertained to be in that form by X-ray powder diffraction.

2.2. EXAFS measurement

Measurements were carried out using the EXAFS facilities at BL-10B of Photon Factory in National Laboratory for High Energy Physics. The Zr-K absorption data were collected in the energy range from about 1000eV on the lower-energy side of the Zr-K absorption edge (about 18keV) to about 1000eV on the higher-energy side. The incident and the transmitted X-ray intensities were measured by the ($75\%N_2$ + 25%Ar) and the 100%Kr ion chambers, respectively.

2.3. EXAFS analysis

EXAFS analysis was performed according to the Teo procedure.[1]

In the extraction of EXAFS oscillation from absorption spectra, the background absorption was calculated by employing the Victreen formula extrapolation, while the smooth K-shell absorption due to an isolated Zr atom was estimated by using the iterative smoothing method.

The energy of absorption edge for Zr-K was taken as 18000eV and the EXAFS oscillation was converted into the k space to obtain $\chi(k)$. Then the $\chi(k)$ weighed by k^3 was Fourier-transformed in the k region from 3.5 to 13.5Å^{-1} to obtain the Fourier transform magnitude, $|F(r)|$. In the Fourier transformation the following Hanning window function was used in order to make cutting smooth

$$
\begin{aligned}
W(k)&=(1/2)[1-\cos\pi(k-k_1)/(k_2-k_1)], & k_1\le k<k_2\\
W(k)&=1, & k_2\le k\le k_3\\
W(k)&=(1/2)[1-\cos\pi(k-k_3)/(k_4-k_3)], & k_3<k\le k_4
\end{aligned}
$$

where k_1, k_2, k_3 and k_4 are 3.5, 4.0, 12.5 and 13.5Å^{-1}, respectively. Both phase shift and backscattering amplitude were not corrected.

2.4. Fourier backfiltering and curve fitting

Fourier backfiltering and curve fitting were also performed according to the Teo procedure[1] and structural parameters about the F coordination environment around Zr were obtained.

In the Fourier backfiltering the Hanning window function used in 2.3 was employed to make cutting smooth. The k_1, k_2, k_3 and k_4 values adopted for each glass are given in the $|F(r)|$ curve. The curve fitting was performed on $k^3\chi(k)$ in the k range from 4.0 to 10.5Å^{-1} by the least squares method. The phase shift and backscattering amplitude values which are necessary in performing the curve fitting treatment were obtained by employing the Li_2ZrF_6 crystal as a reference specimen. As the Zr-F interatomic distance and the F coordination number of Zr for this crystal the X-ray structure analysis values of 2.016Å and 6 were adopted.[2]

2.5. MD simulation

MD simulation was performed on two glasses of compositions $50BaF_2 \cdot 50ZrF_4$ ($BaZrF_6$) and

$33.3BaF_2 \cdot 66.7ZrF_4$ ($BaZr_2F_{10}$). The detailed procedure has already been described elsewhere.[3]

3. RESULTS AND DISCUSSION

3.1. Crystalline fluorozirconates

At first, the reliability of structural parameters determined by the present EXAFS experiment is discussed based on the Zr-EXAFS result of reference crystals Li_2ZrF_6, Cs_2ZrF_6, α-$BaZrF_6$, β-$BaZrF_6$ and β-ZrF_4.

The Zr-K absorption spectra, the EXAFS oscillation curves and the Fourier transform magnitude curves of crystalline Li_2ZrF_6, Cs_2ZrF_6, α-$BaZrF_6$, β-$BaZrF_6$ and β-ZrF_4 are shown in figures 1 and 2.

Structural parameters obtained by Fourier backfiltering and curve fitting are given in table 1, where those determined by X-ray structure analysis[2, 4-7] are also given for comparison. As can be seen from the table, the r_{Zr-F} and CN_F values obtained for Cs_2ZrF_6 and β-ZrF_4 by EXAFS are fairly close to those obtained by X-ray structure analysis, although the EXAFS values are about 0.01Å shorter and about 0.5 smaller than the X-ray values. In α-$BaZrF_6$ and β-$BaZrF_6$, however, the EXAFS values are considerably small, compared with the X-ray values and, in addition, satisfactory curve fittings were difficult for these crystals, as can be seen from figure 3. Such a situation may be due to an extremely wide distribution of Zr-F interatomic distances in the F polyhedra of Zr in these two crystals (figure 4).

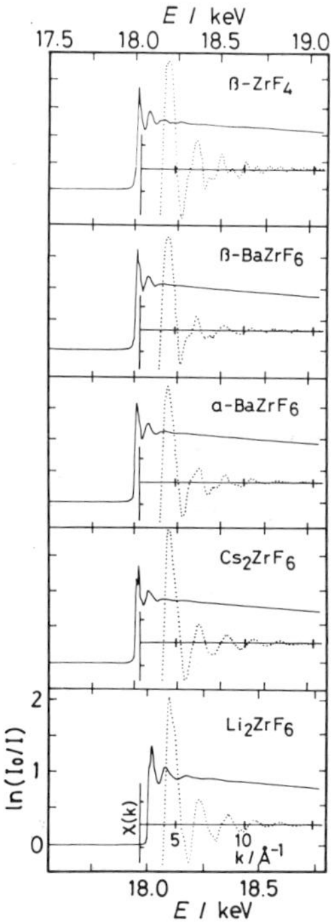

FIGURE 1
Zr-K absorption spectra (solid lines) and EXAFS oscillation curves (dotted lines) of crystalline fluorozirconates

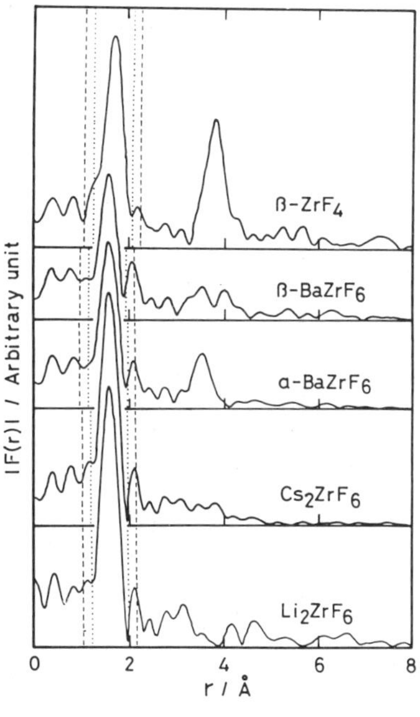

FIGURE 2
Fourier transform magnitude curves of crystalline fluorozirconates (Two broken lines mean the r regions where Fourier backfiltering was carried out, and two dotted lines mean the r regions where the Hanning window function takes a value of unity.)

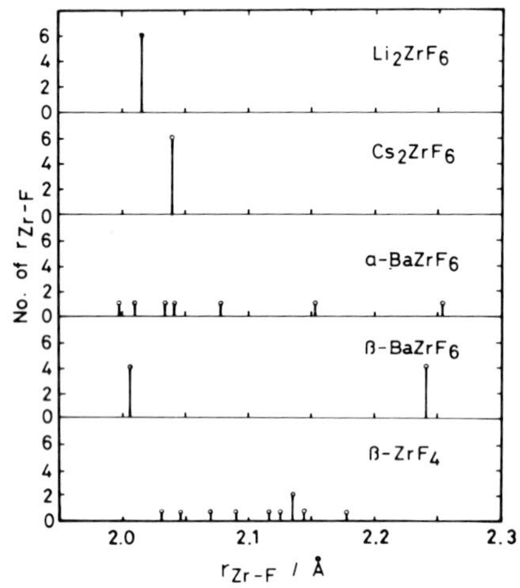

FIGURE 3
Distribution of Zr–F interatomic distances in F polyhedra of Zr in crystalline fluorozirconates

TABLE 1
Structural parameters (Zr–F interatomic distance, r_{Zr-F}, F coordination number of Zr, CN_F, and Debye-Waller factor, σ) obtained for crystalline fluorozirconates by Fourier backfiltering and curve fitting (Numerals in parentheses are values obtained by X-ray structure analysis studies.[2,4-7])

Crystal	r_{Zr-F} (Å)	CN_F	σ (Å)
Li_2ZrF_6	2.016 (2.016)	6.0 (6)	0.071
Cs_2ZrF_6	2.02 (2.029)	5.6 (6)	0.075
α-$BaZrF_6$	2.05 (2.082)	5.4 (7)	0.090
β-$BaZrF_6$	2.06 (2.126)	5.5 (8)	0.104
β-ZrF_4	2.10 (2.11)	7.5 (8)	0.090

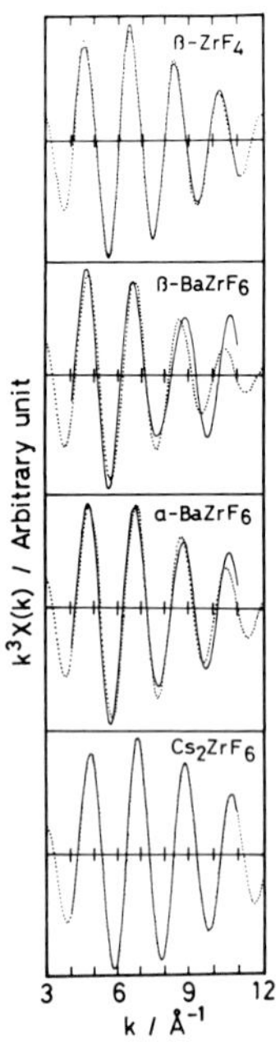

FIGURE 4
$k^3\chi(k)$ curves obtained for crystalline fluorozirconates by Fourier backfiltering (solid lines) and by curve fitting (dotted lines)

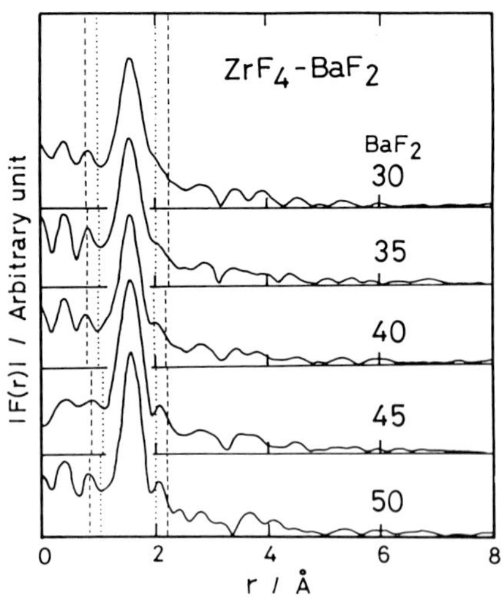

FIGURE 5
Fourier transform magnitude curves of glasses $(100-x)ZrF_4 \cdot xBaF_2$

Thus it may be concluded that, if there is no wide distribution in the Zr-F interatomic distances, the present EXAFS experiment gives structural parameter values very slightly smaller than the true values; about 0.5% short in r_{Zr-F} and about 7% small in CN_F.

3.2. Glasses $(100-x)ZrF_4 \cdot xBaF_2$ (x=30-50)

Figure 5 shows the Fourier transform magnitude curves of the $(100-x)ZrF_4 \cdot xBaF_2$ glasses and table 2 gives the structural parameter values obtained for the glasses. As can be seen from the figure and the table, there are no definitive composition dependences in both the Zr-F interatomic distance and the F coordination number of Zr and that the r_{Zr-F} and CN_F values are 2.06-2.07Å and 6.5-7.5, respectively. Taking into consideration the reliability on r_{Zr-F} and CN_F that was mentioned in 3.1, the real r_{Zr-F} and CN_F values of glasses $(100-x)ZrF_4 \cdot xBaF_2$ (x=30-50) may be regarded as 2.07-2.08Å and 7.0-8.0, respectively.

Figure 6 shows the MD result of simulated glasses $50BaF_2 \cdot 50ZrF_4$ and $33.3BaF_2 \cdot 66.7ZrF_4$. The pair correlation function for a Zr-F pair and the cumulative F coordination number for Zr reveal that the r_{Zr-F} and CN_F values are almost identical in the two compositions. The only difference between these two glasses observed in the simulation data is the shape of the Zr-Zr pair correlation function which provides informations about the linkage between the F polyhedra of Zr. The observed shapes indicate that there are two linkage manners, i.e. edge sharing and corner sharing, and that the edge sharing/corner sharing ratio changes from about 6:4 in the $50BaF_2 \cdot 50ZrF_4$ glass to about 5:5 in the $33.3BaF_2 \cdot 66.7ZrF_4$ glass.[3] On the F coordination environment of Zr, therefore, the MD study gave the same conclusion as the EXAFS study.

3.3. Glasses $55ZrF_4 \cdot (45-x)BaF_2 \cdot xCsF$ (x=0-40)

The Fourier transform magnitude curves of the glasses are shown in figure 7 and the peak maxima and half-widths of Zr-F peaks in the curves are given in table 3. It can be seen from the figure that, in glasses x=27, 34 and 40, small but discernable shoulders emerged around

TABLE 2
Structural parameters obtained for glasses $(100-x)ZrF_4 \cdot xBaF_2$

Glass composition ZrF_4	BaF_2	r_{Zr-F}	CN_F	σ
(Mol%)		(Å)		(Å)
50	50	2.06	7.4	0.119
55	45	2.07	6.5	0.104
60	40	2.07	6.4	0.111
65	35	2.07	6.7	0.115
70	30	2.06	7.4	0.119

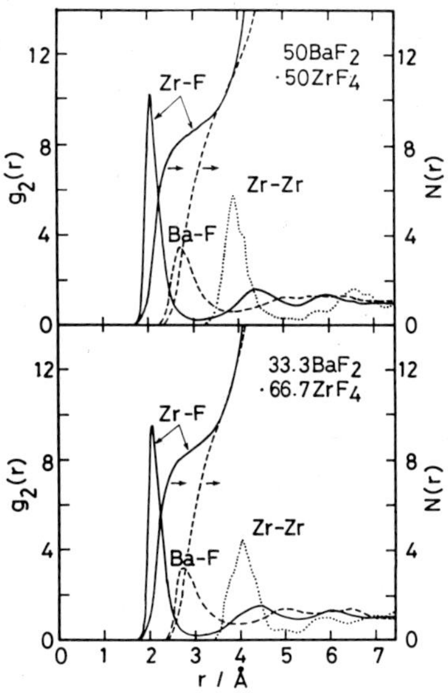

FIGURE 6
Pair correlation functions for pairs Zr-F, Ba-F and Zr-Zr, and cumulative F coordination numbers for ions Zr and Ba in simulated $50BaF_2 \cdot 50ZrF_4$ and $33.3BaF_2 \cdot 66.7ZrF_4$ glasses at room temperature

1.8Å at the right-hand sides of the Zr-F peaks. For these glasses, therefore, the curve fitting was carried out by using a "two shell" model instead of a "one shell" model used for glasses x=0, 7, 14, and 20. The $k^3\chi(k)$ curves obtained for glasses x=0, 20 and 40 by Fourier backfiltering and by curve fitting are shown in figure 8. The structure parameters thus obtained are given in table 4.

The F coordination number of Zr appears to increase slightly with the substitution of CsF for BaF_2. However, if an accuracy of the coordination number evaluation in an EXAFS analysis is taken into account, it will be reasonable to consider that the F coordination number hardly varies with composition.

On the contrary, the Zr-F interatomic distance exhibited the following definitive compositional variation: The Zr-F interatomic distances of glasses x=0, 7 and 14 hardly change, but those of glasses x=27, 34 and 40 split into two different distances, one of which being shorter and the other of which being longer than the Zr-F interatomic distances in the x=0, 7 and 14 glasses. In addition the two Zr-F interatomic distances and the F numbers participating to the respective distances vary, depending on the substituted BaF_2 content.

No "two shell" model was applied to the x=20 glass, but there is no doubt as to the occurrence of a splitting of the Zr-F interatomic distance. The reason is because the half-width of Zr-F peak of this glass is a little wider than those of the x=0, 7 and 14 glasses (table 3).

Thus the followings may be concluded on glasses $55ZrF_4\cdot(45-x)BaF_2\cdot xCsF$: The F coordination numbers of Zr are about 7.0-7.5 in all the glasses. However, when the substitution of CsF for BaF_2 exceeds about 20mol%, a splitting of the Zr-F interatomic distance into two groups with about 0.2Å difference occurrs, accompanying a decrease in average Zr-F interatomic distance.

3.4. Glasses $52.5ZrF_4\cdot5.0LaF_3\cdot42.5MF$ (M: Li, Na, K, Rb and Cs)

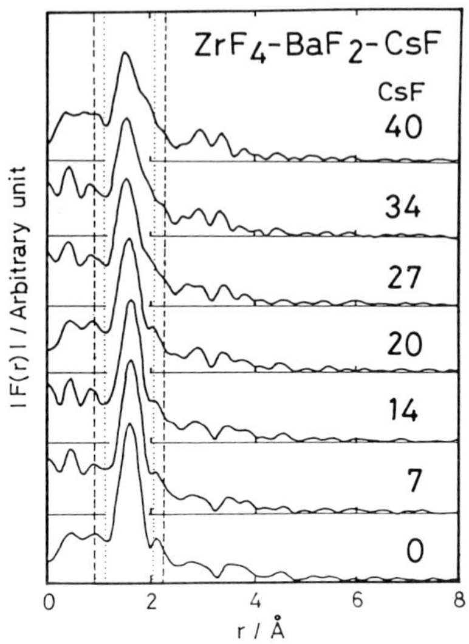

FIGURE 7
Fourier transform magnitude curves of glasses $55ZrF_4\cdot(45-x)BaF_2\cdot xCsF$

TABLE 3
Peak maxima, R_{max}, and half-widths, HW, of Zr-F peaks in Fourier transform magnitude curves of glasses $55ZrF_4\cdot(45-x)BaF_2\cdot xCsF$

Glass composition ZrF_4 (Mol%)	BaF_2	CsF	R_{max} (Å)	HW (Å)
55	45	0	1.59	0.515
55	38	7	1.58	0.495
55	31	14	1.58	0.495
55	25	20	1.57	0.540
55	18	27	1.55	0.545
55	11	34	1.53	0.630
55	5	40	1.50	0.720

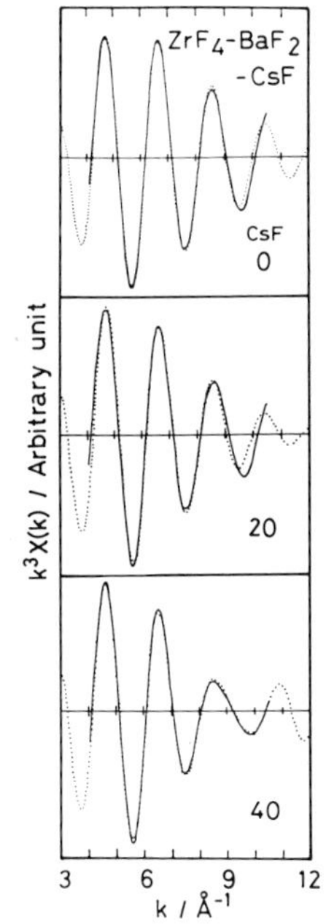

FIGURE 8
$k^3\chi(k)$ curves obtained for glasses $55ZrF_4\cdot(45-x)BaF_2\cdot xCsF$ by Fourier backfiltering (solid lines) and by curve fitting (dotted lines)

TABLE 4
Structural parameters obtained for glasses $55ZrF_4\cdot(45-x)BaF_2\cdot xCsF$

Glass composition ZrF_4 (Mol%)	BaF_2	CsF	r_{Zr-F} (Å)	CN_F	σ (Å)
55	45	0	2.07	6.5	0.104
55	38	7	2.07	6.7	0.109
55	31	14	2.06	6.7	0.114
55	25	20	2.07	7.1	0.118
55	18	27	2.06 2.27	6.4 0.6	0.108 0.002
55	11	34	2.06 2.27	6.5 0.6	0.112 0.006
55	5	40	2.05 2.23	5.1 1.9	0.096 0.063

The Fourier transform magnitude curves of the glasses are shown in figure 9. The peak maxima and half-widths of Zr-F peaks in the Fourier transform magnitude curves are given in table 5. It can be seen from the figure that small shoulders appear at the right-hand sides of Zr-F peaks of the Rb and Cs glasses. Therefore, curve fittings for these two glasses were made by using a "two shell" model and those for the Li,

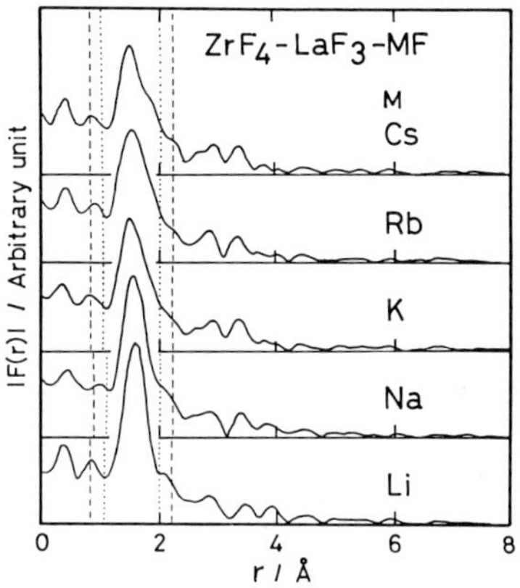

FIGURE 9
Fourier transform magnitude curves of glasses $52.5ZrF_4\cdot5.0LaF_3\cdot42.5MF$

TABLE 5
Peak maxima, R_{max}, and half-widths, HW, of Zr-F peaks in Fourier transform magnitude curves of glasses $52.5ZrF_4\cdot5.0LaF_3\cdot42.5MF$

Glass composition ZrF_4 (Mol%)	LaF_3	MF	R_{max} (Å)	HW (Å)
52.5	5.0	42.5Li	1.61	0.545
52.5	5.0	42.5Na	1.58	0.545
52.5	5.0	42.5K	1.55	0.665
52.5	5.0	42.5Rb	1.54	0.705
52.5	5.0	42.5Cs	1.52	0.725

Na and K glasses were made by using a "one shell" model. The result is given in table 6. Although no "two shell" model was applied to the K glass, there is no doubt as to the occurrence of a splitting of the interatomic distance, since the half-width of Zr-F peak is a little wider than those of the Li and Na glasses (table 5).

TABLE 6
Structural parameters obtained for glasses $52.5ZrF_4 \cdot 5.0LaF_3 \cdot 42.5MF$

Glass composition ZrF_4	LaF_3	MF (Mol%)	r_{Zr-F} (Å)	CN_F	σ (Å)
52.5	5.0	42.5Li	2.06	7.4	0.109
52.5	5.0	42.5Na	2.06	7.2	0.115
52.5	5.0	42.5K	2.07	7.5	0.128
52.5	5.0	42.5Rb	2.07 2.26	6.7 0.6	0.112 0.004
52.5	5.0	42.5Cs	2.04 2.22	5.0 2.0	0.098 0.071

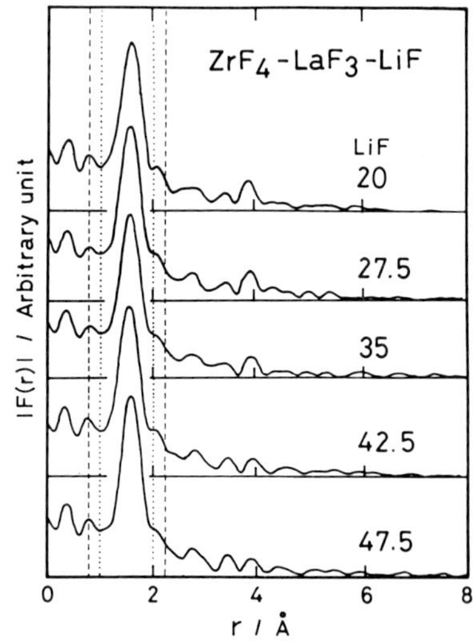

FIGURE 10
Fourier transform magnitude curves of glasses $(95.0-x)ZrF_4 \cdot 5.0LaF_3 \cdot xLiF$

Thus the followings can be concluded on glasses $52.5ZrF_4 \cdot 5.0LaF_3 \cdot 42.5MF$ (M: Li, Na, K, Rb and Cs): The F coordination numbers of Zr are around 8. In the K, Rb and Cs glasses a splitting of the Zr-F interatomic distances into two groups with about 0.2Å differences occurrs, accompanying decreases in average Zr-F interatomic distances.

3.5. Glasses $(95.0-x)ZrF_4 \cdot 5.0LaF_3 \cdot xLiF$ (x=20.0-47.5)

The Fourier transform magnitude curves and the structural parameters obtained for glasses $(95.0-x)ZrF_4 \cdot 5.0LaF_3 \cdot xLiF$ are shown in figure 10 and in table 7. As can be seen from the figure, all the $|F(r)|$ curves are entirely identical even when the composition largely changes from 20.0 to 47.5mol% in the LiF content, i.e. from 75.0 to 47.5mol% in the ZrF_4 content. Consequently the structural parameters also have the same values regardless of the composition of glass. Since the r_{Zr-F} and CN_F values obtained are 2.07Å and 7.0-7.5, respectively, these real values become 2.08Å and 7.5-8.0. It is of much

TABLE 7
Structural parameters obtained for glasses $(95.0-x)ZrF_4 \cdot 5.0LaF_3 \cdot xLiF$

Glass composition ZrF_4	LaF_3	LiF (Mol%)	r_{Zr-F} (Å)	CN_F	σ (Å)
47.5	5.0	47.5	2.07	7.3	0.108
52.5	5.0	42.5	2.07	7.0	0.107
60.0	5.0	35.0	2.07	7.0	0.109
67.5	5.0	27.5	2.07	7.3	0.109
75.0	5.0	20.0	2.07	7.5	0.110

interest that the $(95.0-x)ZrF_4 \cdot 5.0LaF_3 \cdot xLiF$ glasses have such r_{Zr-F} and CN_F values, though the r_{Zr-F} and CN_F values of the Li_2ZrF_6 $(66.7LiF \cdot 33.3ZrF_4)$ crystal are 2.016Å and 6.

Several EXAFS studies of the F coordination environments of Zr in crystalline and vitreous fluorozirconates have already been reported.[8-10] However, some of those results are largely different each other and also from the present result.

ACKNOWLEDGEMENTS

The present authors thank late Dr. H. Morikawa, Tokyo Institute of Technology, for his help in EXAFS experiments. This work was supported by a Grant-in-Aid for Scientific Research (Priority Areas, New Functionality Materials-Design, Preparation and Control) from the Ministry of Education, Science and Culture (No. 63604570).

REFERENCES

1. B.K. Teo, EXAFS: Basic Principles and Data Analysis (Springer-Verlag, Berlin, 1986).

2. G. Brunton, Acta Crystallogr. B29 (1973) 14.

3. Y. Kawamoto, T. Horisaka, K. Hirao and N. Soga, J. Chem. Phys. 83 (1985) 2398.

4. P. Laval, Acta Crystallogr. B34 (1978) 1070.

5. B. Mehlhorn and R. Hoppe, Z. Anorg. Allg. Chim. 425 (1976) 180.

6. H. Bode and G. Teufer, Z. Anorg. Chem. 283 (1956) 18.

7. R.D. Burbank and F.N. Benzey, USA-EC Report K1280 (1956) 14.

8. B. Boulard, A. le Bail, J.P. Laval and C. Jacoboni, J. Physique 43 (1986) C8-791.

9. Z. Shen, F. Ma, L. Ye, K. Lu and Y. Zhao, Extended Abstract 4th Inter. Symp. on Halide Glasses (1987) pp.252-259.

10. F. Ma, Z. Shen, L. Ye, M. Zhang, K. Lu and Y. Zhao, Extended Abstract 6th Inter. Symp. on Halide Glasses (1989) pp.129-143.

Frontiers of Materials Research/
Electronic and Optical Materials
M. Kong, L. Huang (editors)

PREPARATION OF FLUOROZIRCONATE GLASS FIBER

HU Hefang, YI Guanhong, LIN Fengying, YU Yaochu and XU Yongjun

Shanghai Institute of Optics and Fine Mechanics, Academia Sinica, P.O. Box 800-216, Shanghai 201800, PRC

This paper presents the preparing techniques of teflon FEP-cladding fluorozirconate glass fiber. The effect of technological parameters on the scattering loss of the fiber is studied. It is shown that the glass melting temperature, the refining time, and the water content in the environment during the glass-melting and the fiber drawing process are very important parameters in order to reduce the scattering losses. Moreover, the NH_4HF_2 content in the batch, the temperature distribution in the fiber-drawing furnace, and fiber drawing technology also have influence on the scattering loss of the fiber. Based on this study, the FEP-cladding fluorozirconate glass fiber with a minimum attenuation of 0.24 db/m at 2.32 µm is obtained.

1. INTRODUCTION

Heavy metal fluoride glass fiber, such as fluorozirconate glass fiber has been attracted considerable attention as a new generation of optical fiber for telecommunications and as a new fiber laser material. Its minimum intrinsic loss has been estimated to be 10^{-2} to 10^{-3} db/km in the 2.5 to 3.5 µm region. Recently, the attenuation of long fluoride glass fiber has been achieved to 2.6 db/km at 2.55 µm[1]. Fiber doped with Nd^{3+}, Er^{3+}, Ho^{3+} or Tm^{3+} has been made to lase at a variety of wavelength between 0.85 to 2.7 µm. Semiconductor diode pumping close to 795 nm has now been used to drive high efficiency Nd laser at 1.05 and 1.35 µm and its quantum efficiency close to the maximum of 1[2]. Reported attenuation of fluoride glass fiber usually range from 1 to several tens db/km around 2.5 µm exceeding the estimated intrinsic loss by at least two orders of magnitude. These high losses are mainly caused by the absorption of OH group, 3d and 4f elements, and by the scattering of the micro-crystals and bubbles in the fibers. The fluorozirconate melts are characterized by high tendency to crystallize and to react with moisture to form oxides or oxyfluorides. The preform for fiber drawing is usually prepared using a melting -casting method. Therefore, the preparing technology has large influence on the attenuation of the fiber.

In this paper the preparation of Teflon FEP-cladding fluorozirconate glass fiber is presented. The effect of technological parameters, such as melting technology of glass, the water content in the environment during the glass melting and fiber drawing process, fiber drawing technology, as well as the type of raw material on the attenuation of fiber is studied. Based on this study, the FEP-cladding fluorozirconate glass fiber with a minimum loss of 0.24 db/m at 2.32 µm is obtained.

2. EXPERIMENTAL

A fluorozirconate glass with chemical composition (mol%) of $53ZrF_4$-$20BaF_2$-$4LaF_3$-$3AlF_3$-20NaF (ZBLAN) was chosen for study. Chemically pure $(NH_4)_2ZrF_6$, BaF_2, LaF_3, NH_4AlF_4, NaF and NH_4HF_2 were used as starting materials for the study of effect of melting technology on the scattering intensity of glass. The batches were put in a platinum crucible, heated to 850-900 °C and kept at that temperature for some time as required. The melts were cast in a preheated aluminum mold. The annealed glasses were cut and polished to bulk specimens with size of 10x10x20 mm and four polished surfaces. The scattering intensity measurement was carried out at wavelength of 4538 Å by comparison method, and optically pure benzene was used as a standard specimen.

Anhydrous, high pure fluorides with Fe content about 1-2 ppm were used as starting materials to prepare the preforms for drawing fiber. The whole preparing process was carried out in a glove box filled with dry air to avoid hydrolysis of the fluorides. The size of preform is 10 mm in diameter and 200 mm in length. Commercial teflon FEP tube was used as cladding material. The FEP-cladding fluorozirconate glass fiber was drawn under dry nitrogen atmosphere after the preform was chemically etched. The attenuation spectra of fiber in the 1-3 μm wavelength region were measured by cutback method.

3. RESULTS AND DISCUSSIONS

3.1. The effect of melting temperature and refining time

Figures 1 and 2 show the effect of the melting temperature and refining time on the scattering intensity of glass, respectively. It is shown that a suitable melting temperature and refining time are very important technological parameter to reduce the scattering losses. It is well known that the glass melting is a complex physico-chemical process including chemical reaction between components, melting, diffusion, refining and so on. For fluorozirconate glass, this process also includes the volatilization of some components , such as ZrF_4, and the reaction between fluorides and moisture in the atmosphere to form oxides or oxyfluorides. The fluorozirconate glass possesses high tendency to devitrify, a little change in the composition and the existence of oxides or oxyfluorides have obvious effect on its glass-forming ability. Therefore, low melting temperature and short refining time can not ensure the complete dissolve of the compounds with high melting point and the bubble escape from melt. But overhigh melting temperature and overlong refining time increase the volatilization of ZrF_4 and intensify the reaction between fluorides and moisture in the atmosphere, and result in an increment of scattering center.

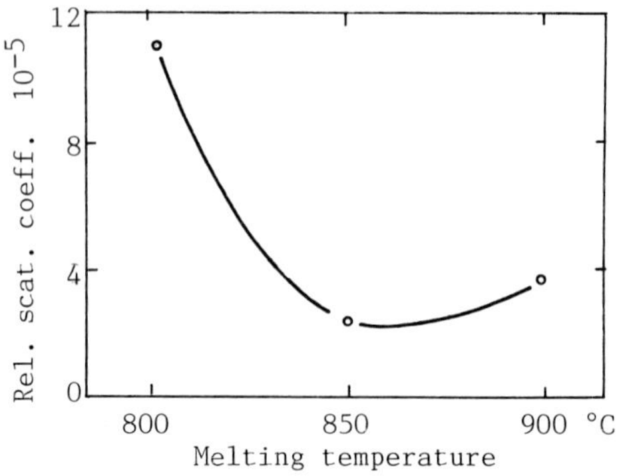

FIGURE 1
Scattering intensity of fluorozirconate glass as a function of glass melting temperature. Refining time: 1.5 hr.

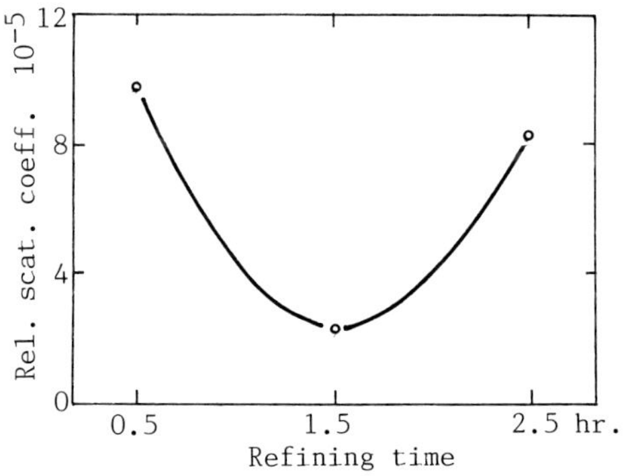

FIGURE 2
Scattering intensity of fluorozirconate glass as a function of glass refining time. Melting temperature: 850 °C.

3.2. The effect of NH_4HF_2 addition

The fluoride starting materials always contain small amount of oxide impurity due to incomplete fluorination and surface hydrolysis. The existence of oxide impurity in the fluorozirconate glass induces part of ZrF_4 to convert to ZrO_2 or ZrO_xF_y, which plays a role of nuclei to promote the crystallization of glass and

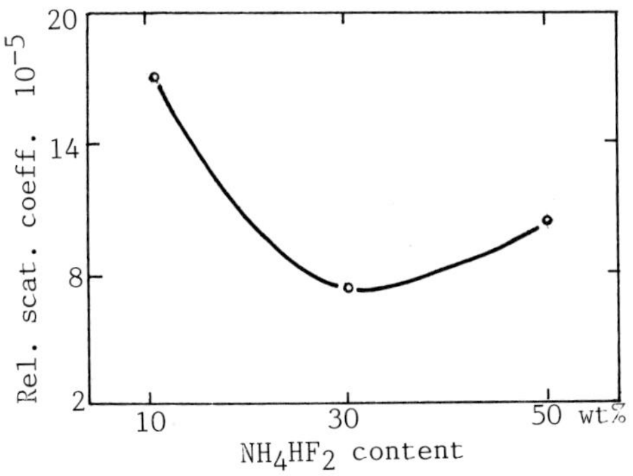

FIGURE 3
Scattering intensity of fluorozirconate glass as a function of NH_4HF_2 content in batch.

causes an additional absorption shoulder at 1350 cm^{-1}[3]. Therefore , it is necessary to introduce certain amount of NH_4HF_2 as a fluorinating agent for the oxide impurity in starting materials. In the case of ZrO_2, fluorination occurs through the following reactions:

$$2ZrO_2(s) + 7NH_4HF_2(s) = 2(NH_4)_3ZrF_7(s) + NH_3(g) + 4H_2O(g) \quad (1)$$

$$(NH_4)_3ZrF_7(s) = ZrF_4(s) + 3NH_4F(g) \quad (2)$$

In addition, the exceeding NH_4HF_2 decomposes to NH_4F and HF at high temperature, which protect the fluoride melts from the reaction with the moisture. Figure 3 shows the scattering intensity of glass as a function of the NH_4HF_2 content in the batch. It is clear that the certain amount of NH_4HF_2 addition is advantageous to reduce the extrinsic scattering loss of glass.

3.3. The effect of water content in environment

Figure 4 shows the attenuation spectra of fluoroziconate glass fiber as a function of water content in the environment during the glass melting process. It is shown as the water content in the environment decreases from 700 ppm to 40 ppm, the minimum attenuation of fiber decreases simultaneously from 1.5 db/m to less than 0.3 db/m, and the intensity of absorption bands at 2.42 and 2.24 μm due to OH^- group becomes small. Figure 5 shows the scattering of the fluorozirconate glass melted in the environment with different water content under He-Ne laser beam. The amount of the scattering center, especially the amount of micro-crystal in the glass decreases obviously with the decrease of water content. The results indicate that the existence of moisture during the glass melting process causes an increment in the intensity of OH^- absorption bands and accelerates the hydrolysis of fluorides to form oxide or oxyfluoride micro-crystals. These increase distinctly the optical loss of the fiber. It is also shown in Fig. 4 that the wavelength dependence of fiber attenuation is varied as the water content in the environment during the glass melting process. When the glass is melted under the atmosphere with high humidity, the attenuation of fiber appears wavelength independent. On the other hand, the attenuation of the glass fiber melted under the dry atmosphere is inversely proportional to the square of wavelength. This illustrates that the mechanism of scattering loss changes from the large particle scattering to Mie's scattering.

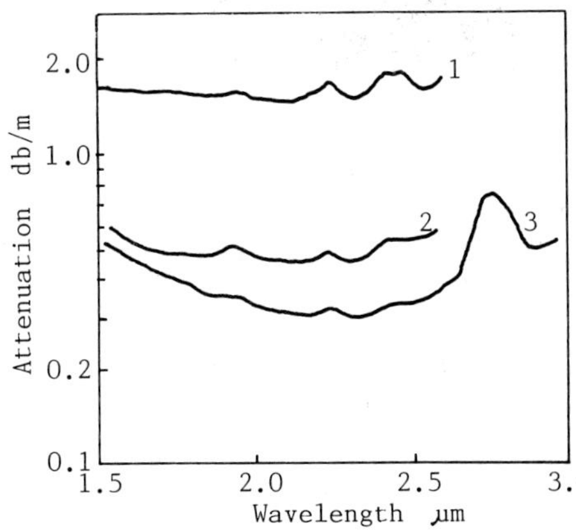

FIGURE 4
The attenuation of fluorozirconate glass fiber as a function of water content in the environment during the glass melting process.
Water content: 1. 700 ppm; 2. 100 ppm; 3. 40ppm.

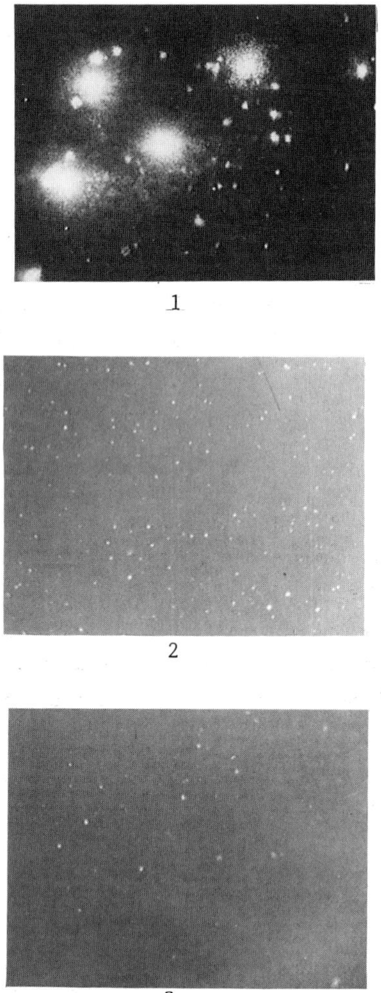

FIGURE 5
The scattering of fluorozirconate glasses melted in the environment with different water content under He-Ne laser beam.
Water content: 1. 700 ppm; 2. 100 ppm; 3. 40 ppm.

In Fig. 6 the attenuation spectra of two fibers drawn under the atmosphere with and without dry nitrogen protection are compared. It is shown that the reaction of the glass surface to the moisture also occurs at the fiber drawing temperature, which causes additional extrinsic loss.

3.4. The effect of temperature distribution in fiber drawing furnace

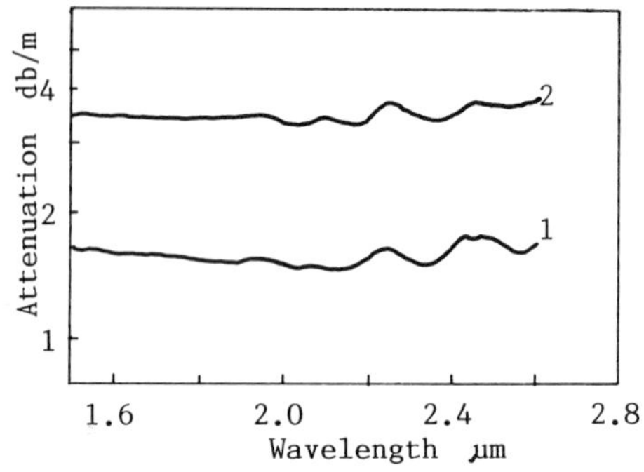

FIGURE 6
The attenuation of fluorozirconate glass fiber drawn with (1) and without (2) nitrogen protection.

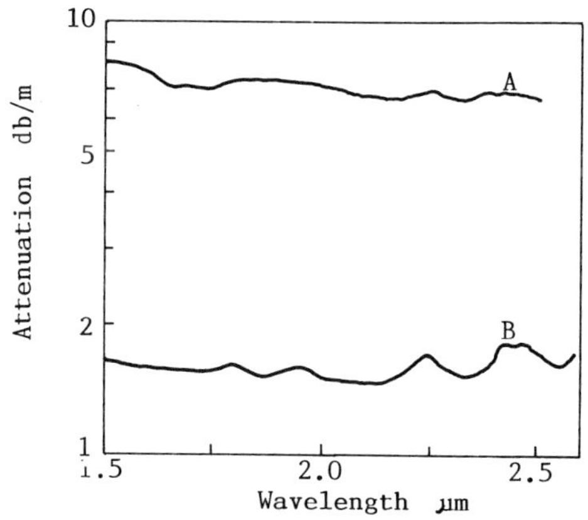

FIRGUE 7
The attenuation of fluorozirconate glass fiber drawn in the furnace with different temperature distribution.

Figure 7 shows the attenuation spectra of fluorozirconate fibers drawn from same preform but in the furnace with different temperature distribution. Fibers A and B were drawn in the furnace with heat zone length about 4 and 1 cm, respectively. The results indicate the furnace with smooth temperature distribution can not used to draw fluoride glass fiber. The smooth temperature gradient offers enough time for crystal nuclei in the glass preform to grow up, thus resulting in an increment of extrinsic loss of the fiber.

3.5. The multiple effect of starting material and preparing technology

The attenuation spectra of fluorozirconate glass fibers, made from different starting materials and by various technology are shown in Figure 8. The fiber 1 was made from commercial, high-pure fluoride and by previous technology. The others were prepared from refined fluorides and improved technology. Among them, sublimated ZrF_4 is used to prepare fibers 2 and 3, while $ZrF_4 \cdot H_2O$ was used to fiber 4. It is pointed out that the FEP-cladding fluorozirconate glass fibers with attenuation about 0.3 db/m can be made from above refined fluorides and by improved technology. The minimum optical loss of the fiber has been reached to 0.24 db/m at 2.32 µm. But the intensity of absorption bands at 2.24, 2.42 and 2.8 µm in fiber 4 is much stronger than those in the others, and its minimum attenuation wavelength also shifts to 2.05 µm. This indicates that more strict control of the processing is required in the preparation of OH-free fluorozirconate glass fibers when $ZrF_4 \cdot H_2O$ is used as raw material.

As indicated in the figures, there are many strong absorption bands in the wavelength region of 2-3 µm. The absorption bands due to Nd^{3+} and Pr^{3+} are found at the wavelength near 2.5 and 2.0 µm, respectively, besides the OH^- absorption at 2.24, 2.42 and 2.8 µm. The existence of these absorption bands cause the minimum attenuation wavelength of most of current fibers shifts to short wavelength about 2.32 µm. The estimated content of Nd^{3+} and Pr^{3+} is about 1-2 ppm. The glass also contains iron of 2-3 ppm and other 3d transition elements. The extrinsic scattering losses are the wavelength-independent scattering and Mie's scattering resulting from the microcrystals, bubbles, as well as the roughness of the interface between core and cladding of the fiber. It should be pointed out that the extrinsic scattering is the major source of the optical loss of current fibers. Therefore, the preparation techniques has to be further improved to obtain low-loss fluorozirconate glass fibers.

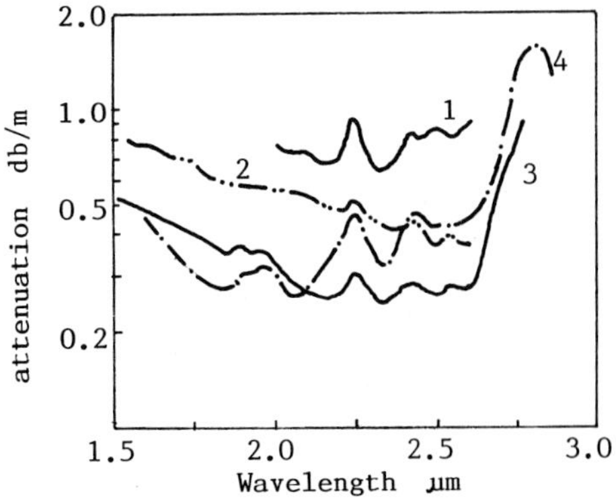

FIGURE 8
The attenuation spectra of fluorozirconate glass fibers prepared by different techniques.

4. CONCLUSIONS

1. In order to reduce the scattering intensity of fluorozirconate glass, it is necessary to select a suitable glass-melting temperature and a optimum refining time. The addition of certain amount of NH_4HF_2 in the batch is also helpful for decreasing the scattering losses.

2. The technological parameters , such as the water content in the environment during the glass melting and fiber drawing process, the distribution of the temperature in fiber drawing furnace, as well as the type of starting material have large enfluence on the attenuation of the fluoride glass fiber.

3. The teflon FEP-cladding fluorozirconate glass fiber with minimum attenuation of 0.24 db/m at the wavelength of 2.32 µm is obtained. The preparation techniques has to be further improved to make low-loss fluorozirconate glass fibers.

ACKNOWLEDGEMENTS

The authors are grateful to the National

Natural Science Foundation of China for the support of this work.

REFERENCES

1. J.R. Williams, S.F.Carter et al., Proceedings of 6th international Sympsium on Halide glasses, Clausthal-Zellerfeld, FRG (1989) pp 521-525.

2. P.W.France, ibid. pp 51-56.

3. HU Hefang and J.D.Mackenzie, J. Non-Cryst. Solids 80 (1986) 495.

Frontiers of Materials Research/
Electronic and Optical Materials
M. Kong, L. Huang (editors)

STUDY ON β-BaB_2O_4 PYROELECTRIC GLASS-CERAMICS

Zhenya DING, Yibing HUANG, Wei WANG and Jianping ZHANG

Beijing Graduate School of Wuhan University of Technology, Beijing 100024, China

Pyroelectric glass-ceramics of crystal structure β-BaB_2O_4, have first been made satisfactorily in this work by crystallizing the glass in a temperature gradient. The microstructure of the materials prepared were investigated by IR, Raman spectra, XRD and SEM. The pyroelectric properties of BBO glass-ceramics were measured and compared with those of BBO single crystal and PZT ceramic. This kind of material has potential applications in the field of pyroelectric detectors.

1. INTRODUCTION

Pyroelectric glass-ceramics or so-called polar glass-ceramics are non-ferroelectric materials with useful piezoelectric and pyroelectric properties due to their oriented polar crystallites. Extensive studies have been carried out by Gardoppe[1] and Halliyal[2] on the preparation, properties and application of polar glass-ceramics in the systems of Li_2O-SiO_2, B_2O_3-GeO_2-TiO_2, BaO-TiO_2-SiO_2, SrO-TiO_2-SiO_2, etc..

Barium metaborate crystal in low temperature forms (β-BaB_2O_4, BBO), discovered and grown succesfully by Chen Chuangtien et al[3], is a new type of optical crystal which offers many attractive properties in addition to its superior non-linear optical property such as high pole stability, good thermal conductivity and thermal shock resistance, unique infrared absroption and bearable to high power laser radiation. It is expected that this crystal would be useful for application in pyroelectric devices, especially for high power infrared detectors.

As known to all, in preparing pyroelectric materials, the growth of high quality, large size crystal is very complicated and expensive, the ferroelectric ceramics need poling and then aging problem is always expected. Compared with single crystal and ferroelectric ceramics, the preparation of polar glass-ceramics is simple and no external poling step needed. As a result, a large size, inexpensive sample can be made easily and aging or depoling can be avoided.

In this study, considering the good properties of BBO crystal and the advantages of polar glass-ceramics, the crystallization technique of temperature gradient is investigated and pyroelectric glass-ceramics of barium metaborate crystalline phase are prepared for the first time by this technique. The experiment results indicate that the BBO glass-ceramics have good pyroelectric properties. So far there is no else report on BBO glass-ceramics and their pyroelectric properties.

2. EXPERIMENTS

The main crystal phase in a polar

glass-ceramics should be polar crystal, so polar crystal β-BaB_2O_4 had been chosen as a main phase of polar glass-ceramics in veiw of its excellent properties. The BBO glasses were prepared with CP grade chemicals in required composition by melting in a platinum crucible in a glober furnace. The melt was maintained at 1280°C for 8-12 hrs. Transparent cylindrical samples were obtained by pouring the melt into a graphite mould and were then annealled at 490°C. The samples in the form of thick diskes were polished and placed on the hot stage for crystallization. Surface grain-oriented glass-ceramics were obtained by high temperature gradient technique with proper heat treatment cycle. The structures of both basic glasses and glass-ceramics were studied by IR and Raman spectra, the main crystal phase and preferred orientation were determined by X-ray diffraction and the pyroelectric coefficient P_3 was measured by modified Glass method[4].

3. RESULTS AND DISCUSSION

3.1 Selection of the main crystal phase

The main crystal phase of glass-ceramics can greatly influence the basic properties of glass-ceramics. To obtain polar glass-ceramics with good pyroelectric properties, the suitable main crystal phase must be selected. As only if a crystal has one axis, can it exhibits pyroelectric effect, so the main crystal phase must be a polar crystal with acentric symmetry.

β-BaB_2O_4 crystalline phase was selected as the main crystal phase of glass-ceramics in this work in view of its excellent pyroelectric properties. This crystal belongs to trigonal system with space group C_3^4 - R_3. There are 6 formul units in one hexagonal cell. The unit cell dimensions referring to hexagonal axes are a=b=12.532 Å, c=12.714 Å. This structure can be considered as a layer step type of built up by Ba^{2+} and $(B_3O_6)^{3-}$ rings alternatively. The anion $(B_3O_6)^{3-}$ group is nearly planar with its plane perpendicular to the threeford axis which passes through the center of the ring. Each barium atom has seven oxygen neighbors.

It is difficulty to prepare homogeneous glass from the stoichiometric composition $2BaOB_2O_3$[5]. According to the principle of glass forming, the composition were modified by adding a proper amount of glass network former SiO_2 to promote the forming of homogeneous glass and the problems like phase seperation and crystallization were avoided.

3.2 Structural characteristic of the BBO basic glasses and glass-ceramics

3.2.1 IR spectra

The IR spectra in this paper were obtained using a Perkin-Elmer 983 spectrometer in the range from 200-4000cm^{-1}. The spectra of the basic glass and polar glass-ceramics are shown in Fig.1. Curve 1 represents the basic glass. The band at 1397 cm^{-1} is the characteristic peak of $(B_3O_6)^{3-}$ hexacycle due to its antisymmetric stretching vibration. The intensity of this band indicates that the $(B_3O_6)^{3-}$ hexacycle is the main structural units in the basic glass. Curve 2 represents the BBO glass-ceramics which is similar to that of basic glass. Curve 2 represents the

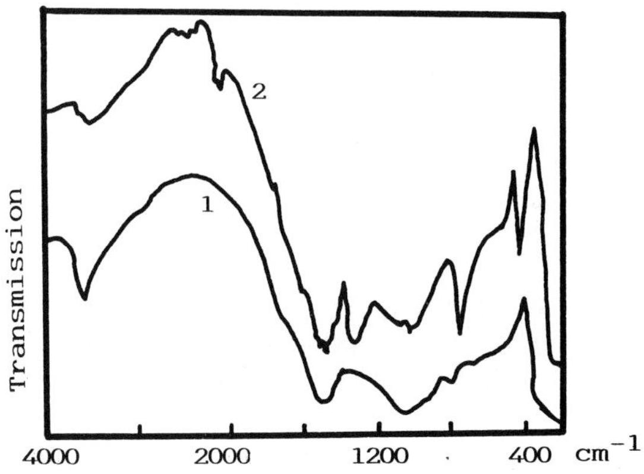

FIG. 1
Infrared spectra of BBO glass and glass-ceramics
(1) BBO glass; (2) BBO glass-ceramics

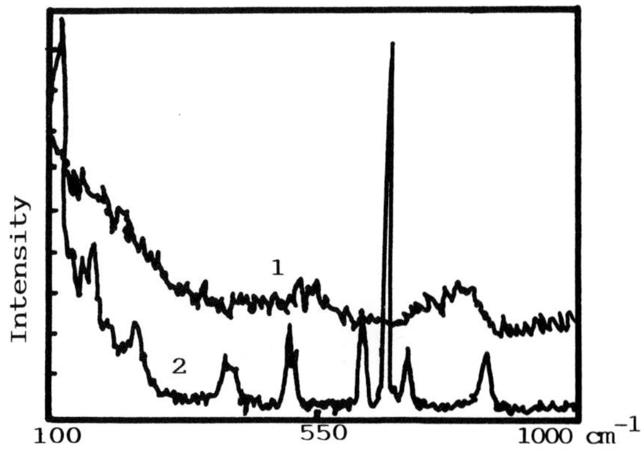

FIG. 2
Raman spectra of BBO glass and glass-ceramics
(1) BBO glass; (2) BBO glass-ceramics

BBO glass-ceramics which is similar to that of basic glass but the characteristic peak of $(B_3O_6)^{3-}$ units is much more sharp. Many less strength sharp peaks appear in the spectrum due to the crystal grains in the glass-ceramics.

3.2.2 Raman spectra

The Raman spectra were obtained using a Spex-Raman Log-IV Laser Raman spectrometer in the range from 100 cm^{-1} to 1000 cm^{-1}. The spectra of the basic glass and the BBO glass-ceramics are showed in Fig.2. Curve 1 represents the basic glass. The band at 738 cm^{-1} is the combined vibration peak of $(B_2O_5)^{4-}$ units and $(B_3O_6)^{3-}$ units in conjunction with two (BO_4) units and the band at 504 cm^{-1} is the characteristic peak of $(B_3O_6)^{3-}$ units with which two boron atoms are linked with one non-bridge oxygen[6]. Compared with Curve 1 the peaks are much sharper in the Raman spectra of the glass-ceramics. (Curve 2). And the Curve 2 is similar to the spectra of BBO single crystal.

3.3 Micro-morphology

The sample were etched by HF on the fresh surfaces and micro-structural photographs were obtained by scanning electronic microscope.

The vertical cross sectional view and the cross-sectional view of glass-ceramics are shown in the Fig.3. The microstructure consists of needle-like crystallites developed from the initial crystallizing surface into the bulk parallel to the temperature gradient, but the grain-orientation of crystallites is dispersed.

3.4 Determination of the main crystal -phase and spatial distribution of grains by XRD

A small section of basic glass was maintained at temperature Tc for 6 hrs. for full-crystallization, and the powder X-ray diffraction pattern was obtained with the full crystallized section Fig.4. Being compared with JCPDS standard diffraction card, the β-BaB_2O_4 was determined the main crystal phase in the full-crystallized section.

(1) Vertical cross-section view (2) Cross-sectional view

FIG. 3
Microstructure of BBO glass-ceramics.

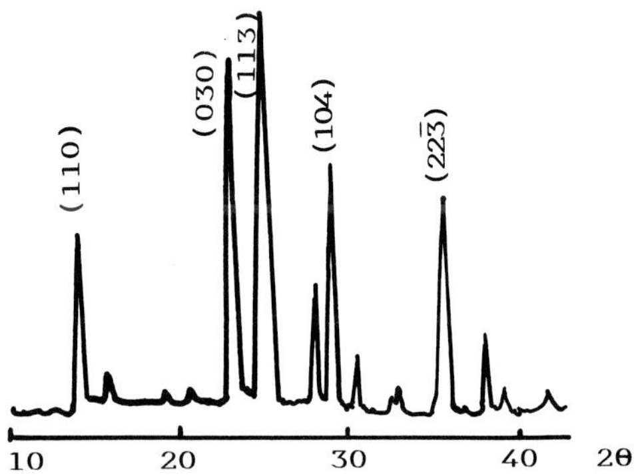

FIG. 4
X-ray powder diffraction of unoriented BBO glass-ceramics

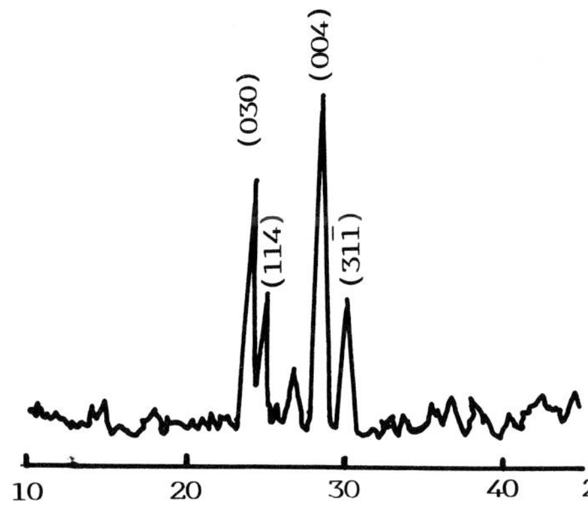

FIG. 5
X-ray diffraction of oriented BBO glass-ceramics

The X-ray diffraction of BBO glass-ceramics is shown in Fig.5. The highest diffraction peak of (004) plane indicated that the (004) plane was fully developed and had the most preferred orientation.

3.5 Pyroelectric properties of BBO glass-ceramics

The pyroelectric properties of glass-ceramics are affected by several factors, such as the degree of preferred orientation of crystallites, the accuracy of the alignment relative to the preferred direction. These factors depend on the orientation crystallization. Pyroelectric properties can be prepared by employing the heat treatment schedule established in this paper. The pyroelectric properties of glass-ceramics are listed in Table 1. From Table 1 it is clear that the pyroelectric constant P_3 of BBO glass-ceramics is lower than that of PZT ceramics, but because of the low dielectric constant K, its figure of merit FM_V is comparative to that of $LiNbO_3$ single crystal and superior to those of PZT ceramics and BBO crystal[7]. This characteristic is of great importantce for applications in high frequency, large area infrared

TABLE 1
Comparison of pyroelectric properties

Property	BBO glass-ceramics	BBO single crystal	PZT ceramic
K	9	10	1400
P_3 (C/cm^3 °C)	1.9×10^{-9}	1.0×10^{-9}	37×10^{-9}
d (g/cm^3)	3.80	3.85	7.50
tgδ(1 KHZ)	0.0075	0.0030	0.0075
C_p (J/cm^3 °C)	1.87	1.91	3.00
$P_3/K_o \cdot K \cdot C_p$ (C·cm/J) (FM_V)	1.44	0.59	0.09

detectors.

Since the glass-ceramics are non-ferroelectric, there is no-poling step involved. Hence, the glass-ceramics can be used at high temperature. The simplicity of preparing method makes it possible to produce large-area element. Moreover, the properties of glass-ceramics can be modified by adjusting compositions of glasses. Therefore, extensive application of this material in pyroelectric detectors is expected.

4. CONCLUSION

(1) non-ferroelectric polar BBO glass-ceramics with good pyroelectric properties can be prepared by high temperature gradient crystallization technique.

(2) The spatial distribution of grains, development of crystallites and pyroelectric properties are dependant upon the technique of orientation crystallization.

(3) BBO glass-ceramics enjoy many advantages such as simple in preparation,low dielectric constant, high applicable temperature, no aging, and seem to be attractive candidate materials for pyroelectric IR detectors.

REFERENCES

1. R.J. Cardopee, Ferroelectric 33 (1981) 155.
2. A. Halliyal, A.S. Bhalla, et al. J. Mater. Sci. 16(1981) 1023.
3. Chen Chuangtian, China Science (B) 7(1984), 598.
4. A. M. Glass, J. Appl. Phys. Vol.40 (1969) 4699.
5. Gan Fuxi, Optic Glasses, Vol.1, 1985, China Science Press.
6. Jin Yifeng, J. Chinese Ceram. Soc. 13(2) 1985, 144.
7. Shi Zikang, ACTA. Physica. SINICA, 34(1) 1985, 140.

Frontiers of Materials Research/
Electronic and Optical Materials
M. Kong, L. Huang (editors)

STUDY ON ELIMINATING IMPURITY ABSORPTION PEAK IN CHALCOGENIDE GLASS FIBERS

Yang Kewu, Li Qiugeng, Wu Peilan, Wei Guosheng

Beijing Glass Research Institute, 1 Dong Da Di, Chong Wen Men Wai, Beijing, 100062, China

This paper described the transmission loss mechanisms in chalcogenide glass fibers. Different absorption peaks for H-M AND O-M bond (M is one of the composed elements in chalcogenide glass) are systematically analyzed. In this paper, a method is presented, in which these impurity absorption peaks are eliminated. An $As_{40}Se_{40}Te_{20}$ glass fibers were prepared, its loss was lower than 5dB/m, at 10.6μm.

1. INTRODUCTION

Chalcogenide glass generally consists of 6th main group elements S, Se, Te in the periodic table and other elements Ge, As, Sb etc. Chalcogenide glass is a very good material for infrared transmission. The cutoff wavelength of glass is longer than that of oxide glass and fluoride glass, for example, As-Se-Te glass transmiting range is in 1~18μm. Chalcogenide glass can be drawn into fiber easily. Chalcogenide glass fibers were used for a wide range of purposes including laser range finding, thermal imaging, sensing and tracking. Most importantly it includes the output waveband of the tunable CO_2 laser. In these applications, how to reduce the loss of the chalcogenide glass fibers is a key problem. It is clear that high-quality glass is prerequisite for producing low loss fibers. Transmission loss in chalcogenide glass fibers is found to be due to the intrinsic absorption and extrinsic absorption caused by fiber imperfections and impurities, which is the main factor to cause the fiber loss.

2. LOSS MECHANISMS IN CHALCOGENIDE GLASS OPTICAL FIBERS

The mechanisms by which losses occur are common to all optical materials and can be classified as either intrinsic or extrinsic. The spectral dependence of intrinsic loss is usually represented as a 'V-curve' as shown schematically in Fig. 1. Included in the figure are the contributions due to the main intrinsic processes of electronic absorption, multiphonon absorption and Rayleigh scattering. Electronic absorption determines the short wavelength limit of transparency, which is caused by excitation of bound electrons into the conduction band. Multiphonon absorption defines the long wavelength limit on transmisson, which is due to coupling of radiation to infrared-active vibrational modes. Rayleigh scattering is caused by intrinsic variations in refractive index. Rayleigh scattering was found to change linearly against a wavelength of the -4 power, its loss decreases as the transmission wavelength increases. The loss in chalcogenide glasses caused by Rayleigh scattering is of the order 10^{-2}dB/km.

A weak absorption tail is observed from the relationship between photon energy and transmission loss. Some scholars think that the tail was caused by electronic transitions associated with intrinsic defects such as unsaturated band, and is, therefore, intrinsic. There is, however, much evidence to suggest that the tail is caused by transitions involving impurity related band-gap states and is, therefore, extrinsic. At present, the argument about this problem is still being carried on. Weak absorption tail is observed in chalcogenide glass fibers and not in oxide glass optical fibers.

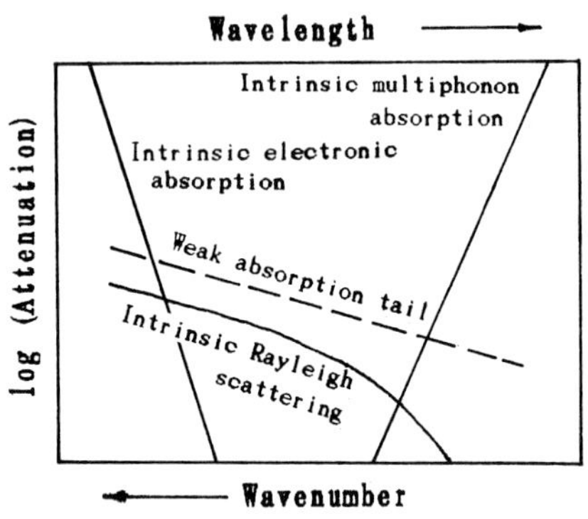

FIGURE 1
Spectral dependence of intrinsic attenuation processes

This indicates that formation energy for a defect in chalcogenide glasses is fairly small, as compared with that for oxide glasses. It is also the primary cause that leads chalcogenide glass structural imperfections [10]. This part of absorption is about 10~100dB/km [1].

The impurities that have the strongest effect upon transmission in chalcogenide glass are oxides, hydrides, and hydroxyls, which have their fundamental vibrational frequencies in the mid-infrared waveband. Even after purification these species are often present at the ppm level and produce peak absorption in the 1~10dB/m range [1].

During chalcogenide glass was melted into shape and was drawn into fiber, extrinsic scattering is caused by impurities and imperfections in the glass, examples of which include crystals, bubbles, inhomogeneities and particulate inclusions. Total extrinsic scattering in the mid-infrared waveband is usually in the range 0.01~50dB/m [1] depending on the quality of the glass. So extrinsic loss is main source of chalcogenide glass fibers loss.

3. IR ABSORPTION CURVES OF CHALCOGENIDE GLASSES

The polycrystalline Ge and five-nine pure As, Sb, Se and Te were put into the silica ampul and sealed in vacuum. After melting at 800~1000℃ in a rocking furnace for 24~36 hours, the melt was cooled to room temperature. The IR absorption curves of the prepared $Ge_{28}Sb_{12}Se_{60}$, $Ge_{30}As_{15}Se_{55}$ and $As_{40}Se_{40}Te_{20}$ glasses are shown in Fig. 2. Fig. 2 shows that the absorption spectrums of (A) and (B) are similar. In order to extend the transmission range of the glasses, a sample (C) is prepared with addition of Te and without Ge. The absorption spectrum of this glass has a slight difference. It is shown in Fig. 2 that the peaks at 2.8, 4.5, 5 and 8μm in such absorption spectrums of chalcogenide glasses can be attributed to bonds vibratiotn of H-O, H-Se, Ge-H, Ge-O respectively [1, 2, 4, 7], and their are also a stronger peak of Ge-O at 12.8μm, and some peaks of Se-O at 11.8μm and 13.5μm, As-O at 12.8, Te-O at 14.5μm. If oxide purity in glass is up to 10^{-4}, these peaks will be extended widely and it could make the glasses untransparent. If the loss 5dB/m in 10.6μm of glasses has achieved, the contents of oxide impurities should be in 10-20ppm range [5]. The attenuation is equal to 1dB/m, the concentration of oxide should be about 1ppm [3]. In addition, the absorption of the energy at IR waveband was also generated by transition metal and carbon impurities.

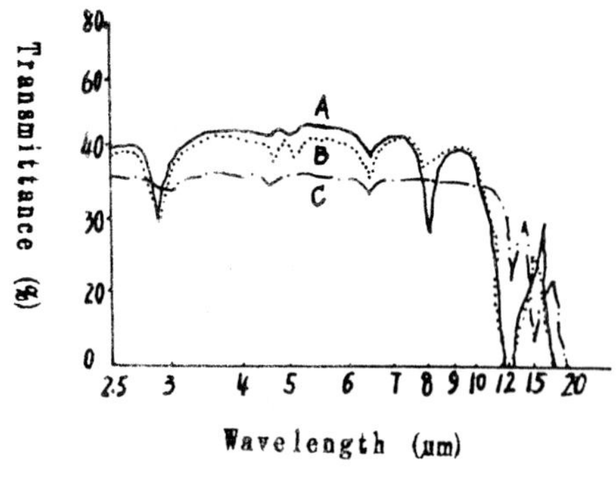

FIGURE 2
Infrared absorption curves for $Ge_{30}As_{15}Se_{55}$ (A), $Ge_{28}Sb_{12}Se_{60}$ (B) and $As_{40}Se_{40}Te_{20}$ (C) glasses

4. STUDY ON ELIMINATING IMPURITY ABSORPTION PEAK IN CHALCOGENIDE GLASS

The important application of chalcogenide glass fibers is to transmit CO_2 laser power. It is a key problem how to remove impure oxide from chalcogenide glass.

Purity raw material by physical or chemical method is the first step to reducing the absorption peaks of M-O bonds (M is one of elements in chalcogenide glass). But in order to eliminating these absorption peaks completely, it is necessary to purify the glass further. It is evidenced by the effect of $Ge_{30}As_{15}Se_{55}$ glass prepared conditions on IR absorption curve (shown in Fig. 3). Although five-nine pure raw material was melted into glass, oxide impurity led into raw material and technology is still considerable.

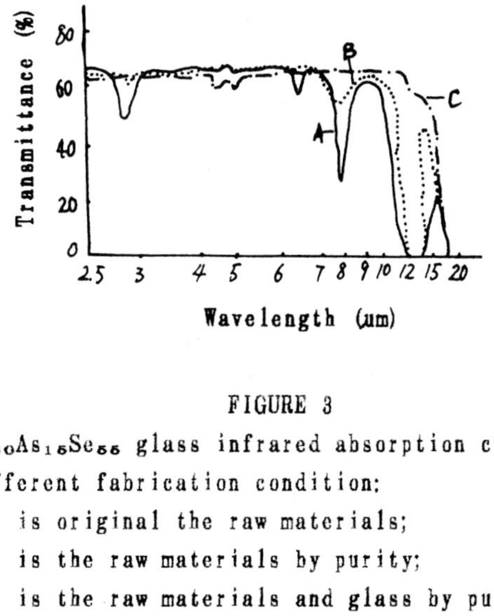

FIGURE 3

$Ge_{30}As_{15}Se_{55}$ glass infrared absorption curves in different fabrication condition:

(A) is original the raw materials;

(B) is the raw materials by purity;

(C) is the raw materials and glass by purity

S can separated from impurities by distillation in vacuum or inert atmosphere, and also the SeO_2, As_2O_3 can be easier to eliminate by this method as their vapour pressure is much larger than that of Se and As. TeO_2 and GeO_2 can be dissolved in some inorganic acid, but Te, Ge can't, so it can remove the oxide from Te and Ge surface by acid-treating.

Putting Al or Zr, the stronger electro redutive metal, into chalcogenide glass as reductive agent, and distillating, the oxide absorption peaks can be eliminated[4, 6, 7]. The result from use this method to $Ge_{30}As_{15}Se_{55}$ and $As_{40}Se_{40}Te_{20}$ glasses is shown in Fig. 4 and Fig. 5. It can be seen in Figs that Zr content about 0.5% is prefered.

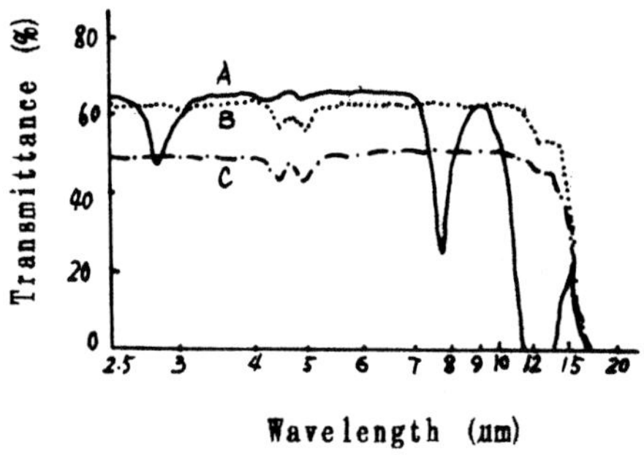

FIGURE 4

$Ge_{30}As_{15}Se_{55}$ glass infrared absorption curves with stiling added Zr:

(A) adding Zr 0.1%;

(B) adding Zr 0.5%;

(C) adding Zr 1%.

It is known that the absorption peaks in the 2.5~7μm waveband in Fig. 2 is caused by H-M bond formed in glass melting. When Zr or Al was added into glasses and distillate, the absorption peaks related oxygen bonds with H and other impurities can be eliminated, but hydride related absorption peaks can't be removed. It is evidenced that the absorption at 5μm is attributed from H_4Ge other than H_2O. Thus elucidated that oxide absorption can be eliminated at the cost of increasing hydride absorption in the 4~6μm waveband.

Chalcogenide glass can also be used to transmit CO laser power (at about 5μm waveband). In this application, the hydride impurities in glasses must be considered. Fig. 6 is taken from U. Saito et al., they considered that hydrogen impurities was produced by adherent to

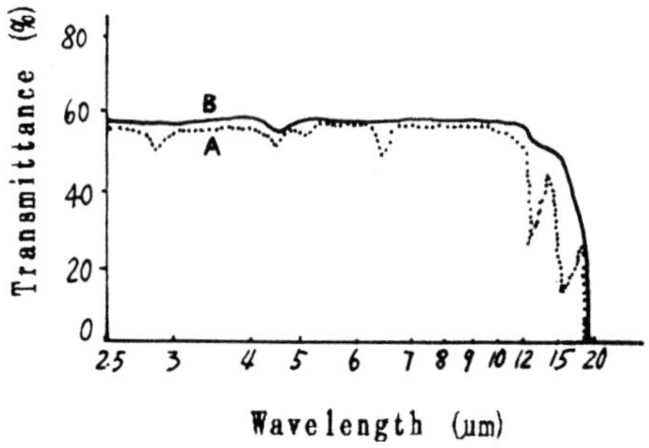

FIGURE 5
$As_{40}Se_{40}Te_{20}$ glass infrared absorption curves with stiling added Zr:
(A) original glass;
(B) adding Zr 0.3%.

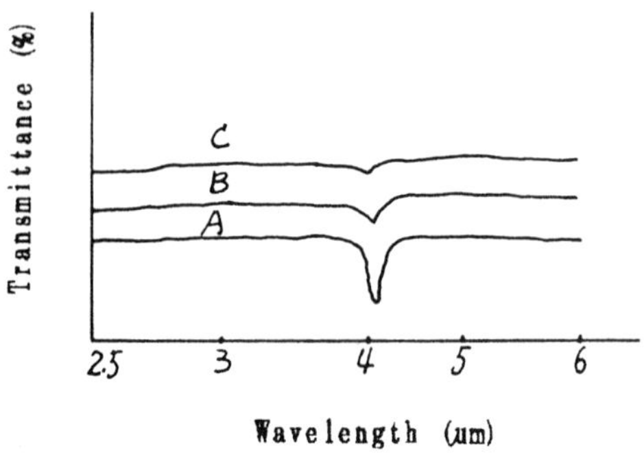

FIGURE 6
As_2S_8 glass absorption curves in different melting temperature and time:
(A) in 800℃, melting 8 hour;
(B) in 800℃, melting 1 hour;
(C) in 600℃, melting 12 hour.

hydrogen on the sureface of silica ampoule, and the glass should be melted in lower temperature and shorter time[8].

In order to reduce O-H and H-S impurities concentration in Ge-S glass, the method presented by Shibata et al. is to mix Ar gas with S_2Cl_2 (or SCl_2) vapour pass through the surface of melted sulphur or bubble from inside to remove impurity hydrogen[9].

5. FIBER-DRAWING AND MEASUREMENT OF THE LOSS

The fibers were drawn in a pressurised crucible under an Ar gas atmosphere, coated with epoxy acrylate. The diameter of fiber was 50~400μm.

The transmission loss of chalcogenide fibers was measured in cut back method at 10.6μm. The CO_2 laser was used as a source. The laser radiation was divided into two beams perpendicular each other by a beam-splitter. One was projected into a power monitor, which consists of a pyroelectric detector and a auto-recorder. The stabilizition of source was monitored by the recorder. And the other was coupled into the fiber in length with 2~3m. The output signal from the fiber was measured by a pyroelectric detector. The loss were measured in the period when the laser power was stabled. The length of the fiber that was cut off is greater than 2/3 of original length. It must be ensured that the ends of fiber are sufficient flat. The ends of fiber before cutting and after cutting were at same place from the detector by means of setting it in a pore.

6. CONCLUSION AND DISCUSSIONS

As-Se-Te glass without Ge can prevent unfavourable effect on optical characteristics caused by Ge-O bond. The fact that the large atomic number of Te in the periodic table makes the shift of glass multiphonon absorption to longer wavelength. All these are helpful to reduce the transmission loss in 10.6μm. Authors had utilized the above-mentioned raw material and glass purifing technique to get a high-quality glass. The loss of the drawn $As_{40}Se_{40}Te_{20}$ glass fiber is less than 5dB/m at 10.6μm.

Chalcogenide glass fibers are suitable waveguide for transmission of mid-infrared radiation and are expected to find a wide range of applications. Theoretical predicts that loss of chalcogenide glass fiber is very small, however the loss of practical used fibers is still several dB/m[1~6]. Further loss reduction is expected by seeking new glass composition, developing purifing method, improving melting, drawing and coating techniques, and further investigation to the weak absorption tail.

References

1. N.J.Pitt, Loss mechnisms in chalcogenide glass optical fibers, 1986
2. Mao Xilai et al., Special Glasses(Chinese), 3, 1986
3. Dimitrij LEZAL et al., J.Non-Cryst. Solids, 90, P557-560, 1987
4. N.J.Pitt et al., Telluride Glass Fibres for Transmission in the 8~12 Micrometres Waveband, 1986
5. Yu Yaochu et al., Glass and Enamel(Chinese), 3, 1988
6. J.P.Parant et al., Glass technology, Vol.24, 3, P161-163, 1983
7. P.Klocek et al, SPIE Vol. 572, Infrared Technology, XI, 1985
8. M.Saito et al., J.Appl.Phys.59(5), 1450, 1986
9. S.Shibata et al., Mat.Res.Bull., Vol.16, P703-714, 1981
10. Terutoshi KANAMORL et al., REVEW of the Electrical Communication Laboratories, Vol.32, 3, P469-477, 1984

Frontiers of Materials Research/
Electronic and Optical Materials
M. Kong, L. Huang (editors)
Elsevier Science Publishers B.V., 1991

NEW FAST Cu^+ ION-CONDUCTING GLASSES IN THE $CuI-Cu_2O-WO_3$ SYSTEM —— PREPARATION, STRUCTURE, AND ELECTRICAL PROPERTIES

Guangming LI and Fuxi GAN

Shanghai Institute of Optics and Fine Mechanics, Academia Sinica, P.O.Box 800-216 Shanghai 201800, P.R.China

New cuprous ion-conducting glasses in the system $CuI-Cu_2O-WO_3$ were prepared by twin-roller quenching technique. The major glass-forming region was found in the binary oxide system $xCu_2O-(1-x)WO_3$ with $0.25 < x < 0.55$ and the pseudobinary system $yCuI-(1-y)Cu_2WO_4$ with $0.57 < y < 0.82$. The IR, Raman, and ESR spectra revealed that these glasses are mainly composed of discrete WO_4^{2-}, Cu^+, and I^- ions as well as condensed anion $W_2O_7^{2-}$ groups, depending on the ratio of Cu_2O/WO_3 in glasses. The ionic and electronic conduction was studied by the complex impedance and the Wagner polarization method, respectively. The mixed ionic-electronic conduction was shown in the binary system, whereas the pure ionic conduction in the pseudobinary system. With the increase of Cu_2O/WO_3 ratio, the glass changes from mixed ionic-electronic conductor to the ionic one. In these new glasses, ionic conductivities at room temperature can reach 10^{-2} Scm^{-1}.

1. INTRODUCTION

Fast ion-conducting glasses have been attracting an increasing attention during recent years due to their potential utilization in solid state batteries and other electrochemical devices[1-5]. Among these materials the glasses containing Ag^+ ion are most studied probably because of the ease of glass preparation, the stability of glasses in the ordinary atmosphere, and the extremely high conductivity. However, the exploring of Cu^+ ion-conducting glasses has nowadays been emphasised since Cu^+ ions have a similar electron configuration in the outmost orbitals and a smaller radius than Ag^+ ions and glasses containing Cu^+ ions may be expected to exhibit higher conductivities than those of glasses containing Ag^+ ions. Although there have occurred several Cu^+ ion-conducting glasses which show very high conductivity, the further works are still urged[6-8].

In this paper, we report the formation of new cuprous ion-conducting glasses in the system $CuI-Cu_2O-WO_3$ by twin-roller quenching technique, the IR, Raman, and ESR spectra of the glasses, and the electrical properties.

2. EXPERIMENTAL

The twin-roller quenching technique was used for glass preparation. Desired amounts of CuI, Cu_2O, and WO_3 powders (> 99%) were fully ground in an agate mortar and sealed off under vacuum in a silica glass ampoule, which was melted at temperatures ranging from 700-900 °C, depending on the composition, for two hours and quenched in water. The solidified materials were then melted for a second time in a silica glass tube on the twin-roller quenching apparatus.Finally the melt was ejected through a nozzle in the bottom of the tube by a proper pressure of N_2 gas and quenched between two copper rollers rotating at 3000 rpm. The samples thus obtained are 20-50 μm thick and several mm across.

X-ray powder diffraction measurements, using Cu-Kα radiation, were made for all the samples prepared to determine whether or not crystalline phases were present.

*The present work was supported by the National Nature Science Fundation of China

The infrared (IR) spectra were recorded on a Hitachi 270-50 IR spectrophotometer in the range 4000-250 cm^{-1} at room temperature; the measurements were carried out by means of glass powder dispersed in KBr pellets (about 1 wt%).

The Raman spectra were obtained on a Spex Ramalog 1403 Raman spectrometer, using the 514.5 nm laser line. The samples used for the measurements were pellets 13 mm in diameter and 0.5 mm thick (pressed from as-quenched glass powder).

The electron spin resonance (ESR) measurement was conducted on a JES-FE2xG ESR spectrometer, working at x band at room temperature.

The total electrical conductivities were determined by the complex impedance method, using a 4192A LF impedance analyzer. The samples used for the measurement also were the pellets prepared under a pressure of $4x10^3$ kg/cm^2. Gold film was evaporated on both surfaces as electrodes.

The electronic conductivity of the glass was measured using the Wagner d.c. polarization technique[9]. Copper and gold were used as the reversible and blocking electrodes, respectively. A potential below the decomposition voltage was applied across both electrodes.

3. RESULTS AND DISCUSSION

3.1. The glass-forming region

The tungstate compositions quenched by the twin-roller technique are shown in Fig.1. Essentially pure glass was produced only around two regions: the binary system $xCu_2O-(1-x)WO_3$ with $0.25 < x < 0.55$ and the pseudobinary system $yCuI-(1-y)Cu_2WO_4$ with $0.57 < y < 0.82$, respectively. It can be seen from Fig.1 that the addition of CuI compound expands the glass-forming ability remarkably. The glass formation in the $CuI-Cu_2O-WO_3$ system is rather similar to that in $AgI-Ag_2O-MoO_3$ or $CuI-Cu_2O-MoO_3$ systems except that the glass-forming region has been extended into the binary system owing to the high cooling rate($>10^5$ °C/s) produced by the twin-roller quenching technique.

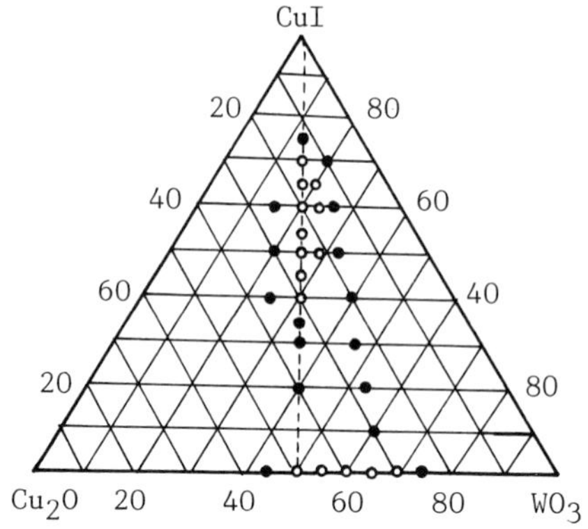

FIGURE 1
Glass-forming region in the system $CuI-Cu_2O-WO_3$ obtained by twin-roller quenching technique
o represents rapidly quenched glass
• partially crystalline

3.2. IR, Raman, and ESR spectra

Figure 2 through 6 presents the IR, Raman, and ESR spectra of these glasses, respectively.

The main IR absorptions of the binary glasses Cu_2O-WO_3 are found in the frequency region 840-820, 630-600, 460-440, and 360-340 cm^{-1}, as shown in Fig.2.

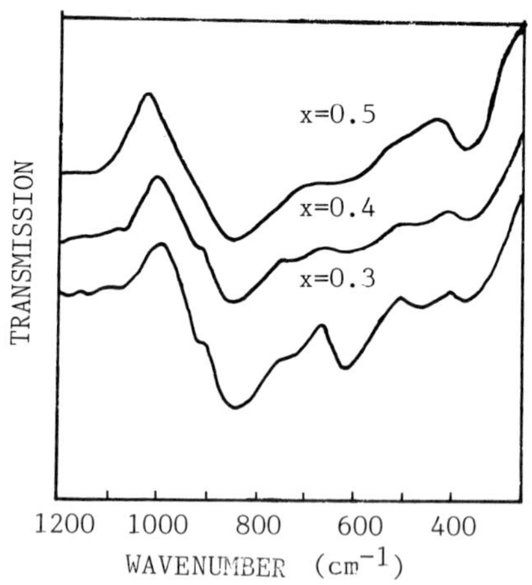

FIGURE 2
IR spectra for $xCu_2O-(1-x)WO_3$ glasses(x= 0.3, 0.4, and 0.5)

With the increase of Cu_2O content, the absorption maximum occurring at 630-600 cm^{-1} is gradually weakened, while the absorption at 360-340 cm^{-1} is strengthened. Since Cu^+ ions and Na^+ ions have a similar ionic radius(0.96 Å for Cu^+ ions and 0.97 Å for Na^+, respectively), it will be helpful to compare the IR spectra given in Fig.2 with those found in corresponding sodium tungstates. Based on Gossink's data[10], the IR spectra of Na_2WO_4 compound exhibit two main absorption bands in the frequency region 840-820 cm^{-1} and 400-300 cm^{-1}. This is consistent with the IR spectra of Cu_2WO_4 glass, which was seen to have two strong absorptions at 840-820 and 360-340 cm^{-1}, respectively. Thus the Cu_2WO_4 glass is expected to mainly consist of discrete WO_4^{2-} and Cu^+ ions only. Besides, it is worth noticing that the absorption maximum occurring at 640-610 cm^{-1}, which is neither shown by WO_3 nor by Na_2WO_4, can be assigned to the vibrating mode of $W_2O_7^{2-}$ groups after Dupuis et al.[11].

With the increase of Cu_2O content, the Raman peak at 910 cm^{-1} is rapidly strengthened, but the 815 cm^{-1} peak weakened. While x= 0.5, two Raman peaks at 910 and 790 cm^{-1} become dominant. According to the literature[12], the peaks at 910 and 790 cm^{-1} can be considered to correspond to those of Na_2WO_4 compound except for a little lower than the Raman peaks of Na_2WO_4, which is perhaps due to the formation of a strong covalence bond between the Cu^+ ions and the WO_4^{2-} groups(see Fig.3).

However, while changing the CuI concentration with Cu_2O/WO_3 = 1, it can be seen from Fig.4 that the IR spectra of CuI-Cu_2WO_4 glasses almost remain unchanged. This implies that the glass structure depends on the ratio of Cu_2O/WO_3 to a great extent and consists of discrete WO_4^{2-}, Cu^+ and I^- ions with CuI dissolved in the matrix glass as strong electrolyte.

In the case of the ratio $Cu_2O/WO_3 < 1$, as shown in Fig.5, a remarkable absorption band at 630-600 cm^{-1} shows up as in the binary system, which is not affected by the CuI content. As discussed previously, this band can be attributed to $W_2O_7^{2-}$ groups.

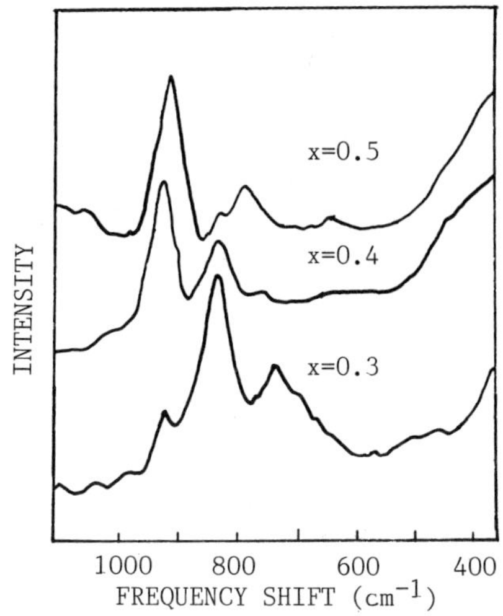

FIGURE 3
Raman spectra for $xCu_2O-(1-x)WO_3$ glasses(x= 0.3, 0.4, and 0.5)

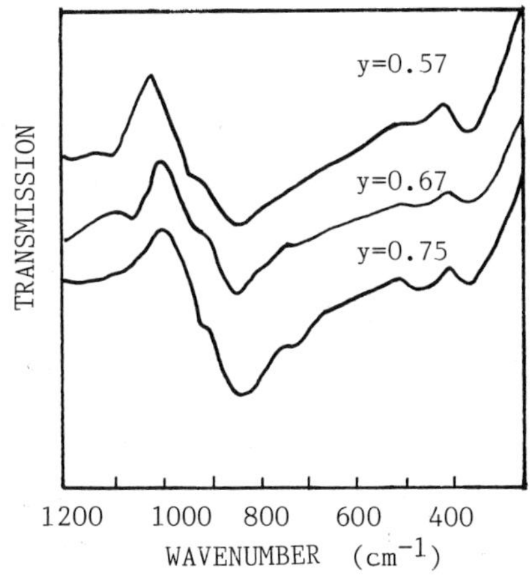

FIGURE 4
IR spectra for the glasses $yCuI-(1-y)Cu_2WO_4$ (y= 0.57, 0.67, and 0.75)

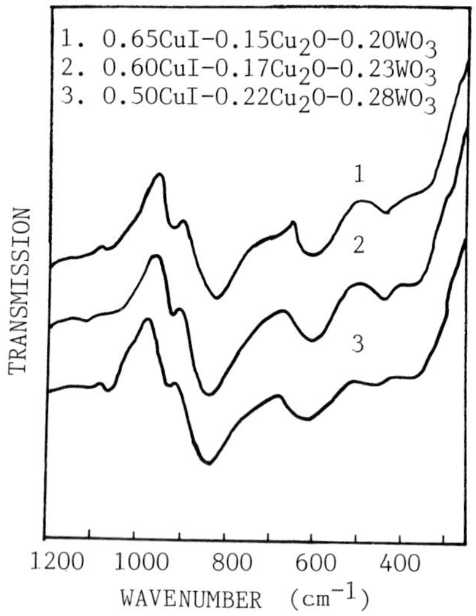

FIGURE 5
IR spectra for the glasses $CuI\text{-}Cu_2O\text{-}WO_3$ with $Cu_2O/WO_3 < 1$

From the above discussions it can be concluded that the structures of $CuI\text{-}Cu_2O\text{-}WO_3$ glasses are basically controlled by the ratio of Cu_2O/WO_3. When the ratio > or = 1, the glasses are composed of discrete WO_4^{2-}, Cu^+ or I^- ions only, called " ionic glass ". However, if the ratio < 1, the glasses will contain other condensed anions such as $W_2O_7^{2-}$ groups.

Figure 6 presents the ESR spectra of glasses in the $CuI\text{-}Cu_2O\text{-}WO_3$ system. It is observable that in glasses of low Cu^+ ion concentration there are three absorptions corresponding to the effective g factors about 4.6, 3.3, and 2.1, respectively. By analysis we consider that the low-field absorptions are probably caused by W^{3+} ions, whereas the high-field one by both of the W^{3+} and W^{5+} ions. Keeping in mind that the addition of Cu_2O tends to convert $[WO_6]$ octahedral units into $[WO_4]$ tetrahedral ones, finally resulting in the discrete WO_4^{2-}, Cu^+ or I^- ions in the glasses. Therefore, it is appropriate to assume that these paramagnetic ions are mainly located in a tetrahedral crystal field. Though the ^4F state of the W^{3+} ions splits in a tetrahedral field so as to leave the $^4T_{1g}$ state lowest, the distortion of the tetrahedral field can not be expected to be the main source for the further splitting of the $^4T_{2g}$ state in the case of the isolated $[WO_4]$ groups, which are expected to possess high tetrahedral symmetry. Hence, the ground state will have a net orbital angular momentum. We must consider the effect of the spin-orbit coupling.

Following Abragam and Pryce' treatment[13], the three degenerate T_{1g} states can be considered as the three components of a state with an orbital angular momentum quantum number L' = 1.(A prime has been used since L applies strickly only to systems with spherical symmetry.) Thus the following modified hamiltonian is required.

$$\hat{H}' = \beta(-\alpha\hat{L}' + g_e\hat{S})\cdot H - \alpha\lambda\hat{L}'\cdot\hat{S}$$

Using the function $|M_{L'}, M_S\rangle$ ($M_{L'} = \pm1, 0$; $M_S = \pm3/2, \pm1/2$) as a basis set, solution of the corresponding secular determinant then yields the three energies

$$W_{\frac{5}{2}} = -\frac{3}{2}\alpha\lambda$$
$$W_{\frac{3}{2}} = \alpha\lambda$$

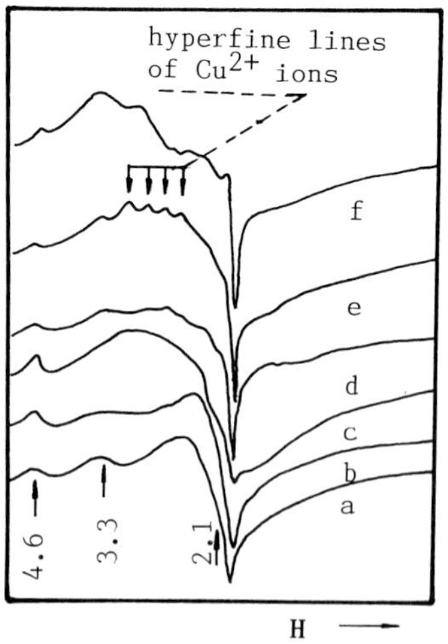

FIGURE 6
ESR spectra of the glasses $CuI\text{-}Cu_2O\text{-}WO_3$ at room temperature
a. $30Cu_2O\text{-}70WO_3$ b. $40Cu_2O\text{-}60WO_3$
c. $50Cu_2O\text{-}50WO_3$ d. $50CuI\text{-}25Cu_2O\text{-}25WO_3$
e. $60CuI\text{-}20Cu_2O\text{-}20WO_3$ f. $70CuI\text{-}15Cu_2O\text{-}15WO_3$

$$W_{\frac{1}{2}} = \frac{5}{2}\alpha\lambda$$

For W^{3+} ions $\lambda > 0$; hence, $W_{\frac{5}{2}}$ will correspond to the state of lowest energy. If $\alpha\lambda >> kT$, then the state with $J' = \frac{5}{2}$ will be the only one which will be appreciably populated. The ground state will be sixfold degenerate in the absence of a magnetic field. Such a ground state may be characterized by a fictitious spin S'= 5/2.

Based on considerations of the symmetry, Lin[14] has shown that in a tetrahedral field there are three possible effective g factors(30/7, 10/3, and 2) for the S-state ions of the half-filled shell, that is, ^6S . This is consistent with our ESR results(see Fig.6).

As for W^{5+} ions the ^2D state splits in a tetrahedral field so as to leave the Eg state lowest. Thus the ground state will retain some orbital angular momentum. For such a nonlinear system with residual orbital degeneracy, the system will be strongly coupled to those lattice vibrations which remove the degeneracy and lower the energy of the ground state due to the Jahn-Teller effect. Therefore, the high-field absorption with the effective g factor near 2 is still expected to occur. This may be the reason why we observe a strong absorption in the high-field range(also see Fig.6).

It is interesting that with increasing the Cu^+ content, the ESR spectra of these glasses become complicated. The absorptions resulting from Cu^{2+} ions are clearly observed, which is characterized by the effective g factor together with four hyperfine lines due to the ^{63}Cu and ^{65}Cu nuclides, as shown in Fig.6.

3.3 The electrical properties

A brief summary of the electrical properties for the $CuI-Cu_2O-WO_3$ glasses is presented in the following table 1.

We can see that in the binary glasses Cu_2O-WO_3, with increasing the Cu_2O concentration, the total electrical conductivity gradually increases, while the electronic conductivity decreases. In these compositional ranges, the glass changes from the ionic-electronic conductor to the pure ionic conductor. It is well known that electron transport in glass comes from transitions of the electron between ions of the variable valence. For example, in the Cu_2O-WO_3 glasses the conducting electrons move between the W^{5+} and W^{6+} ions by means of the bridging oxygen ions as follows,

$$W^{5+} - O - W^{6+} \text{———} W^{6+} - O - W^{5+}$$

As shown in the previous part, addition of Cu_2O into the glasses will convert the $W_2O_7^{2-}$ groups into the discrete WO_4^{2-} groups, resulting in an increase of the non-bridging oxygen.Thus the electronic conductivity will be decreased.

In addition, it is seen from table 1 that the electrical conductivity increases remarkably with CuI concentration. This suggests that the Cu^+ ions bound to the halide ions I^- probably exhibit higher mobility than those bound to the oxygen ions O^{2-}.

In summary, we have prepared glasses in the new $CuI-Cu_2O-WO_3$ system by twin-roller quenching technique. Pure glass can be produced only in a small compositional range. IR, Raman, and ESR spectra show that the glass structure mainly depends on the Cu_2O/WO_3 ratio. In the case of Cu_2O/WO_3 > or = 1, the glasses consist of discrete WO_4^{2-}, Cu^+ or I^- ions, or else there exist other condensed anions such as $W_2O_7^{2-}$ in the glasses. The paramagnetic absorptions in these glasses result from the W^{5+}, W^{3+} or Cu^{2+} ions. Among them the first two are located in a tetrahedral field. The mixed ionic-electronic conduction in the binary Cu_2O-WO_3 glasses has been shown. In the pseudobinary glasses $CuI-Cu_2WO_4$, ionic conductivities at room temperature can reach 10^{-2} Scm^{-1}. Finally it is emphasised that one of the problems in the preparation of Cu^+ ion-conducting glasses is oxidization of the Cu^+ ions which should be overcome before higher conductivity can be achieved.

TABLE 1 The electrical properties of the $CuI-Cu_2O-WO_3$ glasses

Composition	total electrical conductivities (Scm^{-1}, 25 °C)	electronic conductivities (Scm^{-1}, 25 °C)	ion transport numbers	activation energies (eV)
$30Cu_2O-70WO_3$	$2.51x10^{-6}$	$1.14x10^{-7}$	0.9545	0.293
$35Cu_2O-65WO_3$	$8.32x10^{-6}$	$1.08x10^{-7}$	0.9870	0.285
$40Cu_2O-60WO_3$	$2.08x10^{-5}$	$0.97x10^{-7}$	0.9953	0.279
$45Cu_2O-55WO_3$	$5.01x10^{-5}$	$0.63x10^{-7}$	0.9987	0.274
$50Cu_2O-50WO_3$	$1.58x10^{-4}$	$0.38x10^{-7}$	0.9997	0.266
$40CuI-30Cu_2O-30WO_3$	$3.55x10^{-4}$	$0.71x10^{-7}$	0.9998	
$45CuI-27.5Cu_2O-27.5WO_3$	$8.91x10^{-4}$			
$50CuI-25Cu_2O-25WO_3$	$2.14x10^{-3}$			
$55CuI-22.5Cu_2O-22.5WO_3$	$3.98x10^{-3}$			
$60CuI-20Cu_2O-20WO_3$	$7.08x10^{-3}$			
$65CuI-17.5Cu_2O-17.5WO_3$	$1.32x10^{-2}$			

REFERENCES

1. H.L.Tuller, D.P.Button and D.R.Uhlmann, J.Non-Cryst.Solids 40 (1980) 93.
2. D.Ravaine, J.Non-Cryst.Solids 73 (1985) 287.
3. C.A.Angell, Solid State Ionics 18 & 19(1986) 72.
4. T.Minami, J.Non-Cryst.Solids 95 & 96 (1987) 107.
5. M.D.Ingram, Phys.Chem.Glasses 28 (1987)215.
6. C.Liu and C.A.Angell, Solid State Ionics 13 (1984) 105.
7. C.Liu, H.G.K.Sunder and C.A.Angell, Solid State Ionics 18 & 19 (1986) 442.
8. N.Machida and T.Minami, J.Am.Ceram.Soc. 71 (1988) 784.
9. C.Wagner, Int.Comm.Electrochem.Thermodyn. Kinet.Proc.Meet.7th, (1955) 361.
10. R.G.Gossink et al., Philip.Res.Rep.Suppl. 3 (1971) 1.
11. T.Dupuis and M.Viltange, Mikrochim.Acta, (1963) 232.
12. D.Deal et al., J.Non-Cryst.Solids 54 (1983) 207.
13. A.Abragam and M.H.L.Pryce, Proc.Roy.Soc. (London) A205 (1951) 135.
14. Lin Fucheng, Scientia Sinica A 25 (1982) 1298.

Frontiers of Materials Research/
Electronic and Optical Materials
M. Kong, L. Huang (editors)
Elsevier Science Publishers B.V., 1991.

INFLUENCE OF PREPARATION CONDITIONS ON THE PROPERTIES OF a-SiC:H FILMS

CAI Wei, FENG Lianghuan, ZHOU Xinming, ZHENG Jiagui, CAI Yaping, LUO Zhaohe

Department of Material Science, Sichuan University, Chengdu 610064, P.R.China

By changing substrate temperature Ts and fixing the methane content in mixture gases, we research the relationship between Ts and electrical, optical properties of the a-SiC:H thin films prepared from methane-silane mixture gases with RF glow discharge method. The conductivity σ, activation energy Ea, optical energy gap width Eg are measured and the XPS spectra are observed. The results show that as Ts varies from 180°C to 300°C, the σ increases one order in magnitude, while Ea and Eg decrease . The XPS spectra show that there are two bond ways for carbon atom in the films: the C-Si bond and C-H bond. As Ts increases the amount of the C-Si bond reduces. There exists some relation between the variations of the a-SiC:H energy gap width and the amount of C-Si bond component.

1. INTRODUCTION

Since Anderson and Spear[1)] reported in 1977 on the first preparation of hydrogenated amorphous silicon-carbon films by glow discharge decomposition of C_2H_4-SiH_4 mixtures, and Hamakawa[2)] reported in 1981 the first successful application of borondoped a-SiC:H films as window layer in amorphous silicon solar cells, the interest in a-SiC:H film materials has increased rapidly. During these years, many works study the properties of a-SiC:H films, study the influences of ratio x on property of a-SiC:H films, as for x, it is the volume ratio of $CH_4/(CH_4+SiH_4)$ in original mixture gases from which the films are deposited. Now, some experiential relations of influences of x on Eg, σ, Ea have been found, and the effects of x on solar cell parameters, such as Voc, Isc, FF, are researched, too. In addition, some scientists have researched their infrared absorption spectra[3,4)], X-ray photoelectroscopy spectra[5,6)]. Unfortunately, they only focus a great deal of attention upon changing ratio x in original material mixture gases.

In fact, the properties mentioned above may relate not only to the content of carbon atoms actually in solid films, but also to the state and the environment of these atoms. All of these depend strongly on the preparation conditions of the films and on the type of reactor used. In fact, the study of the influences of preparation conditions should be payed much attention for its vital importance. Some papers on this field have been reported. For example, I.Solomon and M.P.Schmidt et al.researched the relations of carbon contents in solid films with respect to the original mixture gases at low power density[8)] and high power density[9)] respectively. Fixing the ration x, we research the influences of different Ts on properties of a-SiC:H films. As for influences of other preparation conditions, they are being researched, too. We find out that the Ts obviously influences the Eg, Ea, σ and doping level of a-SiC:H thin films. As Ts varies from 180°C to 300°C, the σ increases one order in magnitude, while the Ea decreases, so does the gap width Eg. The results from XPS spectra show that carbon atom content in the films are not sensitive with the change of the Ts. And there are two types of carbon bonds in the films: one relates to the C-Si bond, the binding energy of which is 282.4 eV, another relates to the C-H bond, the binding energy of which is 284 eV. As Ts increases, the amount of the C-Si bond reduces. There exists some relationship between the variation of the Eg and the amount of the C-Si bond component.

2. EXPERIMENT

Optical and electrical properties of a-SiC:H thin films depend strongly on the preparation conditions and the type of reactor, we prepare a-SiC:H alloy films from methane and silane mixture gases diluted by hydrogen with the radio frequency glow discharge method. Figure 1 shows a schematic diagram of our glow discharge apparatus.

The mixture gases come from methane and silane. The silane is diluted by hydrogen according as the volume ratio of silane to hydrogen is 1:9. The ratio x is fixed at 0.45. Another type of mixture gases is used for doping: the x is still 0.45, but the diborane B_2H_6 is added into SiH_4+H_2 gases according as B_2H_6 to SiH_4 is 3%.

The Ts is selected at 180°C, 220°C, 260°C, and 300°C, respectively. Several kinds of substrate materials are used: quartz glass for resistivety and transmittivity measurements, high resistivity silicon wafer for infrared absorption (ρ=500Ωcm), and low resistivity silicon wafer for XPS (ρ=0.1Ωcm).

Deposition parameters are as follows:

RF power:	20	Watt
frequency:	13.56×10^6 Hz	
CH_4/CH_4+SiH_4:	0.45	
gas flux:	20	SCCM

The conductivity, optical transmission, infrared absorption and XPS spectra of the a-SiC:H film specimens are measured and analysed. (Cf. Figure 1)

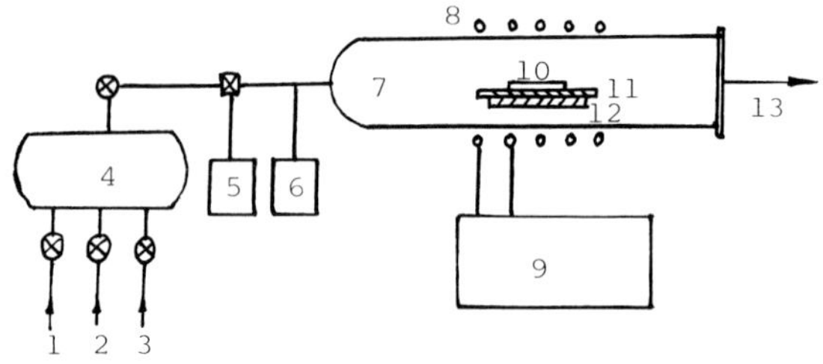

FIGURE 1
Schematic diagram of radio frequency glow discharge deposition apparatus. 1. CH_4; 2. SiH_4+H_2; 3. B_2H_6+SiH_4+H_2; 4. mixture gases; 5. flowmeter; 6. pressure gauge; 7. quartz tube; 8. ring electrode; 9. rf power controller; 10. substrate; 11. substrate frame; 12. heater; 13. pump.

3. RESULTS AND DISCUSSION

3.1. Influence of substrate temperature on electrical property and gap width of a-SiC:H films

The a-SiC:H films used for conductivity measurement are deposited on quartz glass substrate. The Ts is 180°C, 220°C, 260°C and 300°C, respectively. The films are flat and glossy, their thickness is about 400 nm. The film specimens are classed into two groups: undoped, and boron-doped.

The dark conductivity of these films at various temperature is measured. Their activation energy Ea is calculated, too. Our results show: the σ of undoped a-SiC:H is on the order of $10^{-11}\sim10^{-12}(\Omega cm)^{-1}$, the Ea is more than 1eV. The results of 3% diborane doping show: the boron doped a-SiC:H is p-type, its σ is on the order of $10^{-5}\sim10^{-6}(\Omega cm)^{-1}$, the Ea is about 0.46 eV to 0.56 eV. This implies that the doping effect is good. On this respect, our results are consistent with that of Hamakawa[2].

Figure 2 show the temperature dependence of dark conductivity of boron-doped films. When Ts increases from 180°C to 300°C, the σ increases one order of magnitude. Meantime, the slope of these lines decreases. This implies that the activation energy decreases as Ts is increases.

The relations between Ea and Ts, and between resistivity and Ts are ploted in Figure 3. The specimens are boron-doped films. It is obvious that the enhancing slightly substrate temperature is beneficial to boron doping and to

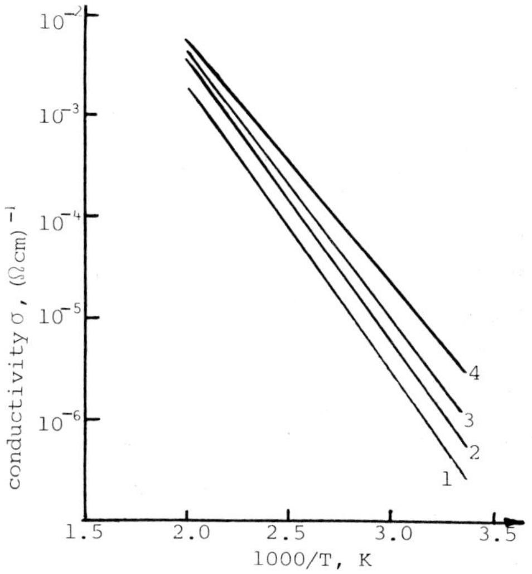

FIGURE 2
Temperature dependence of the dark conductivity of boron-doped films

improving conductivity.

According to the analysis of Tauc

$$(\alpha h\nu)^{\frac{1}{2}} \propto h\nu - Eg,$$

the gap width Eg may be determined.

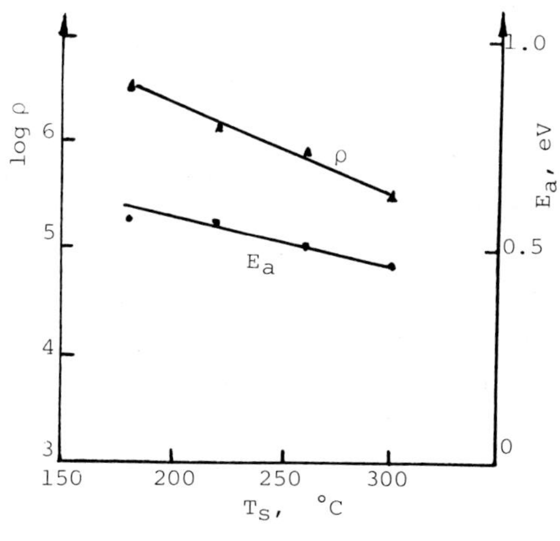

FIGURE 3
Dependence of resistivity ρ, activation energy Ea on substrate temperature Ts, boron-doped a-SiC:H

The $(\alpha h\nu)^{\frac{1}{2}}$ versus hν graphs are drown in Figure 4. When Ts increases from 180°C to 300°C, the Eg drops oppositely from 2.1eV to

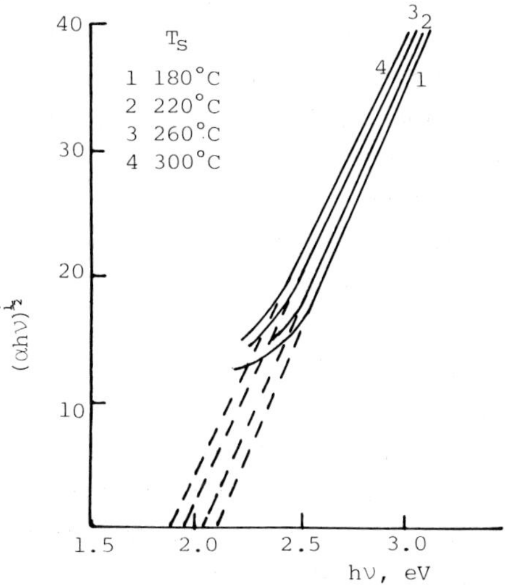

FIGURE 4
Graph of optical absorption versus photo energy

1.85eV. Therefore, high Ts is disadvantageous to widening gap Eg.

3.2. Influence of substrate temperature on carbon concentration in film

In fact, the gap Eg depends on carbon concentration in films and on the coordination of the carbon atom in films. Obviously, the actual carbon concentration in film is not equal to ratio x in gases. When x is fixed, the carbon concentration should depend on the preparation conditions of the film.

The X-ray photoelectron spectroscopy (XPS) may obtain both information of carbon concentration in film and binding energy of these carbon atoms. The XPS spectra are taken with a XSAM 800 spectrometer using Al K_{α} radiation. The vacuum is 10^{-9} torr. Every specimen surface is cleaned by argon ion beam of 4000 eV to eliminate the contaminated carbon and oxygen atoms. The cleaning is important and necessary for a correct spectrum.

All of specimens of a-SiC:H film are prepared under same deposition conditions except different substrate temperature Ts. We find out that the carbon atom concentrations C/(Si+C) in film are near about 22.5% for all our specimens. It is lower than the ratio x=0.45 in the mixture gases. This fact may explained as methane is more stable and more difficult to decompose than silane[7]. Although the Ts varies from 180°C to 300°C, the carbon concentration in the films varies slightly between 22.1% and 23.0% (Table 1). This shows the carbon contents in films are not sensitive to their substrate temperature Ts.

3.3. Influence of substrate temperature on bond-state of carbon atoms in film

Katayama[6] reported that XPS spectrum has two C 1s peaks at 282eV-284eV of binding energy. The peak at higher binding energy (about 284eV) is thought to originate from the adsorbed carbon, and the peak at the lower binding energy is relevant to the carbon atoms within the films. W.Y.Lee pointed out[5] that there are two peaks for C 1s spectrum. The peak centered at 284.6 eV is due to the hydrocarbon or graphite in films. Another peak at about 283 eV belongs to carbon associated with inner carbide in films. After sputter cleaning the peak at 284.6 eV will disappear from spectrum.

Indeed, it is very important that the argon ion beam cleans sufficiently the specimen surface before taking spectrum. We contrast the both spectra before and after ion cleaning. The cleaning proceeds more than 20 minutes. Before cleaning, the composite C 1s peak (the dot curve in Fig.5) has a wide top, and it may be separated into several peaks. The peak at 285 eV binding energy, the dash curve in fig.5, originates from contamination carbon atoms on the film surface. The intensity of this peak implies that the contaminating is serious. After cleaning the composite C 1s peak has very different shape (the dot curve in Fig.6). Its

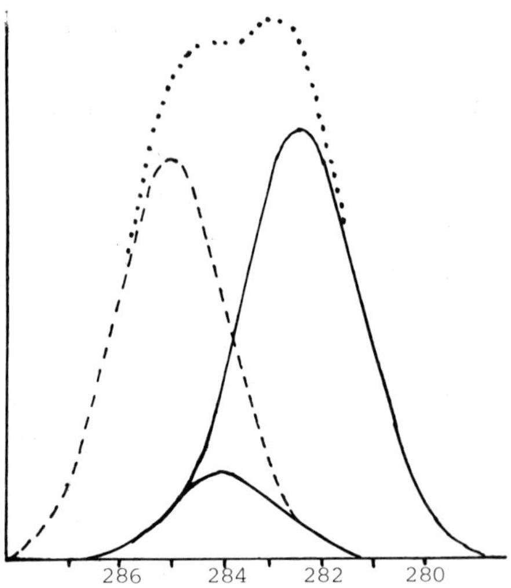

FIGURE 5
C 1s XPS spectrum of a-SiC:H film before cleaning.

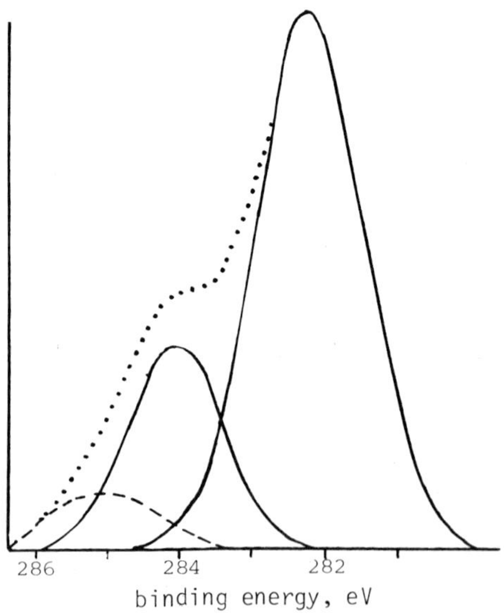

FIGURE 6
C 1s XPS spectrum of a-SiC:H film after cleaning.

higher energy side drops. The peak at 285 eV is a small peak now, which is shown by the dash curve in Fig.6.

Figure 6 shows that the composite C 1s peak may be still separated into two Gaussian type peaks after removing the surface carbon contamination. These two peaks are drawn as the two solid curves in Fig.6. One peak at 284 eV may belong to C-H bond, another big peak at 282.4 eV belongs to C-Si bond in film.

The both peaks imply that inner carbon atoms in films have two types of bond-state. The relative components depend on the ratio x in mixture gases and preparation conditions of the films. When preparation conditions change, the components of both C-Si and C-H bonds change. Therefore, because the intensities of the two peaks vary, the peak shape and the position of the centre of composite peak vary, too. The Si 2p spectrum of same a-SiC:H specimen after cleaning is consist of two peaks. The main peak at 99.2 eV is Si-Si bond, the small one at 100.6 eV should belong to Si-C bond.

Although carbon atom concentration in films is not sensitive to Ts at which the films are prepared, but the C-Si bond component is sensitive to Ts. We can obtain the bond components in films from the areas of these binding energy peaks. Table 1 lists the carbon atom concentrations, C-Si bond concentrations, ratio of $C_{282.4}/(C_{282.4}+C_{284})$, and gap Eg of a-SiC:H films which are prepared at different Ts. When Ts increases from 180°C to 300°C the C-Si bond, binding energy of which is 282.4 eV, decreases. The gap width Eg decreases at the same time. This fact implies that the carbon atom entering a-SiC:H alloy films can widen the Eg because the C-Si bond content increases in films. When other conditions do not change, the more the C-Si bond component is, the wider the Eg is.

4. CONCLUSION

The optical and electrical properties of a-SiC:H films depend not only on methane ratio x in original mixture gases used for films preparation, but also on preparation conditions.

TABLE 1

No.	Ts °C	Eg eV	$\frac{C}{Si+C}$	$\frac{X}{Y}$	$\frac{X}{Si+C}$
1	180	2.10	0.223	0.738	0.165
2	220	2.05	0.230	0.713	0.164
3	260	1.95	0.226	0.688	0.155
4	300	1.85	0.224	0.672	0.150

$X = C_{282.4}$; $Y = C_{284} + C_{282.4}$

By means of RF glow discharge we decompose silane-methane mixture gases to prepare a-SiC:H thin alloy films, fixing methane ratio x and other preparation conditions, but changing substrate temperature Ts only (from 180°C to 300°C). As Ts increases, the conductivity of films increases one order of magnitude, while the activation energy decreases and the optical energy gap width decreases as well. The carbon concentration in the films is not sensitive to Ts, but there are two types of bond states for carbon atom in films: one is C-Si bond, another is C-H bond. The carbon atom amount ratio of both types of bond depends on Ts. As Ts is enhanced, the C-Si bond concentration in films decreases, and the energy gap Eg comes narrow. Therefore, the C-Si bond concentration in films is one of the important factors which influence gap width.

REFERENCES

1. D.A. Anderson and W.E. Spear, Phil. Mag. B 35 (1977) 1.
2. Y. Tawada, H. Okamoto and Y. Hamakawa, Appl. Phys. Lett. 39 (1981) 237.
3. H. Wieder, M. Cardona and C.R. Guarnieri, Phys. Stat. Sol. (b) 92 (1979) 99.
4. Y. Catherine and G. Turban, Thin Solid Films 70 (1980) 101.
5. W.Y. Lee, J. Appl. Phys. 51 (1980) 3365.
6. Y. Katayama, Phil. Mag. B 43 (1981) 283.
7. J.P.M. Schmitt, J. Non-crystall. Solids 59/60 (1983) 649
8. I. Solomon, et al. Phys. Rev. B 38(1988) 9895
9. M.P. Schmitt, I. Solomon, H. Tran-Quoc and J. Bullot, J. Non-crystall. Solids 77/78 (1985) 849.

Frontiers of Materials Research/
Electronic and Optical Materials
M. Kong, L. Huang (editors)

INVESTIGATION ON PROPERTIES OF Si-BASED AMORPHOUS ALLOYS

Jianping ZHOU, Fuxi GAN and Zhishang DIN*

Shanghai Institute of Optics and Fine Mechanics, Academy of Sciences, P.O.Box 800-216,Shanghai,PRC
Zhejiang University, Hangzhou, P.R.China*

Three series of Si-based B, C and P amorphous alloy films were prepared by rf glow discharge method. By means of infrared spectroscopy analysis, we investigated the characteristics of Si-based amorphous alloy matrix and several structural unites, such as B-B two-electron three-center bonds and SiB_4 and/or SiB_6 local enviroment in a-SiB:H, CH_3 radicals attached to Si in a-SiC:H, and PSi_3 configuration and SiP local enviroment in a-SiP:H. According to the structural analyses, the composition dependence of optical energy gap E_{opt} as well as refractive index n of these films were interpreted, and an important conclusion reached. Finally, heat treatment experiments were performed and we give here a preliminary report of the optical characteristic change and crystallization in these samples.

1. INTRODUCTION

Hydrogenated amorphous silicon thin films are today, one of the most interesting materials in scientific research fields, its wide usage in electronic and optical devices is adequate motivation for the continued exploration of its promising properties. For some specific purposes, it would be advantages to have a semiconductor of band gap somewhat smaller or larger than a-Si:H but with substantially the same photoelectronic response[1,2]. Therefore, alloying of hydrogenated amorphous silicon by C, N for the shift of the optical band gap has attracted a lot of interest.Here,we systematically studied a series of Si-based B,C,P amorphous alloys so as to conclude a change role in the network structures which strongly govern their optical properties. Besides, heat treatment experiments were also performed, and the process of optical characteristic change and crystallization discussed in aspect of a potential application for optical storage of information.

2. EXPERIMENTAL

The samples used in these measurements were prepared by rf glow discharge decomposition of SiH_4+B_2H_6,SiH_4+C_2H_2 and SiH_4+PH_5 mixtures under the following conditions: frequency 13.65MHz, total pressure 0.2 Torr and substrate temperture 250°C. The composition, thickness, infrared spectra, transmission and reflection data, as well as internal structure changes of our samples were measured with photoelectron and Auger spectroscopy, a Daktak stylus instrument, IR and UV Perkin-Elmer double-beam grating spectrometers, and a JEM-6AS electron microscope, respectively.

3. RESULTS AND DISCUSSION

3.1. Infrared absorption spectra and E_{opt}

3.1.1. a-SiB:H

One feature of a-SiB:H in fig.1 which grows with increasing boron content is a broad absorption band extending from 900 to 400 cm^{-1}, which is superimposed on the vibrational modes of Si-H bonds near 630 cm^{-1} and 890 cm^{-1}. This broad feature , also observed by Berezin et al[3] between 800 and 550 cm^{-1}, arises from B-B bonds. In rich boron samples, the probability rises for three boron atoms to get close enough to form a two-electron three-center bond. However, the bond lengths can not have their optimal value because the position of each of atoms is also conditioned by its other bonds to

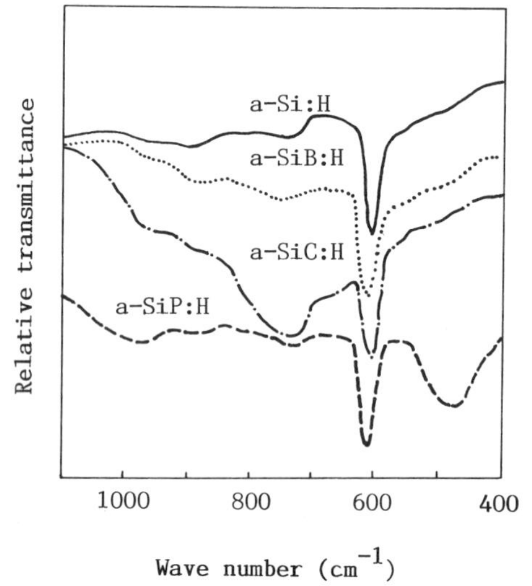

FIGURE 1
Infrared-absorption spectra of a-Si:H,a-SiB:H ($X_{film,B}$=38.9), a-SiC:H ($X_{film,C}$=34.6) and a-SiP:H (X_{film}=31.2) alloy films

the network. as a consequence, it is likely that such three-center bonds have rather large variation in their strengths, this is probably the main cause for the broadening of the infrared absorption bands attributed to B-B vibrations. Besides, there exists a little broadening of 630 cm^{-1} due to Si-B[4] and a slight enhancement in the foundamental vibrations of the Si-Si network. All the factors can bring about a monotonic decrease in E_{opt} as shown in fig.2a. The latter platform at X_{gas}>0.3 may be owing to local chemical ordering in the amorphous films associated with SiB_4 or/and SiB_6, which are the only compound compositions in phase diagram of Si-B binary alloy.

3.1.2. a-SiC:H

The 740 and 690 cm^{-1} bands of a-SiC:H shown in fig.1 can be identified as wagging vibration of CH_3 radicals attached to Si and the stretching mode of Si-C[5], respectively. According to the bond energy among Si, C and H, the bond energy of Si-C (63 kcal/mol) and C-H (98.7 kcal

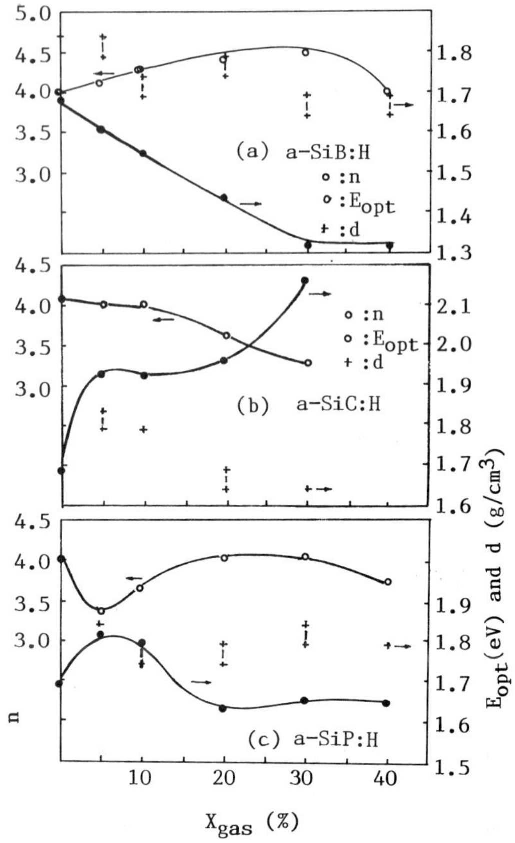

FIGURE 2
Dependence of refractive index n, optical gap E_{opt}, density d on composition $X_{gas,B(C,P)}$

/mol) are larger than Si-Si (42.2 kcal/mol) and Si-H (76 kcal/mol), respectively. Therefore, it can be infered that the main reason for the shift of optical energy gap toward higher energy level with increament of C content in fig.2b is ascribed to the rapid increase of Si-CH_3 and Si-C bond.

Due to the decrease in the deposition rate of carbon atoms, the maximum concentration of C in our samples reached only 34.5%. As suggested in Ref.6, there exists a three-fold coordinated graphitic local enviroment in carbon rich film, whose effect is to increase the overall level of gap states. From our discussion above, we may assume that if the content of H goes up to such a value that the carbon atoms are not easy to gather, but form a polymer-like amorphous structure brought by the high hydrogen content, which might result in a trend of continued increase in E_{opt}, maybe balanced, to some extent, by the effect of some disorder originating from the large amount of bonding hydrogen.

3.1.3. a-SiP:H

The vibrational frequencies of Si-P stretching mode of PSi_3[7] and Si-Si mode located at 470 cm^{-1} and 490 cm^{-1}, respectively, are rather close to each other, so are their atomic weights. From this point of view, we can infer that the difference on bond strengths between them is small. Therefore, in this case, we mainly take the local structural or/and chemical order into account for analysis of optical band gap E_{opt}.

Initially, phosphorus atoms added in the matrix on the Si-rich side of the amorphous alloy system compensate dangling bonds of silicon raising the structure order while reducing the density of band-tail states, and then with increasing phosphorus concentration continually, when all dangling bonds have been compensated, as a result, the phosphorus atoms added in at this moment are 3-fold coordinated and generate P-related defects into the network. Hence, with the increase of phosphorus content, the optical band gap goes up first, and then goes down, reaching its maximum at about $X_{gas,P}=0.05$. When $X_{gas,P}$ is equal to 0.4, E_{opt} does not decrease as expected, but increases a little (fig.2c), maybe, this is the same phenomenon as in a-SiB:H, that is , the increament of E_{opt} is due to local chemical ordering in the amorphous films associated with SiP structural units which is the only compound composition in the phase diagram of Si-P binary alloy.

3.2. Refractive index

Generally speaking, refractive index is approximately in inverse proportion to average energy gap between valence and conduction bands E_g whereas in direct proportion to the square root of the density of valence band electrons per volume N. The density of the sample d enters through N and the band structure E_{opt} through E_g. As shown in fig.2a,2b and 2c, the refractive curves calculated from our measurement accord approximately with this conclusion.

From all the analyses of E_{opt} and refractive index n in these Si-based amorphous alloy films we can conclude that the variation of bond energy in amorphous silicon matrix when adding the second alloy element (B,C,P) in the system plays a decisive role in optical properties , which is more or less influenced by the coincidental variation of short-range ordering negatively or positively.

3.3. Heat treatment

Electron diffraction measurements show that all the samples are in amorphous state before

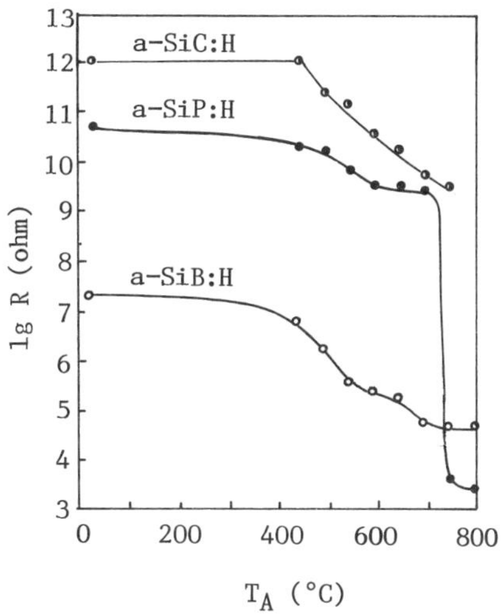

FIGURE 3
Surface resistance R of a-SiB:H, a-SiC:H and a-SiP:H as a function of annealing temperature T_A

heat treatment. When a-SiC:H sample is heated at 600°C, and a-SiP:H at 750°C, their electron diffraction patterns of polycrystallite occur , however, a-SiB:H even heated at 750°C remains noncrystalline. With the help of the electron microscopy results, we can analyse the temperature dependance of surface resistance R of our samples in fig.3.

Apparantly, the cruve of a-SiP:H in fig.3 can be divided into two parts. First the surface resistance goes down slowly with temperature, then seems to keep appoximately constant around 600-700°C and at last reduces drastically. In correspondance with the electric characteristic, its absorption edge also experiences a slight shift to a lower energy level before crystallization shown in fig.4, which is similar to but much smaller

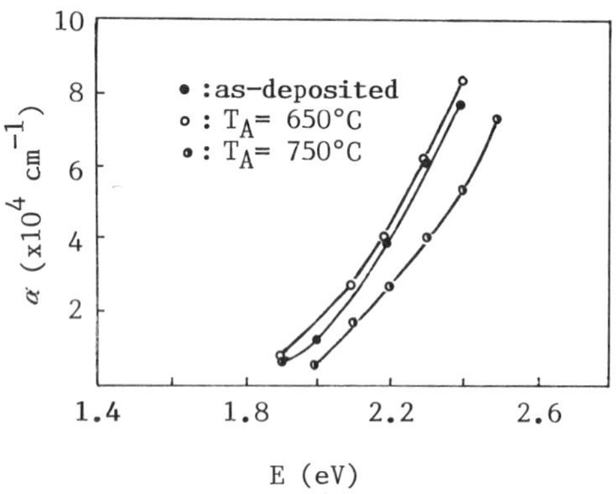

FIGURE 4
Optical absorption spectra for a-SiP:H alloy film with P content $X_{gas,P}$=0.4 in the gas at different annealing temperature T_A

than the optical characteristic change of P-doped a-Si:H reported by Steabler[8], due to desorption of hydrogen and the increase of the density of gap states. Meanwhile, the reflectance all over the spectrum increases approximately by 15% seen in fig.5. However, when crystallization takes place at 750°C, the situation is opposite, the absorption edge has an apparent shift to high energy and the reflectivity drops with occurance of interference

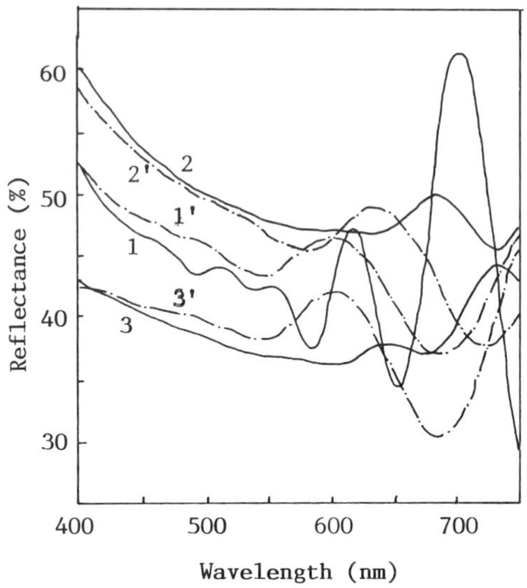

FIGURE 5
Reflectivity of a-SiP:H alloy film (solid line) at annealing temperatue: 1---750°C, 2---650°C, 3---as-deposited, and a-SiB:H alloy film (chain-dotted lines) at annealing temperature: 1'--750°C, 2'--500°C, 3'-- as-deposited

fringes at much lower wavelength. However, the situation in a-SiB:H alloy film is quite different. Although a-SiB:H alloy film, heated even at 750°C is still in amorphous state, its reflectance spectrum drops a little following a considerable enhancement at 500 °C shown in fig.5.

We suggest that the variation of reflectivity and the shift of absorption edge of Si-based amorphous alloy films during optical characteristic change and crystallization be worth further studying for optical information storage.

REFERENCES

1. K.D.Mackenzie, J.H.Burnett, J.R.Eggert, Y.M.

Li and W.Paul, Phys.Rev.B , 38(1988)6120.

2. G.H.Bauer, H.D. Mohring, G.Bilger, A.Eicke, J.Non-Crys. Solids, 77&78(1985)873.

3. A.A.Berezin, O.A.Golikova, M.M.Kazamin, T. Khomidov, D.N.Mirlin, A.V.Petrov, A.S. Umarov, and K.K.Zaitsev, J.Non-Cryst. Solids 16(1974)237.

4. S.C.Shen and M.Cardona, Phys.Rev.B,23(1981) 5322.

5. H.Wieder, M. Cardona and C.R.Guarnieri, Phys. Stat. Sol.(b), 92(1979)99.

6. D.A. Anderson and W.E. Spear, Phil. Mag. 35(1977)1.

7. Wu Zongyan, Shen Yuehua, Ye Yagu, Chinese J. Semiconductors, 5(1984)113.

8. David L. Staebler, J. Appl. Phys. , 50(1979) 3648.

Frontiers of Materials Research/
Electronic and Optical Materials
M. Kong, L. Huang (editors)

METALORGANIC MOLECULAR-BEAM EPITAXY: GROWTH KINETICS AND SELECTIVE-AREA EPITAXY

C.W. TU and B.W. LIANG*,

Department of Electrical and Computer Engineering, University of California at San Diego, La Jolla, California 92093, U.S.A.,

V.M. DONNELLY

AT&T Bell Laboratories, Murray Hill, New Jersey 07974, U.S.A.

Metalorganic molecular-beam epitaxy (MOMBE) or chemical-beam epitaxy (CBE) is interesting because its growth kinetics is different and relatively unexplored compared to that of solid-source MBE and metalorganic chemical vapor deposition (MOCVD) and because its capability of selective-area epitaxy has potential novel applications, quantitatively, in optoelectronics. We propose a new kinetic model for MOMBE of GaAs and InAs. The model can explain the important effects of arsenic overpressure on growth rate for the first time. Group-III and group-V species are equally important in the surface chemical reactions during MOMBE. Excimer-laser-enhanced MOMBE of GaAs was also observed at low substrate temperature with nearly unity efficiency. The effect is due to transient heating during the pulsed laser irradiation.

1. INTRODUCTION

Metalorganic molecular-beam epitaxy (MOMBE)[1] or chemical-beam epitaxy (CBE)[2] has gained increasing interest not only for fundamental studies of growth kinetics[3–5] but also for optoelectronic device applications due to its potential for high throughput[6] and selective-area epitaxy.[7,8]

This paper summarizes our recent results on growth kinetics[4,5,9,10] and laser-enhanced MOMBE of III-V compounds.[7] The kinetic model for MOMBE or CBE that is proposed so far considers only group-III alkyl decomposition pathways[3] and, therefore, cannot explain effects of group-V species. We propose a new kinetic model that includes both group-III alkyls and arsenic in the surface chemical reactions, for the first time.[4,5] The model can fit the growth-rate data for GaAs and InAs very well and leads to new insights into the MOMBE process.

Another interesting aspect of MOMBE is selective-area epitaxy. We have observed excimer-laser-enhanced MOMBE growth of GaAs using triethylgallium (TEGa) and arsenic[4] at substrate temperatures below 450°C, where pyrolytic growth is very slow. From the strong dependence of the enhanced growth rate on laser fluence and X-ray photoelectron spectroscopy of laser-induced decomposition of TEGa molecules,[11] we conclude that at high laser fluences the growth enhancement process appears to be pyrolytic in nature, not photolytic. The enhanced growth is due to rapid transient heating by the laser pulses.

Atomic layer epitaxy (ALE) can be achieved with trimethylgallium (TMGa) more easily than triethylgallium (TEGa).[9,10,14] In addition, using TMGa can result in ultrahigh *p*-type carrier concentration from high 10^{19} cm^{-3} to high 10^{20} cm^{-3} range.[9] Such high *p*-type doping with low diffusivity of carbon may be useful in a variety of device applications. These results have been presented elsewhere.[9,10]

2. EXPERIMENTAL PROCEDURES

The growth was performed in a modified prototype Perkin-Elmer 425A MBE system, as shown in Figure 1. The growth chamber was pumped by a cryopump and an ion pump. TMGa, TEGa, and trimethylindium (TMIn) molecules were directly injected into the growth chamber without hydrogen carrier gas through a leak valve and an

*On leave from the Shanghai Institute of Metallurgy, Shanghai, China.

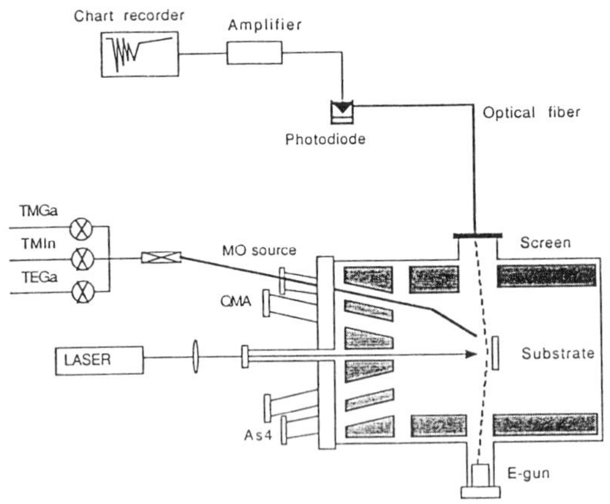

FIGURE 1
A schematic diagram of the MOMBE system.

on-off valve upstream. The reflection-high-energy-electron diffraction (RHEED) intensity was monitored by a photodiode through an optical fiber. GaAs substrates were etched and cleaned in the usual manner and then mounted on molybdenum (Mo) blocks. The substrate was thermally cleaned in vacuum under an overpressure of arsenic. The substrate temperature was calibrated by a thermal couple inserted into a Mo block and also independently by a pyrometer. Nevertheless, the uncertainty in absolute temperature could still be ~20^oC. Fluxes were measured by the ionization gauge behind the substrate holder with the holder out of the way.

ArF excimer laser (Lumonics Hyper EX460) was used for the selective-area-epitaxy experiment. The laser delivered 100-300 mJ, 15 ns pulses at 193 nm. The 0.5×2 cm^2 beam was collimated with a 3-m-focal-length, fused silica lens and directed at the substrate at normal incidence. No mask or additional focusing elements were employed in these studies. The absolute intensity of the laser beam was measured as a function of (x,y) position in the plane of the substrate by reflecting the beam off a mirror and positioning a pinhole and detector at the appropriate distance. This laser-MOMBE system was also used in a separate study of laser-modified MBE growth reported elsewhere.[12] The RHEED-intensity-oscillation measurement was not set up yet during the laser-assisted MOMBE experiment.

3. GROWTH KINETICS

3.1 CBE of GaAs

Because the mean free paths of source molecules are very long compared to the source-to-substrate distance, one can expect that gas-phase reactions are unimportant, unlike metal-organic chemical vapor deposition (MOCVD). However, surface reactions play an important role in determining the MOMBE or CBE growth process. Growth behaviors and growth kinetics mechanisms of GaAs and InAs, grown by CBE or MOMBE, have been studied by several authors.[1–5,13,14] The first kinetic model proposed by Robertson *et al.* considered only triethylgallium (TEGa) decomposition paths.[3] The role of arsenic species was presumed secondary. Subsequently, group-V species, unlike in conventional MBE, were found to have a strong effect on the growth rate.[14,15] This model therefore cannot explain the effects of arsenic species. We propose a new kinetic model, taking into account the effect of arsenic species for the first time, for CBE of GaAs using TEGa and arsine (AsH_3).[4] The model can fit the experimental data of Chiu *et al.* very well,[14] and we gain some more insights into the effect of arsine and TEGa flow rates on the CBE growth process.

The model assumes that diethylgallium (DEGa) decomposition--in the presence of arsenic--is the rate-limiting step for growing GaAs. We expect that monoethylgallium, with two ethyl groups cleaved off, would be very reactive and chemisorb strongly to the surface. Cleavage of the last ethyl group and subsequent incorporation of GaAs would occur very fast due to small activation barrier. This situation is similar to Robertson's model where DEGa decomposition is the rate-limiting step.[3] What differentiates our model is that this DEGa decomposition step includes an arsenic species. Experimentally, the growth rate decreases with increasing arsine flow rate, and this is attributed to the site-blocking effect.[14] In our preliminary study, the site-blocking mechanism does not agree with experimental data. Furthermore, the site-blocking model is more applicable when the TEGa and arsenic fluxes are supplied alternatingly, as in atomic-layer epitaxy (ALE).[16] We therefore propose enhanced desorption of TEGa by excess arsenic on the surface in the continuous CBE or MOMBE process. The proposed chemical reactions on the substrate surface are as follows,

$$(1)\ TEGa(g) \xrightarrow{FS_0} TEGa(ad)$$

$$(2)\ TEGa(ad) \xrightarrow{k_3} DEGa(ad)$$

$$(3)\ TEGa(ad) + As_2\ (ad) \xrightarrow{k} TEGa(g) + \frac{1}{2}As_4(g)$$

$$(4)\ DEGa(ad) \xrightarrow{k_{-2}} DEGa(g)$$

$$(5)\ DEGa(ad) + As_2\ (ad) \xrightarrow{k} GaAs + \text{other products}$$

In reaction (1), F is the TEGa beam flux, and S_0 is the adsorption probability. k's are various reaction rate coefficients. Similar to Robertson's model, we postulate the cleavage of the second ethyl radical to be the rate-limiting step. In our model, the growth rate (R_g) depends not only on the DEGa surface concentration X_2, but also on the arsenic surface concentration C_{As}.

$$R_g = kX_2\ C_{As}. \tag{1}$$

In the steady-state condition, concentrations of adsorbed TEGa (X_3) and DEGa (X_2) are independent of time. Solving for X_2 and X_3 and assuming Arrhenius behavior for the reactions, $k_i = A_i \exp(-E_i/RT)$, we obtain

$$R_g = \frac{FS_0}{(1+\frac{A_1}{C_{As}}e^{-\Delta E_1/RT})\ (1+A_2\ C_{As}e^{-\Delta E_2/RT})} \tag{2}$$

where $A_1 = A_{-2}/A$; $A_2 = A'/A_3$; $\Delta E_1 = E_{-2} - E$ and $\Delta E_2 = E' - E_3$. There are six unknown parameters (A_1, A_2, ΔE_1, ΔE_2, FS_0, and C_{As}). Note that in this model, we have further assumed that (1) there is no Ga desorption because the growth temperature is below that of Ga desorption; and (2) the temperature dependence of S_0 and C_{As} is small in the range studied. Chow and Fernandez have measured As_2 and As_4 sticking probabilities on GaAs (001) in conventional MBE to be independent of substrate temperature up to 540°C.[17] As arsenic species on the surface are depleted during growth, they are replenished by the incoming flux.

We use equation (2) to fit experimental data from reference [14]. First, we fit the growth rate versus substrate temperature at 0.64 sccm of TEGa flow rate and 2.2 sccm of AsH_3 flow rate. We obtain the fitting parameters as follows: $A_1 = 1.6 \times 10^4$ ML, $\Delta E_1 = 22$ kcal/mol, $A_2 = 1.0 \times 10^{-14} ML^{-1}$, $\Delta E_2 = -46$ kcal/mol, $FS_0 = 1.09\ ML/cm^2s$, and $C_{As} = 0.25$ ML. ML stands for monolayers. (A discussion on the reasonableness of these numbers is presented later in this paper.) Then, we fix the values of A_1, A_2, ΔE_1, and ΔE_2, and adjust FS_0 and C_{As} to other growth-rate curves, as shown in Figure 2.

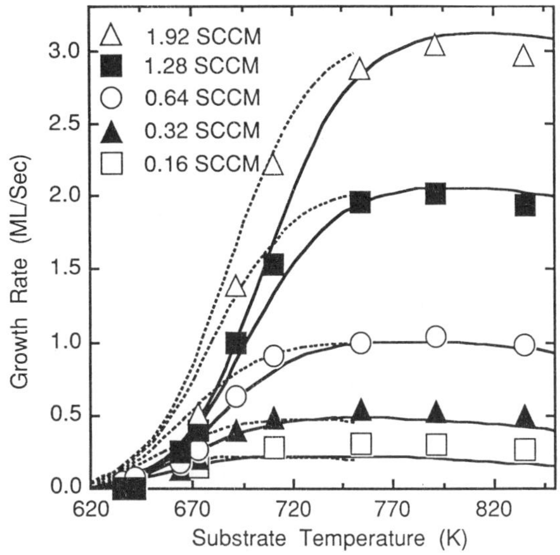

FIGURE 2
Growth rate versus substrate temperature for different TEGa and arsine flow rates. Symbols are experimental data from Ref. [14] for arsine flow rate of 2.2 sccm and various TEGa flow rates (0.16 sccm to 1.92 sccm). Curves are calculation results from our model. Solid lines are for 2.2 sccm arsine flow rate and dashed lines for 1.0 sccm of arsine flow rate.

Here we note that the model fits the low temperature region better than the model without taking into account the effect of arsenic.[6] Figure 3 shows arsenic surface concentration C_{As}, as a function of TEGa flow rate. In the range of lower TEGa flow rate, C_{As} increases linearly with TEGa flow rate. In the range of higher TEGa flow, C_{As} tends toward saturation. This is reasonable because higher TEGa flow rate, resulting in higher growth rate, requires higher arsenic surface concentration. However, when the TEGa flow rate is too high, C_{As} will be limited by the available species.

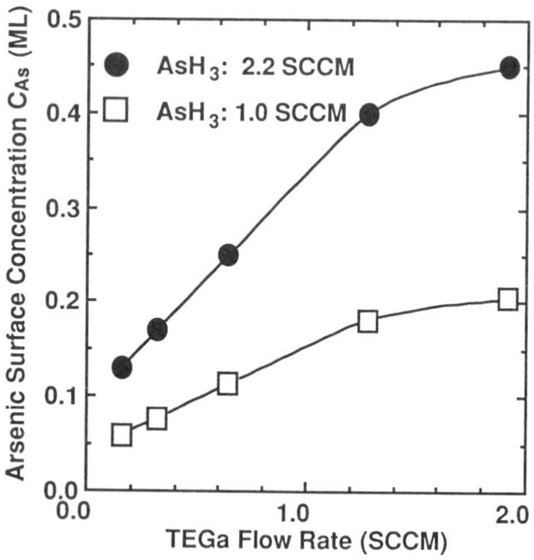

FIGURE 3
Arsenic surface concentration (C_{As}) in units of monolayers (ML) versus TEGa flow rate at two different arsine flow rates.

From the relation between FS_0 and TEGa flow, we can obtain S_0, as shown by the solid line with solid circles in Figure 4. The dashed line is a calculation, with S_0 = 1, for comparison. In our model it appears that the adsorption probability of TEGa is constant, S_0 ~ 0.8,

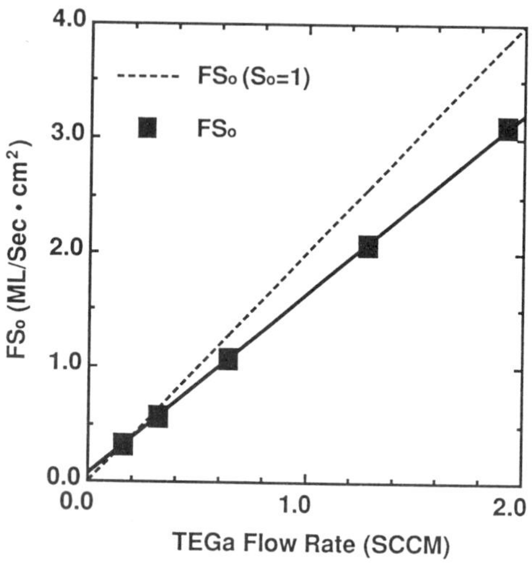

FIGURE 4
FS_0 of TEGa versus TEGa flow rate. The dashed line is a calculation with $S_0 = 1$ for comparison. The data indicates $S_0 = 0.8$.

independent of TEGa flow rate, whereas Robertson *et al.* assume that S_0 is unity.[3]

Using these parameters we calculate the growth rate as a function of C_{As}, as shown in Figure 5 for various substrate temperatures used in Ref. 14. Because we do not know the As_2 flux on the substrate surface, it is not possible to quantify C_{As}, from the fittings, as a function of AsH_3 flow rate. However, qualitatively these curves agree well with data in Ref. 14.

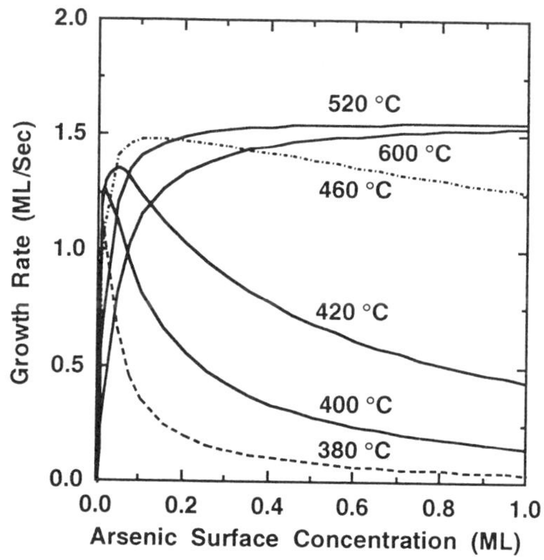

FIGURE 5
Calculations of growth rate versus arsenic surface concentration at different substrate temperatures. The TEGa flow rate is 0.96 sccm. These curves qualitatively agree with data in Ref. [14].

Figure 6 shows a three-dimensional plot of calculated growth rates as a function of substrate temperature and surface arsenic concentration.

At lower substrate temperature, equation (2) can be simplified into $R_g = FS_0 A_2^{-1} C_{As}^{-1} \exp(-|\Delta E_2|/RT)$. The growth rate increases with increasing substrate temperature, as shown in Figures 2 and 6, because higher substrate temperature results in more TEGa decomposition into DEGa. At higher substrate temperature, equation (2) can be simplified into $R_g = FS_0 C_{As} A_1^{-1} \exp(\Delta E_1/RT)$. The growth rate decreases with increasing substrate temperature, but very slowly because ΔE_1 is small. In Figures 5 and 6, at lower

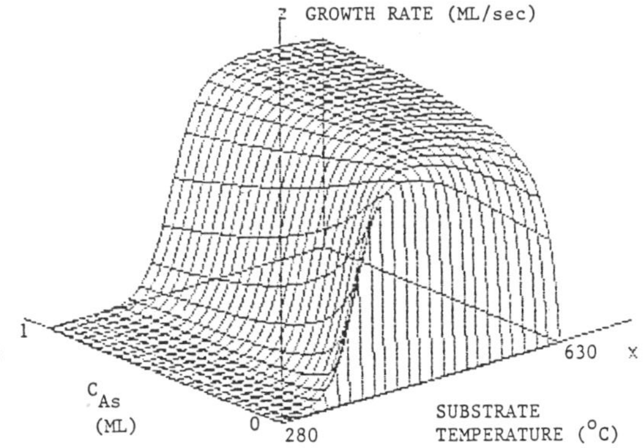

FIGURE 6
Calculated growth rate versus substrate temperature and surface arsenic concentration.

substrate temperature, the growth rate at lower AsH_3 flow rate is higher than that at higher AsH_3 flow rate because too much arsenic on the surface will enhance desorption of TEGa. However, at high substrate temperature the growth rate increases with increasing AsH_3 flow rate because surface reactions become fast. At high AsH_3 flow rate, the growth rate is independent of AsH_3 flow. In this region of high temperature and large AsH_3 flow, the CBE process is similar to solid-source MBE.

We now discuss the parameters obtained from the model. Consider first $\Delta E_1 = E_{-2} - E$, where E_{-2} is the activation energy of DEGa desorption and E is the activation energy for reaction (5). Since DEGa and As_2 are reactive, E is probably small. Therefore, $\Delta E_1 \simeq E_{-2}$, which means $E_{-2} \simeq 22$ kcal/mol. According to Robertson *et al.*,[3] if the activation energies for cleaving the first and second ethyl groups are comparable (~13 kcal/mol), then the desorption energy of DEG would be ~ 26 kcal/mol.[3] This value compares well with our result. Now consider $\Delta E_2 = E' - E_3$, where E′ is the activation energy of arsenic-enhanced desorption, and E_3 is the activation energy for TEGa decomposing into DEGa. The upper limit on ΔE_2 given by Robertson *et al.* is -49 kcal/mol.[3] Our best fitting result is -46 kcal/mol. Since the binding energy for physisorbed large molecules such as TEGa should be in the range of 1-10 kcal/mol,[3] E_3 is in the range of 47 to 56 kcal/mol. This value, close to the gas-phase value of 46 kcal/mol,[18] seems to be somewhat larger than expected, for surface species. This discrepancy may be a result of inaccurate substrate temperature measurement because fitting data depends sensitively on substrate temperature.

In summary, we have developed a growth kinetic model for CBE of GaAs using TEGa and AsH_3. The model can fit the experimental data very well and explain how TEGa and arsine flow rates influence the growth rate. Although our model may not be complete or unique, it does reveal that arsenic plays an equally important role as TEGa in CBE growth of GaAs.

3.2 MOMBE OF InAs

We have applied the same kinetic model to MOMBE of InAs grown with TMIn and As_4.[5] Using equation (3) to fit the growth rate versus substrate temperature at 3.8×10^{-7} Torr arsenic overpressure, we obtained the following parameters: $FS_0 = 1.37$ ML/cm^{2}s, $A_1 = 4.2 \times 10^4$ ML, $\Delta E_1 = 19.5$ kcal/mol, $A_2 = 5 \times 10^{-9}$ ML^{-1}, $\Delta E_2 = -27.8$ kcal/mol and $C_{As} = 0.27$ ML. Then, fixing these parameters, except C_{As}, we fit other experimental growth rates versus substrate temperatures, for different arsenic overpressures, as shown in Figure 7.

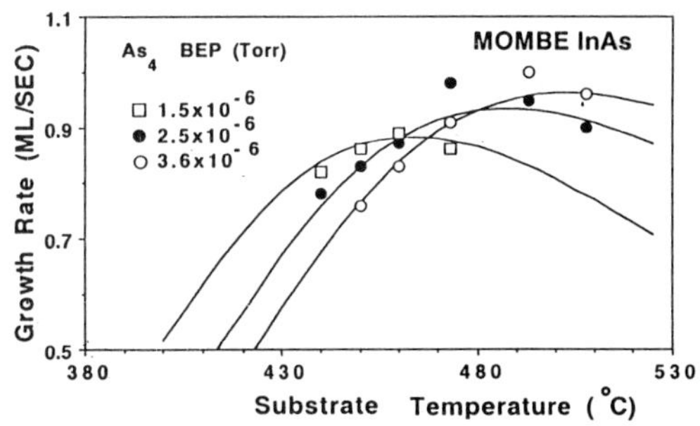

FIGURE 7
MOMBE growth rate of InAs versus substrate temperature at different arsenic surface concentrations.

Figure 8 shows that the best C_{As} for each curves turns out to be linearly dependent on the arsenic overpressure P_{As_4} from the flux measurement.

Using these parameters, we then calculate the growth rate as a function of arsenic surface concentrations for different growth temperatures. The results are plotted in

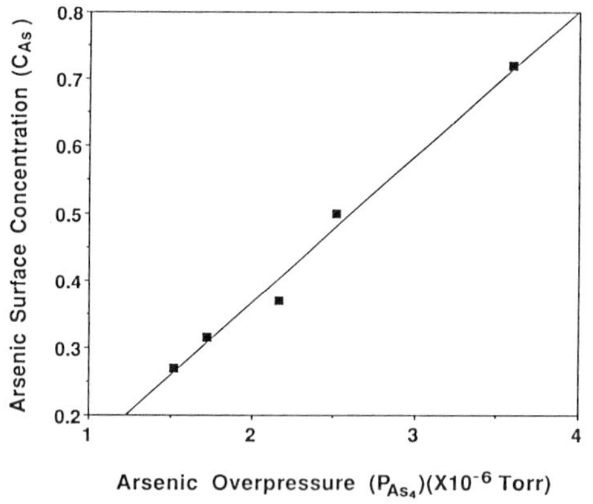

FIGURE 8
The relation between arsenic surface concentration (C_{As}) and arsenic overpressure (P_{As_4}).

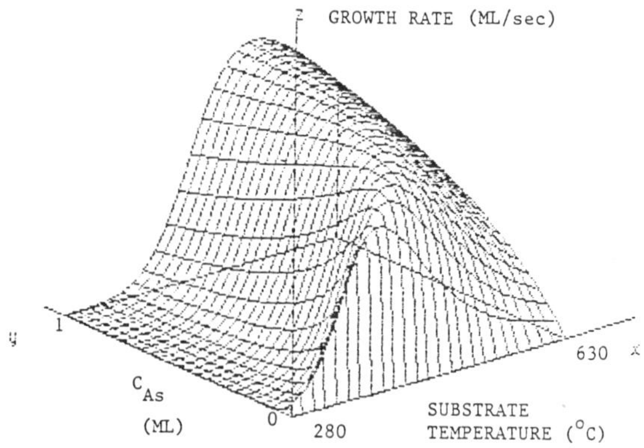

FIGURE 10
Calculated growth rate of InAs versus substrate temperature and surface arsenic concentration.

Figure 9. Various symbols are experimental data, and they agree with calculations remarkably well. Figure 10 shows a three-dimensional plot of growth rates as a function of substrate temperature and surface arsenic concentration.

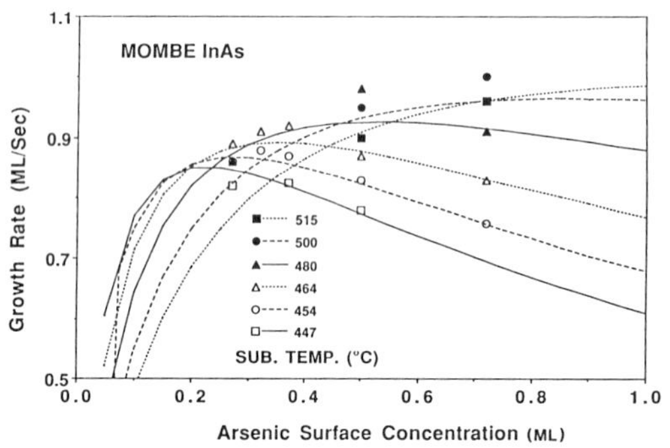

FIGURE 9
MOMBE growth rate of InAs versus arsenic surface concentration for different substrate temperatures.

Due to the surface catalytic effect, E_3, the activation energy for cleaving a methyl radical from adsorbed TMIn, should be lower than that in the gas phase. Because both DMIn and As_2 are reactive, the activation energy of reaction (5), E, should be small, less than 10 kcal/mol. Then E_3 is less than 37.8 kcal/mol, as expected. E_{-2}, the activation energy for DMIn desorption is then less than 30 kcal/mol. This value is similar to the activation energy for TMGa desorption of 25 kcal/mol.[19]

The physical interpretation is similar to that discussed for CBE of GaAs. The growth-rate behavior at high substrate temperature and high arsenic pressure is similar to that in conventional solid-source MBE.

4. LASER-ASSISTED SELECTIVE-AREA EPITAXY OF GaAs

Selective-area growth is attractive because it could make possible the deposition of patterned layers, which could then be covered with layers of a different pattern or composition. Laser-assisted MOCVD and MOMBE have been reported recently. Specifically, Aoyagi *et al.* reported laser-assisted atomic layer epitaxy[20] and MOMBE[21] using argon ion laser. However, we chose ArF excimer laser because the large beam area (~1 cm^2) and short wavelength (193 nm) make possible the growth of chip-size patterns by projection lithography.

Once the surface oxide had desorbed and the growth temperature was established, the laser beam was directed at the substrate, followed immediately by the TEGa beam. Laser-enhanced growth rates were measured with a stylus profiler (Sloan Dektak model II). Excimer laser radiation leads to growth of ~2×10 mm^2 area GaAs films. The growth rate enhancement in the center of the film, in the region of highest laser intensity, is plotted as a function of average substrate temperature in Figure 11. At 520°C the growth of GaAs resulting from pyrolysis of TEGa occurs with maximum efficiency; hence laser enhancement does not occur. However, below ~450°C pyrolysis becomes inefficient, and laser-enhanced growth dominates. The growth rate at low temperature is 0.15 Å $pulse^{-1}$ or 3.0 Å s^{-1} at 20 Hz laser repetition rate.

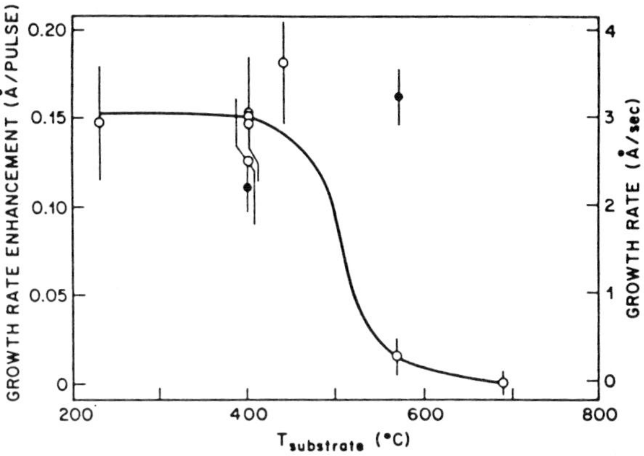

FIGURE 11
Growth rates as a function of steady-state substrate temperature (i.e., neglecting laser heating). (o): Laser-induced growth in Å $pulse^{-1}$ and Å s^{-1}, (•): Total growth rate in Å s^{-1}. TEGa and As_4 beam pressures (uncorrected for ionization guage sensitivity) were 1.5×10^{-6} and 5×10^{-6} Torr, respectively. Laser fluence = 0.01 ± 0.01 J cm^{-2} $pulse^{-1}$. Repitition rate = 20 Hz.

The laser-enhanced growth rate of 3 Å s^{-1} at low temperature can be compared with the absolute TEGa beam flux. We can compute the TEGa flux from the TEGa pressure measured with the ionization gauge. A sensitivity factor of 5.3 for TEGa (relative to N_2) was compted from empirical methods.[22,23] Taking into account geometric factors, we estimate the flux to be 8.4×10^{14} molecules $cm^{-2}s^{-1}$ under the conditions used to obtain the data in Figure 11. From this flux we compute a maximum growth rate of ~3.8 Å s^{-1} for GaAs. Consequently, at T≤500°C, where pyrolytic deposition is inefficient, TEGa molecules impinging on the surface near the region exposed to the most intense laser radiation react to form GaAs with near unity efficiency.

Of the possible mechanisms responsible for enhanced growth, gas-phase photodissociation can be ruled out, since the duty cycle of the laser is so small (3×10^{-7}) and the gas-phase number density is so low (1×10^{-6} Torr). Consequently, the laser must be dissociating adsorbed TEGa. Furthermore, TEGa must stick on GaAs longer than the time between laser pulses (50 ms), otherwise desorption would cause the laser-enhanced growth rate to fall well below the TEGa impingement rate, contrary to observations.

Laser-induced decomposition of adsorbed TEGa could occur by a photolytic (photodissociate TEGa), electronic (create electron-hole pairs in GaAs which aid in TEGa decomposition), or pyrolylitic process (thermally decompose TEGa because of transient heating of GaAs by the pulsed laser). We can gain more insight into these processes by determining the dependence of growth rate on laser fluence (J cm^{-2} $pulse^{-1}$). This was accomplished by measuring the film thickness along the short dimension of the beam with the stylus profiler and then comparing thickness with laser intensities determined from the beam profile measurements. A stylus profile is shown in Fig. 12, along with the laser intensity. Thickness falls off rapidly as a function of laser fluence, indicating a highly nonlinear process.

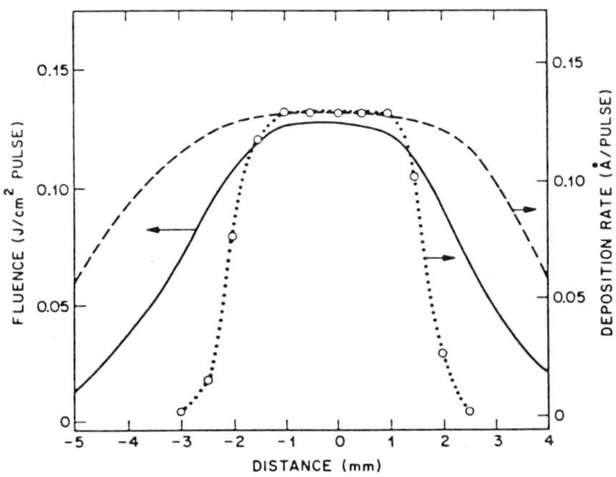

FIGURE 12
Deposition rate and laser intensity as a function of distance from the center of the laser beam across the short dimension. Beam pressures are the same as for Fig. 1. Substrate temperature = 400°C. (o); peak laser fluence = 0.12 J cm^{-2} $pulse^{-1}$, repetition rate = 20 Hz. Solid curve: laser intensity. Dashed curve: predicted deposition rate based on photochemical model.

We have shown elsewhere that the photochemical mechanism would result in a growth-rate enhancement nearly following the laser fluence profile as shown by dashed line in Fig. 12. Since the observed deposition rate falls off much more rapidly, we can exclude the simple photochemical model to explain laser-enhanced growth.[7] From the lack of laser-enhanced deposition at 0.05 J/cm^2 and below, we estimate the photodissociation absorption cross section to be at least ten times lower than the gas-phase value.[7] Likewise, an electron-hole pair mechanism would not depend so strongly on laser fluence and so does not seem likely.

We ascribe enhanced growth to pyrolysis of $TEGa_{(ads)}$ due to transient heating of the GaAs surface from pulsed laser irradiation. Such a process would be much more strongly dependent on peak laser power than the photolysis mechanism, as observed. Based on a numerical solution to the one-dimensional heat flow equation, we estimate a peak temperature rise T_{peak} of ~ 700°C at a fluence of 0.1 J/cm^2. From an Arrhenius plot of log growth rate versus $1/T_{peak}$ across the laser beam, we obtain an activation energy of 22-15 kcal/mol in the region where the film thickness is falling off rapidly with distance from the center. This is in fair agreement with the activation energy for TEGa pyrolysis of 20 kcal/mol derived from data by Pütz *et al.*[25] and with activation energies of 18 and 21 kcal/mol for the growth rate of GaAs from TEGa.[26] One piece of supporting evidence is the lack of wavelength dependence at high fluence for relative carbon coverage versus number of laser pulses. Carbon coverage was measured by x-ray photoelectron spectroscopy on a TEGa-covered GaAs surface as a function of the number of laser pulses at various laser intensities. At low fluence, the carbon coverage depends on laser wavelength, indicating a photolytic effect. At high fluence, the carbon coverage is independent of laser wavelengths, indicating pyrolytic, not photolytic, effects dominate.

Crystalline quality was studied by RBS for samples deposited at 400 and 570°C. In all cases, X_{min}'s were low (3-5%), indicating good crystal quality. Some dechanneling was observed in some cases near the film-substrate interface, indicating defects. However, the best results were obtained at the highest laser intensity (0.13 J pulse^{-1}) and at a substrate temperature of 400°C, where no interface could be found, and the RBS channeling spectrum was indistinguishable from a single crystal substrate. Low-temperature (2 K)-photoluminescence spectra (Fig. 13) revealed the added benefit of layer-by-layer laser annealing on luminescence intensity. the peak at 835 nm is identified as an acceptor band associated with a carbon impurity. The regions exposed to the highest excimer laser intensity during growth exhibited the most intense luminescence, most likely because of laser annealing of defects in the growing epitaxial film, which otherwise acts as a nonradiative recombination center. However, the overall quality of the material is still rather low since the exciton bands are not observed, the most intense peak is broad and due to carbon imputity, and the overall intensity is relatively small. We attribute this to the relatively poor-quality vacuum system in which these preliminary experiments were carried out, compared to state-of-the-art MBE systems. Similar effects were obvserved in GaAs grown from elemental Ga.

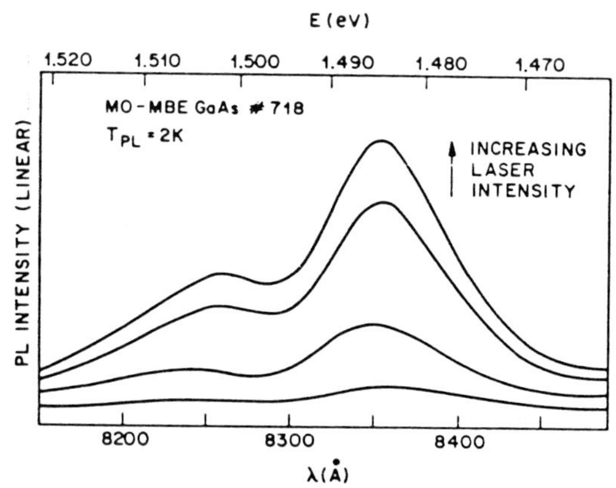

FIGURE 13
Low-temperature photoluminescence spectrum of GaAs deposited at 570°C. TEGa and As_4 beam pressures (uncorrected for ion gauge sensitivity) were 1.3×10^{-5} and 1.7×10^{-5} Torr, respectively. Laser repetition rate = 20 Hz. Laser fluence = 0.095 J cm^{-2} pulse^{-1} for top curve and ≤ 0.01 for bottom curve. The other two curves are for levels of radiation between these two extremes.

In summary, 193 nm ArF excimer laser irradiation enhances the growth of GaAs from triethylgallium and As_4. Below ~ 450°C, growth is controlled by the laser. Film growth is epitaxial, as confirmed by RBS.

Photoluminescence yields are also increased because of laser annealing of the growing film. The mechanism is dominated by laser pyrolysis (as oposed to photolysis) of adsorbed TEGa. Laser-induced growth is very efficient, with ~ 100% of the impinging TEGa being dissociated by the laser to form GaAs. Finally, we have demonstrated the potential for selective area growth, since at low substrate temperature there is no growth in the absence of laser irradiation.

ACKNOWLEDGMENT

The work at UCSD is partially supported by the Air Force Wright Research and Development Center under Contract No. F33615-88-C-1861 and by the Office of Naval Research under Grant No. N00014-89-J-1147. We wish to thank T.P. Chin, L.Y. Wang, and H.Q. Hou for assistance in MOMBE data taking and analysis and V.R. McCrary and J.C. Beggy for assistance during laser-assisted MOMBE.

REFERENCES

1. E. Veuhoff, W. Pletschen, P. Balk and H. Luth, J. Cryst. Growth 55 (1981) 30.
2. W.T. Tsang, Appl. Phys. Letters 45 (1984) 1234.
3. R. Robertson, Jr., T.H. Chiu, W.T. Tsang and J.E. Cunningham, J. Appl. Phys. 64 (1988) 877.
4. B.W. Liang and C.W. Tu, Appl. Phys. Letters, August 13, 1990.
5. B.W. Liang, L.Y. Wang and C.W. Tu, Proc. XII State-of- the-art Prog. on Comp. Semicond., in press.
6. L. Fraas, G.R. Girard, V. Sundaram, C. Master, C. Nelson and R.A. Stall, Mater. Res. Soc. Symp. Proc. 145 (1989) 253.
7. V.M. Donnelly, C.W. Tu, J.C. Beggy, V.R. McGrary, M.G. Lamont, T.D. Harris, F.A. Baiocchi and R.C. Farrow, *Appl. Phys. Letters 52 (1988) 1065.*
8. H. Sugiura, R. Iga, T. Yamada and M. Yamaguchi, Appl. Phys. Letters 54 (1989) 335.
9. C.W. Tu, B.W. Liang, T.P. Chin and J. Zhang, J. Vac. Sci. Technol. B 8 (1990) 293.
10. B.W. Liang, T.P. Chin and C.W. Tu, J. Appl. Phys. 67 (1990) 4393.
11. J.A. McCaulley, V.R. McCrary and V.M. Donnelly, J. Phys. Chem. 93 (1988) 1148.
12. C.W. Tu, V.M. Donnelly, J.C. Beggy, F.A. Baiocchi, V.R. McCrary, T.D. Harris and M.G. Lamont, Appl. Phys. Letters 52 (1988) 966.
13. S. Horiguchi, K. Kimura, K. Kamon, M. Mashita, M. Shimazu, M. Michara and M. Ishii, Jpn. J. Appl. Phys. 25 (1986) L979.
14. T.H. Chiu, Mater. Res. Soc. 145 (1989) 47.
15. C.W. Tu, B.W. Liang and T.P. Chin, J. Cryst. Growth, in press.
16. D.E. Aspnes, E. Colas, A.A. Studna, R. Bhat, M.A. Koza and V.G. Keramidas, Phys. Rev. Letters 61 (1988) 2782.
17. R. Chow and R. Fernandez, Mater. Res. Soc. Symp. Proc. 145 (1989) 13.
18. M. Tirtowidjojo and R. Pollard, J. Cryst. Growth 77 (1986) 200.
19. E. Coltrin and R.J. Kee, Mater. Res. Soc. Symp. Proc. 145 (1989) 119.
20. Y. Aoyagi, A. Doi, T. Meguro, S. Iwai, K. Nagata and S. Nonoyama, Chemtronics 4 (1989) 117.
21. K. Nagata, Y. Iimura, Y. Aoyagi and S. Namba, J. Cryst. Growth 95 (1989) 142.
22. D.K. Sharma and A.N. Pandey, J. Acta Ciencia Indica 1 (1974) 35.
23. K.J. Miller and J.A. Savchik, J. Am. Chem. Soc. 101 (1979) 7206.
24. V.R. McCrary and V.M. Donnelly, J. Cryst. Growth 84 (1987) 253.
25. N. Pütz, E. Veuhoff, H. Heinecke, M. Heyen, H. Lüth and P. Balk, J. Vac. Sci. Technol. B 3 (1985) 671.
26. P. Balk, M. Fisher, D. Grundmann, R. Lückerath, M. Lüth and W. Richter, J. Vac. Sci. B 4 (1987) 1453.

Frontiers of Materials Research/
Electronic and Optical Materials
M. Kong, L. Huang (editors)

OPTICAL PROPERTIES OF ZnSe-ZnS STRAINED-LAYER SUPERLATTICES

X.W.Fan, G.H.Fan, D.Z.Shen, Z.P.Guan, F.Y.Jiang, and J.Y.Zhang

Changchun Institute of Physics, Academia Sinica, Changchun 130021, P.R. China

Good quality ZnSe-ZnS strained-layer superlasttices (SLSs) were grown by atmospheric pressure (AP) MOCVD and identified by X-ray diffraction and photoluminescence (PL) measurements. The PL bands in the SLS were studied and ascribed to transition of n=1,2 subband heavy-hole excitons. The excitonic dispersive optical bistability with ns response was measured in ZnSe-ZnS SLS and its origin was attributed to the excitonic saturating absorption.

1. INTRODUCTION

Superlattice structures can be expected to give entirely new types of semiconductor devices, such as multiquantum well (MQW) laser and optical nonlinearity and bistability devices. Recently, much interest has also been paid to II-VI compound semiconductors, ZnSe and ZnS, as materials for short-wavelength optical devices since they have wide band gaps at room temperature (RT). The ZnSe-ZnS layered structure is one of the most extensively studied superlattices (SLs), and is well known as a type I structure. For ZnSe-ZnS superlattice (SL), the mismatch between their lattice parameters is 5% (ZnS:0.5412nm and ZnSe:0.56676nm) and a strained SL is formed. Fujiyasu et al. have prepared ZnSe-ZnS SLs on (100) GaAs substrates by hot wall epitaxy[1] and observed their photoluminescence associated with the band edge or with excitons at RT and 77K[2]. Yokogawa et al.[3] have successfully fabricated for the first time good-quality ZnSe-ZnS SLS by low pressure metalorganic vapour phase epitaxy (LP MOCVD) using methylalkyls as source materials. Taike et al.[4] have reported the growth of ZnSe-ZnS SLS by metalorganic molecular beam epitaxy (MOMBE) and observed the two sets of satellite in the X-ray diffraction pattern of the SLS. Kawakaini et al.[5] have the first observation of n=1 heavy-hole and light hole excitons of ZnSe-ZnS SLS in the optical absorption spectrum[6]. Yokogawa et al. have reported for the first time the fabrication and characterization of three-dimensional optical waveguides using a ZnSe-ZnS SLS, which exhibits a propagation loss of less than $1cm^{-1}$ at a wavelength of 0.633μm.

In this paper we report that the ZnSe-ZnS SLSs were grown by atmospheric pressure (AP) MOCVD, indentified by X-ray diffraction and photoluminescence measurements. The excitonic emission spectra, and the optical bistability in the ZnSe-ZnS SLSs are measured and studied.

2. GROWTH OF ZnSe-ZnS SLSs

The electronic grade DMZ, H_2S and H_2Se diluted in H_2 carrier gas were used as reactants for the film growth. To avoid the pernicious gas phase premature reaction, a RF-heated reactor with water cooled sheath was used. H_2-carried DMZ went through the innertube and H_2-mixed H_2S and H_2Se went through the central nozzle. They met each other until 2.5 cm apart from the substrates. The reactor design and growth parameters ensure that gas flows of groups II and VI are separated up and down into two steady laminas to prevent the gas phase prereaction between them. The deposition area is situated in steady velocity field and even temperature field. The stagnent spaces in the reactor and pipeline were reduced as fully as possible. The gas flows were automatically controlled by air operated valves using an IBM PC/XT computer.

Typical growth temperature is in the range of 290-420°C. Some SLSs consisting of 20-100 pairs of ZnSe(3-10nm)-ZnS(2-10nm) were grown on (100)GaAs or (111)CaF_2 substrates.

X-ray rocking curves have been obtained in symmetric (400) reflections using a Cu $K\alpha_1$ line (λ=0.15405nm). Fig.1 shows the rocking curves of the SLSs consisting of 100 pairs of ZnSe-ZnS. The zero-order peak exists near the GaAs substrate peak, and the ±1st and ±2nd order satellite peaks can be observed on both sides of the zero-order peak. It is well known that the superlattice periodicity can be obtained from the Bragg angle difference $\Delta\theta$ between the satellite peaks. The superlattice periodicity which corresponds to $\Delta\theta$ agrees well with the estimated periodicity from the growth rates.

PL measurements of the SLSs were carried out with a conventional spectrometer system. The 337.1nm line of a N_2 laser was used for sample excitation. The PL spectrum of the SLS shows an intense and sharp excitonic emission band from the ZnSe wells. The peak position depends on the thickness of the ZnSe well layer. The relationship between the well width L_w and the peak position shift ΔE is $\Delta E \propto L_w^{-2}$. This is in good agreement with the calculation of quantum energy levels using the Kronig--Penney model with the strain effect taken into account. It indicates that this behavior is due to the quantum-size effect of the superlattice.

We have successfully demonstrated that atomospheric pressure MOCVD using DMZ, H_2Se and H_2S can be quite useful for growing good quality ZnSe-ZnS SLS.

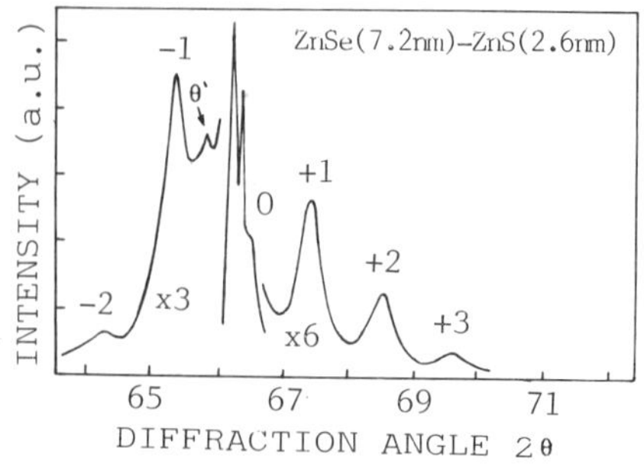

Fig.1 X-ray diffraction satellites of the ZnSe-ZnS SLS.

3. EXCITON EMISSION IN ZnSe-ZnS SLSs

Saito and Shionoya[7] ascribed the E and P bands in PL spectrum of ZnSe single crystal to the exciton-electron (E_x-e) and exciton-exciton (E_x-E_x) collisions, respectively. In our earlier work attention was paid to study the emission of free excitons[8,9], and E_x-E_x interaction[10-12]. Up to now there is no report on the origin of the E_s band in the SLS with ZnSe as well layer.

Fig.2 shows the PL spectra at 77K of a ZnSe(4.8nm)-ZnS(1.5nm) SLS on (100) GaAs under

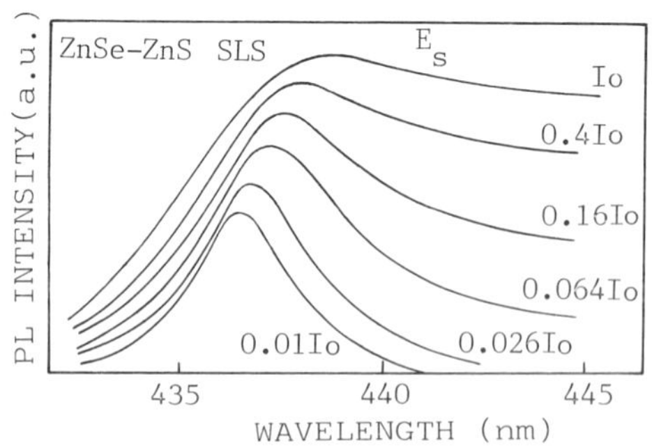

Fig.2 The PL spectra of ZnSe(4.8nm)-ZnS(1.5nm) SLS under different excitation intensities at 77K. (I_o=2MW/cm^2)

different excitation intensities. As seen from Fig.2 there is only one emission band E_s in the spectra. The peak position of the E_s band shifts to lower energy side, and its low energy tail rapidly increases with increasing excitation intensity. One of the most interesting aspects of the work described here concerns with the nature of the processes responsible for the blue emission band E_s in the SLS.

Fig.3 shows the peak energies of the E_s band as a function of temperature for ZnSe epilayer and ZnSe-ZnS SLSs. The curve labelled $E_x(T)$ in Fig.3 describes the temperature dependence of the free exciton energies in ZnSe bulk crystal. As seen from Fig.3, we found that the photon energies of the E_s band for ZnSe(8.7nm)-ZnS(1.1nm) and ZnSe(5.7nm)--ZnS(1.7nm) SLSs are higher than that for ZnSe epilayer at the same temperature due to the quantum size effect in the SLSs. If the curve $E_x(T)$ is shifted vertically, making it overlap with the peak energies of the E_s of two ZnSe--ZnS SLSs at 77K, respectively, the $E'_x(T)$ can be obtained (not shown in the figure). It should be noticed from Fig.3 that the peak engergies of the E_s band shift to the low energy side with temperature by amounts more than the shift of the $E'_x(T)$. The inset of figure 3 shows the variation with temperature of the energy separation $\Delta E(T)=E'_x(T)-E_s(T)$. It is obvious that the $\Delta E(T)$ depends linearly on temperature. This result is in good agreement with the E band obtained by Saito and Shionoya[7], who ascribed the E band in CdS, CdSe and ZnSe to the E_x-e collision. From this consideration, it is reasonable to think that the E_s band obtained here from the ZnSe-ZnS SLSs originated from the E_x-e collision. We carried out comparison of the shape of the E_s band (for $I=0.016I_o$) of a ZnSe(4.8nm)-ZnS(1.5nm) SLS measured at 77k with the theoretical curve calculated according to the theory of Benoit a la Guillaume et al.[13] with $I=0.16I_o$ and effective electron temperature T_e=110K. The fact that the theoretical curve of the E_x-e collision is also in good agreement with the experiment result presents further support for the above conclusion.

As mentioned above, generally only one emission band E_s can be measured in the PL spectra of the ZnSe-ZnS SLS. The E_s band was reasonably considered to be the excitonic emission between the n=1 electron subband

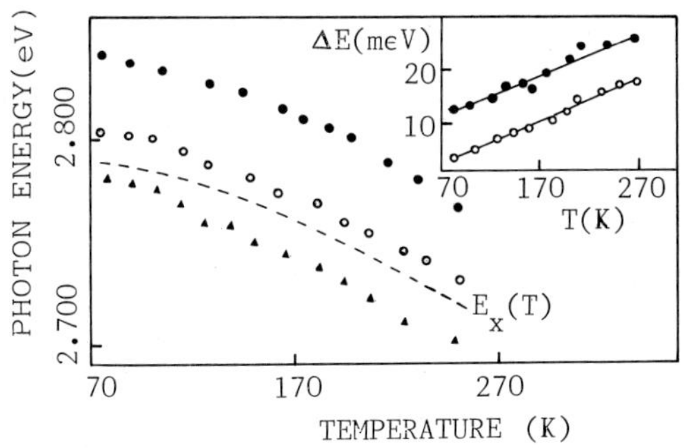

Fig.3 Temperature dependence of photon energies of the PL emission peak for various samples:(▲)ZnSe film, (o)ZnSe(8.7nm)-ZnS(1.1nm) SLS and (•)ZnSe(5.7nm)-ZnS(1.7nm) SLS. The inset shows the variation with temperature of $\Delta E(T)=E'_x(T)-E_s(T)$.

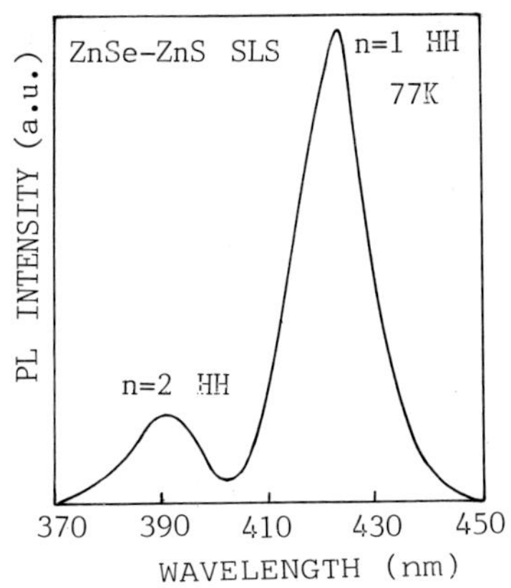

Fig.4 PL spectrum of the ZnSe-ZnS SLS on CaF excited by the 337.1nm line of a N_2 Laser at 77K.

and the n=1 heavy-hole subband. In some selected samples, such as a ZnSe(2.2nm)-ZnS(2.3nm) SLS having 100 pairs on (111)CaF_2 with good quality identified by X-ray diffraction, it is found that the PL spectrum at 77K showed two emission bands in near edge band region, as shown in Fig.4. The two emission bands located at 390 and 421.5nm, respectively. Using the LCAO theory[14], the theories of the stress-induced energy band structure[15,16] and the Kronig-Penney band model[17,18] we have calculated the subband levels of a ZnSe(2.2nm)-ZnS(2.3nm) SLS. The calculated data at 77K are as follows:

E_{1HH}=2.956-3.005eV λ=420.5-413.7nm

E_{2HH}=3.166eV λ=392.6nm

Where E_{1HH} and E_{2HH} are the energy gaps between n=1 electron subband and n=1 heavy-hole subband, and n=2 electron subband and n=2 heavy-hole subband, respectively. It is reasonable to consider that the emission band at 421.5nm could be ascribed to n=1 free excitons, so the difference between the theoretical and experimental values is related to the binding energy of free exciton. The transition between the n=2 heavy-hole subband and the n=2 electron subband is 3.166eV, corresponding to the wavelength 392.6nm, while the experimental peak value in the PL spectrum is 390nm. The difference can be thought to be resulted from the n=2 electron subband level which might be above the minimum of the ZnSe conduction band.

4. OPTICAL BISTABILITY IN ZnSe-ZnS SLS

Since the first observation of optical bistability in GaAs[19] and InSb[20], a variety of semiconductor materials has been considered for possible applications to nonlinear signal processing and optical computing. ZnSe and ZnS interference filters have extensively been studied, but they employ thermal nonlinearities, resulting in inefficient switching with slow switching speeds[21]. Kolodziejski et al.[22] have reported nonlinear excitonic effect at 77K in ZnSe-ZnMnSe SLSs. Gribkovskii et al.[23] have observed optical bistability at 80K in MOCVD-grown ZnSe film. The peculiarities of nonlinear optical response are associated with the polaritonic mechanism of light propagation. Peyghambarian et al.[24] have observed large optical nonlinearities in MBE-grown ZnSe film at RT and 77K. The maximum nonlinear index per excited electron-hole pair at RT is comparable to that of bulk GaAs and GaAs-AlGaAs MQWs. The dominating mechanism for the nonlinearty is ascribed to exciton screening.

A ZnSe(3.8nm)-ZnS(6.2nm) SLS having 100 periods on (100) GaAs substrate was grown by AP-MOCVD. In order to measure the optical absorption spectra, it is necessary to etch a window on GaAs substrate. The excitation source was generated by a dye laser, using Coumarin-440 pumped by a model UV-24 N_2 laser. The dye laser pulse has spectral full width at half-maximum(FWHM) of 4nm with central wavelength of 437nm and duration FWHM of 10ns. The time dependence of incident I_o and transmitted I_t pulses was measured by using a Model 4400 Boxcar.

Fig.5 (a) and (c) show the accumulated and normalized temporal shape of incident I_o and transmitted I_t pulses, respectively. Fig.5 (b) and (d) show the hysteresis loops $I_t=f(I_o)$ as the resultant curve of I_t and I_o in (a) and (c), respectively. In Fig.5 (b) the direction of hysteresis loop is clockwise, but in Fig.5(d) the direction of hysteresis is anticlockwise. The fact mentioned above indicates that the change in absorption coefficient or in refractive index caused by changing absorption coefficient is opposite with increasing the I_o for the different incident light wavelength λ_1 and

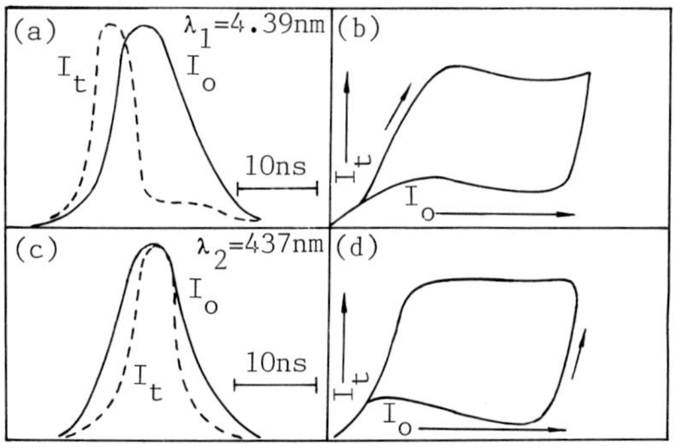

Fig.5 Time dependence of averaged normalized incident (solid) I_o and transmited (broken) I_t pulses (a and c) in the ZnSe-ZnS/GaAs SLS at 77K. The resulting hysteresis loops (b and d).

λ_2 as shown in Fig.5.

Fig.6(a) shows the PL spectrum at 77K in the ZnSe-ZnS SLS on (100)GaAs SLS under the excitation of 337.1nm line of a N_2 laser with pulse width of 10ns and frequency of 20Hz. Fig.6(b) shows the optical absorption spectrum at 77K in the same sample. When we compare the excitonic emission peak position of 438.5nm in Fig.6(a) with the absorption peak position of 438nm, it is reasonable to think that the absorption band at 438nm can be attributed to excitonic absorption. In this case it is reasonable to conclude that the optical bistability obtained here should be attributed to the excitonic absorption.

The contribution to the excitonic optical bistability is due to the change in excitonic

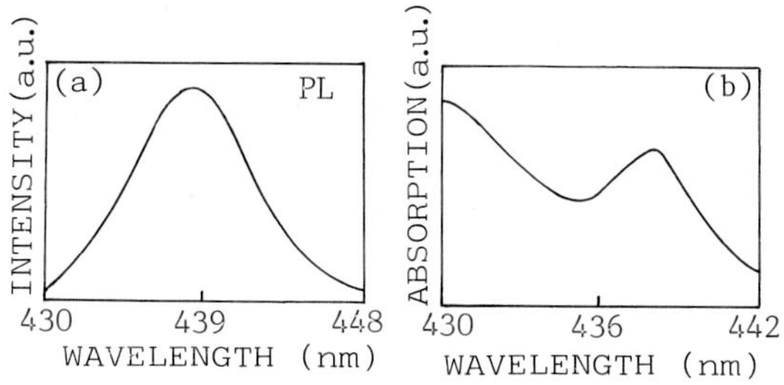

Fig.6 PL spectrum (a) and absorption spectrum (b) at 77K in the ZnSe-ZnS/GaAs SLS.

absorption coefficient $\Delta\alpha$ and refractive index Δn, which is due to the change of excitonic absorption coefficient. On the basis of F-P cavity theorey, the transmission T at the presence of loss is as follows[25]:

$$T=H/(1+F\sin^2\Phi) \quad (1)$$

where

$$H=(1-R_f)(1-R_b)\exp[-(\alpha_o+\Delta\alpha)L]/(1-R_e)^2 \quad (2)$$

$$\Phi=(2\pi/\lambda)(n_o+\Delta n)L \quad (3)$$

and $F=4R_e/(1-R_e)^2$ (4)

with $R_e=\sqrt{R_fR_b}\exp[-(\alpha_o+\Delta\alpha)L]$ (5)

n_o and α_o are the background refractive index and absorption coefficient, respectively. Δn and $\Delta\alpha$ are the changes in refractive index and absorption coefficient, respectively. R_b and R_f are the exit- and entrance-face reflectivity, respectively. λ is the incident light wavelength.

In our case, $\alpha_o + \Delta\alpha$ is about 10^3/cm, the cavity length L is 1 um, $R_f=R_b=R=0.4$. From the above parameters, $\exp[-(\alpha_o+\Delta\alpha)L]$ is about 1, therefore we obtain:

$$T=1/(1+4.4\sin^2\Phi) \quad (6)$$

In this case, the major contribution to the optical bistability is due to the change of refractive index Δn. The fact indicates that the optical bistability in the ZnSe-ZnS SLS should be ascribed to the dispersive optical bistability.

On the basis of the nature of excitonic saturating absorption and the Kramers-Kronig relations, we can obtain the following conclusion: in excitonic absorption region, the relation of $\Delta n<0$ exists when the incident light wavelength λ is longer than the central wavelength λ_o of excitonic absorption, and $\Delta n>0$ exists when $\lambda<\lambda_o$. In this case, the optical bistability with different direction hysteresis loops can be obtained.

5. CONCLUSION

Good quality ZnSe-ZnS SLS on (100)GaAs

and (111)CaF_2 substrates have been prepared by AP-MOCVD for the first time using DMZ, H_2Se and H_2S as source materials. The SLS structure was identified by X-ray diffraction and PL measurements.

A near band edge emission band E_s can usually be observed in the PL spectrum of ZnSe-ZnS SLS, which is attributed to free exciton emission following scattering from electrons in the conduction band. Two emission bands at 421.5 and 390nm were observed in the PL spectrum at 77K for a selected high quality ZnSe(2.2nm)-ZnS(2.3nm) SLS, and can be attributed to n=1,2 subband heavy-hole exciton emissions.

The excitonic dispersive optical bistability with ns response time has been observed in the ZnSe-ZnS SLS on (100)GaAs substrate, and its origin is ascribed to the excitonic saturating absorption.

ACKNOWLEDGEMENTS

This work is supported by the "863" High Technology Research Program in China, the Foundation of Academia Sinica and the National Natural Science Foundation of China.

REFERENCES

1. H. Fujiyasu, H. Takahashi, K. Mochizuki, K. Nakanishi, G. Shimaoka, and H. Kuwabara, in: Proc. Int. Conf. On MBE, San Francisco, (1984).
2. H. Fujiyasu, H. Takahashi, H. Shimisu, and H. Kuwabara, 17th Int. Conf. on the Physics of Semiconductors, San Francisco, (1984).
3. T. Yokogawa, M. Ogura, and T. Kajiwara, Appl. Phys. Lett. 49(1986) 1702.
4. A. Tiake, N. Treraguchi, M. Konagai, and K. Tadahashi, Japan. J. Appl. Phys. 26 (1987) L989.
5. Y. Kawakaini, T. Taguchi, and A. Hiraki, J. Cryst. Growth 93 (1988) 714.
6. T. Yokogawa, M. Ogura, and T. Kajiwara, Appl. Phys. Lett. 52 (1988) 120.
7. H. Saito, and S. Shionoya, J. Phys. Soc. Jpn. 37 (1974) 423.
8. X.W. Fan, and J. Woods, IEEE Tran-ED ED 28 (1981) 428.
9. X.W. Fan, and J. Woods, J. Phys. C. 14 (1981) 1863.
10. D.Z. Shen, and X.W. Fan, Chinese J. Lumin. 9 (1988) 26.
11. D.Z. Shen, and X.W. Fan, Springer Proceedings in Physics, Vol. 38 Electroluminescence, eds. S. Shionoya and H. Kobayashi (Springer-Verlag. Berlin, Heidelberg 1989) pp. 376.
12. L. Ma, Z.K. Tang, and X.W. Fan, J. Lumin. 40 & 41 (1988) 523.
13. C. Benoit a la Guillaume, J.M. Debever, and F. Salvan, Phys. Rev. 177 (1968) 567.
14. W.A. Harrison, "Electronic Structure and the Properties of Solid-The Physics of Chemical Bond", W.H. Freeman and Company, San Francisco, (1980).
15. Chris G. Van de Walle, Phys. Rev. B34 (1986) 5621.
16. M. Chandrasekhar, and F. H. Pollak, Phys. Rev. B15 (1977) 2127.
17. R. Kronig and W.G. Penney, Proc. Roy. Soc. A130 (1931) 499.
18. G. Bastard, Phys. Rev. B24 (1981) 5693.
19. H.M. Gibbs, S.L. McCall, T.N.C. Venkatesan, A.C. Gossard, and W. Wiegmann, Appl. Phys. Lett. 35 (1979) 45.
20. D.A. Miller, S.D. Smith, and A. Johnston, Appl. Phys. Lett. 35 (1979) 658.
21. G. Olbright, N. Peyghambarian, H.M. Gibbs, A. Macleod, and F. Van Milligen, Appl. Phys. Lett. 45 (1984) 1031.
22. L.A. Kolodziejski, R.L. Gunshor, N. Otsuka, S. Datta, W.M. Becker, and A.V. Nurmikko, IEEE J. Quantum Electron. QE-22 (1986) 1666
23. V.P. Gribkovskii, L.G. Zimin, S.V. Gaponenko, I.E. Malinovskii, P.I. Kuznetsov and G.G. Yakushchva, Phys. Stat. Sol. (b) 150 (1988) 761.
24. N. Peyghambarian, S.H. Park, S.W. Koch, A. Jeffery, J.E. Potts, and H. Cheng, Appl. Phys. Lett. 52 (1988) 182.
25. C.D. Poole and E. Garmire, IEEE J. Quantum Electron. QE-21 (1985) 1370.

Frontiers of Materials Research/
Electronic and Optical Materials
M. Kong, L. Huang (editors)

HIGH QUALITY GaAs-AlGaAs FOR HETEROJUNCTION BIPOLAR TRANSISTORS (HBTs) BY MOCVD

Baibiao HUANG, Hongwen REN, Shiwen LIU, Liqiang LIU and Minhua JIANG

Institute of Crystal Materials, Shandong University, Jinan 250100, P.R. China

High electrical and optical quality GaAs and AlGaAs epilayers have been grown by Metallorganic Chemical Vapour Deposition (MOCVD). Preparation of metallurgical and dopant junctions were studied. MQWs with a series of well thicknesses were grown and the thinnest width reached 20A. GaAs/AlGaAs HBTs with current gains of 15-40 at 300K and above 60 at 77K were fabricated, and cut-off frequency of 10.5GHz and maximum oscillation frequency of 5.5GHz were achieved.

1. INTRODUCTION

The use of heterojunctions for high-frequency transistors was first put forward in the late 1940's[1,2,3]. The relatively primitive semiconductor technology, however, prevented any significant experimental studies. The emergence of MBE[4] and MOCVD[5], especially, the recent developments of the two growth technologies have offered the routine growth of heterostructures[6,7], and remarkable improvements of device properties are achieved after the application of heterostructures. One of the most profitable devices is heterojunction bipolar transistor (HBT), which offers the potential of microwave transistors with maximum oscillation frequency of 100GHz[8], switching speed of the order of 10ps at 300K, and a wide range of operating temperatures.

MOCVD growth and characterization of unintentionally , silicon and zinc doped (Al,Ga)As epilayers were studied, and high quality, morphologically smooth and uniform materials were achieved.

GaAs-AlGaAs heterostructures were grown under various growth conditions. Control of dopant and metallurgical junctions were studied. GaAs/AlGaAs multi-quantum wells of a series of thicknesses with different barrier compositions were grown, and the thinnest well width reached 20A. Heterojunction bipolar transistors were prepared and current gains of 15-60 and cut-off oscillation frequency of 10.5GHz has been achieved.

2. EXPERIMENTAL

Growth were performed in a horizontal, atmospheric pressure commercial MOCVD reactor (Epiquip). The main components were TMGa, TMAl and 10% diluted AsH_3 in H_2. Dopants were 100ppm SiH_4 in H_2 and DEZn. The growth temperature ranged between 600°C and 750°C, and growth rates between 0.02 and 0.1 μm/min with V/III ratios between 10 and 30. Carrier gas velocities at the inlet of the cell ranged from 5 to 30 cm/s. Care was taken to minimize the transit time of gases by fast switching and vent/run pressure balance in the manifold.

Hall measurement and photoluminescence spectrum were used to characterize the epilayers and Electrochemical C-V profiles were used to determine the grown structure of the device materials.

3. RESULTS

3.1 GaAs and AlGaAs

n-type unintentionally doped GaAs were grown at 650°C with V/III ratio to be 25. 300K Hall mobility of 5017 cm^2/Vs and electron concentration of $1.39x10^{15}$ cm^{-3} were obtained, and 77K mobility of 23315 cm^2/Vs were achieved. Photoluminescence spectrum showed very sharp band-band emission with FWHM to be 2.95 meV and shallow acceptor emission of Si and C.

Gas flow hydrodynamics was studied by growth of specific epilayers under different carrier gas flow rates. Variation of growth rates along the susceptor were studied as shown in Fig.1 and the experimental results were in good agreement with theoretical calculation.

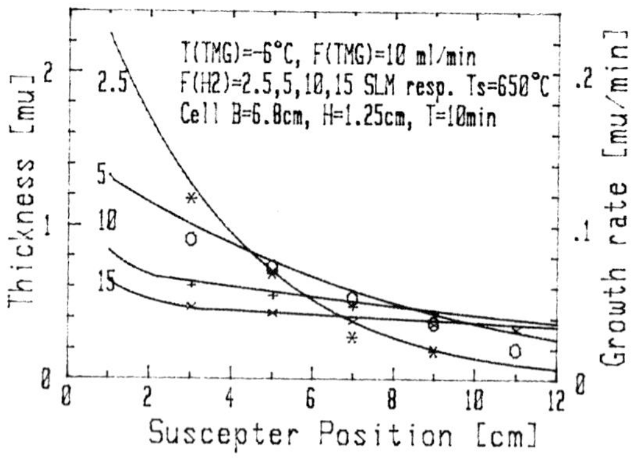

Fig.1

Growth rate variation along susceptor under different H_2 carrier flow rates.

$Al_xGa_{1-x}As$ of different Al composition were grown and very good morphology were obtained. Higher V/III ratio and growth temperature were preferred for higher x values. Al composition were measured by photoluminescence and cathodoluminescence for x<0.35. Measurements of ban dedge adsorption and ellipsometry were used for higher x values. Linear solid composition dependence versus input partial pressure was obtained, as shown in Fig.2.

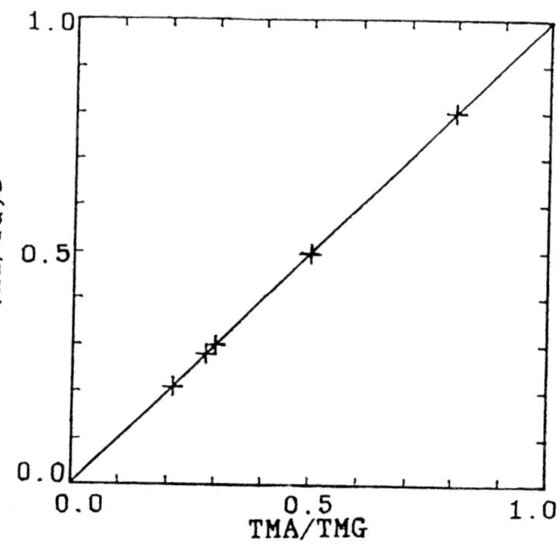

Fig.2

Solid composition dependence of AlGaAs vs. input partial pressure.

3.2 Heterostructures and Quantum Wells

In order to evaluate the width of the (Al,Ga)As/GaAs transition a single heterostructure has been grown with a growth rate of 0.1 µm/min, and the GaAs and $Al_{0.5}Ga_{0.5}As$ thickness were 2 and 1.7 µm, respectively. TEM cross-section study showed the transition width of the heterostructure be less than 50A.

Photoluminescence measurements at 11K

have been made on some quantum well structures. A $GaAs/Al_{0.5}Ga_{0.5}As$ structure with five wells of different well widths has been grown on GaAs substrate by interrupted growth method with a growth rate of 0.02μm/min. The expected thickness of each GaAs layer were 20, 30, 40, 50 and 90Å, respectively, Fig.3 exhibits the photoluminescence spectrum of this structure, the band to band luminescence intensity was as high as 260mV, and the overlap of three emission profiles is caused by layer thickness and compositional fluctuations, the heterojunction transition was estimated to be 10Å.

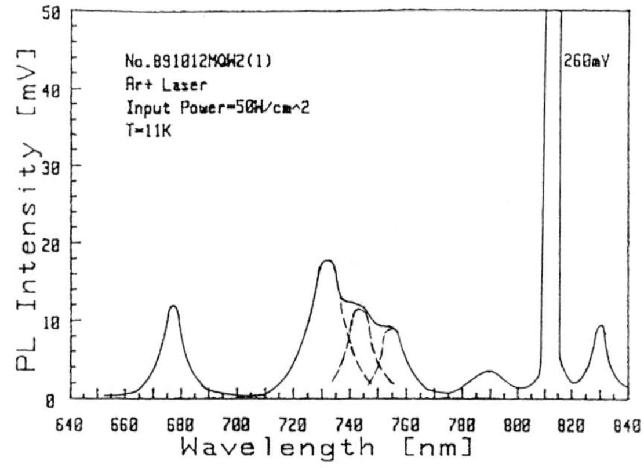

Fig.3

11K Photoluminescence spectrum of five $GaAs/Al_{0.5}Ga_{0.5}As$ quantum wells: 20,30,40,50,90Å.

3.3 HBTs

A n-p-n HBT device structure was designed, and it was composed of the following five layers: n^+-GaAs/n-AlGaAs/p^+-GaAs/n^--GaAs/n^+-GaAs buffer / n^+-GaAs substrate. Special precautions were made to grow the metallurgical and dopant junctions. 100A undoped spacers were inserted at the n-p junctions for preventing the "memory effect" and diffusion of zinc doping. Three technological methods were used for a spacer after the growth of p-type GaAs layer in order to place the AlGaAs/GaAs heterojunction and n-p junction at the same position: interrupted growth for 10 min before the growth of undoped spacer, growth of undoped spacer under reduced growth rate, and growth of the undoped spacer under reduced growth rate within the time to increase the temperature for p-type layer growth to that for n-type layer.

Fig.4 shows the electrochemical C-V measurement of a HBT profile. One can see that uniform doping and abrupt dopant junctions were obtained, which were in good agreement with the theoretical design.

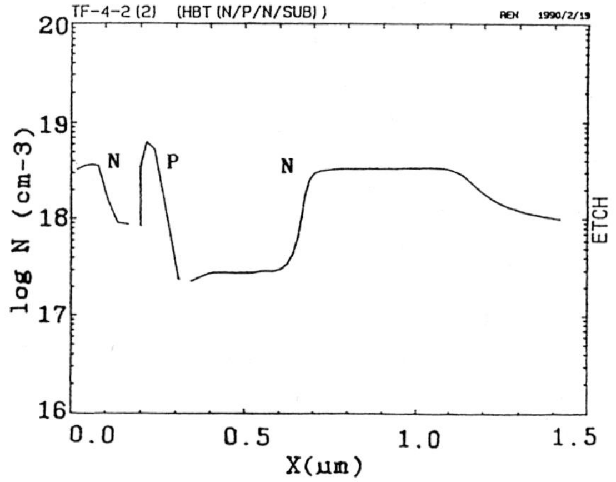

Fig.4

Electrochemical profile of n/p/n HBT

Fabrication of microwave transistors were performed by single and double self-collimated methods. Current gains of 15-40 were normally obtained at 300K. When working temperature dropped to 77K, increased current gains to over 60 were obtained, which may be caused by inactivation of deep levels in the higher band gap AlGaAs materials. Fig.5 shows the collector-emittor I/V characteristics of a HBT at 77K. Microwave performances were tested after they were placed into FET's fixtures. Cut-off frequencies of 10.5 GHz and maximum oscillation frequencies of 5.5 GHz were obtained. Better microwave characteristics may be obtained if fixtures with less parasitic parameters were used.

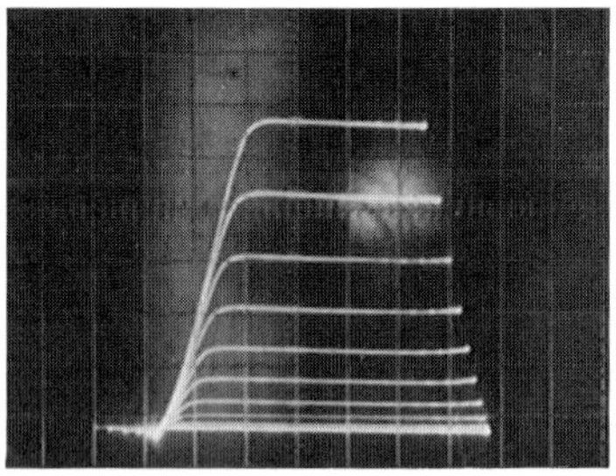

Fig.5

77K HBT collector--emitter I/V character. I_{ce}=2mA/div, V_{ce}=0.5V/div, I_b=50μA/skp, β≈60.

4. DISCUSSION

Qualities of both GaAs and AlGaAs layers are crucial to the device performance. Residual dopants and oxygen concentration should be well controlled during the whole process. High quality substrates, high purity of chemical reagents, vacuum treatment of the growth system and high temperature treatment of substrates before growth are indispensable.

Gas flow instability and delay of gas transition may cause some inconsistency of actual quantum well widths with the design.

5. CONCLUSION

High electrical and optical quality GaAs and AlGaAs have been grown, and abrupt metallurgical and dopant junction were prepared. GaAs/AlGaAs HBTs with current gains 15 40 at 300K and about 60 at 77K were fabricated. Cut-off frequency of 10.5GHz and maximum oscillation frequency of 5.5GHz were achieved for microwave application. These results match the world record in the early 1980's and are the highest in PRC at present.

ACKNOWLEDGMENTS

This work was supported by the "National Natural Science Foundation Committee."

The authors would like to thank Prof. Zhuang Wei-hua and Dr. Xu Qiang in the Institute of Semiconductors of Academica Sinica for PL measurements and Prof. Chen Xiao-jian in the Institute of Nanjing Solid state Devices for the fabrication and test of HBT devices.

REFERENCES:

1. W.Shockley, U.S. Patent 2,569,347 (1951).

2. H.Kroemer Pro.IRE, 1535 (1957) 45

3. H.Ktoemer Proc. IEEE, 70 (1982),13.

4. A.Y.Cho and J.R.Arthur, Prog. Solid State Chem., Pt 3, 10 (1975) 157.

5. R.A. Milano, et al, Appl. Phys. Lett. 39 (1979), 562.

6. R.W.Glew, M.S.Frost J. Crystal Growth 68 (1984) 450.

7. R.A.Milano, et al., Appl. Phys. Lett. 34 (1979) 357.

8. M.Madihian, et al., IEEE Trans. ED. 34 (1987) 1419.

Frontiers of Materials Research/
Electronic and Optical Materials
M. Kong, L. Huang (editors)

MISFIT DISLOCATIONS IN GaAs EPITAXIAL LAYERS ON In-DOPED GaAs GROWN BY MOVPE

DUAN Shukun, WANG Yutian, LI Chengji and LI Xuebin

Institute of Semiconductors, Chinese Academy of Sciences, P. O. Box 912, Beijing 100083, P. R. China

ABSTRACT

The lattice mismatch in GaAs epitaxial layers grown on GaAs:In (001) substrates by Metalorganic Vapour Epitaxy (MOVPE) is reported. When the In content exceeds the critical value for plastic deformation, x_{c2}, a cross-hatch pattern appears on the cathodeluminescence image. Transmission Electron Microscopy (TEM) observation proves that the cross-hatch pattern is an orthogonal dislocation network along [110] and [$\bar{1}$10] direction and is located at the interface between the epitaxial layer and the substrate. X-ray double crystal diffraction technique has been used to study the relationship between lattice mismatch along the growth direction and the In content. The results obtained show that when x less than the critical value for elastic deformation, x_{c1}, the mismatch strain resulted due to lattice mismatch is mainly accomodated by elastic deformation; When x exceeds x_{c2}, the dislocations are generated and the mismatch strain is mainly relaxed by misfit dislocation. The value x_{c2} is not only depending on the mismatch strain, but also relating to the thickness of the epitaxial layer.

1. INTRODUCTION

In recent years, heteroepitaxial structures including lattice matched epitaxial layers and strain-layer superlattices are played a very important role in semiconductor devices. In order to improve the performances and the reliability of such devices, it is necessary to reduce the dislocation density of the substrates and of the epitaxial layers. Many authors[1-4] have reported that high concentration of isoelectronic impurity indium doping strongly reduces the dislocation density in bulk GaAs single crystal. It can be foreseen that growth of GaAs epitaxial layer on such substrates will give rise to mismatch strain and misfit dislocation because of lattice mismatch[5-6]. Weyher and Van de Ven have found, using nomarski microscope, surface morphological "cross-hatch pattern" of GaAs epilayer grown on (100) GaAs:In. According to their deduction this is due to a very small increase of the local growth speed as a result of misfit dislocations.

This paper reports that an orthogonal network has been observed by cathodeluminescence observations and by the misfit dislocation array from TEM observation directly; the relationship between lattice mismatch for epitaxial layers of diverse thickness and various In contents has been studied.

2. EXPERIMENTAL

The unintentionally undoped GaAs epitaxial layers were grown by trimethylgallium (TMG) and arsine (AsH_3) in VP50-RP MOCVD system at atmospheric pressure[7]. High purity hydrogen was used as a carrier gas. The sum of the flow rates for all gases flowing into the reactor was 7.5 l/min. The temperature of TMG bubbler was -5.7°C and the V/III ratio was 20. The substrates were initially heated at 800°C with a flow of H_2 and AsH_3 for 10 minutes prior to growth in order to remove any surface oxides. The growth rate was 2-3 μm/hr at the growth temperature 700°C and the epitaxial layers had a typical carrier concentration of $1x10^{15}cm^{-3}$ and a Hall mobility of 6,000 $cm^2V^{-1}s^{-1}$ at room temperature and 49,500 $cm^2V^{-1}s^{-1}$ at 77K. Surfaces were mirror-like.

The LEC In-doped GaAs ($Ga_{1-x}In_xAs$) substrates used were oriented (001) with 2° off towards the nearest [110]. Three groups epitaxial layers with different thickness (8 μm, 1.5 μm and 0.6 μm)were grown. For each thickness, six substrates with different x-value were used.

The indium content x was determined by JXA-3A electron probe X-ray analyzer and by photoluminescence at 10K. The mismatch between epitaxial layer and substrate along growing direction was measured by X-ray double-crystal diffraction using $CuK\alpha_1$ radiation with the first crystal Si to assume a (224) symmetrical reflection and the diffraction plane of the sample being (004). Misfit dislocations were examined by electron probe X-ray microanalyzer and by EM-430 transmission electron microscope.

3. RESULTS AND DISCUSSION

Unintentionlly doped GaAs epitaxial layers with mirror-like surface morphlogy were grown on GaAs:In (001) substrates. In some samples, however, a cross-hatch pattern along [110] and [$\bar{1}$10] directions appeared in cathodeluminescence photograph as shown in Fig.1. The sample with thickness of GaAs epilayer 8μm and In content of substrate 0.006 was used to take this photograph. It is usually considered that the decrease of

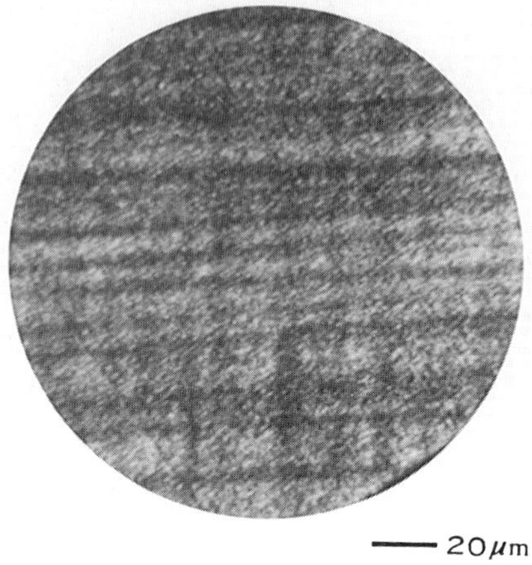

Figure 1
Cathodeluminescence photograph of undoped GaAs epilayer grown on (100) In-doped GaAs substrate.

luminescence efficiency is related to the nonradiative carrier recombination at the dislocation core. This is because the impurites or complexes gettered around the dislocation cores[8].

Fig.2 shows a [110] sectional photograph of same sample by transmission electron microscope. It can be seen that a set of orthogonal misfit dislocation networks exsited at the interface of GaAs epilayer and (001) In-doped substrate. One group of dislocations perpendicular to the surface of the [110] section with a spacing of about 2 μm was observed. A second group parallel with this was also observed. The result is similar to that of the InGaAs/GaAs (001) and GaAlAsP/GaAs (001) composition gradient studied epilayer by Bartels et al[9], using x-ray photography. That is, misfit dislocations

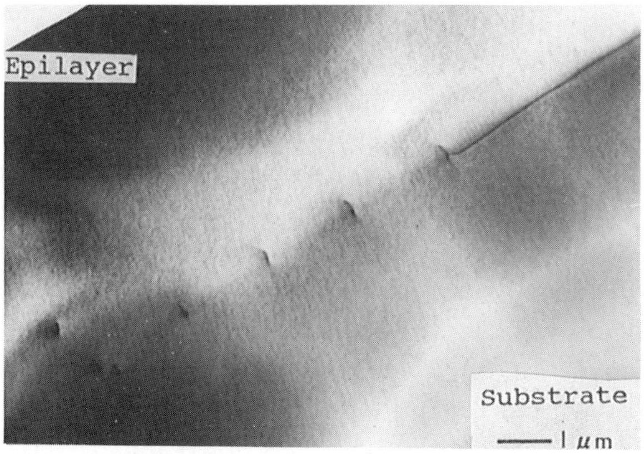

Figure 2
TEM photograph of GaAs undoped epilayer grown on (100) In-doped GaAs substrate.

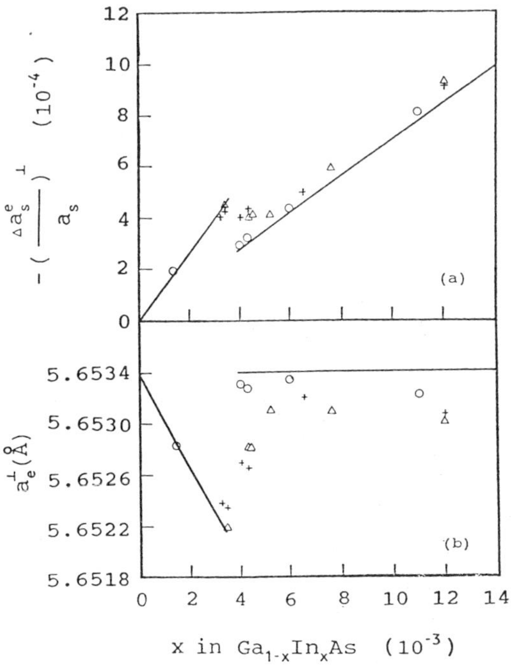

FIGURE.3
Lattice mismatch (a) and lattice constant (b) vs. Indium content, x, of In-doped GaAs substrate. (+) 0.6μm; (**Δ**) 1.5μm; (O) 8μm.

appear in the mutually vertical directions [110] and [$\bar{1}$10] on the (001) plane, and the misfit dislocation densitity as well as its length in parallel with the growth plane is more significant in one direction that in the other.

Fig.3(a) shows that the lattice mismatch between the GaAs epilayer and the GaAs:In (001) substrate is a function of the In content of the substrate. It can be seen that lattice mismatch increases with the increase of the value x, but there is a change when x is about 0.0035. the lattice mismatch increases slowly with the increase of value x when x exceeds 0.004 or 0.005, depends on the thickness of the epilayer is 8 μm or 0.6-1.5 μm.

Fig.3(b) shows the dependence of the lattice parament of GaAs epitaxial layer normal to the growth plane on In content x in the substrate. The lattice constants of various In doped substrates were calculated by wsing Vegard's Law. In the calculation, the lattice constant of GaAs and InAs take 5.6534Å and 6.0584Å respectively. It can be seen that in the range of $0<x<0.0035$, a drops with the increase of value x while after x is larger than 0.004 (or 0.005), $a^{\perp}_{GaAs}$ tends to level off, nearing the lattice constant of GaAs at the free state.

According to the Vegard's law, the relationship between x and the lattice mismatch along the growth direction of the strainless epitaxial layer can be given by the following function:

$$\left(\frac{\Delta a_s^e}{a_s}\right)_0^{\perp} = - \frac{(a_{InAs}-a_{GaAs}) \cdot x}{a_{GaAs}+(a_{InAs}-a_{GaAs}) \cdot x} \qquad (1)$$

where a_s is lattice constant of the bulk

GaAs, Δa_s is the difference between the lattice constant of the epitaxial layer and that of the substrate. The underscript "o" represents strainless state while the superscript "⊥" means the vertical to the growth plane.

While the epilayer is in strain state, the lattice mismatch along the growing direction $(\Delta a_s^e/a_s)_S^{\perp}$ can be expressed[10] as:

$$\left(\frac{\Delta a_s^e}{a_s}\right)_S^{\perp} = -\left(1+\frac{2}{1-\nu}+\frac{8}{1-\nu}\ \frac{t_e}{t_s}\right)\cdot\left(\frac{\Delta a_s^e}{a_s}\right)_o^{\perp} \qquad (2)$$

where ν is Possion's ratio, which is equal to $C_{12}/(C_{11}+C_{12})$, with C_{11}, C_{12} being the elastic moduli and t_e, t_s being the thickness of epitaxial layer and of the substrate.

When $t_e \ll t_s$, expression (2) can be taken as:

$$\left(\frac{\Delta a_s^e}{a_s}\right)_S^{\perp} = -\left(1+\frac{2}{1-\nu}\right)\cdot\frac{(a_{InAs}-a_{GaAs})\cdot x}{a_{GaAs}+(a_{InAs}-a_{GaAs})\cdot x} \qquad (3)$$

From Fig.3(a), it can be seen that when $0<x<0.0035$ (x_{c1}) the experimental data are in agreement with the calculation results from equation (3). This indicates that in this range, the mismatch strain resulted from lattice mismatch is mainly accomodated by tetragonal deformation, the epilayer is in the elastic state and no misfit dislocation is produced. Provided that the volume of the single cell GaAs is unvariable, the parallel lattice constant increase and the vertical lattice constant will surely be compressed, therefore $a_{GaAs}^{\perp}$ decreases along when x increases.

In the case of $x>x_{c2}$ (it is 0.004 for 8 μm or 0.005 for 0.6 - 1.5 μm), the experimental data are approximately equal to the calculation curve of equation (1), indicating that the mismatch strain resulting from lattice mismatch is mainly relaxed by the misfit dislocation produced. Because of the accomodation by the misfit dislocations, the lattice constant $a_{GaAs}^{\perp}$, in the case of a variable x, basically remains unchanged, approaching the lattice constant of GaAs in the free state.

While $x_{c1}<x<x_{c2}$, it seems that there exists a transitional range in which misfit dislocations take place, but the dislocation density is nevertheless small, i.e., the mismatch strain is merely relaxed by the mismatch dislocation in part.

Our experimental data show that the x_{c2} is related to the thickness of the epitaxial layer. The thicker the epitaxial layer, the smaller the value x_{c2} approaching the value x_{c1}, and the more the value $a_{GaAs}^{\perp}$ approaching the lattice constant of the bulk GaAs. This shows that the creation of mismatch dislocation depends on the content of In and is related to the thickness of the epitaxial layer at a fixed growth temperature. Additionally, the growth method, especially the growth temperature of the epitaxial layer also has some influence thereon. From the above stated, it can be seen that in order to decrease the dislocation density of the epitaxial layer, suitable In content of the substrate should be chosen, i.e., as high as decreasing the dislocation density of the ingot and as low as avoiding the formation of misfit dislocation in the epitaxial layer. For a growth on a substrate having a higher In content (e.g. x=0.006 or even higher), superlattice layer[5] or composition gradient layer can be used as the buffer layer to reduce the dislocation density in the epilayer.

4. CONCLUSION

In MOVPE GaAs growth using In doped GaAs (001) substrate, we have found that in the case of x being less than x_{c1}, tetragonal deformation takes place in the epitaxial layer; in the case of x being larger than x_{c2}, misfit dislocation network appears at the interface along [110] and [$\bar{1}$10] directions by CL and TEM. When the In content is between x_{c1} and x_{c2}, there seems to be a transitional range where in the strain is partially relaxed. The magnitude of this transitional range has connection with the thickness of the epitaxial layer.

ACKNOWLEDGEMENTS

The authors wish to thank Fenglian Wang and the Laboratory of Electron Microscopy Chinese Academy of Sciences for taking the TEM photograph, the spectrum research group for PL measurement and Hongja He for providing the GaAs substrates.

REFERENCES

1. G. Jacob, M. Duseaux, J. P. Farges, M. M. B. Van den Boom and P. J. Roksnor, J. Crystal Growth 61 (1983) 417.

2. K. Yamada, H. Kohda, H. Nakanishi and Hoshikawa, J. Crystal Growth 78 (1986) 36.

3. D. L.Barrett,S.Msguigan,H.M.M.Hobgood, G. M. Eldridge and R. N. Thomas, J.Crystal Growth 70 (1984) 179.

4. H. Kimura, C. B. Afable, H. M. Olsen, A. T. Hunter and H. V. Winston, J.Crystal Growth 70 (1984) 185.

5. M. Morioka, T. Mishima, k. Hiruma, Y. Katayama and Y. Shiraki Proc. 12th Intern. Symp. on GaAs and Related Compounds. Karuizawa, Japan 1985, Inst. Phys. Conf. Ser. No. 79. Ed. M.Fujimoto (Adam Hilger Ltd, Bristed and Boston). P121.

6. J. L. Weyher, J. Van de Ven, J. Crystal Growth 88 (1988) 221.

7. S. K. Duan et al., The proceeding of the Second International Conference on Solid State and Integrated Circuit Technology, P.740 (1989).

8. P. M. Petroff. Semiconductors and Semimetals, Vol. 22, Part A. Chap. 6, P.379 (1985). Vol. Ed. W.T. TSANG (ACADEMIC PRESS, INC.)

9. W. J Bartels and W.Nijman, J. Crystal Growth 37 (1977) 204.

10. B. E. Estop, A. Izreal and M. Sauvage, Acta Cryst. A 32 (1976) 627

Frontiers of Materials Research/
Electronic and Optical Materials
M. Kong, L. Huang (editors)
Elsevier Science Publishers B.V., 1991

THE ELECTRICAL AND OPTICAL PROPERTIES OF LPE GaInAsSb WITH RARE EARTH ELEMENTS DOPING

Xiuying GONG, Zhanguo WANG, Fengsheng GAO, Yingdi MA, Wanqiang HAN, Yin YU and Lanying LIN

Institute of Semiconductors, Academia Sinica. P.O. Box 912, Beijing 100083, China

Rare earth elements have successfully been introduced into the LPE GaInAsSb system for the first time, to our knowledge.
The LPE GaInAsSb/InP and GaInAsSb/InAs epilayers were assessed by using Van der pauw measurements, photoluminescence and SIMS analysis. The results show that the purity of the epilayers with Gd doping is considerably improved. Rare earth elements play the catchers of impurities in the quaternary solutions.

1. INTRODUCTION

High purity GaInAsSb epilayers are important material for the detectors in middle infrared region[1]. It is also an attractive material for fabricating high speed devices, such as FETS, HEMT and so on, because of its high electron mobility at room temperature.

In this paper, electrical and optical properties of the LPE GaInAsSb/InP and GaInAsSb/InAs with and without rare earth elements doping were reported. The results show that the purity of the epilayers with rare earth doping are considerably improved. Room temperature electron mobility up to 10900cm^2/v.s for GaInAsSb/InP and 24800cm^2 /v.s for GaInAsSb/InAs have been achieved. The SIMS measurements have been used to investigate the mechanizm of the influnce of rare earth elements on the purity of the epilayers. The results demonstrate that rare earth elements play the catchers of impurities in the melts.

2. EXPERIMENTS

The GaInAsSb epilayers were grown in a conventional sliding glassy graphite boat system by the LPE technique. 7NIn, 6NSb, undoped GaAs and InAs were used as the source materials. The rare earth elements, Gd, Yb or Sm, were introduced as the dopants. The substrates were Fe-doped InP and n-InAs with (100) oriention. The growth temperature and cooling rate are 660-640°c and 0.5°c/min, respectively. The thickness of the epilayers with $\lambda \simeq 2.5$-4.5um are 3-10um.

The electrical properties of the layers were determined by using van der pauw measurements for GaInAsSb/InP, and selective etching / van der pauw technique for GaInAsSb/InAs. The optical properties of the layers were investigated by photoluminescence measurements at 4.2k. The impurity analysis for the epilayers and melts were performed using SIMS analysis.

3. THE INFLUNCE OF RARE EARTH ELEMENTS ON ELECTRICAL PROPERTIES OF LPE GaInAsSb

Van der pauw measurements and selective etching/Van der pauw technique were used to investigate the influnce of rare earth elements on electrical properties of LPE GaInAsSb epilayers. In table 1, the results obtained from the two group samples are shown. It can be seen that carrier concentrations of the epilayers without Gd doping are in the range of 10^{16}-$10^{17}cm^{-3}$, thought the melts, from which the layers were grown, have been baked for 10-18 hours. Whereas carrier densities of Gd dopped epilayers decreased to $(1.01$-$22)x10^{15}cm^{-3}$, room temperature electron mobility up to $10900cm^2$ /v.s for GaInAsSb/InP and $24800cm^2$ /v.s for GaInAsSb/InAs have been achieved. These results show that the purity of the epilayers with Gd doping has been remarkably improved. This is very important for fabricating the middle infrared detectors and high speed electronic devices.

4. OPTICAL PROPERTIES OF GaInAsSb EPILAYERS

Optical properties of GaInAsSb/InP were determined by using photoluminescence measurements at 4.2k.
The experiments were carried out using an H_{25} infrared grating monochrometer at a wavelength range 1-2um. The discernment of this instrument is higher than 20Å. A Ge detector in liquid N_2 used as the receptor of signal has

TABLE1 Electrical properties of LPE GaInAsSb

Sample	Substrate	Baking time (h)	Rare earth elements	n (cm^{-3})	μ_{300k} ($cm^2/v.s$)	$\Delta a/a$(%)
P-409	InP	18	-	$3.60X10^{16}$	8050	
P-705	InP	12	Gd	$1.01x10^{15}$	10900	
As-501	InAs	1	-	$7.0x10^{16}$	14150	-0.09
As-505	InAs	10	-	$20x10^{16}$	19000	<-0.05
As-603	InAs	10	Gd	$1.73x10^{16}$	23690	+0.06
As-607	InAs	10	Gd	$2.20x10^{16}$	22460	+0.08
As-608	InAs	11	Gd	$2.11x10^{16}$	24800	+0.05
As-609	InAs	12	Gd	$7.20x10^{15}$	23690	+0.06

sensitivity $S \simeq 4x10^8$ $-7x10^9$ v/w. The Ar^+ laser was employed as the excitation source. The excitation intensity was kept as low as possible during measurement so that the optimum accuracy can be attined.

In fig.1, the typical photo-luminescence spectra for the two epilayers are shown. It is seen that the FWHM of band A for the layers with Gd doping is as narrow as 6.15Mev, which is much narrower than that of the layers without Gd doping (14.85Mev), and that A/B is increased up to 30.2. These results also indicate the high purity of the epilayers with rare earth elements doping.

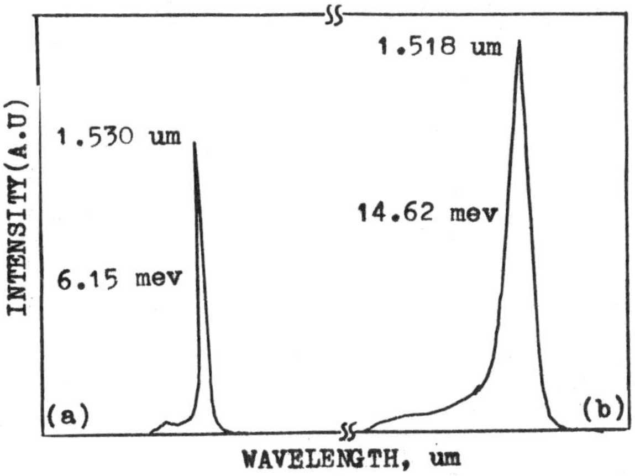

FIGURE 1
Photoluminescence spectra at 4.2k on LPE GaInAsSb/InP with (a) and without (b) Gd doping

5. SIMS MEASUREMENTS

In order to investigate the action of rare earth elements in improving the purity of the GaInAsSb epilayers. SIMS analysis were performed. The experiments were done using an IMA-2A apparatus with the diameter of the beam spot 200um. An Ar^+source was used as the first source. fig.2 shows the SIMS spectra for the two

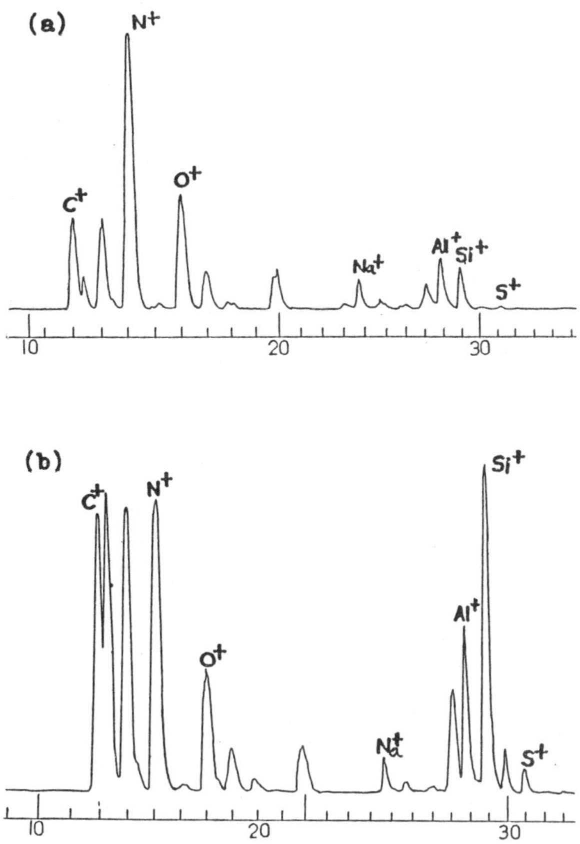

FIGURE 2
SIMS spectra of the main impurities for the two GaInAsSb/InAs epilayers with (a) and without (b) Gd doping.

LPE GaInAsSb samples. It is evident that the main impurity content in the layer with Gd doping is much lower than that in the layer without Gd doping. This result is in agreement with that on LPE InGaAs system[2]. The Gd spectrum can hardly be seen probably due to its very low ionization rate.

SIMS spectra for the melts with and without Gd doping are shown in fig.3 (a) and (b), respectively. It is very interisting that the intensity of the

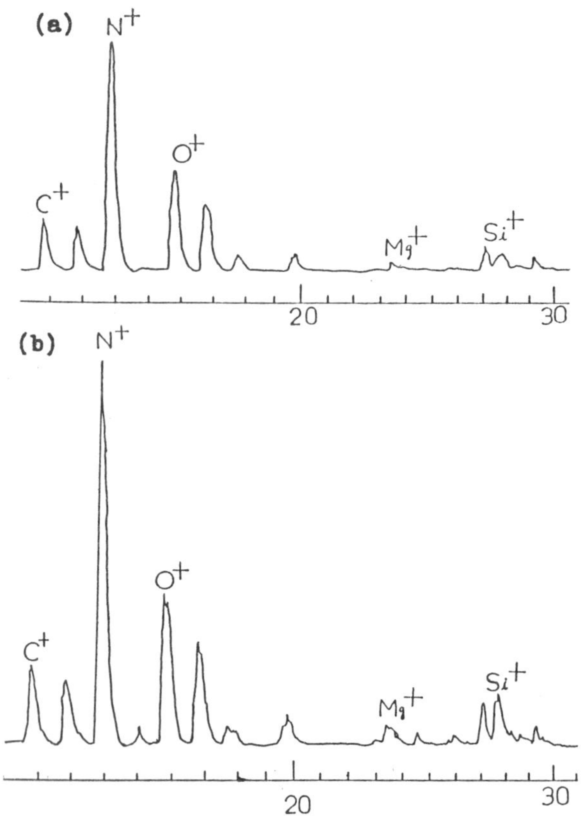

FIGURE 3
SIMS spectra of main impurities of the two melts with (a) and without (b) Gd doping

SIMS spectra of main impurities in the melt with Gd doping (fig.3 a) also lower than that in the melt without Gd doping. It can be deduced that rare earth elements strongly interact with above mationed impurities in the melts, due to their chemical activity, to form large compound molecula which prevent introduction of the residual impurities into the epilayers. Meantime, single elementary impurity molecula decrease in the solutions. Further, the impurity density reduces in the epilayers and the electron mobility increases. Therefore, It seems to us that rare earth elements play the catchers of impurities in the quaternary solutions.

6. CONCLUSION

The rare earth elements, for the first time, were introduced into the LPE GaInAsSb system. Van der pauw measurements and photoluminescence show that electrical and optical characteristics of the epilayers with Gd doping are obviously improved. SIMS measurements indicate that rare earth elements play the catchers of impurities in the quaternary solutions.

ACKNOWLEDGEMENTS

The authors would like to thank Dr Wang Wannian and Li Guang Ping for electrical measurements and SIMS analysis.

REFERENCES

1. J.L. Zyskind et al., J. Appl. Phys. 61 (1987) 2898.

2. N.T. Bagraev et al., Sov. Phys. Semicond. 18 (1984) 49

Frontiers of Materials Research/
Electronic and Optical Materials
M. Kong, L. Huang (editors)

LASER CHARACTERISTICS OF NEW AMORPHOUS LASER MEDIUM $La_{1-x}Nd_xP_5O_{14}$

Jiaji LI, Yanbin LIU, Meizhen ZENG and Sujie ZHAO

Research Institute of synthetic Crystals, P.O.Box 733, Beijing, P.R.China

Recently, interest in amorphous laser medium LNPP has increased due to its advantages over NLPP crystal and Nd-glass. The preparation processes of amorphous LNPP hae been described, spectroscopic measurement of LNPP at room temperature have been performed, both output energy as a function of pump energy, and pump energy as a function of slope efficiency also output energy as a function of ambient temperature have been studied and compared with Nd: YAG and Ne-glass. Some physical and laser output properties are also given.

1. INTRODUCTION:

$Nd_xLa_{1-x}P_5O_{14}$ (NLPP) crystal[1] is a high gain Nd^{3+} laser material[2], which shows only a small decrease in their fluorescence quantum efficiency (QE) as the Nd^{3+} concentration is increased up to the stoichiometric limit. Directly related to this technologically important aspects are severed interesting anomalous luminescence. The weak concentration quenching of the QE in NLPP crystal is linear dependent[2,3] on the fractional Nd^{3+} concentration x; it is contrastive to many other similar system[4], e.g. Nd: YAG, where the quenching goes to as x^2. This anomalous fluorescence property makes NLPP crystal to be an important high gain output energy density laser medium. But the application of NLPP crystal are limited by its poor mechanical property, i.e. very easy to cleave and very difficult to prepare into a large size ets.

Glass medium is one of the important laser material due to its large size and good uniformity, but its thermal conductivity is much poorer than crystal, usually equals to about 2×10^{-3} Cal/cm.s.k. Thus high frequency performance of glass medium is blocked by thermal effect itself. How to improve glass thermal conductivity is still one of the current subject. Some of the works have made encouraging progresses[5,6,7].

Amorphous $La_{1-x}Nd_xP_5O_{14}$ (LNPP) laser medium not only possesses advantages of NLPP crystal and Nd-glass, but also overcomes drawbacks of them as discribed above It is a new kind of laser material. Here we briefly introduce the preparation process, as well as spectroscopy and some basic properties of amorphous LNPP compare with Nd-glass and Nd:YAG.

2. PREPARATION PROCESS

Phosphate is a kind of very complicated compounds, including a great variety and their properties are difference. For keeping crystal advantages, homologous crystal was grown at 550℃ first. Growth this crystals only require their chemical composition, do not require their sizes and qualities, so this processes are much easier than other crystal growth methods. The chemical analysis shows that maximum deviation of composition does not exceed 0.5%, there action formala is:

$$Re_2O_3+15H_3PO_4 = 2ReP_5O_{14}\downarrow +H_3P_5O_{14}+21H_2O$$

(Re is rare earth elements)

For overcoming crystal weaknesses and developing glass advantages, these crystals are translated into amorphous at 1100-1300℃ which temperature is much higher than decomposition temperature. Anticorrosion crucible must be seriously considered, because the crystals have strong corrosion in so high temperature. Composition of this amorphous is as the same as the crystal: $La_{1-x}Nd_xP_5O_{14}$. For improving amorphous properties, it can be doped with other cations to substitute part equivalence of La^{3+}, x can change from 0—1. High Nd^{3+} concentrations are beneficial increased gain and stored energy of LNPP but also increased its self-absorption and effect its thermal properties, common used concentrations of Nd^{3+} are about 7×10^{20} — 12×10^{20} cm^{-3}. In fact, doped with suitable Ce^{3+} in LNPP can improve its antiultravidet property and gives some sensitization. While, sensitization with doped suitable Cr^{3+} is also obviously but the output frequency is decreased.

3. SPECTROSCOPY

Optical properties for amorphous LNPP, including absorption, fluorescence and lifetime of excited states have been investigated. For comparasion, we present optical properties for NLPP crystal, Nd-glass and Nd: YAG as follows:

3.1 Absorption

The absorption spectram of amorphous LNPP in Fig.1 shows that these systems can be pumped by flashlamps, dye or diode lasers.

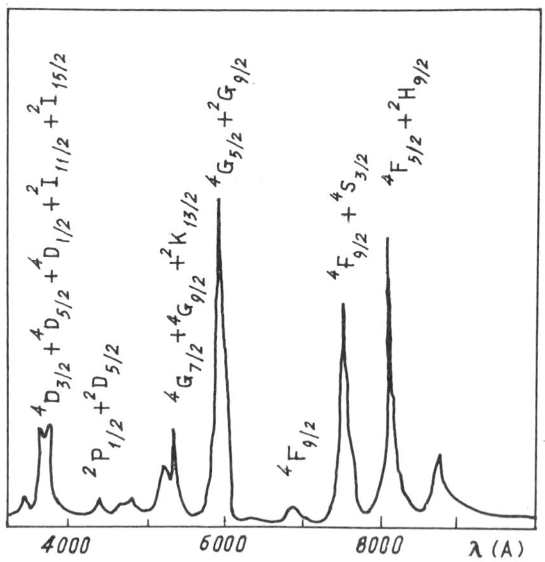

FIGURE 1
Absorption spectrum of LNPP and its energy level

3.2 Fluorescence

The fluorescence spectra of

$^4F_{3/2} \rightarrow {}^4I_{9/2}$ and $^4F_{3/2} \rightarrow {}^4I_{11/2}$ transitions of amorphous LNPP and crystal NLPP are shown in Fig.2 and Fig.3. The LNPP main emission wavelength is 1.053 μm, similar to NLPP crystal.

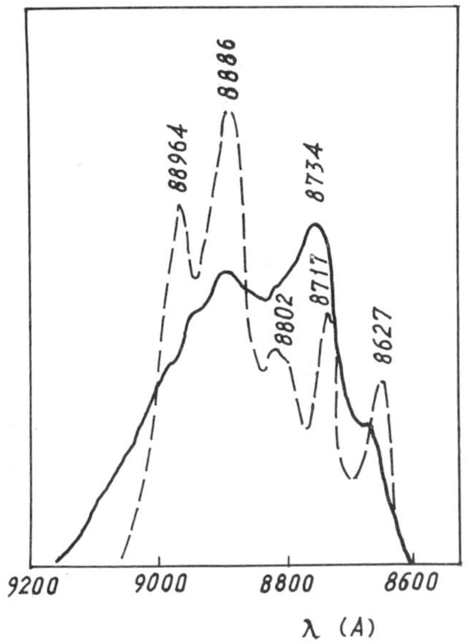

FIGURE 2
Fluorescence spectra of $^4F_{3/2} \rightarrow {}^4I_{9/2}$ transition of LNPP and NLPP
Solid line-LNPP; Dashed line-NLPP

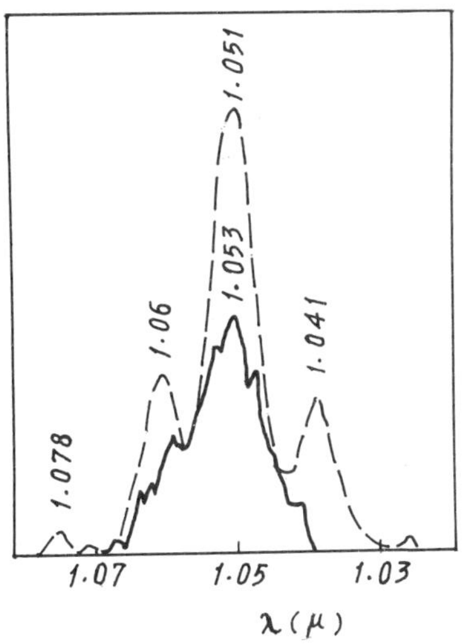

FIGURE 3
Fluorescence spectra of $^4F_{3/2} \rightarrow {}^4I_{11/2}$ transition of LNPP and NLPP
Solid line-LNPP; Dashed line-NLPP

3.3. Fluorescence Quenching Rate

The decay curve of fluorescence $\lambda = 671.8$nm at room temperature of LNPP amorphous is exponential. Fig.4 shows the decay curve. From these decay curves of fluorescence, we have measured the lifetime τ_x for $x = 0.01$, 0.10, 0.20, 0.35, 0.50, 0.70 and 1.00 LNPP samples, The $1/\tau - x$ diagram is a straight line. From the intersection of the straight line with thevertical axis, one can get $1/\tau_0$ (τ_0 is the lifetime of zero concentration sample) . The quenching rate of fluorsecence q(x) is definedas:

$$q(x) = \frac{1}{\tau_x} - \frac{1}{\tau_0}$$

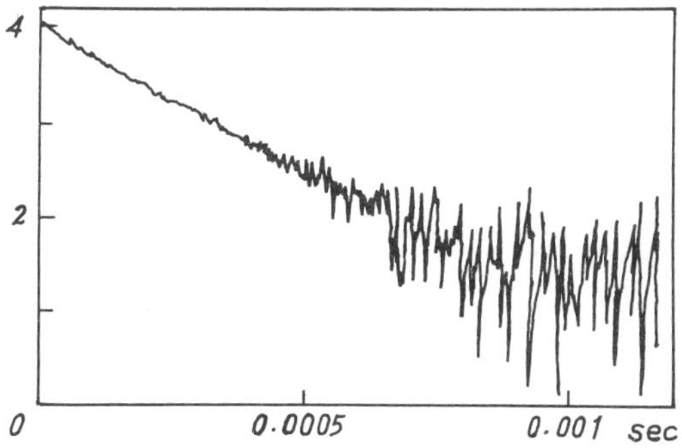

FIGURE 4.
The semilog of decay curve of $\lambda = 671.8$nm fluorescence at room temperature

Fig. 5 is the log q(x) — log x diagram. These are 3 straight lines (a), (b) and (c). The line (a) is the concentration curve of LNPP. The lines (b) and (c) are curve of the same kind for NLPP crystal[8] and

Nd: YAG[8] respectively. The slope of the line (a) for LNPP is equal to 1; it is similar to NLPP crystal (b) for $\nu = 1$. This indicates that for both LNPP and NLPP, q(x) is linearly dependent on the fractional Nd^{3+} concentration x. But the slope of (c) is equal to 2; it means that for Nd: YAG the quenching rate q(x) goes as x^2. From Fig.5 one can easily find that for the same value the quenchingrate LNPP is slower than that of NLPP. The lifetime of LNPP is shorter than that of NLPP.

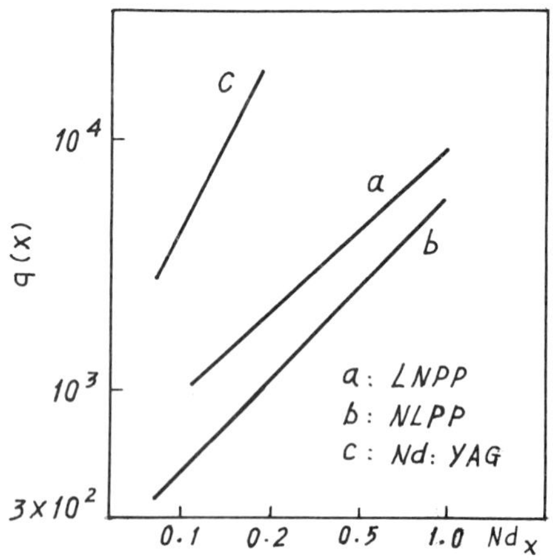

FIGURE 5.
Fluorescence quench speeds for some laser materials

4. LASER CHARACTERISTICS

The pumping and output energy relation for amorphous LNPP is shown in Fig.6. For comparsion, there are 4 curves (a), (b), (c) and (d) which are Cr: LNPP, LNPP, Nd: YAG and Nd-glass respectively. From Fig. 6 one can easily find that one of outstanding characteristic of LNPP is high output energy, that is high gain characteristic too. It is basically in keeping with calculation value of gain coefficient. (The gain of LNPP, Nd: YAG and Nd-glass are 5.9×10^{-2}/cm, 1.5×10^{-2}/cm and 3.7×10^{-2}/cm respectively). The LNPP emission cross is $4.2\times10^{-2}\,cm^2$, although it is not larger than crystal like Nd: YAG and NLPP, but it larger than other glass medium[9]. From Fig.6 one can also find that the LNPP output characteristicis not only obviously better than Nd-glass, but also better than Nd: YAG if pump energy over than 50J.

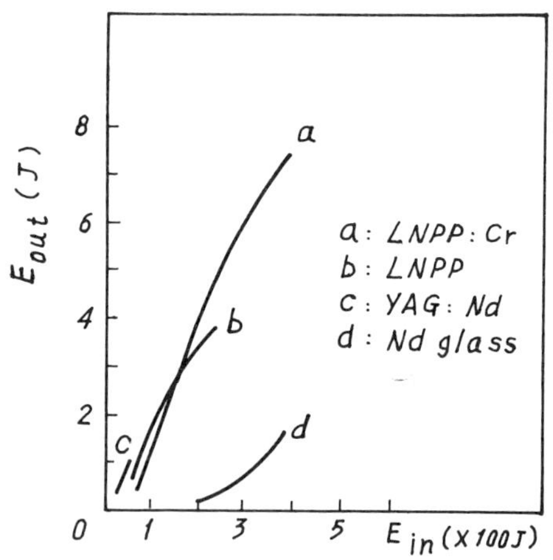

FIGURE 6
Input-output relation for LNPP

Good quality Nd:YAG has rather high slope efficiency on condition of low pump energy, but its slope efficiency tends to saturation quickly if pump energy over than 50J. LNPP does not like Nd: YAG, it is

TABLE 1
Numbers of reversed particles and measured oscillation thresholds for LNPP and some laser materials

medium	LNPP	Nd:YAG	Nd-glass
$\triangle n_t (cm^{-3})$	1.4×10^{18}	1.8×10^{16}	1.4×10^{18}
threshold(J/cm^2)	4.3	2. 6	4.5

advantageous to the output high energy. Nd-glass is for inferior not only output energy, but also slope efficiency to LNPP. From Fig. 7 which is pumping energy and slope efficiency relation for Nd: YAG, LNPP and Nd-glass one can easily find that.

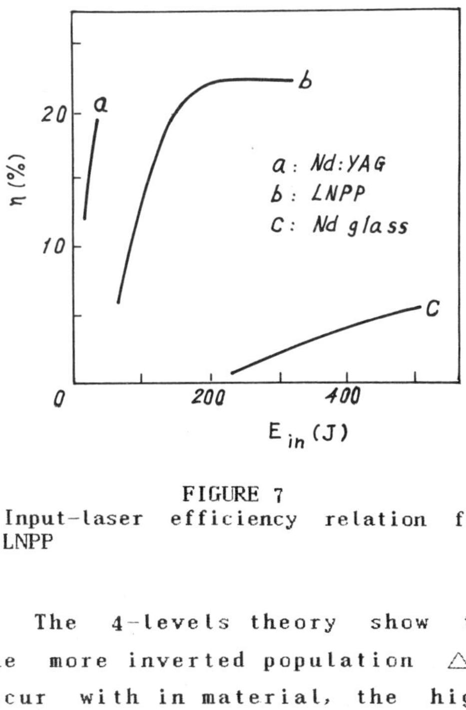

FIGURE 7
Input-laser efficiency relation for LNPP

The 4-levels theory show that the more inverted population $\triangle n_t$ occur with in material, the higher effective pump energy is required and the higher threshold value is[10]. $\triangle n_t$ is smaller in Nd: YAG, than in LNPP and Nd-glass, so its threshold is lowest. $\triangle n_t$ in LNPP is similar to in Nd-glass, their threshold are about the same, but threshold for LNPP is a bit little than that for Nd-glass. Shown in table 1.

LNPP can be Q-switched its output by electrooptics, rotating mirror and dye methods. The output ratio of Q-switching to free running is 0.7, better than Nd: YAG (0. 5) and Nd-glass(0.5). The Q-switch pulse width is closely to Nd: YAG, better than Nd-glass.

In $\triangle S = L\ dn/dt\ \triangle T + L(n-1)\ \alpha\ \triangle T$, [11] (This is thermal effect lead to aberration), since LNPP dn/dt equals 0, so $\triangle S$ is merely effected by the last term. Thus LNPP can bear great thermal load and work on condition of high pumping energy or high frequency in the good forced cooling system because its α is very small. It is shown in Fig.8. that output versus ambient temperature about LNPP and Nd: YAG in water cooling full cavity system. From Fig.8. one can find that Nd:YAG output begin to descend if ambient temperature higher than 35℃, but the descending speed is slowly. LNPP output does not vary if ambient temperature inferior to 40℃. If ambinet temperature higher than 40℃, LNPP output begin to descend steeply, but only that be low 50℃, the LNPP output is better than Nd:YAG, high over 50℃, Nd: YAG output is better than LNPP. LNPP can also work on condition that ambient temperature great varies in the split cooling system.

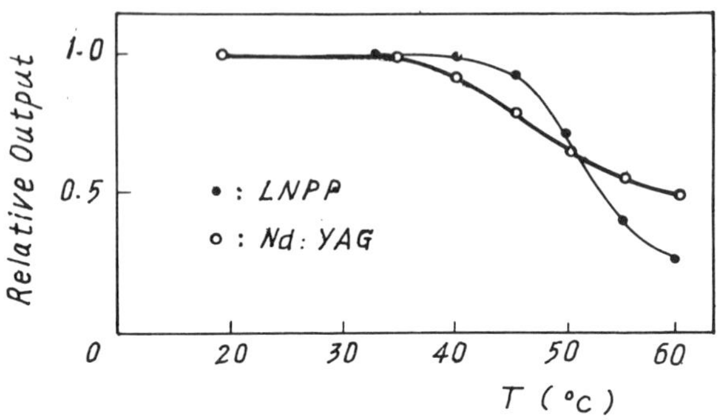

FIGURE 8
Laser outpuut versus ambient temperation

As a result of improve, water-proofing property of LNPP has been increased markedlly. It is about the same as Nd-silicate glass. So LNPP can be apply in various conditions.

5. CONCLUSION

Amorphous LNPP is a kind of new type laser material, which has many advantages properties. Its physical properties and laser output characteristics are shown in table 2 and table 3 respectively. From this two tables one can easily find that LNPP is superior than other materials in pulse laser devices used, specially high power laser devices, such as slab laser device etc.

REFERENCES

1. H.G.Danielmeyer and H.P. Weber. IEEE J. Quant. Electr. EQ-8 (1972) 802
2. S. R. Chinn, H.Y. P. Hong and J. W. Pierce, Laser Focus 12(1976) 64
3. W. Lenth, G.Huber and D. Fay, Phys. Rev. B23 (1981) 3877
4. H.G.Danielmeyer et al., Appl. Phys. 1 (1973) 269
5. Hirotokuroda et al., J.Appl. Phys. 51 (3) (1980) 1351-1356

TABLE 2
Some physical properties of LNPP

refractive index n(1.053 μm)	1.54
n_d	1.56
nonlinear refractive index n_2(10^{-13} esu)	1.12(calculation index)
thermal conductivity (Cal/cm.sec.K)	0.0024
temp. coefficient of optical path length (10^{-6}/℃) (20-40℃)	3.78
linear expansion coefficient (10^{-6}/℃) (20-40℃)	7
softening temp (℃)	550
transition temp. (℃)	514
dn/dt (10^{-6}/℃) (20-40℃)	0
specific density (g/cm³)	3.02
Young's modulus (kg f/mm²)	5400
Poisson's ratio (kg f/mm²)	0.2
knoop number (kg f/mm²)	381

TABLE 3
Laser output characteristics (1.053 μm) of LNPP

central wave	$\triangle\lambda$ (nm)	1053
half width	$\triangle\lambda$ FWHN (nm)	20.5
effective half width	$\triangle\lambda_{eff}$ (nm)	24.8
life time	τ (μs)	150
emission cross section	($10^{-20}cm^2$)	4.2
1.053 μm quantum efficiency	η (%)	43.1
conversion efficiency	slope efficiency (%)	2.8(φ6×100)
	overall efficienvy (%)	2.2(φ6×100)
max. repetition frequency	(Hz)	20(φ5×70)
max. output energy	(J/cm^3)	5.5
max. average output power	(W/cm^3)	14.5
gain	($10^{-2}cm^{-1}$)	5.9
threshold	(J/cm^3)	4.3
destructive threshold	average pump power(W)	> 2000(φ5×70)
	average power density (MW/cm^2)	550
output ratio of Q-switching to free running		0.7
Q-switching pulse width	electrooptics (ns)	10
	rotating mirror (ns)	13
	dye (μs)	22

6. К.Л.Водоцьянов, Б.И.Денкер, Т.В.Максимова, Квантовая Злекроника, 5 (3) (1978) 686
7. Laser Focus (9) 1980. 24
8. M.M.Broer, D.L.Huber and W.M.Yen, Phys. Rev. Lett. 40 (1982) 394
9. Norber Nenroth, Opt. Engin. 26 (2) (1987) 96
10. Zhou Binkun et al., 《Laser Principle》 (Chinese), 214, (1984)
11. Hiroto Kuroda, Harutake Masuko, Shigeru Mackawa, at al., J. Appl. Phys. 51 (3) (1980). 1351

Frontiers of Materials Research/
Electronic and Optical Materials
M. Kong, L. Huang (editors)

THE IN-SITU PRODUCTION OF H_2Se FOR HIGH PURITY EPITAXIAL ZnSe GROWN BY MOVPE

Da-cheng LU

Institute of Semiconductors, Chinese Academy of Sciences, P.O.Box 912, Beijing 100083, People's Republic of China

The MOVPE growth of high quality ZnSe from $Zn(CH_3)_2$ and in situ produced H_2Se is reported. The H_2Se was formed in a small sub-reactor unit by reacting high purity Se and H_2 and by subsequently treating the effluent so that a controlled amount of H_2Se in H_2 was introduced into the growth chamber. A modelling for in situ produced H_2Se process is proposed. This arrangement provide high purity H_2Se while avoiding storing the highly toxic H_2Se.

1. INTRODUCTION

Metal-organic vapor phase epitaxy (MOVPE) has recently attracted much attention as one of the most promising methods for the growth high purity zinc selenide (ZnSe) epitaxial layer, which has a well-recognized potential use in the blue light emitting diodes and lasers. Two kinds of selenium sources have been used for the growth of ZnSe films, one is organometallic selenium compounds[1-2], such as dimethyselenide or diethylselenide, the other is hydrogen selenide (H_2Se)[3-6]. The growth temperature of ZnSe films using H_2Se is around 300°C, which is much lower than that using organic selenium compounds (500°C), so that high quality of ZnSe is expected. But H_2Se is a extreme toxic gas with threshold limit value of 0.05 ppm. The danger is compounded by the fact that it is stored in high pressure cylinders mixed with flammable hydrogen. The risk can be greatly reduced by using hydrogen and selenium as starting materials and producing H_2Se in-situ. The highest value of 77 K electron mobility of ZnSe film has been obtained by using in-situ produced H_2Se and dimethylzinc[7]. This paper describes in detail the control of the molar flow rate of in-situ produced H_2Se.

2. EXPERIMENTAL

A schematic illustration of the MOVPE reactor and the quartz sub-reactor for in-situ producing H_2Se is given in Figure 1. A quadrupole mass spectrometer, with a sampling port immediately down-stream of the susceptor, was used to monitor chemical species in the gas phase. The reactants used for the growth of ZnSe were electronic grade dimethylzinc (DMZ) maintained at -25°C and H_2Se which produced in-situ from Pd-purified hydrogen and 99.9999% pure selenium in the quartz sub-reactor which kept at 0.78 atm and heated by a two zone furnace. A quartz boat containing Se was placed in the low temperature zone kept at 460°C to evaporate Se into gas phase, than these selenium species were carried by a hydrogen flow into the reacting zone kept at 650°C to form H_2Se. The mixture then was passed through a cool trap kept at -78°C to trap excessive H_2Se, non-reacting

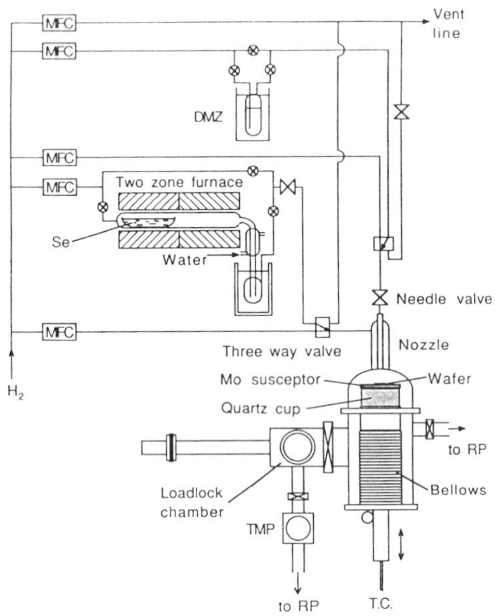

FIGURE 1
Schematic illustration of the MOVPE reactor with the in-situ production H_2Se sub-reactor.

Se and any impurities which have relative low vapor pressure comparing to H_2Se. The hydrogen flow rate were varied from 4 to 70 SCCM in order to change the feed rate of H_2Se in to the growth chamber. The molar flow rate of DMZ and H_2Se were 10-15x10^{-6} mol/min and 20-300x10^{-6} mol/min, respectively. The total flow rate of hydrogen was 1 SLM. The substrates were semi-insulating GaAs (100), oriented 2° towards <110>. The substrate temperature range was 280-480°C, while the reactor pressure was varied between 0.04-0.13 atm. Electrical properties of ZnSe layers were evaluated by the Van der Pauu method.

3. RESULTS AND DISCUSSION

Hydrogen selenide was produced in the reacting zone of the sub-reactor where the reaction, $2H_2+Se_2=2H_2Se$, took place. The partial pressure of H_2Se at the output of the reacting zone is the function of hydrogen flow rate, surface area of liquid selenium, evaporation temperature, reaction temperature and working pressure of the sub-reactor. In order to control the molar flow rate of H_2Se introduced into the growth chamber, the partial pressure of H_2Se should be higher than the vapor pressure of H_2Se at cool trap temperature, so that excessive H_2Se was condensed in the trap and saturated H_2 with H_2Se can be obtained. The equation to calculate the molar flow rate of H_2Se can be derived assuming all of the gases are ideal gases. Thus, we have

$$n_{H2Se}=F_{H2}\ P^o_{H2Se}/[22410\ (P_w-r\ P^o_{H2Se})] \qquad (1)$$

where F_{H2} is flow rate of H_2 introduced to the sub-reactor (SCCM), P_w is the working pressure (atm), P^o_{H2Se} is the vapor pressure of H_2Se and r is a factor defined by

$$r=n^*_{H2Se}/n_{H2Se} \qquad (2)$$

where n^*_{H2Se} is the molar flow rate of H_2Se condensed in the cool trap, which is very difficult to measure. In our case P_w was 0.78 atm and P^o_{H2Se} was 7.57x10^{-2} atm (at -78°C). If r is less than 0.1, Eq.(1) can be simplified to

$$n_{H2Se}=F_{H2}\ P^o_{H2Se}/(22410\ P_w) \qquad (3)$$

with receivable error. That means that the partial pressure of produced H_2Se

should be just slightly higher than the vapor pressure of H_2Se at cool trap temperature.

In order to control the molar flow rate of H_2Se well, a model based on equilibrium concept was proposed. The following assumptions were used because the reaction $2H_2+Se_2=2H_2Se$ proceed very slowly for the attainment of equilibrium, when temperature below 525°C[8]:

1) in the Se boat zone (evaporation zone), usually maintained about 460°C, hydrogen was saturated with selenium vapor, and no chemical reaction took place,

2) in the reaction zone kept above 600°C, the reaction equilibrium was established in the gas phase.

According to Berkowitz and Chupka[9] a series selenium molecules, Se_2, Se_3, Se_5, Se_6, Se_7, and Se_8 were taking into account. The partial pressure of the species can be calculated from the following equations:

$$K_p(1)=P^3_{Se2}/P^2_{Se3}, \tag{4}$$

$$K_p(2)=P^3_{Se2}/P_{Se6}, \tag{5}$$

$$K_p(3)=P^6_{Se5}/P^5_{Se6}, \tag{6}$$

$$K_p(4)=P^7_{Se6}/P^6_{Se7}, \tag{7}$$

$$K_p(5)=P^4_{Se6}/P^3_{Se8}, \tag{8}$$

$$K_p(6)=P_{Se6}, \tag{9}$$

where $K_p(i)$ is equilibrium constant, P_j is partial pressure of species j. The input date of the mole of selenium, n^o_{Se}, for calculating the equilibrium pressure in the reaction zone can be obtained by using these selenium partial pressures:

$$n^o_{Se}=n^o_{H2}(2P_{Se2}+3P_{Se3}+5P_{Se5}+6P_{Se6}+7P_{Se7}+8P_{Se8})/P_H2, \tag{10}$$

where n^o_{H2} is the mole of hydrogen. It is expressed as follows:

$$n^o_{H2}=F_{H2}\ t/22410, \tag{11}$$

where t is 1 minute.

In the reaction zone, the simultaneous equations to be solved are the conservation constraints and mass action equations:

$$n^o_{Se}=n_{H2Se}+(2n_{Se2}+3n_{Se3}+5n_{Se5}+6n_{Se6}+7n_{Se7}+8n_{Se8}), \tag{12}$$

$$n^o_{H2}=n_{H2}+n_{H2Se}, \tag{13}$$

$$K_p(1)=A\ n^3_{Se2}/n^2_{Se3}, \tag{14}$$

$$K_p(2)=A\ n^3_{Se2}/n_{Se6}, \tag{15}$$

$$K_p(3)=A\ n^6_{Se5}/n^5_{Se6}, \tag{16}$$

$$K_p(4)=A\ n^7_{Se6}/n^6_{Se7}, \tag{17}$$

$$K_p(5)=A\ n^4_{Se6}/n^3_{Se8}, \tag{18}$$

$$K_p(11)=A^{1/2}n_{H2}(n_{Se2})^{1/2}/n_{H2Se}, \tag{19}$$

where n_i is the mole of species i and

$$A=P_w/(n_{H2}+n_{H2Se}+2n_{Se2}+3n_{Se3}+5n_{Se5}+6n_{Se6}+7n_{Se7}+8n_{Se8}. \tag{20}$$

The partial pressure of species i in the reaction zone is given by the following equation

$$P_i=n_i\ A \tag{21}$$

Fig.2 shows the partial pressure of Se_n and H_2Se versus the sub-reactor temperature. One can see that the partial pressure of H_2Se drops when temperature increasing. In the range of 600°-650°C

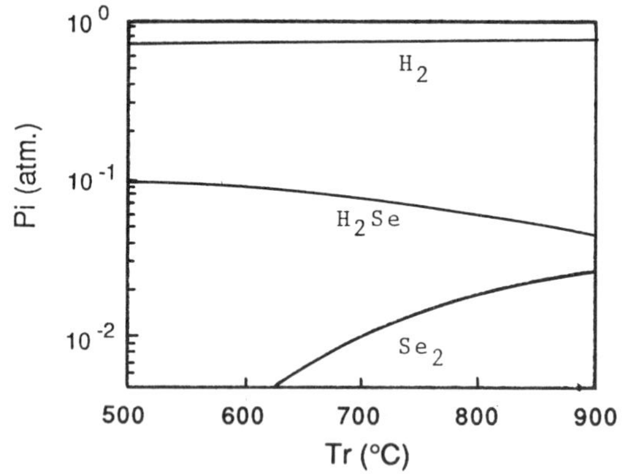

FIGURE 2
Dependence of the partial pressures of Se_n and H_2Se on the reaction temperature. Evaporation temperature (T_e):460°C; P_w:0.78 atm.

high yield of H_2Se can be obtained. The temperature higher than 650°C dose not recommended. The main selenium species are Se_2, Se_3, and Se_5.

The dependence of the partial pressure of these species on the working pressure is shown in Fig.3. The partial pressure of H_2Se decreases when the working pressure decreasing, especially lower than 0.5 atmosphere.

Fig. 4. indicates the relationship between the partial pressure of H_2Se and the evaporation temperature of selenium which is a key parameter. Raising the temperature of Se boat too high, more excessive H_2Se will be formed and condensed on the wall of the cool trap. Consequently, not only more selenium source is consumed, but also makes factor r larger, which increases the error of equation 3. On the other hand, if setting the temperature to low, the partial pressure of H_2Se may be lower than the vapor pressure of H_2Se at the cool trap temperature, and the system may loss the function of controlling the molar flow rate of H_2Se. According to this model a set of conditions for running the sub-reactor was chosen and worked very well.

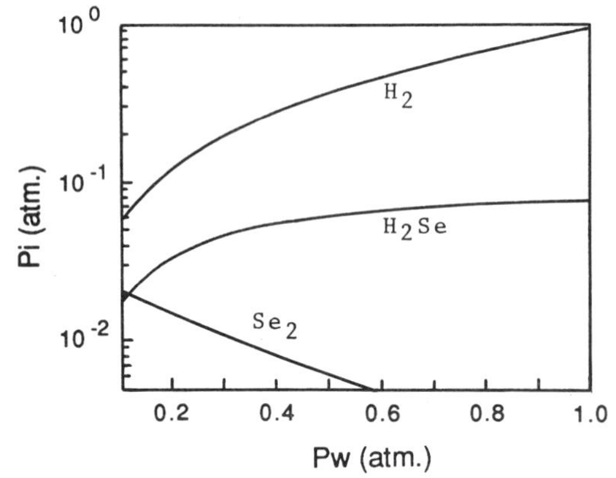

FIGURE 3
Dependence of the partial pressures of Se_n and H_2Se on the working pressure. Evaporation temperature (T_e):460°C; reaction temperature(T_r):600°C.

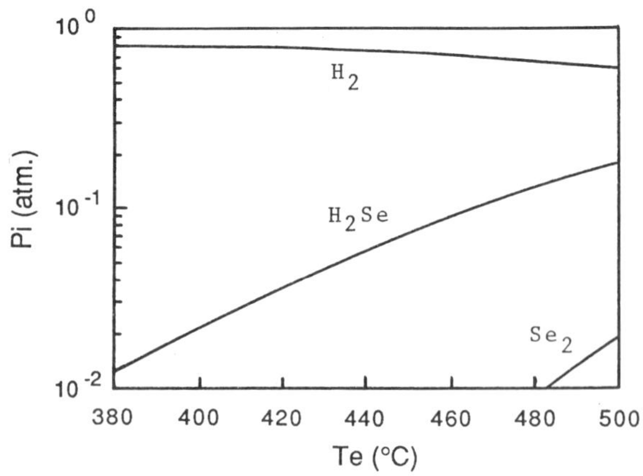

FIGURE 4
Dependence of the partial pressures of Se_n and H_2Se on the evaporation temperature. Reaction temperature (T_r):460°C; P_w:0.78 atm.

The main peak of the mass spectrum of in situ produced H_2Se is mass 80, which consists H_2Se^{78} and Se^{80}. Fig. 5 shows the linear relationship between the height of peak 80 and H_2Se molar flow rate calculated by eq. 3. It means that the molar flow rate of H_2Se introduced to the growth chamber cab be controlled.

All samples investigated were n-type with 300 K net carrier concentrations ranging from $6.4x10^{14}$ to $1.5x10^{16}$ cm^{-3}. The 77K mobility for run No. 62 of 9257 $cm^2/Vsec$ is to our knowledge the highest so far reported for ZnSe grown by any epitaxy method. Fig. 6 shows the electron mobility versus carrier concentration at 300 and 77 K, respectively. Data from other MOVPE and molecular beam epitaxy (MBE)[10] studies of undoped ZnSe are included for comparison. The solid curves have been reproduced from the theoretical work on electron transport in ZnSe by Ruda[11]. The nature of donors and acceptors does not very clear at the present.

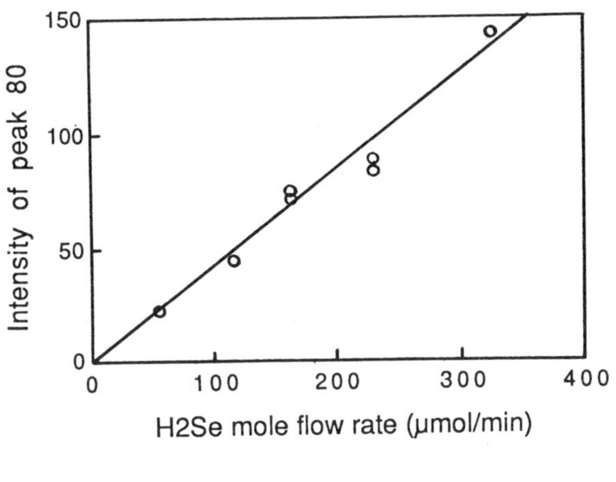

FIGURE 5
Dependence of the height of the peak 80 on the molar flow rate of H_2Se calculated by eq. 3.

The key factor to grow high purity ZnSe is the combination of low growth temperature and the purity of reactants, especially the purity of hydrogen selenide. The formation of H_2Se form hydrogen and selenium is a powerful purification method[12]. In principle, the impurities which have low vapor pressure

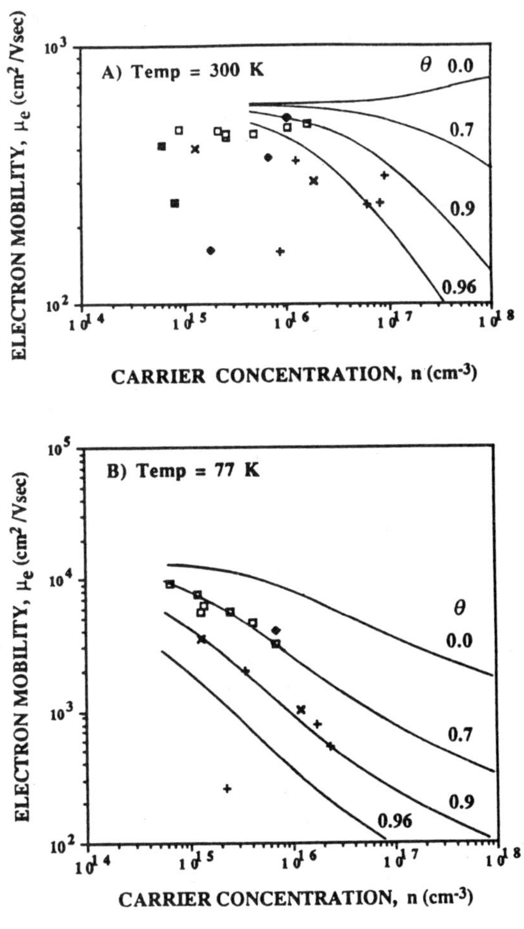

FIGURE 6
Comparison of the Hall electron mobility vs carrier concentration at (a) 300K and (b) 77k for ZnSe. The symbols correspond to experimental data points for nominally undoped films as follows: (□) this study, (+) Ref.4, (■) Ref.5, (×) Ref.6, (♦) Ref.10. The solid curves represent the calculated values of mobility vs carrier concentration for different compensation ratios, reproduced from the work of Ruda (Ref.11).

and/or cannot form volatile hydride at 460°C should be left in the boat. Some impurities and their hydrides will condensed in the cool trap. The impurities which can pass the trap may be H_2S, H_2O, PH_3 and group VI hydrides.

4. CONCLUSIONS

High purity of hydrogen selenide has been produced in-situ utilizing Pd-purified hydrogen and 6N selenium. A model of in-situ produced H_2Se has been proposed. The molar flow rate of H_2Se can be controlled and the danger of using H_2Se cylinder has been greatly reduced by this method. This study indicates that high purity of epitaxial ZnSe films can be grown by low temperature MOVPE with additional purification of the staring materials.

ACKNOWLEDGMENTS

The author would like to thank Professor Klavs F. Jensen for an invitation to University of Minnesota where the experimental work was carrier out and K. P. Giapis for his co-operation.

REFERENCES

1. P.J. Wright et al., J. Crystal Growth 66 (1984) 26.
2. H. Mitsuhashi et al., J. Crystal Growth 77 (1986) 219.
3. W. Stutius, Appl. Phys. Lett. 33 (1978) 656.
4. S. Fujita et al., J. Crystal Growth 72 (1985) 27.
5. G. Fan et al., J.Electr. Mat. 15 (1986) 251.
6. A. Yoshikawa et al., J. Crystal Growth 72 (1985) 13; Jpn. J. Appl. Phys. 23 (1984) L424.
7. K.P. Giapis et al., Appl. Phys. Lett. 54 (1989) 353.
8. J.R. Rawling and J.M. Toguri, Can. J. Chem. 45 (1966) 451.
9. J. Berkowitz and W.A. Chupka, J. Chem. Phys. 48 (1968) 5743.
10. T. Yao, J. Crystal Growth 72 (1985) 31.
11. H.E. Ruda, J.Appl.Phys. 59 (1986) 1220.
12. S. Nilsern and R.J. Heritage, J. Electrochem. Soc. 106 (1959) 39.

Frontiers of Materials Research/
Electronic and Optical Materials
M. Kong, L. Huang (editors)
Elsevier Science Publishers B.V., 1991

FAR-INFRARED REFLECTIVE SPECTRA OF ZnSe-ZnTe STRAINED-LAYER SUPERLATTICES

Jie CUI, Fu-xi GAN, Hai-long WANG and Yun-liang CHEN

Shanghai Institute of optics and Fine Mechanics, Academia Sinica, P.O.Box 800-216, Shanghai, 201800 P.R.C.

Far-infrared reflective spectra have been measured, for the first time, on ZnSe-ZnTe strained-layer superlattices grown by molecular beam epitaxy. The spectra at 300K display the ZnSe and ZnTe transverse optical phonon modes. The reflective spectrum can be explained by the long wavelength superlattice dielectric theory and the reflection theory of multiple absorbing layers. Fitting the experimental data with the computer calculated curve, the transverse optical phonon frequences, mode damping constants, mode oscillation strengths, high frequency dielectric constants have been determined. As far as we know, the present results are the first determination of these parameters.

1. INTRODUCTION

II-VI wide band gap semiconductor strained-layer superlattices (SLSs) are potential materials for application to optoelectronic devices in the short-wavelength visible region. The exciton characteristics of the SLSs such as photoluminescence properties and absorption properties are being studied [1-7]. There are also some studies on the phonon characterisitics by Raman spectra [8-12]. We have studied the TO phonon modes of II-VI compound SLSs by far-infrared reflectivity. Far–infrared reflectivity measurement is a simple way to characterize the quality of the microstructures and it has some important differences to the Raman spectra. First, different phonon modes are active, and the rediation penetration depth is greater in the far-infrared region, so that the entire structure is probed, whereas in Raman spectroscopy only a few periods near the surface are examined. Second, it is simple to write the dielectric response function for the superlattices in the far-infrared region, thus it is easy to make detailed fits to the measured data, which is generally not done for Raman measurements. There are many reports on the infrared reflection measurements of III- V compounds AlAs-GaAs superlattices[13-16]. and some on II-VI nerrow band gap compound HgTe-CdTe superlattice[17-18]. We report, for the first time, the far-infrared reflectivity spectra on II-VI wide band compound ZnSe-ZnTe strained-layer superlattice and characterize the reflectivity by long-wavelength dielectric respones theory and the reflection theory of multiple absorbing layers. Fitting the experimental data with the computer calculated curve, we have determined the phonon parameters such as the transverse optical phonon frequencies, mode damping constants, mode oscillation strengths, high frequency dielectric constants of the II-VI compound ZnSe,ZnTe.

2. EXPERIMENTS

ZnSe-ZnTe strained-layer superlattice samples were grown at 320°C by molecular beam epitaxy on [100] GaAs substrates after growth of ZnSe buffer layers (or [100] InP substrates). The thicknesses of ZnSe and ZnTe layers were determined by x-ray diffraction measurement combining the growth rates of two layers estimated previously. Reflection measurements at near-normal incidence were made in the region 100-500 cm^{-1} at 300K using a Fourier spectrometer, the resolution is 2 cm^{-1}.

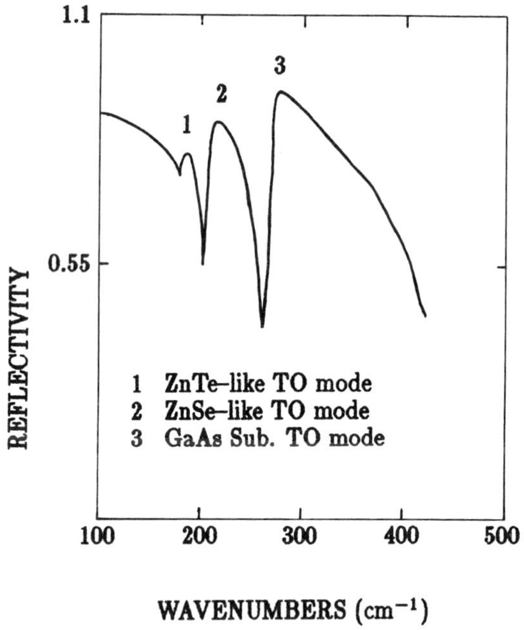

FIGURE 1
The measured far-infrared reflectivity spectrum of ZnSe-ZnTe strained-layer superlattice

3. RESULTS AND DISCUSSION

Figure 1 shows the far-infrared reflectivity spectrum for the sample ZST-22 (the thicknesses of ZnSe, ZnTe layers are 86Å, 86Å respectively; the number of periods is 10; the thickness of ZnSe buffer is 0.12μm; the substrate is [100] GaAs). In the region 150-250 cm^{-1}, there are two peaks near 180cm^{-1} and 210cm^{-1} corresponding to the ZnTe-like TO and ZnSe-like TO modes, respectively. We found that the frequencies of the TO modes from the SLS sample are shifted as compared to the bulk materials and the shifts can be changed according to the structure paremeters[8]. It is confirmed that there are a large strains in the superlattice layers, which shifted the frequences of phonon modes. We also measured the reflectivity of ZnSe-ZnTe mixed crystal system which shows one-mode behavier[19]. The mode is located between ZnTe-like TO mode and ZnSe-like TO mode, as shown in Figure 2 (the peak near 310 cm^{-1} is the InP substrate TO mode). Therefore, by far-infrared reflectivity measurement it is easy to confirm the structure characteristics of the superlattice samples[10].

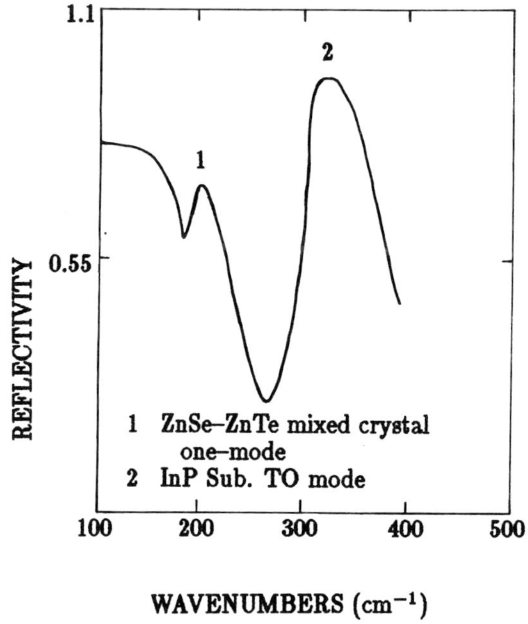

FIGURE 2
The measured far-infrared reflectivity spectrum of ZnSe-ZnTe mixed crystal system

This SLS system has the strong absorption in the far-infrared region. The long wavelength superlattice dielectric response theory combining the reflection theory of multiple absorbing films must be used to study the reflectivity. Though the strains in each layer of SLS are great, the two strain-tensor components parallel to the interface are equall[8]. Therefore, for the normal incidence, the SLS can be considered as a uniaxial crystal system. The dielectric functions of a SLS are[20]

$$\epsilon_{\perp}(\omega) = (\epsilon_1 d_1 + \epsilon_2 d_2)/(d_1 + d_2)$$
$$\epsilon_{//}^{-1}(\omega) = (\epsilon_1^{-1} d_1 + \epsilon_2^{-1} d_2)/(d_1 + d_2)$$

For TO modes, just consider the components perpendicular to the growth direction $\epsilon_{\perp}(\omega)$. The dielectric function of SL at normal incidence ϵ_{SL} is

$$\epsilon_{SL}(\omega) = (\epsilon_1 d_1 + \epsilon_2 d_2)/(d_1 + d_2) \tag{1}$$

here ϵ_1, ϵ_2 and d_1, d_2 are the bulk dielectric response functions and thicknesses of two layers in a SL unit cell, respectively. The buffer layer and the substrate can be treated as bulk materials which are represented by a Lorentzian oscillator of the form

$$\epsilon(\omega) = \epsilon_\infty + \frac{S\omega_{TO}^2}{\omega_{TO}^2 - \omega^2 - i\gamma\omega} \tag{2}$$

where ϵ_∞ is the high frequency dielectric constant, S is the mode strength, ω_{TO} is the transverse optical phonon frequency, and γ is the mode damping. These parameters are unknow for II-VI compounds.

The layers are absorbing, with complex indices $N = n - ik$. The dielectric function is written as an additive form

$$\epsilon = \epsilon' + i\epsilon'' = (n^2 - k^2) - i(2nk) \tag{3}$$

we have

$$n = \sqrt{\frac{1}{2}\sqrt{\epsilon'^2 + \epsilon''^2} + \frac{\epsilon'}{2}} \tag{4}$$

$$k = \sqrt{\frac{1}{2}\sqrt{\epsilon'^2 + \epsilon''^2} - \frac{\epsilon'}{2}} \tag{5}$$

from (1) to (5), we can get the n, k for the SLS layer, buffer layer and substrate repectively. Furthermore, to get the reflectivity of the SLS system we proceed as follows using the reflection theory of multiple absorbing films[21].

With normal incidence, the Fresnel coefficient for the light travelling from $j-1$ to j layer is given by

$$r_j = g_j + ih_j = \frac{N_{j-1} - N_j}{N_{j-1} + N_j}$$

where

$$g_j = \frac{n_{j-1}^2 + k_{j-1}^2 - n_j^2 - k_j^2}{(n_{j-1} + n_j)^2 + (k_{j-1} + k_j)^2}$$
$$h_j = \frac{2(n_{j-1}k_j - n_j k_{j-1})}{(n_{j-1} + n_j)^2 + (k_{j-1} + k_j)^2}$$

we can obtain the effective Fresnel coefficient for the jth layer $\rho_j \exp(i\Delta_j)$

$$\rho_j^2 = (G^2 + H^2)/(G'^2 + H'^2)$$
$$\tan\varsigma_j = H'/G'$$
$$\tan\eta_j = H/G$$
$$\Delta_j = \eta_j - \varsigma_j$$

where

$$H = h_j + \exp(-\alpha_j)(h_{j+1}\cos\theta_j - g_{j+1}\sin\theta_j)$$
$$G = g_j + \exp(-\alpha_j)(g_{j+1}\cos\theta_j + h_{j+1}\sin\theta_j)$$
$$H' = \exp(-\alpha_j)[(h_j g_{j+1} + h_{j+1} g_j)\cos\theta_j - (g_j g_{j+1} - h_j h_{j+1})\sin\theta_j]$$
$$G' = 1 + \exp(-\alpha_j)[(g_j g_{j+1} - h_j h_{j+1})\cos\theta_j + (h_j g_{j+1} + h_{j+1} g_j)\sin\theta_j]$$

and

$$\alpha_j = -4\pi k_j d_j/\lambda$$
$$\theta_j = -4\pi n_j d_j/\lambda$$

The effective Fresnel coefficient for the $j-1$th layer

is then given as

$$\rho_{j-1}\exp(i\Delta_{j-1}) = \frac{r_{j-1}+\rho_j\exp(i\Delta_j - 2i\delta_{j-1})}{1+r_{j-1}+\rho_j\exp(i\Delta_j - 2i\delta_{j-1})}$$
$$\delta_{j-1} = 2\pi N_{j-1}d_j/\lambda$$

The process is repeated successively until the final coefficient ρ_1 is obtained for the whole system. The reflection coefficient of the system is then given by ρ_1^2, which has a maximum near ω_{TO}. Fitting the reflectivity curve with computer, the phonon parameters in the equation such as ϵ_∞, S, γ and ω_{TO} can be determinted.

Figure 3 shows the fit curve, (*) are the measured data, (—) is the curve calculated by the equation list above. The phonon parameters we determinded list in Table 1.

TABLE 1.
Phonon parameters giving best fit to the measured reflectivity curve for the ZnSe–ZnTe SLS sample (ZST-22).

layer	ϵ_∞	$\omega_{TO}(cm^{-1})$	S	γ	$d(\text{Å})$
ZnTe	8.3	178.2	4.63	24.7	86
ZnSe	5.8	206.0	1.91	27.4	86
ZnSe buffer	6.0	210.5	1.8	27.0	0.12μ
GaAs sub.	11.5	268.2	5.9	0.46	

Because the entire structure of sample is probed in the infrared region, the inhomogeneous thicknesses of layers made the peaks widen in the reflectivity spectrum, which is a consequence of the interference effects of the layers. As we see from Fig.3 the measured peak of ZnSe-TO mode is wider than the calculated one, since we assume the layer thickness remain unchanged through the whole sample. From Table 1, we find the ZnSe-like TO mode frequency is shifted toward low frequency compared to the bulk sample (209cm^{-1}). In contrast to this, ZnTe-like TO mode frequency is shifted toward high frequency compared to the bulk

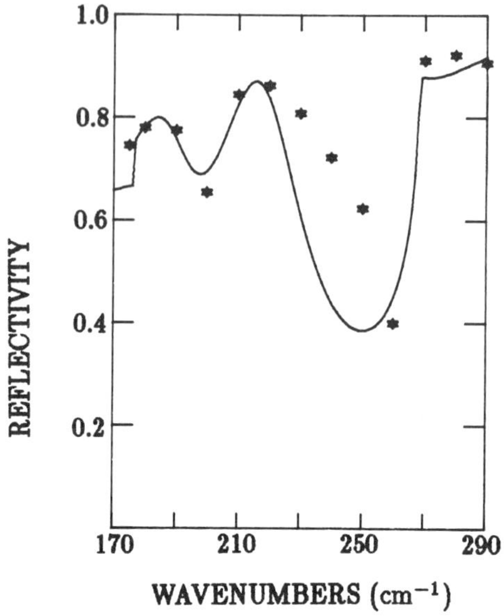

FIGURE 3
Measured (*) and fitted (—) far-infrared reflectivity spectrum of the ZnSe-ZnTe SLS sample ZST-22 at 300K.

ZnTe sample (177cm^{-1}). This phenomenon had been observed by Raman scattering[10]. The ZnSe-ZnTe SLS is zincblende crystals, and the lattice constants of ZnSe, ZnTe are 5.669Å, 6.104Å respectively. When the SLS is grown on a (100) substrate, in the (100) plane ZnSe layers are under biaxial tensile stress and ZnTe layers are under biaxial compressive stress. The displancements for the TO mode lie in the (100) plane also. From measured curve we can conclude that the tensile stress made the TO mode shifted to lower frequency while compressive stress made the TO mode shifted to higher frequency.

4. CONCLUSION

In conclusion, Far-infrared reflectivity measurement is a simple way to check the quality of SLS and to determine the TO mode frequencies. We have successfully measured the TO modes of II-VI compound ZnSe-ZnTe

SLS. Fitting the curve, we have determined the phonon paremeters of these materials.

ACKNOWLEDGEMENTS

The authors would like to thank Dr. L. Xu, Dr. A. D. Shen, and Ms. Y. H. Shen for assistance in the preparation of the SLS samples.

REFERENCES

1. L. A. Kolodzieski, R. L. Gunshor, N. Otsuka, S. Datta, W. M. Becker and A. V. Nurmikko, IEEE J. Quantum Electronics QE-22 (1986) 1666.

2. Masakazu Kobayashi, Naoki Mino, Hironori Katagiri, Ryuhei Kimura, Makoto Konagai, and Kiyoshi Takahashi, Appl. Phys. Lett. 48 (1986) 296.

3. Hiroshi Kuwabara, Hiroshi Fujiyasu, Masaki Aoki, and Shoi Yamada, Japan. J. Appl. Phys. 25 (1986) L707.

4. Toshiya Yokogawa, Motosugu Ogure, and Takao Kajiwara, Appl. Phys. Lett. 49 (1986) 1702.

5. Akira Taike, Nobuaki Tewraguchi, Makoto Konagai, and Kiyoshi Takahashi, Japan. J. Appl. Phys. 26 (1987) L989.

6. Shigeo Fujita, Yoshinobu Matsuda, and Akio Sasaki, Appl. Phys. Lett. 47 (1985) 955.

7. K. Mohammed, D. J. Olego, P. Newbury, D. A. Cammark, R. Dalby, and H. Comelissen, Appl. Phys. Lett. 50 (1987) 1820.

8. S. Nakashima and Y. Nakakura, Appl. Phys. Lett. 48 (1986) 236.

9. S. Nakashima and A. Wada, J. Appl. Phys. 62 (1987) 2009.

10. Masakazu Kobayashi, Makoto Konagai, and Kiyoshi Takahashi, J. Appl. Phys. 61 (1987) 1015.

11. Y. H. Wu, H. Yang, A. Ishida, and H. Fujiyasu, Appl. Phys. Lett. 54 (1989) 239.

12. J. Menendez, A. Pinczuk, and J. P. Valladares, R. D. Feldman, and R. F. Austin, Appl. Phys. Lett. 50 (1987) 1101.

13. M. S. Durschlag and T. A. Detemple, Solid State Commun. 40 (1981) 307.

14. K. A. Maslin ,T. Parker, N. Ra ,D. R. Tilley, P. J. Dobson, D. Hilton and C. T. B. Foxon, Solid State Commun. 60 (1986) 461.

15. R. Sudharsanar, S. Perkowitz, S. S. Yom and T. J. Drummond, SPIE 794 (1987) 197.

16. A. S. Barker, J. L. Merz and A. C. Gossard, Phys. Rev. B17 (1978) 3181.

17. S.Perkowitz, D. Rajavel, I. J. Sou, J. Reno, J. P. Faurie, C. E. Jones, T. Casselman, K. A. Harris, Jr. Cook, J. W., J. F. Schetina, Appl. Phys. Lett. 49 (1986) 806.

18. S. Perkowitz, R. Sudharsanan, and S. S. Yom, J. Vac. Sci. Technol. A5 (1987) 3157.

19. S. Nakashima, T. Fukumoto and A. Mitsuishi, J. Phys. Soc. Japan 30 (1971) 1508.

20. N. Raj and D. R. Tilley, Solid State Commun. 55 (1985) 373.

21. O. S. Heavens, Optical Properties of Thin Solid Films (Butter Worths Scientific Publications, London,1955) pp.40–46.

Frontiers of Materials Research/
Electronic and Optical Materials
M. Kong, L. Huang (editors)

INFRARED STUDY OF PHONON AND IMPURITY ABSORPTIONS IN AlSb

D. S. JIANG*, C. T. LIN, E. SCHÖNHERR and A. BREITSCHWERDT

Max-Planck-Institut fur Festkorperforschung, Heisenbergstr. 1, 7000 Stuttgart 80, F. R. Germany
* and Institute of Semiconductors, Academia Sinica, Beijing 100083, China

Infrared absorption spectra due to impurities and two-phonon processes in AlSb single crystals are investigated at 10K, 77K and room temperature, utilizing high resolution Fourier transform spectroscopy. The residual donor impurity which induces electronic excitation peaks is identified as substitutional sulfer. A strong local vibrational mode absorption caused by carbon has first been observed with a frequency at 590. 5 cm^{-1}. The structure of two-phonon spectra is analysed in terms of the combination of phonons at critical points as well as at boundary planes of the Brillouin zone.

1. INTRODUCTION

Aluminium Antimonide is a promising semiconductor material for high temperature electronic devices and photon detectors. Recently it has been used to constitute the strained-layer GaSb/AlSb superlattice for optoelectronic device application. [1,2,3] It has a relatively wide energy gap of 1. 61 eV at room temperature. The perfect AlSb crystal is difficult to prepare since AlSb reacts with H_2O, air and many crucible materials. Recently, Lin et al[4] have reported to successfully grow high quality AlSb single crystals in Al_2O_3 or vitreous carbon crucibles by the Czochralski technique.

AlSb has zinc-blende structure with an indirect energy band gap. Till now only a few data are available about shallow impurity levels in AlSb. [5,6]. It was found that the donor impurities, Se and Te, have relatively large ionization energies than acceptors due to the splitting of the sixfold-degenerated donor ground states. The two-phonon band absorptions in AlSb were reported and analyzed by four characteristic phonons. [7]

In this paper, the absorption of a local vibrational mode (LVM) in AlSb is reported for the first time. The spectra of electronic excitations of donor impurities are detected in different crystals. Sulfur is assigned as a main residual donor impurity with an ionization energy close to Se donor. The spectral features of the two-phonon absoption bands at 10K are measured and analysed in comparison with those of other III-V compounds.

2. EXPERIMENTAL

Single AlSb crystals were grown from the melt in Al_2O_3 or vitreous carbon crucibles. The single crystal growth and selective doping with impurities have been described in detail elsewhere. [4] The nominally undoped samples show p-type conductivity with free hole concentrations between 10^{16}- 10^{17} cm^{-3}, The samples doped with isoelectronic impurities Yb or Pr show also weak P-type conductivity at room temperature, while the Selenium-doped AlSb is n-type in agreement with the well-known donor behavior of Se in AlSb.

The AlSb samples are cut in 1 mm thick plates and then polished to mirror-like surfaces. They are kept in vacuum for preventing oxidation before IR measurements.

3. RESULTS AND DISCUSSIONS

I. IR absorption related to impurities

The IR absorption spectra of AlSb samples are measured in the wavenumber region between 300 and 4000 cm^{-1}. Above 700 cm^{-1} the Se-doped AlSb samples show the well-known absorption features at 10K, which are related to the optical transitions of Se donors between the split ground 1S states induced by the valley-orbital splitting, and between the ground state to excited states, i. e. the 1S(A1)-1S(T1), 1S(A1)-$2p_{\pm}$ and other transitions. [5] In the Yb-doped AlSb samples, the similar absorption features appear as well at 10K, with two strong peaks at 942 and 1116 cm^{-1}, and small peaks at 1259, 1320 and 1450 cm^{-1}, respectively. A broad band is superposed on the high energy side, as shown in Fig. 1. Their intensities decrease with increasing temperature. At 77K the absorption intensities of

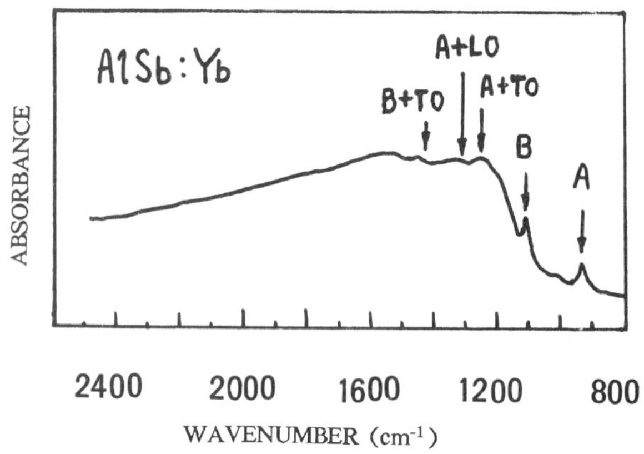

FIGURE 1

The absorption spectrum of a Yb-doped sample, measured at 10K.

A represents the transition

1S (A_1) — 1S (T_1).

B represents the transition

1S (A_1)— $2P_{\pm}$

the separate peaks are too weak to distinguish from the background of broad band absorption. At room temperature, the broad band absorption becomes also much weaker than at low temperatures.

The main peaks in Yb-doped samples should correspond to the donor excitations, as in Se-doped AlSb. The additional peaks are optical phonon side-bands, as denoted in Fig. 1.

The ionization energy of this donor is calculated to be 146. 5 meV, assuming that the energy separation between $2P_{\pm}$ and the bottom of conduction band can be treated by the effective mass theory.[7] The absorption should not be caused by the acceptor impurities because the acceptor ionization energies in AlSb are normally much smaller.[8] Within the effective mass approximation the acceptor level in AlSb is only about 42. 5 meV above the top of valence band.

The isoelectronic Yb impurity in Ⅲ - V compounds behaves as a deep center showing 4f ion transitions. Yb has been tentatively suggested to induce acceptors in AlSb,[9] but not donors.

The sharp donor-acceptor-pair (DAP) line luminescence from these AlSb samples indicated that both donor and acceptor impurities take substitutional sites in the Sb sublattice, and according to the bound exciton energy the donor is more likely sulfur instead of selenium.[10] Another group Ⅵ element, Te, can be excluded from the candidate list because its ionization energy of 78 meV[7] is much smaller than the measured value. Sulfur do really exsist in our samples as a residual impurity. A small quantity (near 1 PPM) of sulfur but no selenium was detected in our Yb-doped AlSb samples by chemical analysis of inductively coupled plasma (ICP) atomic emission spectroscopy. Therefore, we assign the residual donor impurity responsible for the absorption peaks to be substitutional S on Sb sublattice.

Between 350 and 700 cm^{-1}, most features in the absorption spectra are similar for all samples. They are attributed to lattice phonon bands (details in the next paragraph). It is noted, however, that at 10K a sharp peak at 590. 5cm^{-1} appears only in some samples. Fig. 2 shows two IR spectra from an undoped sample and a Se-doped sample, in the latter the peak at 590. 5 cm^{-1} is especially sharp, having a half width smaller than 2 cm^{-1}. The half width of this peak becomes broader at the high temperatures but is still well-resolved from the background. Apparently its integral absorption does not change with temperature, indicating that the absorption is not due to the electronic transition. The temperature dependence of the peak frequency is different from the lattice phonon features; The frequency shift between 10K and 300K is 0. 5 cm^{-1}, which is small in comparison to the nearly 5 cm^{-1} of the neighboring peaks. We suggest that carbon gives rise to the LVM peak at 590. 5 cm^{-1}. The carbon atoms may be incorporated from the carbon crucible or from the graphite heater during the crystal growth. Similar LVM absorptions exist also in GaAs and GaP, e. g. at 582 cm^{-1} due to C_{As} in GaAs. Considering the similarity of the chemical nature of Ga and Al, it is reasonable to expect that the carbon-induced LVM frequencies are close to each other in GaAs and AlSb. The presence of carbon acceptor on Sb site was proposed by Ennen et al.[9]. This is consistent with our assignment.

II. TWO-PHONON ABSORPTION

The absorption spectra of AlSb show prominent features above the restrahlen band (centered at 318 cm^{-1}), in the wavenumber range between 400 and 700 cm^{-1}. The ear-

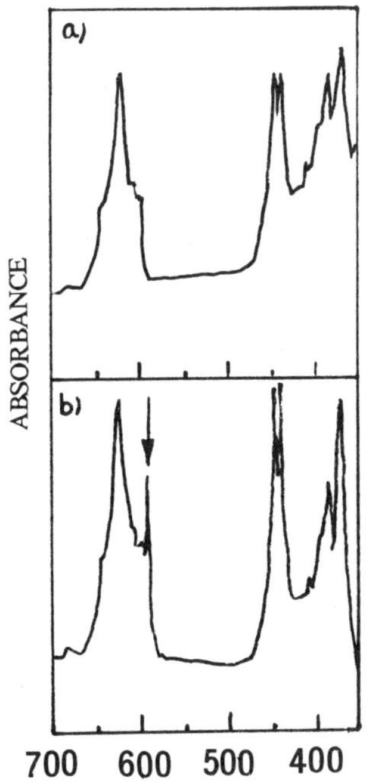

FIGURE 2

The absorption spectra of a) undoped and b) Se-doped samples at 10K. The LVM absorption at 590. 5 cm-1 appears in Fig. 2b and is marked by an arrow.

ly work of Turner and Reese[6] identified the features in this region as two-phonon sum absorption bands by four characteristic phonons. Our better resolved spectra provide more detailed informations of the two-phonon bands. The main spectral features are shown in Fig. 2b and listed in Table I .

For multi-phonon absorption processes both energy and momemtum conservation rules should be observed. The particular types of discontinuities on the two-phonon density-of-states depends on the critical points of the phonon dispersion curve. For AlSb, it is difficult to assign the optical features because no neutron scattering data of the phonon dispersion curve is yet available. Only the phonon energies at Γ and X points of the Brillouin zone were reported. However, the two-phonon absorption spectra in different Ⅲ-V compounds often show many similar structures when the frequency coordinate is normalized in terms of their respective restrahlen frequencies.[12] Such similarity is helpful for the assignment of the features appearing in the infrared spectra of AlSb.

Two-phonon absorption spectra of AlSb and GaAs are shown in Fig. 3 with the abscissa normalized to ν / ν_{TO}, where ν_{TO} is the frequency of the transverse optical mode frequency at the center of Brillouin zone. (ν_{TO}= 269 cm^{-1} for GaAs and ν_{TO}=323 cm^{-1} for AlSb). There are

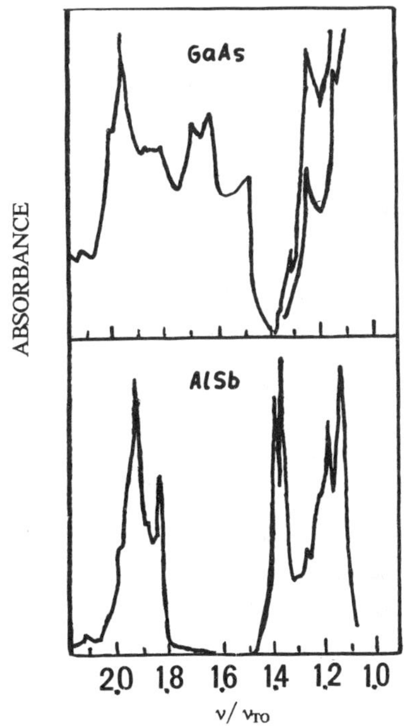

FIGURE 3

The absorption spectra of a) GaAs and b) AlSb with normalised abscissa ν/ν_{TO}. For GaAs the absorption spectrum in the low frequency region was measured using two samples of different thicknesses.[12]

remarkable differences between the two normalized spectra caused by the different phonon dispersion curves of AlSb and GaAs. In particular, between the group of peaks centered at ν/ν_{TO}=1. 93and ν/ν_{TO}=1. 4 a broad gap in the density of states appears in AlSb. A similar gap was found in InP,[13] suggesting that there is a similarity between the phonon dispersion curves of InP and AlSb. However, several important features are similar and characteristic. Both spectra exhibit a sharp peak at ν/ν_{TO}= 1. 93 and a group of peaks in its vicinity. The peaks in AlSb can be assigned to be overtones and combinations of optical mode phonons on the edge of the Brillouin zone, according to their counterparts in GaAs (and many other Ⅲ - V compounds). Several strong peaks near $\nu/\nu_{TO}= 1.2$ are assigned to the LO+TO and TO +TA in a similar way. A small peak is present at $\nu/\nu_{TO}= 2.1$, it is common to nearly all Ⅲ - V compounds and is suggested to be due to saddle-point critical point phonons near the zone center.[12] It should be noted that the critical points of the multi-phonon joint density of states are not limited to the high-symmetry points in the Brillouin zone, but may also occur on the hexagonal faces of the Brillouin zone, as in GaAs. For example, the strong peaks near 662.5 cm^{-1} are attributed to the combinations or overtones of optical mode phonons on the hexagonal planes. They can not be well fitted by the summation of two phonon energies at the high-symmetry points as Γ and X points. Since these phonons do not always possess a simple polarization character, the assignment of longitudinal or transverse modes is not strict. It can also be confirmed from the IR data that the phonon energies of LO(X) and TO(X) is lower than that of TO(Γ), 323 cm-1. This is consistent with luminescence measurements[14] but in contrast with the theoretical prediction.[16]

ACKNOWLEDGEMENT

we are grateful to J. Weber for the critical reading of the manuscript, to G. Hofmann for stimulating discussions. The support and encouragement of prof. H-J Queisser for this work is deeply acknowledged.

Table I. The main optical features in the wavelength region of 400-7000 cm^{-1} and the tentative assignment, where (h) represents hexagonal face, (*) represents the critical points other than Γ and X. The phonon energies at Γ and X points are refered from literatures.[14,15]

position (cm^{-1})	shape	strength	assignment
675	peak		LO+TO (near Γ)
640	shoulder		LO+TO (h)
622.5	peak	strong	2LO (h)
608	peak	weak	2LO (*)
602	peak	weak	2LO (*)
595	peak	weak	2LO (near X)
590.5	sharp peak		LVM
457.5	shoulder		LO+LA (*)
445	peak	strong	LO+LA (h)
440	peak	strong	LO+LA (h)
408	peak		TO+TA (*)
382.5	peak		TO+TA (*)
367.5	peak	strong	LO+TA (h)
319			TO (Γ)
340			LO (Γ)
328			TO (X)
291			LO (X)
67.7			TA (X)
144.4			LA (X)

1. R. J. Stirn and W. M. Becker, Phys. Rev. 148, 907 (1966)
2. J. H. Yee, S. P. Swierkoski and J. W. Sherohman, IEEE Trans. Nuc. Sci. 24,1962 (1977)
3. Y. Ohmori, Y. Suzuki and Okamoto, Jap. J. Appl. Phys. 24, L657 (1985)
4. C. T. Lin, E. Schoenherr, H. Bender and C. Busch, J. of Crystal Growth, 94, 955 (1989)
5. C. T. Lin, E. Schoenherr and H. Bender, ICCG-9, Sendai, Japan, Sept. (1989)
6. W. J. Turner and W. E. Reese, Phys. Rev. 127, 126 (1962)
7. B. T. Ahlburn and A. K. Ramdas, Phys. Rev, 167,717 (1968)
8. B. T. Ahlburn and A. K. Ramdas, Phys. Rev. 187, 932 (1969)
9. H. Ennen, G. Pomrenke and A. Axmann, J. Appl. Phys. 57, 2182 (1985)
10. G. Hofmann, C. T. Lin, E. Schoenherr and J. Weber, J. Appl. Phys. 67, 1478 (1990)
11. E. S. Kotteles, W. R. Datars and G. Dolling, Phys. Rev. B. 9, 572 (1974)
12. E. S. Koteles and W. R. Datars, Can. J. of. Phys. 54, 1676 (1976)
13. E. S. Koteles and W. R. Datars, S. S. Commun. 19, 221 (1976)
14. R. Linnebach and K. W. Benz, J. Crys. Growth, 55, 531 (1981)
15. A. Mooradian and G. B. Wright, S. S. Commun, 4, 431 (1966)
16. H. Matsuo Kagaya ant T. Soma, Phys. Stat. Solidi (b), 127, 89(1985)

Frontiers of Materials Research/
Electronic and Optical Materials
M. Kong, L. Huang (editors)
Elsevier Science Publishers B.V., 1991

APPLICATION OF GaAs, GaSb AND InSb FOR PRESSURE SENSOR DESIGN

Wojciech B. WLODARSKI, Robert A. SHANKS

Victoria University of Technology, Microelectronics and Materials Technology Centre & Department of Communication and Electrical Engineering, GPO Box 2476V, Melbourne, Victoria, 3001, AUSTRALIA.

Results are given of studies on the Gallium Arsenide and Gallium Antimonide p-n junctions as well as Indium Antimonide thin film for designing static and rapidly varying pressure sensors in non-conducting liquid media, in the range up to 50MPa. The static transfer characteristics have been determined. It is shown that transfer characteristics of the GaAs and GaSb p-n junctions as direct pressure sensors are linear, for both diffusion and recombination currents (forward bias) and for both Zener and avalanche breakdown mechanisms (reverse bias). Theoretical results were compared with experimental values and good agreement was found. The static characteristics of the InSb thin films as direct pressure sensors are also linear. Finally results are presented of studies on the application of GaAs and GaSb junctions as well as InSb thin films, as rapidly varying pressure sensors in liquid media. These sensors have been used for measuring rapidly varying pressure in fuel pumps of Diesel engines.

1. INTRODUCTION

Why gallium arsenide instead of silicon for pressure sensor design? At least three reasons. Firstly GaAs has higher pressure sensitivity. The pressure coefficient of the energy band gap is $+ 9 \times 10^{-11}$ eV/Pa while that for Si is less ($- 1.5 \times 10^{-11}$ eV/Pa). Secondly it has a higher temperature operation range, up to 350 ^{0}C (for Si, $n_i = 1.45 \times 10^{10}$ cm^{-3}; for GaAs, $n_i = 1.79 \times 10^{6}$ cm^{-3}). Thirdly it has a faster response (for Si,μ_e = 1500 cm^2/Vsec; for GaAs,μ_e = 8500 cm^2/Vsec). GaSb and InSb are even more sensitive to pressure. The pressure coefficient of energy band gap is $+12.0 \times 10^{-11}$ eV/Pa and $+ 15.0 \times 10^{-11}$ eV/Pa respectively.

Pressure may significantly change effective electron mobility in GaSb as a result of intervalley transitions in the conduction band. The reason is that the energy minimum Γ_1 in GaSb (located at point k(000) of the Brillouin zone) being the bottom of the conduction band shifts with pressure towards higher energies with pressure coefficient $\partial E_{\Gamma_1}/\partial p$ = 12–16 x 10^{-11} eV/Pa. The energy minimum L_1 (located at point k (111) the Brillouin zone) is only 0.08 eV higher than minimum Γ_1 (at T = 300 K)[1]. The pressure coefficient of this minimum is much smaller than that of Γ_1 and amounts to $\partial E_{L_1}/\partial p = 5 \times 10^{-11}$ eV/Pa.

Essential differences appear also in the values of the effective masses of electrons filling these minima. The effective mass of electrons filling minimum Γ_1 is relatively low and amounts to $m^*_{e\Gamma_1} = 0.047\ m_0$ (where m_0 is rest electron mass) while $m^*_{e\Gamma_1}$ is several times greater than $m^*_{e\Gamma_1}$. At atmospheric pressure the majority of carrier fills minimum Γ_1. With pressure increase, the two minima shift at different rates and consequently part of the electrons move to minimum L_1. In view of differences of the effective masses of light and heavy electrons, the effective electron mobility changes markedly. With further pressure increase the electrons in GaSb move from minimum L_1 to minimum X_1[k(100)] since the pressure coefficient of the latter $\partial E_{X_1}/\partial p$ is negative.

TABLE I:

PROPERTIES OF SOME III–V SEMICONDUCTOR COMPOUNDS AND SILICON[1,2,3]

Properties	Si	GaAs	GaSb	InSb
Crystal* structure	D	Z	Z	Z
Band**	I	D	D	D
Energy gap (eV)	1.12	1.42	0.72	0.17
Intrinsic carrier conc. (cm^{-3})	1.45×10^{10}	1.79×10^{6}	$\sim 10^{13}$	$\sim 10^{16}$
Mobility (drift) (cm^2/Vs)	1500 450	8500 400	5000 850	80000 1250
Pressure coeff. of the energy gap (eV/Pa)	-1.5×10^{-11}	$+9.0 \times 10^{-11}$	$+12.0 \times 10^{-11}$	$+15.0 \times 10^{-11}$

* Crystal structure: (D=diamond, Z=zincblende)
** Band: (D=direct, I=indirect)

The effect of intervalley electron transitions under pressure occurs in a number of semiconductors. However in materials such as Si, Ge, GaAs the energy differences between individual minima are much greater than in GaSb.

The pressure coefficients of individual conduction band minima do not differ so markedly. Therefore changes of effective mobility are correspondingly lower.

2. STATIC PROPERTIES OF P–N JUNCTIONS USED AS PRESSURE SENSORS

2.1 Operation at Forward Bias

In the first order theory transfer characteristic of p–n junctions forward biased has the form:

$$\Delta U_F = \frac{a_p \Delta p}{q} \qquad (1)$$

where: $a_p = \partial E_g/\partial p$ is the pressure coefficient of bandgap, Δp is the pressure change and, q is the electron charge. This characteristic is linear in principle and independent of current value, manufacturing technology, junction profile, metallurgical depth and doping level of both junction areas. The change of forward voltage U_F depends, first of all, on the change of bandgap. More accurate analysis should take into account mean mobilities and life times of electrons and holes, energy of impurity ionization, effective masses of state densities, dielectric constant and junction area. However these factors change ΔU_F value only by few percent (at Δp = 100 MPa).

These factors introduce also a slight nonlinearlity of the transfer characteristic. Pressure sensitivity $\alpha_F = \Delta U_F/\Delta p$ is of the order to 10^{-11} – 10^{-10} V/Pa. It depends markedly on semiconductor materials. The absolute value of pressure sensitivity decreases according to following sequence: InSb, GaSb, GaAs, Ge, InAs, Si, GaP, AlSb (Fig. 1).

The experimental investigations were performed on:

- diffused and epitaxial gallium arsenide p–n junctions with silicon, zinc, manganese and cadmium as acceptor impurities,
- liquid phase epitaxy gallium antimonide p–n junctions with zinc and cadmium as acceptor impurities.

Practical application of a forward biased p – n junction as a pressure sensor is limited by large value of temperature sensitivity.

2.2 Operation at Reverse Bias

Operation at reverse bias assures better metrological properties, provided that the operating point is in the breakdown region. Two types of junctions should be here considered: junctions with avalanche and with Zener breakdown.

Avalanche breakdown voltage changes with pressure according to:

$$\frac{\Delta U_{BR}}{U_{BR}} = 2\,\frac{a_p\, k_1\, \eta_n}{E_i\left[\,1 + \frac{1}{m}\left(\frac{q\, L_{po}\, K_{max}}{E_i}\right)^m\right]}\,\Delta p \qquad (2)$$

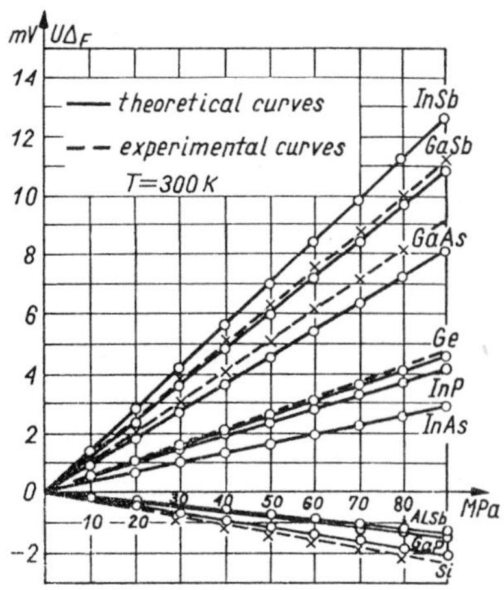

Figure 1 – Characteristic $\Delta U_F = f(\Delta p)$ for some p–n junctions in forward mode of operation.

where m = 1 for Si,GaSb junctions (Chynoweth ionization model) and m = 2 for GaAs junctions (Wolf ionization model). The following symbols have been used in eq. (2): L_{po} – mean free path for scattering on optical phonons, E_i – impurities ionization energy, K_{max} – maximal electric field intensity in the junction, k_1 – coefficient relating the pressure changes of threshold ionization energy to the changes of energy band gap (for Si – k_1 = 0.63; for Ge and electron ionization – k_1 = 0.59 or k_1 = 0.63 for hole ionization; for GaSb, GaAs, – k_1 = 1.5)[4], η_n – coefficient equal to one for planar junctions, but for cylindrical and spherical ones depending on metallurgical junction depth r_j, initial impurities concentration N_d and field intensity K_{max}.

Direct analysis of the relation between junction parameters and pressure sensitivity based on eq. (2) is impossible. These parameters influence also K_{max} and U_{BR} and interrelation of the last quantities is very complicated. Due to these reasons the avalanche ionization condition

$$\frac{\ln\xi}{\xi - 1}\int_0^w \alpha\,[\,K(x)]\,dx \geq 1 \qquad (3)$$

seemed useful in following analysis. In eq. (3) $\xi = \alpha_h/\alpha_e$; α_h and α_e – impact ionization coefficient for both hole and electron ionization (for Si – $\xi = 0.2$, for GaSb, GaAs, – $\xi = 1$), K(x) – electric field distribution in the analyzed junction.

It has been assumed, that ionization coefficient α depends on field intensity in junction according to Albrecht and Miller[5].

Computations, fully confirmed experimentally, lead to the conclusion about linearity of transfer characteristics for p–n junctions with avalanche breakdown (Fig. 2.). Nonlinearity error is lower than 0.5% for GaSb, GaAs and Si diodes (for $\Delta p = 250$ MPa). Such experimental results have been obtained for all tested diodes, irrespective of manufacturing technology, crystallographic orientation and junction depth.

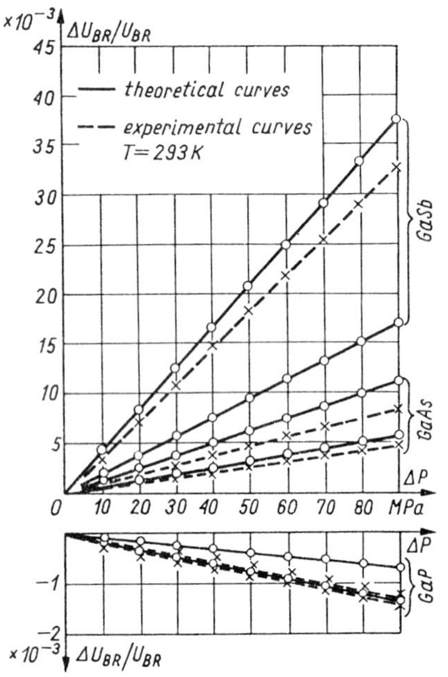

Figure 2 – Characteristics $\Delta U_{BR}/U_{BR} = f(\Delta p)$ for some p–n junctions in avalanche breakdown mode of operation.

Pressure sensitivity $\alpha_R = (1/U_{BR})/(\partial U_{BR}/\partial p)$ depends on junction profile, metallurgical junction depth (for cylindrical and spherical junctions), doping level and ionization carriers. Most sensitive are planar junctions the least sensitive ones – spherical junctions. Sensitivity of Si junctions is in the range ($|-0.46 \times 10^{-11}\ Pa^{-1}| \le \alpha_R \le |-1.04 \times 10^{-11}\ Pa^{-1}|$). Very interesting are more sensitive GaAs junctions ($6.50 \times 10^{-11}\ Pa^{-1} \le \alpha_R \le 12.54 \times 10^{-11}\ Pa^{-1}$ and especially GaSb junctions ($19.02 \times 10^{-11}\ Pa^{-1} \le \alpha_R \le 41.45 \times 10^{-11}\ Pa^{-1}$).

Absolute changes of breakdown voltage ΔU_{BR}, for changing pressure, depend linearly on voltage value. For the pressure values in the range of 100 MPa they are between several tenths of a volt to several volts in GaSb junctions, several times smaller in GaAs junctions, dropping to tenths of a millivolts in Si junctions.

Similar almost linear, transfer characteristics may be observed in the junctions with Zener breakdown. Linearity depends on the change of Zener effect probability with changing pressure. The transfer function is described by equation:

$$\frac{\Delta U_{BR}}{U_{BR}} \approx 3\,\eta_n \frac{a_p}{q} \Delta p \qquad (3)$$

Pressure sensitivity is about one order of magnitude larger than for a forward biased junction and several times larger than for reverse biased junction with avalanche breakdown. Maximal values of pressure sensitivity are for Si junctions – $\alpha_R \approx -4.5 \times 10^{-11}\ Pa^{-1}$, for GaAs $\alpha_R \approx 27 \times 10^{-11}\ Pa^{-1}$ and for GaSb – $\alpha_R \approx 36 \times 10^{-11}\ Pa^{-1}$.

It has been determined experimentally for reverse biased junctions that temperature coefficients of breakdown voltage $\beta_R = \partial U_{BR}/\partial T$ do not depend on pressure. Similarly pressure sensitivity does not depend on temperature and thus makes temperature compensation very easy.

2.3 Temperature Compensation

Temperature compensation method relies on selction of piezojunctions with opposite sign of temperature coefficients. Pressure sensitivity of GaAs and GaSb junctions independent on breakdown mechanism and bias polarisation is positive.

The temperature coefficient value, however, depends, on breakdown mechansim and bias polarisation. Series connection of several diodes and proper polarisation helps to get a sensor with almost zero value of temperature coefficient.

Compensation of avalanche breakdown diode with one forward biased piezodiode (or more if needed) is easiest in practical realisation.

The plot of breakdown voltage change versus the pressure for temperature compensation sensor (1), made of an avalanche piezodiode (2) connected in series with two compensating forward biased piezodiodes (3,4) is shown in Fig. 3.

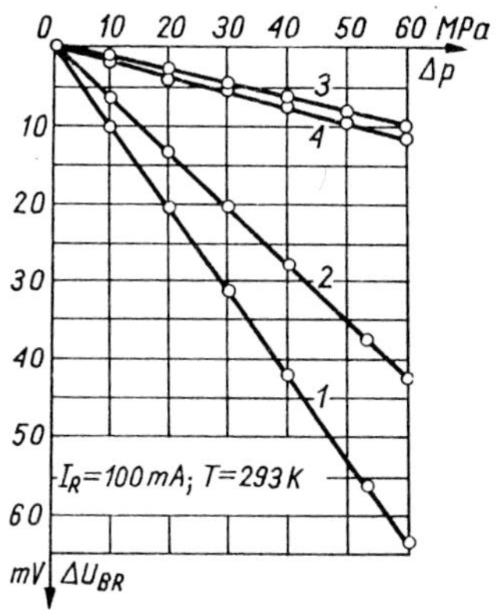

Figure 3 – Characteristics $\Delta U_{BR} = f(\Delta p)$: 1 – temperature compensated sensor, 2 – piezodiode in avalanche breakdown mode, 3,4 – compensating forward biased piezodiodes.

The temperature coefficient of breakdown voltage for compensated sensor, may be postitive or negative and its absolute value depends on current value. The operating point of every sensor should be selected experimentally. For carefully compensated sensors the temperature coefficient may be as low as $1 \times 10^{-6}\ K^{-1}$ for a temperature range 213 K to 523 K (GaAs); the temperature error of the pressure value has been lower than 10 kPa/K. The temperature coefficient of a compensated sensor for pressure up to 100 MPa is practically constant.

3. STATIC PROPERITES OF InSb THIN FILMS USED AS PRESSURE SENSORS

The properties of InSb thin films recrystallized from the liquid phase have also been studied. The electron mobility of samples at room temperature was shown to be 3 to 4 x $10^4\ cm^2\ V^{-1}\ s^{-1}$.

Films were deposited on ceramic substrates. In the pressure range up to 60 MPa the resistance of the thin films examined varied linearly with pressure. The measurements were carried out with a deadweight–piston manometer of 0.2 class. The pressure sensitivity is about 600 mΩ/MPa. The temperature coefficient of this sensor in the temperature range 20–40^{0}C is $-1 \times 10^{-3}\ {}^0C^{-1}$ and the pressure error due to the temperature of the uncompensated sensor has the value: ~330 KPa/^{0}C.

4. DYNAMIC PROPERTIES

To investigate the possibility of applying the GaAs and GaSb p–n junction, or InSb thin film as a sensor for measuring rapidly varying pressures in the range up to 50 MPa, a comparative method was used in which the variation of pressure at the examined sensor output was compared with that at the quartz pressure sensor output.

Bosch and Friedman – Maier stands for testing fuel pumps for Diesel engines were used as the sources of rapidly varying pressures. These stands provide for obtaining periodical, determined pulse cycles under vibration, acceleration and disturbance conditions. i.e. under which conditions other known pressure sensors must operate. The spectrum of measured pressure reached 10 kHz.

Figure 4 shows measuring set–up. At the outlet of the fuel pump, 2, connected to the Bosch or Friedman – Maier stand, 1, was fixed a two cavity adapter, 3, on which the sensor, 4, to be tested and the quartz pressure sensor, 5, were placed. The voltage signal from the tested sensor which together with the decade resisitor constituted a half bridge, 6, was amplified in the amplifier, 8, filtered through a low–pass filter, 9, and fed to one of the inputs of a two channel oscilloscope, 11. The signal from the quartz pressure sensor was amplified in the charge amplifier, 7, filtered through a low–pass filter, 10, and fed to the second input of the oscilloscope.

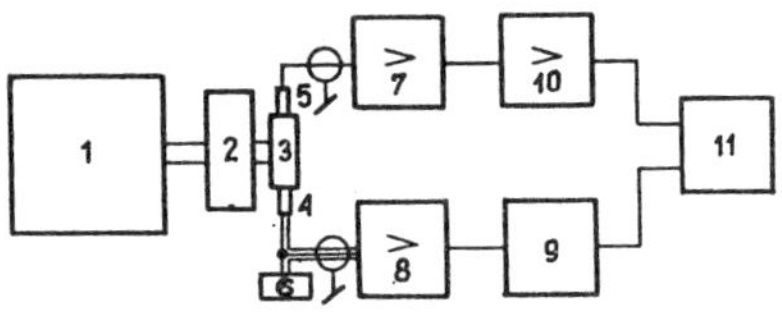

Figure 4 – Measuring set–up for rapidly varying pressure.

Some results are presented on Fig. 5.

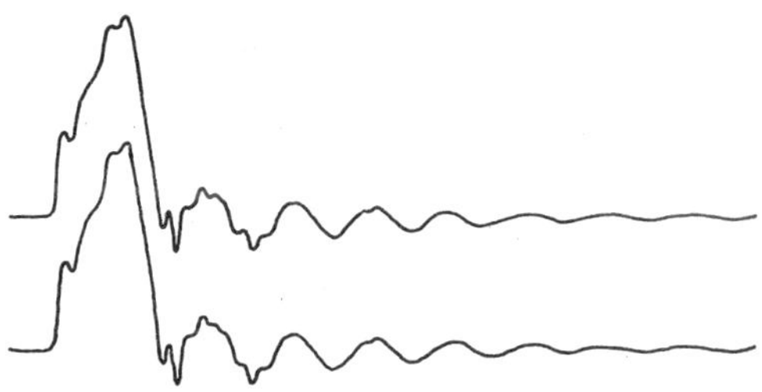

Figure 5 – Comparison of the GaAs p–n junction rapidly varying pressure sensor with the quartz pressure sensor.

The upper curve presents ouput signal from the tested sensor, the lower curve from the quartz sensor. There is good agreement between the two.

The main sources of error in the investigted sensors are the effects of the dynamic temperature of the pressure transmitting medium, which are due to the low temperature time constant of the sensor, noise in the sensor itself and in the electronic circuit and various electric disturbances permeating to the measuring set–up.

In the proposed measuring set–up the detectability of the pressure signal is about 0.3 MPa (3 atm). Difficulties in detecting signals below this value are due to the noise level.

The dynamic error of the measuring set–up including the tested sensors was < 3%.

5. CONCLUSIONS

From the results of measurements it can be concluded that the GaAs and GaSb p–n junctions as well as thin InSb films, used as direct pressure sensors are promising for application in non–conducting liquid media, especially in the range of small pressures.

Their main advantage is their significant sensitivity, without the need for use of an intermediate elastic elements such as membranes, springs and bellows, etc.

These sensors also feature: linearity of transfer characteristic, simple design, small size, simple temperature compensation method, high resonant frequency, high acceleration free limit. Further investigations into the metrological properties, technology and construction of these sensors are in progress.

REFERENCES

1. Paul, W. and Warschauer, D.M., Solid under Pressure, McGraw–Hill, New York, 1963.

2. Sze, S.M., Physics of Semiconductor Devices, John Wiley & Sons, New York, 1981.

3. Gowar, J., Optical Communication Systems, Prentice Hall Int., New York, 1984.

4. Hauser, J.R. and Wortman, J.J., J. Appl. Phys., 37 (1966) 3884–3892.

5. Albrecht, H. and Miller, R., AEU, 32 (1978) 239–247.

6. Moeschke, B. and Wlodarski, W., Bull de l'Acad. Pol. Sci., 28 (1980) 63–69.

7. Wlodarski, W. and Moeschke, B., Bull de l'Acad. Pol. Sci, 28 (1980) 17–29.

Frontiers of Materials Research/
Electronic and Optical Materials
M. Kong, L. Huang (editors)

GROWTH, PERFORMANCE AND APPLICATION OF $Li_2B_4O_7$ CRYSTALS IN LARGE DIAMETER

Shi-ji FAN and Wen WANG

Shanghai Institute of Ceramics, Academia Sinica, Shanghai 200050, P.R. China

T. SHIOSAKI

Department of Electronics, Kyoto University, Kyoto Japan

Large diameter (3 inch diameter X 4 inch) $Li_2B_4O_7$ (LBO) crystals with high quality (core-free, no inclusion and no crack) have been grown by the Bridgman method (BR) for the first time. The piezoelectric, optical, thermal and mechanical properties etc of BR LBO crystals have been measured. The performance of SAW resonator filters with low insertion loss (1.6 dB) and high image attenuation (60 dB) in UHF fabricated on BR LBO substrates indicates that BR LBO crystal is an industrially piezoelectric material for SAW devices.

1. INTRODUCTION

Lithium tetraborate ($Li_2B_4O_7$: LBO) single crystal was found to be a non-ferroelectric piezoelectric material in 1981[1]. It has fairly large SAW coupling coefficient ($k^2 \simeq 1\%$) and the orientations of zero temperature coefficient of delay (TCD). The surface acoustic wave (SAW) reflectivity per aluminum strip for interdigital transducers (IDTs) on the (110) wafer of LBO is at least 5 times larger than that of the wafer of LT, LN, or quartz[2]. The technology of photolithography and preparation of wafer of LBO for SAW devices have been developed well[3,4]. LBO raw materials are abundant and inexpensive. So LBO crystal is a very useful substrate material with temperature compensation for SAW and bulk acoustic wave (BAW) devices of high frequency.

LBO crystal has been grown mainly by CZ method[5-8]. The most serious problem hindering industrial application of LBO crystal is coring in CZ growth of LBO and very low growth rate causes low yield of LBO wafers, specially for large diameter such as ϕ3 inch. The piezoelectric, elastic and dielectric properties etc have been investigated in details[9,10], and the SAW resonator filters have been fabricated on CZ LBO substrates for mobile communication[3,4,11]. S.J. Fan et al used the modified Bridgman method to grow LBO crystal for the first time[12] and have developed the technology of LBO BR growth successfully.

In this paper, we report on the results of BR growth of large diameter LBO crystals and the measurements of some properties which have not been studied as well as the application of BR LBO for fabrication of SAW resonator filters of UHF.

2. EXPERIMENTAL

The procedure of LBO crystal growth by the Bridgman method is about the same as that in reference [12]. The Bridgman furnace and the platinum crucible were designed for ϕ3 inch LBO growth. <100>, <001> or <110> seed and the blocks of raw material of LBO are put in the crucible in sequence and then the crucible is placed in the Bridgman furnace for LBO crystal growth. The analysis of content of impurities in the raw materials and crystals of LBO was made by atomic absorption spectrometer (AAS) (type ICP-5000). The core-free, no inclusion and water-clear transparent LBO crystals have been chosen for the measurements of optical, thermal and mechanical properties etc. The SAW

TABLE 1. The contents and k of impurities in BR growth

Impurities		Fe_2O_3	MgO	CaO	Na_2O	K_2O
Content $\times 10^{-4}\%$	in melt	4.5	1	7	3.5	1
	in crystal	1	0.3	0.4	1	0.5
k		0.2	0.3	0.06	0.3	0.5

resonator filters of BR LBO were fabricated by photolithographic procedure with lift-off process and vacuum evaporation of Al strips of IDTs.

3. RESULTS AND DISCUSSION

3.1. The segregation of impurities in BR growth of LBO

The AAS analysis indicates that the impurities of Fe, Mg, Ca, Na and K in LBO melt enter into the crystal lattice of LBO and the distribution coefficients k of them in BR growth are less than 1 shown in Table 1.

For LBO growth and properties no obvious influence appears by the contents of these impurities.

3.2. Coring and the morphology of core regions

For the growth of large diameter LBO (ϕ2 or 3 inch), because of the poor thermal conductivity of the melt and the crystal, the concave S-L interface appears much often, therefore coring is easier than that in growth of small diameter.

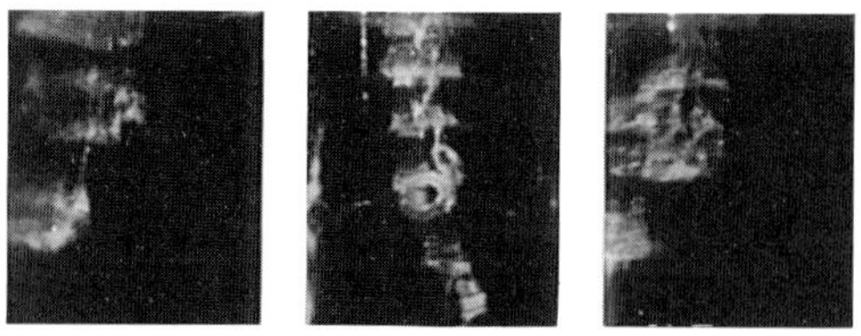

a. <110> b. <001> c. <001>

FIGURE 1
The morphology of the core regions in BR growth of LBO (x0.37)

The morphology of core regions shown in Fig. 1 indicates the change of them with the positions of the bottoms of concave S-L interfaces and vibration of temperature field.

After we optimize the conditions of growth for ϕ3 inch LBO, core-free, no inclusion and no crack of ϕ3 inch LBO shown in Fig. 2 have been grown stably by the Bridgman method.

3.3. The optical, mechanical properties etc of BR LBO

The piezoelectric, elastic and dielectric properties etc of CZ LBO have been reported in details by T. Shiosaki et al. The data of these properties for BR LBO which were measured by us are the same as that of CZ LBO. We fill up the measurements of optical, mechanical and thermal properties of BR LBO (Fig. 3) and summarize them in Table 2.

3.4. The performance of SAW resonator filters of BR LBO

FIGURE 2
ϕ3 inch LBO boule, wafers and SAW devices

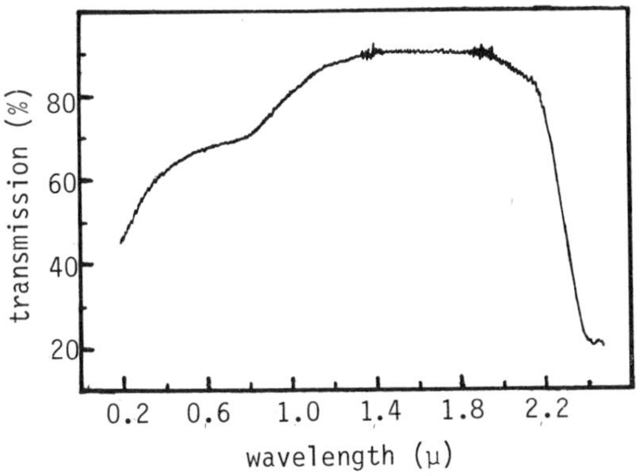

FIGURE 3
The spectrum of transmission of the LBO crystal

TABLE 2. The properties of LBO

Crystallographic system		4mm
Melting point (°C)		917
Density (g/cm^3)		2.44
Specific heat (cal/g·°C) (60 — 500°C)		0.3287
Transmission (0.3-2.2μ)%		55 — 90
Hardness (Mohs)		7
Compressive strength (Kg/cm^2)	<a>	2878
	<c>	5750
	<110>	4496
Bending strength (Kg/cm^2)	<a>	1886
	<c>	950
	<110>	1946
Dielectric const.	ε_{11}^T	9.33
	ε_{33}^T	10.1
Piezoelectric const. (c/m^2)	e_{33}	0.93
Elastic const. ($x10^{11}$ N/m^2)	c_{11}^E	1.35
Coupling coeff.	K_{33}	0.42

The SAW resonator filters of 380 MHz (Fig.2) have been fabricated on (110) wafers of BR LBO crystals. The insertion loss of the devices is very low in so far as 1.6 dB and the image attenuation is high up to 60 dB. The uniformity of performance of the devices is very good therefore BR LBO substrates are suitable to miniaturized SAW devices of high frequency.

ACKNOWLEDGEMENT

The authors wish to express their many thanks to Mr. G.S. Shen, J.L. Li and X.H. Le for LBO crystal growth and Mrs. R.Y. Sun for optical observation and Mr. J.C. Pan for measurements of electric properties etc.

REFERENCES

1. R.W. Whatmore, N.M. Shorrocks, F.K. Ainger and I.M. Young, Elect. Lett. 17 (1981) 11.

2. Y. Ebata and M. Koshino, Jpn J. Appl. Phys. 26 Suppl. 26-1 (1987) 123.

3. H. Abe, H. Saitou, M. Ohmura, T. Yamada and K. Miwa, Proc. IEEE 1987 Ultrason. Symp. (1987) 91.

4. S. Matsumura, T. Omi, N. Yamaji and Y. Ebata, Proc. IEEE 1987 Ultrason. Symp. (1987) 247.

5. J.D. Garrett, M.N. Iyer and J.E. Greedan, J. Crystal Growth 41 (1977) 225.

6. D.S. Robertson and I.M. Young, J. Mater. Sci. 17 (1982) 1729.

7. M. Adachi, T. Shiosaki and A. Kawabata, Jpn J. Appl. Phys. 24 Suppl. 24-3 (1985) 72.

8. R. Komatsu, T. Suetsugu and M. Ono, Jpn J. Crystal growth 15 (1982) 12 (in Japanese).

9. T. Shiosaki, M. Adachi, H. Kobayashi, K. Araki and A. Kawabata, Jpn J. Appl. Phys. 24 Suppl. 24-1 (1985) 25.

10. M. Adachi, T. Shiosaki, H. Kobayashi, O. Ohnishi and A. Kawabata, Proc. IEEE 1985 Ultrason. Symp. (1985) 228.

11. T. Shiosaki, Ferroelectrics 91 (1989) 39.

12. Fan Shi-ji, Shen Guan-shun, Wang Wen, Li Jin-long and Le Xiu-hong, J. Crystal growth 99 (1990) 811.

Frontiers of Materials Research/
Electronic and Optical Materials
M. Kong, L. Huang (editors)

POWDER SECOND HARMONIC GENERATION OF 4,4'-DISUBSTITUTED BENZYLIDENE-ANILINE DERIVATIVES*

Daoyu LIU, Chaoyang DAI, Jingui QIN

Department of Chemistry, Wuhan University, Wuhan 430072, P.R.China

Chuangtian CHEN, Baichang WU

Fujian Institute of Research on the Structure of Matter, Chinese Academy of Science, Fuzhou 350002, P.R.China

Forty 4,4'-disubstituted benzylidene-aniline derivatives were synthesized, and their second harmonic generation (SHG) efficiency was measured by the Kurtz powder technique. The effect of the electronic property and the positions of the substituents as well as the crystallization media on the powder SHG efficiency was studied.

1. INTRODUCTION

Study on organic nonlinear optical (NLO) materials is currently of great interest because they often exhibit much higher second harmonic generation (SHG) effect than traditionally used inorganic materials, and hence have shown potential applications in integrated optics[1,2].

There has been, however, little study thus far on the NLO property of 4,4'-disubstituted benzylidene-aniline (X-⟨◯⟩-CH=N-⟨◯⟩-Y)[3]. It is noteworthy that two benzene rings of these molecules are not fully coplanar but twist a little in some degrees. However, a conjugated π-electron system still extends throughout the whole skeleton, and an intramolecular charge-transfer can take place from one benzene ring to the other if X and Y are electron donor (D) and acceptor (A) respectively[4]. Moreover, the symmetry of these molecules is lowered due to the torsion. This would favour the formation of noncentrosymmetric crystal structure, which is a basic requirement for a material to exhibit SHG effect. Therefore, in the view of the "Molecular Engineering"[5], 4,4'-disubstituted benzylidene-aniline derivatives are good candidates for new NLO materials.

This paper describes the synthesis,

* Supported by the National Foundation for Natural Science of China.

powder Second Harmonic Generation (SHG) of the 4,4'-disubstituted benzylideneaniline derivatives. In particular, the effect of the electronic property and the positions of the substituents as well as the crystallization media on the powder SHG efficiency is discussed.

2. RESULTS AND DISCUSSION

Forty 4,4'-disubstituted benzylideneaniline compounds (where either X or Y is an electron donor such as Me_2N-, Et_2N-, H_2N-, MeO-, HO-, or halogen, while the other is an electron acceptor such as -NO_2, -CO_2Me, -COMe, or halogen) were synthesized by the condensation reaction of corresponding para-substituted benzaldehyde with para-substituted aniline. The compounds were characterized by microelemental analysis. The SHG effect was measured by Kurts powder technique using a nanosecond pulsed Nd:YAG laser (1.064 μm) as a source and KDP (KH_2PO_4) as a reference.

Eleven compounds showed SHG effect with a maximum value of 100 times that of KDP (Table 1).

Table 1 Structure and SHG Efficiency of Some Compounds

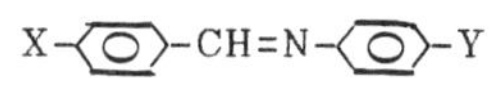

No.	X	Y	SHG (× KDP)
1	MeO	NO_2	11
2	NO_2	MeO	0
3	HO	NO_2	50
4	NO_2	HO	0
5	HO	CO_2Me	22.5
6	CO_2Me	HO	1.6
7	NO_2	Br	95
8	Br	NO_2	0
9	NO_2	I	100
10	I	NO_2	0
11	NO_2	Cl	12
12	Cl	NO_2	0
13	Me_2N	NO_2	0
14	NO_2	Me_2N	0
15	Me_2N	CO_2Me	0
16	Et_2N	CO_2Me	0
17	H	Br	0.2
18	MeO	Br	8
19	n-$C_5H_{11}O$	Br	2
20	n-$C_5H_{11}O$	COMe	2

It is interesting to note that all compounds which showed high SHG values (i.e. $>10 \times$ KDP) are those bearing a relatively weak donor (e.g. MeO-, HO-, I-, or Br-) and a strong acceptor (e.g. -NO_2). On the other hand, the other four compounds containing both strong donor and strong acceptor (see Table 1, No. 13 - 16) did not show SHG effect at all. There may be two reasons for this. Firstly, it was well established that the microscopic second order polarizability of molecules is proportional to the difference between the excited state and the ground state dipole moment $|\mu_e - \mu_g|$[6]. The molecules containing both strong D and strong A usually possess both large μ_e and μ_g, however,the difference may be small. In the case of the molecules containing relatively weak D and strong A, μ_g is relatively small while μ_e is still quite large, giving rise to a relatively larger difference. Secondly, the dipole-dipole interaction of the molecules with relatively weak D and strong A is relatively weak due to the relatively small ground state dipole moments, which favours the formation of noncentrosymmetric crystals. As a result, these compounds showed large SHG effect. This kind of structure-property relationship was also discovered in stilbene derivatives[7]. In addition, fourteen compounds bearing both relatively weak D(e.g. MeO-, HO-, I- or Br-) and relatively weak A (e.g. halogen or -COMe) were synthesized, which showed either relatively small (see Table 1, No.17-20) or no SHG effect. Apparently, this is partly due to the weak intramolecular charge transfer.

There are structural isomers of this type of compounds if the two substituents are the same but their positions are reverse. The compounds 1-12 in Table 1 are such six pairs of the isomers. It is shown in Table 1 that the isomers may exhibit very different SHG efficiency. In most cases, one isomer showed large SHG signal while the other did not show the effect.

It was reported[3b] that the donors would better be attached to the aniline fragment so that the benzylidene aniline derivatives may have more chance to show SHG effect. However, Our results, as shown in Table 1, indicated that this is not necessary. In the cases that the donors are halogens (Cl, Br or I, see Table 1, the compounds 7-12), the above principle is valid. But in the cases that the donors are hydroxyl ormethoxyl (see Table 1, the compounds 1-6), the compounds in which the donor is attached to aniline ring showed no or small SHG effect. In contrast, those where the acceptor is attached to aniline fragment did show high SHG effect.

The big difference of the isomers in SHG effect may be understood in two reasons. Firstly, the isomers may possess different molecular dipole moments due to the polarity of C=N bond, leading to different extent of the intramolecular charge transfer. This can be proved by the UV-visible

Table 2. Polymorphs and Their SHG Effect

Crystal Growing Media	SHG (× KDP)				
	Compd.1	Compd.5	Compd.7	Compd.9	Compd.11
Ethanol	0.3	22.5	95	60	2.5
Cyclohexane	11	0.8	0	100	1.2
Toluene	1	3.2	10	11	12
Ether	5	0	2.5	0.7	1.3
Dichloromethane	0.25	2.5	75	18.5	3.5
Acetone		3	40	6	0
Ethyl acetate	0.25	0	20	3.5	10
DMF	0		45	10	2

spectra of the isomer pairs. For example, the UV-visible spectra of the compounds 1 and 2, both in acetone, showed λ_{max} of 346 and 372, respectively. Secondly, the two benzene rings may twist in different angles for the isomers, leading to the different molecular configuration and different packing mode in their crystals, which then gave rise to the different SHG property

Five compounds which showed large SHG effect have been recrystallized from various organic solvents. It was demonstrated that the crystals of the same molecule but grown from different solvents exhibited very different SHG values (Table 2). For example, the compound 7 may shown either very high SHG efficiency of 95× KDP (grown from ethanol) or no SHG effect at all (recrystallized from cyclohexane). This phenomenon clearly indicated the different molecular packing in the crystals and the existance of polymorphism [8]. The single crystal X-ray diffraction study is in progress.

3. EXPERIMENTAL

The chemicals are used as purchased without further purification. The microelemental analysis was carried out on a Carlo Erba 1106 Analyser. UV-visible spectroscopic study was conducted using Shimadzu UV-240 spectrophotometer. SHG value was measured using modified version of Kurtz powder technique with Nd-YAG laser emitting at 1.06 μm as a source, and KDP as a reference.

Synthesis of the compounds: The equimolar solutions in absolute ethanol of corresponding para-substituted benzaldehyde and para-substituted aniline were mixed up. Two drops of glacial acetic acid (or solution of $AlCl_3$ in absolute ethanol) were added. The mixture was heated at 40-50°C with stirring for 2 hours. The yellow or white precipitates were filtered, washed twice with ethanol, once with ether, and dried in air. The compounds were then recrystallized from organic solvents. Microelemental analytical results were consistent with the calculated values.

4.CONCLUSIONS

Forty 4,4'-disubstituted benzylideneaniline derivatives were synthesised. Eleven compounds showed SHG by powder measurement with the maximum values of 100 times that of KDP. The effect of the electronic property and the positions of the substituents as well as the crystal growing media on powder SHG efficiency was observed as the following:

a) The compounds which showed higher SHG values were those which contain a relatively weak donor and a strong acceptor .

b) The structural isomers which contain the same substituents but in reverse positions showed very different SHG values.

c) The crystals of the same molecules recrystallized from different solvents exhibited very different SHG values, indicating the existance of polymorphism.

REFERENCES:

1.D.J.Williams, ed., Nonlinear Optical Properties of Organic and Polymeric Materials , Am.Chem.Soc.Symp.Ser.233, (ACS, Washington, 1983).

2.D.S.Chemla and J.Zyss, eds. Nonlinear Optical Properties of Organic Molecules and Crystals , (Academic Press, Orlando, 1987).

3.a) J.F.Nicoud and R.J.Twieg, in Ref.2, Vol.2, p.221.
b) T.Tsunekawa, T.Gotoh, and K.Egawa, Eur. Pat., Appl. EP 250,099.

4.K.Maeda, K.A.Muszkat and S.Sharafi-Ozeri, J.Chem.Soc.,Perkin II,(1980) 1282.

5.a) D.S.Chemla and J.L.Oudar, Phys. Rev., B12 (1975) 4534.
b) C.T.Chen and G.Z.Liu, Ann. Rev. Mater. Sci., 16 (1986) 203.

6.J.L.Oudar and D.S.Chemla, J.Chem. Phys., 66 (1977) 2664.

7.Y.Wang, W.Tam, S.H.Stevenson, R.A. Clement and J.Calabrese, Chem. Phys. Lett., 148 (1988) 136.

8.J.Bernstein, Mol.Cryst.Liq.Crystd, 50 (1979) 233.

Frontiers of Materials Research/
Electronic and Optical Materials
M. Kong, L. Huang (editors)

DEVELOPMENT OF ORGANIC NONLINEAR OPTICAL CRYSTAL 3-METHYL-4-NITROPYRIDINE-I-OXIDE

Li Songxian, Chen Jianzhong, Dong Meibin, Wang Jiahe, Li Junqian

(Center for Research in Solid Materials, Fuzhou University, Fuzhou 350002, Fujian, China)

Wu Bochang

(Fujian Institute of Matter Structures, Academia Sinica, Fuzhou 350002, Fujan, China)

In this paper, the proper conditions of synthesis of 3-POM have been explored. The growth kinetics of the single crustal has been investigated. The single crystals of optical quality up to $18\times12\times6mm^3$ in size have been successfully grown by solution cooling method. We determined the transmission wave length range to be 0.480-2.100 μm. The optical damage threshold measurement, using a YAG laser funsamental and doubled output (10 ns pulse duration), was found to be $1073MW/cm^2$ and $173MW/cm^2$ on 1.06μm and 0.532μm, respectively. The type I phase-matching curve locus of 3-POM single crystal has been calculated and verified by our experiments. We have also measured the nonlinear optical properties at the direction ($\theta=54^o$, $\phi=0^o$) of type I phase-matching, the frequency doubling efficiency d^{3-POM} is 18 times larger than d_{14}^{KDP}, its conversion efficiency is about 36%. Based of their molecular and crystal structure, we discussed the obvious difference for frequency doubling properties and molecular dipole moment between 3-POM and 2-POM.

1. INTRODUCTION

A new organic molecular crystal 3-methyl-4-nitropyridine-1-oxide (POM) with excellent nonlinear optical properties was reported by J.Zyss and D.S.Chemla in 1981[1]. During studying the relation between electric structure and frequency doubling effects of some organic molecular crystals, we also discovered POM has the possibility to be an excellent frequency doubling crystal material[2]. Consequently, we continued to investigate the growth kinetics of the single crystal (reported in another paper), the conditions of single crystal growth by solution cooling method, and also measured its various optical properties. POM single crystal has many advantages: wide transparent region from visible range to near infrared range, high optical damage threshold, fine nonlinear optical coefficient and high conversion efficiency. Thus it is rather promissing to apply it to second-harmonic generation, electro-optical modulation, optical parametric oscillation, frequency mixing in the future.

In this paper, the work in our laboratory, the proper conditions of synthesis of POM, screen of the solvent and the growth conditions of single crystal are reported. We determined the metastable region of POM in acetone solution, and was successed in obtaining the POM single crystals with optical quality by solution cooling method. We measured its transmissive wave length range and transmittance, the frequency doubling efficiency at the direction of type I phase matching and its conversion efficiency, calculated typeI phase matching curve of POM single crystal and verified by our experiments.

2. SYNTHESIS OF POM CRYSTALS

POM was synthesized by two steps[3]: oxidation, adding 30% hydrogen peroxide to a

The Project Supported by National Natural Science Foundation of China.

mixture of glacial acetic acid and 3-methyl pyridine at 70°C with stirring; nitrotion, adding fuming nitric acid and concentrated sulfuric acid at 100°C. In each step the products were extracted, distillated under reduced pressure and then repeatly crystallized.

POM crystallizes are yellow prisms from acetone solution. Its melt point 136±0.5°C, its IR spectra identify with those in the references.

Single crystal structure of POM was determined by Shiro et al. in 1977[4]. POM is orthorhombic system with space group $P_{2_1 2_1 2_1}$, unit cell parameters a=21.359, b=8.111 and c=5.132Å. Our experiment results agree quite well with these parpmeters. The typical habit of a large defect-free crystal is shown in Fig. 1. The most developed faces are the slowly growing faces of the dome (201), (201), (201), (201), the prism (201), (210), (210), (210) and the pinacoid (100), (100).

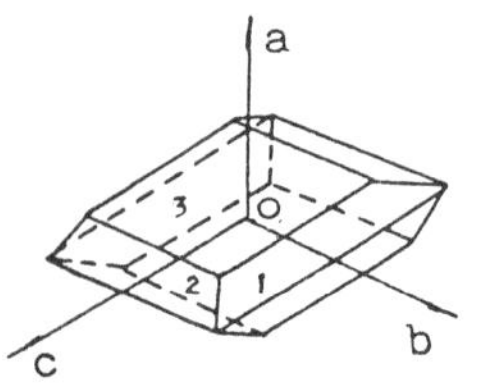

Fig. 1. Appearance of POM single crystal

3. SINGLE CRYSTAL GROWTH OF POM

3.1 Screen of the solvent

Several solvents had been selected as growth meduim of POM, for example, chloroform[1], 1,2-dichloroe-thane, methylacetate, acetonitrile[5] etc. But there are still some problems, when we used these solvents to grow the POM single crystal. The crystals growing in methylacetate are thin, in acetonitrile are imcomplete, and chloroform and 1,2-dichloroethane not good to operators for their high noxiousness. So that we have tried looking for some other solvents: alcohol, acetone, water and some mixed solvents etc, as proper growth meduim of POM, at last we find that acetone is the most suitable one. The solubility of POM in acetone is modest, that single crystal growing in it are integrated and transparent.

3.2 Determination of metastable region

The solubility curves of POM in acetone were determined by optical effect method. The snowslide of one and two once nucleation of different concentration were measured. The solubility curves and supersaturation curves are shown in Fig.2.

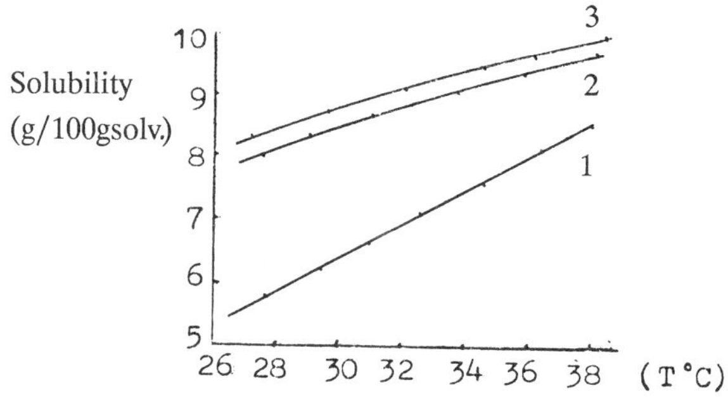

Fig. 2. Solubility and supersaturation curves
1. Solubility Curse
2. Supersaturation Curse at without seed crystal
3. With seed Crystal

This Fig. 2 shows that while the temperature increasing to more than 34°C, the metastable region become narrow, the stability of the solution decreases.

3.3 Control of growth condition

In order to obtain single crystals with high quality, the raw material must be completely purified. The traces of reactant and by-product in synthetic process obviously effect single crystal growth, so that extraction, distillation and repeat crystallation should be strictly carried on, then the solution of compound need be purified by adsorption. The saturative solution should be carried on superheat treatment for several hours, then the temperature drops naturally to higher

0.05 than saturation point, stays at this three hours, until the seed crystal slightly solves, then the temperature drops to saturation point, normal growth begins.

In normal conditions, the quantity of droping temperature need be strictly controled, it is very small at the first 3-5 days, then increases gradually to about 0.1°C/day. The C axial cut section of seed crystal is used, the surface of the crystal is carried on a polishing treament and cleaned with acetone, then absorted immediately by filter paper.

The defect which appears commonly for POM single crystals could be produce additional crystals, once the temperature fiuctuates considerable or the quanlity of drop temperature is too larger, additional crystals should be produced easily. Another defect is wedge , two facts which account for the occurrence are existence of impurity and too law super-saturation. The single crystals of good optical quality up to 20 × 12 × 6mm^3 in size have successful grown by solution cooling method.

4. DETERMINATION OF PROPERTIES OF POM SINGLE CRYSTAL

4.1 Transmittance waveband and transparency

Transparency experiment adopted Lampda-9 type UV/VIS/NIR spectrometer. It shows

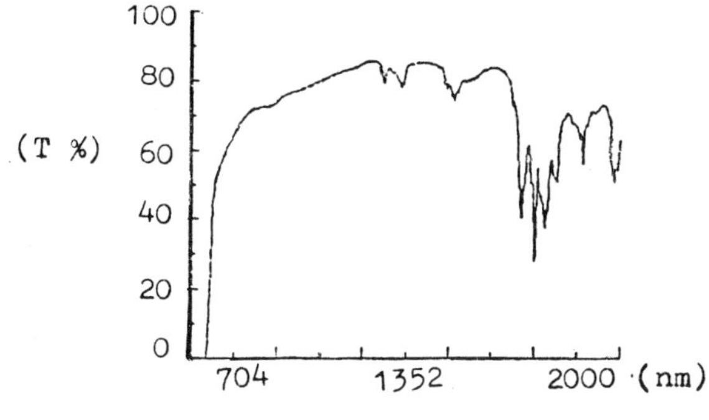

Fig. 3. Transmissive curve of a 2.7mm thick plate of POM

transmittance waveband ranges from 0.47 to 2.0μm, the result is displayed in Fig. 3.

Transparences are 89% and 67% respectively for 1.06 and 0.53μm, 87% and 85% respectively for 1.30 and 1.55μm. These experimental values are higher slightly than literate values, and the thickness is 2.7mm, it is more thicker than the values is reference[1]. It is clear thus that the transmittance of tested crystals is rather fine.

4.2 Calculation and provableness of phase matching properties

In order to improved frequency doubling efficiency, the phase matching technology is adopted generally. We explored proper conditions with which the effective nonlinear optical coefficient reaches a maximum.

We have calculated the phase matching curve and effective nonlinear optical coefficient d_{eff} of POM at 1.064μm by the procedure of literature[6]. The results are displayed in Fig. 4.

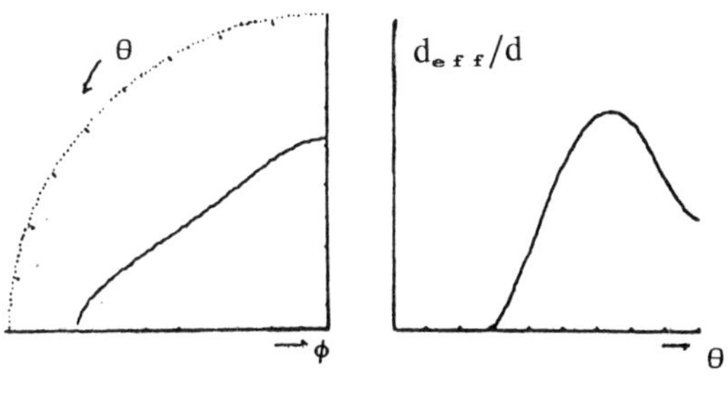

Fig. 4. Type I phase matching angle and d_{eff}/d of POM

The best values of the type I phase matching angle are $\theta=54.0^\circ$, $\phi=0^\circ$, its relative nonlinear optical coefficient $d_{eff}/d=-0.9504$, this calculated value agrees with literary value.

The best values of the type II phase matching angle are $\theta=63.1^\circ$, $\phi=34.4^\circ$, corresponding, $d_{eff}/d=0.8559$, as shown by Fig. 5. A 2.0×6.3×4.0mm^3 of POM single crystal has been processed according to $\theta=54.0^{\circ}$, $\phi=0^{\circ}$. The type I phase matching curve has been proved in Shanghai Jiao Tong University. The experimental results given in Table 1. The experimental points are discribed on Wolf's net. They are in good agreement with the calculated results. See Fig. 6.

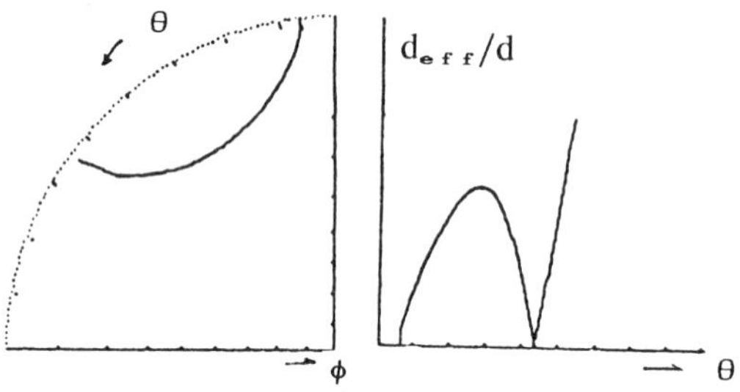

Fig. 5. Type II phase matching angle and deff/d of POM

Tab. I. Experimental datas of type I phase matching angle of POM

$\theta(^{\circ})$	54.0	54.0	54.1	53.9	53.6	53.4	53.2	52.9	52.6
$\phi(^{\circ})$	0	0.6	1.1	1.7	2.9	4.0	5.1	6.3	7.4

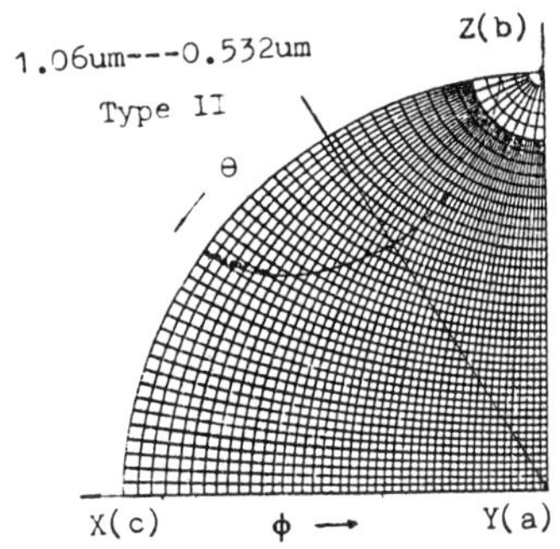

Fig. 6. Tested curve of type I phase matching angle

POM single crystal has also been processed according to $\theta=63.1^{\circ}$, $\phi=34.4^{\circ}$, for type II phase matching. While fundamental frequency to be 1.064μm, the frequency doubling experiment of this single crystal produces stronger green light, so the calculated results were vertified.

4.3 Determination of frequency doubling coefficient

POM is orthorhombic with point group 222, correspondingly, there are three different three-step tensor units. When the Kleiman approximation is considered, there are $d_{14}=d_{25}=d_{36}$.

The nonlinear optical coefficient of POM had been measured with the Maker fringe method in reference[1], and the value of d^{POM} is $(23\pm3)10^{-9}$ esu. When the section of POM and of KDP are selected according to type I phase matching $\theta=54^{\circ}$ $\phi=0^{\circ}$ and $\theta=41^{\circ}$, $\phi=45^{\circ}$ respectively, their length that the light passing through L are 3.1mm and 3.3mm respectively, the experimental data of intensity $I^{2\omega}$ are 160 and 0.8 respectively. Thus, the relative second-order nonlinear optical coefficient is given by

$$d_{eff}^{POM}/d_{eff}^{KDP}=(L_K/L_P)\sqrt{(n_{p,2}n_{p,1}I_p^{2\omega})/n_{k,2}n_{k,1}I_p^{2\omega}}.$$

The corrective facts have been introduced: the corrective factor of refractive index is 1.1, the effective factor and the absorptive factor are 0.66 and 0.59, respectively, the absorptive coefficient of fundamental wave and harmonic wave of POM are 0.77 and 1.88 cm^{-1}, respectively. The corrected calculation result is $d^{POM}=14d_{14}^{KDP}$, it accords with reference[1].

4.4 Determination of conversion efficiency

POM single crystal had been cut a plate at $\theta=52.5^{\circ}$, $\phi=0^{\circ}$ (it drifted off a little type I phase matching angle), the conversion efficiency η are determined to be 32.2%, 35.7% in both Shanghai Institute of optics and Fine Mechanics and Fujian Institute of Research on the Structure of Matter. Basides, crystal plate has been cut according to type I phase matching angle, determined η is 52.0% with following conditions: the diameter of facula and silt are 5.0mm and 1.0mm, respectively, 20 ns pulse duration, the length and area of passing light are 3.7mm and 8.0×3.0mm^2, respectively, while optical power density is 40Mw/cm^2.

4.5 Measurement of optical damage threshold

The optical damage threshold D_τ has been measured using the fundamental and doubled output of Q-switch YAG laser (10 ns pulse duration, P.D.), the D_τ values of the experiment and of the reference as shown in table 2.

Tab. 2. Optical damage threshold value of POM

Wavelength (μm)	P.D. (ns)	Experimental D_τ (Mw/cm^2)	P.D. (ns)	Literal D_τ (Mw/cm^2)
0.53	10	1.70×10^2	20	50
1.06	10	1.07×10^3	20	2×10^3

4.6 Measurement of volume-bearing width of temperature

The volume-bearing width of temperature is 5-6°C, which is the bearable temperature change of the crystal when $I^{2\omega}$ drop to the original half intensity.

5. CONCLUSION

We have synthesized high pure POM crystals and screened a satisfactory solvent. The rational range of temperature of POM single crystal growth could be defined from the determination of metastable region. The speed of temperature decreasing in the initial stage is controlled at smaller than 0.01°C/day. The crystal seeds which are out into slices along C axis do not easily product cuneatity.

The transmittance waveband range has been determined, the transparency is from 0.47 to 2.1μm. The optical damage threshold measurement, using a YAG laser fundamental and doubled output (10 ns pulse duration), is found to be 1073Mw/cm^2 and 173Mw/cm^2 on 1.06μm and 0.532μm, respectively. The type I phase matching curve of 3-POM single crystal have been calculated and vertified by our experiments. We have also measured the nonlinear optical properties at the direction of type I phase matching, the frequency doubling efficiency d^{POM}_{PM} is 14 times large than d^{KDP}_{14}, Its conversion efficiency is about 36%, so that POM crystal would be a promising nonlinear optical meterial.

REFERENCES

1. J.Zyss, D.S.Chemla and J.F.Nicond; J. Chem. Phys., 74, (1981) 4800.

2. Li Songxian et al.; Chin. J. Lasers 14 (1987) 662.

3. N.Rabjohn; Organic Synthses, I (1963) 114.

4. M.Shiro, M.Yamakawa and T.Kubota, Acta crystallogr. Sect., B33 (1977) 1549.

5. R.Hierle, J.Badan and J.Zyss, J. Cryst. Growth, 69(1984) 545.

6. Xie Shengwu et al.; J. Shanghai Jiao dong University, No.1 (1982) 37.

Frontiers of Materials Research/
Electronic and Optical Materials
M. Kong, L. Huang (editors)

STUDY OF ALLYLTHIOUREA COMPLEX TYPE NEW NONLINEAR OPTICAL MATERIALS

Duorong YUAN, Nan ZHANG, Xutang TAO, Dong XU, Minhua JIANG and Zongshu SHAO

Institute of Crystal Materials, Shandong University, Jinan 250100, P.R. China

In this paper, a series of new nonlinear optical materials i.e. ATCC, ATCB, ATMC and ATMB are reported. Synthesis, crystal growth and characterizations have been carried out. The results obtained have shown that the SHG intensities of ATMC and ATMB are three times higher than that of urea.

1. INTRODUCTION

The rapid development of optical communications has led to the searching of more efficient components for the processing of optical signals. The potential of organometallic materials for this purpose has been demonstrated recently [1-3]. Based on the theoritical "double-radical" model [4] in which the distorted octahedron combined with the organic conjugated moleculars (or groups), we have found several organometallic complex type NLO materials. They all take allylthiourea as ligand and Zn^{2+}, Cd^{2+} and Hg^{2+} as center ions, their potential applications are confirmed by powder method measurements. The SHG intensities of ATMC and ATMB are three times higher than that of urea. In this paper, we reported the synthesis, crystal growth and properties of these new NLO materials, i.e. Tri-allylthiourea cadmium chloride (ATCC), tri-allylthiourea cadmium bromine(ATCB), tri-allylthiourea mercury chloride (ATMC), and tri-allylthiourea mercury bromine (ATMB).

2. EXPERIMENTAL

The raw materials used are : allylthiourea (c.p.), $CdCl_2\cdot 2.5H_2O$(A.R.), $CdBr_2\cdot 4H_2O$(c.p.), $HgCl_2$ (A.R.) and $HgBr_2$ (A.R.). They reacted in ionized water according to the following equation

$$3(\text{R-NH}\underset{\underset{\text{S}}{\|}}{\text{C}}\text{NH}_2) + MX_2 \longrightarrow M(\text{R-NH}\underset{\underset{\text{S}}{\|}}{\text{C}}\text{NH}_2)_3X_2$$

where M represents Cd^{2+} and Hg^{2+}, X represents Cl^- and Br^-, and R is $CH_2{=}CHCH_2-$.

The element-composition analysis of ATCC, ATCB, ATMC and ATMB have been carried out by using gravimetry method. The results obtained are listed in Table 1.

The crystals of ATCC, ATCB and ATMC/have been grown in an improved Hobden crystallizer by temperature cooling method.

We have carried out the structure analysis for these three crystals, and the IR spectrum of them are shown in Figure 1 . From the comparison of the spectrum of these crystals to that of allylthiourea we can see that there are three absorption bands in the spectrum of ATCC, ATCB and ATMC which are closed to one another at the region of 570-440 cm^{-1}, whereas no absorption band appeared in this region on IR spectrum of pure allylthiourea. Therefore , it is evident that these absorption bands are the complex absorption bands of S-Cd and S-Hg, and the three crystals seem to be coordinated by sulphur to form complexes.

We have measured the structures of these crystals by 4-circle diffractrometer. The diffraction intensity data were collected on a Nicolet R3M/E diffractrometer. The results have shown that ATCC, ATCB and ATMC crystals all belong to trigonal system and the space

TABLE 1

Data of the element-composition analysis

Formula	Theoretical value (%) C	N	S	H	Cl(Br)	Experimental value (%) C	N	S	H	Cl(Br)
$Cd(C_4H_8N_2S)_3Cl_2$ (ATCC)	27.09	15.80	18.08	4.55	13.30	27.00	15.50	17.90	4.42	13.30
$Cd(C_4H_8N_2S)_3Br_2$ (ATCB)	23.22	13.54	15.49	3.9	25.70	23.55	13.61	15.20	4.12	25.45
$Hg(C_4H_8N_2S)_3Cl_2$ (ATMC)	23.24	13.55	15.51	3.90	11.40	23.10	13.70	15.45	4.15	11.30
$Hg(C_4H_8N_2S)_3Br_2$	20.34	11.85	13.57	3.39	22.54	20.25	11.80	13.55	3.35	20.60

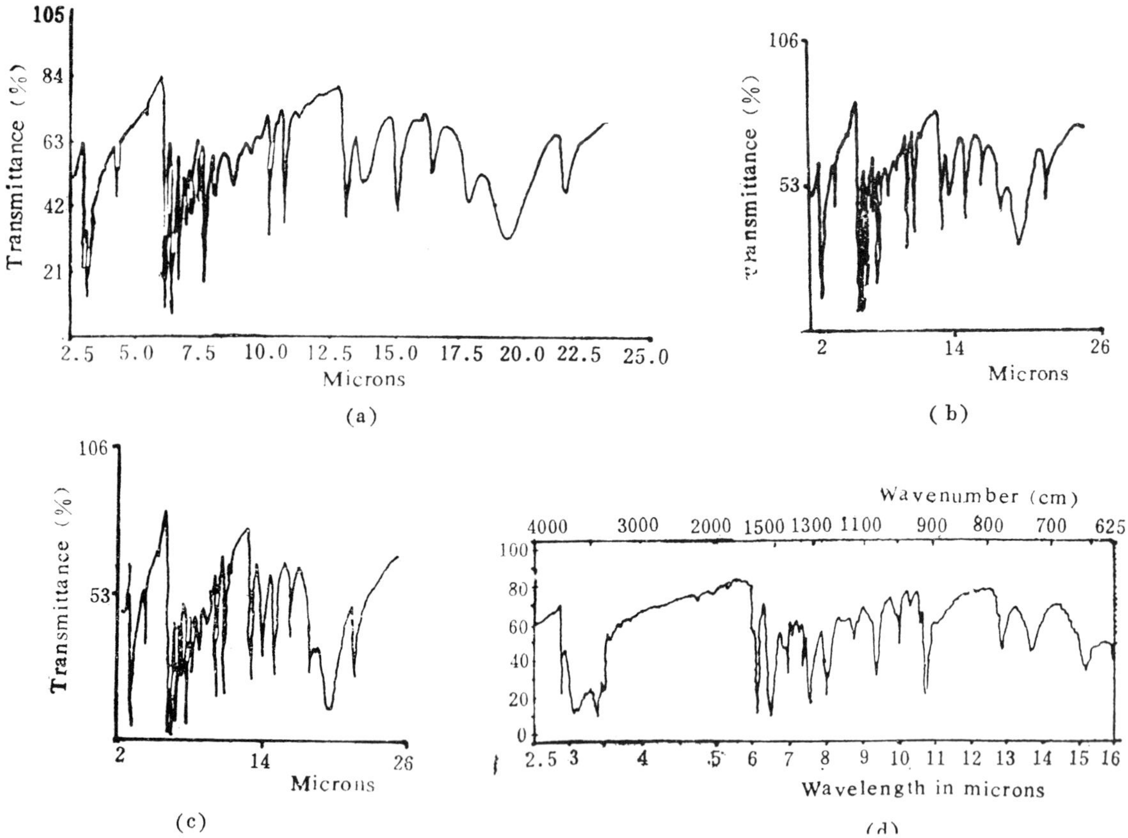

FIGURE 1

IR spectra of the crystals

(a) ATCC crystal (b) ATMC crystal

(c) ATCB crystal (d) Allylthiourea

TABLE 2
Structural Data

Crystal	Point group	Space group	Unit cell parameter	Dc (g/cm^3)	Dm (g/cm^3)
ATCC	C_{3v-3m}	R_{3c}	a=b=11.527(4) A^O c=27.992(6) A^O $\alpha=\beta=90^O$ =120° Z=6	1.65	1.63
ATCB	C_{3v-3m}	R_{3c}	a=b=11.621(2) A^O c=28.569(7) A^O $\alpha=\beta=90^O$ =120° Z=6	1.85	1.82
ATMC	C_{3v-3m}	R_{3c}	a=b=11.436(3) A^O c=28.106(8) A^O $\alpha=\beta=90^O$ =120° Z=6	1.94	1.91

group is R_{3c} . The structural data are listed in Table 2, and Figure 2 is the typical molecular configuration.

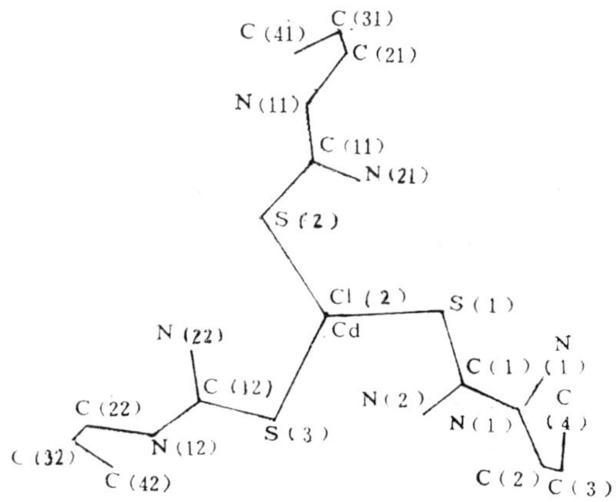

FIGURE 2
Typical molecular configuration of complex

The differential scanning calorimetry (DSC) and thermogravimetry (TG) were carried out by using Pekin Elmer's DSC-2S and TGS-12 thermoanalysis equipments. The results show that these crystals have no crystalline water and their melting points(i.e.ATCC=101°C, ATMC=133°C and ATMB=125°C, respectively) are all higher than that of allylthiourea(i.e. 73°C).

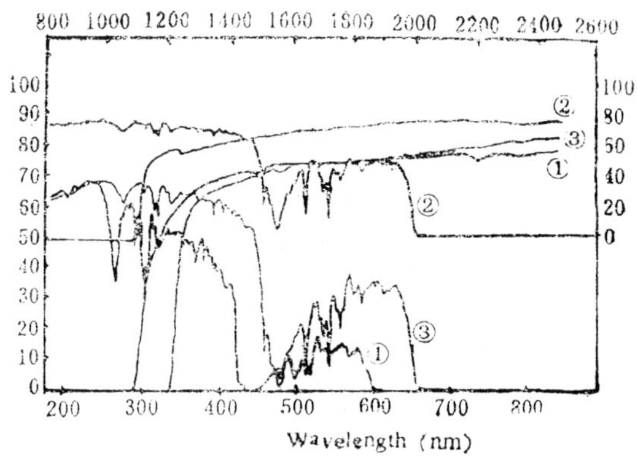

FIGURE 3
The transmission spectra of the crystals
(1) ATCC (L=16.5 mm)
(2) ATCB (L=2 mm)
(3) ATMC (L=4 mm)

The transmission spectra of these crystals determined by using Hitachi Model 340 spectrophotometer are shown in Figure 3. The coordinate of ATCB is at the right side in this figure. From the figure, we can see that these crystals have wide range of transmission (i.e. 300-1500 nm).

The nonlinear coefficients of these crystals were measured by using Maker Fringe method, and the results obtained are listed in Table 3.

TABLE 3
Data of NL coefficients of the crystals

	d_{22}	d_{31} ($x10^{-9}$ esu)	d_{33}
ATCC	0.7	1.5	1.9
ATCB	2.73	0.5	0.9
ATMC	8.2	0.27	2.7

3. CONCLUSION

ATCC, ATCB, ATMC and ATMB are new organometallic NL frequency-doubling materials. They all have wide range of transmission, and their UV cutoff wavelength is about 300 nm.

These four NLO materials have high efficient SHG effects. The SHG intensities of ATMC and ATMB are 3 times higher than that of urea.

The discovery of these allylthiourea complex type NLO materials has once again proved that the organometallic complex field is a potential field for the exploration of new NLO materials possessing useful properties for wide applications.

ACKNOWLEDGEMENTS

The authors would like to thank Mr. W. T. Yu and Mr. Z.H. Yang for the providing of the results of crystal structure measurements and thermoanalysis.

REFERENCES

1. G.C. Xing, M.H.Jiang and Z.S. Shao, Chinese Phys. Lett. 14 (1987) 1.

2. X.T. Tao, M.H. Jiang, D. Xu and Z.S. Shao , Kexue Tongbao(in Chinese)16(1987) 1234.

3. N.Zhang, M.H. Jiang, Chinese Phys. Lett. 6 (1989) 280.

4. D.Xu,Synthetical Crystals 16 (1987) 1.

Frontiers of Materials Research/
Electronic and Optical Materials
M. Kong, L. Huang (editors)

GROWTH OF $LaAlO_3$ SINGLE CRYSTAL BY VERNEUIL METHOD

Hong WANG, Mengkai Lü, Zhuo WANG, Shuxia SHANG and Xiaolin WANG

Institute of Crystal Materials, Shandong University, Jinan 250100, P.R. China

Single crystal of $LaAlO_3$ have been grown by Verneuil method. The powders of Al_2O_3 (99.90%) and La_2O_3 (99.99%) were used as starting materials. The crystals with size up to ϕ15 X 25 mm^3 and light brown colour were obtained. The XRD analyses show that the $LaAlO_3$ crystal have hexagonal structure with lattice parameter of a=b=5.365Å and c=13.114Å. The dielectric constant and dissipation were measured by using HP 4192A LF Impedance Analyzer at temperature from 77K to 300K and frequency from 1 KHz to 13 MHz. At room temperature the dielectric constant is 24.3 and the dissipation is less than 1 X 10^{-3}.

1. INTRODUCTION

After the discovery of high Tc superconductor $YBa_2Cu_3O_x$ (YBCO), the growth of YBCO thin film has been extensively studied[1,2]. Due to the anisotropic structure of the YBCO unit cell, the properties of YBCO thin film depend critically on the grain structure and materials. Accordingly, the properties of YBCO thin film are largely determined by the characteristics of the substrate on which they are formed. $LaAlO_3$ is of interest as a possible substrate material for thin films of YBCO because of its superior dielectric properties (compared with $SrTiO_3$) and its samll lattice mismatch with YBCO ($\angle$1%). We report in this paper the growth of single crystal $LaAlO_3$ by Verneuil method. The size of the as-grown $LaAlO_3$ crystal is up to ϕ15 X 25 mm^3. The dielectric constant and dissipation of as-grown crystal were measured at temperature between 77K to 300K and frequency from 1k Hz to 13 MHz. At room temperature the dielectric constant is 24.3 and dissipation is less than 1 X 10^{-3}.

2. EXPERIMENTAL AND RESULTS

The single crystal of $LaAlO_3$ were grown by Verneuil method. Powders of high purity Al_2O_3 (99.90%) and La_2O_3 (99.99%) were used as starting materials. They were weighted in the atom ratio of La:Al=1:1, then ground and thoroughly mixed. The mixture was calcined at 1200--1300°C for 2 days. The powders obtained were used for crystal growth. The crystal growth apparatus is shown in Figure 1. The typical growth rate is about 2-3 mm/h. The crystal obtained has light brown colour. The size is up to ¢15 X 25 mm^3. Figure 2 shows a transparant crystal sheet and three crystal pieces cut from the bulk crystal. The crystal structure was measured by XRD analysis using a Japan Rikaku D/Max RA X-ray diffractometer. The results show that $LaAlO_3$ crystal has hexagonal structure with lattice parameters a=b=5.365Å, c=13.114Å.

The dielectric properties were measured by using a HP 4192A LF Impedance Analyzer. The thin samples with paralleled face were cut from the as-grown bulk $LaAlO_3$ crystal.

The gold electrodes were sputtered on the opposite surfaces of sample to give a parallel plate capacitor configuration. The dielectric constant were obtained using the equation:

$$\frac{\varepsilon}{\varepsilon_0} = \frac{cd}{A\varepsilon_0}$$

where d,A,c are the thickness, the area, and the measured capacitance of the sample, respectively.

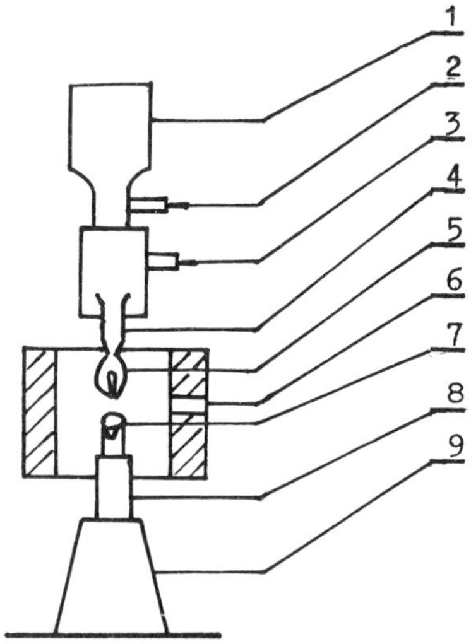

FIGURE 1
Apparatus used for growth of $LaAlO_3$ single crystal: (1) Hopper, (2) O_2 inlet, (3) H_2 inlet, (4) Nozzel, (5) Flame, (6) Window, (7) Seed crystal, (8) Pedestal, (9) Elevator

Temperature between 300K and 77K were attained in a conventional glass double Dewar system. The temperature of the sample was measured by a chromel-alumel thermocouple mounted close to the sample. The temperature dependence of dielectric constant measured at 10k Hz is shown in Figure 3. The dielectric constant decreased slightly with the decreas of the temperature. The relative change is within 4%. The dissipation is less than 1 X 10^{-3}

FIGURE 2
A transparent sheet and three crystal pieces of $LaAlO_3$ single crystal

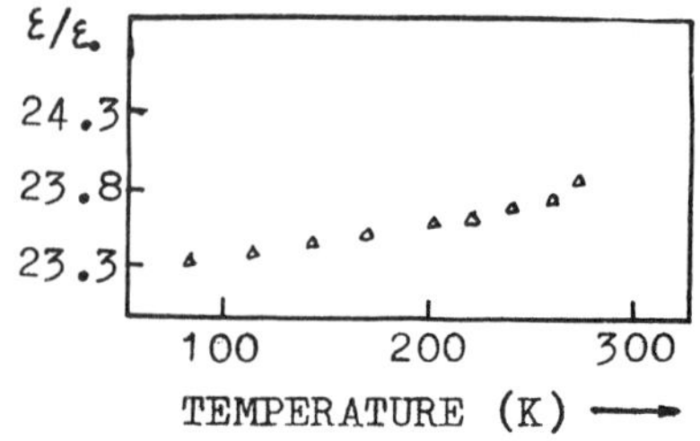

FIGURE 3
Temperature dependence of dielectric constant of $LaAlO_3$ crystal measured at 10k Hz

when the temperature is above 260K, and less than 1 X 10^{-3} when it is below 260K. The dependence of dielectric constant on frequency measured at room temperature is shown in Fig.4. From 1k Hz to 13 MHz, the dielectric constant decreased from 24.3 to 23.9, and the relative change is less than 2%. The disspation measured in this frequency range is less than 1 X 10^{-3}.

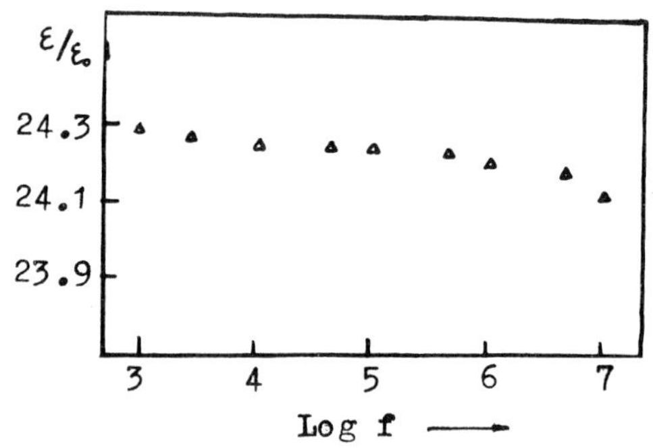

FIGURE 4
Dielectric constant as a function of measurement frequency at 296K

3. CONCLUSION

Single crystal of $LaAlO_3$ has been grown by Verneuil method. The measurements of dielectric constant and dissipation show that the as-grown $LaAlO_3$ crystal has superior dielectric properties compared with that of $SrTiO_3$. The dissipation is less than 1×10^{-4} at low temperature and the dielectric constant is about 24 at frequency between 1k Hz and 13 MHz and temperature between 77K and 300K. These results show that $LaAlO_3$ crystal grown by Verneuil method is a superior substrate material for thin film of high Tc YBCO superconductor.

REFERENCES

1. R.B. Laibowitz, R.H. Koch, P. Chaudhari and R.J. Gambino, Phys. Rev. B 35 (1987) 8821

2. A. Hohler, D. Guggi et al., Appl. Phys. lett. 54 (1989) 1066

Frontiers of Materials Research/
Electronic and Optical Materials
M. Kong, L. Huang (editors)

OPTICAL CHARACTERISTICS OF FERROELECTRIC KIO_3 SINGLE CRYSTALS*

Wang Xiu, Xiao Dingquan and Li Heping

Department of Materials Science, Sichuan University, Chengdu 610064, China

Lu Mengkai

Institute of Crystallography, Shandong University, Jinan 250100, China

The optical diffraction bands whose polarization plane rotates 90° with respect to that of the incident light were experimentally observed in the temperature region from -183° to 240°C when laser beam passes through KIO_3 single crystals, and the change of the intensity I(T) of the diffraction light with temperature was measured. The relation between the pattern and intensity of the diffraction light and the electric domains of the crystals was discussed. The optical absorption, anomalous photovoltaic effect and non-linear optical effect of KIO_3 crystals were also investigated.

INTRODUCTION

Potassium iodate (KIO_3) single crystals(space group P1 at room temperature) are piezoelectric, pyroelectric, electro-optic and non-linear optic crystals with the non-linear optical coefficient being the largest among all iodate crystals. The thermal stability of the crystals is very good, which indicates that the crystals can be used in a wider temperature region. These crystals possess peculiar ferroelectric behaviour and have several reversible phase transitions from low temperature up to 250°C, and therefore they are peculiar crystals for fundamental study.

In this paper, the experimentally observed optical diffraction bands whose polarization plane rotates 90° with respect to that of the incident light when a laser beam passes through KIO_3 single crystals are reported, the dependence of the change of the intensity and the pattern of the diffraction light on temperature, and the phase transition and electric domains of the crystals are discussed, and the characteristics of optical absorption, anomalous photovoltaic effect and non-linear optical effect of KIO_3 crystals are also investigated.

90° POLARIZATION CHANGED LIGHT DIFFRACTION BANDS

The colorless and transparent KIO_3 crystals grown from solution are hexagonal prism crystals [1]. The polar direction of crystals is parallel to crystal growth direction which is the direction of the hexagonal prism edges. According to the crystals shape, we choose the Cartesian coordinate system O-XYZ with Z-axis parallel to the crystals growth direction, Y-axis normal to the prism surface of as grown crystals, and define the planes perpendicular to X, Y and Z axes as X, Y and Z planes respectively. The samples used in our experiments were rectangular parallelepipeds with their surfaces being X-, Y- and Z- planes respectively. All the samples were mirror-polished. An intracavity He-Ne laser (λ =6328Å) was used as the light source, and the polarizer and analyzer were placed in front of and behind the sample respectively.

1. The light diffraction bands at room temperature[2]

When a laser passes through the crystal, the

*The project supported by the National Foundation of Natural Science of China

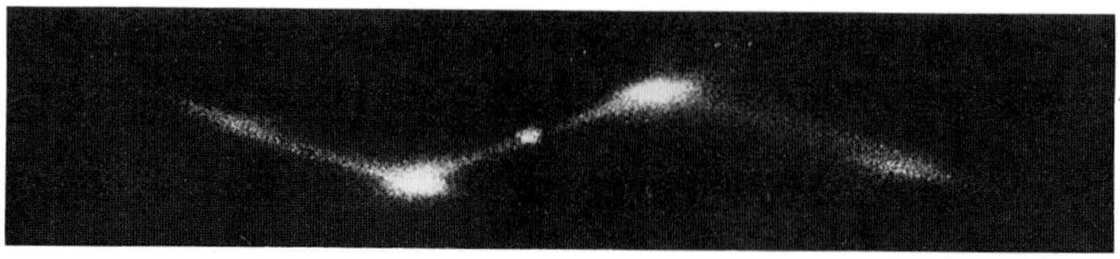

FIGURE 1 $E_{\parallel} \rightarrow E_{\perp}$ diffraction bands

light diffraction bands shown in Fig.1 were observed with Y-plane being transmission plane and the analyser perpendicular to the polarizer. The plane of the polarization of the diffraction light rotates 90° with respect to that of the incident one, i.e., when the electrical vector of the incident light is parallel to Z-axis, the electrical vector of the diffraction light is perpendicular to this direction (noted as $E_{\parallel} \rightarrow E_{\perp}$ mode), on the contrary, when the electrical vector of the incident light is perpendicular to Z-axis, the electrical vector of the diffraction light is parallel to this direction (noted as $E_{\perp} \rightarrow E_{\parallel}$ mode). The intensity of the $E_{\parallel} \rightarrow E_{\perp}$ diffraction band is almost as large as that of the $E_{\perp} \rightarrow E_{\parallel}$ bands. The pattern of the diffraction bands is essentially centrosymmetric. These diffraction bands are easily observed when a laser passes through the different parts of a sample, and the diffraction spots will also appear when a laser passes through some parts of a sample. With X or Z-plane being transmission plane the 90° polarization changed diffraction bands were observed as well, however the patterns of these diffraction bands were different from that shown in fig.1.

2. The change of the intensity of the diffraction light with temperature

The intensity I(T) of the 90° polarization changed diffraction bands was measured in the temperature regions from 20° to 240°C and 20° to -183°C respectively. The samples were put into a special heating furnace with light window or a cryostat. A segment of the $E_{\parallel} \rightarrow E_{\perp}$ diffraction bands was measured to determine the change of the intensity I(T) of the diffraction light with temperature by using a photomultiplier as a detector. The measured results were shown in Fig. 2.

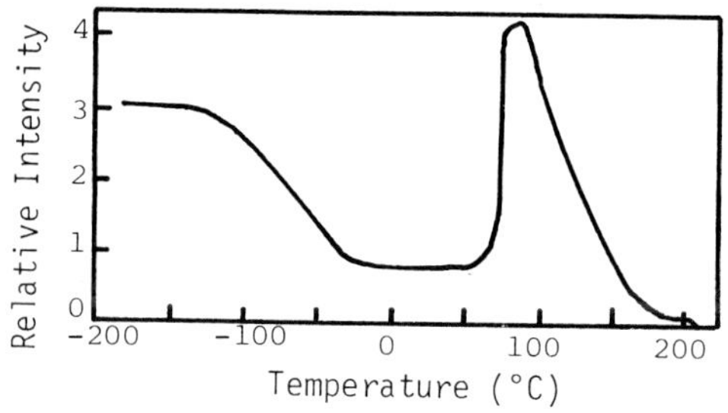

FIGURE 2
The change of the intensity I(T) of $E_{\parallel} \rightarrow E_{\perp}$ diffraction bands with temperature (Y-plane was the transmission plane).

In the heating process from room temperature, the pattern of the diffraction bands essentially remains the same, and at the temperature of 212°C, the pattern of the diffraction bands vanishes from sight. In the cooling process from room temperature down to -183°C, the pattern of the diffraction bands changes obviously.

3. The dependence of the change of the intensity I(T) of the diffraction bands on the phase transition and electric domains

From Fig.2 it can be seen that the intensity I(T) of the diffraction bands reaches its maximum at 72°C and becomes to zero at 212°C in heating processes. In literature[3-4] the phase transitions of KIO_3 crystals take place at 72°C and 212°C respectively. Therefore, the measurement of the dependence of the intensity I(T) of 90° polarization changed diffraction light on temperature provides an additional method for determining the phase transitions of crystals.

In cooling processes, the intensity I(T) of the diffraction light increases obviously with the decrease of temperature in the temperature region below -50°C, but there is no dramatic change in the intensity I(T). According to the literature[4], the electric domains in KIO_3 crystals will spit into laminar ones in the temperature region from -75°C to -90°C. It is thought that the increase of the intensity I(T) of the diffraction bands is related to the change of domains with temperature in the low temperature region.

4. Discussion

The KIO_3 crystals grown from solution are polydomain crystals. One can observe the stripe-distributed 120° domains of the crystal in the cross section perpendicular to the growth direction[2]. The near-regular stripe-distributed domains have the crystals form volume phase grating[5]. The volume phase gratings resulted from domains cause the light diffraction resulting in the 90° polarization changed diffraction bands. The phase transition or the change of domains in KIO_3 crystals cause the change of the parameters of the volume phase grating resulting in the increase or decrease of the diffraction efficiency of the phase grating, thus it cause the change of I(T) as shown in Fig.2.

The experimental results show that the measurement of the pattern and the intensity of the 90° polarization changed diffraction bands provides a simple new method for the investigation of phase transition and domains structure of crystals[2,6].

OPTICAL ABSORPTION PROPERTIES[7]

The optical absorption properties of KIO_3 crystals were studied from ultraviolet to infrared ranges. The optical transmissivity of various KIO_3 samples was measured by using Hitachi 340 Recording Spectrometer and Nicolet MX-1E FT Spectrometer. The results were shown in Fig.3 for incident light perpendicular to the polar axis. The thickness d of the sample is 7.0 mm. By using the experimental results in the visible range, the wavelength dependence of the absorption coefficient α and its root (α-hν, $\alpha^{\frac{1}{2}}$-hν) were calculated and the characteristics of the absorption edge were discussed. The forbidden energy gap Eg of KIO_3 crystals, which corresponds to intrinsic absorption related to direct transition, is 4.13ev, slightly higher than that reported in the literature[8]. The energy gap Eg′ (3.88 ev) corresponding to the indirect transition and the energy Ep(806.7 cm^{-1}) of phonons taking part in the indirect transition at room temperature were calculated from the $\alpha^{\frac{1}{2}}$-hν curve, the latter was verified from the Raman spectrum of KIO_3 crystals measured in our experiments.

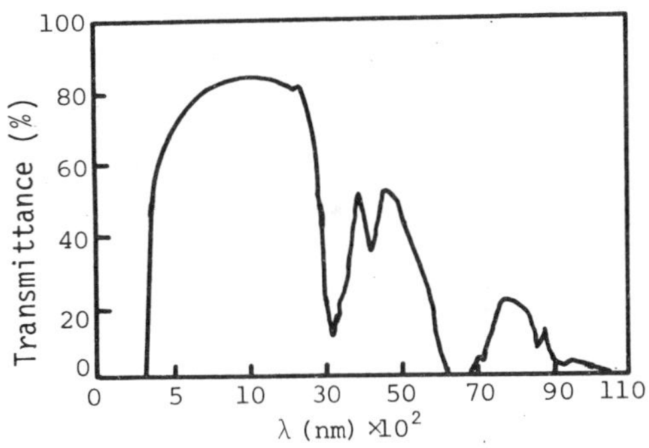

FIGURE 3
The transmissivity curves of KIO_3 crystals

The anisotropy of the transmissivity of KIO_3 crystals was also observed, that is, the transmissivity of KIO_3 with incident light perpendicular to the polar axis is different from that with incident light parallel to the polar axis. It becomes more obvious in the long wavelength end of the visible range. It could be regarded as the effect of the special domains structure of KIO_3 crystals on transmissivity [See Ref.(7) for detail].

OTHER OPTICAL EFFECTS

1. The anomalous photovoltaic effect

By using the method of current-voltage characteristics we studied experimentally the anomalous photovoltaic effect (APV) of KIO_3 crystals with good optical quality and large size of 18×18×40mm^3. The sample used were cut perpendicularly to the polar axes from different batches. The light sources were xenon lamp (500W), nitrogen laser (3371Å), and the blue (4416Å), green (5337 Å) and red (6355Å) lines from white light.

The preliminary experimental results show that KIO_3 crystals possess apparent APV effect, and that the anomalous photovoltaic current is linear with the light intensity. The photo-current is intrinsic and extrinsic in nature, the impurity energy levels may be due to the defects in the crystals. The dynamic measurement of the photo-current indicates that the relaxation time of the crystals is about 27 to 30 minutes, which is longer than that of strontium-barium niobate (SBN) crystals measured in our experiments.

2. The non-linear optical effect

KIO_3 single crystals possess good second harmonic generation (SHG) property. By applying KIO_3 crystals in Nd:YAG laser stronger light rings of second harmonic wave was observed. The shape and strength of the rings almost remain the same in the temperature region from RT to about 220°C. It is believed from the experiment that the frequency doubler made of KIO_3 crystals can be used in higher temperatures.

In conclusion, the 90° polarization changed light diffraction, optical absorption, anomalous photovoltaic effect and second harmonic generation of KIO_3 crystals grown from solution were experimentally investigated. The experimental results have revealed some new phenomena of the crystals, provided fundamental data for further research and a new method for study of phase transition, and verified that KIO_3 crystals are useful non-linear single crystals especially for SHG in higher temperatures.

REFERENCES

1. M.K.Lu, K.C.Zhang, Scientia Sinica A, 1986, 957,
2. X.Wang, D.Q.Xiao, W.C.Zheng, G.Y.Luo and M.K.Lu, Acta Physica Sinica, 35(1986), 136
3. E.Salje, Z.Kristallogr., 134(1971), 107
4. U.Helg, Z.Kristallogr., 131(1970), 241
5. H.G.Yang, C.X.Li and Y.Z.Huang, Acta Physica Sinica,33(1984), 706
6. H.G.Yang, Y.Z.Huang, Acta Optica Sinica, 5(1985), 926
7. D.Q.Xiao, X.Wang, W.C.Zheng and M.K.Lu, Ferroelectrics 100(1989), 213
8. V.M.Fridkin, Ferroelectrics Semiconductors, Consultants Bureau, New York (1980), p.105

Frontiers of Materials Research/
Electronic and Optical Materials
M. Kong, L. Huang (editors)

GROWTH INSTABILITIES OF FACETED CRYSTAL FROM MELT

Tao HUANG *, Yili LYU *, Yaohe ZHOU *, Xishen CHEN * *

* *National Lab. of solidification Processing, and Dept. of Materials Science & Engineering, Northwestern Polytechnical University, Xi'an 710072, CHINA*
* * *Institute of Physics, Academia Sinice, Beijing 100080*

The instabilities and pattern formation of faceted crystal growth have not been definitively investigated either experimentally or theoretically up to now. In situ observation and measurements of faceted crystal growth have been carried out with salol based transparent organic material during constrained melt growth. The studies clearly underscore the importance of anisotropic kinetics effects and control parameter on growth instabilities.

1. INTRODUCTION

It is well known that the stability of growing interface makes the controlling of growth processing possible and it dominates the distribution of solute or impurities in crystal, that is ,only the stabilized growth of planar interface can produce qualified crystal. Faceted cellular growth has recently become of interest because of its importance to the understanding of the growth of semiconductor materials[1,2]. Faceted dendrite is the common pattern forming in cast polycrystalline Si and Ge which is of potential usefulness for fine micro–processing or solar cells[3]. More generally, the question of growth shapes of faceted solids when the solidification dynamics is , at least partly, controlled by diffusion is with the noticeable exception of the work of Chernov and collaborators[4]on freely growing crystals. Faceted cellular growth has been observed[5]and the time – dependent faceted cellular array growth has been modelled numerically[6].

The study of interfacial instabilities and pattern formation, and particularly the transition from one type of growth to the other, requires the knowledge of the position and the morphology of the interface as a function of time. In this paper, a systematic quantitative experimental study on constrained growth morphologies of transparent faceted materials is developed for measuring the spatial and temporal instability of the faceted planar interface and the value of threshold velocity Vc. The hysteresis phenomena of interfacial growth velocity with the imposed pulling velocity has been reported. A series of pattern transitions with increasing growth velocity have been revealed, such as, from planar to ladder – shaped wave pattern, then to triangle wave pattern, coupling with transition of growth crystallography planes. During growth, the instability is caused by transport phenomena giving rise to faceted cellular, and subsequently to arrays of faceted dendritic shapes.

2. EXPERIMENTAL

2.1.Apparatus

Directional solidification experiments were carried out with a Hele – Shaw cell by using a temperature gradient stage described by Hunt and Jackson[7]which was very similar in principle. The motor – screw combination with microcomputer control system give velocities range from 0.01 to 300μm / sec. A special installation was designed to adjust the crystal orientation.

2.2. Preparation of materials

In situ observations of the growth in a high anisotropic material–salol (phenyl salicylate $C_{13}H_{10}O_3$) have been carried out. Salol is orthorhombic with space group Pbca. Because its melting point closes to room temperature (Tm = 43.7℃) and faced growth form corresponds high value of L / KTm = 7, salol was often selected as a model system. The material used in our experiment is impure with a few percent of O_2 content so that the chosen material salol – O_2 is a dilute binary system.

2.3. Preparation of perfect seed crystals and determination of growth orientation in direction of temperature gradient.

A seed crystal of rhombic form with four (112) faces was bred in the sample cell with its (001) face parallel to glass plane of the cell. The seed crystal was grown under steady state to occupy all space of the cell. The (012) face was adjusted in parallel to isothermal line in the cell on temperature gradient stage corresponding to the 55 ° angle between (012) and (112).

3. RESULTS AND DISCUSSION

. 3.1. Faceted planar interface stability

The time taken for the interface to reach the destabilized velocity Vd was found to be inversely proportional to Ve, and it is shown in Fig.1. The experimental data fall on a smooth surve. It is shown that there is a definate relation between Ve and t_d. The regression equation of the curve in Fig. 1 has been found by using computer.

$$V(t) = 7.29250e^{-1.08 \times 10^{-3} t} + 2.0787 \qquad (1)$$

and regression coeffidcient r = 0.9947368

According to the existence of Vc, there is a horizontal asymptotic line of V(t) inevitably[8] and

$$Vc = \lim_{t \to \infty} V(t) \qquad (2)$$

therefore, for salol which contains a few

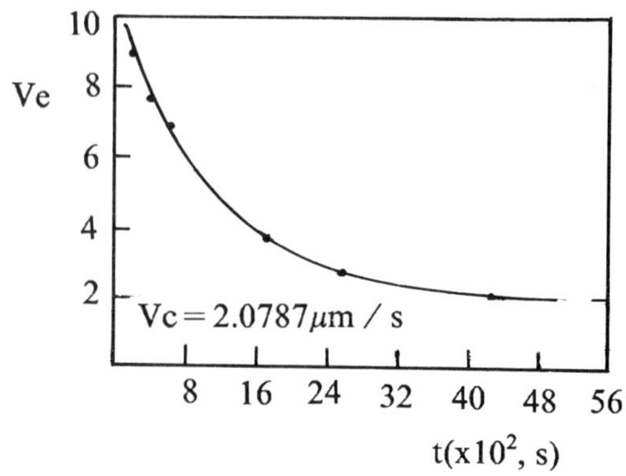

FIGURE 1
The variation in velocity with the destabilized time t_d at temperature gradient G = 2.14K / mm for Salol–O_2 system

percent of O_2, at G = 2.14K / mm

$$Vc = 2.0787 \mu m / s \qquad (3)$$

and the criterion of planar stability:

$$G / V \geqslant 1.029 \times 10^{9} K \cdot Sec / m^{2} \qquad (4)$$

3.2 Instability of the crenellated front.

The interface undercooling equation can be written:

$$\triangle T = \triangle T\kappa + \triangle Tc + \triangle Ts \qquad (5)$$

where $\triangle T\kappa$ is kinetic undercooling

$\triangle Tc$ is curvature undercooling

$\triangle Ts = mC_{1I}$ is solute undercooling (m is liquid slope, and C_{1I} is the liquid solute concentration at the interface)

The interface undercooling at any point Pi on interface is temperature gradient G multiplies interface distance Zi,

$$\triangle Ti = G \cdot Zi \qquad (6)$$

One can assume that there is no curvature undercooling for a facet, and $\triangle Tk$ equals to constant for a face, therefore, for flat region

$\triangle Tc=0$, but at the breakdown point $\triangle Tc \neq 0$, and $\triangle Ts$ at breakdown point is larger than that for flat region. According to eq. (6). it is sure that there is a groove observed at breakdown point. The anisotropic interface kinetics and diffusion give rise to the instability phenomena and process of pattern formation observed as shown in Fig. 2. It takes quite a long time from a to c as seen from Fig.2. This should result in a very strong hysteresis of the front morphology, related to the depth of the cusp in the Wulff plot, i.e. to the magnitude of the step energy for the relevant facet and also in a relatively weaker effect of diffusion.

3.3 Faceted cell to faceted dendrite transition and faceted dendritic instability.

Experiment results show that faceted cellular growth is slow in relating to external perturbation, noise and weak fluctuation and largely depend on the early stage of the breakdown of planar interface just because of the faceting effects. It has been found again[6] that the cell tip splitting and loss of cells (overgrowth) are the two main mechanisms for cell spacing adjustment. The fast adjustment mechanism has been observed that the quasi–concave or quasi–convex would be generated and propagated to creat new tips. The time scale of the tip splitting taken for isotropic material SCN system is of the order of seconds, except for anistropic salol is minutes or hours. Cellular interactions in the array are the direct result of the different growth rates of facets in the array. It is suggested that the cell tip splitting occurs when the local kinetic undercooling at cell tip become zero. In practice this would lead to a flat non–faceted cell tip which would become unstable and split as shown in Fig. 2–b.

Fig.3 shows the array of growth morphology of faceted cells and faceted dendrite formation in two different experiments, the external velocities of which closed to each other. It has been pointed out that the variation of solute concentration along the crystal face is destabilizing. It has been proposed that growth becomes unstable when the solute supersaturation becomes zero at one point on the face. When the transition from cell to dendrite would occur, the higher solute or impurity concentration in groove of cells

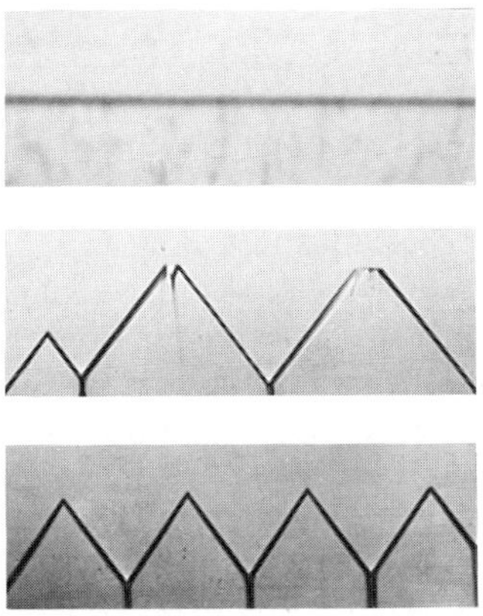

a. t=0

b. t=220s

c.t= 1680s

200μm

FIGURE 2

Transition of faceted planar–to–faceted cell G= 2.14K / mm, Ve=9.182μm / s

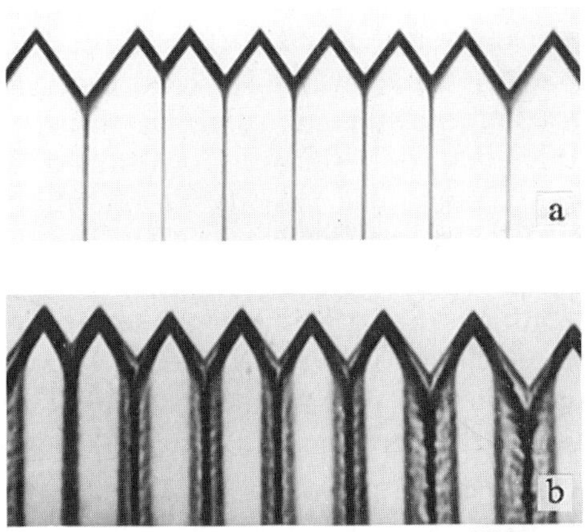

500μm

FIGURE 3

Morphology of faceted cell (Ve=23.13μm / s) and faceted dendrite (Ve=24.56μm / s); G=2.14K / mm made the (102) face of cellular sides destabilized and the O_2 gaseous inclusion–formed on the cellular sides (as seen in Fig. 3–b). It is believed that a shift in growth mechanism is responsible for the transition from inclusing free growth to inclusion – forming growth along (112) face. It is quite different from the systems of rough interface where the mechanism is responsible for coherent side branching and therefore for the cell to dendrite transition a periodic array of deep cells undergoes an oscillatory instability at sufficiently large growth velocity. Faceted dendritic growth was focused on the dependence of growth rate on orientation and processes of step generation and propagation on (112) face, and the interaction of a sequence of steps, as well as the solute or impurity concentration field about the crystal (112) face.

4. CONCLUSION

In–situ observation and measurements of growth instabilities and pattern formation of faceted crystal growth have been carried out with a high anisotropic salol–O_2 transparent organic matarial during constrained melt growth. The critical velocity Vc, at which instability of planar interface is observed, has been determined by defining to be equal to the horizontal asymptotic line of the experimental curve of relation between V and t_d, Therefore, the criterion of planar interface salol which contains a few percent of O_2 for (012) face is found to be

$$G / V > 1.029 \times 10^9 \mathrm{K} \cdot \mathrm{Sec} / \mathrm{m}^2$$

The dynamic study has shown that the coupling between solute or impurity diffusion and faceting plays a role in interface instability. The initial stage of breakdown of faceted planar interface demonstrated the perturbing propagation of moving growth step along interface. The moving step was squeezed at a random point because of concentration of impurity there and this point became the breakdown point at interface. The stability of the crenelleted front related to the magnitude of the step energy for the relevant facet and diffusion has a relatively weaker effect. The mechanism of faceted cell to faceted dendrite transition is responsible for a growth mechanism transition of impurity inclusion from free growth to growth on {112} faces.

ACKNOWLEDGEMENTS

This work was supported by National Natural Science Foundation of China under contract No. 8588010 and Foundation of National Education Committee. One of us (Huang) acknowledges receipt of support from FOK YING TUNG Education Foundation.

REFERENCES

1. M. W. Geis, H.I. Smith, B.Y. Tsaur,J.C.C. Fan, D.J. Silversmith and R.W. Mountain, J.Electrochen. Soc. 129 (1982) 2812.
2. D. Dutarte, Mater.Res. Soc. Symp. Proc. 107 (1987) 157.
3. T.F. Cised, in" Crystal Growth of Electronic Materials" edited by E. Kaldis, Elsevier Science Publishers B.V. (1985) P. 210.
4. A.A. Chernov, T. Nishinaga,in " Morphology of Crystals", Part A, Ed. I, Sunagawa (Terra, Scient, Publ. Co.) 1987, P. 207
5. D.K. Shangguan and J.D. Hunt, in "Solidification Processing 1987"Eds. J. Beech and H. Jone (Institute of metals, London) PP.175 – 178 (1988).
6. D.K. Shangguan and J.D. Hunt, J. Crystal Growth 96 (1989) 856
7. J.D. Hunt, K.A. Jackson and H. Brown, Rev. Sci. Instrum. 37 (1966) 805
8. Huang Weidong , Zhou Yaohe, to be published (1990)

Frontiers of Materials Research/
Electronic and Optical Materials
M. Kong, L. Huang (editors)

PREFERRED GROWTH DIRECTION OF $Nd_2Fe_{14}B$ DENDRITE DURING DIRECTIONAL SOLIDIFICATION

Guangrong ZOU,Zhongmin CHEN,Leyi WANG,Zhengxing SHI and Hengzhi FU
Teaching and Research Section 403, Northwestern Polytechnical University,Xi'an,710072,China

Directional solidification process of $Nd_{17}Fe_{75}B_8$ magnets has been studied. Unidirectionally aligned structure of $Nd_2Fe_{14}B$ dendritic crystal has been obtained.Its primary arm spacing is 45 – 120μm.It is found that the $Nd_2Fe_{14}B$ matrix crystallizes directly out of the melt and grows directionally in dendrite formation.The preferred growth direction of $Nd_2Fe_{14}B$ dendrite is [100],which is perpendicular to its easy direction of magnetization [001].

1. INTRODUCTION

Directional solidification technique has been used successfully to produce workable Sm – Co rare earth magnets[1]. In this technique, magnet to be get its easy magnetization direction oriented alignment is with the help of controlling the single direction heat flow during directional solidification processing. Because of the principal motivation for the development of directional solidification is a desire to control the grain morphology of magnetic alloys, and also, to get a sharp crystallographic texture due to the dominance of the fast growing crystal orientation parallel to the direction of solidification and the dendrites are similarly aligned. Therefore,to produce anisotropic oriented magnets using directional solidification technique, it is essential that the preferred growth direction of matrix should be parallel to its easy nagnetization direction.

As for NdFeB magnets, it is well known that the matrix $Nd_2Fe_{14}B$ has tetragonal structure and its easy magnetization direction is [001][2]. In order to investigate the possibility of producing anisotropic NdFeB magnets by directional solidification technique and to detect the relationship between $Nd_2Fe_{14}B$ crystal growth direction and its easy magnetization direction, the crystal growth morphology and preferred growth direction of $Nd_2Fe_{14}B$ in directional solidification have been studied experimentally in this paper.

2. EXPERIMENTS

Alloys used in this study were $Nd_{17}Fe_{75}B_8$ in constitution and were prepared from materials 99.5% Nd, 99.9% Fe and 20% B–Fe. They were melted and poured into rods with 8 mm in diameter and 100 mm in length by using the vacuum induction furnace in an argon atmosphere. The rods were then set vertically in a Liquid–Metal–Cooling directional solidification furnace, remelted and solidified directioly by being pulled down to the cooling zone in predetermined withdrawal rate. The temperature gradient in front of the solid–liquid interface is about 200 ℃ / cm. The pulling rate of crystallizer is 5 to 15 mm / min. The melting and solidifying were carried out under the vacuum and then protection of argon gas.

Microstructures were observed in optical microscopies. Bitter technique was used to display domain patterns. Chemical composition was determined with JEOL–35c SEM scanning electron microscopy and its accessory EDAX9100 energy spectrum analyzer. To determine phase structure and orientation,x–ray diffraction and electron diffraction were performed in D / max–γA x–ray diffractometer and H–800 TEM, respectively.

3. RESULTS AND DISCUSSION

3.1. $Nd_2Fe_{14}B$ crystal growth morphology

Fig.1 shows the solid–liquid interface morphology during directional solidification by suddenly quenching the solidifying ingots into cooling zone. It can be seen that the alloy crystallizes in typical dendrite structure. The dendrite crystal grows from solid zone to liquid zone against the direction of heat flow. The chemical composition microanalysis, as shown in table 1, indicates that the dendrite is $Nd_2Fe_{14}B$ matrix phase. The composition in every parts of dendrites is well–distributed. Nd–rich liquid phase is appeared in front of and among the dendrites, and Nd / Fe proportion changes greatly in Nd–rich phase. This indicates that $Nd_2Fe_{14}B$ is firstly crystallized as primary crystal from the liquid when the alloy is cooled down from the temperature above the melting point. $Nd_2Fe_{14}B$ crystal forms and grows into unidirectional dendrite columnar crystal in directional solidification. The orientation of $Nd_2Fe_{14}B$ columnar crystal is mainly depended on the controlling of unidirectional heat flow. The various microscopic solidification morphologies can be achieved by adjusting the solidifying parameter. The most important parameter influencing the microscopic solidification morphology is the values of freezing rate V_1. In practice conditions, although it can not be assumed that the rate of mould withdrawal V is identical to the velocity of the alloy melt interface freezing rate, V_1, the major differences will be transient effects at the beginning and end of solidification, for most of the slidification length where the temperatures on either side of the solid / liquid interface attain steady state profiles, $V_1 = V$. For our experiment, the temperature gradient in front of the solid / liquid interface, G_L, is about 200℃ / cm, and the directional solidifying temperature is about 1200℃ . When different withdrawal rate V is selected, the various of grain morphology can be obtained. With the increasing of the values of V, the value of G_L / V_1 decrease and the value of $G_L \cdot V_1$ (cooling rate) increase, the grain size is getting small and the alignment diverging. The equiaxed grain will be appearing when V is more than 20 mm / min.

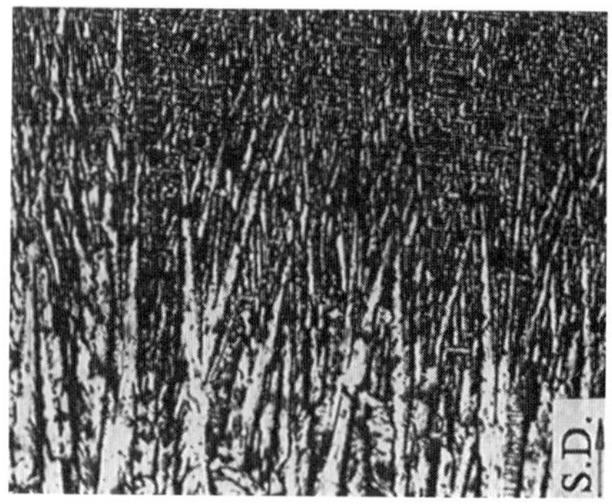

Fig.1 The solid / liquid interface morphology in directional solidification.X50 (S.D.–solidifying direction)

Tabe 1 ESA microanalysis in solid / liquid interface

element	tip of dendrite		inter–dendrite		front of dendrite
Nd	32.32	31.75	33.25	37.79	33.08
Fe	67.68	67.98	66.58	61.32	66.64
Si		0.27	0.17	0.20	0.28
Ca				0.69	

Well–orientating alignment of $Nd_2Fe_{14}B$ unidirectional columnar crystal can be obtained when the crystallizer withdrawal rate is small than 15 mm / min. The primary arm spacing is 45 to 120 μm.

3.2. Preferred growth direction of $Nd_2Fe_{14}B$ dendrite

In directional solidification, the growth of the dendrite possesses a preferred growth direction due to the anisotropy of solid–liquid interface energy and growth kinetics[3]. Fig.2 and Fig.3 show the domain patterns and the X–ray diffraction spectra of the directionally solidified samples, respectively. In section parallel to the solidifying growth direction, both stripe domain and maze domain are observed in matrix and $Nd_2Fe_{14}B$ crystal planes of (006), (204) and (004) show strong diffraction intensity. It means that there is preferred orientation of c axis in the section parallel to the growth direction. While in section perpendicular to the

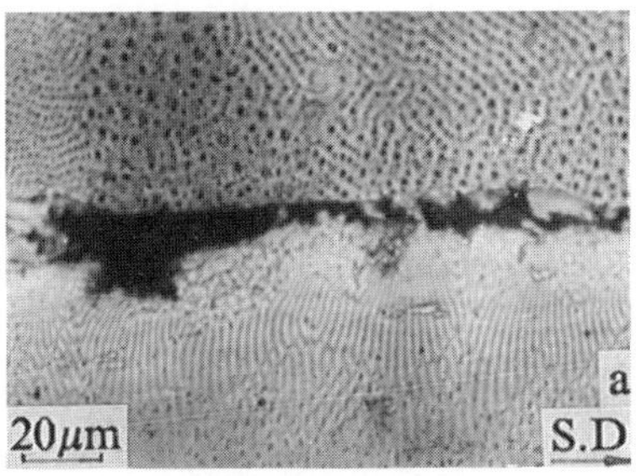

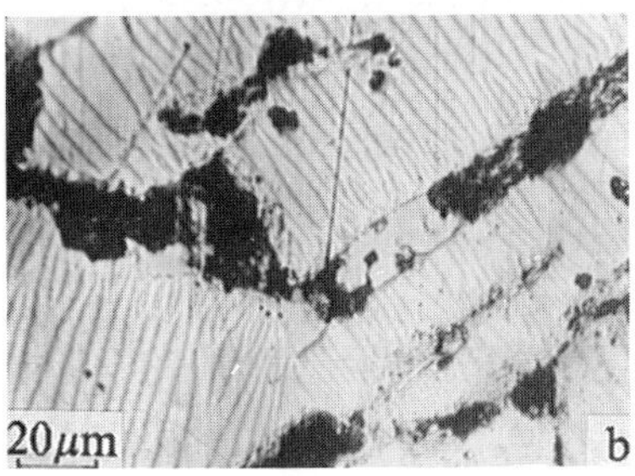

Fig.2. Domain patterns in section (a) parallel and (b) perpendicular to the solidifying direction. (S.D.–solidifying direction)

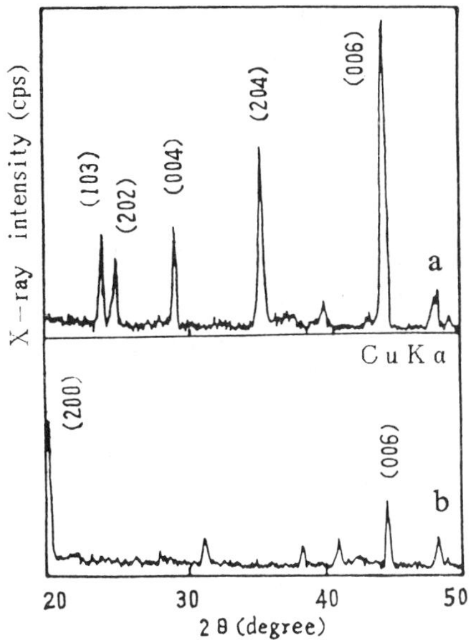

Fig.3. X–ray diffraction spectra in section (a) parallel and (b) perpendicular to the solidifying direction.

solidifying direction, as shown in Fig.2b, only the stripe domain is examined, the diffraction intensity of (006) plane is weak but that of (200) plane is strong(Fig.3b). It can be concluded that there exist preferred orientations of a axis in the section perpendicular to growth direction. The x–ray diffraction results performed on powder samples of directionally solidified rods show the same as that on sintered magnets. They all contain three phases, $Nd_2Fe_{14}B$ matrix, Nd–rich and B–rich phases. They are representative of no preferred orientating diffraction spectra.

All of these reveal that the preferred growth direction of $Nd_2Fe_{14}B$ dendrite is perpendicular to its easy magnetization direction [001]. To determine the exact growth direction, electron selected area diffraction is performed on the film sample prepared from perpendicular to growth direction. The electron beam is kept perpendicular to the film surface in operation, so that the direction of crystal zonal axis of diffraction patterns is parallel to the solidifying growth direction. Fig.4 is electron SAD patterns. The direction of crystal zonal axis is calculated to be [100]. It means that the alloy

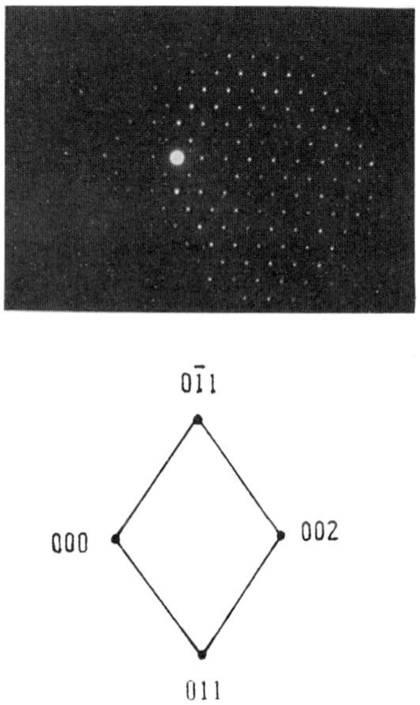

Zonal Direction:[100]

Fig.4 Transmission electron selected area diffraction patterns and its indexing showing the [100] zonal axis direction.

grows in [100] direction. As we know, directional freezing clearly encourages the production of long grains and transverse grain boundaries do not occur in the appropriate condition of solidification, in general, because the elongated grains grow freely into the melt and do not impinge on other grains growing in some other dirction[4]. The development of a preferred growth texture, [100] in the case of NdFeB permanent magnetic alloys, occurs because crystals with this orientation grow at a faster rate than others and spread transversely to eliminate the slowly growing orientations. Our experiment results clearly show that, during directional solidificaion of NdFeB permanent magnetic allays, competitive growth in an initially random population of crystals eliminating unfvourable grains and leaving the fast growing [100] crystals. The final product therefor consists of a series of long crystals oriented with [100] directions parallel to the solidification axis but misoriented with respect to each other by various rotations.

Therefor, the preferred growth direction of $Nd_2Fe_{14}B$ dendrite in directional solidification is [100], which is perpendicular to its easy magnetization direction.

The fact that $Nd_2Fe_{14}B$ crystallizes directionally as primary crystal from the melt of NdFeB permanent magnetic alloys and grows preferentially along [100] direction indicates that it is impossible to produce anisotropic NdFeB magnets only by normal directional solidification process. The research of controlling directional solidification processing and magnetic orientation is now under investigation.

REFERENCES

1. Hengzhi Fu, Xincai LiU, Zhengxing Shi, Leyi Wang and Wei Tang. The studies of Directionally Solidified Sm–Co magnets. Digests of Intermag'89 Conference,GC–12, Mar, 28–31,(1989), Washington, D.C.

2. J.F.Herbst, J.J.Croat, F.E.Pinkerston and W.B.Yelon, Phys. Rev. 29 (1984) 4170.

3. W.Kurz, D.J.Fihil. Principle of Solidification. Trans. by Mao Xiemin, The Publish House of Northwestern Polytechnical University, (1987), P65.

4. M.Mclean, Directionally Solidified Materials for High Temperature Service (Teddington, Middlesex, 1983).

Frontiers of Materials Research/
Electronic and Optical Materials
M. Kong, L. Huang (editors)

PREPARATION AND PROPERTIES OF A NEW COMPOSITE MATERIAL----PVDF-TGS FILM

Changshui FANG, Min WANG and Hongshen ZHUO

Institute of Crystal Materials, Shandong University, Jinan 250100, P.R. China

A new composite material, PVDF-TGS film, has been investigated. The spectroscopy, dielectric constant and pyroelectric coefficient of this film have been measured. Its figure of merit $M(p/\varepsilon)$ was calculated. This composite film has already been used to make IR detectors. All the results obtained have shown that PVDF-TGS composite is a prospect and important pyroelectric material.

1. INTRODUCTION

With the development of science and technology, composite material becomes an important field for the research of material science. Especially, the ferroelectric composite formed by ceramic and organic polymer has already been used in piezoelectric and pyroelectric fields. Newnham et al.[1-3] have made great contribution to the development of composite materials. This kind of material has not only good pyroelectric and piezoelectric properties, but also the advantages of good flexibility which makes it easy to be processed into homogeneous large area thin films and low cost. For all these reasons, composite materials have attracted wide scale of studies. This paper reports, for the first time, a new composite i.e. the composite film of triglycine sulfate (TGS) single crystal powders and poly vinylidene fluoride (PVDF). As we know, TGS single crystal is an IR detection material with good pyroelectric property. It has large pyroelectric coefficient and low dielectric constant. The IR devices made by it has sensible IR response and can be used at room temperature. But, TGS crystal has high fragileness, easy to crack and difficult to be processed into large area thin slices. Besides, the growth period is long and the cost is high. These disadvantages have limited its applications. PVDF is an organic polymer having good flexibility and easy to be prepared into homogeneous large area thin film. It has low dielectric constant. After phase transition and polarization, it will have certain extent of pyroelectricity. But, its pyroelectric coefficient is still one order of magnitude smaller than that of TGS. Preparing composite film by combining these two materials together, a new kind of pyroelectric thin film possessing the merits of them both can be obtained. We have made investigations on the preparing method and physical properties of PVDF-TGS film. This paper mainly reports its spectroscopy, dielectric and pyroelectric properties.

2. PREPARING METHOD

Different magnitudes of TGS crystal grains(45-75 μ) and PVDF powders were mixed according to different concentrations proportions. These materials were dissolved in an organic solvent N-dimethylformamide and were stirred sufficiently. Different solutions with different PVDF-TGS ratio concentration were obtained. Then these solutions were poured into several special-made glass containers in such a way that they formed homogeneous thin films usually of 50-100 μ thick. In order to prevent some crystal grains from baring on the upper surface of the film during the solidification procession, a PVDF layer of 1-2 μ thick should be coated on the upper surface. The PVDF-TGS composite films obtained were observed by using electron microscope and the results are shown in Fig.1 (a) and (b).

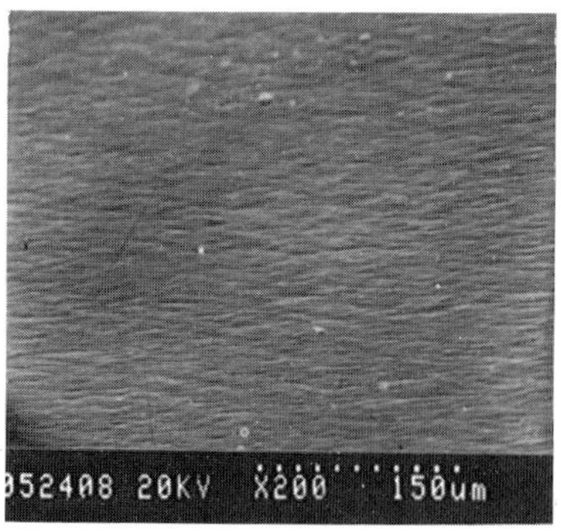

(a) pure PVDF film

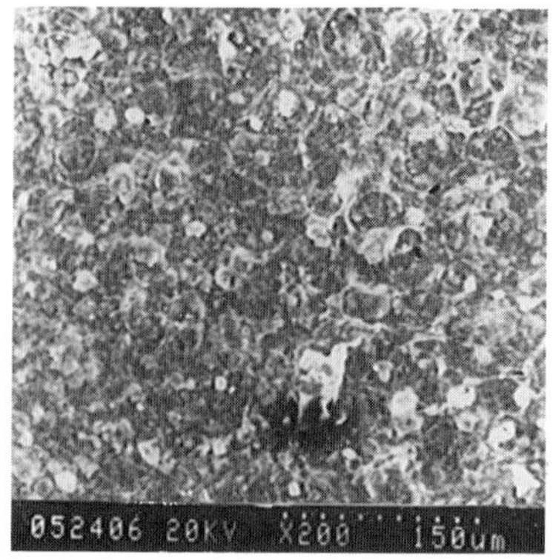

(b) PVDF-TGS film

FIGURE 1
The structures of pure PVDF and PVDF-TGS films

Fig.1 (a) is the structure of pure PVDF film and (b) is that of PVDF-TGS composite film in which a lot of crystal grains can be observed.

3. CHARACTERISTICS OF IR SPECTROSCOPY

The optical transmission of pure TGS crystal measured is 0.23-2 μ . It has been found that above 2 μ , there is no transmissivity. The optical transmission of pure PVDF(α phase, 70 μm thick) is within 0.2-6 μ , while above 6 μ, the transmissivity reduced obviously and its characteristic absorption peak appeared. For PVDF-TGS composite film, its NIR optical transmission reduces from that of pure PVDF (i.e. 60%) to 10%. Of course, the characteristic absorption peak of PVDF still existed, only its position moved slightly. These results are shown in Fig.2.

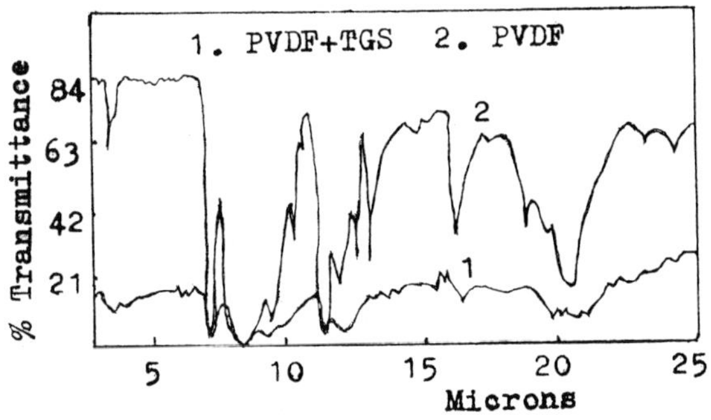

FIGURE 2
IR spectroscopy

From Fig.2, we can see that the composite film inherits the merits of TGS, i.e. it reduces the transmission of PVDF in IR waveband considerably, so that the sensibility of the devices made from the composite films is greatly improved. The detailed data of the spectroscopy of the composite film will be reported in our another paper.

4. DIELECTRIC AND PYROELECTRIC PROPERTIES

The two faces of the sample for measuring were evaporated with Al electrode in vacuum. One face was partially coated with Ø8 mm electrode and the other face was completely coated. The dielectric constant and dielectric loss were measured by using Model HP427A LRC meter at 1 kHz frequency. The intensity of the electric field applied is 5 V/cm.[4] The results of dielectric constant measurements are shown in Fig.3. The dielectric constant of pure PVDF is also given.

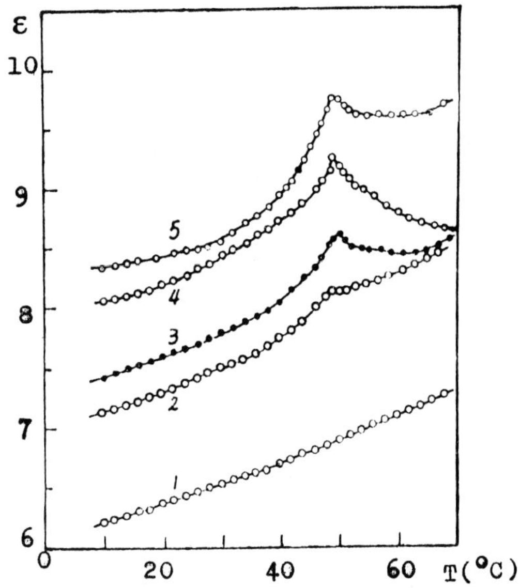

FIGURE 3
Temperature dependence of dielectric constant
Curve 1--pure PVDF; Curve 2-- W_{PVDF}/W_{TGS}=1:1; Curve 3--1:2; Curve 4--1:3; Curve 5--1:4.

The results shown in Fig.3 indicate clearly that the dielectric constant of the composite film increases with the increasing of TGS component, and there appears a distorted peak in the vicinity of the temperature corresponding to Curie temperature of TGS crystal (T_c=49°C). This phenomenon has revealed that TGS plays an important role in the composite film. The dielectric constant of the composite film is smaller than ε of TGS crystal. This shows that the PVDF component in the composite can lower the dielectric constant.

The dielectric constant ε of PVDF-TGS film is lower than those of the other composite materials. This depends mainly on the processing technology. By coating a PVDF layer on the upper surface of the composite film and forming a series-type condensor in the film, the dielectric constant can be reduced. Otherwise, it will become higher.

The results of loss tan δ measurements are shown in Fig.4. From this figure, we can see that the loss tan δ of PVDF-TGS film is one order of magnitude higher than that of TGS crystal. Because this composite film is a mixture having many crystal grain boundaries and defects, it is necessary for us to improve the preparing method so as to make the dielectric constant reduce.

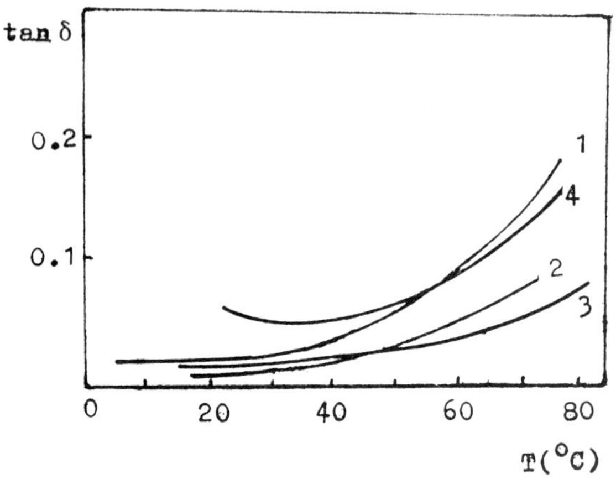

FIGURE 4
Temperature dependence of dielectric loss tan δ

Curve 1—W_{PVDF}/W_{TGS}=1:1; Curve 2—1:2; Curve 3—1:3; Curve 4—1:4; Curve 5—1:5.

The pyroelectric coefficient was measured by using charge integral method[4]. The temperature-rising rate is 3°C/min. Because the sample was freely suspended, the results obtained are the sum of the first and the second pyroelectric effects. Owing to the fact that the spontaneous polarization orientation of TGS grains is random, the magnitude of pyroelectric coefficient depends on the completeness of polarization. For this reason, the sample should be poled sufficiently before the measurement. Usually, we apply different polarization electric fields (generally, 4-6 kV/cm) at 70-80 °C for 2-3 hours. After the temperature was lowered to below 10°C, then remove the electric field. The results obtained (see Fig.5) have shown that each sample has its own optimal polarizing electric field. For comparison, the ½ value of the pyroelectric coefficient of TGS crystal is also shown in Fig.5.

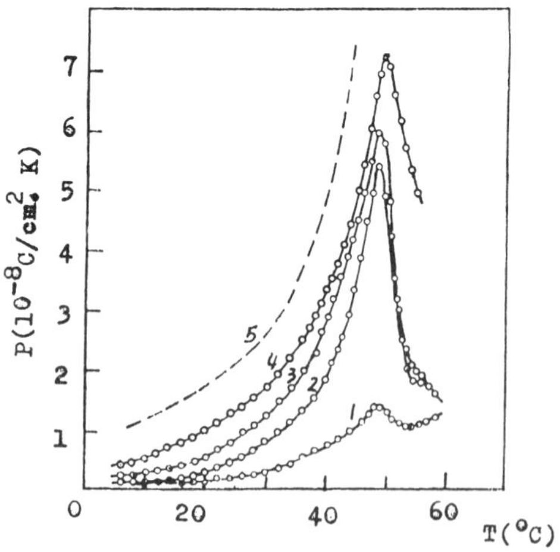

FIGURE 5
Temperature dependence of p:W_{PVDF}/W_{TGS}—
Curves 1(2:1);2(1:2);3(1:4);4(1:1)5(½TGS).

It can be seen from Fig.5 that the pyroelectric coefficients of PVDF-TGS film are all lower than that of pure TGS. The regularity of the pyroelectric coefficient p of the composite film varied with temperature is similar to that of TGS crystal, i.e. it increases nonlinearly with the increase of temperature. It reaches to a peak in the vicinity of the temperature corresponding to the Curie temperature of TGS (T_c=49°C). Because there is correlation between the grains, the corresponding temperature range of p can even be extended to 60°C. The results shown in Fig.5 indicate that p becomes higher when TGS grain content reaches to about 50%. This has proved that there is an optimum mixing ratio between the two components.

The figure of merit (p/ε) has been calculated based on the data mentioned above and the results are shown in Fig.6.

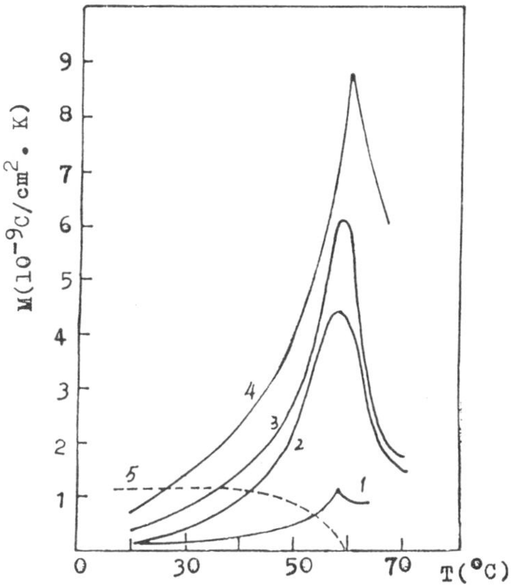

FIGURE 6

Figure of merit of PVDF-TGS film W_{PVDF}/W_{TGS}—Curve 1(2:1);Curve 2(1:2); Curve 3(1:4); Curve 4(1:1); Curve 5 (pure TGS).

From Fig.6, we can see that the M value of PVDF-TGS film at high temperature surpasses that of TGS crystal. When the content of TGS grains in the film reaches 50%, then the value of M is the largest. If the content of TGS is too high, the M value will be reduced and the structure of the composite material will become relatively loose, which results in the failure of using it to process into homogeneous large area film. By improving the preparing method, we have prepared homogeneous PVDF-TGS composite films and they have already been used to fabricate IR detectors. The size of the sensor is Ø3.5 mm and its thickness is 50 μm. The device is easy to fabricate and has good performance. Its D^* value is (5-7) x 10^7 cm·$H^{\frac{1}{2}}W^{-1}$. We are sure that if the size of TGS grains and the thickness of the composite film can be minimized, then D^* value can certainly be improved further.

REFERENCES

1. R.E. Newnham, D.P. Skinner and L.E. Cross, Material Research Bulletin 13 (1978) 525.

2. J. Mendiola et al., Ferroelectrics 53 (1984) 159.

3. M. Amin et al., Ferroelectrics 59 (1984) 233.

4. M. Wang and C.S. Fang, J. Shandong University, Natural Science Edition (in Chinese) Vol. 21 (1987) 94.

Frontiers of Materials Research/
Electronic and Optical Materials
M. Kong, L. Huang (editors)

STUDY OF CRYSTALLIZATION KINETICS IN NEW OPTICAL GLASSES

Hu Lili and Jiang Zhonghong

Shanghai Institute of Optics and Fine Mechanics, Academia Sinica, P.O.Box 800216, Shanghai, 201800, P.R.China

With thermal analysis method and nonisothermal Johnson-Mehl-Avrami equation $\ln T_p^2/\alpha = \ln E/R + E/RT_p - \ln k_0$, kinetic parameters of crystallization are resolved in La_2O_3-Nb_2O_5-B_2O_3, La_2O_3-BaO-B_2O_3, fluoride and fluorophosphate glasses. Characteristics of crystallization in these optical glasses are described and compared with $k(T_p)=k_0\exp(-E/RT_p)$ values. It is concluded that most stable glasses could be obtained after proper composition adjustment in each glass systems. As a new stability criterion $k(T_p)$ has provided a convenient approach to determine the stability of optical glasses.

1. INTRODUCTION

There are two main directions in development of optical glasses. These glasses containing La_2O_3, Ta_2O_5 and Nb_2O_5 are developed to satisfy design requirement for high refraction and low dispersion. Glasses that satisfy other requirement for low refraction and high dispersion are fluoride and fluorophosphate glasses. The improtment problem of new optical glass development is the tendency toward crystallization in melting and forming process[1]. It is necessary to establish a stability criterion and to evaluate the most stable glasses in these optical glass systems.

In order to assess the stability level of glasses, different quantitative criteria such as Hurby[2], Iqbal[3], $k(T_g)$[4,5] criteria have been suggested. These criteria have their limitation in application though in some cases they give good evalution of glass stability.

Based on differential thermal analysis method and nonisothermal Johnson-Mehl-Avrami equation, rate constant of crystallization of glass can be easily achieved. In present paper, analysing from the variation of $k(T_p)$, rate constant of crystallization at crystallization peak temperature, with composition in these optical glasses, it is considered that $k(T_p)$ is a good criterion to classify the glass stability. The main glass forming ability criteria are shown in Table 1.

Table 1: Main glass forming ability criteria

Hurby[2]	$HR=(T_x-T_g)/(T_m-T_p)$
Iqbal[3]	$ST=(T_x-T_g)/R_c$
Surinach[4,5]	$k(T_g)=k_0\exp(-E/RT_g)$

From the development of optical glasses, different compositions of La_2O_3-Nb_2O_5-B_2O_3, La_2O_3-BaO-B_2O_3, fluoride and fluorophosphate glass systems are chosen. The variation of crystallization characteristics of these optical glasses with compositions will be researched with our new criterion. Position of these four glass systems in n_D-ν_D figure is shown in Fig. 1.

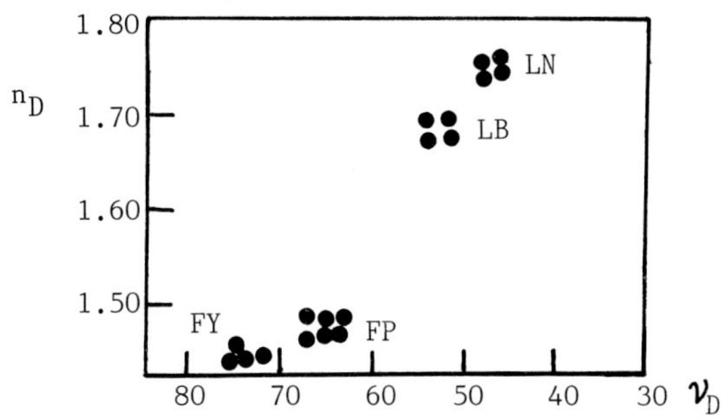

FIGURE 1
n_D-ν_D feature of optical glasses

2. EXPERIMENTAL

All glasses were made from c.p. grade raw materials in platinum crucible. The fluoride glasses are melted in protective atmosphere. Differential thermal analysis curves of all these bulk glass samples at four heating rates were measured by Rigaku PTC-10A device. The temperatures at each heating rates were caliberated by Pb sample. Reference material of DTA is α-Al_2O_3 power. Compositions of all these glasses and their critical temperature T_g, T_x, T_p and T_m are shown in Table 2-4. The defination of T_g, T_x, T_p and T_m are shown in Fig.2.

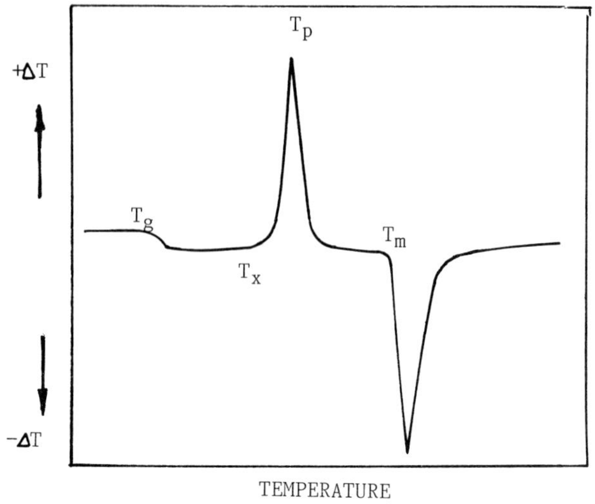

FIGURE 2
Typical DTA curve of glass

Table 2 Compositions and critical temperatures of fluoride glasses

glass	composition(mol%)										
	AlF_3	MgF_2	CaF_2	SrF_2	BaF_2	YF_3	YbF_3	T_g(K)	T_x(K)	T_p(K)	T_m(K)
FY0	40	10	30	10	10			687	750	767	1023
FY5	40	10	30	10	10	5		692	798	829	1005
FY10	40	10	30	10	10	10		690	803	834	1015
FY15	40	10	30	10	10	15		702	815	847	1013
FY20	40	10	30	10	10	20		706	826	852	1015
FY25	40	10	30	10	10	25		701	811	847	1024
FYb5	40	10	30	10	10		5	699	793	821	1011
FYb10	40	10	30	10	10		10	693	799	844	1018
FYb15	40	10	30	10	10		15	701	803	837	1012
FYb20	40	10	30	10	10		20	704	815	849	1012
FYb25	40	10	30	10	10		25	698	819	848	1013

Table 3 Compositions and critical temperatures of borate glasses

glass	compositions(mol%)							
	B_2O_3	La_2O_3	BaO	Nb_2O_5	T_g(K)	T_x(K)	T_p(K)	T_m(K)
LB1	74	22	4		915	1041	1062	1265
LB2	74	18	8		919	1071	1092	1255
LB3	74	14	12		909	1055	1069	1255
LB4	74	10	16		892	1040	1061	1253
LN1	64	20.5		15.5	912	1099	1135	1290
LN2	65	28		7	924	1108	1138	1283
LN3	70	25		5	930	1103	1139	1295
LN4	60	30		10	922	1103	1138	1295

Table 4 Compositions and critical temperatures of fluorophosphate glasses

glass	compositions(mol%)										
	AlF_3	MgF_2	CaF_2	SrF_2	BaF_2	LiF	$M(PO_3)_x$	T_g(K)	T_x(K)	T_p(K)	T_m(K)
FP1	27	17	16	12	12	10	$Al(PO_3)_3$ 6	693	865	888	-
FP2	33	17	7	12	12	10	$Ca(PO_3)_2$ 9	673	915	925	-
FP3	33	17	16	3	12	10	$Sr(PO_3)_2$ 9	676	873	895	-
FP4	33	17	16	12	3	10	$Ba(PO_3)_2$ 9	673	858	885	-

3. EXPERIMENTAL RESULTS

According to the nonisothermal Johnson-Mehl-Avrami equation[6]

$$\ln T_p^2/\alpha = \ln E/R + E/RT_p - \ln k_0 \qquad (1)$$

where E is activation energy for crystallization, k_0 is frequency factor, R is gas constant, α is heating rate and T_p is peak temperature of crystallization curve. In a certain temperature range, $\ln T_p^2/\alpha$ has linear relation with $1/T_p$. E and k_0 can be resolved from this relation according to the measured T_p values at different heating rates. Typical $\ln T_p^2/\alpha$ versus $1/T_p$ curve of glass is shown in Fig.3.

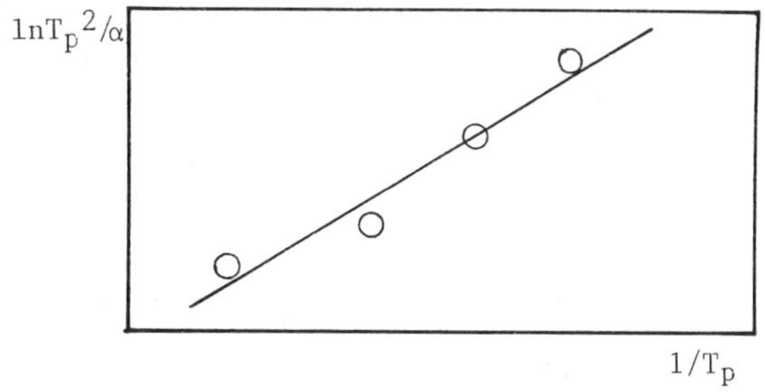

FIGURE 3
$\ln T_p^2/\alpha$ versus $1/T_p$ curve of glass

With least-square method, E, k_0 and rate constant of crystallization $k=k_0exp(-E/RT)$ can be obtained. The kinetic parameters E, k_0 and $k(T_p)$ for all glasses are shown in Table 5-7. It is known that E, k_0 and $k(T_p)$ values of glasses vary with compositions. Comparing $k(T_p)$ values of these glasses in each glass systems, it is found that FP1, LN4, LB2, FY15 and FYb10 glasses have smallest $k(T_p)$ values in corresponding systems. These glasses do not behave top HR values in corresponding systems as shown in Table 5-7.

Table 5 E, k_0 and data obtained from different criteria in fluoride glasses

glass	HR	$k(T_g)(min^{-1})$	$k(T_p)(min^{-1})$	E(kJ/mol)	$k_0(min^{-1})$
FY0	0.25	$4.61*10^{-3}$	0.674	273	$2.64*10^{18}$
FY5	0.60	$1.20*10^{-3}$	0.53	212	$1.21*10^{13}$
FY10	0.62	$5.64*10^{-4}$	0.387	217	$1.51*10^{13}$
FY15	0.68	$2.73*10^{-3}$	0.222	150	$3.96*10^{8}$
FY20	0.74	$2.31*10^{-4}$	0.471	261	$4.73*10^{15}$
FY25	0.62	$3.84*10^{-4}$	0.478	241	$3.49*10^{14}$
FYb5	0.49	$2.10*10^{-3}$	0.428	208	$7.33*10^{12}$
FYb10	0.61	$1.23*10^{-3}$	0.374	184	$9.13*10^{10}$
FYb15	0.58	$3.52*10^{-4}$	0.468	258	$5.91*10^{15}$
FYb20	0.68	$3.69*10^{-4}$	0.499	247	$7.85*10^{14}$
FYb25	0.73	$6.71*10^{-5}$	0.556	296	$9.52*10^{17}$

Table 6 E, k_0 and data obtained from different criteria in borate glasses

glass	HR	$k(T_g)(min^{-1})$	$k(T_p)(min^{-1})$	E(kJ/mol)	$k_0(min^{-1})$
LB1	0.62	$4.07*10^{-4}$	0.432	383	$2.98*10^{18}$
LB2	0.93	$7.88*10^{-4}$	0.336	292	$3.12*10^{13}$
LB3	0.78	$7.75*10^{-4}$	0.339	307	$3.40*10^{14}$
LB4	o.77	$2.24*10^{-4}$	0.386	347	$4.68*10^{16}$
LN1	1.21	$1.79*10^{-4}$	0.281	284	$3.31*10^{12}$
LN2	1.27	$6.62*10^{-5}$	0.360	350	$4.05*10^{15}$
LN3	1.11	$3.57*10^{-4}$	0.255	278	$1.47*10^{12}$
LN4	1.15	$7.11*10^{-4}$	0.211	230	$7.63*10^{9}$

Table 7 E, k_0 and data obtained from different criteria in fluorophosphate glasses

glass	HR	$k(T_g)(min^{-1})$	$K(T_p)(min^{-1})$	E(kJ/mol)	$k_0(min^{-1})$
FP1	-	$4.48*10^{-4}$	0.248	146	$9.62*10^{7}$
FP2	-	$1.27*10^{-5}$	0.367	211	$3.02*10^{11}$
FP3	-	$2.73*10^{-6}$	0.452	276	$5.81*10^{15}$
FP4	-	$3.72*10^{-6}$	0.482	275	$8.22*10^{15}$

4. DISCUSSION

In nonisothermal heating process, the volume fraction of crystallization x can be expressed[6]

$$x=1-\exp[-(\int k(t)dt)^n] \quad (2)$$

where n is a constant related to crystallization mechanics. From this equation, in crystallization temperature range the larger the k value, the more likely the glass tends to crystallize. While the smaller the k value, the more stable the glass is. Combining two kinetic parameters E and k_0, k value is more reasonable than other criteria in assessing the stability of glasses. Because $k(T_p)$ is typical k value in glass, for simiplicity, $k(T_p)$ is chosen as our new criterion. The larger the $k(T_p)$ value, the more likely the glass crystallizes. HR criterion give only rough evalution of glass stability. In some glasses the T_m temperatures can not be measured or T_m can not be measured exactly. HR criterion can not be obtained, and it is meanless. Iqbal criterion contains critical cooling rate, which can not be conveniently obtained from simple DTA curves. Surinach proposed $k(T_g)$ criterion which infered from metallic glasses is not reasonable because this criterion neglected the temperature interval of E and k_0. Both E and k_0 values are achieved in crystallization temperature range. Obviously, $k(T_g)$ criterion is not appropriate in comparing crystallization behavior of glasses in the crystallization temperature ranges. The results obtained from Hurby, Surinach and $k(T_p)$ criteria are shown in Table 5-7. It is observed that HR and $k(T_g)$ values vary randomly with composition in each glass systems. While $k(T_p)$ value is in minimum in FP1, FY15,FYb10, LB2 and LN4 glasses. From above discussion, it is confirmed that crystallization properties of glasses can be quantitatively described by $k(T_p)$ criterion. The $k(T_p)$ criterion provides a convenient method for new stable glass system determination. From $k(T_p)$ criterion, FP1, FY15, FYb10, LB2 and LN4 glasses are most stable glasses in corresponding glass systems. The results deduced from $k(T_p)$ is in good consistant with experiment results.

5. CONCLUSIONS

After proper composition adjustment, La_2O_3-Nb_2O_5-B_2O_3, La_2O_3-BaO-B_2O_3, fluoride and fluorophosphate glasses have good glass forming ability. The glass forming ability and stability of glass can be quantitatively described by rate constant of crystallization at T_p, i.e. $k(T_p)$. It is proved that $k(T_p)$ is a good criterion to classify crystallization tendency of glasses, and it provides a simple method to search for new stable optical glasses.

REFERENCES

1. W.Vogel "Glass Chemistry" Light Industry Publication (1988) in Chinese pp3-17

2. Hurby,A Czech.J.Phys. B22 (1972) 1187

3. T.Iqbal, M.R.Shahiari, G.H.Sigel,Jr. Proceeding of 6th International Symposium on Halide glasses, West Germany, (1989) pp225-261

4. S.Surinach, M.D.Baro et al J.Mat.Sci. 19 (1984) 113-119

5. S.Surinach, M.D.Baro et al J.Non-Cryst. Solids 111(1989) 113-119

6. N.P.Bansal, R.H.Doremus and C.T.Moynihan J.Am.Ceram.Soc. 66(4) (1983) 233-238

Frontiers of Materials Research/
Electronic and Optical Materials
M. Kong, L. Huang (editors)

DIRECT ELECTRON MICROSCOPIC OBSERVATION OF THE CRYSTALLITES IN A FLUOROALUMINATE GLASS

Ruihua LI, Haobing WANG and Peizhen DENG

Shanghai Institute of Optics and Fine Mechanics, Academy of sciences, P. O. Box 800-216, Shanghai 201800, P. R. China.

With the help of Leitz optical microscopy, scanning electron microscopy, energy dispersive spectroscopy and wavelength dispersive spectroscopy, the original crystallites in a MeF_2-YF_3-AlF_3 (Me=Mg+Ca+Sr+Ba) fluoroaluminate glass have been studied. Crystallites forming in the process known as homogeneous nucleation and heterogeneous nucleation respectively were found in the glass. Two crystalline phases, $CaMg_2Al_2F_{12}$ and CaY_4F_{14}, were preliminerily determined. The relative merits of the analysis techniques used will be discussed.

1. INTRODUCTION

Extrinsic scattering centers such as microcrystallies, inclusions and bubbles in fluoride glasses have been recognized as one of the major causes of the fiber losses that exceed the theoretical minimum loss value by several orders magnitude. It is of great importance therefore to have a complete understanding of their crystallization behaviors in order to achieve full potential of these glasses as optical waveguides and for other applications. Developmental researches on fluoride glasses have been mostly focussed on fluorozirconate glasses especially ZBLAN system[1,2]. The ZrF_4-based glasses, however, seems to be poor in chemical durability and mechanical strength. It has been reported that the fluoroaluminate glasses appear to be less susceptible to these problems. But the fluoroaluminate glasses have shown great tendency to devitrify during preparation. Little work has been reported on identifying the original crystallites in the fluoroaluminate glasses which have not been heat-treated.

The aim of this paper is to report our study results of the crystallies in a MeF_2-YF_3-AlF_3 (Me=Mg+Ca+Sr+Ba) fluoroaluminate glass by using Leitz optical microscope (OM), scanning electron microscope (SEM), energy dispersive spectrometer (EDS) and wavelength dispersive spectrometer (WDS). The relative merits of the analysis techniques used are discussed.

2. EXPERIMENTAL

2.1. Glass preparation

The composition of the glass examined is $35AlF_3$-$20CaF_2$-$15YF_3$-$10MgF_2$-$10SrF_2$-$10BaF_2$ (mol%). The glass was prepared by HU Hefang et al. from hydrous chemical pure materials. Appropriate mixture of fluorides with ammonium bifluoride in a Pt crucible was heated at 400°C for 2 hr to fluorinate any oxide impurities. The temperature then rose up to 1050°C and held for 1 hr in dry N_2 atomsphere. The melt cooled to 750°C and cast into a non-rust steel mould preheated and annealled by cooling slowly to room temperature.

2.2. OM observation

Initial characterization of the crystallite was performed with a transmitted light polarizing OM equipped with ultra-long working distance objectives. This permits observing the appearance and distribution of all crystallites throughout the sample and locating the crystalline particles of interest.

2.3. SEM image

Samples for SEM examination was prepared by first carefully polishing the glass until the individual crystallite was exposed at the polished surface and then making several marks around the crystallite with a Leitz micro-hardness tester. The samples were carbon coated and the crystallites were imaged in SEM. Backscattered electron images can give us preliminery chemical information of the crystallite in fluoride glasses, since the brightness of backscattered image is concerned with the atomic number of examined material.

2.4. EDS and WDS analysis

The chemical information about the crystallite was obtained by EDS and WDS. EDS can give us elemental information conveniently and rapidly. Certain elements, such as F and O, however, cannot be detected in the fluoride glass system. The L lines of Sr and Y overlap. These two problems can be alleviated by using a WDS though WDS analysis is more time-consuming.

3. RESULTS

Fluoroaluminate glasses are characterized by more ionic bonding than conventional glasses. As a result, such glasses are relatively unstable and tend to devitrify. Figure 1 exhibits the transmitted light micrographs of three kinds of crystalline particles which are commonly found in this glass under OM.

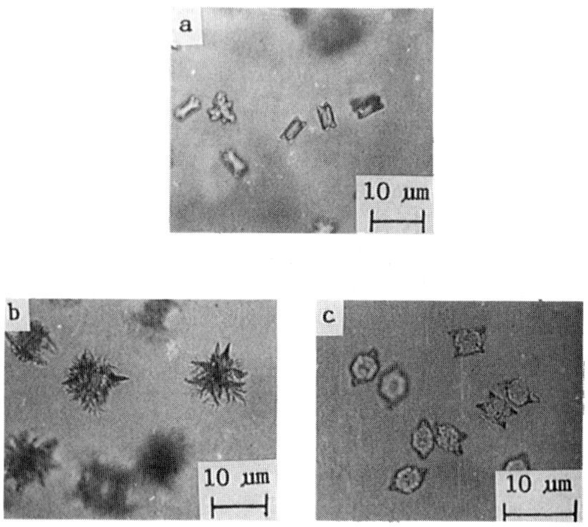

FIGURE 1
Three kinds of crystalline particles commonly found in the fluoroaluminate glass.

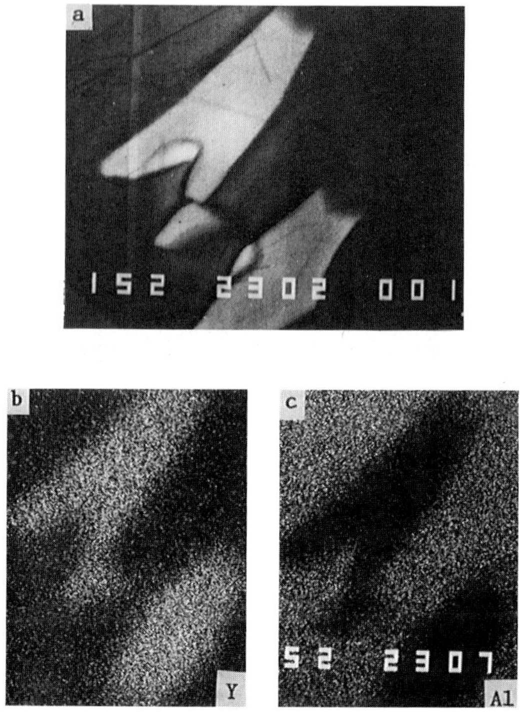

FIGURE 2
a. Backscattered electron image of the crystallite in figure 1a.
b, c. Two dimensional distribution of element Y and Al at the polished surface corresponding to this crystallite.

The crystallite in figure 1a (crystal A) at the expolished surface was imaged by SEM (shown in figure 2a). Figure 3 and figure 4 where the EDS spectra corresponding to this crystallite and the glass matrix are shown reveal striking difference of chemical composition between the crystallite and matrix. It is clearly seen that the peaks belong to Mg(k), Al(K) and Ba(L) have a marked drop while the peaks of Sr,Y(L) and Ca(K) become much higher in the crystallite as compared to that of the glassy surrounding. This is consistent with the results of plane

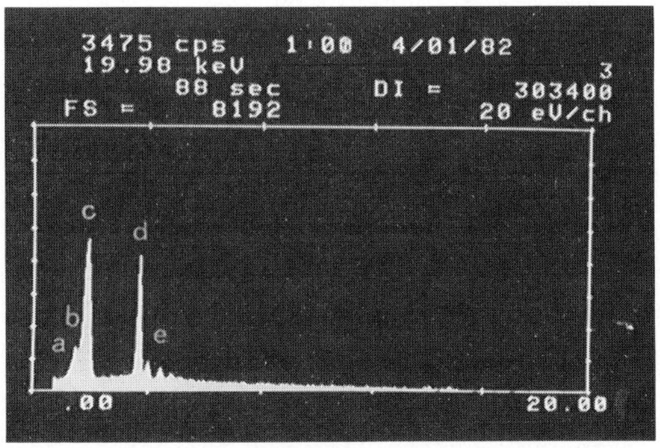

FIGURE 3
EDS spectrum of the crystallite in figure 1a.
a, b, c, d,and e stand for Mg(K), Al(K), Sr,Y(L), Ca(K) and Ba(L) respectively.

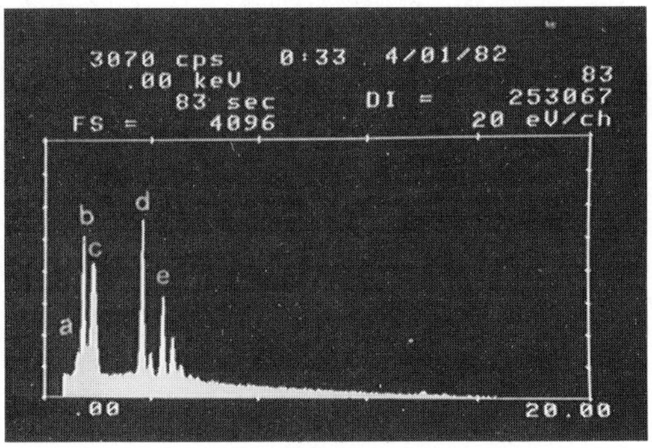

FIGURE 4
EDS spectrum of the glass matrix.

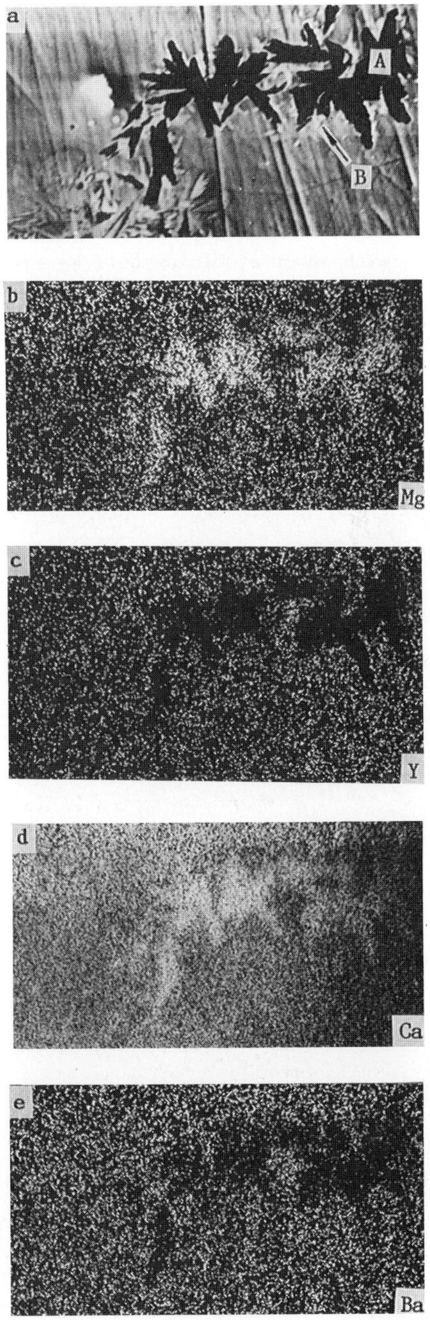

FIGURE 5
a. Backscattered electron image of the crystallite in figure 1b.
b, c, d and e. Two dimensional distribution of element Mg, Y, Ca and Ba at the polished surface corresponding to this particle.

scan of every element contained in the glass. For example, figure 2b and 2c is the distribution of Y and Al. WDS analyses make clearly that the ascendent of Sr,Y(L) peak is due to the enrichment of element Y and the anion in this crystallite is F^-. Based on the above experiment results, crystal A may be preliminerily determined to be $CaF_2 \cdot 4YF_3$.

The branched crystalline particles in figure 1b were prevalent in the as-cast glass. Figure 5a is its backscattered electron image. From this picture we can see clearly that the particle consists of two kinds of crystals. The backscattered image of crystal B, the inside part of the particle, appears much darker than that of crystal C, the outside part of the particle. This means the mean atomic number of crystal B is substantially lower than that of the bulk glass and crystal B is mainly composed of the elements with lower atomic number. We can find that the nuclei of crystal B form at random throughout the volume of liquid, i.e, homogeneous nucleation. It is also obvious that the nuclei of crystal C form directly on crystal B, in a process known as heterogeneous nucleation.

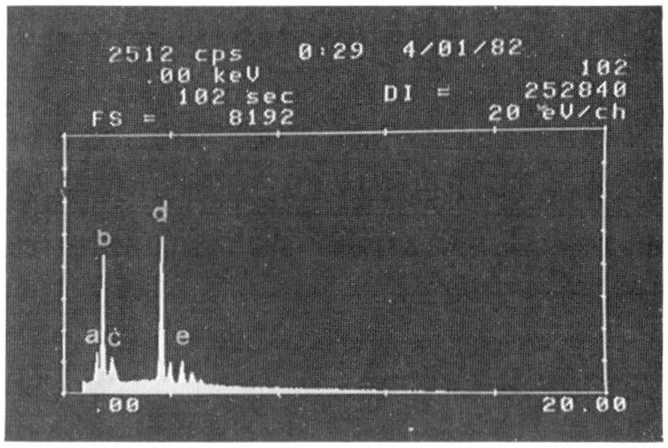

FIGURE 6
EDS spectrum of a point on crystal B.

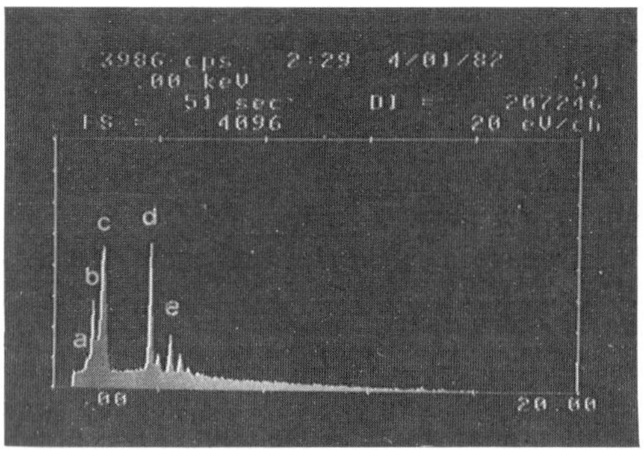

FIGURE 7
EDS spectrum of a point on crystal C.

Crystal B may be the nucleation site of crystal C. The EDS spectra corresponding to the crystallites are shown in figure 6 and figure 7, indicating that the crystals and matrix glass have different chemical composition. Crystal B is enriched in Mg, Al and Ca. This behavior is prominently displayed in figure 5, where the results of two dimensional distribution of element Mg, Y, Ca and Ba at the polished surface corresponding to this particle are shown.The anion in this crystallite was characterized to be F^- by EDS analysis, implying crystal B may be $CaMg_2Al_2F_{12}$. It is difficult to identify crystal C. The particle of crystal C was smaller (<1μm) than the electron beam spot size, so EDS spectrum of crystallite necessarily included some contributions from the matrix. However, the occurrence of strong Sr,Y(L) and Sr,Y(L) peaks in the crystalline regions, along with Al(K), Mg(K) and Ba(L) peaks which are greatly reduced compared to the glass matrix, indicated that crystal C was probable the same kind of crystal as crystal A, i.e CaY_4F_{14}.

The same techniques are used to examine the crystallite in figure 1c. Figure 8 is the backscattered electron image of it. The results of EDS and WDS analysis indicate that it is the same crystal as crystal B, i.e $CaMg_2Al_2F_{12}$.

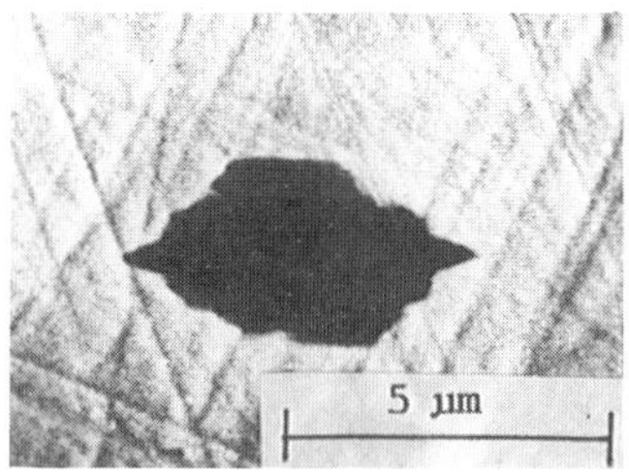

FIGURE 8
Backscattered electron image of the crystallite in figure 1c.

4. DISCUSSION

Two kinds of crystals, $CaMg_2Al_2F_{12}$ and CaY_4F_{14}, are preliminerily determined. Crystallization in the fluoride glasses is a complex process. Since most of the stable compositions contain at least four components, many crystalline phases can form. There exists a tendency for the formation of metastable speceis and the composition of the crystalltes may be stoichiometric or non-stoichiometric. Moreover, as a result of their relative ease of devitrification, their crystallization products and kinetics maybe quite sensitive to minor impurity inclusions, which are related to glass preparation precedures. Besides, the small size and relatively small amount of these particles on a volume basis, it is not easy to reveal the exact phase of an individual crystallite present in the glass.

The crystallization behavior of some fluoride glasses as a function of time and heat-treatment has been studied by means of DSC, X-ray diffraction [3,4,5]. But DSC cannot distinguish between the many crystals which have been observed in fluoride glasses. Concurrent growth of several types of crystals will lead to composite kinetic data which can be misleading. X-ray diffraction is often used to obtain crystallization information about the heat-treated glass. Heat-treating the glass, however, may not produce the same type of the crystals as those which occur upon cooling only. During heat-treatments, the glass goes through the high nucleation rate temperatures twice. This can provide a possibility for introducing a large number of different nuclei and thus different crystals. The techniques we used allow direct analysis of the individual crystalline particles in the glass as compared with other techniques such as DSC, X-ray diffraction.

Fluoroaluminate glasses exhibit characteristic temperatures higher than those of fluorozirconate glasses by 100 to 200°C. Their transmission is reduced at the longer wavelength[6]. Intrinsic optical losses lie between those of silica and fluorozirconate glasses [7] making them potential materials for low loss optical fibers. The present problems encountered originate from devitrification. The results we obtained are only elemental composition information about the crystallites. In the light of these findings, we are now conducting more systematic study to clarify mechanism of crystallization in fluoroaluminate glasses based on Raman microprobe, scanning transmission electron microscope, electron energy loss spectrum and selected area electron diffraction.

ACKNOWLEDGEMENTS

We would like to thank prof. HU hefang , YI Guanghong et al. of Shanghai Institute of Optics and Fine Mechanics, Academy of Sciences for providing the examined glass.The research work leading to this paper was subsidized by National Natural Science Foundation of China.

REFERENCES

!. M. poulain et al. Mat. Res. Bull,12 (1977) 151-156.

2. D.C.Tran et al. Electron. Lett. 18 (1982) 1046-1048.

3. J.M.Parker,A.G.Clare and A.B.Sedden, Material Science Forum, 5 (1985) 257-262.

4. G.F.Neilson, G.L.Smith and M.C.Weibeig, Mat. Res. Bull. 19 (1984) 279-282.

5. N.P.Bansal,R.H.Doremus,A.J.Bruce and C.T. Moynihan, Mat. Res. Bull. 19 (1984) 577-590.

6. M.R.Shahriari, T.Iqbal, G.H.Sigel and G. Merberg, extended abstracts of 5th. Int.Symp. on Halide Glasses, (1988) 540-547.

7. S. Shibata, M. Horignshi, K. Jinguji, S. Mitachi, T. Kanamori and T. Manabe, Electron. Lett. 17 (1981) 775.

Frontiers of Materials Research/
Electronic and Optical Materials
M. Kong, L. Huang (editors)

LASER-INDUCED PHOTOLUMINESCENCE DECAY EFFECT OF CdS(Se) MICRO-CRYSTALLITES IN GLASS

Haiyan CHEN, Guangzhao WU and Fuxi GAN

Shanghai Institute of Optics and Fine Mechanics,Academy of Sciences, P.O.Box 800-216, Shanghai, P.R.China 201800

A series of glass samples with CdS(Se) micro-crystallites were prepared and the spectro-luminescence properties have been investigated. The phenomenon of luminescence decay of CdS(Se) micro-crystallites in glasses excited by high-power-density laser beam has been reported in detail.

1. INTRODUCTION

CdS(Se) doped glasses, with its high non-linear effect, its good transparence and its similar way to the spectrum of CdS(Se) single crystal, have been paid great attention [1-4]. The emission and absorption spectra of CdS(Se) microcrystallites were studied in detail. In case of the excitation by CW Argon ion laser, the authors observed a unusual luminescence decay(fatigue) phenomenon which has not been paid attention before. In order to study this phenomenon, a series of glass samples were prepared and the transient emission spectra were studied with the OMA-2 system and Argon ion laser.

2. EXPERIMENT

A series of glass samples with different CdS(Se) doping contents were prepared. The composition of host glass is as follows:
$64SiO_2 \cdot 3B_2O_3 \cdot 13ZnO \cdot 11K_2 \cdot 9NaO$
The doping concentration of samples are shown in Tab.1. The treatment of A1 and B1 are diffrent.

Table 1
CdS(Se) doping concentration of glasses

sample	A1,B1	A2	A3
doping CdS wt%	1.4	1.05	0.15
Doping CdSe wt%	0.002	0.07	0.15

An Argon ion laser and OMA-2 system with wavelength range from 3000Å to 6000Å were used to record luminescence spectra. The exciting laser beam was focused to obtain high power density. In order to avoid the absorption of the sample, a very small incident angle of exciting beam was applied in the experiment.

3. RESULT AND DISCUSSION

A new luminescence decay phenomenon was observed, when the sample was excited by a focused laser beam. A similar phenomenon has been reported in amorphous Si film [5] and as so called "fatigue". After the sample excited by CW Argon ion laser , a very strong "luminescence trace" appeared at the begning of the excitation. But in following seconds, the intensity of the luminescence decreased rapidly, and kept at very low intensity, although the intensity of the exciting beam was not changed. In order to record the process of luminescence decay of the samples investigated here, the wavelength of spectrometer was set at peak of emission and an Argon ion laser with 4880Å was used as exciting source. The process of this luminescence decay was shown in Fig.1. It can be seen that the stronger the exciting power is, the more obvious the luminescence change will be. Therefore, luminescence decay behavior can not be observed under low power excitation, for example, under excitation of xenon-lamp.

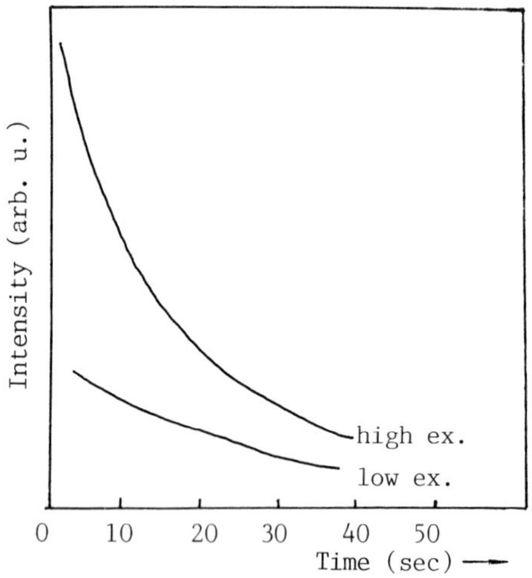

Figure 1
Luminescence decay of CdS(Se) doped glass

A broad and low power beam was used to excite the sample after laser irradiation, it was observed that a luminescence "dark trace" appeared in sample. That is, the position irradiated by laser beam has low photoemission than the other position in the sample. Thus, "dark trace" of luminescence was formed.

In order to study the mechanism of this luminescence decay(fatigue) process, the transient spectra of four glass sample were recorded by OMA-2. The data of luminescence intensity were processed by computer for studying the dynamic process. The results are express as the ratios of two luminescence intensity recorded at different time interval. In Fig.2, it is shown that the luminescence intensity of sample A1 increased between 210ms and 420ms, and the increment in the side of short wavelength was greater than that in long wavelength side, a maximum can be seen in the intensity ratio curve(Fig.2b). Between the interval of 420ms and 630ms, the luminescence intensity continue to increase at short wavelength side, while that at peak wavelength decreased. As the time went on, the ascendand peak position in intensity ratio curve gradually

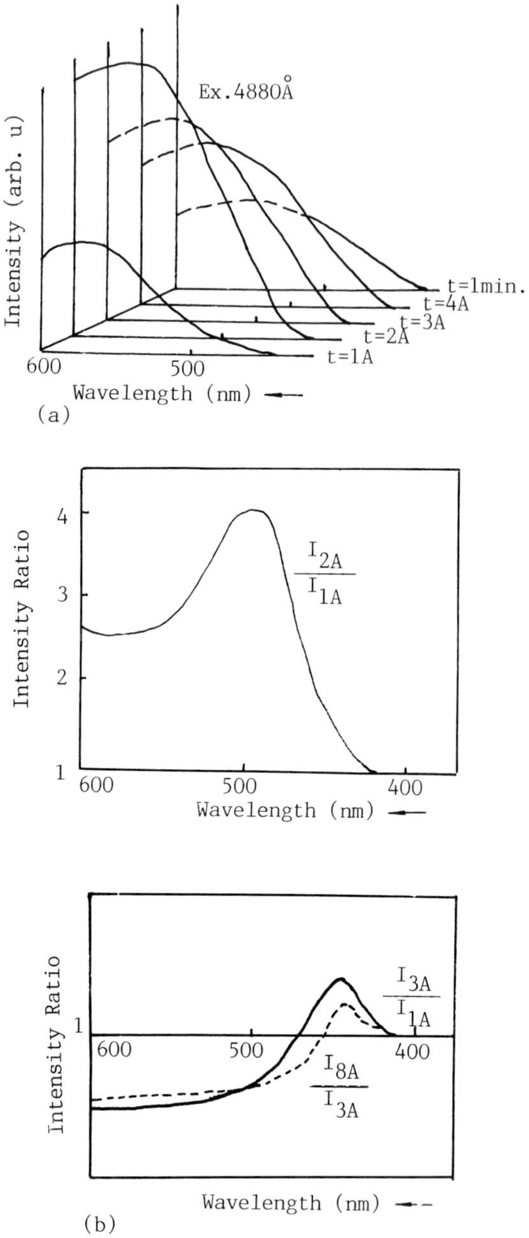

Figure 2
(a) Transient luminescence spectra of sample A1 at the first minute, A=210ms
(b) Intensity ratio curve of A1

move to short wavelength and the intensity gradually decreased. Till after 1.5 seconds, the

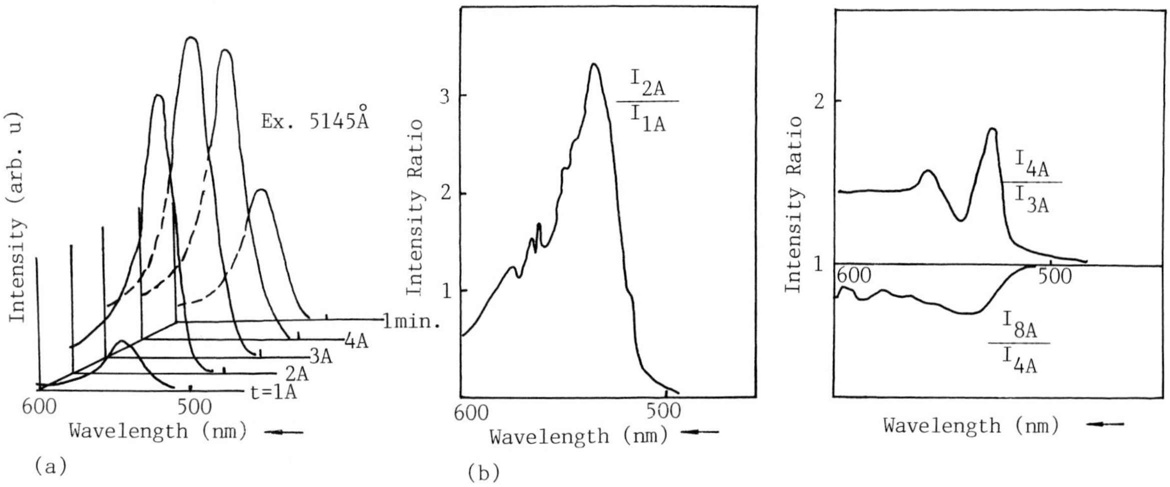

Figure 3
(a) Transient luminescence spectra of sample A3 at the first minute, A=210ms
(b) Intensity ratio curve of A3

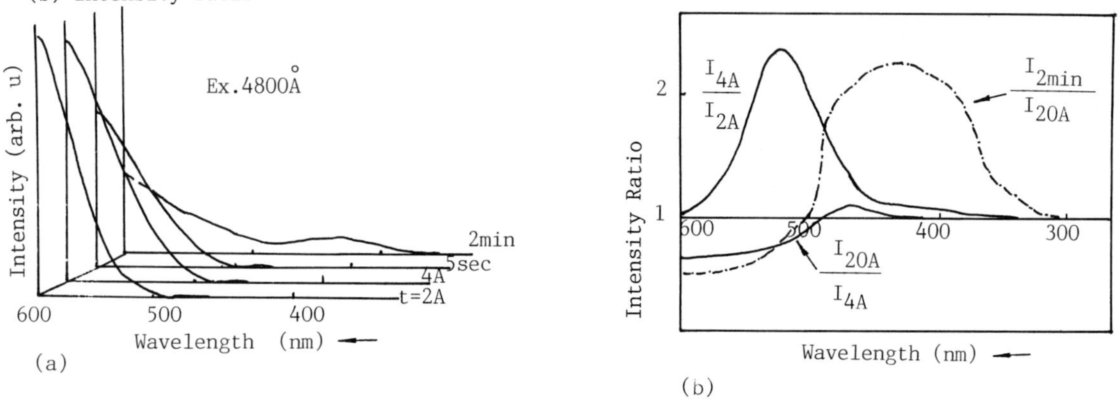

Figure 4
(a) Transient luminescence spectra of sample B1 at first two minutes, A=210ms
(b) Intensity ratio curve of B1

intensity of whole wavelength range decreased at the same rate. The same process occurred to sample A2. Sample A3 had an almost the same process as that of A1, as shown in Fig.3, but the difference between A1 and A3 is that there is a minimum in intensity ratio curve. It is considered that the luminescence ascension is limited by some absorption mechanism, because this minimum corresponds to the excitation peak in exciting spectra [4], which is an emission from acceptor to valence band. Sample B1 had a similar luminescence behavior to A1 before the 4th second. After 4-second laser irradiation, a new broad emission band appeared at 4775Å and then was getting more intensified and broader, although the main emission peak was decreasing, as shown in Fig.4.

From the experimental results mentioned above, it can be concluded that the luminescence of CdS(Se) doped glasses is a rather complicated process. The luminescence decay process induced by irradiation of high power density laser can be considered mainly as follows:

1. After excitation of intense laser beam, electrons transferred from valence band to conduction band and captured by donor level. As we know, the energy of donor(D) and acceptor(A) combination luminescence can be expressed as

$$h\nu=E_g-(E_A+E_D)+\frac{e^2}{\epsilon R} \quad (1)$$

where E_g is band gap, E_A and E_D are the A and D energy level respectively, the third term is Columb energy, R is the distance between D and A ions. There is a broad band in D-A pairs luminescence spectra because of the different R. The intensity of D-A pairs combination luminescence is of course related to the numbers of electrons on D level. The main emission peak of the glasses are the D-A pairs luminescence [4]. Thus, we consider that at initial time of excitation, because of the accumulation of a great number of electrons on D level, luminescence intensity increased. In this period, the intensity is expressed as

$$I\propto 1-e^{-(k+a)t} \quad (2)$$

where a is probability of transition, and k is the probability of excitation. It is assumed here that k is a constant to all of the D-A pairs. Because transition probability a is proportional to e^{-R/R_o}, the combination of the D-A pairs with short distance is faster than that with long distance. Thus, it turns out from eq.2, that the luminescence intensity will ascend fast to short distance D-A pairs. This effect can be seen in luminescence spectra, namely, the maximum of intensity ratio first appeared at short wavelength and then moved to rather short wavelength gradually in the process of luminescence decay. For sample A3, due to the absorption, the electrons transfer from acceptor to conduction band, which cause a minimum in intensity ratio curve.

2. The bound between D and A is mainly due to the Columb's force. When the sample was excited by high power density laser, the various interaction in microcrystallites, such as lattice vibration, were strengthened. Meanwhile, D and A gained a large amount of kinetic energies. In semiconductor the diffusion of impurity atoms is much fast, the rate of the diffusion is about 10^{-14}-10^{-15} m/sec. Therefore, if D or A in microcrystallites get enough energy, they will overcome the potential barrier and diffuse from one place to another. Departing away, A and D can get rid of the bound and cause the break of D-A pairs. Once the D-A pairs broke, the intensity of luminescence decreased obviously. D-A pairs with long distance will break easily. This can be proved from transient luminescence spectra, in which luminescence was ascending at high energy side, while luminescence was descending at peak position. In microcrystallites, D and A recombinations are always existed through the luminescence decay process, however, the probability of recombination is so small that break of D-A pairs is the main process at the begining of irradiation. As the time went on, the numbers of D-A pairs decreased and the numbers of independent D and A increased, so the process of break and recombination tended to be balanced. After the balance was gained, the intensity no longer changed. In this process, independent D or A at microcrystallite surface could also overcome the potential barrier and enter into the surrounding glass medium. Therefore, the number of D-A pairs in irradiated range was much fewer than that of unirradiated range. This caused the luminescence "dark trace".

3. As to the new broad emission band of B1, it is shown that the transition energy of new luminescence band is larger than that of exciting laser. The appearance of this band is the result of decrease of D-A pairs number and accumu-

lation of electrons on conduction band, which may cause the transition on high energy. This kind of transition appears in several way. The first, the electron energy below the band edge may be caused by transition from conduction band to acceptor or from donor to valence band. The second, the electron energy above the band edge may be produced by double electron process or quantum energy level in microcrystallites.

4. CONCLUSION

The phenomenon of unusual luminescence decay process(fatigue) was observed in CdS(Se) doped glasses excited by high power density laser beam. This phenomenon is caused by breakdown of D-A pairs under high density excitation.

REFERENCES

1. J. Warnock and D.D. Awschalom, Appl. Phys lett. 48(16) (1986) 425.

2. N.F. Borrelli et al., J. Appl. Phys. 61(12) (1987) 5399.

3. A.I. Ekimov, Proc. XV Int. Congs. on Glass, leningard 1989 P.190.

4. Zhang Jinshan, Gan Fuxi and Qi Changhong Acta Optica Sinica 8(5) (1988) 385.

5. Wang Fucao et al., Chin. J. Infrared Res. 7A (1988) 275.

Frontiers of Materials Research/
Electronic and Optical Materials
M. Kong, L. Huang (editors)

LIQUID PHASE EPITAXIAL GROWTH OF InAsPSb FOR LASER EMISSION IN THE 2-4um WAVELENGTH REGION

Jin Changchun, Wang Yongzhen, Lu Guijin, Wang Jifeng and Jiang Jinxiu

Changchun Institute of Physics, Academia Sinica, Changchun 130021, CHINA

Wang Hailong

Shanghai Institute of Optics and Fine Mechanics, Academia Sinica, Shanghai 201800, CHINA

High quality Zinc doped quaternary InAsPSb epitaxial layers have been directly grown on InAs substrate without buffer layer using both the gradual temperature lowering and step-cooling liquid phase epitaxial techniques. The quaternary layer are lattice matched to InAs 1.6 X 10^{-3} . X-ray diffraction spectra half width of the quaternary epilayer is about 200sec. Laser emission with wavlength 3.09μm was observed from P-N junction made of P-$InAs_{0.82}P_{0.12}Sb_{0.06}$ and N-InAs substrate at 12K.

1. INTRODUCTION

Semiconductor lasers operating in the 2-4μm range are of considerable interest for future optical communications using low loss fluoride glass fibres. InAsPSb quaternary alloy is one of important III-V compounds used for making semiconductor lasers in the 2-4μm range. Laser emission have been observed in this wavelength region using quaternary InAsPSb double heterostructure(1-2). There has been no report so far concerning single heterostructure made from InAsPSb epilayer and InAs substrate. Generally InAsPSb was grown on InAs buffer epitaxial layer in order to improve the quality of InAsPSb epitaxial layer. In this paper, we describe the characteristics of the InAsPSb was directly grown by liquid phase epitaxy and of the P-N junction made of the InAsPSb and InAs substrate.

2. EXPRIMENTAL PROCEDURE

The quaternary epitaxial layers have been directly grown on InAs substrate without buffer layer by two kinds of liquid phase epitaxial growth methods, the gradual temperature lowering and step-cooling methods.

Undoped InP as a P source, undoped InAs as a As source, and Sb were dissolved in In at higher temperature of 650°C. A S-doped n-InAs substrate with a (100) orientation was used. Prior to growth, the undersaturated quaternary melts was held on an undoped InAs source wafer for a hour to achieve saturation at a given temperature. The epitaxial layers were grown at a temperature of 530°C. The quaternary epilayers were grown by the gradual temperature lowering LPE method at 0.3°C/min, and using step-cooling method under 5-10°C the saturation level. The solid composition of epitaxial layer was determined by EPMA analysis. The lattice-mismatch ratio $\Delta a/a(=(a_{InAs}-a_{Epi})/a_{InAs})$ was estimated from (004) X-ray diffraction pattern of CuKα1 line.

3. RESULTS AND DISCUSSIONS

Both methods yielded very smooth epitaxial layer surfaces. We found that the surface morphology is dependent on the growth temperature and saturation level. The epitaxial layers were lattice match to InAs about 1.6 X 10^{-3}. Fig.1 shows a SEM photograph of a cross-section of single heterostructure made from quaternary $InAs_{0.82}P_{0.12}Sb_{0.06}$ and InAs substrate.

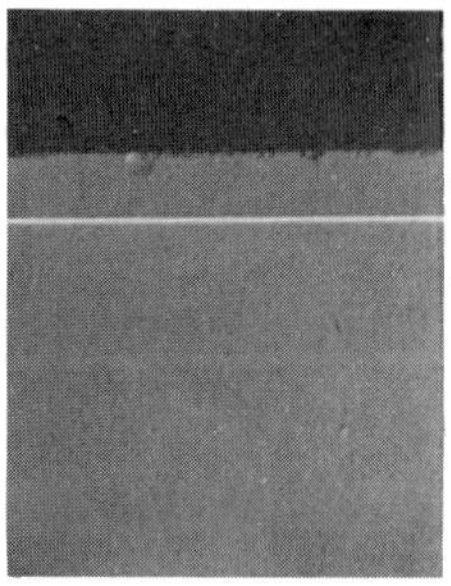

FIGURE 1
A SEM Photograph of a cross-section of the InAsPSb/InAs SH wafer.

The interface between the InAsPSb epitaxial layer and InAs substrate is very flat, and it is suggested that there was no interface problem. The solid composition of epitaxial layer determined by the EPMA analysis can be controled within 0.04-0.14. X-ray diffraction spectra half width of the quaternary epilayer is about 200 sec. It is smaller than that reported by Kobayashi et al[1]. Fig.2 shows a typical result for the compostional variation in the epitaxial layer along the growth direction. The phosphorous content has maximum value at the interface, and decreases gradually

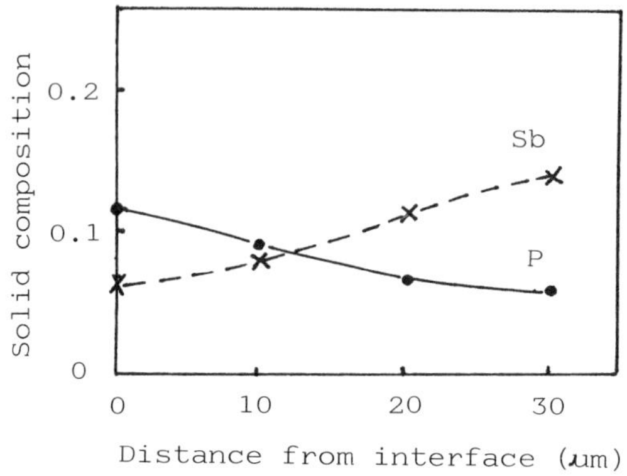

FIGURE 2
Typical result of compositional variation in epitaxial layer along the growth direction.

along the growth direction[3]. On the contrary, the antimony content increases gradually along the growth directin for the epitaxial layer grown by gradually temperature lowering method. But the phosphorous and antimony content have almost a constant value along the growth direction for epitaxial layer grown by the other

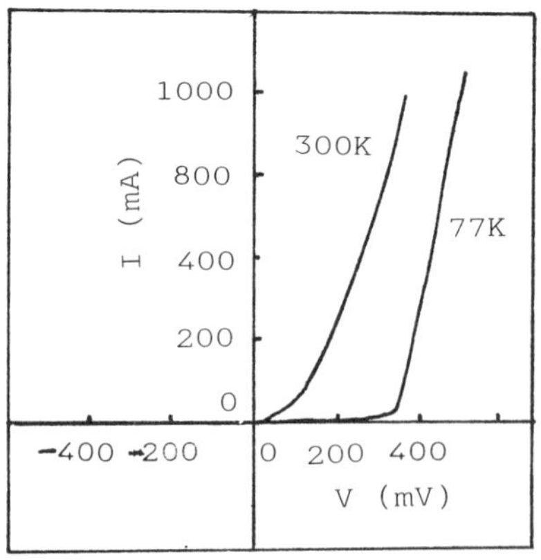

FIGURE 3
The characteristic of P-N junction made of P-InAsPSb and N-InAs substrate.

method. Fig.3 shows the characteristic of a P-N junction made of P-$InAs_{0.82}P_{0.12}Sb_{0.06}$ and N-InAs substrate at 77K and 300K respectively. It is shown that the slope of I-V relationship changes with temperature, in other words the forward current component is due to different mechanism at different temperature. A 200µm X 350µm diode chips was formed. The diodes were driven by current pulses 10us duration and 500HZ-5KHZ pulses/s repetition. The laser emission with wavelength 3.09µm was observed from single heterojunction made of P-$InAs_{0.82}P_{0.12}Sb_{0.06}$ and N-InAs substrate at 12K for the first time.

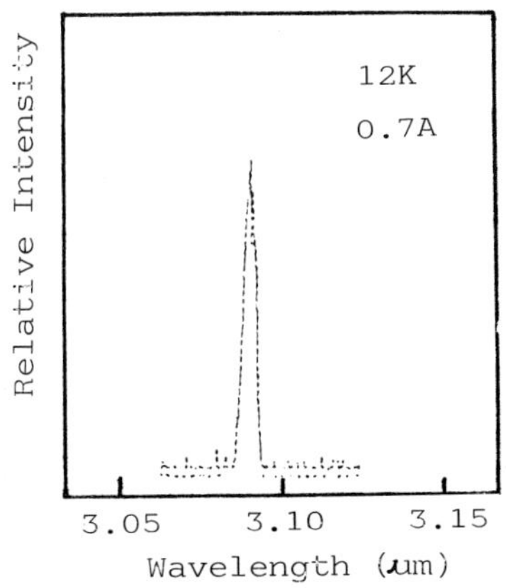

Figure 4
Typical lasing spectrum of the SH laser at 12K.

Typical lasing spectrum of the SH laser at 12K is shown in Fig.4. The threshold current of this laser is 0.7A and the half width is about 48 angstrons.

REFERENCES

1. N.Kobayashi and Y.Horikoshi, Jpn.J. Appl.Phys., 19(1980)L641.

2. S.Akiba, Y.Matsushima, T.Iketani, M.Usami, Electronics Letters, 24(1988)1069.

3. N.Kobayashi and Y.Horikoshi, Jpn.J. Appl.Phys., 20(1981)2301.

Frontiers of Materials Research/
Electronic and Optical Materials
M. Kong, L. Huang (editors)
Elsevier Science Publishers B.V., 1991

EMBEDDED GROWTH AND CHARACTERIZATION OF GaAs ON Si BY MBE

Jiben LIANG and Meiying KONG

Institute of Semiconductors, Academia Sinica,
Beijing, 100083, China.

J.De BOEK, R.Merterns and G.Borghs

Interuniversity Micro-Electronics Center, Leuven. Belgium.

A GaAs on Si coplanar technology by molecular beam epitaxy suitable to combine GaAs and Si circuits on chip was reported. The wells for embedded growth were accomplished by wet chemical etching with an isotropic solution of $HF+HNO^3$. Metallization without loss of continuity was performed over the GaAs to Si border with evaporated metal lines, 150nm thick and 1.25μm wide. The results of measurements with SEM, TEM, Raman scattering, X-ray double crystal diffraction and PL demonstrated that the quality of GaAs epilayers grown on Si in the wells is equal to those on large unetched areas.

1. INTRODUCTION

In recent years, there has been growing interest in the heteroepitaxy of GaAs on Si. The monolithic GaAs on Si (MGS) takes the advantage of optoelectronics and high speed GaAs technology and high density Si technology [1]. However, the 4.1% lattice mismatch and large thermal mismatch between GaAs and Si introduce a large density of misfit dislocations and strain in the GaAs epilayer. Rapid progress in the growth of GaAs on Si has enabled the fabrication of devices with performances comparable to those of devices on GaAs substrate. It was reported that GaAs and Si devices such as GaAs MESFET's and Si MOSFETs as well as GaAs/AlGaAs double heterostructure LEDs and Si MOSFETs were fabricated on the same MGS wafers. In spite of the inital success, however, the surface planarity essential for digital IC process was not considered. A serious problem is a step height of several microns created by the growth of GaAs layer on Si substrate which posed an extreme difficulty in subsquent processing. One way to avoid this problem is to selectively recess the Si region to a depth equal to the thickness of GaAs epilayer such that the GaAs surface is level with the Si surface after the GaAs growth. Some problems concerning the embedded growth need to be solved. The etched Si surface in the wells preserved the flatness, damage free and intentional misorientation of several degree from (100) that is employed to suppress the formation of antiphase domains[2]. The growth of GaAs on the sidewall in the wells exists different nucleation behavior due to exposure of different crystallographic orientations and geometrical constraints[1].

In this paper, we present a GaAs on Si heterocoplanar technology and the characterization of the GaAs epilayer in the wells.

2. EXPERIMENTS

Light P-doped 10 ohm-cm 3 inch (100) Si wafers oriented 4° off towards[110] were used. The preparation for growth involves forming a 400nm thick SiO_2 layer and exposing the Si misorientation with different widths(20 ,50, 100, and 150µm) and spacing (5, 10, 15µm). The wells were etched 2-3 µm deep into Si substrate by an isotropic solution of $HF:HNO_3$(1:19), finally cleaning the wafer with the solution of $HCl+H_2O_2+H_2O$ to form a thin and volatile oxide film.

The Si substrate was then loaded into a Riber 2300 system using an indium free mount and heated at low temperature(300°C) to remove the moisture and gases. Consequently heated in high temperature (850°C) to desorb the thin oxides on the Si. The RHEED showed a very sharp (2x1)reconstruction of Si surface. To avoid the formation of islands , the first few hundred angstrums of GaAs was grown at T=350°C and heated to a temperature to turn the GaAs thin layer from amorphous into polycrystalline then monocrystalline that could be monitored by RHEED images. Thermally strained buffer layers[1] were used in the initial procedure then GaAs layers were grown on the buffer layers at a conventional growth temperature of 570-600°C and a growth rate of 1µm/hr .

Nomarski optical microscope, SEM, TEM, double crystal X-ray diffraction, PL, Raman scattering were used to characterize the GaAs epilayer and the heterocoplanar technique.

3. RESULTS AND DISCUSSIONS

Typically the tested sample with

different microareas mentioned above, consist of an undoped GaAs buffer layer with the thickness of 300nm and a top 2μm thick GaAs layer Si doped (7×10^{15} cm^{-3}). The surface morphology of GaAs epilayer on Si show well-oriented wavy surface structure and mirror like both in the microareas and on the large unetched areas observed by Nomarski contrast optical microscope[3]. The major concerns of GaAs epitaxial growth on Si(100) is the reduction of the threading dislocation density and the suppression of antiphase domain formation. FIG.1 shows a TEM cross section of GaAs epilayer by the sidewall. Most of the meander type dislocation encapsulated during growth were restricted to a narrow region parallel to the heterointerface not thread to the top surface. A little of threading dislocation lying in (111) plane can be observed, and there is a low defect density in the GaAs epilayer by the sidewall as well as in the wells. The double crystal X-ray diffraction was used to monitor the quality of GaAs epilayer. The rocking curve with FWHM of 270arcsec from a 2μm thick GaAs epilayer as grown indicates the low dislocation density in the epitaxial films.

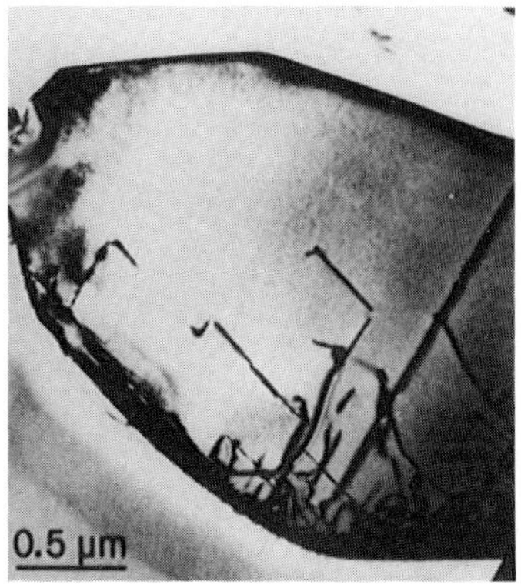

FIG.1. Bright field, high voltage electron micrograph of the dislocation network originating at the GaAs/Si interface.

Raman scattering has been useful to examine the GaAs quality on Si at room temperature as a function of the intensity of TO phonon relative to LO phonon that is 0.06 in our sample The 5145Å line from an Ar ion laser was focused to a spot size of 3um. FIG.2 shows a spetrum of GaAs on Si with different size of microareas. A very small To phonon and a sharp LO phonon Raman peak were observed. The same results have been obtained in different microareas and large unetched areas. Since the TO phonon is forbidden by a selective rule. we have interpreted this peak as due to some surface roughness by the interface or disorder with exposed faces from which TO scattering is allowed. The optical properties of GaAs layers were characterized by PL measurement at low temperature. The PL spectrum of the GaAs/Si sample with linewidth of 8mev at 4K and a downward shift of 10meV in the peak energy of the PL which indicate band gap narrowing, due to the biaxial tensile stress in the GaAs layer caused by the difference in the coefficents of thermal expansion between Si and GaAs [4]. All results indicate that the quality of the epitaxial GaAs layer on Si in the embedded wells are comparable with that on the large unetched area.

The GaAs layer grown on bare Si in the wells is single crystal material during MBE growth, while that on the SiO_2 is polycrystalline, GaAs devices are to be fabricated in the wells, the realization of the coplanar GaAs on Si surface, thus requires precise control the thickness of the epilayer in order to equalize the depth of the etched wells . After the "lift off" polycrystalline GaAs with the SiO_2, a GaAs/Si coplanar surface and a good step coverage of metallization were performed. FIG.3 shows a cross sectional SEM photo demonstrating how the GaAs epilayer level with Si surfaces [4]. Although there

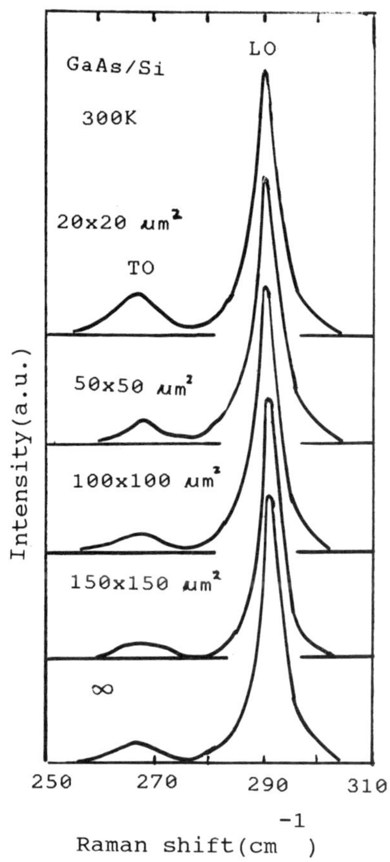

FIG.2. Raman scattering of spectrum of GaAs on Si with different areas. All regions are on the same subtrate

exists a GaAs surface step of 0.6µm on GaAs to Si border, it is a gentle slope and acceptable for MGS interconnection. The shape and height of hillock depend on the geometry of the sidewall. The nonplanar heteroepitaxial GaAs on Si MBE growth mechanism that is responsible for this hillock formation will be discussed in detail elsewhere. Successful metallization was performed on the GaAs on Si coplanar surface without an intermediate planarizing layer[2,4]. Metal lines 150nm thick with 1.25µm wide cross the perimeter barrier without loss of continuing. This interconnection indicates the feasability of the monolithic integration of MBE grown GaAs on Si.

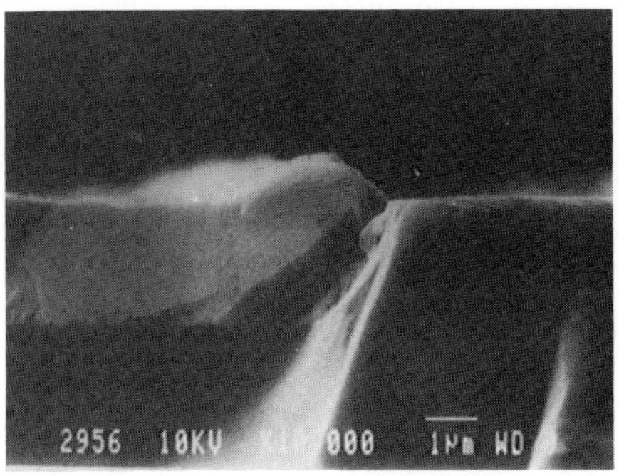

FIG.3. A SEM photo showing the cross section of GaAs on Si coplanar

4. CONCLUSION

Embedded growth of GaAs on Si in the wells is a viable technique for the preparation of substrates for MGS. A controlled and isotropic solution of HF and HNO_3 were used to recess the wells several microns deep into Si substrate. The etched surface meets the requirements of flatness and retains the intentional misorientation of several degree from(100). The results of measurements indicated the same high quality for the GaAs layer grown on Si both in the wells and on the large unetched areas. The GaAs on Si coplanar technology which is suitable for high resolution metallization is realized.

ACKNOWLEDGEMENTS

The authors would like to thank the colleagues of Interuniversity Microelectronics Center(Belgium) and Institute of Semiconductors(China) who help us in this work.

REFERENCES

1. M.R.S Symp. Proc. Vol 67 (1986), Vol 91 (1987) and Vol 116 (1988).
2. J.B.Lang, J.De Boeck, R.Merterns and G.Borghs, Journal de Physique. Tome 49, c4-701, 1988.

3. J.De.Boeck, J.B.Lang, K.Deneffe, J.Vanhellemont, D.J.Arent, C.Van Hoof, R.Merterns, and G.Borghs. Appl. Phys. Lett. 53(12) 1071, 1988

4. J.B.Lang, J.De Boeck, K.Deneffe, D.J.Arent, C.Van Hoof, J.Vanhellemomnt and G.Borghs. J.Vac.Sci.Technol.B7(1) 116, 1989

Frontiers of Materials Research/
Electronic and Optical Materials
M. Kong, L. Huang (editors)

IMPURITIES AND DEFECTS IN AS-GROWN InP STUDIED BY POSITRON ANNIHILATION AND OTHER MEASUREMENTS

YU-JIE HAN

HEBEI SEMICONDUCTOR RESEARCH INSTITUTE, P.O.BOX:179,SHIJIAZHUANG,HEBEI,050051,P.R.CHINA, Fax:011-86-311-745695

This paper reports experimental results of a positron-lifetime study on the interaction and relationship among point defects,microdefects, dislocations and impurities in InP.

1. INTRODUCTION

InP crystal has been the most useful substrates for DH semiconductor lasers and p-i-n detectors operating in the wavelength range between 1.1 to 1.7 μm. In order for such devices to be operated reliably,their substrates must be free from any crystalline defects. In particular,native threading dislocations in the substrates must be eliminated.

A convenient method for obtaining dislocation-free InP was reported by Seki et al.[1] They found that doping with a high concentration of donor impurities(n>1.0E+18/cm^3) could lead to substantial reduction in dislocation density. But the mechanism of dislocation reduction by doping impurities is still unclear.

Positron annihilation has higher sensitivity for point defects.Most of the previous work has been directed toward the investigation of bulk defect structures, little is known for Indium Phosphide(InP). This paper reports experimental results of a positron-lifetime study on the interaction and relationship among point defects, microdefects,dislocations and impurities in InP.

2. EXPERIMENTAL DETAILS

The standard positron-lifetime measurement technique was used.The experiment was performed at room temperature by using a fast coincidence device with a time resolution of 380ps (FWHM). Each lifetime spectrum contains 1.0E+6 counts.A ^{22}NaCl positron source of 6.0E+5Bq(15 μCi) deposited and sealed between two 1 μm thick nickel foils, was sandwiched by two identical pieces of the sample under study.

Van der Pauw method was used to obtain free carrier concentration and resistivity combined with Chemical etching to obtain Etching Pit Density (EPD).

The samples used were polished(100)-oriented and (111)-oriented crystal wafers. The specification of the crystals investigated is shown in Table I. The specific positron trapping rate is 2.0E-9cm^3/s. Here Nd denotes the concentration of vacancy defects (corresponds to the lifetime of τ_2). τ_b and τ_d denote the lifetimes of positron annihilation from free state (bulk lifetime) and defects respectively. K is positron trapping rate. n is the concentration of the free carrier. σ is the conductivity of InP crystals, logσ is the logarithm of electrical conductivity of the InP crystals. The HN1 to HN6 samples are all none doped. HU1 and HU3 and HU7 are S-doped and co-doped with Ga and Sb. HU1 and HU7 were cut from the same ingot. The former was cut from the middle and the latter was cut from the tail. HU2 was doped with (Sn+Sb +Ga+As). HU4 was S-doped and co-doped with (Ga +Sb+As).HU5 was S-doped and HU6 was S-doped and co-doped with Ga.

Table I A specification of the samples investigated and results of positron lifetime, chemical etching as well as electrical measurements.

	τ_1 (ps)	τ_2 (ps)	τ_3 (ps)	I_1 (%)	I_2 (%)	I_3 (%)	τ (ps)	τ_b (ps)	K (1.0E +9/s)	EPD (1.0E 4/cm³)	τ_d/τ_b	Nd (1.0E+ 18/cm³)	n (1.0E+ 16/cm³)	logσ
HN1	163	273	2303	54.2	45.5	0.3	213.2	199.7	1.13	4.00	1.37	0.57	0.410	0.45
HN2	170	273	1602	56.3	43.5	0.3	214.9	203.5	0.97	3--4	1.34	0.49	0.683	0.65
HN3	172	289	1650	62.0	37.3	0.7	216.2	203.2	0.88	1.00	1.42	0.44	2.104	1.09
HN4	170	278	2132	58.6	41.2	0.3	214.6	202.5	0.94	1--2	1.35	0.47	1.124	0.86
HN5	174	323	1750	66.9	31.5	1.7	221.7	204.2	0.85	0.80	1.58	0.43	2.148	1.10
HN6	171	276	2205	57.3	42.4	0.3	215.7	204.0	0.95	3.00	1.35	0.48	1.149	0.85
HU1	189	300	1852	72.4	27.2	0.4	219.3	210.3	0.54	0.20	1.43	0.27	1165.	3.22
HU2	175	279	1956	60.6	39.1	0.3	215.8	205.0	0.84	0.80	1.36	0.42	352.1	2.99
HU3	175	279	1897	60.3	39.4	0.3	216.1	205.2	0.84	0.89	1.36	0.42	162.3	2.67
HU4	188	306	1919	73.2	26.4	0.4	219.3	209.4	0.54	0.20	1.46	0.27	468.7	3.10
HU5	176	274	2290	51.0	48.8	0.2	223.9	213.3	0.99	4--6	1.28	0.50	510.2	3.03
HU6	178	286	2917	60.9	38.9	0.2	220.1	208.7	0.83	0.78	1.37	0.42	844.6	3.17
HU7	187	281	1972	61.0	38.7	0.2	223.5	214.9	0.69	0.60	1.31	0.35	969.6	3.22

3. RESULTS AND DISCUSSIONS

The PA spectra are dissociated to three components by multi-exponential fitting method. τ_1 is the positron lifetime of free state, monovacancy,complexes of monovacancy with impurities, dislocation,interstitial and interstitial cluster.τ_2 is the positron lifetime of multi-vacancy cluster and microcavity, I_1 and I_2 is relative intensity with lifetime τ_1 and τ_2 respectively. The long-lifetime component τ_3 is most likely due to annihilations at the surfaces of the sample and source rather than in the bulk of the sample, therefore it is left as a free parameter in subsequent analysis.

We use

$$\tau(\text{average lifetime})=(\tau_1*I_1+\tau_2*I_2)/(I_1+I_2) \quad (1)$$

$$K=I_2/I_1*(1/\tau_b -1/\tau_d) \quad (2)$$

$$(\tau_b)^{-1}=(I_1/\tau_1+I_2/\tau_2)/(I_1+I_2) \quad (3)$$

$$K= \mu*Nd \quad (4)$$

where μ is the specific positron trapping rate of defect. Nd is concentration of the vacancy defects(corresponds to the lifetime of τ_2).

We got positron-lifetime τ_1, τ_2 and their relative intensity I_1 and I_2 dependence with EPD. It is noted that τ_1,τ_2 and I_1 decrease with the increase of dislocation density.During EPD increasing, some of interstitial atoms released were absorbed by dislocations and vacancies,others were merged into interstitial clusters, the concentration of interstitial atoms increase, the monovacancy's decrease. On the other hand, with the EPD increasing, more dislocations,isolate interstitial atoms and impurities would interact with vacancies and clusters. Furtherly,we propose vacancy clusters is the source of the dislocation, while monovacancy (we think V_{In}) resist the creation and motion of the dislocations.The higher the concentration of the complexes of the impurities with monovacancy,the lower the dislocations are. As seen in Table I when doping concentration is around 4.0E+18/cm^3,EPD is lowered remarkbly. (Sn+Ga+Sb+As)-doping and (S+Ga+Sb+As)-doping and (S+Ga+As)-doping samples aren't better than (S+Sb+Ga)-doping crystals. EPD of HU3 is also low,it seems Sb-doping produce bigger reduction. As-doping seems to produce a smaller reduction

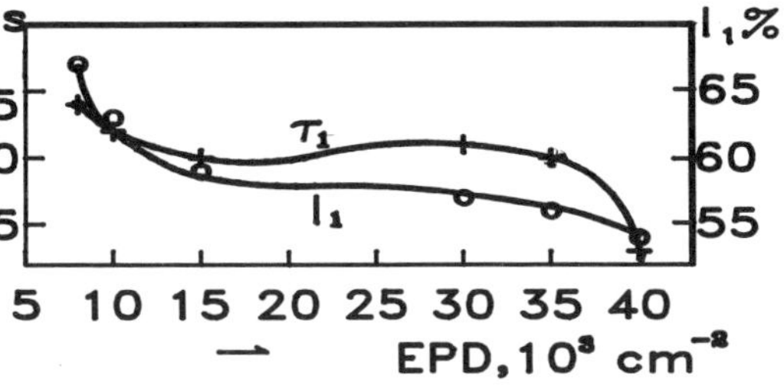

FIGURE 1 (a)
undoped InP τ_1 and I_1 versus EPD.

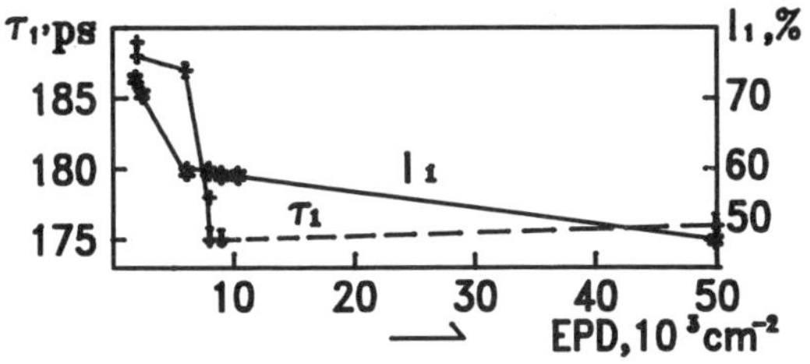

FIGURE 1 (b)
multi-doped InP τ_1 and I_1 versus EPD.

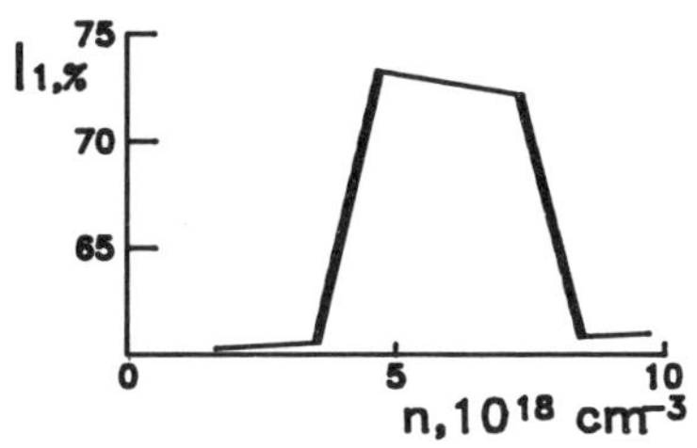

FIGURE 1 (c)
multi-doped InP I_1 versus free carrier concentration n

in EPD. Dislocation density of (Ga+Sb+S)-doped and (Ga+Sb+Sn)-doped InP crystals are nearly the same. The results also indicate that the dislocation density of both (Ga+Sb+S)-doped and (Ga+Sb+Sn)-doped InP crystals reduced remarkbly, nearly one order magnitude lower than the dislocation density of InP crystals doped with S or Sn or (S+Sn) or (Sn+Ga). Normally,EPD of InP doped with S or Sn or (S+Ga) or (Sn+Ga) is lower in the front part of the ingot,however higher in the middle or tail of the ingot. S and Sb occupy P sublattice, as a result the concentration of V_P lowers while the concentration of V_{In} increases. From Fig. 1 I_1 is bigger when the concentration of free carrier varies between 4.0E+18/cm^3 and 8.0E+18/cm^3. When the doping concentration exceeds the concentration of native point defects,the effect of EPD reduction by impurities lowers. Dislocation reduction effect of Ga and Sb combined with the mutual compensation of their effective segregation coefficient was used.EPD of InP crystals was reduced remarkbly by doping with (Ga+S+Sb) and (Sn+S+Sb). Doping together with Ga and Sb can make up for the lattice mismatch each other and reduce the distortion energy of InP lattice.

Most of the grown-in dislocations were not observed in the central zone of an ingot. They were blocked in the peripheral zone of an ingot by forming helical dislocations. These dislocations were probably formed by absorbing interstitial atoms or emitting vacancies.The mechanism of dislocation reduction by impurity was understood by taking into account the combination of the impurity effect on dislocation pinning and the helical dislocation formation metioned above.

In addition,the impurity doping causes tensile or stretching dislocations of near-neighbor atoms around impurities which may contribute to reduce dislocations by delocalizing strains.

ACKNOWLEDGEMENTS

The author thank cordially Prof. M.R. Huang and Ms. J.H. Yang from the Institute of High Energy Physics,Academia Sinica.

REFERENCE

1. Y. Seki et al., J.Appl.Phys. 49 (1978) 822.

Frontiers of Materials Research/
Electronic and Optical Materials
M. Kong, L. Huang (editors)

SUBSOLIDUS PHASE EQUILIBRIA IN THE SYSTEM $MgO-Nb_2O_5-Li_2O$

Yanfei ZHOU, Peiling WANG and Haoran TAN

Shanghai Institute of Ceramics, Academia Sinica. Shanghai 200050, China

Forty four compositions were selected to study the subsolidus compatibility phase relationships in the system $MgO-Nb_2O_5-Li_2O$. Conditions for solid state reaction and resulting phase as determined by X-ray powder diffraction method were described. The subsolidus phase diagram of this ternary system was obtained, containing 5 subsolidus compatibility triangles, 4 binary phase regions including $LiNbO_3$ (s.s.) and a broad solid solution region of single phase $MgO:LiNbO_3$ at room temperature, MgO forms solid solution with $LiNbO_3$ up to 25 mol%. The existence of two ternary compounds, $Li_2MgNb_2O_7$ and $Li_3Mg_2NbO_6$ was determined. The former can coexist with the solid solution of $LiNbO_3$, but the latter can not.

1. INTRODUCTION

In recent years lithium niobate has been the subject of extensive study because it is a versatile crystal material. But its optical applications were severely limited by its photorefractive effects which induces birefringence changes at high laser intensities. In 1980 it was first reported that lithium niobate doped with certain amount of MgO can remarkably enhance its resistance to optical damage[1]. Since then, the effects of MgO dopant have aroused widespread interest in lithium niobate research.

The first phase diagram of the system Li_2O-Nb_2O_5 was given by Reisman and Holtzberg[2]. Lerner et al. made a substantial improvement, showing that in the system there is a fairly large $LiNbO_3$ solid solution range, its maximum melting temperature does not occur at the stoichiometric composition but rather the composition of 48 to 49 mol% Li_2O[3]. Carruthers et al. used NMR and dielectric constant measurements to construct a phase diagram of Li_2O-Nb_2O binary system which showed the congruent composition to be 48.6 mol% Li_2O[4].

There are at least four compounds in the system $Li_2O-Nb_2O_5$, namely $Li_2Nb_{28}O_{71}$, $LiNb_3O_8$, $LiNbO_3$ and Li_3NbO_4. The crystal structure of $LiNbO_3$ at room temperature belongs to the rhombohedral (trigonal) space group R3C, the point group is 3m. This form is stable up to the ferroelectric transition temperature around 1100 to 1200°C, where a transition to the nonpolar point group $\bar{3}m$ occurs. Abbattista et al. recently reported that in the part of Li_2O-Nb_2O_5 binary system ranging from Li_3NbO_4 to Li_2O, there are other meta-stable compounds[5].

In the system $MgO-Nb_2O_5$, the binary diagram has not been reported. But at least three compounds, namely $Mg_4Nb_2O_9$, $Mg_5Nb_4O_{15}$ and $MgNb_2O_6$ in the binary system have been determined.

No study of the binary system $MgO-Li_2O$ has been reported.

The ternary phase diagram $MgO-Nb_2O_5-Li_2O$ was investigated by Grambaier[6]. Subsolidus phase equilibria in the system $MgO-Nb_2O_5-Li_2O$ has been studied and a solid solution region of the single phase $MgO:LiNbO_3$ was found.

In the present paper, the subsolidus phase relationships in the system $MgO-Nb_2O_5-Li_2O$ and solid solution range of MgO dopant in lithium niobate were studied in detail. The subsolidus phase diagram of this ternary system was established and two ternary compounds, $Li_2MgNb_2O_7$ and $Li_3Mg_2NbO_6$ were identified.

2. EXPERIMENTAL PROCEDURE

(1) Materials and Equipments

The starting materials used were reagent

regrade MgO (99.99%, Li_2CO_3 (99.99%) and Nb_2O_5 (99.95%). These chemicals were weighted in required molar ratios and mixed in an agate mortar with alcohol as media. The mixture obtained (about 5g) were pressed into disks (15 mm in diameter and 10 mm in thickness) with a handjack. Then the disks were put into an Al_2O_3 crucible with Pt sheet and prefired in an electric muffle furnace at 900°C for 4-10h. The prefired samples were smashed up, grinded and repressed as above. The final firing was made in air at a certain temperature below liquidus. The phases present in reacted samples were routinely identified by an X-ray diffractometer with Cukα radiation. For determination of the single phase compounds X-ray powder photographs were also taken in a focusing camera of Guinier-Hägg type, using Cukα radiation. SiO_2 (a=5.430880 Å) was added as an internal standard. The films were analyzed with a film scanner system and the compounds were indexed with TREOR program[7].

(2) Methods

44 compositions were selected to establish subsolidus phase diagram in the $MgO-Nb_2O_5-Li_2O$ system. The subsolidus compatibility relations in the system were determined by the phase present in 44 samples fired at high temperature. X-ray diffraction patterns which suggested non-equilibrium conditions were followed by further heat treatments until no further change was observed in the phase assemblages. For determination of phase relationships one or two composition were selected on each possible tieline and in each phase region in the $MgO-Nb_2O_5-LiNbO_3$ subsystem. But in the $MgO-LiNbO_3-Li_2O$ subsystem the phase relationships are more complicated, a hypothetical diagram illustrating all possible tielines was constructed and the compositions corresponding to the points of intersection of tielines were chosen. Results which established the existence of a true tieline allowed the elimination of conflicting tielines and thus simplify the procedure of experiment. Evidence suggesting ternary compounds were further identified by preparation of several samples of specific compositions.

3. RESULTS AND DISCUSSION

(1) The System $MgO-Nb_2O_5-Li_2O$

The 44 compositions in $MgO-Nb_2O_5-Li_2O$ system for determination of the subsolidus compatibility relations, their heat treating conditions and resulting phases as determined by X-ray powder diffraction method are listed in Table 1. Solid state reaction of compositions within the $MgO-Nb_2O_5-Li_2O$ ternary system resulted in the identification of 5 subsolidus compativility triangles, 4 binary phase regions and a broad solid solution region of single phase $MgO:LiNbO_3$ at room temperature. The subsolidus relations, as illustrated in Fig. 1, show that the maximum of solid solubility of MgO in $LiNbO_3$ at room temperature is about 25 mol% (as revealed by a sample with MgO 25.0, Nb_2O_5 41.3 and Li_2O 33.7 mol% which was identified to be $LiNbO_3$(s.s.) after treated at 1120°C for 7h).

The tielines in the $Mg_4Nb_2O_9-Li_3NbO_4-Li_2O-MgO$ subsystem appears as dashed lines because the samples fired usually contain MgO phase, which can not be fully reacted at high temperature and prolonged heat treating time due to the significant volatilization of lithia. For example, samples 24 and 31-36 were calcined at 1150°C for 100h, their loss of weight was found to be about 0.5 wt%. Besides, the Li_2O-rich composition samples react with Pt disk at high temperature and change their color into yellow-green after firing. Their diffraction patterns contain some unidentified peaks. Consequently, equilibrium state could not be established in these samples.

(2) Ternary Compounds

The existence of a new compound, $Li_2MgNb_2O_7$ (denoted by X_1 in Fig. 1) was identified. It explained the phase relationships in the central

Table 1. Results of Heating Experiments on Composition in the System $MgO-Nb_2O_5-Li_2O$

No.	Composition (molar) $MgO:Nb_2O_5:Li_2O$	Calcining Conditions Temp. (°C)/Time(h)	Phases Present
2	0.0:85.0:15.0	1100/12	$N(s)+LN_3(s)$
3	10.0:75.0:15.0	1100/12	$N(s)+MN(s)+LN_3(s)$
4	13.7:68.0:18.3	1100/12	$MN(s)+LN_3(m)$
5	5.0:56.5:38.5	1120/7	$LN(s)+LN_3(m)$
6	15.0:55.0:30.0	1120/7	$LN_3(m)+LN(s)+MN(w)$
7	22.7:47.3:30.0	1120/7	$LN(s)+MN(m)$
8	0.0:57.0:43.0	1120/7	$LN(s)+LN_3(w)$
9	9.0:49.7:41.3	1120/7	$LN(s)+MN(w)$
10	25.0:41.3:33.7	1120/7	$LN(s)$
11	17.0:45.5:37.5	1120/7	$LN(s)+MN(w)$
12	5.0:53.0:42.0	1120/7	$LN(s)+LN_3(w)$
13	15.0:43.7:41.3	1120/7	$LN(s)+X_1(w)$
14	7.0:45.7:47.3	1120/7	$LN(s)+X_1(w)+L_3N(vw)$
15	0.0:49.5:50.5	1150/7	$LN(s)$
16	35.0:40.0:25.0	1120/7	$LN(s)+M_4N(m)+MN(m)$
17	35.0:33.4:31.6	1120/7	$LN(s)+M_4N(w)+X_1(s)$
18	33.3:33.3:33.3	1150/7	$X_1(s)$
19	70.0:15.4:14.6	1120/7	$M_4N(s)+M(w)+X_4(m)$
20	60.0:10.0:30.0	1120/7	$M(m)+X_4(s)$
21	25.0:35.0:40.0	1180/30	$LN(s)+X_1(s)$
22	20.0:35.0:45.0	1120/7	$LN(s)+X_1(s)+L_3N(w)$
23	32.5:28.3:39.2	1150/7	$X_1(s)+M_4N(w)+M(m)$
24	30.0:27.0:43.0	1150/7	$X_1(s)+M(m)+L_3N(w)$
25	50.0:25.0:25.0	1160/20	$M_4N(s)+X_1(m)+L_3N(w)$
26	20.0:28.0:52.0	1180/60	$X_4(s)+M(m)$
27	37.5:12.5:50.0	1180/20	$X_4(s)+M(m)$
28	21.0:21.0:58.0	1180/20	$L_3N(s)+M(m)+X_4(vw)$
29	50.0:12.5:37.5	1180/20	$X_4(s)+M(m)$
30	50.0:17.0:33.0	1160/20	$M_4N(s)+X_4(m)$
31	40.0:20.0:40.0	1180/30	$M(s)+M_4N(m)+L_3N(w)$
32	35.0:25.0:40.0	1150/100	$X_1(s)+M_4N(w)+M(m)$
33	35.0:20.0:45.0	1150/100	$M(s)+L_3N(vw)$
34	34.5:24.0:41.5	1150/100	$X_1(s)+M(w)+L_3N(w)$
35	34.0:;2.5:43.5	1150/100	$M_4N(s)+L_3N(m)+X_4(w)$
36	34.0:22.0:44.0	1150/100	$M_4N(s)+L_3N(m)+X_4(w)$
37	44.0:12.0:44.0	1160/20	$X_4(s)+M(m)$
38	33.3:16.7:50.0	1180/30	$M(s)+X_4(w)$
39	20.0:20.0:60.0	1180/20	$M(s)+L_3N(m)+X_4(vw)$

Continued

No.	Composition (molar) $MgO:Nb_2O_5:Li_2O$	Calcining Conditions Temp. (°C)/Time(h)	Phase Present
40	2.5:37.5:60.0	1120/7	LN(s)+L_3N(m)
M_4N	80.0:20.0:0.0	1500/7	M_4N(s)+M_5N(vw)
MN	50.0:50.0:0.0	1150/7	MN(s)+M_4N(vw)
L_3N	0.0:25.0:75.0	1150/7	JCPDS(16-459)
LN_3	0.0:75.0:25.0	1150/7	JCPDS(26-1175)

Note: N = Nb_2O_5, M = MgO, LN = $LiNbO_3$, L_3N = Li_3NbO_4, LN_3 = $LiNb_3O_8$, M_4N = $Mg_4Nb_2O_9$
MN = $MgNb_2O_6$, M_5N_2 = $Mg_5Nb_4O_{15}$, X_1 = $Li_2MgNb_2O_6$, X_4 = $Li_3Mg_2NbO_6$
s: Strong, m: Middle, w: Weak

Table 2. X-ray Powder Diffraction Data of $Li_2MgNb_2O_7(X_1)$

hkl	d(Å)	I/I_0
120	5.2805	100
201	4.8172	25
211	4.4682	5
002	400408	97
030	3.9264	81
410	2.8997	18
041	2.7690	8
312		
132		
411	2.7368	78
420	2.6709	66
113	2.5665	27
421	2.5415	30
241	2.5120	5
213	2.4038	5
500		
123		
142	2.3362	13
150	2.3083	16
431	2.2852	11
432	2.0534	9
114	1.9630	5
034	1.7950	31

portion of the system $MgO-Nb_2O_5-Li_2O$ very well. Two tielines between $LiNbO_3$ (s.s.) and X_1 were established. The $Mg_4Nb_2O_9-X_1$ and $Li_3NbO_4-X_1$ tielines were also determined. The diffraction data of this new compound indexed by TREOR program are presented in Table 2. This compound has an orthorhomic unit cell with a=11.999 Å, b=11.761 Å and c=8.082 Å.

In 1982 Gard reported the existence of a ternary compound $Li_3Mg_2NbO_6$ (denoted by X_4 in Fig. 1) and its indexed powder diffraction data[8]. This compound has a composition of sample 29 (50 MgO, 37.5 Li_2O, 12.5 mol% Nb_2O_5). In our experiment, sample 29 was first prepared with chemicals MgO, Nb_2O_5 and Li_2CO_3 and fired at 1050°C for 100h and 1180°C for 20h respectively. After that, its powder diffraction showed a mixture phases $Mg_4Nb_2O_9$, MgO and $Li_3Mg_2NbO_6$. Then, the sample with the same composition but mixed from Li_3NbO_4 and MgO in a molar ratio 1:2 was prepared and fired at 1050°C for 100h and 1180°C for 20h respectively. Under this condition the reaction product contained only a little MgO phase with $Li_3Mg_2NbO_6$ as the major phase. The MgO phase was not able to vanished even at prolonged firing time. As the MgO content was reduced to Li_3NbO_4:MgO =1:1.98 (molar ratio), the sample still contained a little MgO phase after fired at 1180°C for 60h. These re-

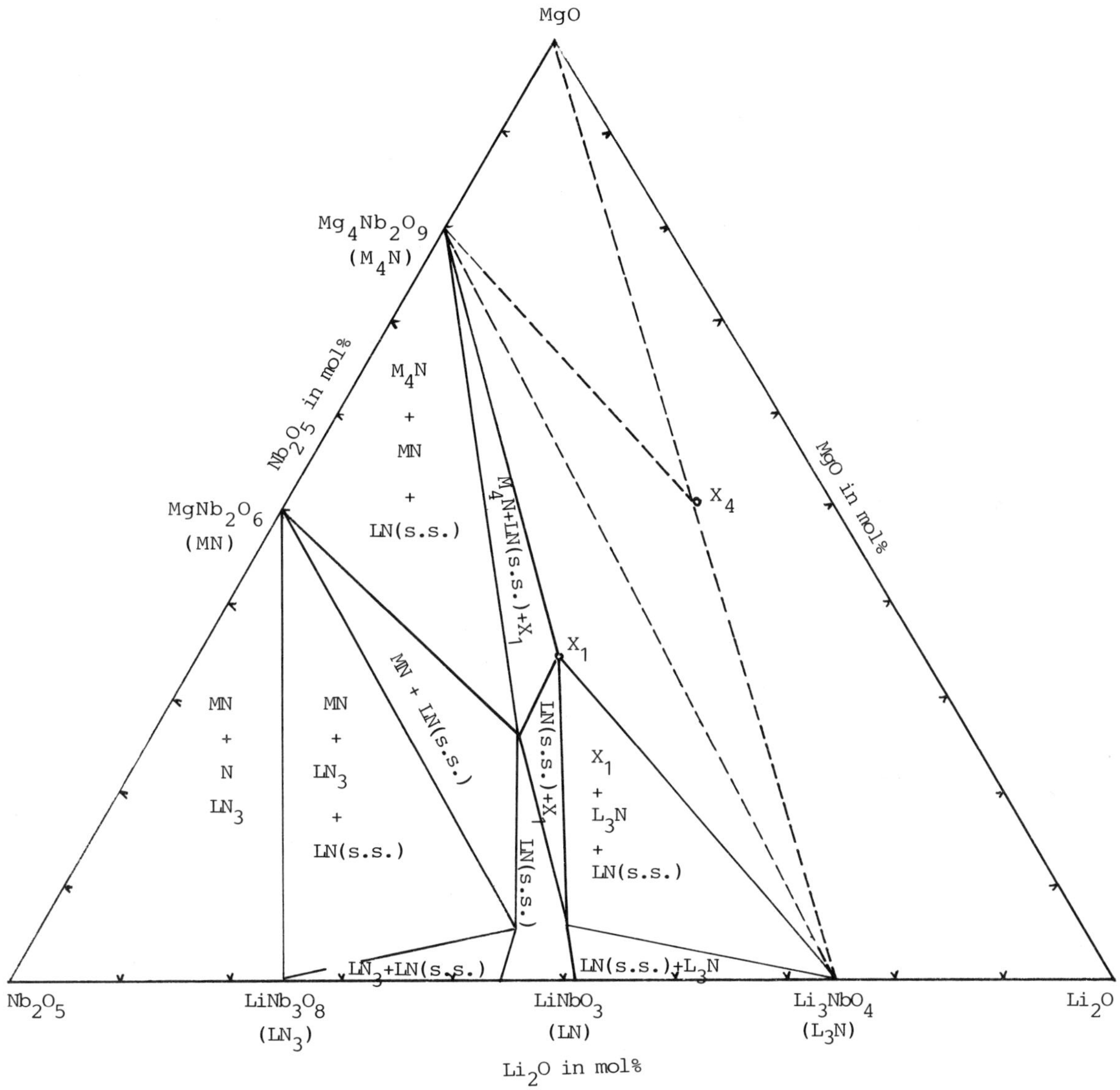

FIGURE 1
Subsolidus phase relationship of $MgO-Nb_2O_5-Li_2O$ system

sults suggested that the compound should be a nonstoichiometric composition.

4. CONCLUSIONS

Study on the subsolidus phase relationships in $MgO-Nb_2O_5-Li_2O$ system is essential for an understanding of the effect of MgO dopant. Fig. 1 shows the experimental results obtained by means of high temperature solid state reaction and X-ray diffraction technique for 44 samples of different compositions. From this figure we could arrive at the following conclusions:

(1) MgO forms solid solution with $LiNbO_3$ up to 25.0 mol% MgO.

(2) MgO:$LiNbO_3$(s.s.) can coexist with $LiNb_3O_8$, Li_3NbO_4, $Mg_4Nb_2O_9$ and $MgNb_2O_6$ to form several subsystems.

(3) Two ternary compounds $Li_2MgNb_2O_7$ and $Li_3Mg_2NbO_6$ have been confirmed. The former can coexist with $LiNbO_3$(s.s.) and there is no tie-line between the latter and $LiNbO_3$(s.s.).

REFERENCES

1. Zhong Giguo, Jin Jian, Wu Zhongkang, 11th International Quantum Electronics Conference June 23-26 (1980) 631.
2. A. Reisman and F. Holtzberg, J. Am. Chem. Soc. 80 (1958) 6503.
3. P. Lerner, C. Legrab and J. P. Dugas, J. Cryst. Growth 3 (1968) 231.
4. J. R. Carruthers, G. E. Peterson, M. Grasso and P. M. Bridenbaugh, J. Appl. Phys.
5. F. Abbattista, M. Vallino and D. Mazza, Mat. Res. Bull. 22 (1987) 1019.
6. B. C. Grabmaier, F. Otto, J. Cryst. Growth 179 (1986) 682.
7. P. E. Werner, L. Eriksson, M. Westdahl, J. Appl. Crystallogr. 18 (1985) 367.
8. J. A. Gard and A. R. West, J. Appl. Cryst. 15 (1982) 116.

Frontiers of Materials Research/
Electronic and Optical Materials
M. Kong, L. Huang (editors)
Elsevier Science Publishers B.V., 1991

GROWTH OF LITHIUM TRIBORATE (LBO) SINGLE CRYSTAL FIBERS BY THE LASER-HEATED PEDESTAL GROWTH METHOD

GROWTH OF LITHIUM TRIBORATE (LBO) SINGLE CRYSTAL FIBERS BY THE LASER-HEATED PEDESTAL GROWTH METHOD

HUO Yujing

Tsing Hua University, Beijing, P.R. CHina

ZHAO Shuqing JI Yangyang※ZHANG Hongwu HUANG Chaoen

Research Institute of Synthetic Crystals, Beijing, P.R. China

The growth of LBO single crystal fibers by the laser-heated pedestal growth method (LHPG) is reoprted. Some LBO single crystal fibers were obtained by the seeds with different orientations, different source rods and different fluxes. The LBO single crystal fibers obtained by the seeds with phase matching dirction give an output of 0.532um green light when it is used as a SHG device of Nd: YAG laser.

1. INTRODUCTION

Lithium Triborate (LBO) is a newly discovered nonlinear optic crystal[1]. This crystal has a sufficient large nonlinear coefficient, good chemical stability and excellent moisture resistance. Its hardness is reasonable. Phase matching can be carried out at room temperature. In addition, it has high damage threshold, wide transparency range, large acceptance angle for phase matching and high conversion efficiency. All these made LBO a conspicuous crystal in nonlinear optical field. Many LBO nonlinear optical devices have been made.

Recently, single crystal fibers have become of a great interest because of its potentiality for application in a variety of electro-optical and nonlinear optical devices such as phase and amplitude modulators, frequency shifters, harmonic generators, sum and difference frequency generators and optical parametric oscillators. Because they combine properties of bulk crystals and of glass fibers with each other, LBO single crystal fibers have many advantages. For example, they would not be overheated when a laser beam with high power density is focused on to their small crosssection. Besides the superiority of bulk LBO crystal, LBO single crystal fibers do not

---

※ Present address: Institute of Physics, Academia sinice, Beijing, P.R.China suffer from the same problem of walk-off angle as in bulk nonlinear optical devices, so that the length of interaction and hence the coherent length could the much longer then ever before. A input laser beam of high power density is allowed to get a high conversion efficiency. Devices made from LBO single crystal fibers can be effectively coupled with other glass fibers, decreasing the coupling loss and on the whole improving the system performance. They can be widely used in optical communication, fiber sensors and other fields. Many works to research on LBO single crystal fibers have been done in our lab for above reasons.

2. GROWTH METHOD

Various methods to grow single crystal fibers have been reported. The laserheated pedestal growth method (LHPG)[2] was widely used because of its advantages such as the high growth speed, the small diameters of fibers and the posibility to grow fibers from either congruent or incongruent melting materials. For the first time the LHPG method was chosen to grow LBO single crystal fiber.

The LHPG method is shown schematically in fig. 1. A ring-shaped CO_2 laser beam produced by the optical system shown in fig. 2 was focused onto the top of a source rod, creating a molten zone used for fiber growth.

The source rod can be prepared from either bulk single crystal or starting materials with the same composition . To initiate the growth of a single crystal fiber, an oriented seed crystal was dipped into the molten zone and then slowly withdrawn up. To keep the volume of the stationary molten zone constant, the source rod was slowly pushed into the molten zone as the growing fiber crystal was withdrawn from it. According to the law of conservation of mass in steady-state, the ratio of feeding rate to pulling rate is easily calculated by the equation:

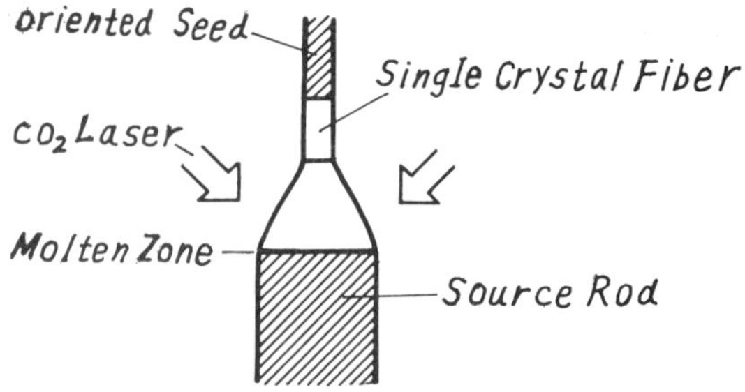

Fig. 1. Schematic diagram of fiber growth

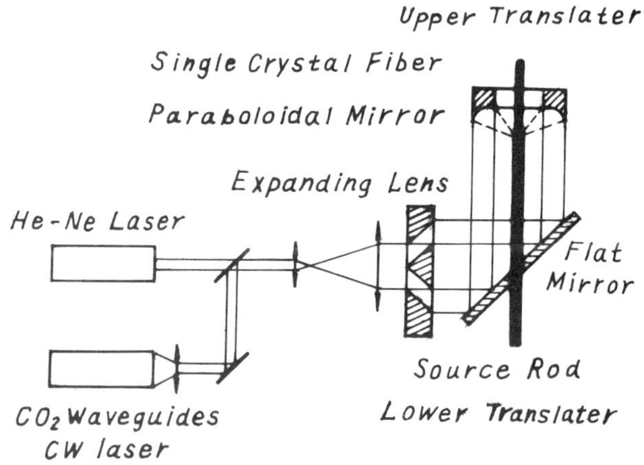

Fig. 2. Optical system of LHPG

$$V_s / V_f = (D_f / D_s)^2 \cdots\cdots\cdots\cdots (1)$$

Where, V_s is the feeding rate of the source rod, V_f is the growth speed of the fiber, and D_f and D_s are the diameter of the fiber and source rod respectively. In the LHPG growth method, the diameter ratio of fiber to source rod for stable growth is usually between 1/4 and 1/2. Another important for growth stability is the zone length. Empirically[3], the stable growth occurs for zone length is:

$$L = 3/4(D_f / D_s) \cdots\cdots\cdots\cdots (2)$$

Where D_f is the diameter of fiber, D_s is the diameter of source rod. A more detail description about the LHPG method is givev by Feigelson in ref.4.

3. GROWTH PROCESS

A LBO bulk single crystal starting materials with the same composition were used as the source rod and oriented LBO single crystal was used as the seed. The source rod was precisely processed as a rod that has a square cross section in a size of 0.4×0.4 mm or a circle cross section with a diameter of 0.4 mm.

In the growth, fluxes such as B_2O_3, LiF, $B_2O_3+Li_2B_4O_7$ etc. have to be added into growth area because LBO decompose easily at high temperature. The amount of fluxes added is very important for LBO fiber growth. If too much fluxes are added, it is hard for LBO fiber to crystallize from the molten zone of a low LBO saturation in the melt. Otherwise if a too small amount of fluxes is used, it is also difficult fo grow the fiber because the molten zone is vevy unstable.

According to the $Li_2O-B_2O_3$ binary phase diagram, the $Li_2O.3B_2O_3$ decomposed at 834±4℃, therefore the growth temperature must be controlled below this point. Because of a high temperature gradient in LHPG, the temperature difference between the center and the surface of the molten zone may be more than 100℃ , solidification in this case may start from the surface and the fluxes may flow into the center of the molten zone. The growing fiber in fact is shellshaped with high stress and often cracked while growing. So for a successful LBO fiber growth by this method it is very important to use a suitable flux and to made the horizontal temperature gradient as small as posible.

The growth rate was often controlled between 0. 1mm/min and 0. 2mm/min. The fiber will be not transparent or easily cracked if it is growing too fast, and its surface will be not homogeneous because of the anisotropy of LBO growth if the growth rate is too small.

4. RESULTS

LBO single crystal fibers are successfully pulled out by LHPG method. Different fluxes, different oriented seeds, different source rods made from bulk LBO single crystals or other materials with the same composition as LBO were used. The fiber pulled in the phase matching direcion is 10mm in length, 200um in diameter. The two end-faces of the fiber in phase matching direction were polished and an output of 0.532um green light was obtained from one end if a 1. 064um Nd: YAG laser beam was incident on the other end. The double frequency effect is preliminarily testified. Other work on measurements of LBO fibers is in progess.

REFERENCES

1. Zhao Shuqing, Zhang Hongwu and Huang chaoen, J. of Synthetic Crystals (in Chinese) 18 (1989) 9
2. M. M. fejer, J. L. Nightingale, G.A.Magel and R.L.Byer, Rev. Sci. Instrum. 55 (1984) 1791
3. D. Y. Tang, R. K. Route and R. S. Feigelson, J. Cryst. Growth 91 (1987) 81
4. R.S.Feigelson, J.Cryst. Growth 79 (1986) 669
5. H.Koing and R.Hoppe, Z.Anorg . Allg. Chem. 439 (1978) 71

AUTHOR INDEX

and

SUBJECT INDEX

AUTHOR INDEX

SUBJECT INDEX